D0643808

Springer Advanced Texts in Chemistry

Charles R. Cantor, Editor

Springer Advanced Texts in Chemistry

Series Editor: Charles R. Cantor

Robert B. Gennis

Biomembranes

Molecular Structure and Function

With 142 Figures in 236 Parts

Springer-Verlag
New York Berlin Heidleberg London Paris
Tokyo Hong Kong Barcelona Budapest

Robert B. Gennis
School of Chemical Sciences
University of Illinois at Urbana-Champaign
Urbana, IL 61801, USA

Series Editor:
Charles R. Cantor
Department of Human Genetics and Development
College of Physicians and Surgeons
Columbia University
New York, NY 10032, USA

Library of Congress Cataloging-in-Publication Data
Gennis, Robert B.
Biomembranes: molecular structure and function/ Robert B. Gennis.
p. cm.—(Springer advanced texts in chemistry)
Bibliography: p.
Includes index.
ISBN 0-387-96760-5 (alk. paper)
1. Membranes (Biology) I. Title. II. Series.
QH601.G435 1989
574.87'5—dc19 88-26558

Printed on acid-free paper

Typeset by Precision Graphics, Champaign, Illinois.
Printed and bound by R.R. Donnelley & Sons, Harrisonburg, Virginia.
Printed in the United States of America.

9 8 7 6 5 4 3

ISBN 0-387-96760-5 Springer-Verlag New York Berlin Heidleberg
ISBN 3-540-96760-5 Springer-Verlag Berlin Heidelberg New York

Dedicated to my family for all their encouragement:

To Joe and Sylvia, my parents,
to Christine, my wife,
and
to Emmalyn Rose, my daughter,
for all the happiness she has brought us.

Klaatu berada nikto

Series Preface

New textbooks at all levels of chemistry appear with great regularity. Some fields like basic biochemistry, organic reaction mechanisms, and chemical thermodynamics are well represented by many excellent texts, and new or revised editions are published sufficiently often to keep up with progress in research. However, some areas of chemistry, especially many of those taught at the graduate level, suffer from a real lack of up-to-date textbooks. The most serious needs occur in fields that are rapidly changing. Textbooks in these subjects usually have to be written by scientists actually involved in the research which is advancing the field. It is not often easy to persuade such individuals to set time aside to help spread the knowledge they have accumulated. Our goal, in this series, is to pinpoint areas of chemistry where recent progress has outpaced what is covered in any available textbooks, and then seek out and persuade experts in these fields to produce relatively concise but instructive introductions to their fields. These should serve the needs of one semester or one quarter graduate courses in chemistry and biochemistry. In some cases, the availability of texts in active research areas should help stimulate the creation of new courses.

New York, New York CHARLES R. CANTOR

Preface

The study of membranes has become a meeting ground for a number of diverse scientific disciplines ranging from biophysics to molecular biology. Students can enter the realm of membrane research from almost any direction, as physicists or genetic engineers or almost anything in between. To write a textbook that will be useful to such a wide audience is a challenge. There are, however, a body of knowledge and a set of guiding principles which should be comprehended by anyone who wants to appreciate the current state of research in biomembranes. I have focused on these fundamentals. My perspective is that of a biochemist and I have organized the text primarily around molecular structure and structure-function correlations. The book should be useful for graduate-level courses or for self-guided reading in the broad area of membrane structure and function, or it could be used to provide background information required for courses on special topics such as transport, receptors, signal transduction, or membrane biogenesis. The citations in the text cover through the end of 1987.

I have tried to create a text that I would have liked to have as a student first entering the field as well as one that I will find useful as an active research scientist. With these goals in mind, I have tried to bring some perspective to an awesome body of work by showing how research studies in diverse areas are related to each other and to a common conceptual framework. I have included ample documentation to make it easy for the reader to go directly to the research literature to probe more deeply into areas that can only be briefly discussed in a text of this sort. There may already be more detail given in the text to suit some readers, but I feel the density of information provided is suitable for an advanced textbook. It is always an easy matter to skip over sections in which one is not interested.

I would like to thank a number of individuals who have been helpful and encouraging during the long, seemingly endless, process of writing this book.

Many friends and colleagues were kind to read portions of the manuscript and offer encouragement and advice as well as point out errors. I would like to thank the following: Richard Anderson, Vyto Bankaitis, Lewis Cantley, Charles Cantor, Tony Crofts, John Cronan, Pieter Cullis, Tom Ebrey, Don Engelman, Gerry Feigenson, Sidney Fleischer, George Fortes, Michael Glaser, Neil Green, Lynne Guildensoph, Ari Helenius, Rick Horwitz, Wayne Hubbel, Ken Jacobson, Ron Kaback, Jim Kaput, Steve Kaufman, David Kranz, Vishnawath Lingappa, Mark McNamee, Chris Miller, Eric Oldfield, Elliot Ross, Ted Steck, and John Whitmarsh. Special thanks go to J. Keith Wright, who has been particularly helpful in efforts to improve the manuscript, to Ann Dueweke, who was a great help in collecting the copyright permissions and in putting the list of references in order, and to Karen Shannon at Precision Graphics (Champaign, Illinois) for all the figures. Thanks are also due to all those who were helpful in proofreading the final manuscript: Kathe Andrews, Rose Beci, Visala Chepuri, Tom Dueweke, Hong Fang, John Hill, Tamma Kaysser, Kiyoshi Kita, Laura Lemieux, Gail Newton, Kris Oden, Petr Pejsa, Jim Shapleigh, Steve Van Doren, Cecile Vibat, Melissa White, and Chris Yun.

Finally, my deep gratitude goes to the excellent secretarial service provided within the physical chemistry office at the University of Illinois by Evelyn Carlier, Jan Williams, Karen McTague, and Betty Brillhart.

Urbana, Illinois

ROBERT B. GENNIS

Contents

Chapter 7 Interactions of Small Molecules with Membranes: Partitioning, Permeability, and Electrical Effects 235

Chapter 8 Pores, Channels, and Transporters 270

Chapter 1
Introduction: The Structure and Composition of Biomembranes

1.1 The Importance and Diversity of Membranes

Membranes play a central role in both the structure and function of all cells, prokaryotic and eukaryotic, plant and animal. Membranes basically define compartments, each membrane associated with an inside and an outside. If this were all they did, membranes would be considerably less interesting than they are. But, membranes not only define compartments, they also determine the nature of all communication between the inside and outside. This may take the form of actual passage of ions or molecules between the two compartments (in and out) or may be in the form of information, transmitted through conformational changes induced in membrane components. In addition, attached to membranes are many cellular enzymes. Some of these enzymes catalyze transmembrane reactions, involving reactants on both sides of the membrane or molecular transport. Others are involved in sequential reactions involving a series of enzymes which are concentrated in the plane of the membrane, thus facilitating efficient interactions. Still other enzymes have membrane-bound substrates and/or are involved in the maintenance or biosynthesis of the membrane. Most of the fundamental biochemical functions in cells involve membranes at some point, including such diverse processes as prokaryotic DNA replication (e.g., refs. 807, 777, 803), protein biosynthesis, protein secretion, bioenergetics, and hormonal responses.

Electron micrographs of mammalian cells reveal the wealth of membranous organelles which comprise a large part of the intracellular volume. It is now clear that the structural principles for all these membranes are basically the same. Furthermore, these structural similarities apply also to plant cell membranes and bacterial membranes. These common features, recognized by Robertson in the late 1950s (1231), allow us to apply lessons learned in one membrane system, such as the erythrocyte membrane, to other systems, tempered with a reasonable degree of caution. This caution is necessary because, paradoxically, one of the most salient points to be made about membranes is their remarkable diversity. This diversity is due primarily to the different functions of the proteins present in each membrane and to the way in which these proteins interact with each other as well as with cytoplasmic components. These interactions result in distinct morphologies, such as in the microvilli of the intestinal epithelium or the tubular endoplasmic reticulum, and may result in lateral inhomogeneities within a given membrane (see Section 4.5). The main point is that there is a common ground for studying membranes in general, but that an appreciation for the subject lies in large measure in the comprehension of the molecular and biological basis for the diversity in membrane structure and function.

Progress in the study of membranes has come from exploiting the advantages for studying the membranes from a variety of organisms. Bacteria have relatively simple envelopes containing one or two membranes, which can be manipulated genetically or by altering the growth conditions. Enveloped viruses enter animal cells by membrane fusion (Section 9.52) and exit by budding (Section 4.53). The maturation of viral proteins provides an excellent experimental system for studying membrane protein biosynthesis (Section 10.2).

Eukaryotic cells have numerous membranous organelles, and each membrane is unique in composition, structural detail, and function. In order to understand the motivation behind many of the studies described in later chapters it is important to have some background in the biological functions of these various membrane systems. Figure 1.1 shows a schematic indicating the various membranes as they appear in a generic animal and plant cell. Note that the appearance of the organelles will be different in other cell types, and, in addition, some cells, such as the rod cell of the retina or the skeletal muscle cell, have highly specialized membranes which have unique functions.

(1) *Plasma membrane*: The plasma membrane defines the boundaries of the cell and is the point of contact between the cell and its environment. As such, the plasma membrane contains specialized components involved in intercellular contacts and communication, hormonal response, and transport of both small and large molecules into and out of the cell. However, the plasma membrane is itself divided into specialized regions in those cells which are simultaneously in contact with different environments. Figure 1.2 shows the location of the *apical* and *basolateral* plasma membrane domains for a hepatocyte and for a polarized epithelial cell. The apical membrane is that which is in contact with the "external" environment, such as the bile canaliculus in the case of the liver cell or the gastrointestinal lumen for an epithelial cell in the gut. The apical membrane can contain specialized structures such as the *microvilli*, which can be organized to form the *brush border membranes* in some absorptive cells. Microvilli greatly increase the effective surface area of the membrane and facilitate efficient transport. The basolateral membrane is that which is in contact with other cells (lateral or contiguous membrane) or blood sinusoids (sinusoidal membrane). In the hepatocyte, the lateral and sinusoidal membranes are morphologically and biochemically separable (402).

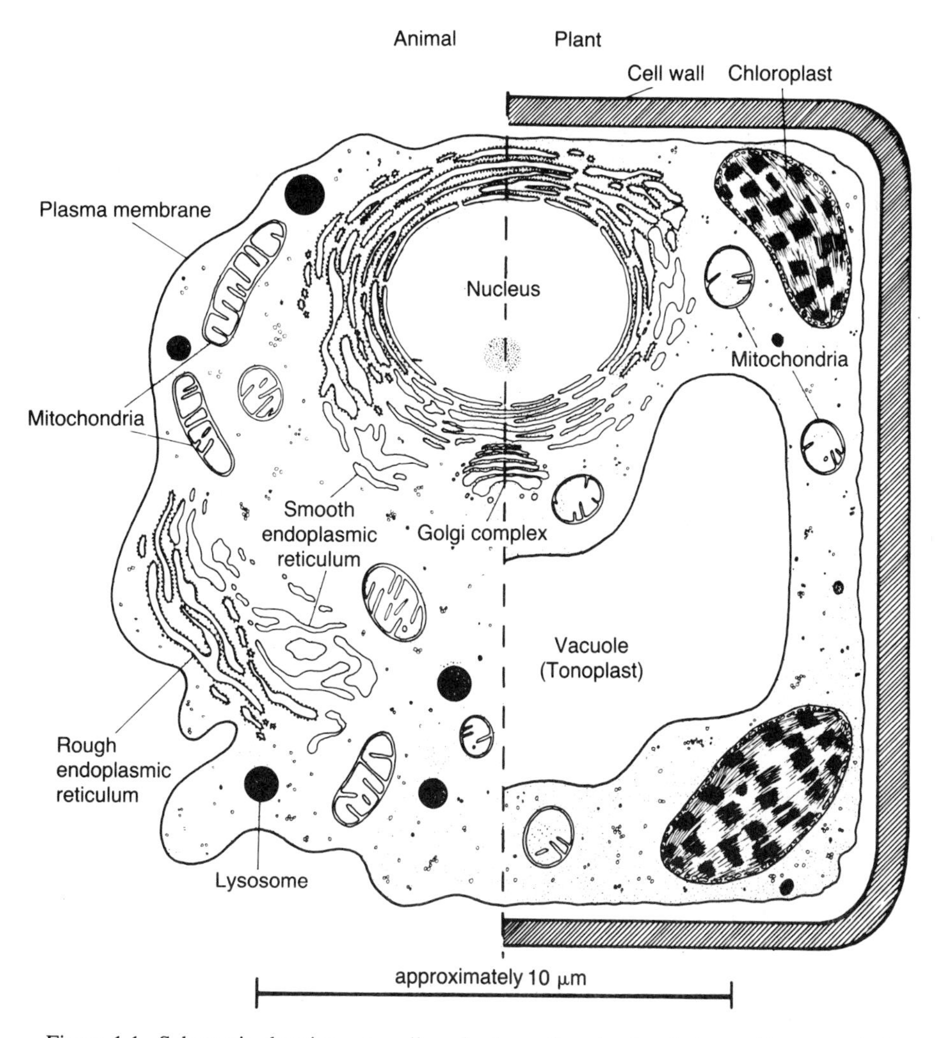

Figure 1.1. Schematic showing organelles of eukaryotic animal and plant cells as revealed by electron microscopy. Adapted from ref. 425a.

The basolateral membrane in the hepatocyte contains several specialized structures for cell–cell adhesion and intercellular transport.

Tight junctions seal the contacts between cells to prevent mixing the contents of the bile and blood vessels.

Gap junctions contain a regular array of pores that allow small molecules to pass through the plasma membranes of two adjacent cells. Electron microscopic and biochemical studies have revealed some molecular detail of these pores, showing each to contain a hexagonal array of protein subunits (see Section 8.21).

Desmosomes also function as adhesion sites between cells and are involved in contacts between the plasma membrane and cytoskeletal elements (see Section 4.3).

The apical, lateral, and sinusoidal portions of the plasma membranes are morphologically distinct and have unique compositions and functions. If the cells are disrupted gently, these specialized regions of the plasma membrane can be physically separated and purified (402). It is not understood on a molecular level how these specialized domains of the plasma membrane are maintained in the cell, but, clearly, there cannot be free diffusion of all membrane components between them (see Section 4.51).

(2) *Nuclear membrane*: The nuclear envelope which is present in interphase cells appears in electron micrographs as a double membrane, with a narrow space in between called the perinuclear space (459). The nuclear envelope appears to be formed from portions of the endoplasmic reticulum (see below) and these two systems may, in fact, be physically continuous. The most prominent morphologi-

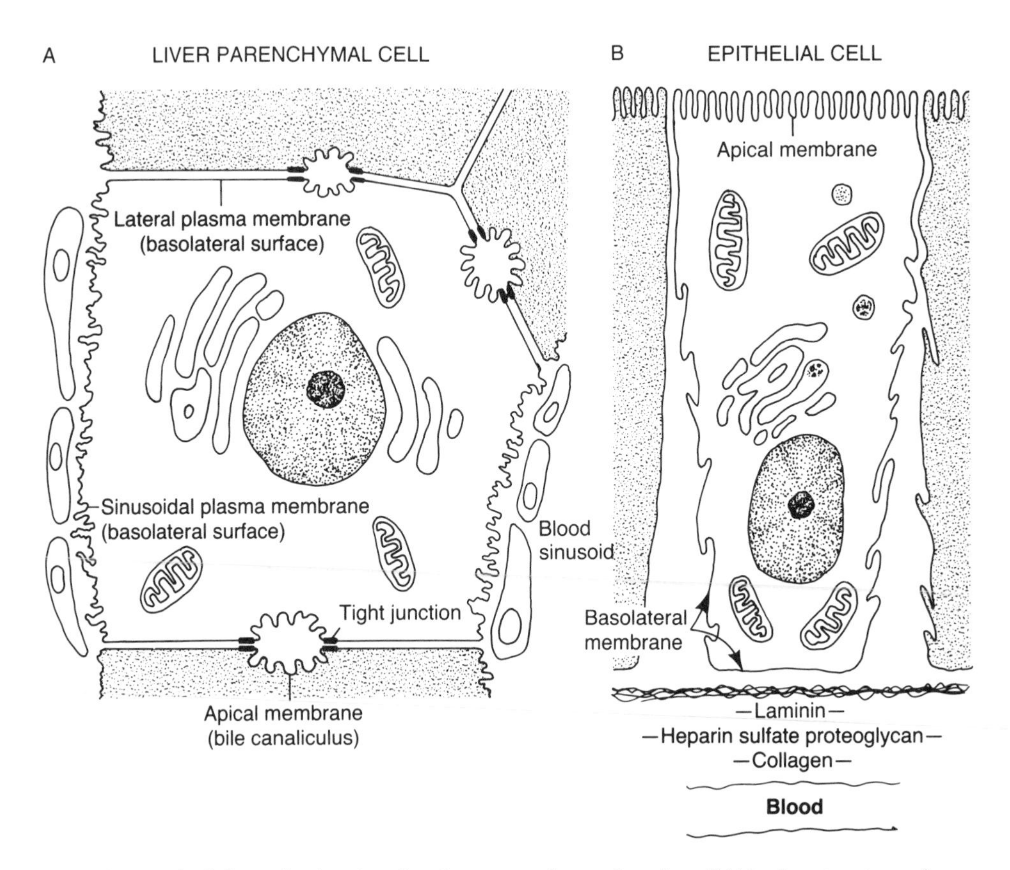

Figure 1.2. Schematic showing the plasma membrane domains of (A) a hepatocyte, and (B) a polarized epithelial cell.

cal features of the nuclear envelope are the pore structures. These nuclear pore complexes have a diameter of about 600 Å and appear to be assembled by morphologically distinct constituents arranged in octagonal symmetry (1483) (see Section 8.22). They are located in regions where the inner and outer nuclear membranes appear to fuse. The pores presumably allow the passage of mRNA–protein complexes from the nucleus to the cytoplasm and of regulatory proteins from the cytoplasm into the nucleus. Little biochemical work has been reported on the nuclear envelope.

(3) *Endoplasmic reticulum* (ER): This is a complex network of cisternae or tube-like structures which occupies a considerable portion of the internal volume of a typical animal cell. The primary purpose is to provide a site for the biosynthesis of proteins destined for secretion, for internalization into lysosomes, or for incorporation into the plasma membrane. Potentially lethal hydrolytic enzymes which are to be secreted outside the cell or sequestered in the lysosome are processed to mature forms inside the ER. Ribosomes are frequently associated with the ER membrane, giving it a rough appearance in electron micrographs (rough ER). The complex processes by which membrane, secretory, or lysosomal proteins are synthesized, matured, and properly delivered to their destinations are described in Chapter 10.

Other portions of the ER are devoid of ribosomes (smooth ER), and these are apparently sites for reactions involving sterol biosynthesis, detoxification reactions, and fatty acid desaturation, all of which involve a complex interactive electron transport system containing cytochrome b_5 and the cytochrome P450 enzymes (see Chapter 6).

(4) *Golgi apparatus*: This organelle appears as a series of tubules and stacked, disk-shaped structures called cisternae. The primary function is the post-translational modification of glycoproteins synthesized initially in the endoplasmic reticulum and destined eventually for secretion or for incorporation in the plasma membrane or for delivery to the lysosome. The organelle contains a number of glycosidases and glycosyltransferases which act sequentially as the protein being processed is passed, probably via vesicles, from one end of the Golgi stack where it enters (*cis*-Golgi), to the other end where it exits (*trans*-Golgi). The Golgi, therefore, is really composed of a series of distinct membranes making up the cisternae, In fact, different sub-fractions can be physically separated which are enzymatically distinct (1248) (see Section 10.2; Figure 10.4). The manner in which membrane and secretory proteins are transported through this system is discussed in Chapter 10.

(5) *Lysosome*: This organelle is responsible for macromolecular degradation, and contains a number of hydrolytic enzymes such as proteases and lipases (316). Materials taken into the cell by endocytosis and phagocytosis which are to be degraded are delivered via vesicles to the lysosome. Also, the breakdown involved in the normal turnover of cellular components is accomplished within the lysosome. The manner in which lysosomal enzymes are synthesized, marked for delivery to the lysosome, and then delivered is one in which some of the steps are understood reasonably well, and is discussed in Chapter 10.

(6) *Peroxisome*: This organelle contains oxidative enzymes involved in the breakdown of small molecules, such as amino acids, xanthine, and, in particular, fatty acids (933). The name derives from the presence of catalase, which breaks down peroxide, a byproduct of the oxidative reactions.

(7) *Mitochondrion*: This organelle is the site of oxidative phosphorylation where ATP is produced at the expense of the oxidation of substrates such as NADH and succinate. The mitochondrion contains two membranes and an intermembranous space. The interior is called the matrix (see Figure 10.1). The inner membrane is invaginated to form a series of septa called cristae, and contains the enzymes involved in electron transport and ATP synthesis. The role of diffusion within the plane of the membrane of the components of the electron transport chain and its functional significance are discussed in Chapter 6. The manner in which proteins which are synthesized in the cytoplasm but end up within one of the mitochondrial compartments or membranes is covered in Chapter 10.

(8) *Chloroplast*: This is the organelle containing the photosynthetic apparatus. It has an outer envelope consisting of two membranes and an interior called the stroma. Within the stroma are the thylakoid membranes, in which the photosynthetic components reside. The thylakoid membranes are closely stacked or appressed in some regions and are unstacked and exposed to the stroma in other places (see Figure 4.8). The compositions of the appressed and stroma-exposed domains of the thylakoid membrane are different, demonstrating lateral heterogeneity (see Section 4.52). The enzymology of the photosynthetic electron transport chain is discussed in Section 6.6.

Each of the membrane systems mentioned above, along with other specialized membranes from animal cells, plant cells, or bacteria, poses a different set of important and interesting research questions and provides opportunities for biochemical research. Other systems will be described in subsequent chapters in the text.

1.2 Historical Perspective

It was recognized in the mid-nineteenth century that the plasma membrane at the surface of cells is a discrete structure. At the turn of the century, Overton noted a correlation between the rate at which various small molecules penetrate plant cells and their partition coefficients between oil and water (1112), leading him to speculate on the lipid nature of the membrane. In 1925 Gorter and Grendel (532) proposed that lipids in the erythrocyte membrane are arranged in the form of a bimolecular leaflet, or lipid bilayer. This conclusion resulted from elegantly simple experiments. The erythrocyte lipids were extracted in acetone and then dispersed in water in a Langmuir trough (see Figure 2.23) so as to form a thin layer at the surface of the water. A thin thread was drawn across the surface, thus compressing the lipid molecules at the air–water interface. At a well defined point, the surface layer offered resistance, and this was interpreted as the point where the lipid layer was a closely packed monomolecular layer. When the measured area occupied by the lipids was compared with the computed area of the erythrocytes from which the lipids were extracted, a 2:1 ratio was found. Thus, it was concluded that the membrane consists of lipids arranged in two layers. Although it is likely that the conclusions of Gorter and Grendel were correct only due to fortuitously offsetting errors (64), this work is historically significant since the concept of the lipid bilayer as the structural basis of the membrane has been dominant ever since and is certainly correct.

The bimolecular lipid membrane was further elaborated in 1935 in the Davson–Danielli or "paucimolecular" model, in which it was postulated that proteins coat the surfaces of the lipid bilayer [279] (Figure 1.3). This was a remarkably successful model, and during the next 30 years numerous experimental results, notably from X-ray diffraction and electron microscopy (see next section), provided overall support. During the same period of time, however, the enormous diversity of membrane functions was becoming clear, and the basic Davson–Danielli model was modified by numerous other workers to account for this functional diversity (e.g., 1387, 425, 880, 1229, 1500).

The rapid development leading to our current view of membranes has largely been due to the progress in characterizing membrane proteins. Freeze-fracture electron microscopy (see next section) revealed globular particles apparently embedded within the membrane (135, 136, 162). Biochemists, meanwhile, were successful at using detergents to dissociate membranes into functional "particles" (542, 543, 1501). Spectroscopic evidence also indicated that membrane proteins had an appreciable amount of α-helix and that they were likely to be globular rather than spread out in a monolayer on the lipid bilayer surface (836, 1543). The non-polar characteristics of membrane proteins (e.g., 543, 1216) also encouraged speculation on hydrophobic contacts between the proteins and the lipid bilayer interior. At the same time, experimental techniques were developed which revealed the fluid nature of the lipid bilayer (468). Singer and Nicolson amalgamated these ideas into the fluid mosaic model, which basically pictures the membrane as a fluid-like phospholipid bilayer into which freely diffusing globular proteins are embedded to varying degrees (1348, 1349). The earlier Davson–Danielli model was a static structural model largely successful at explaining the rather low-resolution structural data available at the time (see next section). In contrast, much of the focus in membrane research since 1970 has been directed at the dynamics of the membrane, and the relationship between this dynamics and membrane function. The emphasis of the fluid mosaic model has undergone modification and will continue to do so. In particular, it is now clear that membrane proteins do not all diffuse freely in the fluid lipid bilayer (690; see Chapter 5), and that there is evidence for differentiated lateral domains within membranes (693; see Chapter 4). The role of the cytoskeleton is under increasing scrutiny (see Chapter 4). There is also a growing belief that some regions of biological membranes may not be arranged in the traditional bilayer (265). Nevertheless, in various modified forms the fluid mosaic model will clearly continue to provide the conceptual backdrop for much of membrane research in the foreseeable future.

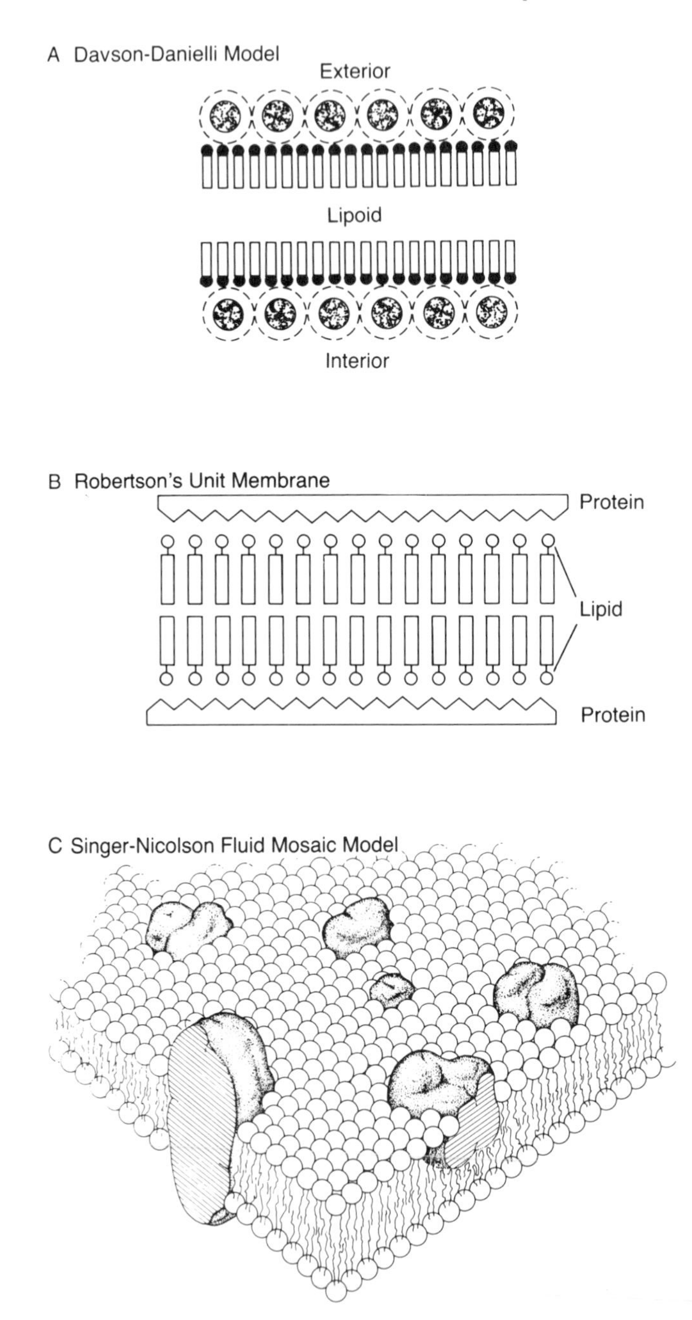

Figure 1.3. Three influential models proposed for biomembrane structure. (A) Davson-Danielli, (B) Robertson's unit membrane, and (C) Singer and Nicolson's fluid mosaic model. See text for details. Adapted from references 279 (Part A), 1230 (Part B) and 1349 (Part C). Drawing for part C provided by Dr. Singer.

1.3 Membrane Morphology

Two methods which have been of historic importance in defining membrane morphology have been X-ray diffraction analysis and electron microscopy. Both techniques have been used to confirm the bilayer model, but both methods are very limited in revealing molecular detail.

1.31 X-Ray Diffraction

X-ray diffraction techniques are capable of yielding high resolution structural information when applied to highly ordered crystalline samples. For less ordered samples, the power of this method is considerably limited. Some specialized membrane systems are naturally stacked in regular arrays and provide useful systems for applying X-ray techniques. Most notable is the myelin sheath of peripheral nerve, which is a membranous system which wraps around the axon many times, providing a regular concentric array of membranes. X-ray diffraction studies on myelin dating back to the 1930s are consistent with the bilayer model (1299, 427). Similar results have been obtained with the rod outer segment of vertebrate retinal cells (108), which also have a naturally stacked membrane system (disks), as well as with artificially stacked membranes formed by centrifuging (and collapsing) vesicular membrane preparations from sources such as the mitochondrion (1449) and the erythrocyte (426). In all cases, the data yield a similar electron density profile across the membrane, pictured in Figure 1.4.

In order to interpret the X-ray data, it is necessary to obtain the phases of the X-ray reflections in addition to the measured amplitudes. This is greatly simpli-

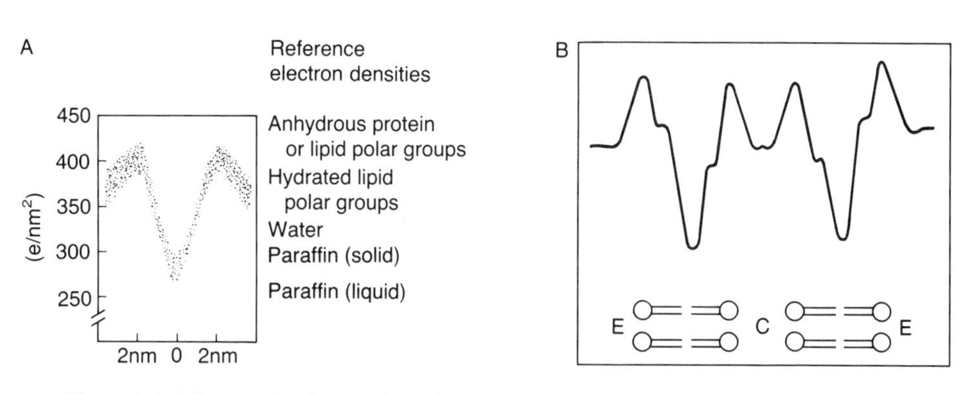

Figure 1.4. Electron density profiles of membranes from X-ray diffraction analysis. (A) Schematic of an electron density profile with calibration standards indicated. (B) The electron density profile of myelin. This represents a centrosymmetric pair of apposed membranes, with "C" indicating the apposition of the pair of cytoplasmic surfaces and "E" the extracytoplasmic surfaces. A schematic of the pair of membranes corresponding to the electron density profile is also indicated. Adapted from references 425a and 1057b.

fied for the stacked membrane systems because they all have repeating units with centrosymmetric symmetry (see 109). The data indicate that all the membranes have a similar structure, with a hydrocarbon interior (low electron density) and polar groups on either side (high electron density). There are only relatively subtle variations between different membranes seen in the X-ray data despite large differences in the protein composition of the various membranes examined, ranging from 20% to almost 80%. It is possible to obtain some information about the relative position of the bulk of the protein mass with respect to the lipid bilayer (e.g., embedded vs peripheral), but basically this X-ray technique does not yield molecular detail.

Wilkins et al. (1586) pointed out in 1971 that X-ray diffraction analysis can also be applied to dispersions of membranes and phospholipids. Reflections arise from the two polar layers on either side of the bilayer, yielding the thickness between polar headgroups (about 36 Å for pure phospholipid) and, from the ordered hydrocarbon chains, yielding the spacing between paraffin chains (about 4.2 Å when highly ordered). Again, the membrane samples from different sources yielded a similar pattern, confirming the universality of the bilayer model.

The lack of molecular detail obtainable from this technique has limited its application to biomembranes. However, these methods have been particularly useful for analyzing the ordered structures of lipid-water systems (1334).

1.32 Electron Microscopy

Transmission electron microscopy of thin sections of myelin and, in fact, virtually all membranes, shows a "trilamellar image," with two electron-dense bands separated by about 80 Å. This image is, in large part, dependent on the commonly used treatment of the samples by osmium tetroxide (956). Although the image, termed the "unit membrane image" by Robertson (1231, 1230), to stress its universality, has been viewed as confirmation of the bilayer model, the molecular basis of the osmium staining pattern is not known. It is clear, however, that the techniques used to prepare membrane samples for transmission electron microscopy can be damaging. In particular, osmium tetroxide treatment is known to cause extensive loss of protein (956) from the erythrocyte membrane. The trilamellar image in some way reflects the basic bilayer structure of membranes, but further molecular details of the location of the proteins are not obtainable.

The more recent (by now "classical") techniques of freeze-cleaving and freeze-etching yield some information about membrane proteins. The samples are rapidly frozen and not subjected to the deleterious procedures used for preparing thin sections. As pictured in Figure 1.5, the procedure involves the following steps (see 432):

1. After freezing, the sample, which may consist of a suspension of cells or membranes, is fractured with a knife at low temperatures (-100°C) and under high vacuum. This creates shear forces which result in a shear plane passing

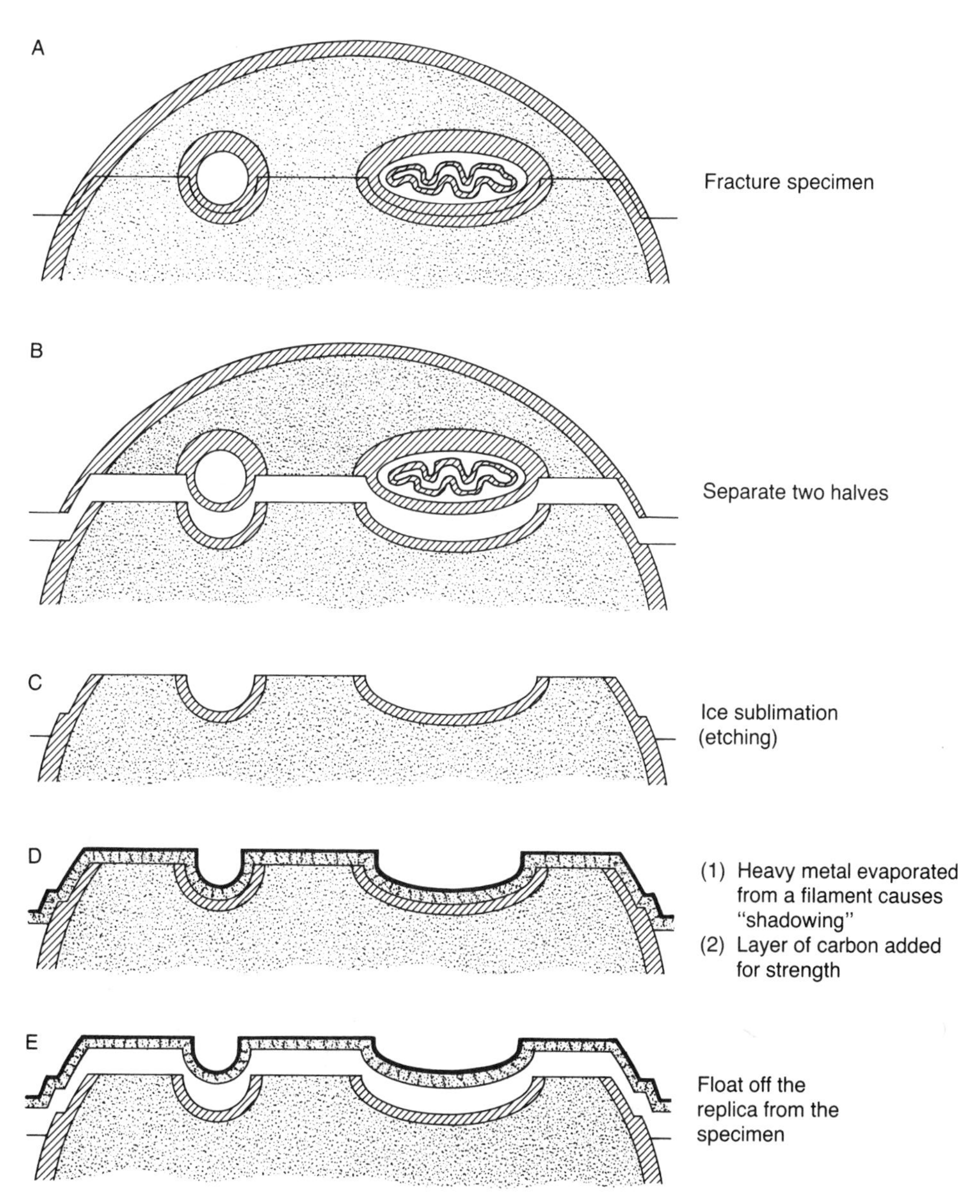

Figure 1.5. Schematic illustrating the freeze-cleavage technique. (A) A frozen cell is split by a cleavage plane which traverses, in part, through the middle of various membranes. (B) The two halves are separated. (C) The sample may be etched, exposing surface features. (D) A platinum layer is shadowed onto the sample, followed by a layer of carbon. This creates a replica of the surface of the sample. (E) The replica is removed and then examined by electron microscopy. Adapted from ref. 698. Introduction to Biological Membranes, by M. K. Jain and R. C. Wagner, Copyright © 1980. Reprinted by permission of John Wiley & Sons, Inc.

through the sample. Experience has shown that when the shear plane passes through a membrane it preferentially passes through the middle, thus dividing the membrane in half. The interior of the membrane is now exposed in these places.
2. If desired, the sample can be freeze-etched by simply allowing ice to sublime under vacuum. This results in exposing the surface structures of the cellular membranes.
3. A replica of the surface is now constructed. It is this replica which is actually observed in the electron microscope. The replica is produced by first shadowing the sample with platinum at about a 45° angle to highlight topological features. The platinum replica is given mechanical strength by depositing a layer of carbon on the sample. The sample is now thawed and the replica is floated off and picked up on an appropriate grid for examination.

The most significant features visualized by the freeze-cleavage studies are numerous intramembranous particles, about 80 Å to 100 Å in diameter, in the plane of the cleaved membranes. They are usually randomly dispersed but may also be aggregated in groups. Considerable study has shown that these particles correlate with membrane proteins. There are no corresponding features found in thin section electron microscopy. The topological features seen in the two halves of the cleaved membrane are not necessarily complementary, indicating that some of the particles are associated primarily with one or the other half of the bilayer. Freeze-fracture studies were prominently cited by Singer and Nicolson in their formulation of the fluid mosaic model, since these studies strongly implied that globular proteins reside in the interior of the bilayer and are not constrained to the bilayer surface (1349).

Box 1.1 An Example of Particles Visualized by Freeze-Cleavage

Figure 1.6 shows the results of freeze-cleavage of liposomes composed of egg yolk lecithin reconstituted with a crude preparation of Band 3, a membrane protein from the human erythrocyte (1625, 1626). Band 3 is a major protein component of the red blood cell membrane and it is known to function as an anion transporter (see Section 8.33). In the absence of this protein, the phospholipid vesicles present a smooth surface in freeze-cleaved preparations. Reconstitution of Band 3 in the phospholipid results in the appearance of particles which are indistinguishable from those observed in the red blood cell membrane (1625, 1626). Furthermore, it is known that at pH 5.5, the particles in the red cell aggregate within the plane of the membrane and that this aggregation is dependent on an interaction between the Band 3 protein with two other proteins, spectrin and actin. These latter two proteins are part of the cytoskeletal network found on the inner surface of the erythrocyte membrane (see Chapter 4). The reconstituted Band 3-lecithin system manifests similar behavior, and clumping of the particles is observed in the presence of added spectrin and actin at pH 5.5, but not at pH 7.6 (1625, 1626).

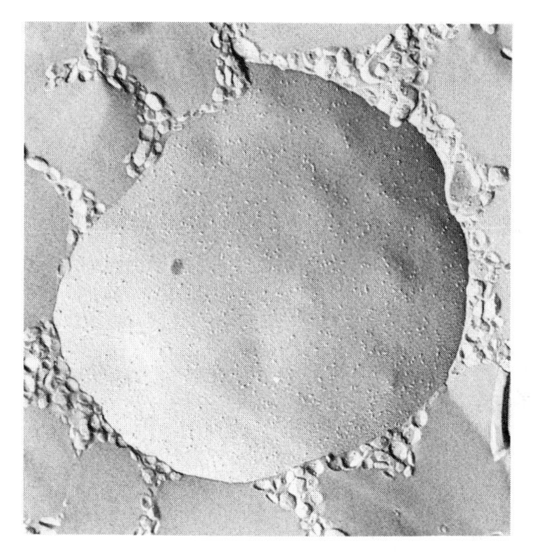

Figure 1.6. Freeze-fracture electron micrograph of reconstituted vesicles composed of egg phosphatidylcholine and a crude preparation of Band 3 from the human erythrocyte. The particles observed here resemble those seen in similar preparations of erythrocyte membrane. From ref. 1625.

These kinds of data helped to strengthen the concept that membrane proteins are globular units free to diffuse laterally in the plane of the membrane. Ironically, the static electron micrographs from freeze-cleave studies contributed to the appreciation of the dynamic potential of the membrane. We shall see later in Chapter 5 that there are many proteins which are not freely diffusing in the "sea of lipid," including Band 3.

1.4 Membrane Isolation

There has been a growing appreciation over the past 3 decades of the enormous number of cellular functions which are membrane-dependent. Both plant and animal cells are highly compartmentalized, and many of the cytoplasmic organelles are membranous, as outlined in Section 1.1. In addition to the organelles indicated for the generalized case, there are specialized membranous systems such as the sarcoplasmic reticulum of muscle cells, the myelin sheath around peripheral nerve axons, the chloroplast thylakoid membrane, and the disk membranes of the rod cells in the retina. Prokaryotic organisms also have membranes, although the internal elaborations found in the eukaryotic systems are generally not present. Gram-positive bacteria such as *Bacillus subtilis* have a single cytoplasmic membrane, whereas gram-negative bacteria such as *Escherichia coli* have, in addition, an unusual outer membrane external to a thin peptidoglycan cell wall (see Section 4.42). Some specialized organelles are also found in prokaryotic organisms, notably the chromatophore containing the photosynthetic apparatus in purple non-sulfur bacteria such as *Rhodobacter sphaeroides*. Some animal cell viruses, known as enveloped viruses, also are surrounded by true membranes, and these have been quite useful systems for study.

Most studies on membranes require as a prerequisite the purification of the particular membrane to be examined. Each system presents unique preparative problems. For example, if one is interested in studying the plasma membrane of a particular cell population (e.g., hepatocytes of the liver) it is obviously advantageous to first isolate these cells from the whole tissue. Then, one must consider the optimal procedures for cell disruption and for physically separating the membrane of interest from the other cellular components. An important consideration is the criterion used to assess the purity of the membranes obtained.

1.41 Cell Disruption

It is desirable to select a procedure which will effectively disrupt the cell without destroying the membrane structure to be isolated. In the case of many animal cells, relatively gentle procedures can be used, such as a Dounce or Potter–Elvehjem glass–Teflon type of tissue homogenizer (e.g., 436). This disrupts the cells by shear forces by forcing the suspension through a narrow gap between a Teflon plunger and the glass wall of the apparatus. This should strip off the plasma membrane and sever the connections between the various organelles but still maintain the integrity of the individual organelles. Specialized regions of the plasma membrane, such as the basolateral and apical membranes of epithelial cells, can also be severed by these procedures. It is usually desirable to work under conditions in which the organelles remain intact to minimize release of hydrolytic enzymes (e.g., from lysosomes) and to optimize subsequent separation procedures.

Harsher procedures are required to disrupt cells which have walls, such as bacteria, fungi, and plant cells. Sometimes the cells are pre-treated with degradative enzymes prior to physical disruption to assist in breaking the cell wall. For example, Tris-EDTA and lysozyme treatments can be used when disrupting *E. coli* (1102). The more harsh disruption techniques rely on grinding, sonication, and extrusion. Grinding is usually done in the presence of an abrasive such as sand, alumina, or glass beads. Small-scale work can be done with a mortar and pestle, but mechanical devices can also be used. Sonication is often used for breaking bacterial cells. Presumably, this technique works by creating shear forces in solution produced by cavitation. Shear forces are also produced by extruding the cell suspension through a small orifice, e.g., as occurs with the French press. There are many variations on these techniques and the choice will depend on the particular system being examined.

It should be noted that disrupted and fragmented membranes will usually spontaneously form vesicles. Examples (see 435) include (1) microsomes derived from plasma membrane, endoplasmic reticulum, or specialized systems such as the sarcoplasmic membrane of muscle cells; (2) submitochondrial particles, from inner mitochondrial membrane; (3) synaptosomes, derived from pinched-off nerve ends at synaptic junctions; and (4) bacterial membrane vesicles (Kaback vesicles) from the cytoplasmic membrane of *E. coli*. Other membrane systems such as

Golgi also vesiculate. In most cases, the size of the vesicles is critically dependent on the method used to disrupt the cells. Since the vesicle size in large part determines the sedimentation rate (see next section) and behavior in subsequent purification steps, the disruption step is of obvious importance. Some membranes do not form vesicles, notably the lateral or contiguous membranes of animal cells (see Figure 1.2), which are stripped off as pairs of adjacent membranes derived from neighboring cells, held together by junctions. The presence of these junctions prevents vesicle formation, and these membranes are isolated as sheets or ribbon-like structures (402).

The choice of medium used for cell disruption can also be important. For example, in order to maintain the structure of sealed membranous organelles, it is important to use a breakage medium which is iso-osmotic with the organelle interior. Sucrose (0.25–0.30 M) is most commonly used for this purpose, but sorbitol and mannitol are also utilized and in some cases favored (e.g., 52). It should be noted that the subsequent preparative steps for intact organelles are usually critically dependent on the maintenance of isotonic conditions.

1.42 Membrane Separations

By far the most widely used technique for membrane separations is centrifugation. (See Box 1.2; refs. 1209, 1459.) Particles can be separated from each other on the basis of their sedimentation rate or on the basis of differences in their buoyant density. The former is called S-value or zonal centrifugation, whereas the latter is equilibrium density or isopycnic centrifugation. In practice, separations are actually based on a hybrid of these two methods. Figure 1.7 shows the location of a number of subcellular particles in "S-ρ space." On the axis are plotted the sedimentation coefficients of the "particles," and on the ordinate is the density. Separations based on sedimentation rate are clearly pictured by comparing S-values along the axis. For example, nuclei have a relatively high S-value, indicating that the sedimentation rate is substantially higher than those of most other subcellular organelles. Nuclei can be pelleted by differential sedimentation of cellular homogenates leaving the other organelles in the supernatant. On the other hand, smooth and rough endoplasmic reticulum cannot be separated by zonal sedimentation.

In order to separate different membrane fractions from a cellular homogenate, it is often necessary to take advantage of the differences in density. This is done by centrifuging in a density gradient of a centrifugation medium (1178). Most often, sucrose density gradients are utilized. There are, however, serious drawbacks to the use of sucrose. In order to attain the density required to separate various membrane fractions, it is necessary to use high concentrations of sucrose which are both highly viscous and hypertonic. Exposure of subcellular organelles to hypertonic sucrose solution results in dehydration and, often, readjustment of the solution afterwards to isotonic conditions results in lysis and damage of the organelle (e.g., 1020). Another problem is that many membrane organelles are

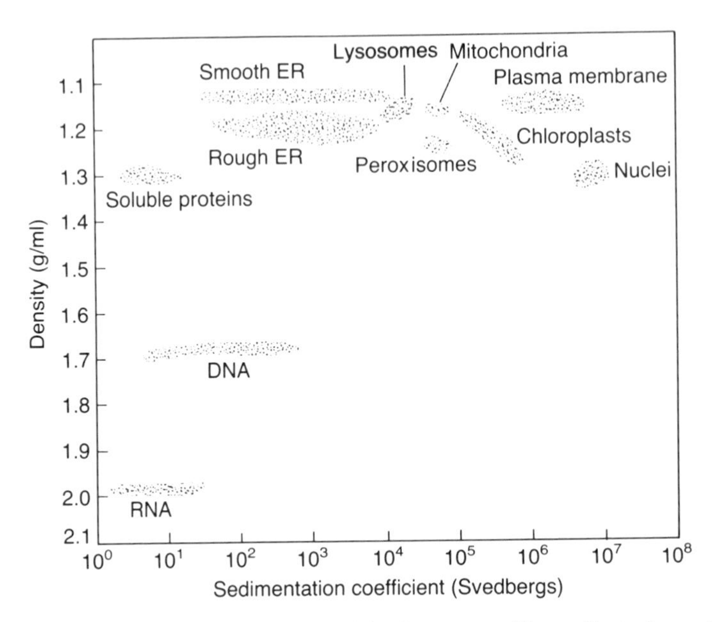

Figure 1.7. Subcellular particles displayed in S-ρ space. The ordinate is equilibrium density and the abscissa is the apparent sedimentation coefficient on a logarithmic scale. Note that the actual values will vary in different gradient media. The values for RNA and DNA apply to CsCl gradients. Adapted from ref. 1178.

permeable to sucrose. This can also result in osmotic disruption of the organelle. Penetration by sucrose can also alter the effective density of the particles being separated (1140).

To overcome some of these drawbacks, other density gradient media are becoming increasingly utilized. Some are listed in Table 1.1.

(1) *Ficoll*: A high molecular weight (ca. 400,000) hydrophilic polymer of sucrose which can be used to cover a density range up to about 1.2 g/ml. A major advantage is low osmotic pressure compared to the equivalent concentration of sucrose (w/v %). Density gradients can, thus, be constructed which are isotonic throughout by including sucrose (0.25 *M*) or physiological salts in the medium. Disadvantages include high viscosity and very non-linear dependence of the viscosity and osmolarity on concentration (1039).

Table 1.1. Physical properties of gradient media

	Concentration (% w/v)	Density (g/ml)	Viscosity (cP)	Osmolality (mOs/kg H_2O)
Sucrose	20	1.06	30	700
Metrizamide	30	1.16	2	260
Ficoll	30	1.10	49	130
Percoll	26	1.13	10	10

Data from ref. 1140.

(2) *Metrizamide*: A triiodinated benzamide derivative of glucose (mol. wt. 789) (1039). At equivalent concentrations, metrizamide solutions are more dense than Ficoll solutions. A major advantage is very low viscosity to facilitate rapid separation. At a concentration of 35% metrizamide, the solution has approximate physiological osmolarity, so most separations can be performed without exposing the membranes to hypertonic solutions. *Sodium metrizoate* is a related compound with similar properties but with the difference that it is isotonic at about 20% (w/v). This is used primarily for isolating intact cells. *Nycodenz* is also a derivative of triiodobenzoic acid with three hydrophilic side chains. Upon centrifugation, it will form its own density gradient rapidly, and is used for the preparation of subcellular organelles (e.g., 403) (Accurate Chemical and Scientific Corp., Westbury, N.Y.)

(3) *Percoll*: A colloidal suspension of polyvinylpyrrolidine (PVP)-coated silica (1140). The coating mitigates many of the toxic properties of the silica gel. Major advantages are that Percoll does not penetrate biological membranes, and solutions have low viscosity and low osmolarity. Because of the large particle size, centrifugation at moderate speeds results in self-generation of the Percoll density gradient (e.g., 30,000 × g for 30 minutes). Hence, separations are usually very rapid. Centrifugation medium can be made isotonic throughout by inclusion of salts or sucrose. Shallow gradients are easily generated which result in very high resolution of membrane fractions based on buoyant density (e.g., 38, 914, 680).

(4) *Sorbitol and mannitol*: Sometimes used in place of sucrose because they reportedly do not penetrate some biological membranes as readily as does sucrose (1179).

Note that glycerol is not used as a density gradient medium because sufficiently high densities cannot be attained. Alkali salts such as CsCl are used only when high densities are required. The salt concentrations required for equilibrium density are often deleterious.

Other techniques are also used for separating membranes from cell homogenates, though much less frequently than centrifugation.

(1) *Phase partitioning* (14, 15): Separates membranous particles according to their surface properties. Two (or three) immiscible aqueous layers are formed by mixing different water-soluble polymers with water. Examples are polyethylene glycol–dextran and dextran–Ficoll. The particles separate according to their relative affinities for the phases. The phases can be designed to separate species based on such properties as surface charge or hydrophobicity.

(2) *Continuous free-flow electrophoresis* (583, 584, 403): Separates on the basis of electrical charge. The sample is added continuously to a thin film of buffer flowing vertically. An electric field is applied perpendicular to the direction of flow. Particles are separated by electrophoresis across the flowing sheet of buffer, which is collected in a series of fractions at the bottom of the separation chamber.

(3) *Affinity adsorption*: Separation is based on a biospecific interaction between a membrane component and a solid phase. The advent of monoclonal antibodies, in particular, has made it possible to design preparative protocols based on the presence of a particular antigenic component in the membrane of interest. Once a monoclonal antibody preparation is available, the antibodies can

be covalently attached to a solid support and used to specifically bind to the membrane of interest. More frequently, this method is used to isolate a particular membrane protein (e.g., 1613). Problems encountered include the necessity of eluting the membranes without denaturation.

(4) *Silica microbeads* (553): Plasma membranes can represent as little as 1% of the total membrane mass of a eukaryotic cell. Consequently, the isolation of plasma membrane free of contaminating membranes can be a major problem. One approach specifically utilized for plasma membranes is to coat the intact cell (or protoplast) with a layer of cationic microbeads. These bind tightly to the outer surface of the plasma membrane and the plasma membrane sheets bound to the beads are easily separated on a sucrose gradient from all other contaminants by virtue of the high density of the beads. An additional feature is that the cytosolic face of the plasma membrane is exposed to solution in the final preparation.

Box 1.2 Sedimentation Velocity and Sedimentation Equilibrium

The rate at which a particle sediments in a centrifugal field is given by its sedimentation coefficient or S-value. This essentially is a measure of the steady-state velocity of the particle per unit of applied force:

$$S = \frac{dr/dt}{\omega^2 r} \qquad [1.1]$$

where S = sedimentation coefficient, usually given in Svedberg units (1 Svedberg = 10^{-13} sec)

r = distance from the center of the rotor

ω = circular velocity of the rotor, rotor rpm × ($2\pi/60$) sec^{-1}

The S-value is determined by the size, shape, and density of the sedimenting particle and by the density and viscosity of the medium.

$$S = \frac{M(1 - \bar{V}\rho)}{Nf} \approx \frac{M(1 - \rho_{solution}/\rho_{particle})}{Nf} \qquad [1.2]$$

where M is the molecular weight of the particle

$\bar{V}$ is the partial specific volume of the particle or approximately the inverse of the particle density

$\rho_{particle}$ is the density of the particle

$\rho_{solution}$ is the solution density

N is Avogadro's Number

f is the frictional coefficient, which measures the amount of friction due to the particle moving through the solution

Note that the tabulated S-values, $S_{20,w}$, are all given for standard solution conditions: water at 20°C. This equation says that the rate of sedimentation is directly proportional to the mass of the particle which has been corrected for buoyancy by subtracting the mass of displaced solution, and is inversely proportional to the frictional resistance encountered. This frictional coefficient contains information about size and shape of the particle. For spherical particles, the frictional coefficient $f = 6\pi\eta R$, where η is the solvent viscosity and R is the radius of the particle. Everything else being equal, any shape other than a sphere will have a higher frictional coefficient and, therefore, sediment more slowly.

Differential sedimentation takes advantage of the differences in S-values of particles to be separated. If these differences are large, one class of particles will pellet at the bottom of the centrifuge tube, and the other particles will remain in the supernatant. The differences in S-values may result from differences in particle mass, shape, or density or, as is usual, a combination of all three. For example, whole cells and nuclei are removed from disrupted membranes by pelleting. Mitochondria can similarly be separated from microsomes (see Figure 1.7).

Density gradient centrifugation is required when simple pelleting will not suffice. In this case, the density of the medium is varied from most dense at the bottom of the tube to least dense at the top. The gradient may be in any form, and the most commonly used gradients are either linear or step gradients. The gradient serves several purposes. It stabilizes the solution in the centrifuge tube against convective disturbances. This allows one to separate stable zones of material which have sedimented at different rates by stopping the centrifuge before pelleting has occurred or equilibrium has been reached. Importantly, density gradient centrifugation also allows one to do isopycnic centrifugation. In this case the particles sediment to a point in the tube where $\rho_{particle} = \rho_{solution}$, at which the sedimentation velocity is zero (see Equation 1.2).

Note that if the particle is more dense than the surrounding medium, it will sink ($S > 0$) in the centrifugal field. If the particle, however, is less dense, then $S < 0$, and the particle will rise. This phenomenon is called *flotation* and is frequently used to isolate membranes. When membranes are separated by isopycnic centrifugation, one is exploiting the differences in membrane density, which are usually determined by the protein/lipid ratio. For example, rough endoplasmic reticulum with attached polysomes is more dense than the smooth endoplasmic reticulum. Bacterial inner and outer membranes from gram-negative organisms also are clearly separated on the basis of density due to compositional differences.

Often membrane purification protocols are empirically optimized and represent a hybrid separation due to S-value differences and density differences. In a single density gradient some particles (or bands) may be at equilibrium (isopycnic) whereas others may not be equilibrated at the time the centrifugation is stopped.

1.43 Criteria of Membrane Purification

By far the most critical assessment of membrane purity is to assay for particular components known to be present uniquely or predominantly in the membrane being purified. Usually, these are enzymes, called marker enzymes. Examples of marker enzymes which have been used to monitor membrane purification are shown in Table 1.2. In performing these assays one must be aware of the potential problem of enzyme latency, which can result if an enzyme is inside of a vesicular membrane and unable to interact with its substrate. Other potential problems are discussed in ref. 436, but in many cases the procedures have been standardized.

Table 1.2. Markers used to monitor membrane purification from mammalian cells.[l]

Cell fraction	Marker enzyme activity
Plasma membranes	5'-Nucleotidase
	Alkaline phosphodiesterase
	Na^+/K^+-ATPase (basolateral in epithelial cells)
	Adenylate cyclase (basal in hepatocytes)
	Aminopeptidase (brush border in epithelial cells)
Mitochondria (inner)	Cytochrome *c* oxidase
	Succinate-cytochrome *c* oxidoreductase
Mitochondria (outer)	Monoamine oxidase
Lysosomes	Acid phosphatase
	ß-Galactosidase
Peroxisomes	Catalase
	Urate oxidase
	D-Amino acid oxidase
Golgi	Galactosyltransferase (see Figure 10.4)
Endoplasmic reticulum	Glucose-6-phosphatase
	Choline phosphotransferase
	NADPH-cytochrome *c* oxidoreductase
Cytosol	Lactate dehydrogenase

[l]From refs. 436, 698, and 404.

In some cases, the most convenient membrane markers are not enzymes but receptors for specific lectins, hormones, toxins, or antibodies. For well characterized systems, the protein profile on sodium dodecyl sulfate (SDS)-polyacrylamide gel electrophoresis can be used to indicate purity. For example, the outer membrane of gram-negative bacteria has a distinctive set of polypeptides not found in the cytoplasmic membrane (885). Other features used to evaluate a membrane preparation include morphology, as determined by electron microscopy, and gross compositional characteristics. For example, subcellular membrane preparations containing plasma membrane, Golgi, or mitochondria can be distinguished based on morphology. Cholesterol content can, in some cases, be used to evaluate a preparation. For example, mitochondria have a much lower cholesterol content than does Golgi or plasma membrane.

1.5 Composition of Membranes

The major components of membranes are proteins and lipid. Carbohydrate may comprise as much as 10% of the weight of some membranes, but the carbohydrate is invariably in the form of glycolipid or glycoprotein. The relative amounts of protein and lipid vary significantly, ranging from about 20% (dry weight) protein (myelin) to 80% protein (mitochondria). Tables 1.3 and 1.4 summarize the com-

Table 1.3. Composition of subcellular membranes from rat liver[1]

	Percentage of total phospholipid					
	Mitochondria	Microsomes	Lysosomes	Plasma membrane	Nuclear membrane	Golgi membrane
Cardiolipin	18	1	1	1	4	1
Phosphatidylethanolamine	35	22	14	23	13	20
Phosphatidylcholine	40	58	40	39	55	50
Phosphatidylinositol	5	10	5	8	10	12
Phosphatidylserine	1	2	2	9	3	6
Phosphatidic acid	—	1	1	1	2	<1
Lysophosphoglycerides[2]	1	11	7	2	3	3
Sphingomyelin	1	1	20	16	3	8
Phospholipids (mg/mg protein)	0.175	0.374	0.156	0.672	0.500	0.825
Cholesterol (mg/mg protein)	0.003	0.014	0.038	0.128	0.038	0.078

[1]Data from ref. 284. Additional tables of lipid compositions are found in ref. 1570. Endosomes are reported to have a composition similar to that of the plasma membrane (404).

[2]High values of lysophosphoglycerides should be viewed with caution, since this could result from breakdown during preparation.

Table 1.4. Protein and lipid composition of some animal cell and bacterial membranes. *L/P* is the ratio of lipid/protein (dry weight).[l]

Membranes	Major proteins	*L/P* (w/w)	Major lipids
Myelin (human)	Basic protein Lipophilin (proteolipid)	3–4	PC 10% PE 20% PS 8.5% SM 8.5% Ganglioside 26% Cholesterol 27%
Disk membranes (bovine)	Rhodopsin	1	PC 41% PE 39% PS 13% Trace of cholesterol
Erythrocytes (human)	Band 3 Glycophorin Spectrin Glyceraldehyde-3 phosphate dehydrogenase	0.75	PC 25% PE 22% PS 10% SM 18% Cholesterol 25%
Rectal gland plasma membrane (dogfish)	Na^+/K^+-ATPase (I)		PC 50.4% PE 35.5% PS 8.4% PI 0.5% SM 5.7% Cholesterol
Cholinergic receptor membranes (*Torpedo marmorata*)	Acetylcholine receptor	0.7–0.5	PC 24% PE 23% PS 9.6% Cholesterol 40%
Sarcoplasmic reticulum (rabbit)	Ca^{2+}-ATPase	0.66–0.7	PC 66% PE 12.6% PI 8.1% Cholesterol 10%
E. coli (inner membrane)		0.4	PE 74% PG 19% CL 3%
Purple membrane (*Halobacterium halobium*)	Bacteriorhodopsin	0.2	Phosphatidyl-glycerophosphate 52% Glycolipids 30% Neutral lipids 6%

[l] The major proteins in these membranes have been intensively studied and are discussed in other portions of the book. Abbreviations used: PC, phosphatidylcholine; PE, phosphatidylethanolamine; PS, phosphatidylserine; SM, sphingomyelin; PI, phosphatidylinositol; PG, phosphatidylglycerol; CL, cardiolipin. Data from ref. 330.

position of a number of membranes. The density of a membrane is directly proportional to the amount of protein in the membrane. Higher protein composition results in increased density as determined by isopycnic centrifugation.

To some extent, the protein components associated with a membrane will depend on the procedures used to isolate the membrane. A number of proteins are not strongly associated with membranes and can easily be removed by such procedures as washing at high or low ionic strength, washing at alkaline pH, or including a chelator such as EDTA in the buffer. In some cases, it is difficult to distinguish proteins which should properly be considered as membrane components from cytoplasmic proteins which may bind adventitiously to the membrane surface during the isolation procedures.

1.51 Membrane Lipids

The most striking feature of membrane lipids is their enormous diversity. The reason for the diversity is not at all clear, although there is an increasing awareness of the multiple roles of lipids in membranes (see Section 1.52). Certainly the major role of membrane lipids is to form the bilayer matrix with which the proteins interact. The major lipid classes are pictured in Figure 1.8 and are briefly discussed below.

Glycerophospholipids

These are the most commonly found membrane lipids. One of the glycerol hydroxyls is linked to a polar phosphate-containing group and the other two hydroxyls are linked to hydrophobic groups. Glyceride nomenclature is often in terms of the stereospecific numbering (*sn*) system. When the glycerol is drawn in a Fischer projection, with the hydroxyl in the middle drawn to the left, the positions are numbered as shown in Figure 1.9, and the prefix *sn*- is used before the name (e.g., *sn*-3 position). Several different stereochemical conventions are used: *sn*, D/L, and *R*/*S*. Figure 1.9 also illustrates the stereochemistry about carbon atom C-2 in the three conventions (see ref. 604). Natural phospholipids generally have the *R* (or D) configuration.

Most phosphoglycerides have the phosphate at the *sn*-3 position of glycerol. The phosphate is usually linked to one of the several groups as indicated in Figure 1.10, including choline, ethanolamine, *myo*-inositol, serine, and glycerol.

The long-chain hydrocarbons attached to *sn*-1 and *sn*-2 positive may be attached through ester or ether linkages. The chains themselves vary widely in terms of length, branching, and degree of unsaturation.

(a) *1,2-Diacylphosphoglycerides or phospholipids*. These fatty acid esters of glycerol are the predominant lipids in most eukaryotic and prokaryotic membranes, excluding archaebacteria (1296). Phosphatidylcholine is a major component in animal cell membranes, and phosphatidylethanolamine is often a major component in bacterial membranes. Table 1.5 lists a number of the more common

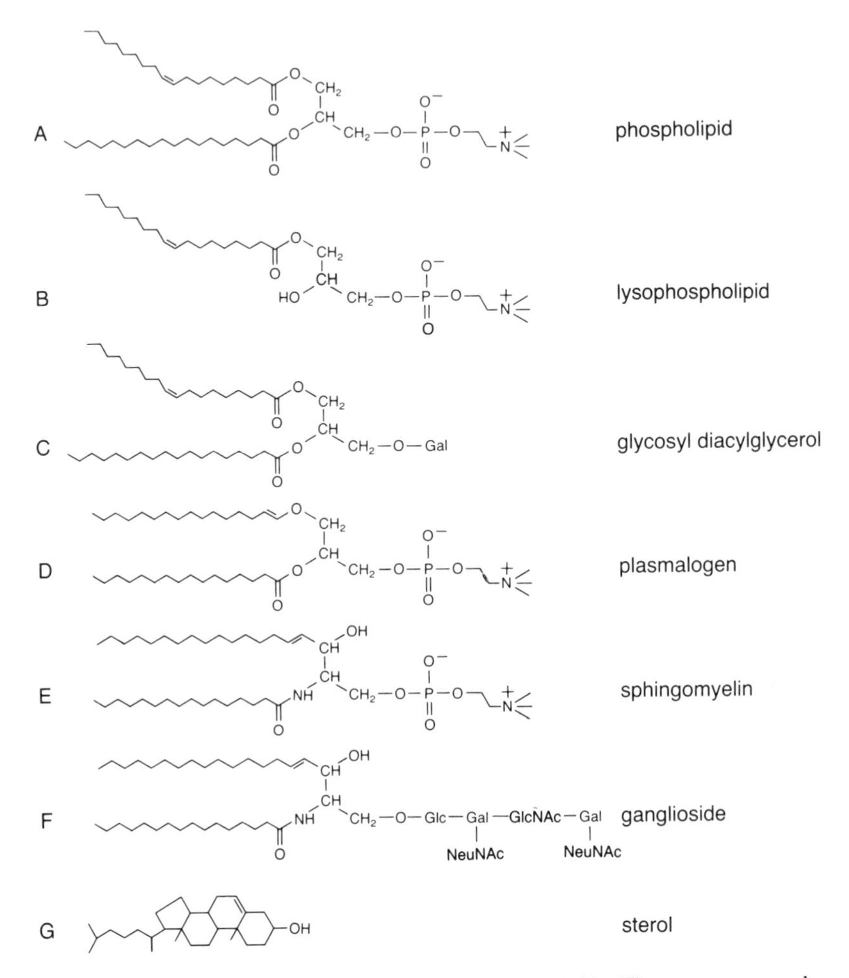

Figure 1.8. The structures of some classes of membrane lipids. The structures are drawn so as to emphasize the amphipathic nature of the lipids, with nonpolar groups on the left and polar moieties on the right. Gal, galactose; Glc, glucose; NeuNAc, *N*-acetylneuraminic acid (sialic acid); GlcNAc, *N*-acetylglucosamine. Adapted from ref. 658. Copyright 1982, Reprinted by permission of John Wiley & Sons, Ltd.

Figure 1.10. Structures of membrane lipids, illustrating the variety of polar head groups. At neutral pH, the amino group of the ethanolamine moiety will be protonated.

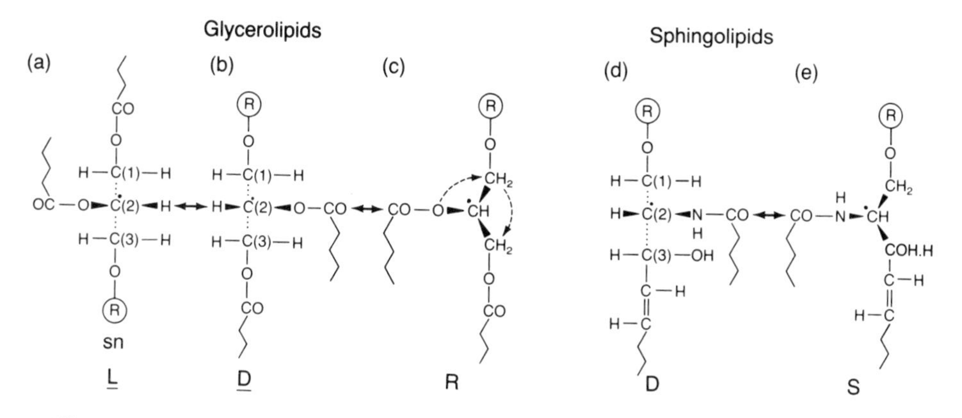

Figure 1.9. Stereochemical conventions for glycerolipids and sphingolipids. The *R*/*S*, D/L, and *sn* nomenclatures are illustrated for the C(2) positions. Adapted from ref. 604.

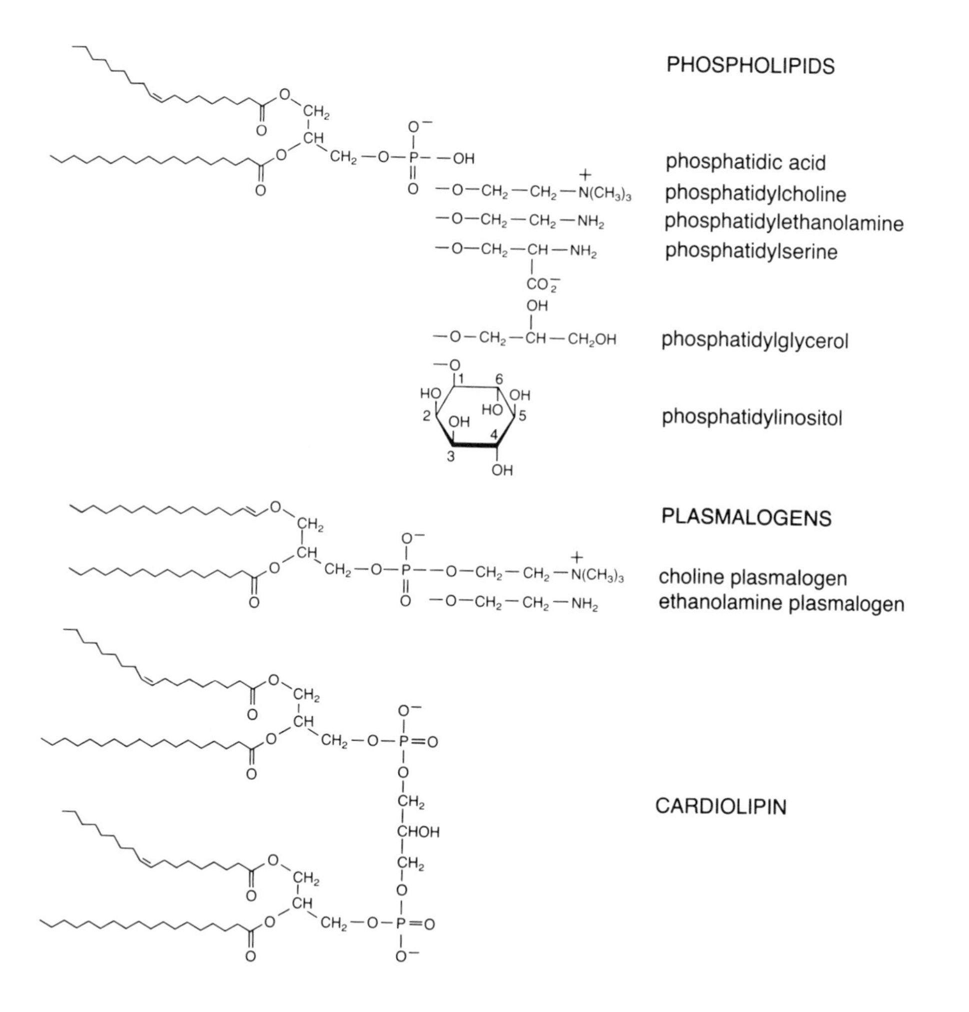

Table 1.5. Some fatty acids commonly found in membrane lipids. The common names are given.

Common name	Chain length: unsaturation
Lauric	12:0
Myristic	14:0
Palmitic	16:0
Palmitoleic	16:1 (9-*cis*)
Stearic	18:0
Oleic	18:1 (9-*cis*)
Vaccenic	18:1 (11-*cis*)
Linoleic	18:2 (9-*cis*, 12-*cis*)
γ-Linolenic	18:3 (6-*cis*, 9-*cis*, 12-*cis*)
α-Linolenic	18:3 (9-*cis*, 12-*cis*, 15-*cis*)
Arachidic	20:0
Behenic	22:0
Arachidonic	20:4 (5, 8, 11, 14-(all)*cis*)

fatty acids found in phospholipids, and Table 1.6 gives the fatty acid compositions of organelle membranes from rat liver mitochondria. The acyl chains nearly always have an even number of carbons, ranging from C14 to C24. Most common are C16, C18, and C20. The degree of unsaturation varies widely, but the most common unsaturated species are 18:1, 18:2, 18:3, and 20:4. In this notation, the first number indicates the chain length and the second figure is the number of double bonds. Nearly all naturally occurring double bonds are *cis* rather than *trans*. This places a kink in the molecule which is generally disruptive of ordered packing of the chains in the bilayer (see Chapter 2). Many phospholipid molecules have one saturated and one unsaturated chain. In animal cells, the unsaturated chain is usually found esterified to the *sn*-2 position of glycerol, and the same is true in *E. coli* (see Figure 10.16). Polyunsaturated chains are generally not conjugated. Branched chains, cyclic chains (e.g., cyclopropane-containing chains), and ß-OH groups are found in some bacterial membranes. Figure 1.11 gives the structures of some of these less common fatty acid structures (1262).

(b) In *archaebacteria* the stereoconfiguration of the glycerophospholipids is reversed, with the phosphoryl groups on the *sn*-1 position of the glycerol (1296). In many of these bacteria the hydrophobic constituents are isopranyl glycerol ethers rather than fatty acid esters (1296, 719, 1007, 307, 888) (see Figure 1.11).

(c) *Cardiolipids or diphosphatidylglycerols* (see Figure 1.10). These are essentially dimeric phospholipids. They are a significant component of the mitochondrial inner membrane, chloroplast membrane, and some bacterial membranes, but are rare in other membranes.

(d) *Plasmalogens*. These are phosphoglycerides where one of the hydrocarbon chains is linked via a vinyl ether linkage (see Figures 1.9 and 1.10). Ethanolamine plasmalogens are an important component of myelin and of the cardiac sarcoplasmic reticulum (554).

Table 1.6. Fatty acid compositions of some membranes from rat liver[1]

Membrane fraction	Fatty acids (as % total by weight)														
	14:	15:	16:	16:1	17:	18:	18:1	18:2	18:3	20:	20:1	20:2	20:3	20:4	22:6
Mitochondrial (outer)	0.4	27.0	4.1	21.0	13.5	13.5							1.1	15.7	3.5
Mitochondrial (inner)	0.3	27.1	3.6	18.0	16.2	15.8							1.0	18.5	3.8
Plasma membrane	0.9	36.9		31.2	6.4	12.9	tr	tr						11.1	
Smooth ER	0.4	28.6	3.1	26.5	10.6	14.9							1.4	14.0	0.7
Rough ER	0.5	22.7	3.6	22.0	11.1	16.1							1.8	19.7	2.9
Golgi	0.9	34.7		22.5	8.7	18.1	tr	tr						14.5	

[1]Data from ref. 1570. ER denotes endoplasmic reticulum.

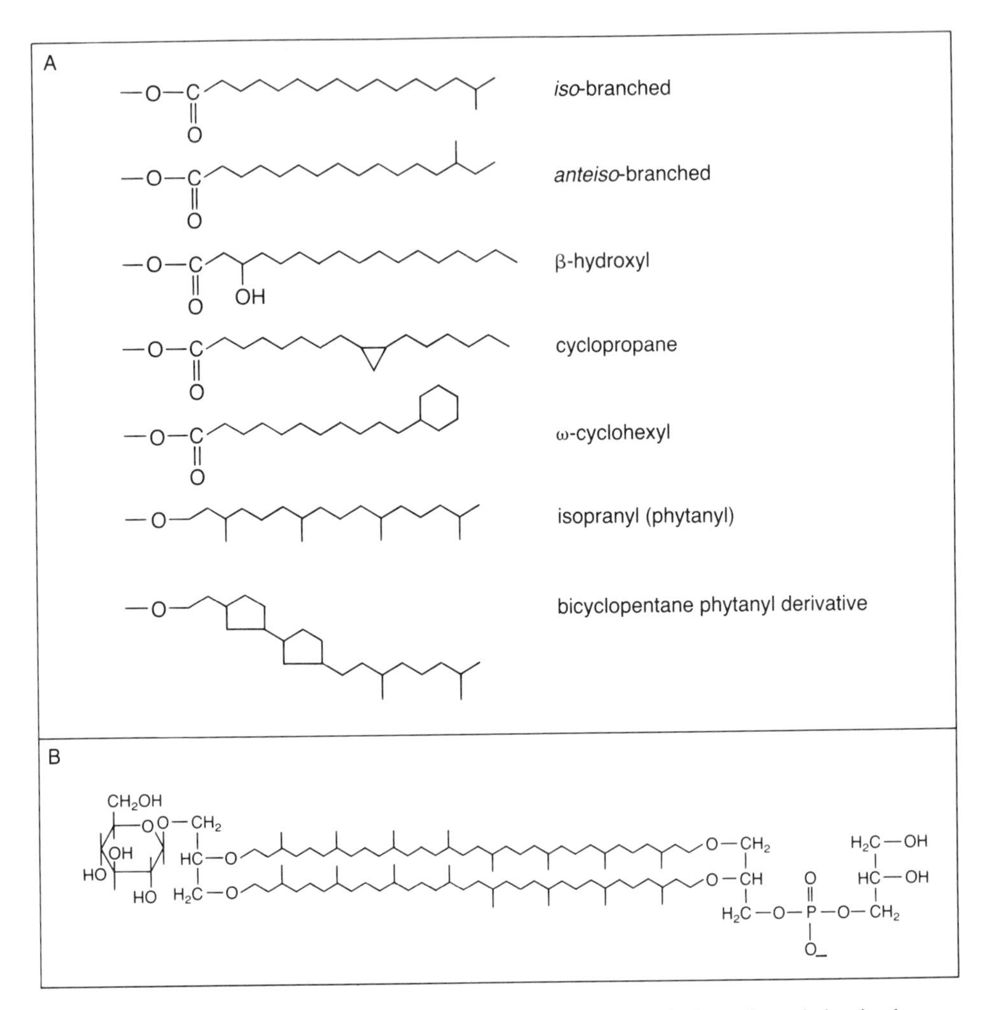

Figure 1.11. (A) Some less common acyl- and ether-linked hydrocarbon chains in the membrane lipids of bacteria. Many of the bacterial species containing these lipids are adapted for growth under extreme environmental conditions, e.g., thermophiles, acidophiles, and halophiles (see 1262, 307, 888 for reviews). Adapted from ref. 1262. (B) An example of an unusual 40-carbon tetraether that spans the bilayer of *Thermoplasma acidophilum* membranes. From ref. 111.

Phosphosphingolipids

These contain the same kinds of polar substituents (e.g., phosphorylcholine) as do the glycerophospholipids, but the hydrophobic group is a ceramide. Sphingomyelin (ceramide l-phosphorylcholine) is widely found in animal cell plasma membranes (Figure 1.8). In myelin, the predominant fatty acids are 24:1 and 24:0. Phosphosphingolipids are rarely found in plants or bacteria. Other phosphosphin-

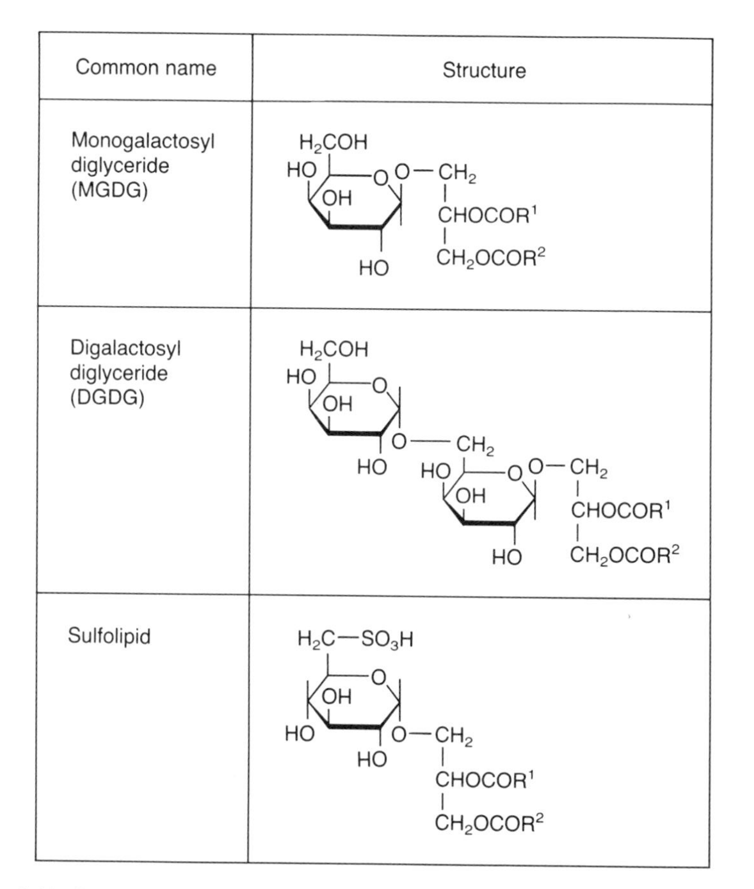

Figure 1.12. Representative structures of several glycerol-based glycolipids. Note that the sulfolipid has a carbon-sulfur bond. The R groups represent fatty acid hydrocarbon chains. These lipids are primarily found in plant leaves and in algae.

golipids are also widely distributed, such as ceramide l-phosphorylethanolamine, ceramide l-phosphorylinositol, and ceramide l-phosphorylglycerol (e.g. see 590).

Glycoglycerolipids (see 97 for review)

These are polar lipids in which the *sn*-3 position of glycerol forms a glycosidic link to a carbohydrate such as galactose. Glycoglycerolipids are predominant in the chloroplast membrane and are also found in substantial quantity in blue-green algae and bacteria. Monogalactosyldiacylglycerol (Figure 1.12) has been termed "the most abundant polar lipid in nature" since it comprises half of the lipid in the chloroplast thylakoid membrane (533). Gram-positive bacteria, in particular, have glycoglycerolipids with a variety of sugars. Archaebacteria also contain similar lipids, but as with the glycerophospholipids, the stereoconfigurations are reversed, with the glycosidic linkage at the *sn*-1 position (307, 888). Glycoglycerolipids are rare in animals.

A. Structures of Some Glycosphingolipids

Glucocerebroside	Cer—Glc
Ceramide lactoside	Cer—Glc—Gal
Globoside	Cer—Glc—Gal—Gal—GalNAc
Sulfatide	Cer—Gal(3)—OSO_3^-
Galactocerebroside	Cer—Gal

Gangliosides

GM_3	Cer—Glc—Gal—NeuNAc
GM_2	Cer—Glc—Gal—GalNAc (Gal | NeuNAc)
GM_1	Cer—Glc—Gal—GalNAc—Gal (Gal | NeuNAc)

B. Detailed Structure of Ganglioside GM_1

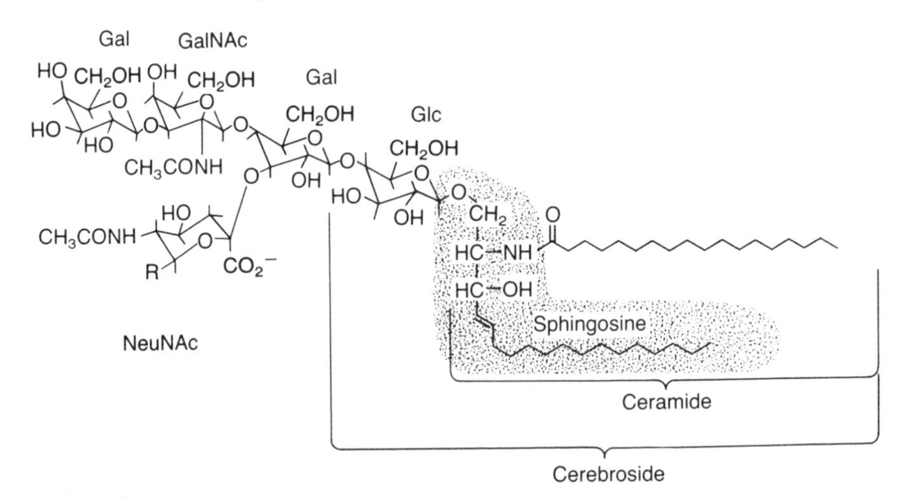

Figure 1.13. Structures of some glycosphingolipids. (A) Shorthand summary of several examples. Abbreviations used: Cer, ceramide; Glc, glucose; Gal, galactose; NeuNAc, sialic acid or N-acetylneuraminic acid; GalNAc, N-acetylgalactosamine. Note that GM_3 is also called hematoside and GM_2 is called Tay-Sachs ganglioside. (B) A more detailed structure of ganglioside GM_1, also showing the names applied to various parts of the structure.

Glycosphingolipids (see 271 for review)

These lipids have a glycosidic linkage to the terminal hydroxyl of ceramide. The classification scheme is according to the carbohydrate moiety, which can range from a single sugar to very complex polymers (see Figure 1.13). Monoglycosyl

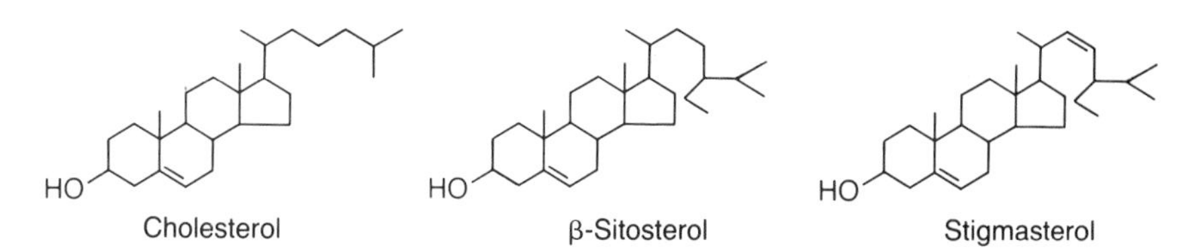

Figure 1.14. Structures of three sterols found in the membranes of eukaryotic cells.

ceramides are generally known as cerebrosides. Gangliosides are a class of anionic glycosphingolipids that contain one or more molecules of sialic acid (*N*-acetylneuraminic acid, NeuNAc) linked to the sugar residues of a ceramide oligosaccharide. Globosides refer to neutral glycosphingolipids which do not contain the negatively charged sialic acid residues.

The glycosphingolipids are found on the outer surface of animal cell plasma membranes, usually as minor components, but occasionally they can be a significant lipid class [e.g., epithelial plasma membrane (849)]. Monogalactosyl ceramide is the largest single component of myelin sheath of nerve. In some cases, the glycosphingolipids are present in intracellular membranes rather than in the plasma membrane (1416).

The glycosphingolipids in the erythrocyte membrane carry blood group antigens (494, 320). Human adenocarcinomas result in the accumulation of novel glycosphingolipids, which are fucosylated and which can be used to recognize these cells and to monitor the oncogenic progression (e.g., 573).

Sterols

These are found in many plant, animal, and microbial membranes. Cholesterol is, by far, the most commonly found sterol. This molecule is a compact, rigid hydrophobic entity with a polar hydroxyl group. Cholesterol is found in animal cell plasma membranes, lysosomes, endosomes (404), and Golgi. It constitutes about 30% of the mass of the membrane lipids of many animal cell plasma membranes. Whereas cholesterol is the major sterol found in animal cells, other sterols, notably sitosterol and stigmasterol, are found in higher plants. These plant sterols (phytosterols) frequently have an additional side chain at position C-24 and/or a double bond at position C-22 (see Figure 1.14). Ergosterol is often found in yeast and other eukaryotic microorganisms. Hopanoids are sterol-like lipids which are found in bacteria and some plants (see 1181 for review).

Minor Components

There are a number of other lipid components found in membranes which can be considered minor in terms of the amounts present. Free fatty acids and lysophospholipids are usually present, but at very low levels. One possible exception is the chromaffin granule membrane, where an extraordinarily high amount of free fatty acids has been reported (668). Monoacyl- and diacylglycerides are also minor components. Diacylglyceride serves an important function as a second messenger in signal transduction in a mechanism by which a variety of biologically active substances can activate cellular responses. This signal response system will be discussed in more detail in Chapter 9.

Polyisoprenoid lipids are also commonly found in membranes. These include ubiquinones and menaquinones, which function in membrane-bound electron transport chains. Other examples are undecaprenol and dolichol, which function respectively as lipid carriers for intermediates in prokaryotic cell wall biosynthesis and eukaryotic glycoprotein biosynthesis in the Golgi. These lipids can be stretched out to lengths considerably beyond the thickness of the bilayer and it is not known how they reside in the bilayer. It is also not known why the polyisoprenoid structure is apparently preferred for these various lipid carrier molecules.

1.52 Lipids Play Multiple Roles Within Membranes

Although the distribution of lipids in various membranes does not seem to be random (1257), there is no satisfying explanation for the observed patterns. Any single membrane can contain well over 100 unique lipid species. Why are there so many and why does each membrane have a unique distribution of lipids? The biosynthesis of membrane lipids and the mechanisms by which they are distributed to different membranes are discussed in Section 10.4. However, the reasons for the heterogeneity are not known, although lipids are increasingly being recognized as active participants in membrane-associated processes. Several factors can be considered.

1. Minimally the lipid mixture must form a stable bilayer in which the proteins can function. This is discussed in the next chapter.
2. Some lipids may be required because their shapes favor packing configurations that may be necessary to stabilize regions of high curvature, junctions between membranes, or optimal interactions with specific proteins (see 303, 267, 272). This polymorphic aspect of membrane lipids is discussed in the next chapter.
3. Some lipids are important as regulatory agents. Most notable are the derivatives of phosphatidylinositol in the plasma membranes of eukaryotic cells (see Section 9.73).
4. Some lipids participate in biosynthetic pathways. For example, in *E. coli* phosphatidylglycerol provides the glycerol phosphate moiety in the biosynthesis of periplasmic oligosaccharides (see Section 10.43).
5. Specific lipids may be required for optimal enzyme activity of particular enzymes. This topic is addressed in Chapter 6.
6. Gangliosides, in particular, have been implicated as playing a role in the regulation of cell growth (1368), in binding to specific receptors in the plasma membrane (210), and in adhesion (812).
7. Other lipid components are also known to play specialized roles. These include the polyisoprenoids, such as dolichol, ubiquinones, menaquinones, and carotenoids, and the platelet activating factor (576).

It must be remembered that it has been demonstrated that organisms can often tolerate quite drastic changes in their membrane lipid composition without deleterious effect. For example, by genetic manipulations, strains of *E. coli* can be made in which the membranes contain 34% phosphatidic acid, which is not normally found in the wild-type strains (1365; see Figure 10.16). Obviously, the exact lipid composition found in wild-type strains is not required for viability of the organism, at least under laboratory growth conditions.

1.53 Membrane Proteins

As shown in Tables 1.3 and 1.4, membranes contain between 20% and 80% (w/w) protein. It is the proteins, of course, which are the biochemically active components of the membrane and provide the diversity of enzymes, transporters, receptors, pores, etc. which distinguishes each particular membrane. Progress in our understanding of membrane proteins was initiated when biochemists learned to use detergents to solubilize these proteins from membranes in biochemically active forms. Initially, success was with the enzyme complexes of the mitochondrial inner membrane. The realization that membrane proteins were not predominantly ß-pleated sheet, as postulated to best fit the Davson–Danielli–Robertson "unit membrane" model, but contained significant amounts of α-helix, was a significant step forward. Also important was the insight that membrane proteins extended deeply into or completely through the lipid bilayer and were stabilized by hydrophobic interactions. This thermodynamic argument was essentially an extension of the principles of the "hydrophobic force" being developed to understand protein structure, i.e., nonpolar hydrophobic interior and polar hydrophilic exterior in contact with water.

As techniques for membrane protein purification were developed, more membrane proteins were obtained in homogeneous form. The insoluble characteristics of most membrane proteins and of the hydrophobic peptides derived from them made primary structure determination difficult. Progress on two membrane proteins, glycophorin and cytochrome b_5, helped to establish the structural themes which have been dominant since the mid-l970s. The amino acid sequence of glycophorin, a sialoglycoprotein from the erythrocyte membrane, indicated a short stretch of 23 nonpolar amino acids near the middle of the molecule (1461) (see Figure 3.17). Topological and other studies indicated that glycophorin extended completely through the erythrocyte membrane and that this hydrophobic stretch was in an α-helical form and was buried in the membrane (475). This work contributed to the now firmly entrenched concept of membrane-spanning α-helical domains in proteins. Electron microscopy image reconstruction studies on bacteriorhodopsin in the purple membrane of *Halobacterium halobium* (619) and X-ray diffraction studies on bacterial photosynthetic reaction centers (319) have provided the highest resolution data available for membrane-spanning proteins. These proteins consist of a series of α-helical segments traversing the bilayer (see Chapter 3).

Sequence studies on the intact form of microsomal cytochrome b_5 were also very suggestive, showing a relatively short stretch of hydrophobic amino acids near the carboxy-terminus (1113, 437) (see Section 4.22). This "hydrophobic anchor" could be removed by proteolysis, releasing the heme-binding domain in a water-soluble form. The membrane-binding hydrophobic domain or "anchor" has been another dominant theme in the analyses of the structure of membrane proteins.

The major point is that membrane proteins are now generally viewed as being folded so as to present a nonpolar hydrophobic surface which can interact with the nonpolar portions of the lipid bilayer. Polar or charged regions of the protein can interact with the lipid headgroups at the surface of the bilayer. Many membrane proteins are transmembranous and extend through the bilayer. Other membrane proteins are probably bound to the membrane exclusively through interactions with other proteins.

Membrane proteins are generally bound to the membrane through noncovalent forces, such as the hydrophobic force or electrostatic interactions (see Chapter 3). There is, however, a small but growing number of examples of membrane proteins which are covalently bound to lipids (see Section 3.8). Many of the proteins in plant or animal plasma membranes are glycoproteins, such as glycophorin. The carbohydrate residues are always located on the extracytoplasmic side of the membrane.

Operationally, membrane proteins are classified as extrinsic (or peripheral) or intrinsic (or integral). This generally denotes the degree of harshness of the treatment required to release the protein from the membrane. Extrinsic proteins are dislodged by washing the membranes in low ionic strength buffer, in a buffer at low or high pH, and/or in the presence of a divalent cation chelator such as EDTA (1438). Such proteins are thought to be weakly bound to the membrane surface by electrostatic interactions either with the lipid headgroups or with other proteins. It is often hard to distinguish an extrinsic membrane protein from a cytoplasmic protein which has become adventitiously bound to the membrane during the isolation procedure. As much as 30% of the proteins associated with the erythrocyte membrane is solubilized by treatment at low ionic strength (922, 1348). Somewhat harsher treatments to release extrinsic proteins involve the use of chaotropic agents such as ClO_4^- or SCN^- (see 1438). These reagents are, in some cases, strong enough to disrupt some protein-protein interactions but are not sufficently strong to denature the individual polypeptides. Chaotropics or "water structure breakers" act by effectively reducing the magnitude of the hydrophobic force (278) (see Chapter 2).

Treatments involving detergents or, occasionally, organic solvents are required to release intrinsic membrane proteins. The detergents disrupt the lipid bilayer and presumably bind to the membrane proteins at the nonpolar binding sites normally in contact with the bilayer interior. Intrinsic membrane proteins require the continued presence of detergents to remain in a soluble, monodisperse form. Removal of the detergent invariably results in the formation of high-molecular-weight protein aggregates and, usually, precipitation. Further information on protein–detergent interactions and the structure and analysis of membrane proteins is in Chapter 3.

1.6 Chapter Summary

Historically, X-ray diffraction and electron microscopy have contributed substantially to our current view of biological membranes. It is now clear that the lipid bilayer forms the structural framework of virtually all biomembranes, and that structural similarities underlie the functional diversity one finds in comparing different membranes. There are a large number of chemically distinct lipids in any given membrane. The reasons for this diversity are unknown, although the unique biological functions of particular lipids are receiving increasing attention. Proteins comprise between 20% and 80% of biomembranes. Many of these proteins extend across the lipid bilayer and are usually solubilized by the use of detergents. Other proteins, called peripheral membrane proteins, are more easily dislodged from the membrane, such as by changing the pH, ionic strength, or chelating divalent cations.

Biochemical techniques have been developed to isolate and characterize distinct membrane populations from prokaryotic organisms, from animal cells, and, to some extent, from plant cells. These techniques most often take advantage of differences in the size and/or density of the different membrane populations in disrupted cell homogenates. Differences in surface properties and in electrophoretic behavior can also be exploited to separate different membranes from a cell homogenate. The isolation of pure membranes is an essential first step prior to biochemical characterization.

Chapter 2
The Structures and Properties of Membrane Lipids

The previous chapter briefly described the historical development of the concept that the lipid bilayer forms the structural basis of all biomembranes. It is important to understand the detailed structure of the lipid bilayer as well as the thermodynamic principles underlying its stability in order to approach an understanding of biological membranes. In addition, some phospholipids spontaneously organize in structures which are not bilayers, such as the inverted hexagonal H_{II} phase, and these lipids have been postulated to play specialized roles within membranes (see 303,267). In this chapter we will survey the structures and thermodynamics of lipid–water systems, with the emphasis on those features which give some insight into the properties of biological membranes.

2.1 Lipid Crystals (see 6 for review)

High-resolution structures from X-ray crystallography have been reported for several membrane lipids: lysophosphatidylcholine (605), dimyristoyl phosphatidic acid (594), dimyristoyl phosphatidylcholine (1133), dilauroyl phosphatidylethanolamine (636), dimyristoyl phosphatidylglycerol (1126), and a cerebroside (1125). These crystals contain very little water. However, it appears that the lipid structures in these crystals are similar to those adopted in the fully hydrated forms. Figure 2.1 shows the conformations adopted by some of these membrane lipids within the crystals. The glycerol carbon atoms, and the corresponding atoms in sphingosine, have been blackened for clarity. Figure 2.2 shows some of the structural notation used to describe the lipid conformation.

Let us start by considering the crystal structure of dilauroyl phosphatidylethanolamine (Figure 2.1A). The most salient features of these crystals are

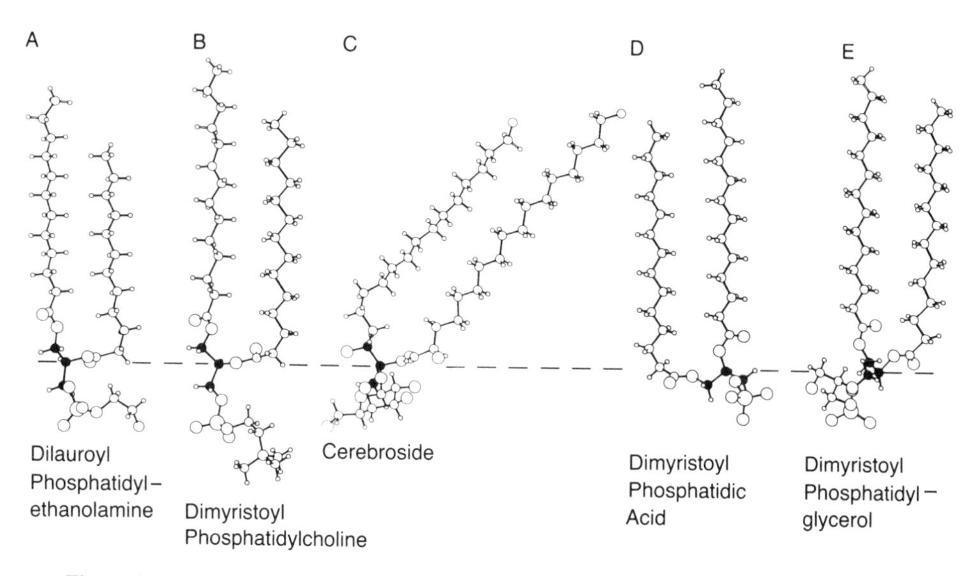

Figure 2.1. Structures of five membrane lipids as determined by X-ray crystallography. The three glycerol and sphingosine carbon backbone atoms have been colored black. Note A, B, and C have a similar conformation for the glycerol or sphingosine moiety, whereas D and E are each different. Adapted from references 604 and 636 (A); 1133 (B), 1125 (C), 594 (D) and 1126 (E). Figures kindly provided by Dr. I. Pascher.

1. The molecular area (*S*, as in Figure 2.2) is 39 Å^2.
2. The polar headgroup is virtually parallel to the plane of the bilayer. In fact, the amino group is hydrogen bonded to the unesterified phosphate oxygens of an adjacent molecule. The glycerol group is essentially oriented normal to the plane of the bilayer.
3. The *sn*-2 fatty acid chain extends parallel to the plane of the bilayer for the first two carbons and then is directed down into the bilayer. The *sn*-1 fatty acid extends directly into the bilayer.
4. The acyl chains are directed perpendicular to the bilayer surface and, with the exception of the initial part of the *sn*-2 fatty acid, are in a fully extended, i.e., all-*trans*, configuration.

While similar in many respects, the crystal structures of the phosphatidylcholine and the cerebroside have important differences from that of phosphatidylethanolamine. The most significant and obvious difference is the tilt of the acyl chains, most evident in the structure of the cerebroside. This is rationalized as being due to a simple packing problem. The bulky polar headgroups in these lipids do not allow the relatively simple structure seen with dilauroyl phosphatidylethanolamine. The cross-sectional areas (*S*) required by these headgroups are larger than the 39 Å^2 occupied by the acyl chains of each molecule (2Σ, as in Figure 2.2). In the case of the cerebroside, the packing problem is eliminated by the tilting of the acyl chains with respect to the bilayer normal. This effectively increases

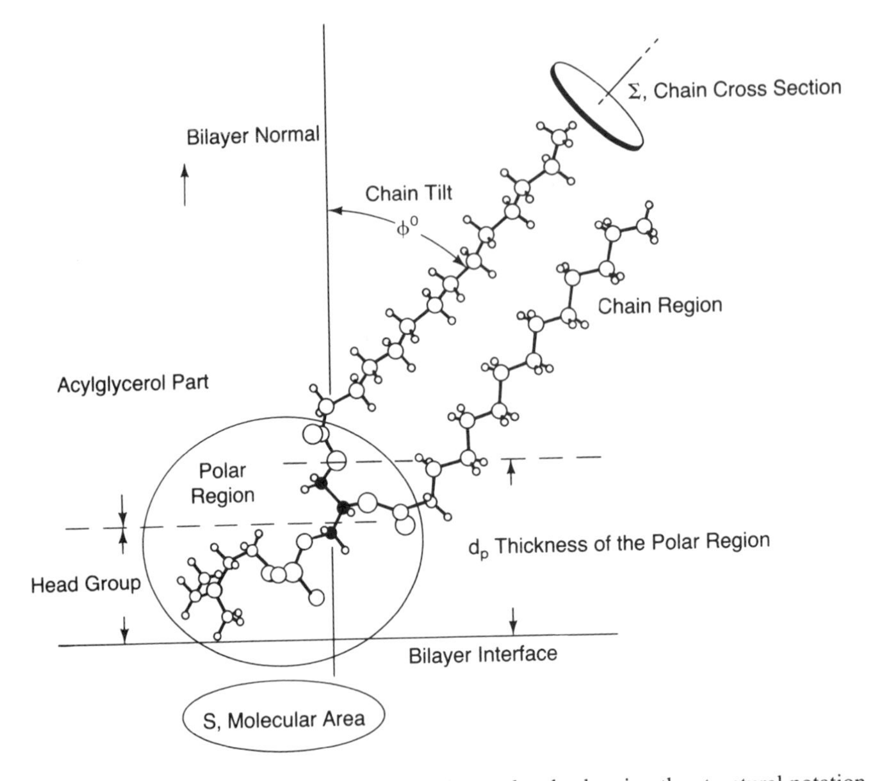

Figure 2.2. Schematic of a phosphatidylcholine molecule showing the structural notation defining various regions. Note that the cross-sectional area of the hydrocarbon chain, Σ, is taken perpendicular to the chain axis. The cross-sectional area parallel to the plane of the membrane will be, in this case, significantly larger due to the tilt of the chains. Adapted from ref. 604.

the cross-sectional area of the acyl chains with respect to the plane of the bilayer. Figure 2.3 shows schematically how the chain tilt allows for stabilizing interactions between the acyl chains of adjacent molecules, while maintaining the large molecular area due to the bulky headgroup. The acyl chains in the dimyristoyl phosphatidylcholine are tilted only about 12° away from the bilayer normal, in contrast to 41° in the cerebroside (1133). The diacyl phosphatidylcholine solves the problem of packing the bulky headgroups by having adjacent molecules alternately displaced along the bilayer normal, schematically shown in Figure 2.3D. There is substantial evidence that, in some cases, fully hydrated lipid bilayers in the gel phase also have substantial chain tilting (e.g., see 604, 1320; Section 2.2). This is one example where simple steric arguments, i.e., taking into account lipid "shapes" (S vs. 2Σ), are very useful in rationalizing lipid structures. Further examples will be described in Section 2.34.

In all the crystal structures, except that of phosphatidic acid, the initial part of the *sn*-2 fatty acyl chain is directed parallel to the bilayer surface. This feature

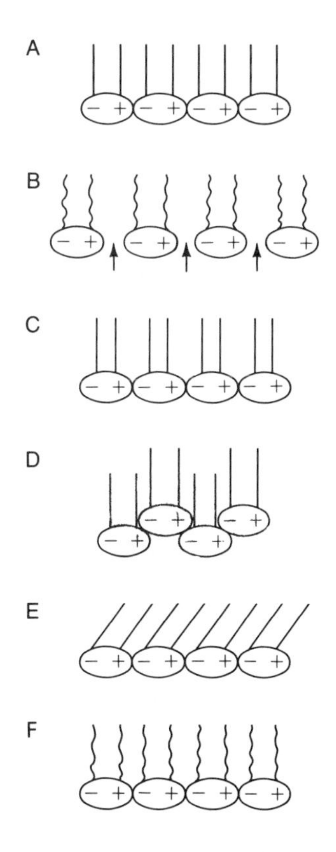

Figure 2.3. Schematic showing packing adaptations to differences in the cross-sectional areas of the polar group and the acyl chains (S and Σ, in Figure 2.2). These figures depict the behavior of phosphatidylethanolamine and phosphatidylcholine. (A) Phosphatidylethanolamine in a crystal, where the space requirements of the headgroup and the two acyl chains are the same. (B) Phosphatidylethanolamine in a liquid crystalline lamellar phase, where the area per molecule requires that the headgroup lattice be disrupted. This will occur only if water or other polar molecules can bridge the gap between lipid neighbors and stabilize the structure. (C) Phosphatidylcholine in a hypothetical arrangement illustrating that the space required by the headgroup (50 Å^2) is substantially larger than required by the acyl chains (38 Å^2). This requires some packing adjustment. (D) One adjustment of phosphatidylcholine observed in crystals is for the headgroups to be displaced in an overlapping fashion. (E) An adjustment observed in the lamellar gel phase of phosphatidylcholine is for the chains to be tilted such that they accommodate a larger cross-sectional area (50 Å^2). (F) In the lamellar liquid crystalline phase, the size of the headgroup of phosphatidylcholine does not require the headgroup lattice to be disrupted. Adapted from ref. 604.

has also been confirmed by nuclear magnetic resonance (NMR) of phosphatidylethanolamine and phosphatidylcholine bilayers and for the phospholipids in membranes from *E. coli* (487). The physiological relevance of this feature is not known. However, it has been noted that in egg phosphatidylcholine, the average length of the *sn*-2 fatty acid is 18, whereas the average length of the *sn*-1 fatty acid is 16 carbons. This may be in compensation for the configuration of the 2-fatty acid, so that the two acyl chains extend about the same depth from the bilayer surface.

In summary, there are five major features of the crystal structures which have importance for considering the structure of the lipid bilayers:

1. The structures observed are all lamellar, with the polar and nonpolar groups organized as in a bilayer.
2. There are packing problems resulting from having a bulky headgroup such as in phosphatidylcholines and cerebrosides ($S > 2\Sigma$). Such packing considerations are important determinants of the properties of membrane lipids, not only in crystals but also in model membranes and, possibly, within biomembranes.
3. The polar headgroups essentially lie flat in the plane of the bilayer, with intermolecular hydrogen bonding where possible.
4. Acyl chains (saturated) are in an all-*trans* configuration.

5. In most cases, the *sn*-2 fatty acyl chain does not extend into the bilayer until after the C-2 position.

Each of these features has been confirmed for lamellar structures formed by lipid-water mixtures for the gel phase (points 1, 2, 3, 4, 5) and/or the liquid crystalline phase (points 1, 3, 5) (see next section). The structures of lipids in crystals are an excellent starting point for understanding the conformations of lipids within biological membranes.

2.2 Lipid–Water Mixtures

Mixture of lipids and water are polymorphic. Even for single purified lipids there is more than one kind of organized structure when hydrated. The particular form which predominates depends on such parameters as the lipid concentration, temperature, pressure, ionic strength, and pH. X-ray diffraction techniques have been particularly valuable in defining the kinds of structures in lipid–water systems, and this is most often studied as a function of lipid concentration and temperature, with the data presented in the form of a phase diagram indicating the structure in various regions of the "temperature–concentration" plot. Differential scanning calorimetry is often used along with X-ray diffraction to define the phase boundaries of the lipid–water phase diagrams. These studies are usually performed at high lipid concentrations [>40% lipid (w/w)]; however, many of the structures which have been characterized at high lipid concentrations also exist in lipid dispersions in a large excess of water.

The major organized forms of lipid–water systems (see 1334, 263 for reviews) are depicted schematically in Figure 2.4. The major forms are

1. *Lamellar liquid crystalline phase* (L_α): This form is what is usually thought of as representing the bulk of the lipids in the biological membrane. There is two-dimensional order, but there is considerable disorder in the acyl chains, as indicated by the X-ray diffraction data.
2. *Lamellar gel phase* (L_β): This is formed at low temperatures in those lipids which form the lamellar structure. The molecules are packed more tightly together (smaller surface area per molecule) and the acyl chains are much more highly ordered, corresponding to the all-trans configuration found in the structure of lipid crystals. Because the chains are maximally extended in the gel phase, the bilayer thickness is greater than in the liquid crystalline phase. The density of the gel phase is slightly greater than that of the liquid crystalline phase. In lipids which have bulky polar headgroups, such as dipalmitoyl phosphatidylcholine, the acyl chains are tilted with respect to the bilayer normal (700, 461) similar to structures seen in some lipid crystals (Figure 2.1). The chain tilt is often denoted by a prime ($L_{\beta'}$). It is interesting to note that dispersions of phosphatidylcholine in solvents containing some alcohols (1346) or glycerol (1084) form an unusual gel phase in which the acyl chains from opposing halves of the "bilayer" are fully interdigitated. The biological significance of this is not known.

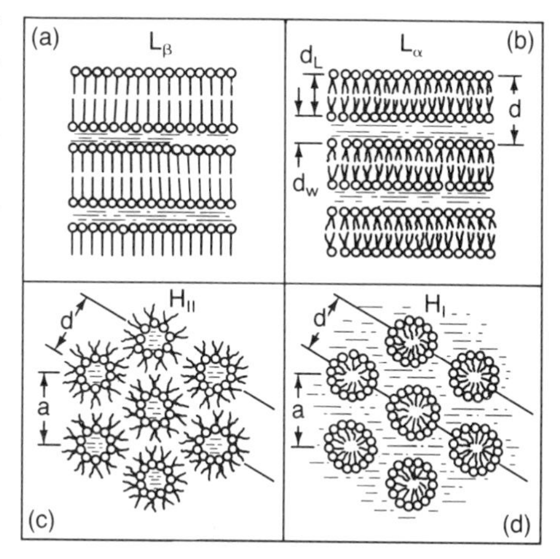

Figure 2.4. Schematic representations of lipid–water phases: (a) lamellar gel; (b) lamellar liquid crystalline; (c) hexagonal type II; (d) hexagonal type I. Various dimensions that can be measured by X-ray diffraction are indicated. Adapted from ref. 1334.

3. *Hexagonal I phase (H_I)*: In this form, the lipids are organized in the form of cylinders with the polar groups on the outside, in contact with water. The cylinders are packed in a hexagonal pattern.
4. *Hexagonal II phase (H_{II})*: The lipids are in the form of cylinders, but in this case the polar groups face the inside, where there is a column of water. Again, the cylinders are packed in a hexagonal array.

One major point is that a number of lipids form nonbilayer structures. In fact, many purified membrane lipids do not form stable bilayers, but prefer the H_{II} hexagonal phase. Examples are unsaturated phosphatidylethanolamines and the glycolipid monogalactosyldiglyceride (see Figure 2.5). The reasons for this, and the possible biological relevance, is discussed in subsequent sections.

2.21 Lipid Hydration

X-ray diffraction methods (see 1320, 910) can yield the dimensions indicated in Figure 2.4. As a general rule, the dimension determined primarily by the length of the acyl chains remains nearly constant as the percent water in the system is increased. The lipid headgroups bind to water and become hydrated. This has been studied extensively using NMR techniques, either ^{1}H-NMR or ^{2}H-NMR. Results with phosphatidylglycerol, phosphatidylcholine, and phosphatidylethanolamine using ^{2}H-NMR indicate a primary hydration shell of 11–16 water molecules per lipid, in rapid exchange with the bulk water (123). Other measurements indicate that the phosphatidylethanolamine headgroup binds fewer water molecules than the headgroup of phosphatidylcholine, and it has been speculated that this lower hydration favors the formation of the nonlamellar hexagonal H_{II} phase by unsaturated phosphatidylethanolamines (1616).

Many lipids swell in water. Lipids which are neutral or isoelectric (e.g., phosphatidylcholine) show no or limited swelling, as indicated by the limited thickness of the water layer between lamellae (e.g., see 195, 600). In excess water,

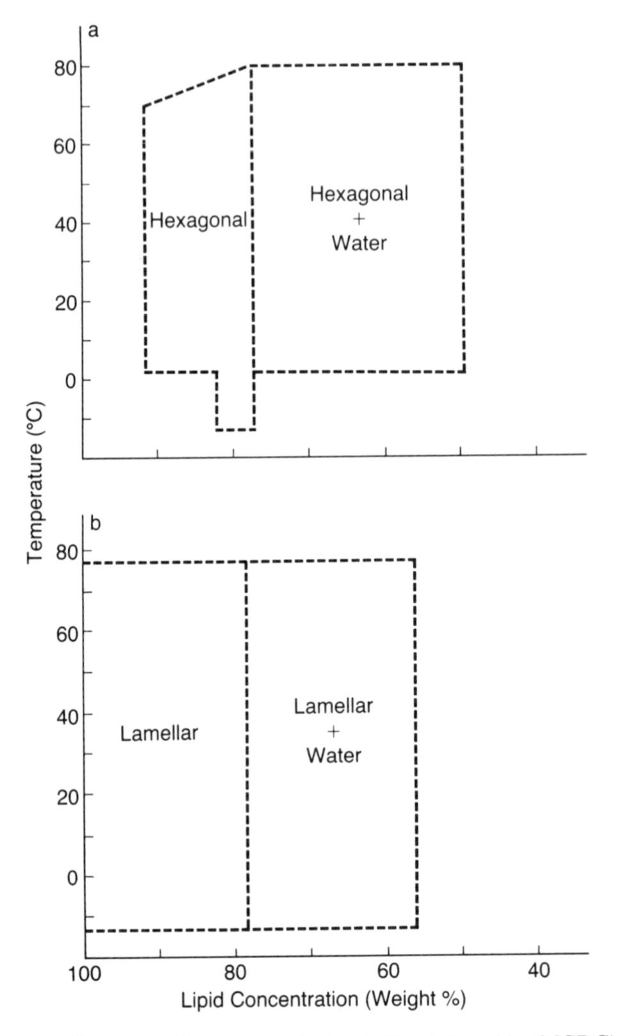

Figure 2.5. Phase diagrams of (a) monogalactosyl diacylglyceride (MGDG) and (b) digalactosyl diacylglyceride (DGDG) extracted from leaves. The acyl chains in both cases are predominantly unsaturated. Adapted from ref. 1334a.

two phases exist, a multilamellar lipid structure in equilibrium with bulk water. Charged lipids show continuous swelling with water added between lamellae up to a certain threshold at which point two phases form, a fully hydrated unilamellar vesicle with bulk water (600). The swelling and relative stability of the multilamellar and unilamellar forms are determined by electrostatic interactions. Low ionic strength results in destabilizing the multilamellar forms. Lipid mixtures with only a few percent of a charged lipid can exhibit continuous swelling behavior.

The polarization of water molecules by the lipid polar headgroups results in a strong repulsive interaction when two bilayers are brought close together (see 1193, 195). This "hydration" force keeps hydrated bilayers at least 30 Å apart. It is this hydration force which provides the major energetic barrier which must be overcome to obtain membrane fusion (see Chapter 9). Phosphatidylethanolamine vesicles tend to aggregate, possibly because the hydration of this headgroup is relatively low (see Section 9.51).

Studies using electron spin resonance (ESR) probes designed to monitor the polarity at different distances from the bilayer surface indicate that water is able to partially penetrate into the hydrocarbon core in the liquid crystalline phase (551). Studies using neutron scattering show that water does not extend beyond the glycerol backbone of the polar headgroup in bilayers in the gel phase (1606).

2.22 Examples of Phase Diagrams of Single Component Lipid–Water Systems

Figure 2.5 shows the temperature–concentration phase diagrams of two common glycolipids isolated from plants (see 1334, 1041, 533). Digalactosyl diglyceride (DGDG) forms a stable lamellar phase. Only a single hydrated phase is present at low water content. Beyond about 20% water, two phases are present, a multilamellar hydrated lipid phase and bulk water. This lipid does not exhibit continuous swelling.

Box 2.1 Some Lipids Isolated from Biological Membranes Do Not Form a Stable Bilayer

The phase diagram of monogalactosyl diglyceride (MGDG) is also shown in Figure 2.5. Note that the removal of a single sugar residue from the lipid headgroup has a drastic effect on the structure of the lipid–water system. Only hexagonal phase H_{II} is formed, and the lipid does not form a stable bilayer by itself. This highly unsaturated galactolipid makes up about 20% of the dry weight of the chloroplast thylakoid membrane and about half of the lipid in the membrane (533). It has been suggested (1041, 533) that MGDG plays an important role in stabilizing the regions of high concave curvature in the thylakoid membrane system and may possibly play other specialized roles.

Another example is provided by phosphatidylethanolamine isolated from the bacterium *Pseudomonas fluorescens*. This lipid is about 75% of the total phospholipid of the membrane (262). It has a heterogeneous and unsaturated acyl chain composition and forms a stable hexagonal H_{II} phase in excess water at room temperature (1334, 262). Clearly, the presence of other components in the membrane stabilizes this lipid in the bilayer form. Saturated phosphatidylethanolamines also show complex phase behavior but do not form H_{II} phase near physiological temperatures (1315). Changes in phase that occur as a function of water composition at constant temperatures are called *lyotropic transitions*.

Figure 2.6 shows a phase diagram of dipalmitoyl phosphatidylcholine (see 1269). This lipid exists in a lamellar form under most conditions. It "swells" upon addition of water until a maximum amount of water is absorbed between the bilayers, at which point a two-phase system with multilamellar liposomes is formed. Note that there is a distinct phase in between the gel ($L_{\beta'}$) and liquid crystalline (L_{α}) lamellar forms. This is the so-called "ripple phase" ($P_{\beta'}$). In this phase, the surface of the bilayer is rippled and presents a wave-like appearance in electron micrographs (Figure 2.7). The thermotropic phase transition $P_{\beta'} \rightarrow L_{\alpha}$ is termed the main transition, whereas the transition $L_{\beta'} \rightarrow P_{\beta'}$ is called the pre-transition (see Section 2.4).

Lipids that have a bulky polar headgroup, such as choline, generally show the pre-transition, indicating a phase which is intermediate between the gel and liquid crystalline phases (604). In dipalmitoyl phosphatidylcholine the acyl chains are probably tilted in the $P_{\beta'}$ phase (461), but Raman studies (181) indicate that the chains are still predominantly all-*trans*, as in the gel phase (see Section 2.4).

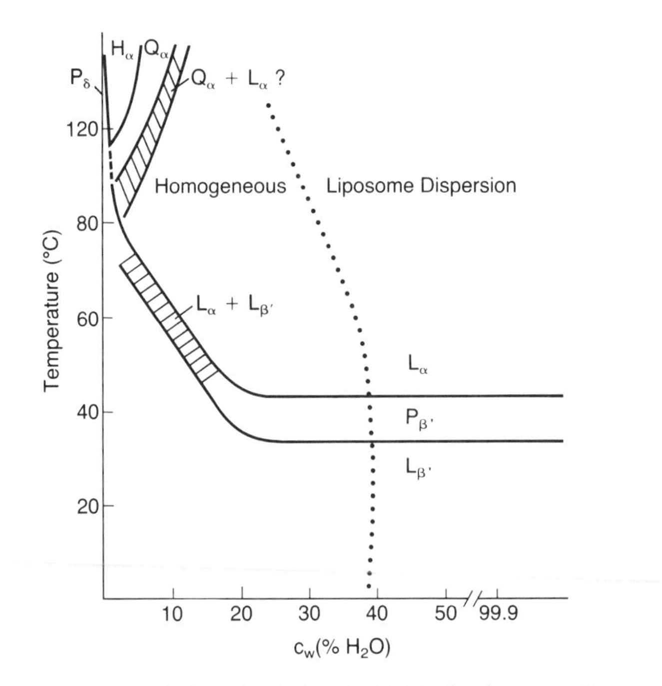

Figure 2.6. Phase diagram of dipalmitoyl phosphatidylcholine in water. The percentage of water is indicated on the abscissa with pure lipid on the left and a dilute lipid suspension on the right. Indicated are the lamellar gel phase ($L_{\beta'}$), lamellar liquid crystalline phase (L_{α}), and intermediate ripple phase ($P_{\beta'}$). At high temperatures and low hydration, other phases can form (Q_{α}, cubic phase and H_{α}, hexagonal phase). The dotted line indicates the maximum absorption of water by the homogeneous lipid-water mixture. Adapted from ref. 1269.

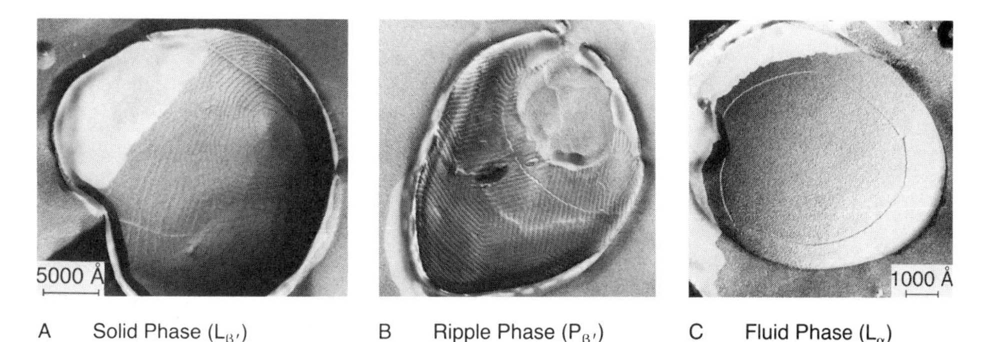

Figure 2.7. Freeze-etch electron micrographs showing the textured appearance of gel ($L_{\beta'}$), ripple ($P_{\beta'}$), and liquid crystalline (L_α) phase of pure phosphatidylcholines. (A) Dipalmitoyl phosphatidylcholine at 25°C; (B) dipalmitoyl phosphatidylcholine at 35°C; (C) dimyristoyl phosphatidylcholine at 25°C. Adapted from ref. 1269. Photographs kindly provided by Dr. E. Sackmann.

2.23 Two Techniques Used for Examining Lipid Polymorphism

Besides X-ray diffraction, a number of techniques have been used to characterize the properties of lipid phases. One that has been particularly useful is freeze-fracture electron microscopy. A second technique, ^{31}P-NMR, has been used more recently, with particular application to detecting nonbilayer structures, which are felt by some to play specialized roles in biomembranes (e.g., 1507, 264).

(1) Electron Microscopy of Freeze-Etched Lipids. One of the techniques which has been very useful in examining the architecture of the various lipid phases is freeze-etch electron microscopy (see Chapter 1 and refs. 263, 313, 1509, 783). Examples are shown in Figure 2.7. The liquid crystalline phase (L_α) always presents a smooth surface, whereas the $P_{\beta'}$ phase appears rippled. The gel phase ($L_{\beta'}$) can appear smooth or, depending on the manner of sample preparation, can exhibit a spiral pattern resulting from temporary defects in the close-packed structure. The hexagonal phase (H_{II}) formed by lipids such as unsaturated phosphatidylethanolamines can also be identified by freeze-etch electron microscopy (Figure 2.8), looking like a stack of cylinders. In preparing samples, the lipids are equilibrated at the appropriate condition to form a particular structure and then rapidly frozen before the lipids can rearrange.

Finally, electron microscopy of lipid bilayers has been used to examine "lipidic particles" (Figure 2.9). These are frequently observed in binary mixtures of lipids where one lipid prefers H_{II} and the other a bilayer configuration (see 263, 1507, 1508). These particles are observed in pure lipid preparations and are different from the "particles" associated with proteins which are observed in biomembranes and in reconstituted protein–lipid systems (see Figure 1.6). It has been speculated that lipidic particles represent regions of inverted micelles within the bilayer and

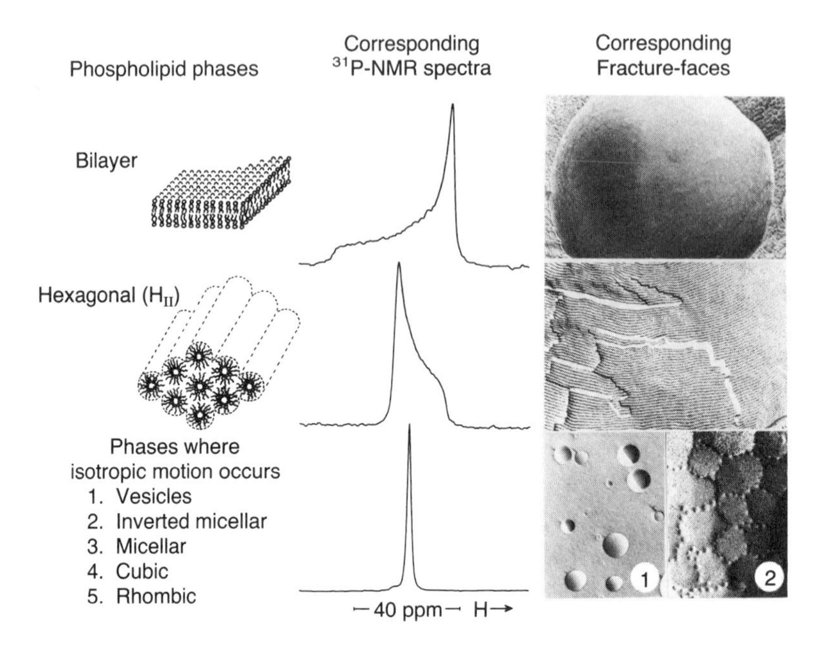

Figure 2.8. ^{31}P-NMR and freeze-fracture characteristics of phospholipids in various phases. Notice that the lamellar liquid crystalline phase (L_{α}) and hexagonal H_{II} phase have very different and diagnostic ^{31}P-NMR spectra as well as quite different appearances in freeze-fraction electron micrographs. The bilayer spectrum was obtained with egg yolk phosphatidylcholine and the hexagonal phase spectrum was obtained with phosphatidylethanolamine prepared from soya beans. The sharp ^{31}P-NMR spectrum, characteristic of rapid motion, is most often observed from either small unilamellar vesicles (photo 1) or large lipid structures containing "lipidic particles" (photo 2). Adapted from ref. 263. Photographs kindly provided by Dr. P. Cullis and Dr. M. Hope.

that they play a function in biological processes such as facilitating membrane fusion or stabilizing regions of high curvature, as in the thylakoid membrane (263, 1507, 1041, 533). However, solid evidence concerning any biological significance of lipidic particles is, so far, lacking.

(2) ^{31}P-NMR of bilayers (see 1454). This technique is also used to characterize the structural properties of hydrated lipids (see 263, 1318, 264, 266, 487, 748). For example, phospholipids in the H_{II} hexagonal phase give rise to a distinctly different ^{31}P-NMR spectrum than do phospholipids which are in a lamellar phase (266, 487) (Figure 2.8). This method has been used to detect lamellar $\rightarrow$ H_{II} transitions in lipids and lipid mixtures (e.g., 487, 748). This technique is subject to ambiguities, especially when interpreting spectra which indicate "isotropic" averaging caused by relatively rapid motions. One structure presumed to be consistent with such a spectrum is the "lipidic particle." However, this is not a unique interpretation of such a spectrum. ^{31}P-NMR has been shown to be reliable in detecting the presence of hexagonal H_{II} phase in pure lipid dispersions, but

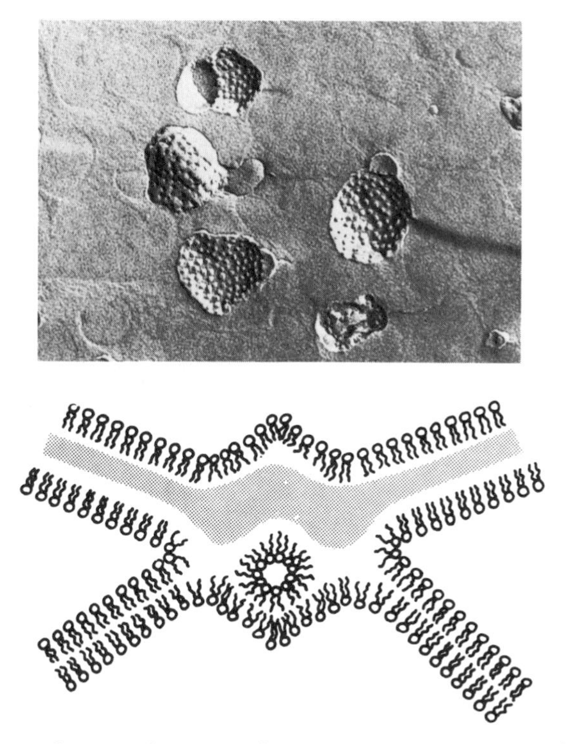

Figure 2.9. Freeze-fracture micrograph of lipidic particles induced by Ca^{2+} in mixture of soya phosphatidylethanolamine and cardiolipin (4:1 molar ratio). A hypothetical model depicting a lipidic particle as an inverted micelle is shown below the micrograph. The shaded region indicates the fracture region. From ref. 263. Figure kindly provided by Dr. P. Cullis and Dr. M. Hope.

such interpretations must be viewed with caution when applied to biomembranes, unless other techniques are also used.

^{31}P-NMR has also been used to study the orientation and dynamics of the phospholipid polar headgroups, as well as the perturbations caused by integral membrane proteins on the bilayer (see next section and Chapter 5).

2.24 Lipid Headgroup Orientation in the Bilayer (for reviews, see 604, 1320)

A number of techniques indicate that in lamellar phospholipid-water dispersions, the lipid polar headgroups are oriented approximately parallel to the plane of the bilayer, as observed in lipid crystals (Section 2.1). For phosphatidylcholines, this has been shown for both the gel and liquid crystalline phases by X-ray and by neutron diffraction. ^{2}H-NMR studies are also consistent with this orientation, but cannot rule out alternate interpretations. Results indicate similar structures for phosphatidylglycerol, sphingomyelin, and phosphatidylserine. ^{2}H-NMR studies on intact mouse fibroblast cells as well as on isolated membranes indicate that both the phosphatidylcholine and phosphatidylethanolamine components have their headgroups oriented parallel to the membrane surface (1293). In contrast, neutron diffraction studies show that phosphatidylglycerol isolated from *E. coli* has its headgroup oriented about 30° from the membrane surface, making its negatively charged phosphate moiety accessible for interaction with cations (995).

The orientation and dynamics of lipid headgroups may be influenced by intermolecular hydrogen bonds at the membrane surface (for review, see 115). Clearly, lipids such as phosphatidylserine, phosphatidylethanolamine, and various glycolipids can participate as donors and acceptors of hydrogen bonds. Studies on model membrane systems indicate headgroup hydrogen bonding can be important, even in the aqueous environment at the membrane surface, but the relevance to the structure of biomembranes is not known.

2.25 Acyl Chain Configuration and Packing in the Bilayer

Let us consider saturated chains first. There is free rotation about each C-C bond, with preferred energy minima, most easily seen in a Newman projection (Figure 2.10). The *trans* configuration is most stable and there is an estimated energy barrier of 3.5 kcal/mol to rotate past the eclipsed configuration to the *gauche* form. The all-*trans* configuration allows the chain be maximally extended, whereas a *gauche* bond alters the direction of the chain. A sequence of *gauche-trans-gauche* for three consecutive C-C bonds results in a kink in the chain which effectively displaces the portions of the chain above and below the kink, as seen in Figure 2.11. Note that each gauche configuration can be designated g^+ or g^- depending

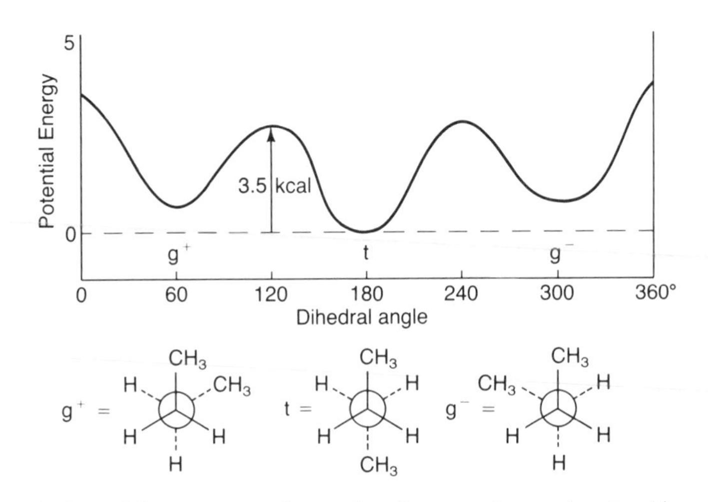

Figure 2.10. Potential energy curve for rotation about a carbon–carbon bond in an alkane. Below is the Newman projection diagram of the minimum energy *gauche* and *trans* conformations of butane: g^+, g^-, and *t*. Adapted from ref. 535.

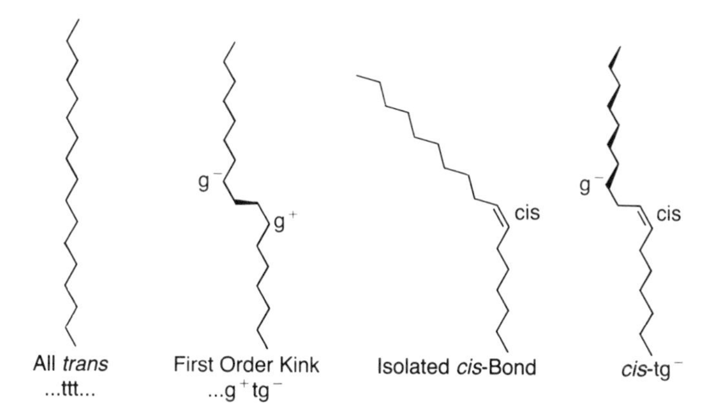

Figure 2.11. Illustration of several alkyl chain configurations. Adapted from ref. 808.

on the sense of rotation in going from C_1 to C_4 (see Figure 2.10). A kink which results in a simple displacement can be either g^+tg^- or g^-tg^+. Almost all double bonds found in membrane lipids are *cis* and introduce the same kind of change of direction in the chain as does a *gauche* configuration. The presence of kinks, *cis*-double bonds, cyclopropyl groups (352), or other alterations from the simple all-*trans* chain configuration results in increasing the cross-sectional area of the hydrocarbon chain from the minimum of about 19 Å^2, and this has important consequences. The packing of lipids in the bilayer can be rationalized by considering the relative space requirements of the hydrocarbon chains and polar headgroup, using the same principles inferred from the lipid crystal structures (Section 2.1). These same principles will be discussed in Section 2.3 in predicting micelle shapes.

Numerous techniques including X-ray diffraction, neutron diffraction, and Raman and IR spectroscopies indicate that in the gel phase, the hydrocarbon chains of saturated diacyl phospholipids are predominantly in the all-*trans* configuration (see 604, 1320). The minimal cross-sectional area per molecule required by diacyl phospholipids is about 38 Å^2 (2Σ in Figure 2.2). This is about the same as the area required by the phosphatidylethanolamine headgroup, and saturated phosphatidylethanolamines in the gel phase can pack with their acyl chains parallel to the bilayer normal as in the lipid crystals. However, as observed in crystals of phosphatidylcholines (605, 1133), the minimal packing requirement for the headgroup is in the range of 50 Å^2 and, hence, dipalmitoyl phosphatidylcholine in the gel phase cannot pack in the same way as does phosphatidylethanolamine. In the gel phase, the acyl chains of dipalmitoyl phosphatidylcholine are tilted about 30° with respect to the bilayer normal, effectively increasing their cross-sectional area to be compatible with that of the headgroup; the chains remain in the all-*trans* configuration. In the liquid crystalline phase, the introduction of *gauche* configurations increases the effective chain cross section to at least 50 Å^2 for the diacyl phospholipid, and in aqueous dispersions the effective liquid crystalline molecular area is typically in the range 60 to 70 Å^2 (see 604). Hence, in the liquid crystalline phase, the hydrocarbon chains are not tilted. The headgroups are sufficiently far apart under these conditions, and they require water or other polar molecules to act as spacers or to bridge between neighboring headgroups. ^{2}H-NMR indicates the thickness of the hydrocarbon domain of dipalmitoyl phosphatidylcholine is 35 Å in the liquid crystalline phase, compared to 45 Å expected if the chains were all-*trans* and oriented along the bilayer normal (see 1320). The shorter distance is due to *gauche* configurations (chain disorder), and the chains are basically aligned and perpendicular to the plane of the bilayer and are not coiled. Figure 2.12 illustrates the linear variation of bilayer thickness of fluid phase vesicles of diacyl phosphatidylcholines on the acyl chain length, determined by X-ray scattering (848).

2.26 Techniques Useful for Characterizing the Interior of the Bilayer

Two techniques which have been particularly useful for obtaining a detailed picture of the interior of the bilayer have been ^{2}H-NMR and vibrational spectroscopy (infrared and Raman spectroscopies). Proton and ^{13}C-NMR have been hindered in their application to membranes and lipid-water systems because of the demonstrated need to use ultrasonic dispersal to yield small vesicles or membrane fragments. The application of the technique of magic angle sample spinning eliminates this need, however, and this is likely to open the way for the application of state-of-the-art NMR methodologies to the study of model membranes and biological membranes (1093).

^{2}H-NMR and vibrational spectroscopy are nonperturbing and do not involve the use of probes embedded in the bilayer, which might influence the structure

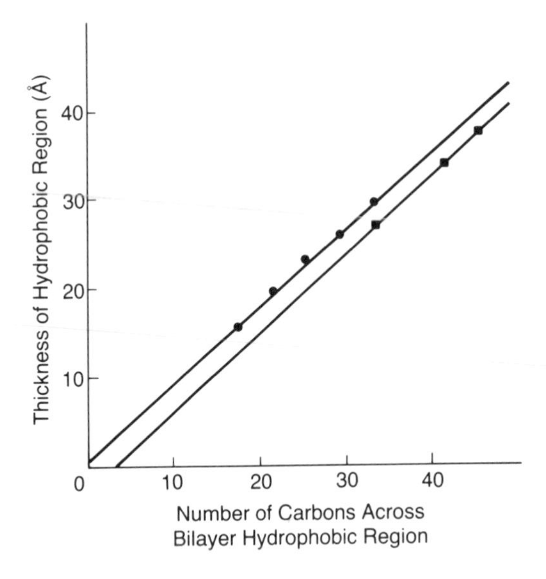

Figure 2.12. The measured thickness by X-ray scattering of the bilayer hydrocarbon region as a function of the number of carbon atoms in the hydrocarbon chain. The number of carbons is taken starting at C-2 of each acyl chain. Measurements were made on liquid crystalline diacyl phosphatidylcholines. Circles, saturated chains; squares, monounsaturated acyl chains. The molecular area was measured to be 65 to 70 Å^2. From ref. 848.

of the surrounding lipids. A brief description of these techniques and a summary of results relating to acyl chain configuration follow. The use of membrane probes for either ESR or fluorescence measurements will be discussed in Chapter 5 in connection with measurements of membrane dynamics and lipid-protein interactions.

Raman Spectroscopy

Laser Raman spectroscopy can be used to probe the physical state of model membranes and biomembranes (for reviews see 1513, 1544). The technique measures the energy differences between scattered photons and the incident photons caused by interaction with the vibrational modes of the sample. The vibrational modes and their intensities are very dependent on the physical state of the lipids, and the spectra of gel and liquid crystalline phospholipids are quite different (e.g., 477). Of particular use in monitoring these differences are the C-C stretch modes. For example, the appearance of *gauche* bonds at the expense of *trans* segments as lipid chains melt results in an increase in a band at ~1080 cm^{-1}. This technique indicates that some *gauche* bonds persist in the gel state until the temperature is decreased to very low temperatures, near -200°C. It is worth noting that the vibrational modes often involve the entire molecule, so quantitative interpretation in terms of particular conformations is not simple.

Typically, samples are about 1 mg/ml lipid suspensions. Membranes with chromophores (e.g., hemes) or impurities which result in fluorescence cannot be used. Background fluorescence makes measurement of the relatively small Raman signal impossible, and chromophores will cause heating problems, due to laser light absorption.

Infrared Spectroscopy (IR)

This technique also monitors the vibrational modes of the sample, but until recently the application of IR to biological samples was limited due to the inability to use aqueous suspensions. However, the recent development of Fourier transform infrared (FT-IR) instrumentation has eliminated many of these problems. Hence, lipid dispersions and biomembranes can now be examined by FT-IR (for reviews see 190, 32). This technique has the advantage over Raman spectroscopy of significantly improved sensitivity and the fact that fluorescence impurities or chromophores do not interfere with the measurement. As with Raman spectroscopy, the FT-IR spectrum is sensitive to changes in phase of polymorphic lipids. Hence, FT-IR has been used to monitor the pretransition in phosphatidylcholine bilayers (181), the main gel-to-liquid crystalline transition (180), as well as the transition from lamellar to hexagonal (H_{II}) exhibited by egg yolk phosphatidylethanolamine (919).

Changes in lipid chain conformation can be monitored by the frequency shifts of the CH_2 absorption bands, and these can be interpreted in terms of changes in the number of *gauche* isomers in the chains. For example, cholesterol causes the number of *gauche* configurations in dipalmitoyl phosphatidylcholine to decrease above the main phase transition temperature (250), consistent with changes in the order parameter monitored by ^{2}H-NMR (1092). The incorporation of intrinsic membrane proteins in the bilayer (see Chapter 5), however, has quite a different effect, resulting in little or no change in the number of *gauche* isomers in the liquid crystalline phase but increasing the *gauche* isomers in the gel phase. The proteins interfere with the ability of the acyl chains to pack in the all-*trans* configuration (250).

^{2}H-NMR

By far the most detailed picture of the structure of the interior of the lipid bilayer has emerged from ^{2}H-NMR (for reviews see 1319, 291). Deuterium can be chemically substituted for hydrogen at specific places in the lipid molecule. This is a relatively benign substitution and is generally considered nonperturbing. The spectra of several deuterated derivatives of dimyristoyl phosphatidylcholine are shown in Figure 2.13. The separation between the two peaks is the quadrupolar splitting, $\Delta\upsilon_Q$, and this is dependent on the time-averaged orientation between the C-D bond vector and the bilayer normal (Figure 2.14). This time-averaged orientation is usually quantified in terms of an order parameter by the following equation:

$$S_{CD} = 1/2\ (3\langle\cos^2\theta\rangle - 1)$$

where $\langle\cos^2\theta\rangle$ implies a time average and S_{CD} is the bond order parameter. Note that what is measured is the average of this value over all the molecules.

For $\theta = 0°$, $\langle\cos^2\theta\rangle = 1$ and $S_{CD} = 1$
For $\theta = 90°$, $\langle\cos^2\theta\rangle = 0$ and $S_{CD} = -1/2$
Random orientation, $\langle\cos^2\theta\rangle = 1/3$ and $S_{CD} = 0$

Frequently, the *molecular order parameter* (S_{mol}) is reported, which represents the orientation of the vector perpendicular to the plane formed by the CD_2 group (Figure 2.14). This indicates the average orientation of this segment of the acyl chain.

$$S_{mol} = -2S_{CD}$$

The order parameter obtained from ^{2}H-NMR reflects the average orientation and says little about the dynamics of the system or the range of motion.

Of critical importance is the fact that the local magnetic field sensed by a particular deuterium depends on the orientation of the C-D bond with respect to the external magnetic field. The motions due to molecular vibrations and rotations which influence the orientation of the C-D bond in the bilayer, in general, occur at rates sufficiently rapid ($>10^6$ sec^{-1}) so that any particular deuterium senses a single average magnetic environment. The particular environment sensed depends on the surrounding atoms as well as on any constraints on the range of motion. This can be contrasted with Raman or infrared spectroscopies where *trans* and *gauche* rotamers interconvert at far slower rates (10^9 sec^{-1}) than the frequency

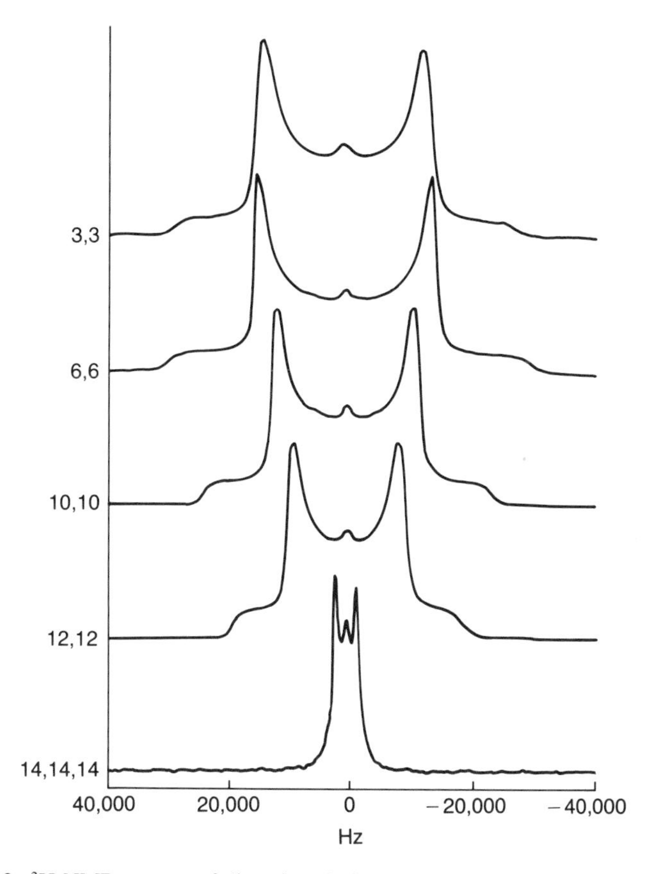

Figure 2.13. ^{2}H-NMR spectra of dimyristoyl phosphatidylcholine deuterated at different positions on the acyl chain. The numbers on the left indicate the position of the 2 (or 3) deuterium atoms in each chain. Note that the spectrum from the sample deuterated at the terminal methyl residue (14, 14, 14) is much more narrow, indicating considerable disorder in the middle of the bilayer. The figure was kindly provided by Dr. E. Oldfield.

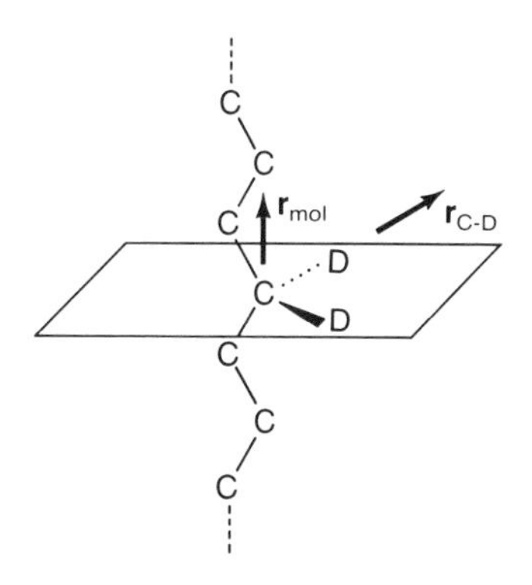

Figure 2.14. An illustration of the molecular vector ($\mathbf{r}_{mol}$) and the C-D bond vector ($\mathbf{r}_{C-D}$) which are used for computing order parameters obtained from ^{2}H-NMR.

difference separating the vibrational bands associated with each form (~10^{12} Hz). Hence, IR and Raman spectroscopies give what amounts to an instantaneous snapshot, summing spectroscopic contributions for *trans* and *gauche* rotamers. Other techniques, in particular ESR and fluorescence, are also used to obtain order parameters, and the use of these probe-dependent techniques will be discussed in Chapter 5.

Figure 2.15 shows the ^{2}H-NMR order parameters for a series of specifically deuterated derivatives of various phospholipids where the deuterium is present in specific methylene groups on the *sn*-1 palmitic acyl chain. The measurements are made on lipids or membranes in the liquid crystalline phase, since in the gel phase the spectra are very broad due to the rigidity of the lipids and are difficult to analyze. Two general remarks can be made about these results:

1. The order parameter is relatively constant for C-2 through about C-8 or C-10. The methylenes located toward the middle of the bilayer show considerably more disorder than those near the surface.
2. The same order parameter profile is obtained for synthetic lipids of various types, including phosphatidylcholine, phosphatidylserine, and sphingomyelin, and also for biological membranes which have incorporated deutrated probes. Hence, the profile is basically not sensitive to lipid chemical structure or membrane composition so long as the bilayer is in the liquid crystalline phase.

Quantitative interpretations of these data require molecular modeling using statistical mechanical techniques (e.g., 921; for review see 1161). For example,

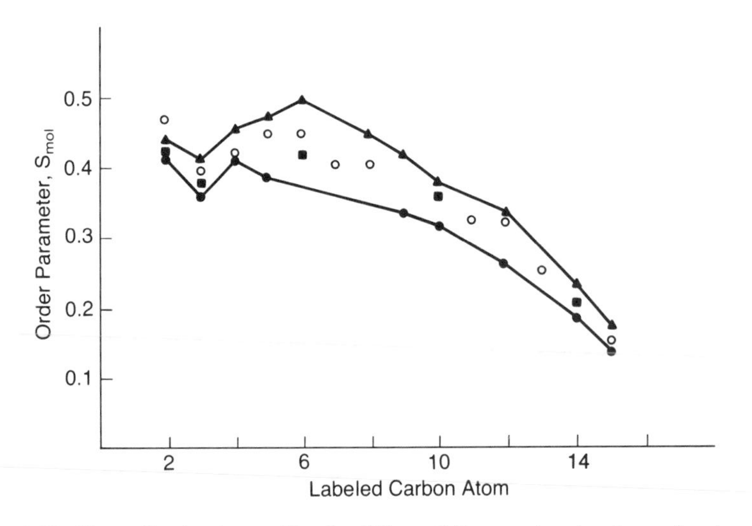

Figure 2.15. Normalized order profiles for different bilayers, showing the molecular order parameter as a function of position in the acyl chain. Closed circles, dipalmitoyl phosphatidylcholine; closed triangles, 1-palmitoyl-2-oleoyl phosphatidylcholine; closed squares, dipalmitoyl phosphatidylserine; open circles, *Acholeplasma laidlawii* membranes. Adapted from ref. 1321. Reprinted by permission of the publisher from "General Features of Phospholipid Conformation," by J. Seelig and J. L. Browning, FEBS Lett. 92, pp. 41-44, Copyright 1978 by Elsevier Science Publishing Co., Inc.

the data are consistent with the presence of about four or five *gauche* rotamers in each chain in dipalmitoyl phosphatidylcholine, but very few kinks (0.5/chain) (see 1320). Because each chain is essentially tethered at the bilayer surface, there is greater order in the portion of acyl chain nearest the surface. Note that the bilayer is obviously a highly cooperative system. An acyl chain cannot change direction without compensating changes in the neighboring chains. Hence, groups of adjacent chain segments must move in a cooperative manner. Deviations in chain segment direction from the bilayer normal will accumulate as one proceeds from bilayer surface to the interior. Hence, the disorder is maximal at the center of the bilayer, and in this region the chains appear similar to those in a paraffin liquid. Other techniques show that molecular motions are also maximized at the bilayer center. However, note that disorder is a static property and implies nothing about motions. For example, one can have a highly disordered structure, such as window glass, which displays little motion (see Chapter 5).

Neutron Diffraction

The specifically deuterated phospholipids can also be studied by neutron diffraction (see 1320). Whereas X-ray scattering reflects the electron distribution, neutrons are scattered by atomic nuclei. The scattering properties of ^1H and ^2H are very different, which permits the localization of deuterated sites in the scattering density profile. For example, the C-5 position can be localized 15 Å from the bilayer center in dipalmitoyl phosphatidylcholine in the gel phase, consistent with an all-*trans* chain configuration (161). By using D_2O as solvent, the location of water in the gel phase has been determined, demonstrating penetration up to the level of the glycerol moiety (1606, 622). The location of ions such as Ca^{2+} has also been studied using neutron scattering (622) (see Chapter 7).

2.3 The Thermodynamics of Lipid Polymorphism

The structural data presented in previous sections show that hydrated lipids exhibit polymorphism. In all the lipid structures, the nonpolar hydrocarbon portions of each molecule are aggregated and the polar headgroups are in contact with water. In this section, we will briefly describe the thermodynamic principles of micelle formation by amphiphilic lipids. The thermodynamic arguments will be summarized diagrammatically in terms of molecular shapes, providing a useful rationale for such diverse questions as why detergents disrupt membranes and how cholesterol affects the phospholipid bilayer.

2.31 The Hydrophobic Force (for review, see 1432)

The major thermodynamic driving force stabilizing hydrated lipid aggregates is the hydrophobic force. Other stabilizing factors are

1. *van der Waals forces*: Short, weak attractive forces between adjacent hydrocarbon chains. The attraction results from interactions between polarizable electrons (induced dipoles).

2. *Hydrogen bonding*: Between polar headgroups of some lipid molecules such as phosphatidylethanolamine (see 115). Intermolecular bridging by divalent cations can also be important in some circumstances with anionic lipids.

These are relatively minor stabilizing factors, however, compared to the hydrophobic force.

The hydrophobic force is the thermodynamic drive for the system to adopt a conformation in which contact between the nonpolar portions of the lipids and water is minimized. This so-called "force" is entropic in origin and results from the unfavorable constraints placed on water as it packs around a nonpolar hydrocarbon.

The structure and dynamics of pure water are complex but are clearly dominated by intermolecular hydrogen bonds (see 1392 for review). When an ion such as Cl^- is placed in water it becomes solvated and water molecules form a hydration shell around the ion. The orientation of these water molecules is unfavorable entropically, but this is more than compensated by a large favorable electrostatic interaction, so the overall free energy change in going from a salt crystal to the dissolved salt is favorable. When a nonpolar substance is dissolved in water, it also causes an unfavorable organization of the water around each molecule. In essence, the water molecules orient themselves to maintain intermolecular hydrogen bonds (each worth about 5-7 kcal/mol), but since those solvent molecules in direct contact with the non-polar solute molecule have fewer water molecules as neighbors, there are substantial configurational constraints on the system. Hence, there is a decrease in the entropy of the system. However, there is no large compensating electrostatic interaction as in the case of ionic or polar solutes. As a result, the net free energy change upon transferring a nonpolar solute from a nonpolar solvent (e.g., heptane) to water is unfavorable due to this entropic effect on the water solvent. Possible models for the way in which water molecules orient about nonpolar solutes are exhibited by crystals of hydrated nonpolar molecules or atoms (e.g., argon) showing water cages or clathrates surrounding the "solute" (see 460).

The unfavorable interaction between nonpolar solutes and water is what is termed the "hydrophobic force," and thermodynamic measurements can quantify the tendency for nonpolar materials to minimize their contact with water. This is a major stabilizing factor for virtually all biological macromolecular structures, including globular proteins as well as the phospholipid bilayer. The "hydrophobicity" of simple molecules such as alkanes can be quantified simply by measuring the equilibrium distribution of the solute (e.g., ethane) between two solvents, e.g., water and heptane.

Expressing the solute concentration in mole fraction units in water $[X]_{H_2O}$, or hydrocarbon, $[X]_{HC}$, we define an equilibrium constant K:

$$\frac{[X]_{H_2O}}{[X]_{HC}} = K \qquad [2.1]$$

and $-RT \ln K = \Delta G^\circ_{trans} = (\mu^\circ_{H_2O} - \mu^\circ_{HC})$ [2.2]

The standard state transfer free energy, ΔG°_{trans}, is a measure of hydrophobicity (see 1432). Hydrophobicity has been shown to be proportional to the surface area of contact between water and the nonpolar solute (1212). Larger molecules (e.g., long-chain alkanes) cause a larger perturbation on the water structure because there is more area of contact. Figure 2.16 shows that for a series of alkanes, the hydrophobicity (ΔG°_{trans}) increases in proportion with surface area.Using van der Waals radii to compute the surface area of contact between water molecules and the alkanes, it was computed that ΔG°_{trans} changes by about -25 cal/Å^2. For straight chain alkanes the hydrophobicity changes by about -800 cal/mole of $-CH_2-$. In other words, for every increase in chain length of two methylenes, the equilibrium constant changes by about a factor of 10 in favor of the hydrocarbon solvent.

2.32 Micelle Formation

Consider what happens when a long-chain alkane is dissolved in water. Because of the very unfavorable "hydrophobic" interaction described in the previous section, the solubility will be very low. Up to a point, alkanes such as dodecane (C_{12}) can be dissolved in water, but beyond a certain concentration the dodecane will form a separate phase. Further additions of dodecane will simply increase the amount in the separate phase of dodecane and not increase the concentration of dissolved hydrocarbon in the aqueous phase (see Figure 2.17).

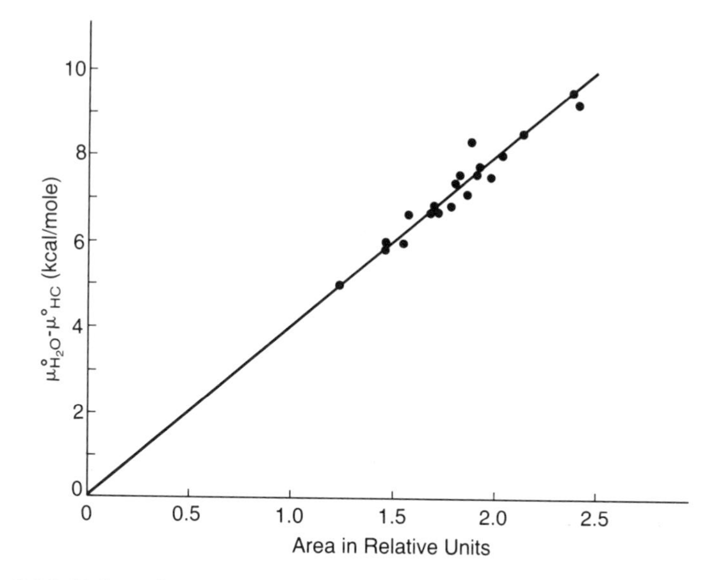

Figure 2.16. Unitary free energy of transfer of hydrocarbons from the pure liquid to aqueous solution at 25°C plotted as a function of relative surface area. The area of isobutane on this scale is 1.45. Adapted from ref. 1212.

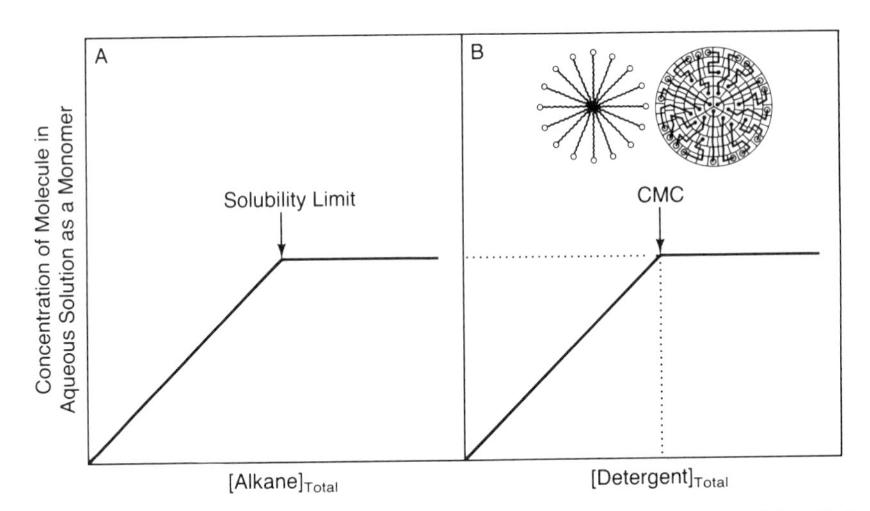

Figure 2.17. Schematic to illustrate the conceptual relationship between a solubility limit and the critical micelle concentration (CMC). Alkanes will dissolve in water up to a defined concentration, at which point a separate phase forms (part A). Amphiphiles, such as detergents, will dissolve up to a point where micelles start to form, defined as the CMC (part B). Also shown in part B are two drawings showing the cross section of a spherical micelle. The drawing on the left is the common way of indicating a micelle, but it is not realistic because the density of material in the center would be much greater than near the edge and the chains cannot pack together in this way. The drawing on the right shows a more realistic distribution of conformations of the hydrocarbon chains in a micelle, based on a statistical theory (ref. 336).

Now consider what happens when an amphiphilic molecule such as dodecyl sulfate (SDS) is added to water. This common detergent has both a nonpolar portion (dodecyl chain) and a highly charged polar group (sulfate) at one end. When the solubility limit for the monomeric form of this detergent is reached, it will also form a separate phase. However, the "phase" in this case is in the form of aggregates, called micelles, dispersed throughout the solution. Because of the highly favorable interactions between the polar headgroup (sulfate in this case) and water, it is preferable to maintain this part of the molecule in contact with water while still excluding water from contact with the nonpolar portion of the molecule. The result is a globular aggregate. The concentration at which 50% of the detergent is in the form of micelles is called the critical micelle concentration, or CMC. Operationally, it is often convenient to define the CMC as the concentration where micelles first appear. The CMC is similar to a solubility limit for the monomeric form of the molecule. Additions of dodecyl sulfate beyond the CMC essentially increase the concentration of micelles.

Lipid aggregates or micelles come in many sizes and shapes. Dodecyl sulfate in water forms spherical micelles with about 60 molecules per micelle. Other detergents or amphiphiles form globular or rodlike aggregates. Phospholipids spontaneously aggregate to form bilayers, and the bilayer is essentially another micelle form. The reason why biological phospholipids form stable bilayers is not difficult to understand and is described in the next section. First, however, we will quantify the relationship between molecular hydrophobicity and the CMC.

An aqueous solution of an amphiphile might consist of a mixture of species, including monomers ($N = 1$) and various aggregates containing many molecules each. At equilibrium, the chemical potential of the amphiphile in each of the various forms will be the same (see 683, 682).

$$\mu_N = \mu^\circ_N + \frac{kT}{N} \ln (X_N/N) = \text{constant, same for all } N \qquad [2.3]$$

where μ°_N is the standard state chemical potential of species with N molecules, X_N is the mole fraction of the amphiphile which is in the aggregated species with N molecules, k is Boltzmann's constant, and T is temperature. For simplicity we will assume a monodisperse system with M molecules per aggregate. That is, there is only one aggregated form which exists ($N = M$) in equilibrium with the monomer. Although this is clearly an approximation, it is not an unreasonable assumption for molecules that form spherical micelles or small bilayer, single-walled vesicles. Now we apply the condition for equilibrium at the critical micelle concentration, and define the CMC as the concentration where $X_1 = X_M$ ($\equiv X_{CMC}$)

$$\mu_1 = \mu_M$$

$$\mu^\circ_1 + kT \ln (X_{CMC}) = \mu^\circ_M + \frac{kT}{M} \ln \left(\frac{X_{CMC}}{M}\right)$$

$$\text{or } \Delta G^\circ_{mic} = (\mu^\circ_M - \mu^\circ_1) \cong kT \ln X_{CMC} \qquad [2.4]$$

A major component of ΔG°_{mic} is the hydrophobic transfer free energy due to excluding water from the nonpolar portion of the amphiphile when burying it inside the micelle. Note that a more negative ΔG°_{mic} results in a smaller value of X_{CMC}. That is, very hydrophobic molecules tend to aggregate at lower concentrations. In fact, for simple amphiphiles with single alkane chains (e.g., alkyl sulfates such as dodecyl sulfate) the chain length dependence of ΔG°_{mic} is very similar to the chain length dependence of ΔG°_{trans} for alkyl chains (equation 2.2). Thermodynamically, transferring a nonpolar moiety from water to a liquid hydrocarbon is similar to transferring it to the interior of a micelle. Quantitatively, this means that for each increase in chain length of two methylene groups, there will be decrease in the CMC of about an order of magnitude.

The importance of this to membranes is that for biological phospholipids, which typically have two long alkyl chains per molecule, the hydrophobic component of ΔG°_{mic} very strongly favors the aggregated state (the bilayer). The CMC for such lipids is $< 10^{-10}$ M. In other words, for most purposes, the concentration of monomeric phospholipids in equilibrium with the membrane is negligible. Nature has designed special proteins for binding and transferring the monomeric forms of lipids within or between cells, and these are discussed in Chapter 10.

2.33 Micelle Shapes: Why Does a Bilayer Form?

As developed in the previous section, it is clear that biological phospholipids spontaneously aggregate in aqueous solution. We now must consider the problem of what micelle form is most favored. Here it should be noted once again that some biological phospholipids, such as phosphatidylethanolamine with unsaturated fatty acyl chains, do not form stable bilayers when dispersed in water. In order to understand the stability of the bilayer and the possible role of "non-bilayer-forming" membrane components, it is necessary to develop further the thermodynamics of these systems and to discuss the concept of lipid shapes in relation to the constraints of packing lipids together in micelles. Readers interested in a more qualitative, pictorial view can skip directly to the next section.

Why don't membrane phospholipids form globular micelles? In considering the problem of how amphiphilic molecules pack into a particular micelle geometry (e.g., a sphere), it is convenient to consider the packing requirements of the molecule in two parts. First, the nonpolar portion of the molecule has a fixed molecular volume (v) and a maximal length (605). Without any other considerations, this will determine the maximal radius of a spherical micelle, for example, as well as the number of molecules that can fit into the micelle. The second factor that must be considered, however, is the optimal surface area required by the polar headgroup (S_o). For biological phospholipids, the area per molecule in a hypothetical spherical micelle is much larger than the optimal value for headgroup packing and, hence, these amphiphiles do not form stable spherical micelles. In what follows, we will first consider the factors which determine the *optimal surface area per molecule* for an amphiphile at a micelle surface and then see how this value can be incorporated in a *critical packing parameter* which can be used to determine what micelle shape is favored for any amphiphile.

Optimal Surface Area per Molecule

By definition, the distribution of lipids in various aggregated forms (e.g., spherical micelles, bilayers, rods...) is determined by the relative standard state chemical potentials (or mean free energies) of the molecules in each structure, μ°_N. In order to proceed we need to assign a reasonable form for μ°_N, taking the geometry of the micelle into account. Following the development of Israelachvili and his colleagues (683, 682), we will postulate contributions from three terms for the chemical potential of a molecule with average surface area, S, in a micelle with aggregation number N. It is the dependence of μ°_N on the average molecular surface area which allows us to account for micelle geometry.

$$\mu^\circ_N = \underset{1}{\gamma S} + \underset{2}{C/S} + \underset{3}{H_N} \qquad [2.5]$$

Terms 1 and 2 are interfacial attractive and repulsive terms, respectively, and term 3 is a bulk energetic term. The origin of these terms is not hard to see. The H (term 3) is the free energy associated with the alkyl chains. To a first approximation, this will be similar in all micelles in which the alkyl chains are buried in a hydrocarbon-like environment from which water is excluded. It is this term, or more precisely $(H_N\text{-}H_1)$, which changes with the chain length of the nonpolar part of the molecule and determines the chain length dependence of the CMC. $(H_N\text{-}H_1)$ is the hydrophobicity.

Terms 1 and 2 are energetic contributions due to intermolecular interactions at the water–hydrocarbon interface. These terms vary depending on how close together the lipids are packed in the micelle and, thus, are dependent on micelle shape. The micelle shape which is most favored is that which minimizes the free energy of the system, and it can be, at least qualitatively, understood in terms of the following simple thermodynamic expressions.

Term 1: γS, interfacial surface tension. An attractive term equivalent to the surface tension stabilizing liquid–liquid interfaces in water–hydrocarbon systems. γ is the surface tension expressed as energy per cm^2 and can be estimated as about 50 erg/cm^2. This is equivalent to the work required to change the interfacial area per cm^2 with a lateral pressure of 50 dyne/cm. Surface tension can also be thought of as a "negative pressure" due to the various attractive molecular interactions at the interface and expressed in units of dyne/cm (50 dyne/cm in this case).

Term 2: C/S, repulsive intermolecular interactions. A crude way of lumping together all sorts of repulsive interactions at the interface, including electrostatic and steric interactions. The essence of the term is that these repulsive interactions vary inversely as the average area per molecule at the micelle hydrocarbon surface. In other words, when the molecules are close together (small area per molecule) these interactions become very large and unfavorable.

These two terms represent what Tanford has called the principle of opposing forces (1432). The drive for the molecules to associate is counteracted by repulsive interactions, lumped into the constant C, which ultimately determines the optimal packing density in the micelle.

The optimal value of S is obtained by setting $(d\mu^\circ_N/dS) = 0$, i.e., minimizing the free energy with respect to molecular surface area. The result gives S , the optimal surface area per molecule:

$$S_0 = \sqrt{C/\gamma} \qquad [2.6]$$

Even in this oversimplified model, one can see that the molecular constant C determines S_0. For dodecyl sulfate, for example, one would expect a large electrostatic repulsive interaction between the charged sulfates at the micelle surface (e.g., large C) to result in a large value of S_0, especially at low ionic strength. The polar headgroup of this molecule essentially requires a large surface area at the micelle surface to keep the charged groups far apart. This requirement dominates and determines the preferred spherical micelle shape. To see how, consider the following.

Micelle Geometry and the Critical Packing Parameter

Three molecular parameters must be considered in determining the most stable micelle geometry:

1. S_o, the optimal surface area occupied by the molecule at the hydrocarbon interface. This will be, in part, dependent on the solution conditions, especially ionic strength in the case of charged molecules.
2. l, the maximum length of the alkyl chain for simple single-chain amphiphiles and for phospholipids. This will determine the upper limit on micelle size, such as the radius of a spherical micelle or thickness of a bilayer. In no case are micelles considered as having holes or gaps, so the radius of the spherical micelle cannot exceed l, though it could be shorter. Usually this distance is slightly shorter than the fully extended, all *trans*, configuration.
3. v, the molecular volume of the hydrocarbon portion of the amphiphile. The micelle volume enclosed within the envelope of the hydrocarbon–aqueous interface is considered to be equal to Mv, where M is the number of molecules in the micelle.

The surface area available per unit volume is purely a function of micelle geometry and it is this which basically determines what kind of micelle is formed by different amphiphiles. Let us briefly consider some possible micellar forms:

1. *Spherical*: Given the dimensional constraints imposed by the length of the lipid chain, the sphere has the highest area/volume ratio of any form and is favored by lipids with a large value of S_o, such as dodecyl sulfate in water.
2. *Distorted spheres:* Have a lower surface/volume ratio than spheres:
   (a) *Ellipsoids*: considered to be unlikely (682) because packing would be highly unfavorable in many places (e.g., the edge of oblate ellipsoid)
   (b) *Globular*: bi-lobed, like two spheres merged. A likely form (682).
3. *Rods and Cylinders*: Even lower surface/volume ratio. The ends would likely be hemispherical so as to exclude water from the nonpolar portions yet maintain reasonable packing.
4. *Bilayer*: Smallest surface/volume ratio, favored by lipids with a large molecular volume, such as lipids with two alkyl chains. Note that disks or flat sheets would be highly disfavored due to contact by water at the disk edge. Spherical bilayer vesicles (liposomes) eliminate this edge and are also favored by being smaller and, hence, entropically preferable to very large bilayer sheets. Some proteins and peptides can stabilize phospholipid disks (see Chapter 3).

The parameter (v/lS_o) can be used to predict (or rationalize, in retrospect) which micelle form will be favored for a particular molecule. This is called the *critical packing parameter*, and it is a number containing the packing requirements of the amphiphile, including the volume and length of the nonpolar portion and the optimal surface area of the polar headgroup.

For example, consider a spherical micelle with radius R and containing M molecules.

$$\text{Total micelle area} = MS_o = 4\pi R^2$$

$$\text{Total micelle volume} = Mv = 4/3\pi R^3$$

so the micelle radius

$$R = \frac{3v}{S_o}$$

But the micelle radius must be less than or equal to l ($R \leq l$), the maximum length allowed for the lipid chain, so a criterion for a lipid being able to pack into a sperical micelle is

$$\frac{v}{lS_o} \leq 1/3 \qquad [2.7]$$

The same calculation is easily done for cylinder and planar bilayer shapes with the critical values being

$$\text{cylinder: } (v/lS_o) = 1/2 \qquad [2.8a]$$

$$\text{bilayer: } (v/lS_o) = 1 \qquad [2.8b]$$

This leads to the following predictions, given values for v, l, and S_o. If the value of (v/lS_o) is less than 1/3, a spherical micelle is expected; if between 1/3 and 1/2, the micelles will be globular or cylindrical; and if the value is between 1/2 and 1 the lipid should form a stable bilayer. It is the two long acyl chains in biological phospholipids which increase the bulk packing requirement (large molecular volume, v) which results in the stable bilayer. Single-chain phospholipids, like most synthetic detergents, have values of (v/lS_o) between 1/3 and 1/2 and do not form stable bilayers. Similarly, diacyl phospholipids with very short chains (e.g., $n = 6$) do not form stable bilayers for the same reason.

Of particular interest is that some biological lipids have values of $(v/lS_o) > 1$ and, hence, do not form stable bilayers. These pure lipids, which have relatively small polar headgroups, form inverted hexagonal phase (H_{II}) aggregates, as we have already seen. The role that such lipids play in biomembranes is uncertain but subject to considerable speculation (e.g., 263, 1041, 533, 303, 267).

2.34 Lipid Shapes

The previous discussion began with a consideration of the thermodynamics of lipid aggregation, but it is clear that an excellent qualitative understanding is possible just by considering the packing requirements of different lipids and the constraints imposed by simple geometric considerations. This is pictorially represented by considering the gross shapes of lipid molecules, in particular, comparing the cross-sectional area of the hydrocarbon portion (crudely, v/l) to the optimal surface area required by the polar headgroup (S_o). The same considerations were used in rationalizing differences in phosphatidylethanolamine and phosphatidylcholine in crystals and in the gel phase bilayer (Sections 2.1, 2.2). Lipids can be simply classed as cones, cylinders, or inverted cones depending on the relative packing requirements of these two regions of the lipid. Figure 2.18 summarizes this modeling with examples. This is essentially a pictorial representation of the thermodynamic consequences as discussed in the previous sections,

LIPID	PHASE	MOLECULAR SHAPE	CRITICAL PACKING PARAMETER (v/ℓS.)
Lysophospholipids Detergents	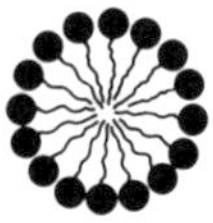 Micellar	 Inverted Cone	<⅓ (Sphere) ⅓ to ½ (Globular Shapes; Rods)
Phosphatidylcholine Sphingomyelin Phosphatidylserine Phosphatidylinositol Phosphatidylglycerol Phosphatidic Acid Cardiolipin Digalactosyldiglyceride	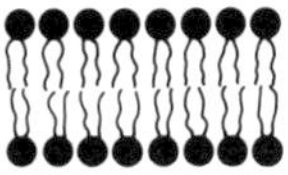 Bilayer	Cylindrical	½ to 1
Phosphatidylethanolamine (Unsaturated) Cardiolipin - Ca^{2+} Phosphatidic Acid - Ca^{2+} (pH < 6.0) Phosphatidic Acid (pH < 3.0) Phosphatidylserine (pH < 4.0) Monogalactosyldiglyceride	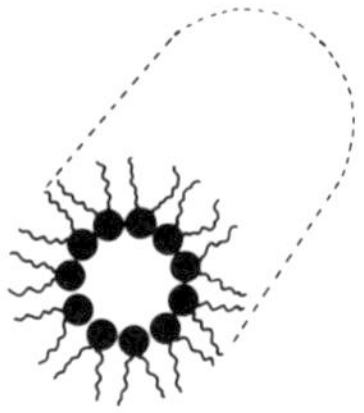 Hexagonal (H_{II})	 Cone	>1

Figure 2.18. Polymorphic phases, molecular shapes, and the critical packing parameter for some membrane lipids. Adapted from ref. 263. Drawing kindly provided by Dr. P. Cullis and Dr. M. Hope.

but it allows one to rationalize quickly large amounts of experimental data, at least qualitatively. Quite possibly, these same simple concepts of lipid shape can be used to understand the roles specific lipids play within the bilayer, such as stabilizing regions of high curvature or packing around membrane proteins (263, 1041, 533, 303, 267).

2.4 Lipid Phase Transitions

The thermodynamics of lipid phase transitions has been the subject of numerous experimental (see 829) and theoretical (see 1048, 1163, 1008) studies. Although thermodynamic parameters (e.g., transition temperature, $\Delta H°$, $\Delta S°$) do not give structural information, they must be consistent with any physical chemical model of the lipid. The ultimate goal of these studies is the quantitative and qualitative description of the biomembrane, including the behavior of the phospholipids in the bilayer and the effects of "perturbants" such as proteins or cholesterol. The goal is far from being realized, but considerable insight derives from concepts and models based on studies of simple systems, starting with aqueous dispersions of pure, homogeneous lipids and proceeding to binary mixtures of different phospholipids as with cholesterol or with specific proteins reconstituted with phospholipids. It must be realized, however, that, generally, biological membranes do not exhibit phase transitions. Hence, much of the material discussed in this section is not directly pertinent to biomembranes. However, the basic physical chemistry of lipids and lipid–protein mixtures is relevant to biomembranes, especially as regards possible lateral inhomogeneities (see Section 4.5) and the way in which components interact in the bilayer.

The most studied lipid phase transition is between the lamellar gel and liquid crystalline phases. However, lipid phase transitions have also been examined between the lamellar and the inverted hexagonal (H_{II}) phases, as well as between other mesomorphic lipid forms. Phase transitions can be induced in several ways, included changes in pressure (215), temperature (e.g., 1444, 1120), ionic strength, or pH (e.g., 194, 1467), depending on the lipid being examined. It is most common to measure the thermally induced or thermotropic transition, not only because it may be biologically meaningful in some cases as in organisms that do not control their own temperature (Section 10.5) but also because the measurements can be made relatively easily and with great precision, and the results directly yield useful information on heat capacities and $\Delta H°$ values. These values can be used for comparison with theoretical models. The technique commonly applied is differential scanning calorimetry (DSC).

2.41 Differential Scanning Calorimetry (DSC) (see 56, 949 for reviews)

This technique is of primary importance in obtaining information about the thermodynamics of model membranes and biomembranes. It is used to monitor and characterize changes in physical state in polymorphic lipids and also to characterize the perturbations on pure lipids by the interactions with other materials, such as other lipids, proteins, ions, or small hydrophobic molecules. In DSC, a sample and inert reference are heated independently to maintain an identical temperature in each. The heat for the endothermic gel-to-liquid crystalline bilayer transition, for example, would be required in excess over the heat required to maintain the same temperature in the reference. Differential heat flow is then plotted as a function of temperature. Highly sensitive instruments allow one to use samples of dilute aqueous suspensions of lipids (1 mg/ml, 1 ml sample size). The parameters reported from this technique are:

1. Transition temperature, T_c: the temperature marking the beginning of the transition.
2. Transition midpoint, T_m: where the transition is 50% complete.

3. Transition enthalpy, ΔH: the actual heat required for the entire transition normalized per mole or per unit weight.
4. Heat capacity, C_p: the amount of heat (per gram or per mole) required to raise the temperature of the sample by 1 degree.

Box 2.2 Information Is Obtained from Both the Midpoint and Width of the Transition

For a two-state transition, T_m is defined as the point where $\Delta G° = O$, so

$$\Delta G° = O = \Delta H° - T_m \Delta S°$$

$$T_m = \Delta H/\Delta S° \qquad [2.9]$$

which can be used to obtain $\Delta S°$, the transition entropy, from measured values of T_m and $\Delta H°$.

The width of the transition can be an important parameter. This is quantified in terms of the slope of a van't Hoff plot. The data are analyzed in terms of a two-state model (e.g., gel and liquid crystalline phases) and the extent of the transition is used to calculate the fractions in the gel and liquid crystalline state at any temperature. For example, at T_m, the fraction of each state is by definition 0.5. An equilibrium constant, K, is defined as

$$K = \frac{\text{[fraction in liquid crystalline state]}}{\text{[fraction in gel state]}} \qquad [2.10]$$

The van't Hoff plot is ln K vs. $1/T$ and yields as a slope $\Delta H°_{vH}/R$, giving the transition enthalpy (calories/mol). However, this is normalized per mole of the cooperative unit that is actually undergoing the transition. If lipids were to melt in units of 100 molecules, representing a "cooperative unit," the $\Delta H°_{vH}$ would be 100 times the "calorimetric" $\Delta H°_{cal}$, obtained directly by DSC and normalized per mole of lipid molecule.

Thus, a large cooperative unit or highly cooperative thermal phase transition is characterized by a steep slope of the van't Hoff plot (large $|\Delta H°_{vH}|$) and a very sharp transition. Smaller values of $|\Delta H°_{vH}|$ result in broader (less cooperative) thermal transitions.

The cooperative unit is defined as follows:

$$\text{Cooperative unit} = \frac{\Delta H°_{vH}}{\Delta H°_{cal}} \qquad [2.11]$$

A major problem is that a broad transition can result from other causes, most notably the presence of even small amounts of impurities in the lipid bilayer (16). It is quite common, however, to interpret changes in the transition breadth in phospholipid bilayers upon the addition of "perturbants" (e.g., proteins, cholesterol) as being due to changes in intermolecular lipid chain interactions resulting in a decrease in the size of the cooperative unit. Often, the addition of such perturbants will also result in a lower value for $\Delta H°_{cal}$, and this is usually interpreted as due to a fraction of the lipid being "sequestered" in some manner and not participating in the "bulk" phase transition. Some hydrophobic intrinsic membrane proteins (e.g., Ca^{2+}-ATPase) have this effect, for example (see 526).

Figure 2.19 shows some DSC scans for several phospholipids. For each lipid, the phase transition between the gel and liquid crystalline phases is evidenced by a sharp peak in the heat capacity over a narrow temperature range. The excess heat is required to convert the lipid from the gel to the liquid crystalline phase and the process is usually likened to a simple melting of lipid such as the liquid water/ice transition. The midpoint of the thermal transition is often referred to as a melting temperature. This transition is thought to be a first-order transition which, in theory, is characterized by an infinite heat capacity at the transition temperature. In practice, this is not the case, and the transitions are also often spread over a width of several degrees. It appears that the major cause for the breadth of many of the reported transitions of single-component lipids is the

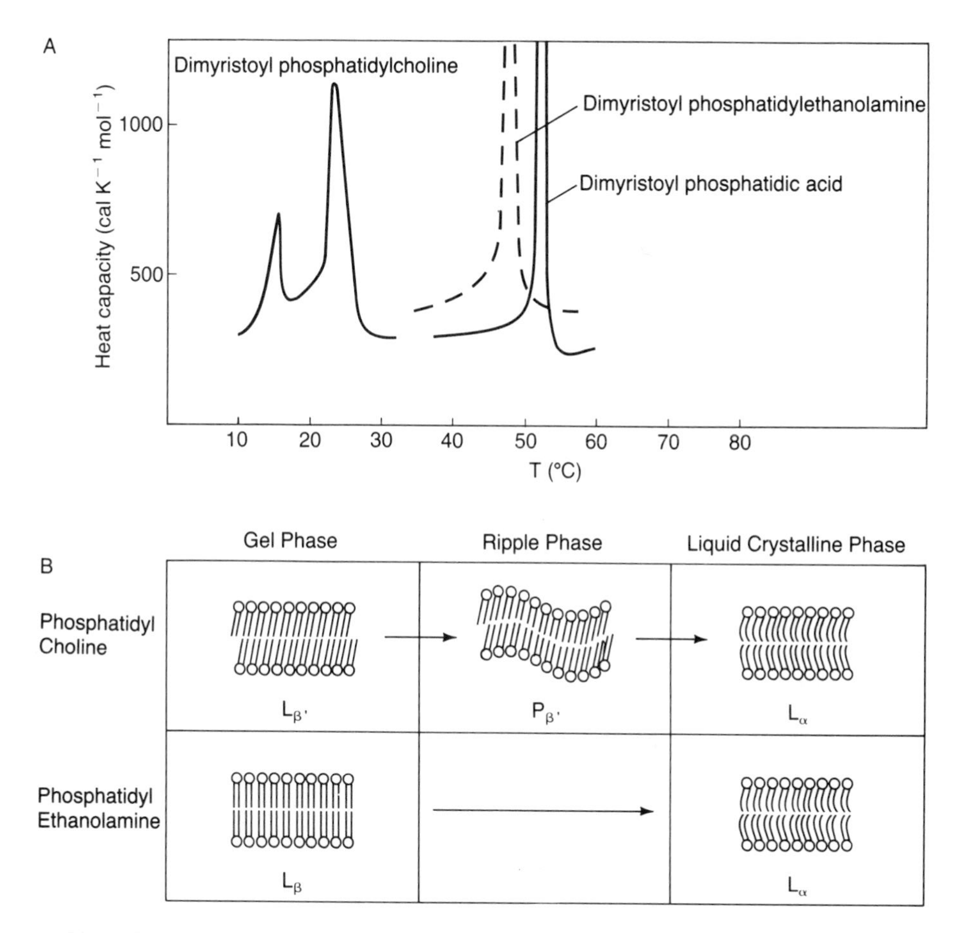

Figure 2.19. (A) Differential scanning calorimetry profiles of three phospholipids. Adapted from ref. 112. Reprinted with permission from "Apparent Molar Heat Capacities of Phospholipids in Aqueous Dispersion," by A. Blume, Biochemistry 22, pp. 5437-5438, Copyright 1983, American Chemical Society.(B) Schematic showing the molecular organization of phosphatidylcholine and phosphatidylethanolamine as a function of temperature. Adapted from ref. 112.

presence of minor impurities. For example, the phase transition of highly purified dipalmitoyl phosphatidylcholine is very sharp and is quite sensitive to impurities (16).

The melting temperature of a lipid is determined by a balance of competing factors. Entropically, disordered chains in the liquid crystalline state, characterized by *gauche* conformations, are favored over the highly ordered all-*trans* chain configuration in the gel state. However, attractive chain contacts (van der Waals interactions) are favored by the more ordered gel phase. In addition, the lower cross–sectional area required by the acyl chains in the ordered state results in altering the distance between nearest neighbor polar headgroups. Close interactions may be favorable, as in the case where there are intermolecular hydrogen bonds or bridging divalent metal cations (e.g., Ca^{2+}), or they may be unfavorable, as in the case where bulky groups interact sterically or charged lipids interact electrostatically. In these latter cases, the state of protonation (pH), the ionic strength, and the presence of divalent cations can have dramatic effects on the T_m value for a particular lipid. As the temperature is increased, eventually the entropic effect dominates to stabilize the state with the *gauche* rotamers.

Changes in polar headgroup interactions (e.g., by changing ionic strength) will alter the T_m, since these are primarily changes in $\Delta H°$, and $T_m = \Delta H°/\Delta S°$. For example, high ionic strength will reduce the repulsive electrostatic interactions between the phosphate groups in phosphatidic acid bilayers. However, the effect will be greater (more stabilization) for the gel phase since the polar group density, i.e., charge density, is larger than in the liquid crystalline state. High ionic strength, therefore, results in a larger magnitude for $\Delta H°$, and an increase in T_m. Thus, one can induce phase transitions isothermally under certain conditions by altering the ionic strength or other parameters which affect polar group interactions (see 194).

Table 2.1 and Figure 2.20 summarize some data obtained with pure single component phospholipid systems. Some qualitative remarks are pertinent.

1. The phase transition temperature is most dependent on the fatty acyl chain because of the importance of van der Waals forces in determining the relative stability of the gel and liquid crystalline phases.

Table 2.1. Thermodynamic data for thermotropic phase transitions of a series of diacyl-saturated phosphatidylcholines determined by DSC[/].

Lipid	Tm_1 (°C)	Tm_2 (°C)	ΔH_1 (kcal/mol)	ΔH_2 (kcal/mol)	ΔS_2 (cal/°K mol)
DMPC (C_{14})	15.3	24.0	1.3	6.5	21.9
DPPC (C_{16})	35.5	41.5	1.6	8.7	27.7
DSPC (C_{18})	51.0	54.3	1.8	10.4	33.3
DAPC (C_{20})	62.1	64.1	1.7	12.3	37.6

[/]From ref. 112. Abbreviations: DMPC, dimyristoyl phosphatidylcholine; DPPC, dipalmitoyl phosphatidylcholine; DSPC, distearoyl phosphatidylcholine; DAPC, diarachidoyl phosphatidylcholine. The subscripts (1, 2) refer to the pretransition (1) and main transition (2).

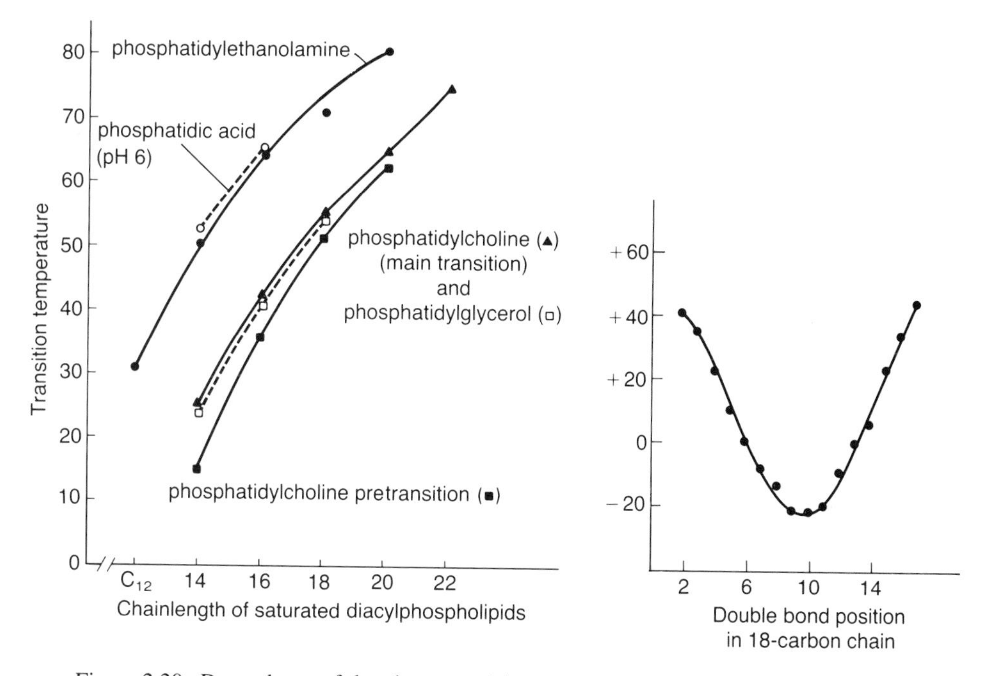

Figure 2.20. Dependence of the phase transition temperature on the chain length of saturated diacyl phospholipids and on the position of the double bond in mono-unsaturated 1,2-dioctadec-*cis*-enoyl phosphatidylcholines. At pH 12, phosphatidic acids have the same transition temperatures as do phosphatidylcholines and phosphatidylglycerols of the same chain length. Data from refs. 112, 73, 698.

(a) Longer chain lengths result in higher T_m values, because of the increased van der Waals interaction for the longer chains.
(b) A *trans* double bond reduces the T_m as this will disrupt the ability of the chains to interact optimally in the gel state.
(c) A *cis* double bond has an even larger affect than a *trans* double bond. The magnitude of the effect is dependent on the position of the double bond in the chain, with the maximal effect when the *cis* double bond is in the middle of the chain. Prokaryotic organisms have a wide variety of hydrocarbon chains containing structures such as cyclopropyl or methyl groups which also destabilize the gel state and lower the T_m (1262) (see Chapter 1).
(d) Fluorine, often used to replace hydrogen in phospholipids for the purpose of providing an NMR probe, has an influence on T_m similar to a *cis* double bond.

2. Lipids with a large area requirement for the polar headgroup, such as phosphatidylcholines, show a pretransition between two gel states ($L_{\beta'}$ and $P_{\beta'}$) (see 855). These states and the chain tilt exhibited by such lipids are discussed in Section 2.2.

3. Saturated phosphatidylethanolamines melt about 20° higher than do equivalent phosphatidylcholines, presumably due to stabilizing hydrogen bonding possible in the gel state of phosphatidylethanolamine (see 115).
4. The presence of a *cis* double bond in the fatty acyl chains largely removes the difference between T_m values for phosphatidylethanolamine and phosphatidylcholine.
5. Phospholipids with various headgroups (including cardiolipin) have all been shown to exhibit similar phase transitions, as have glycolipids.
6. Charged lipids are very sensitive to ionic strength, pH (if the lipid has a pK in the range being examined), and divalent cations (194). The binding of ions to phospholipids is discussed in Chapter 7.
7. Pressure can also be used to induce a phase transition in phospholipids and alter the T_m. High pressure increases the stability of the gel phase, increasing the T_m of dipalmitoyl phosphatidylcholine for example, by 22°C/kbar (215). The gel phase is stabilized at high pressure because it is more dense than the liquid crystalline phase.
8. At the T_m, the lipid is partially in a gel state and partially in a liquid crytalline state. Experimentally, this has been observed to result in leakiness in vesicles, and the size of the apparent "pores" has been measured in multilamellar liposomes of dimyristoyl phosphatidylcholine to be as high as 18 Å diameter (1489). This is interpreted in terms of packing defects at the transient boundaries between macroscopic gel and liquid crystalline domains of the bilayer. The presence of such "defects" is discussed further in Chapter 9 in relation to membrane fusion.
9. Small vesicles generally exhibit a broad thermotropic transition, presumably due to packing difficulties of the gel state lipids in vesicles with a small radius of curvature.

Finally, there are numerous theoretical approaches to phase changes in lipid bialyers (see 1048, 1163, 1008). Successful models have in common the assumption than chain segments can be described in simple terms as either *trans* or *gauche* (g^+ or g^-). By taking into account hard core short-range repulsive interactions between chains and longer-range attractive interactions, the "melting" of the phospholipid bilayer can be adequately described in terms of the introduction of *gauche* conformers. To some extent these models can be extended to describe the behavior of binary lipid mixtures (1008).

2.42 Lipid Mixtures

The next level of complexity beyond the single lipid is a binary lipid mixture. Many such mixtures have been examined experimentally (see 829). The thermodynamics and structure of such mixtures are relevant to biomembranes since they are informative about the miscibility of different kinds of lipids. A question of particular interest is whether the lipid mixtures found in biomembranes might spontaneously organize in separate domains of different lipid composition and, hence, with different physical and chemical properties. The concepts of lateral phase separations in biomembranes (see Section 4.54), though far from being established experimentally, derive directly from physical chemical studies of simple lipids and lipid mixtures where such phase separations have been clearly demonstrated (e.g., 783).

Just as the phase transition of homogeneous lipids has been considered analogous to the melting transition of a normal three-dimensional fluid, binary lipid mixtures can be characterized using regular solution theory (829, 830). This has been most successful when applied to mixtures of dissimilar phospholipids.

Phospholipid Mixtures

Thermodynamic studies have clearly demonstrated that dissimilar phospholipids do not mix ideally. In the gel phase, packing requirements may prevent two lipids from being miscible, resulting in clustering or lateral phase separations. Even in the liquid crystalline phase, two lipids may be miscible but the mixture often behaves nonideally. That is, interactions between lipids of the same type are different from interactions between unlike lipids, resulting in preferential arrangements of nearest neighbors.

Nonideal behavior can be characterized by comparing the experimental phase diagram with that predicted theoretically using regular solution theory. The phase diagram is experimentally determined by monitoring the extent of the transition from gel to liquid crystalline phases, often by using differential scanning calorimetry. However, any method which reports the fraction of lipid in each state can be used. Whereas single-component systems show sharp melting profiles, mixtures exhibit considerably broader thermal melts. For mixtures of different composition, the temperatures at which melting initiates and is finished are recorded and used to construct a phase diagram (Figure 2.21A). In between the lines connecting the temperatures where melting is started and where it is terminated is a region where the lipid has partially melted, and liquid crystalline (fluid) and gel (solid) domains are in co-existence. The shapes of these lines depend on the thermodynamic characteristics of the melting of the individual components and of the mixing of the two components. Ideal behavior implies zero enthalpy and entropy of mixing of the two lipids. Equations describing nonideality are relatively easy to derive (see 829, 830), assuming nonideal free energy of mixing, and phase diagrams derived from theory can be compared to experiment. For a mixture of dimyristoyl and dipalmitoyl phosphatidylcholine, the data fit reasonably to curves derived assuming nonideal mixing (nonzero $\Delta H°$ for mixing) of these two lipids in the gel phase but ideal mixing in the liquid crystalline phase. These lipids differ by only two methylene groups. More dissimilar lipids behave less ideally. For example, mixtures of dilauroyl and distearoyl phosphatidylcholine (C_{12} and C_{18}) appear to mix nonideally in both gel and liquid crystalline phases. Wilkinson and Nagle (1587) showed that dimyristoyl and distearoyl phosphatidylcholine (C_{14} and C_{18}) are nearly immiscible in the gel

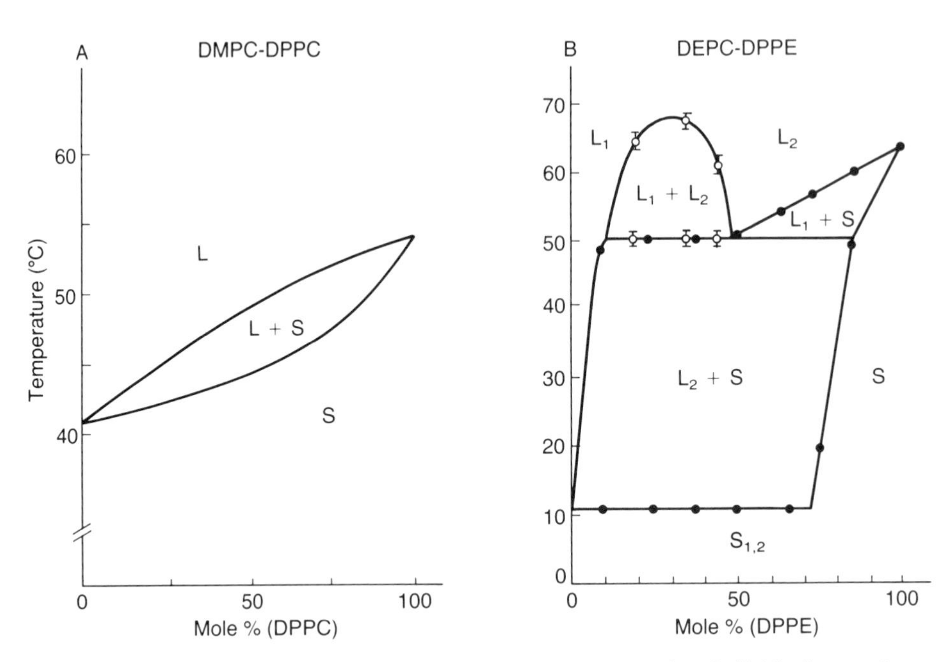

Figure 2.21. Representative phase diagrams of two-component phospholipid mixtures. L, L_1, and L_2 refer to liquid crystalline phases. S refers to solid-like or gel phase. In part A, the lipids differ only slightly in chain length, and there is complete miscibility in both the gel and liquid crystalline phases. In part B, the lipids are quite different and there is immiscibility in both the gel and liquid crystalline phases. Adapted from ref. 644. Reprinted with permission from "Phase Separations in Phospholipid Membranes," by S. Hong-wei Wu and H. M. McConnell, Biochem. 14, pp. 850-851, Copyright 1975, American Chemical Society. Abbreviations: DMPC, dimyristoyl phosphatidylcholine; DPPC, dipalmitoyl phosphatidylcholine; DEPC, dielaidoyl phosphatidylcholine (18:1, *trans*-9); DPPE, dipalmitoyl phosphatidylethanolamine.

phase and that dimyristoyl and dielaidoyl phosphatidylcholine (C_{14} and $C_{18:1,trans}$) are completely immiscible in the gel phase.

Finally, mixtures containing an anionic lipid as one component are dramatically influenced by divalent cations such as Ca^{2+}. The cations apparently form intermolecular cross bridges, resulting in the clustering and lateral phase separation of the anionic lipid component. This can be seen by DSC by the appearance of distinct melting transitions associated with each lipid component (e.g., 1458). This phenomenon can also be directly visualized by fluorescence microscopy (607).

In summary, phospholipids, not surprisingly, behave nonideally in mixtures and show preferences in their nearest neighbors both in the gel and liquid crystalline phases. The significance of these model studies to the distribution of lipids in real biomembranes is not known.

Phospholipids and Cholesterol

The interactions between cholesterol and phospholipids have been extensively studied. Most of the work has used phosphatidylcholines, but other lipids such as phosphatidylethanolamine or sphingomyelin have also been studied (e.g., 398). There is reasonable agreement on the phenomenological description of cholesterol–phospholipid mixtures, but there is no consensus on the interpretation of data in terms of specific structural models. X-ray and neutron scattering show that cholesterol inserts normal to the plane of the bilayer with the -OH group near the ester carbonyl of the lipid (1606) (see Figure 2.22). However, Raman spectroscopy indicates that no actual hydrogen bond is formed with these carbonyls (171). The presence of cholesterol has a substantial effect on the order parameters measured along the lipid hydrocarbon chain by ^{2}H-NMR (1092) and on the phase transition of the phospholipid (890). FT-IR studies (see 250) show that above T_m, cholesterol decreases the fraction of *gauche* rotamers in the phospholipid hydrocarbon chain, whereas just the opposite is seen below T_m. The ordering phenomenon is due to the difficulty in packing the hydrocarbon chains adjacent to the rigid

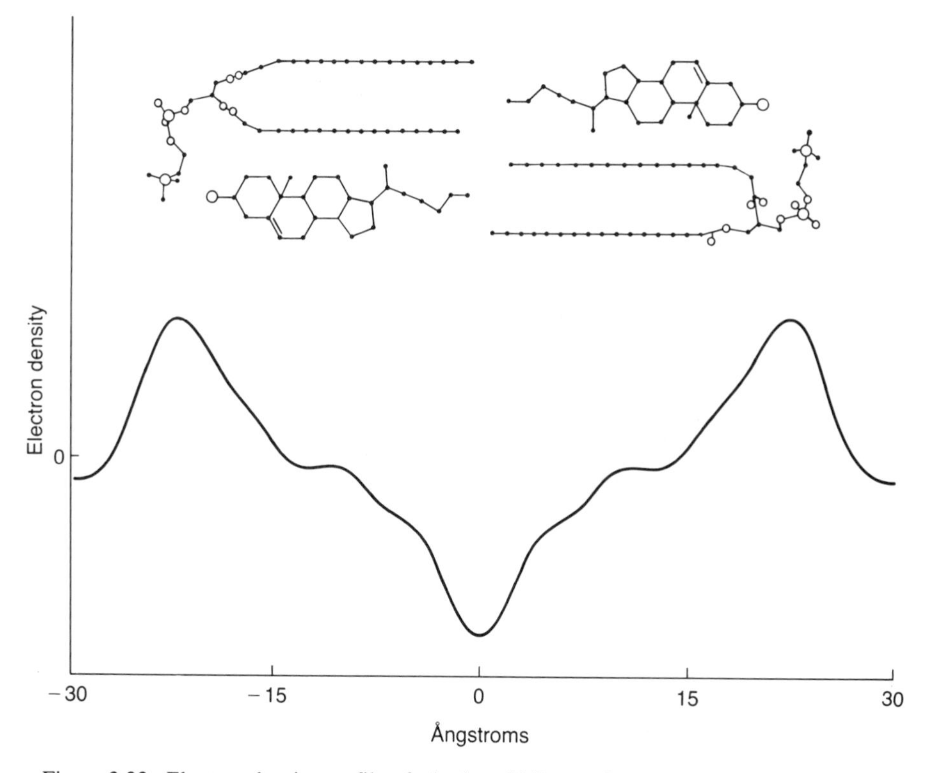

Figure 2.22. Electron density profile of a hydrated bilayer of egg phosphatidylcholine and cholesterol derived from X-ray diffraction analysis. A molecular model consistent with the data is also shown. Adapted from ref. 461.

sterol moiety. In the liquid crystalline state, the sterol results in conformational constraints on the phospholipid chain, whereas in the gel state the sterol inhibits optimal packing of the all-*trans* chain configuration. The result is that lipid–cholesterol mixtures behave in some ways (e.g., disorder) as intermediate between the gel and liquid crystalline states of the pure phospholipid. Basically, cholesterol acts as a "spacer" and reduces the attractive forces between the lipid hydrocarbon chains. In contrast, cholesterol has little effect on the polar headgroups.

A key question of biological relevance is whether cholesterol in biomembranes is distributed randomly in the bilayer or whether it clusters or is concentrated in distinct domains. No simple picture has emerged from the model studies on binary mixtures, but it is clear that phospholipid–cholesterol mixtures do not behave ideally with randomly distributed components. DSC studies on cholesterol mixtures with dipalmitoyl phosphatidylcholine are consistent with the existence of two separate phases below 20 mol% cholesterol and a single phase between 20% and 50% cholesterol (890). Above 50%, no phase transition is observed as the cholesterol completely obliterates the cooperative interactions between the lipid hydrocarbon chains. Various other physical techniques have indicated several phase boundaries in mixtures near 20%, 33%, and 50% cholesterol (839). The structural arrangements within each phase are not known. On the basis of freeze-fracture electron microscopy, Copeland and McConnel (242) have proposed that in the concentration range 0-20% cholesterol with dipalmitoyl phosphatidylcholine, alternating strips containing pure phospholipid and 20% cholesterol exist, resulting in a rippled appearance. It is possible that stable phospholipid–cholesterol molecular complexes may exist but there is no strong evidence to support this. There is evidence, however, that cholesterol interacts preferentially with some phospholipids (1557). Neutron scattering studies (765) indicate that mixtures of cholesterol and dimyristoyl phosphatidylcholine above the phase transition (35°C) are fully miscible, while several distinct phases were observed below the phase transition of the mixture (7°C). Theoretical models (1163) have been moderately successful at duplicating some of the effects of cholesterol on the lipid phase transition parameters (T_m and $\Delta H°$) without including any specific chemical interactions.

In summary, cholesterol–lipid mixtures are complex and polymorphic and there is some evidence for lateral phase separations under certain conditions. How this applies to the role of cholesterol in biomembranes is not clear (see Section 10.4).

Other Mixtures

The influences of numerous hydrophobic compounds on the thermotropic properties of membrane lipids have been studied (see 829). These include numerous drugs, anesthetics, alkanes, alcohols, and fatty acids, as well as natural membrane components such as carotenoids and polyisoprenoids (dolichol, bactoprenol, ubiquinone). Compounds which do not mix well with phosospholipids have less effect on the bilayer phase transition, either because they form a separate phase within the bilayer, even in the liquid crystalline phase, or because they do not become significantly incorporated in the bilayer. Cholesterol esters, for example, are not miscible beyond a few percent with lamellar phase lipids (1453). Physical chemical studies with ubiquinone to locate this molecule within the bilayer have not been conclusive, in part because of uncertainty about whether the quinone is substantially incorporated in the phospholipid bilayer (e.g., 1477, 1391). Whereas many compounds such as polyisoprenoids cause the T_m to decrease because they cannot pack well with gel phase lipids (1477, 1516, 809), other compounds, such as long chain fatty acids and alcohols cause the T_m to increase (889). Some small hydrophobic molecules such as short-chain alkanes appear to form a separate phase in the middle of the bilayer and result in a decrease of the T_m (950).

2.5 Model Membrane Systems

Numerous model membrane systems have been developed for studying the properties of pure lipids, lipid mixtures, and reconstituted lipid–protein mixtures. These model systems can be grouped as (1) monolayers, (2) planar bilayers, and (3) liposomes or vesicles. Each of these systems and their many variants have advantages and disadvantages, and information from each kind of system has been valuable in our developing concepts of the biomembrane. In this section, some of the model systems and their uses are briefly summarized.

2.51 Monolayers at an Air–Water Interface

Many kinds of molecules with substantial non-polar character will be adsorbed at the air–water interface (see 53). This can take the form of a layer only one molecule thick and can be studied as such at the air–water interface or the monolayer can be transported to other supports. Phospholipids and other amphiphilic molecules form an oriented monolayer with the polar portions in contact with the aqueous phase and the hydrocarbon chains extended above. Phospholipids form an insoluble monolayer, since the concentration of the lipid in the aqueous subphase is essentially negligible. This is studied traditionally in a Langmuir film balance or trough, which has a movable float on one side used to control the surface area available to the monolayer (see Figure 2.23). One can use a known amount of lipid to form the insoluble monolayer by, for example, dropping an aliquot of lipid dissolved in volatile solvent on the aqueous surface. The Langmuir film balance allows one to accurately measure the surface area (A) and lateral pressure (π) required to maintain the monolayer. The major advantage of monolayers is the ability to vary the surface density and pressure, and the system has been especially useful in studying the physical chemistry of "surface-active" lipids as well as the enzymology of soluble enzymes such as lipases (324) which act at the lipid–water interface. Although proteins can be incorporated into monolayers, the system is less suited for most studies of intrinsic membrane proteins. Monolayers are usually characterized by "pressure–area" curves. Examples are shown in Figure 2.24. These experimental plots show two important characteristics:

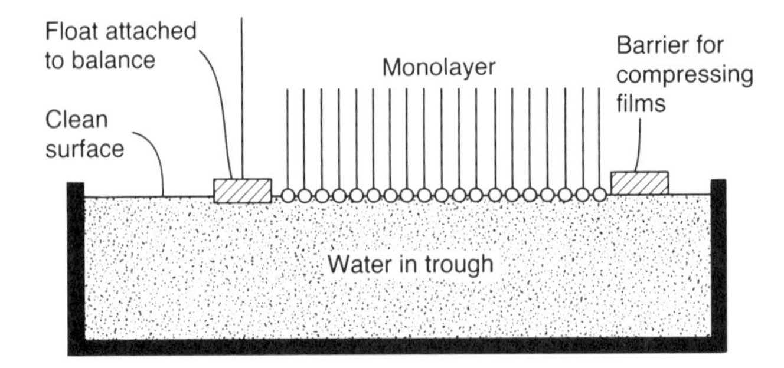

Figure 2.23. Schematic of a monolayer in a Langmuir trough. The polar headgroups are depicted as being in the aqueous phase and the nonpolar chains sticking up. Adapted from ref. 482. "Insoluble Monolayers at Liquid-Gas Interfaces," by G. L. Gaines, Jr., Copyright © 1966 by Interscience Publishers. Reprinted by permission of John Wiley & Sons, Inc.

1. *Discontinuities*, which are due to apparent phase transitions from expanded "liquid-like" or "gas-like" phases at low monolayer densities to more "solid-like" compressed forms at high monolayer densities. The physical properties of monolayers are related to those of bilayers (896, 1160, 505).
2. *Collapse point*, at which the molecules are packed to their maximum density and any further compression results in breakdown of the monolayer. This gives the minimum surface area per molecule.

Saturated alkyl chains occupy approximately 20 Å^2, as seen also in the X-ray structural analyses. *Cis*-unsaturated chains have a larger minimum surface area due to chain packing. The minimum surface area required by a disaturated phospholipid would therefore be expected to be 40-45 Å^2, which would correspond to a gel phase in the bilayer. The minimum area required by a lipid with *cis* unsaturation would be closer to 60 Å^2. Large polar headgroups, such as in phosphatidylcholines or in gangliosides, result in increased minimal surface area requirements.

The phase transition from the fluid "liquid-expanded" phase to the "liquid-compressed" phase has been subject to considerable theoretical and experimental investigation (see 1160, 53). At a given pressure, monolayers can be induced to undergo a thermotropic phase transition which can be directly compared to that observed in the bilayer. For example, at 15 dynes/cm, a monolayer of dipalmitoyl phosphatidylcholine will undergo a phase transition with T_m = 27°C and $\Delta H°$ = 8.7 kcal/mol. This compares with the corresponding bilayer transition of T_m = 41°C and $\Delta H°$ = 8.7 kcal/mol. Pink (1160, 505) has shown that a relatively simple model, which takes into account a very small attractive interaction between the two monolayers comprising the bilayer, can account for the difference in T_m between the monolayer and bilayer. This is an important point and suggests that from a thermodynamic viewpoint, the bilayer behaves largely as two nearly independent monolayers. Hence, it is conceivable that if the two monolayer leaflets making up a biomembrane were of very different composition, the physical properties could also be different and, to a limited extent, uncoupled. Of course,

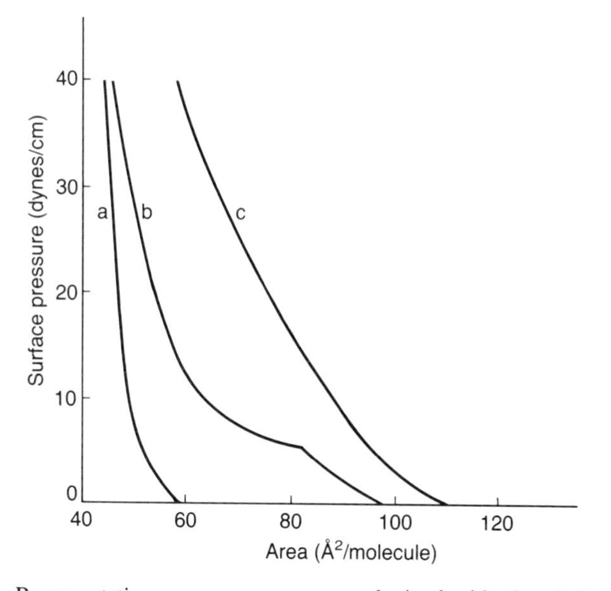

Figure 2.24. Representative pressure–area curves obtained with phospholipid monolayers: (a) Dibehenoyl phosphatidylcholine (22:0), which remains in a solid-like phase; (b) dipalmitoyl phosphatidylcholine (16:0), which exhibits a phase transition from liquid-like to solid-like upon compression; (c) egg phosphatidylcholine, which remains in a liquid-like phase. Adapted from ref. 698. "Introduction to Biological Membranes," by M. K. Jain and R. C. Wagner, Copyright © 1980. Reprinted by permission of John Wiley & Sons, Inc.

transmembrane proteins or other components could modify this. Membrane asymmetry is discussed in more detail in Chapter 4. Finally, the model is consistent with the internal pressure of the bilayer as being 2×15 dynes/cm = 30 dynes/cm. The thermodynamic forces stabilizing the bilayer result in an internal pressure due to the steric repulsions of both the hydrocarbon chains and polar headgroups. This has been estimated (see 1160) between 12.5 and 50 dynes/cm, balanced formally by the attractive surface tension (683).

It should be noted that the pressure, π, measured for a monolayer is actually the difference in surface tension between the surface with the monolayer (γ) and the pure air–water interface (γ_o):

$$\pi = (\gamma_o - \gamma)$$

The spontaneous formation of a monolayer at the air–water interface invariably results in a reduction of the surface tension. Surface tension can be viewed as a negative pressure due to the attractive interactions of the molecules at the interface, and this is lowered by the "surfactant" making up the monolayer. Dipalmitoyl phosphatidylcholine, for example, is a major component of lung surfactant, which reduces the work required to change the surface area of the lung during breathing to near zero ($\gamma = 0$) (598).

Most exciting visually are the data from a technique called epifluorescence in which the fluorescence of a lipid probe doped into a monolayer is monitored by fluorescence microscopy (946, 1145). Probes are used which partition favorably into either the gel or liquid crystalline phase. When both phases coexist, this is visualized by light and dark areas due to differences in the fluorescence intensity from the probe in the different regions. Very striking periodic arrangements of solid–phase regions begin to appear in monolayers of dipalmitoyl phosphatidylcholine at pressures as low as 5 dynes/cm (20°C), and as the pressure is increased, these regions grow at the expense of the fluid-like regions. Figure 2.25 shows an example where the monolayer has an average molecular area of about 60 Å^2. Studies on these monolayers not only show the coexistence of phases but also demonstrate long-range order, possibly stabilized by electrostatic interactions (946).

2.52 Monolayers on a Solid Support

Monolayers formed at the air–water interface can be transferred to a solid support such as an alkylated glass coverslip simply by bringing the glass into contact with the monolayer (946). The alkylation of the glass provides a nonpolar surface. The lipid polar headgroups remain in contact with water. These monolayers can be transferred to the solid support at different surface pressures (π) and then exam-

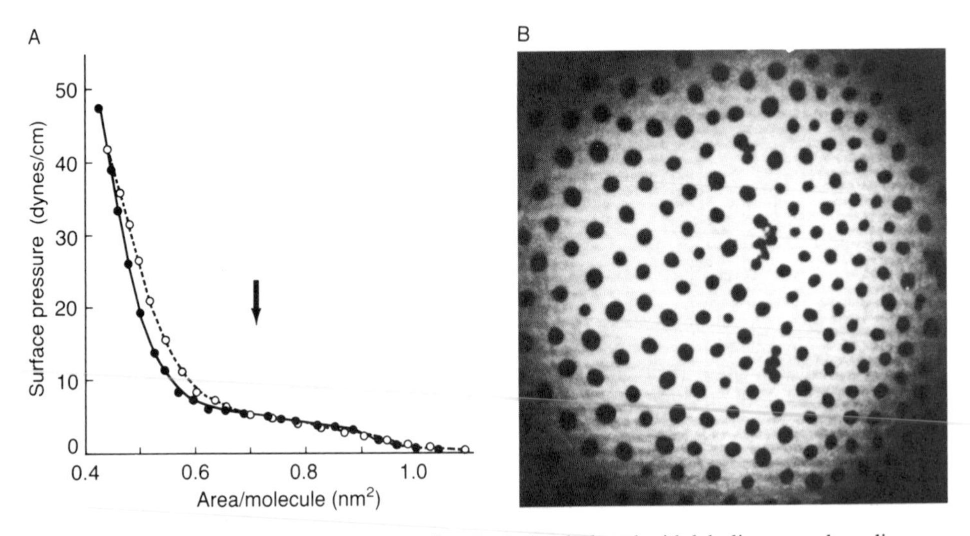

Figure 2.25. (A) Pressure–area curve for dipalmitoyl phosphatidylcholine spread on distilled water at 20°C. The sample contained 2% mol of a fluorescent phospholipid, NBD-egg phosphatidylethanolamine (see Table 5.1). At the point of compression indicated by the arrow in part A, the epifluorescence of the monolayer was photographed (part B). Dark regions are solid-like, and the strongly fluorescent regions are liquid-like domains in which fluorescent probe is preferentially dissolved. The regular pattern is thought to be stabilized by electrostatic forces. The field of view is about 200 μm across. From ref. 946. Photograph kindly provided by Dr. H. McConnell.

ined. Several studies (946, 1145, 1565, 1450) have demonstrated that the dynamic and thermodynamic properties of these supported monolayers as well as monolayers at the air–water interface are similar to those observed for bilayers. Fluorescence methods have also been developed to examine the orientation of molecules in the supported monolayer (1450).

2.53 Planar Bilayer Membranes

Traditionally, planar membranes are produced by painting a concentrated solution of phospholipid in a solvent such as decane over a small hole (~1 mm diameter) in a nonpolar partition (e.g., polystyrene) separating two chambers containing aqueous buffers. Much of the excess solvent disperses in the medium and, under appropriate conditions, the lipids spontaneously form a bilayer across the small hole. These membranes are frequently called bimolecular lipid membranes or BLMs. Because of the lack of light reflectance, they are also called black lipid membranes. Planar membranes formed in this way can be used to study membrane proteins (e.g., ion channels, 987, 988), but they have the disadvantage of containing an unknown amount of residual solvent and they are also unstable, particularly in the presence of small amounts of detergents or other impurities. An alternate procedure for making planar membranes has been developed to avoid the difficulties with the solvent and also to facilitate the incorporation of intrinsic membrane proteins (1009, 1010, 1011, 246). The planar membrane is made across a small orifice from a monolayer at an air–water interface by a simple dipping procedure using either a small pipet (Figure 2.26) or a nonpolar partition with a small hole. Monolayers containing purified membrane proteins can be formed from vesicular protein–phospholipid dispersions and these monolayers used to form the planar membrane. In addition, membrane vesicles containing ion channels can be fused with a preformed planar membrane, incorporating the protein to be studied into the model membrane (see Section 8.14).

The overwhelming advantage of using planar membranes is for making electrical measurements, and this system is particularly valuable for studying pores, channels, or carriers that facilitate or catalyze the transfer of charge across the bilayer from one compartment to the opposing compartment (see Chapter 8). Electrodes are easily utilized in the aqueous chambers, solutions easily changed, and the measurement of current and/or voltage is very accurate and sensitive. One variation of the use of planar membranes is to form them over the tip of a small pipet (diameter < 1 μm) (Figure 2.26). This small patch of bilayer in the "patch pipet" can be used for very sensitive electrical measurements with reconstituted purified membrane proteins, such as the acetylcholine receptor, yielding data on the properties of individual molecular events (1402) (see Section 8.14).

2.54 Planar Bilayer Membranes on Solid Support

Phospholipid bilayers can also be formed on solid hydrophilic supports such as oxidized silicon wafers by the sequential transfer of two monolayers from an air–water interface (1430). The bilayer probably has a water layer between the

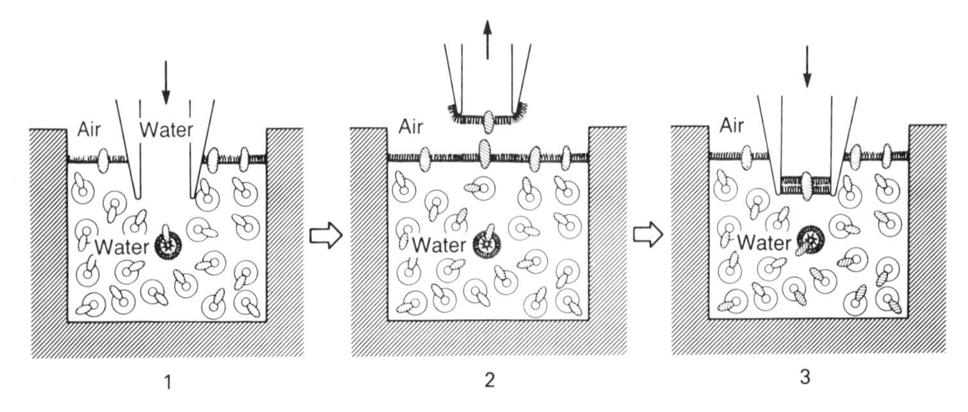

Figure 2.26. Schematic showing the process of forming a bilayer at the tip of a patch pipet. A monolayer is derived from reconstituted phospholipid vesicles containing a protein of interest (see Chapter 3). The pipet is dipped into the solution under positive pressure (Step 1). Upon withdrawing the pipet (Step 2), the monolayer adheres to the pipet by interactions with the lipid headgroup, leaving the hydrocarbon tails in contact with air. The pipet is then re-immersed into the solution (Step 3), resulting in the formation of a bilayer. From ref. 1402. Reprinted with permission from "Single-Channel Recordings from Purified Acetylcholine Receptors Reconstituted in Bilayers Formed at the Tip of Patch Pipets," by B. A. Suarez-Isala, et al., Biochemistry 22, p. 2320, Copyright 1983, American Chemical Society.Figure kindly provided by Dr. M. Montal.

solid support and the lipid polar headgroups. By starting with a monolayer with a reasonable average molecular area, a stable bilayer can be made in this fashion. These bilayers are convenient for physical chemical studies of pure lipids and, possibly, eventually for studying membrane proteins. Bilayers of dipalmitoyl phosphatidylcholine made in this way exhibit sharp thermal phase transitions corresponding to the main and pre-transitions observed in lamellar lipid dispersions. These bilayers are especially well suited for measuring the lateral diffusion of membrane-bound molecules using fluorescence techniques (see Chapter 5), and the results are consistent with data from other systems.

2.55 Liposomes

The term "liposome" can be defined as any lipid bilayer structure which encloses a volume (for reviews see 312, 1417, 649). Many phospholipids when dispersed in water spontaneously form a heterogeneous mixture of vesicular structures which contain multiple bilayers forming a series of concentric shells. These were the first "liposomes" to be characterized and are now termed multilamellar vesicles or MLV. Of greater utility are single-walled or unilamellar vesicles, which can be prepared by a variety of methods. These can be generally characterized as small unilamellar vesicles (SUV), with diameters in the range 200 Å to 500 Å, and large unilamellar vesicles (LUV) with diameters from 500 Å to 5000 Å. Very large or cell-sized phospholipid vesicles can also be prepared and used as model membranes, with diameters as large as 300 μm (1031).

The primary uses of liposomes are (1) as model membranes in which proteins are incorporated and studied and (2) to encapsulate solutes for such uses as drug delivery systems. Liposomes are characterized by their lipid composition, their average diameter, and the extent of size heterogeneity in the population. Sizing is performed by (1) gel filtration chromatography, (2) light scattering, (3) ultracentrifugation, or (4) electron microscopy (see 1306, 1214). Of particular interest to those interested in the ability of liposomes to encapsulate solutes are the (1) *captured volume*, or volume of entrapped solute per mole of lipid, and (2) *encapsulation efficiency*, or the percentage of the aqueous volume inside the vesicles. The former parameter increases with liposome diameter, and the latter is obviously proportional to lipid concentration.

Some of the characteristics and preparative techniques for liposomes are summarized below (see 312, 1417, 649).

(1) *Small unilamellar vesicles (SUV)*: These are usually prepared by sonication of aqueous dispersions of phospholipid (659). The vesicles can be further sized by gel filtration chromatography or by glycerol gradient centrifugation (529) to yield a highly homogeneous distribution with diameter ~250 Å. Alternatively, these can be prepared by quickly injecting an ethanolic solution of lipids into the aqueous phase. A French press can also be used to prepare SUVs.

Although the population homogeneity is advantageous, the small size can be a disadvantage or, at the least, a factor to be considered. The small radius of curvature of SUVs results in packing difficulties of lipids. The surface area of the outer monolayer of lipids is almost twice that of the inner monolayer, and about 70% of the lipids on SUVs are in the outer leaflet. Lipids with an "inverted cone" shape (see Figure 2.18), such as phosphatidylcholine or sphingomyelin, will preferably partition into the outer leaflet, resulting in lipid asymmetry. This packing phenomenon is also likely responsible for the asymmetric orientation of some membrane proteins reconstituted in vesicles (both SUVs and LUVs). The thermotropic phase transition of SUVs is broadened (841), presumably as a result of the effect of the small radius of curvature on the transition cooperativity. Typical capture volumes are quite small for SUVs, in the range of 0.5 to 1 liter/mol (312). Some properties of SUVs are summarized in Table 2.2 and Figure 2.27.

(2) *Large unilamellar vesicles (LUV)*: The need for larger vesicles became obvious in the 1970s and since then numerous procedures have been devised to produce LUVs. The method of choice may depend on the lipid composition but certainly depends on the intended use for the liposomes. For example, a procedure which results in a dilute liposome solution is not suitable for drug encapsulation. On the other hand, techniques involving organic solvents are not suited for protein reconstitution. Techniques employed include the following (see 312, 1417, 649).

(a) *Detergent dialysis or dilution* (390, 993, 1475, 25): By far the most popular method utilized for biochemical studies because it is relatively easy and is compatible with the inclusion of proteins for reconstitution in the resulting LUVs. Excess detergent is codispersed with the lipid (or lipid plus protein) and the detergent is then removed. Detergents with a high CMC, i.e., in equilibrium with a high concentration of monomer (mM range), can be removed almost

Table 2.2. Parameters of SUVs prepared from egg phosphatidylcholine[l].

Parameter	Value
Molecular weight (hydrodynamic)	1.88×10^6
Partial specific volume	0.9848 ml/g
Outer radius	99 Å
Inner radius	62 Å
Outer monolayer thickness	21 Å
Inner monolayer thickness	16 Å
Number of lipid molecules	
outer monolayer	1,658
inner monolayer	790
Surface area per lipid headgroup	
outer monolayer	74 Å^2
inner monolayer	61 Å^2
Acyl chain cross section at bilayer center	
outer monolayer	46 Å^2
inner monolayer	97 Å^2
Anhydrous bilayer thickness	37 Å

[l]Data from ref. 660.

entirely by dialysis (e.g., cholate, deoxycholate, octylglucoside). Detergents with a low CMC (e.g., Triton X-100) can be removed in part by using a hydrophobic resin such as BioBeads, which specifically sequesters the detergent (e.g., 1475). In some cases, simple dilution of the mixture will suffice to reduce the final detergent concentration to a point where the lipids spontaneously vesiculate. The liposome size depends not only on the lipid and detergent utilized, but also on the kinetics of detergent removal (1475). Typically, cholate gives smaller LUVs, 500-800 Å diameter, and octylglucoside gives somewhat a larger population, 1,000-2,000 Å. Residual impurities can be a problem (1120). Capture volumes are much larger for LUVs than SUVs, being in the range 5 to 20 liter/mol (312).

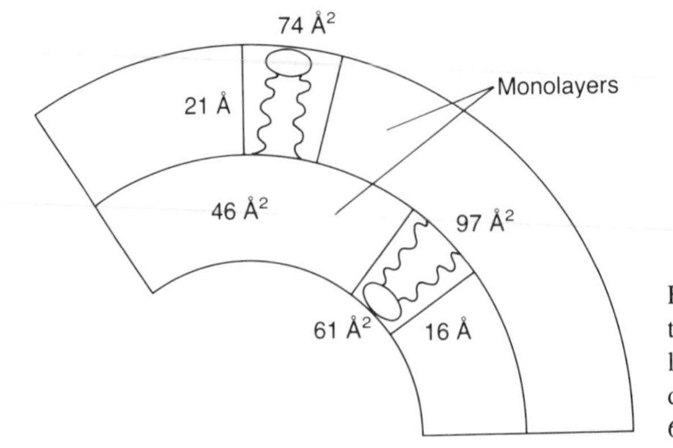

Figure 2.27. Cross section of the bilayer in a small unilamellar vesicle, showing various dimensions. Adapted from ref. 660.

(b) *Infusion and reverse phase evaporation* are separate techniques which involve the use of nonpolar solvents to form LUVs (see 312, 1417). These techniques are not well suited for making model biomembranes with proteins.

(c) *Fusion methods* include several techniques in which SUVs are caused to fuse to form larger liposomes. These techniques include freeze/thaw cycling (898), which is generally applicable and can be used with some proteins, and the use of Ca^{2+} to cause fusion of SUVs made of phosphatidylserine (1119).

(d) *Addition of short-chain phosphatidylcholines* (479, 478) to about 20% of the total lipid has been shown to convert multilamellar bilayers to stable LUVs.

(e) *Addition of fatty acids or detergents* under some circumstances will cause SUVs to fuse to form LUVs (724, 1310). The addition of fatty acids can also result in the incorporation of proteins in the LUV bilayer (1310).

(f) *Rapid extrusion* of multilamellar dispersions through polycarbonate filters can also yield LUVs with diameters in the range 600-1000 Å (648, 649). This method has many advantages.

(g) *Transient increase in pH* causes vesiculation of phosphatidic acid and causes mixtures containing phosphatidic acid to form both SUVs and LUVs (606).

(3) *Cell-size unilamellar vesicles* are formed by simple hydration of lipids or lipid—protein mixtures at low ionic strength (1031). Sizes range from 0.1 to 300 μm, with a capture volume up to 300 liter/mol. They are reported as sufficiently stable to be sorted by centrifugation according to size and can be impaled by microelectrodes for electrical measurements.

(4) *Multilamellar vesicles* are generally not used because of the advantages of single-walled vesicles. Multilamellar vesicles are osmotically active (1621), but the complexity of the many internal volumes makes interpretation difficult.

2.6 Chapter Summary

It is essential to understand the physical chemical behavior of the lipid bilayer in order to appreciate many of the properties of biological membranes. To a large extent, the properties of individual lipids when studied in isolation or in lipid mixtures can be understood qualitatively in terms of shape theory. This is a pictorial way of summarizing information about the way in which the lipid molecules interact. Such considerations are likely to be important even in highly complex biological membranes, and for this reason the study of pure lipids and simple mixtures is relevant to biological systems. Lipids are polymorphic and can exist in structurally distinct phases depending on temperature, pressure, ionic strength, and composition. The lipid bilayer in biological membranes most closely resembles the lamellar liquid crystalline phase which has been characterized for simple lipids and lipid mixtures.

The lipid bilayer is stabilized by the hydrophobic force, which favors structures which minimize the surface area of contact between water and the nonpolar acyl chains. Intermolecular interactions within the polar headgroup region can also be important, including hydration and hydrogen bonding between lipids. A number of spectroscopic techniques are useful for characterizing the structure and dynamics of the bilayer, including NMR and Raman spectroscopies.

There are several model membrane systems which have been experimentally useful. These include monolayers, planar bilayers, and liposomes. Each has advantages and disadvantages, and each has been useful in extending our understanding of biological membranes.

Chapter 3
Characterization and Structural Principles of Membrane Proteins

3.1 Overview of Membrane Protein Structure

Whereas the primary role of membrane lipids is to provide the structural framework of the biomembrane in the form of a stable bilayer, the proteins provide the active components of the biomembrane. In this chapter we discuss some of the operational principles which have proved useful in understanding the structure of membrane proteins. Specific examples which have provided paradigms are discussed in this chapter, and further examples of membrane proteins are covered in Chapters 6, 8, and 9, which deal with major functional classes of membrane proteins. The interactions between membrane lipids and proteins are summarized in Chapter 5, along with a discussion of the motions of both the lipid and protein components.

Many years ago, it was believed that membrane proteins were structurally rather homogeneous, consisting of ß-sheets arranged along the bilayer surface. There is now a tendency to assume that, at least for transmembrane proteins, the membrane-embedded portions are largely, if not uniquely, composed of α-helical segments. It is important to have some perspective on the factual basis for the generalizations which are so tempting. Considering the enormous diversity, as well as the structural themes, found in soluble proteins (1305), it is difficult not to believe that there are structural complexities to intrinsic membrane proteins that go well beyond our present knowledge. The structural classifications of soluble proteins (1305) were only clear once the high-resolution structures were available for over 100 different proteins. By contrast, the high-resolution structure of only a single transmembrane protein class is presently available, the bacterial photosynthetic reaction center protein (see Section 3.51). Along with the lower-resolution data available from electron microscopy image reconstruction of bacteriorhodopsin (Section 3.52), these provide the only models for most other transmembrane proteins.

It is also important to realize the great variety of modes of attachment of membrane proteins to the membrane. Figure 3.1 schematically illustrates some of these types.

1. Interactions with other proteins which are embedded in the bilayer. Examples are the F_1 portion of the H^+-ATPase which binds to the F_0 portion embedded in the membrane (see Section 8.42) and several cytoskeletal proteins (see Section 4.34)

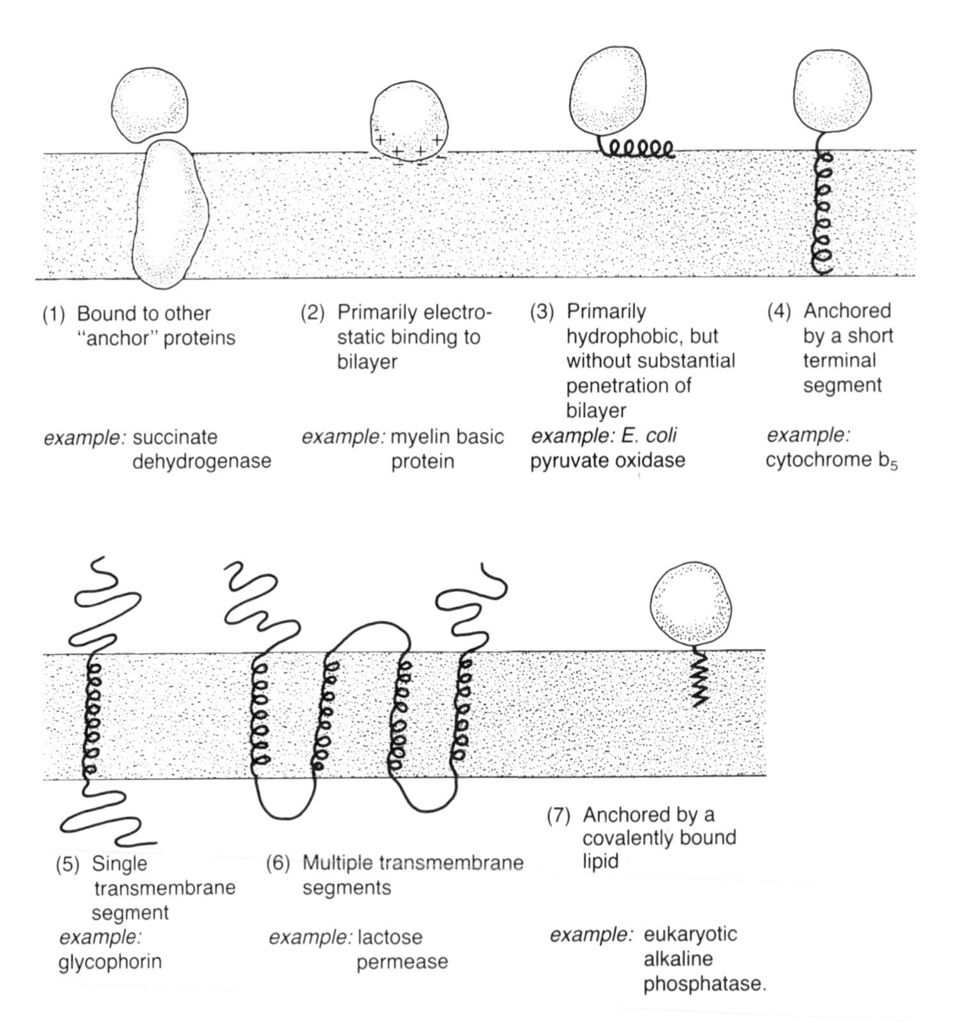

Figure 3.1. Schematic illustration of the variety of ways by which membrane proteins can be attached to the membrane. Peptide anchors (no. 4) can be located at either the amino or carboxyl terminus. Transmembrane proteins (nos. 5 and 6) can be oriented with their amino and carboxy termini on the cytoplasmic or extracytoplasmic surface (see Figure 10.3).

2. Interactions with the surface of the bilayer. This interaction might be primarily electrostatic [e.g., myelin basic protein (201)] or hydrophobic [e.g., surface active peptides (385) and, possibly, phospholipases (324)]. Some membrane proteins may have hydrophobic domains on their surfaces as a result of either secondary or tertiary folding patterns of the polypeptide chain. These types of surface interactions may be used primarily to supplement the membrane-binding properties of proteins which are attached to the membrane by other means, such as by transmembrane "anchors."
3. Interactions with the membrane via a hydrophobic "anchor" or tail, usually identifiable from the primary sequence as a stretch of residues with nonpolar character [e.g., cytochrome b_5 (1420; Section 4.22)]. There are also membrane proteins covalently attached to fatty acids or phospholipids which serve as an anchor (see Section 3.8).
4. Proteins which are transmembranous. These may pass completely through the membrane only once [e.g., glycophorin (1323; Figure 3.17) or many times [e.g., lactose permease (453; Figure 8.11), bacteriorhodopsin (620; Figure 3.8)].

The distinction between an extrinsic (or peripheral) and intrinsic (or integral) membrane protein does not clearly define the mode of attachment to the bilayer, just the relative strength of the attachment.

3.2 Purification of Membrane Proteins

The purification of an intrinsic membrane protein in a biochemically active form almost invariably requires detergents to solubilize the protein and to maintain it in solution. This requirement for detergents and their proper manipulation introduces additional problems to those usually encountered in protein purification. Although there are numerous separation techniques, such as hydrophobic phase separation (905, 1184) and hydrophobic chromatography (1308), specifically designed for resolving intrinsic membrane proteins, most purification protocols rely on the same chromatographic and hydrodynamic procedures used for soluble proteins, i.e., DEAE cellulose or Sepharose, gel filtration chromatography, hydroxylapatite, sucrose density gradients, etc. The critical feature is the proper choice of detergent, since it is the detergent which disrupts the biomembrane and replaces the lipids surrounding a particular protein (614), and thus has a substantial effect on protein stability. The mechanism of detergent solubilization of membranes is reviewed in ref. 614.

3.21 Detergents

During the past two decades, a large variety of detergents have become available for the purification of integral membrane proteins. Ideally, what is desired is a detergent which leaves the tertiary and quaternary structures of the membrane proteins unaltered, but merely replaces most or all of the membrane lipid molecules in contact with the hydrophobic portions of the protein by detergent molecules. One wants to essentially convert a protein embedded in a lipid bilayer to one embedded in a detergent micelle. The purification strategy is then to resolve these protein–detergent complexes.

The first challenge is to find the optimal conditions for solubilizing the membrane protein of interest. Detergents which invariably denature proteins, such as SDS, are termed "harsh" and are obviously unsuited for this delicate task. On the other hand, many detergents are not effective at disrupting membranes and forming protein-containing mixed micelles. Such detergents may be either too hydrophobic or too hydrophilic to effectively mix with the membrane lipids and, at sufficiently high concentration, convert the bilayer to globular mixed micelles. At one point, it was hoped that the selection of the proper detergent could be systematized with a single parameter, called the hydrophilic lipophilic balance or HLB number. This scale, ranging from about 1 to 20, is used in the surfactant industry (see 293) as a measure of relative hydrophobic character, and, indeed, there are correlations showing that the HLB value of a detergent is useful in predicting some detergent effects in biological systems (e.g., 1478). Generally, detergents with HLB values in the range 12.5 to 14.5, are the most effective at solubilizing intrinsic membrane proteins. However, finding the optimal detergent for a particular membrane protein involves many considerations and always requires empirical screening.

The primary considerations are

(1) Maximizing the solubilization of the desired protein. Solubilization can be defined as rendering the protein in the supernatant following centrifugation which will pellet the membranes, e.g., $100{,}000 \times g$ for 1 hour.

(2) Solubilization of the protein in the desired form, usually defined in terms of enzyme activity, but this can also be defined by spectroscopic criteria or in terms of protein associations (1221). This criterion also implies stability of the protein after solubilization. In some cases, exogenous phospholipids are added along with the detergent in order to maintain biochemical activity, as exemplified by one of the preparations of *E. coli* lactose permease (1062) and by the sodium channel protein (70). In other cases, the addition of glycerol or another polyol may be stabilizing after solubilization, as also observed for soluble proteins. It is always prudent to include protease inhibitors upon detergent extraction and to use conditions minimizing the potential for proteolytic damage.

(3) Detergent compatibility with the preparative protocol. The most important properties to be considered are detergent charge (e.g., when using ion exchange chromatography), pH compatibility (e.g., bile salts often precipitate unless maintained above pH 8), and the CMC and micelle size of the detergent. These last properties are critically important because detergents with low CMC values which form large micelles cannot be eliminated by dialysis or ultrafiltration since the concentration of the monomeric species of detergent, which is the form small enough to pass through the membrane pores, is too low. From a practical stand-

point, this means that if one concentrates a protein by ultrafiltration, the concentration of the detergent with a low CMC will also go up, often resulting in protein denaturation. For this reason, many workers prefer to use detergents which have high CMC values such as octyglucoside, bile salts, or some of the newer zwitterionic detergents. Polystyrene resins, such as BioBeads SM-2, preferentially bind to detergents such as Triton X-100 and, thus, remove them from solution, by a procedure which does not involve dialysis, and these have proved to be quite valuable (473).

Another property to consider is optical absorbance of the detergent. Some detergents, such as Triton X-100, absorb in the near ultraviolet and make it impossible to monitor protein concentration by simply measuring the A_{280} of the solution. This can be very inconvenient.

Considering all these factors, it is not surprising that many preparative protocols for intrinsic membrane proteins require the detergent to be changed during the procedure. For example, solubilization may be best obtained with Triton X-100, but the separation obtained during DEAE chromatography may be better if eluted in the presence of octylglucoside. Detergents may be changed during a chromatography step, during density gradient centrifugation, or, in some cases, by dialysis. It is useful to keep in mind that a detergent that may be ineffective at solubilizing a particular protein from the membrane may be quite effective at maintaining the protein in solution following a detergent exchange. The mechanism of solubilization (see 614) is very different from that of maintaining solubility. It is almost always necessary to keep excess detergent in solution during a preparative protocol. If this is not done, in most cases, the equilibrium will favor the membrane proteins aggregating together rather than remaining as separated protein–detergent complexes. In some cases, this aggregation may be acceptable or even desirable, and the final step in a protocol may involve detergent removal (e.g., preparative protocols for glycophorin, cytochrome b_5, cytochrome *c* oxidase). But in most cases, the result is irreversible precipitation and loss of the protein.

The requirement to maintain detergent concentration, usually at or above the CMC, results in problems beyond those normally encountered in protein purification, some of which have already been alluded to, such as difficulties in protein concentration by ultrafiltration. Additional problems are encountered in using the standard technique of salting out by high concentrations of ammonium sulfate. In many cases, the protein precipitates in an aggregate with detergent and lipid. Since the salt solution has a high density and since the detergent in the aggregate has a relatively low density, the result is that the precipitate will float upon centrifugation (see Box 1.2). Depending on the nature of the material, this can be difficult to cleanly remove. It is always important to recall that the species being purified is a protein–detergent complex, often with significant amounts of bound phospholipid. This has profound effects on the resolution obtained during chromatographic procedures as well as in the procedures used to characterize the final product (see next section). Tables 3.1 and 3.2 list some of the most commonly used detergents and some relevant properties. Empirically, those which have

Table 3.1. The structures of some detergents used to solubilize and purify membrane proteins

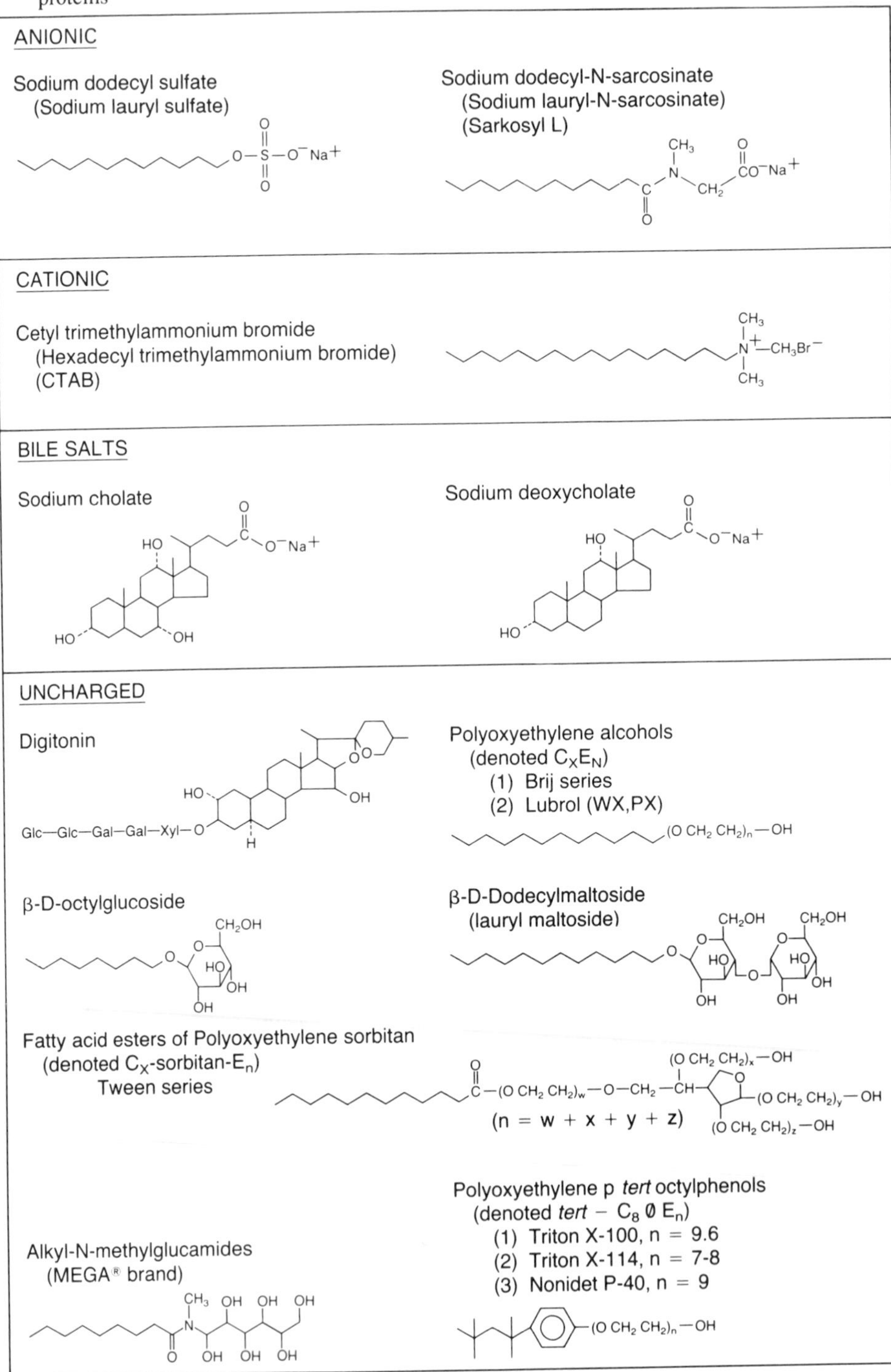

Table 3.1 continued

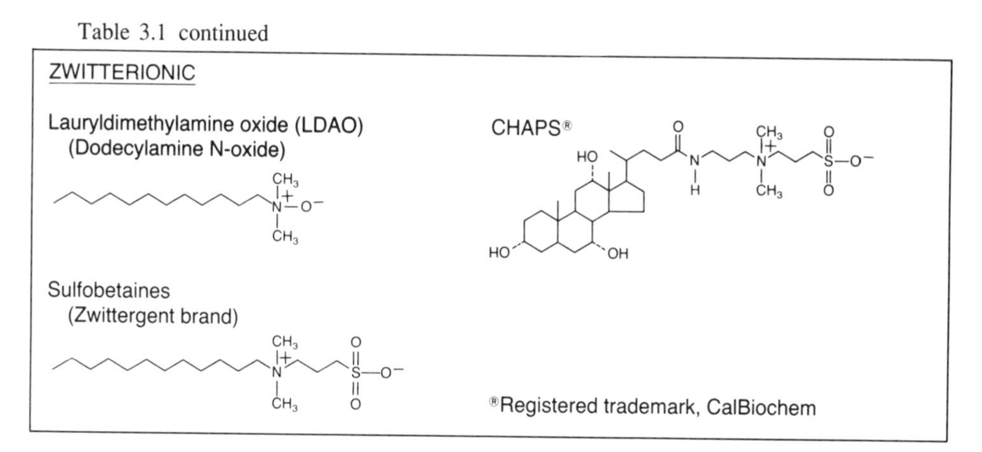

proved most effective are (1) nonionic detergents (e.g., Triton X-100, octylglucoside), (2) bile salts (e.g., cholate, deoxycholate), and (3) zwitterionic detergents (e.g., CHAPS, Zwittergent). Finding the detergent best suited for solubilizing and purifying a particular membrane enzyme is still a matter of trial and error.

3.3 Characterization of Purified Intrinsic Membrane Proteins

Once a membrane protein has been purified, characterization of even the most apparently simple sort can present difficulties. As with a soluble protein, fundamentally, one wants to know the number and molecular weight of the different polypeptide subunits and their stoichiometry, the size and possibly the shape of the molecule, and, if relevant, information about biochemical activity.

3.31 Subunit Molecular Weight (SDS-PAGE Analysis)

The use of dodecyl sulfate-polyacrylamide gel electrophoresis (SDS-PAGE) is routine, but intrinsic membrane proteins present special problems. In this procedure, dodecyl sulfate binds to the polypeptide chains and the SDS-protein complexes are separated in a polyacrylamide gel according to Stokes radius, which is, in most cases, a function of molecular weight. Molecular weight is estimated from the relative electrophoretic mobility in comparison to known standards. However, if the binding of SDS is qualitatively different for the unknown protein than for the standards, an incorrect molecular weight will be obtained. This is commonly found for intrinsic membrane proteins with a high percentage of nonpolar residues. Whereas SDS forms similar complexes with most soluble proteins (glycoproteins excepted) with 1.4 g SDS/g protein (1211), more detergent can bind to proteins with highly nonpolar character. The addi-

Table 3.2. Properties of selected detergents

Detergent	CMC (mM)	Mol. wt.	Micelle size	Aggregation number	Specific volume (ml/g)	References
Sodium dodecyl sulfate	1.33	288	24,500	85	0.864	612, 1383
Sodium cholate[1]	3	408	2,100	5	0.778	612, 1383
Sodium deoxycholate[1]	0.91	392	23,000	55	0.771	612, 1383
$C_{12}E_8$	0.11	538	68,000	12	0.973	612, 1383, 1434
Triton X-100[2]	0.24	628	90,000	140	0.908	612, 1383
Tween 80[2]	0.012	1300	76,000	60	0.896	612, 655
Lauryl dimethylamine oxide	2.2	229	17,000	75	1.112	612
ß-D-Octylglucoside	25	293	8,000	27	0.820	1242, 1213, 612
ß-D-laurylmaltoside	0.16	510	50,000	98	0.820	1242
CHAPS	8	615	6,150	10	0.802	637
Zwittergent 3-12	3.6	335	—	–	0.957	527, 1383

[1]Solution properties of bile salts depend strongly on temperature, pH, counterion type, and ionic strength. Small and large micelles can be formed, depending on conditions. Data are given for 0.15 M NaCl, 20°C, pH 9. Bile salts tend to precipitate about 1 pH unit above the pK_a, which can lead to gel formation, especially with deoxycholate. The pK_a value for cholate is 5.2 and for deoxycholate 6.2 (see 612).

[2]These detergents are polydisperse mixtures. Values in the table are average values. Properties are likely to vary with different samples.

tional negative charge due to the excess bound anionic detergent results in an anomalously high electrophoretic mobility, indicating a smaller molecular weight in comparison to the standards. In other cases, the membrane protein may not be completely unfolded by the interaction with SDS, resulting again in an anomalously high electrophoretic mobility due to the compact shape of the protein—SDS complex. These are not a small effects. For example, the lactose permease has an apparent molecular weight of 33,000 by SDS-PAGE analysis, but is actually calculated to be 46,000 from the gene sequence (159). In many cases, a more accurate molecular weight can be obtained by doing a "Ferguson plot" in which the dependence of the electrophoretic mobility on the percent acrylamide is determined for both the standard proteins and the membrane protein (see 611). This parameter is dependent on the Stokes radius and less so on the charge of the complex (i.e., amount of bound SDS). For example, one of the subunits of an *E. coli* cytochrome complex has an apparent molecular weight of 28,000 in a 12% polyacrylamide gel, but a Ferguson plot analysis indicates 43,000 (990), the correct molecular weight deduced from DNA sequencing.

Another source of trouble is residual quaternary structure. Some membrane proteins aggregate, even in the presence of SDS. Dimers of glycophorin A (25, 3200) or of the M13 (or fd) bacteriophage coat protein (1082), for example, are predominant forms observed on SDS-polyacrylamide gels. In some cases, the aggregation is enhanced when the protein–SDS sample mixture is heated. This is observed for subunits of both mitochondrial and bacterial terminal oxidases (e.g., 990). It is important to compare the SDS-PAGE patterns obtained for samples which have been heated and those which have not in order to assess the tendency of the protein to aggregate irreversibly. A related problem is sometimes observed due to the presence of the detergent used in the purification of the membrane protein. This detergent must be diluted out and displaced by SDS, and, in some cases, clear differences in electrophoretic mobility are observed depending on the detergent in which the enzyme is solubilized prior to the addition of SDS.

In short, subunit molecular weights obtained by SDS-PAGE analysis can be unreliable for highly nonpolar intrinsic membrane proteins. Unfortunately, no simple alternative exists, and the correct value is often derived from the complete primary sequence (usually from the DNA sequence of the gene) or from a rigorous hydrodynamic analysis.

3.32 Molecular Weight of the Native Protein by Hydrodynamic Measurements

This can be very difficult for a membrane protein because of the complication of detergent binding. To appreciate this fully, let us first consider the simple case of a *soluble* protein for which one knows the subunit molecular weight by SDS-PAGE and wants to know if it is a monomer, dimer, or higher oligomer in the nondentured, active form. Although gel filtration chromatography is often used to obtain molecular weight by comparison to standard proteins, this is insufficient because the standard proteins are all globular in shape, and the wrong molecular weight will be obtained if the unknown protein is not globular but is even slightly elongated. A protein of 50,000 molecular weight might easily elute at an apparent molecular weight of 100,000 due to this shape effect. To get around this problem, the gel filtration column should be calibrated in terms of Stokes radius, the size of the "equivalent hydrodynamic sphere," and a second technique used in combination. Usually, the S-value or sedimentation velocity is measured, either in an analytical ultracentrifuge or, more commonly, by sucrose density gradient ultracentrifugation.

$$S^\circ = \frac{M(1 - \overline{V}\rho)}{N6\pi\eta R_s} \quad [3.1]$$

where M = molecular weight of the protein
$\overline{V}$ = partial specific volume of the protein (reciprocal density)
η = solution viscosity (~1 centipoise)
ρ = solution density (~1 g/ml)

Since one knows ρ and η, and R_s is obtained by gel filtration chromatography, there are only two unknowns, $\overline{V}$ and M. But for water-soluble proteins, $\overline{V}$ can be estimated from the amino acid composition (231), or directly measured, or just estimated to be ~0.72 to 0.75 ml/g. Hence, by measuring S°, M can be calculated.

Now, consider the same situation with a membrane protein. Additional problems arise because the hydrodynamic particle is a protein-detergent complex, so M and $\overline{V}$ are the molecular weight and specific volume of the complex, M_{cx} and $\overline{V}_{cx}$. Unfortunately, $\overline{V}_{cx}$ cannot be simply estimated in the absence of any information about the composition of the complex. Two procedures are used to get the protein molecular weight:

(1) Directly measure the amount of bound detergent per gram of protein. This is usually done spectroscopically or with radioactive detergent, using various ways to isolate the complex, such as gel filtration chromatography. Once the relative masses of protein and detergent are known for the complex, the value of $\overline{V}_{cx}$ is estimated as a weighted average of the values for the pure protein and pure detergent. When this has been done, M_{cx} is easily obtained, and since one knows the percentage of the complex which is protein, the protein molecular weight is obtained (see refs. 567, 1281 for examples).

(2) Measure S° in media with different values of solution density, ρ. This is usually obtained by using H_2O and D_20 mixtures. By plotting S° vs. ρ, one can solve for both M_{cx} and $\overline{V}_{cx}$. One then assumes that $\overline{V}_{cx}$ is a weighted average of the value for pure protein and pure detergent.

$$\overline{V}_{cx} = (\text{fraction protein}) \times \overline{V}_{protein} + (\text{fraction detergent}) \times \overline{V}_{detergent} \quad [3.2]$$

By estimating $\overline{V}_{protein}$ and looking up $\overline{V}_{detergent}$ (Table 3.2), one can obtain the molecular weight of the protein portion of M_{cx}.

The easiest way to measure S° vs. ρ is to use an analytical ultracentrifuge (see 567, 1401, 1281; review in 1213). Alternatively, one can use a sucrose density gradient, using H_2O and D_2O mixtures, but in this case, the analysis is much more complicated, though conceptually identical (1357).

An alternative procedure to obtain the molecular weight of the native form of a membrane protein is to use equilibrium sedimentation, either in an analytical centrifuge or even in an airfuge. The distribution of material at equilibrium is such that the slope of a plot of ln(concentration) vs. r^2 is

$$\text{slope} = \frac{\omega^2 M(1 - \overline{V}\rho)}{2RT} \qquad [3.3]$$

where r = position in the tube measured from the center of the rotor
ω = circular frequency of the centrifuge

One again has a simple task if $\overline{V}$ is known or easily estimated, as for most soluble proteins. For membrane proteins, the procedure again is to measure the slope at different values of ρ, obtained by H_2O/D_2O combinations. As seen previously with sedimentation velocity, M_{cx} and V_{cx} are simultaneously obtained and can be used to calculate the protein molecular weight (e.g., 567, 1401, 1281).

Additional complications are introduced if a third component is present in the complex, such as substantial amounts of residual bound lipid (e.g., 567). The major point is that these measurements are far from trivial, and there are many potential errors. The amount of bound detergent can be substantial for purified intrinsic membrane proteins, often 0.3 to 1.5 × the mass of protein. Small errors in this value will result in significant errors in the estimate of the protein molecular weight. Table 3.3 lists some of the values obtained for bound detergent for several proteins. Note that soluble proteins do not bind to these detergents (613, 909, 223), emphasizing that it is the nonpolar portion of the protein, normally in contact with the membrane lipids, which is responsible for detergent binding.

3.33 Radiation Inactivation (see 742, 1512 for reviews)

A procedure which is being applied with increasing frequency to membrane proteins is radiation inactivation or target size analyses (741, 742, 1512, 524). This can be performed with purified proteins or impure preparations, including intact biomembranes. What is required is a specific assay to be able to selectively monitor the percentage of the protein being examined which has been damaged by the incident radiation. This can be done by using an enzymatic assay, hormone or other ligand binding essay, or even spectroscopic criteria. The idea behind this technique is relativly straightforward. The sample, usually frozen, is placed in a beam of high-energy radiation, such as electrons from a synchrotron. Samples are removed at various times, thawed, and assayed. Any interaction between the protein of interest and the beam particles will result in substantial damage to the protein, including covalent bond breakage, which can be monitored by SDS-PAGE. For single subunits, experience shows that a hit anywhere in the polypeptide is sufficient to destroy any biological activity. Essentially, a larger protein will be hit proportionately more frequently than a smaller protein and, thus, be

Table 3.3. Detergent binding to selected membrane proteins.

Protein	Detergent	Bound detergent (mg/mg protein)	References
Na^+/K^+-ATPase	Triton X-100	0.28	558
Ca^{2+}-ATPase	Triton X-100	0.20	558
Band 3	Triton X-100	0.77	558
Acetylcholine receptor	Triton X-100	0.70	558
Rhodopsin	Triton X-100	1.10	558
ADP/ATP carrier	Triton X-100	1.5	567
Insulin receptor	Triton X-100	0.15, 0.31, 0.54	1169,7
Insulin receptor	Deoxycholate	0.01, 0.03	7
Cytochrome *c* oxidase	Triton X-100	0.6	1281
Cytochrome *c* oxidase	Laurylmaltoside	0.55	1401

inactivated at a faster rate. This can be calibrated in terms of the molecular weight of the protein. The technique is sensitive not to the shape of the molecule but to its mass. Typically, internal standards of known molecular weight are examined along with the unknown in order to facilitate the analysis. Problems in interpretation arise when the protein contains more than one subunit (see 1512). A hit in one subunit does not necessarily cause covalent damage in other subunits. As a result, multisubunit enzymes in which different subunits have different activities may give different target sizes depending upon the assay used to monitor inactivation.

This technique is unique in that it can be used with intrinsic membrane proteins *in situ*, within the biomembrane. The various artifacts and interpretive difficulties have been described (741). One obvious difficulty is that one needs a high-energy radiation beam, so most of these experiments are done collaboratively with laboratories which have both access to the beam and expertise in proper analysis.

3.34 Spectroscopic Methods and Secondary Structure

Several methods can be used to measure the amount of α-helix and ß-sheet in a membrane protein. In the absence of three-dimensional structures, this can be valuable in modeling efforts. By far the most widely used method is circular dichroism (CD). Infrared and Raman spectroscopies are also being used more frequently for this purpose and solid-state NMR methods have also been used.

(1) *Circular dichroism* (see 183) measures the differential absorption of left- and right-handed circularly polarized light, and this *optical activity* measures molecular chirality or asymmetry. In the far ultraviolet region of the spectrum (190 nm to 240 nm), CD is primarily due to the absorption of the amide carbonyl groups in the polypeptide backbone. Regular secondary structures, such as the α-helix, result in defined spectroscopic shapes due to the distinct electronic environments sensed by the amide groups within these structures. Most of the algorithms for the analysis of CD spectra of proteins represent the protein spectrum as the sum of component spectra arising from the portions of the protein which are α-helix,

ß-sheet, and "random coils." There are several ways to obtain spectra for each of the components, and, once this is done, this spectroscopic basis set is used to obtain a best fit to the experimental spectrum of the sample protein. The parameters which are varied in the fitting procedure are the weighting coefficients, which are interpreted as the fraction of the protein represented by each kind of secondary structure.

These techniques have been developed with soluble proteins, but there is no reason why the same algorithms should not apply equally well to membrane proteins. This assumes the secondary structures are basically the same, and no evidence suggests otherwise. The same uncertainties and sources of error which apply to soluble proteins also apply to membrane proteins. In addition, membrane proteins can present unique problems, depending on the nature of the sample. Some proteins can be examined *in situ*, using membrane suspensions. Examples are bacteriorhodopsin, located in the purple membrane of *Halobacterium halobium*, and the Ca^{2+}-ATPase within the sarcoplasmic reticulum membrane. Purified membrane proteins can also be examined by CD in detergents, as long as the detergent absorption in the far UV is not excessive, or in reconstituted vesicles. Problems arise from two sources: (1) differential light scattering when the membrane particles are large relative to the wavelength of light and (2) absorption flattening due to the concentration of protein in the membranes or vesicles, i.e., not homogeneously distributed in solution. These artifacts can be very substantial and procedures exist for evaluating and correcting for them (1539, 518, 1442, 1538, 1047).

Because of the general lack of high-resolution structural data on intrinsic membrane proteins, there is no way to evaluate accurately the interpretations of CD spectra. Except for a few cases, different spectroscopic techniques have not been utilized and quantitatively compared. Interestingly, bacteriorhodopsin has been examined by CD, IR, and solid-state NMR methods and all three techniques have been claimed to be consistent with substantial ß-sheet (see Section 3.52). There are substantial uncertainties with each technique, however. For example, the CD interpretation of a relatively large proportion (46%) of ß-sheet in bacteriorhodopsin (518) is very dependent on the way in which the raw data are corrected for optical artifacts (1047, 1526), emphasizing the need for care in such measurements with large membrane fragments. The relatively low resolution data from electron microscopy image reconstruction (see Section 3.42) have been interpreted in terms of 80% α-helical structure with no ß-sheet (620). A final understanding of these inconsistencies will have to await atomic resolution structural analysis of this protein. Other examples of membrane-spanning proteins which appear to have substantial ß-structure are the porins in the outer membrane of *E. coli* (737; Section 3.53) and the α-toxin protein of *Staphylococcus aureus* (1457), both of which form pores through the bilayer (see Chapter 8).

(2) *Infrared and Raman spectroscopies*, in addition to yielding information on membrane lipid conformation (see Chapter 2), can also be used to measure protein secondary structure (1590). Vibrational bands from the polypeptide backbone are sensitive to secondary structure and can be used to quantify the amount of α- and ß-structures. These methods can be applied to air-dried films, aqueous membrane

suspensions, as well as purified proteins, either in detergent or in reconstituted vesicles (see 1513, 32). For example, FT-IR spectroscopy indicates that the Ca^{2+}-ATPase has primarily α-helical and random coil structures in the native membrane (832), whereas the hydrophobic myelin protein (lipophilin) has both α- and ß-structure in reconstituted vesicles (1414). Future application of these techniques will require systematic studies on purified membrane proteins.

(3) *NMR spectroscopy* can also be used to examine membrane proteins. However, there are severe limitations, largely due to the relatively slow motions undertaken by integral membrane proteins in situ and usually also in detergent complexes. Thus, powerful NMR techniques, such as two-dimensional NMR, which can yield precise conformational information about relatively small proteins in solution, cannot yet be applied to membrane proteins. It is more appropriate to apply methods of solid-state NMR to membrane proteins. Deuterium and carbon-13 NMR seem to offer the most potential, though their application to date has been limited. Information about average backbone conformation (850) as well as side chain dynamics (see Chapter 5) is available. Solid-state NMR techniques are not widely used, nor are they applicable in most cases. Nevertheless, for obtaining detailed information in a few select cases, such as bacteriorhodopsin (850), these methods can be very powerful.

3.35 Enzymological Characterization

Certainly one of the most important methods used to characterize purified membrane proteins is to measure biochemical activity. The criteria are essentially the same as for soluble proteins, but there are special problems in working with membrane proteins. One major problem is that the biochemical activity of membrane proteins is often highly dependent on the lipids and detergents bound to the protein. Loss of activity may be either reversible or irreversible if the enzyme is placed in a detergent environment in which it is inactive. It is always best to have some measure of what the specific activity is in vivo or in the membranes prior to solubilization. The presence of excess detergent may be inhibitory, for example, by causing nonpolar substrates to effectively become diluted within the micelle population, thus reducing enzyme activity. The lipid environment which is optimal for activity must be considered in assaying any membrane protein. A second type of problem is presented by proteins whose activity is transbilayer, such as channel or transport proteins (Chapter 8). In these cases, the assay must follow solute movement or exchange across the bilayer between separate compartments (e.g., inside vs outside for liposomes). Chapter 6, on Membrane Enzymology, contains a more detailed discussion of these problems.

3.36 Quaternary Structure and Chemical Crosslinking

Many membrane enzymes are multisubunit complexes. Examples are the H^+-ATPase, Na^+/K^+-ATPase, mitochondrial electron transfer complexes, and the photosynthetic reaction centers. In some cases, intrinsic membrane proteins are strongly bound to soluble proteins via noncovalent protein-protein interactions, as is the case with the F_0 and F_1 components of the mitochondrial H^+-ATPase (476). In the case of the *E. coli* enzyme, the F_0 portion, containing three kinds of subunits (a, b, c) by SDS-PAGE analysis, forms a proton channel, and the F_1 portion, with five kinds of subunits (α, ß, γ, δ, ε), has the active site for ATP hydrolysis (476). An obviously important goal is to determine the subunit identity, stoichiometry, and nearest neighbor relationships of the components of the functional complex. This is not easy, even once the protein complex has been isolated. The problems and solutions are essentially the same as with soluble protein complexes, but there are additional difficulties with the membrane proteins.

First, it is important to realize that the strength of the protein association may be very dependent on the lipids and detergents to which the proteins are bound (1221, 238). For example, the succinate dehydrogenase of *E. coli* appears to contain four subunits when solubilized in Lubrol PX, but only two subunits in most other detergents, including Triton X-100 (238). In this case, the *sdh* operon is known to encode all four polypeptides, and the two-subunit form has an abnormal EPR spectrum, so it is clear that the four-subunit form exists in vivo. Both forms, however, have succinate dehydrogenase activity, so the biochemical criteria one uses are obviously important in judging whether one has solubilized the correct form.

A related problem is that one may have complexes between membrane proteins in the bilayer due to the high concentration of the proteins in the membrane (540). Solubilization, regardless of the detergent used, will effectively dilute the membrane proteins and disperse them. By mass action, this can cause dissociation of the components of a complex where the interactions are weak. It is often very difficult to determine the extent of complex formation in situ in the membrane in relation to what one observes upon solubilization and purification. This is the kind of problem being addressed in numerous milticomponent systems, such as the ß-adrenergic receptor–adenylate cyclase system, the mitochondrial electron transfer chain, and the microsomal cytochrome P450 and cytochrome b_5 systems (see Chapter 6).

In order to address the question of subunit stoichiometry and associations in a purified complex, there are limited experimental tools: (1) chemical crosslinking, (2) quantitative amino-terminal amino acid analysis, and (3) determination of the mass ratio of the subunits on SDS-polyacrylamide gels by the intensity of staining with a dye, autoradiography, or immunoblotting. Each method has limitations, but all have been utilized. For example, the subunit stoichiometry of the five subunits in the nicotinic acetylcholine receptor (2:1:1:1:1 = α/ß/γ/δ/ε) was determined by quantitative amino-terminal analysis (1189) (see Section 8.24). The stoichiometry of the 3 subunits of the F_0 portion of *E. coli* H^+-ATPase was found to be a/b/c = 1:2:10 by mass ratio on SDS-polyacryalamide gels (452) (see Section 8.42). It is important to note that Coomassie Brilliant Blue, the usual dye used to visualize proteins following SDS-PAGE, binds more to proteins with basic residues (1424), and there are cases of very nonpolar intrinsic membrane proteins which barely stain at all.

Chemical crosslinking is one technique that has been used with purified protein complexes as well as with intact membranes in order to determine nearest neighbor relationships (see 480). Several hydrophobic crosslinkers are designed specifically for use with membrane proteins. The structures of several are shown in Table 3.4. The techniques employed are the same as those used with soluble systems.

Table 3.4. Some crosslinking reagents used to study the quaternary structure of membrane proteins

Crosslinking Reagent	Comments
(1) Dithiobis(succinimidylpropionate), (DSP) 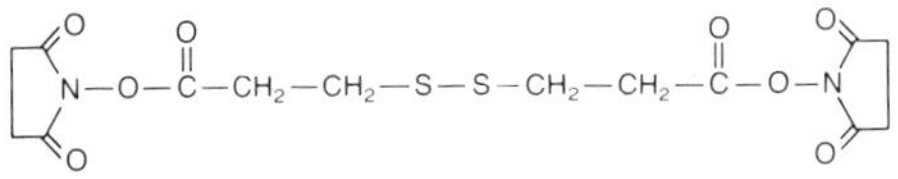	Homobifunctional, cleavable, amino-reactive at physiological pH. Moderately polar, 12Å span. (49, 1143)
(2) Dimethyl 3,3'-dithiobispropionimidate · 2HCl, (DTBP) 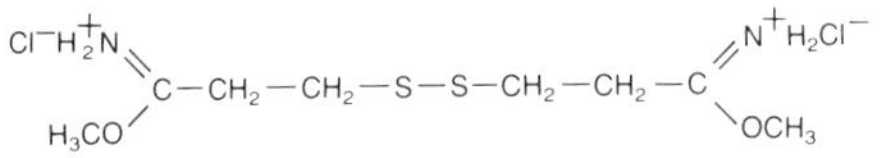	Homobifunctional, cleavable, amino-reactive at pH9. Charged, but penetrates membranes. 12Å span. (49, 1143)
(3) Disuccinimidyl tartrate, (DST) 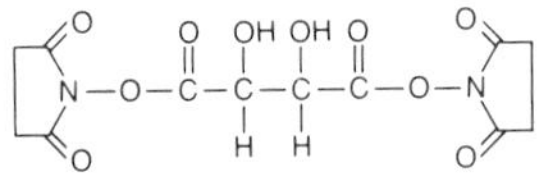	Homobifunctional, periodate cleavable, amino-reactive at physiological pH. Polar, uncharged, 6Å span. (49)
(4) N-Succinimidyl(4-azidophenyl)1,3'-dithiopropionate (SADP) 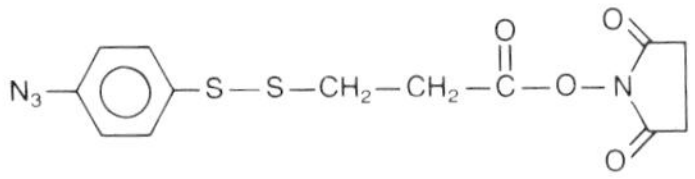	Heterobifunctional, photoactivatable, cleavable. (1143)
(5) Disuccinimidyl suberate (DSS) 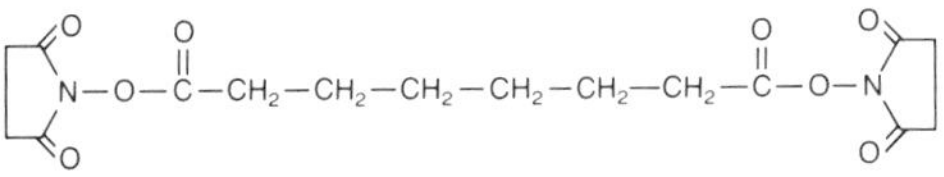	Homobifunctional, amino-reactive at physiological pH. (1159)
(6) 4,4'-Dithiobisphenylazide (DTBPA) 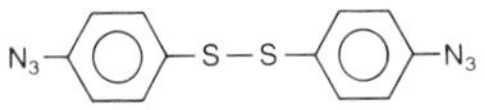	Non-polar, photoactivatable, cleavable. (49)

[1] See Pierce Chemical Company Handbook and Catalog, and reference (480) for additional reagents and information.

The crosslinked products are usually analyzed by polyacrylamide gel electrophoresis, often with cleavable crosslinkers, to allow the polypeptides to be analyzed. Antibodies directed toward the individual polypeptides can also be used for immunoblotting following SDS-PAGE, to identify the components of each product formed. Conceivably, the relatively long lifetime of the reactive species could allow proteins within biomembranes to crosslink even if they just casually collide by random diffusion in the bilayer. However, there are several reports where this does not seem to be a problem, as judged by the fact that the crosslinked products formed in membranes represent specific protein associations and not a random array. An example is the crosslinking observed in the photosynthetic membrane of *Rhodobacter capsulata*, showing crosslinks between subunits of the reaction center components, but also between the reaction center and the B870 antenna complex involved in the energy transfer to the reaction center (1143).

Finally, it should be noted that chemical crosslinking has often been used to identify intrinsic membrane proteins which bind to known soluble proteins or components. Examples are the observed crosslinking of (1) the a and b subunits of the F_0 portion of the H^+-ATPase with the ß-subunit of the soluble F_1 portion (49), (2) cytochrome *c* with subunits of mitochrondrial cytochrome *c* oxidase (see 480), and (3) peptide hormones, such as insulin, with the hormone receptors (see 1159).

3.4 Three-Dimensional Structures from Diffraction Studies and Image Reconstruction

3.41 Crystallization of Membrane Proteins

The most detailed structural information on purified membrane proteins is potentially available through X-ray diffraction studies of three-dimensional protein crystals. Unfortunately, intrinsic membrane proteins have proved to be extremely difficult to crystallize. Once removed from their native lipid environment, the nonpolar regions of the proteins tend to promote aggregation in nonordered forms unsuitable for crystallographic studies. Special techniques are clearly required to minimize these problems, and progress has been made. Michel has pointed out two types of crystals which can be formed (Figure 3.2). Type I crystals resemble stacks of membranes. Such crystals can have points of contact between nonpolar portions laterally, and polar portions of the protein connect the membrane-like sheets. Crystals of this type have been obtained for several proteins, but none has been suitable for high-resolution diffraction studies. Type II crystals have contact points between polar portions of the protein, and the small amphiphile or detergent basically fills in the gaps. Note that the size, charge, and other properties of the detergent are critical, since unfavorable detergent interactions can destabilize the crystal structures. The crystals of the photosynthetic reaction center of *Rhodopseudomonas viridis* (980, 319), which have led to a high resolution structure, are of this type. Table 3.5 lists intrinsic membrane proteins which have been crystallized. Not all these crystals, however, are suitable for structural studies.

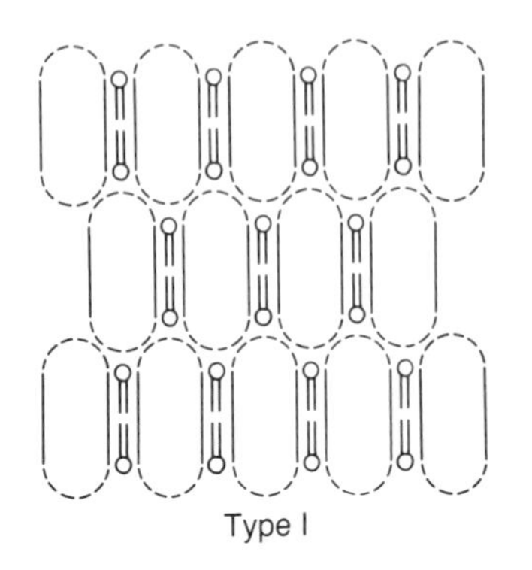

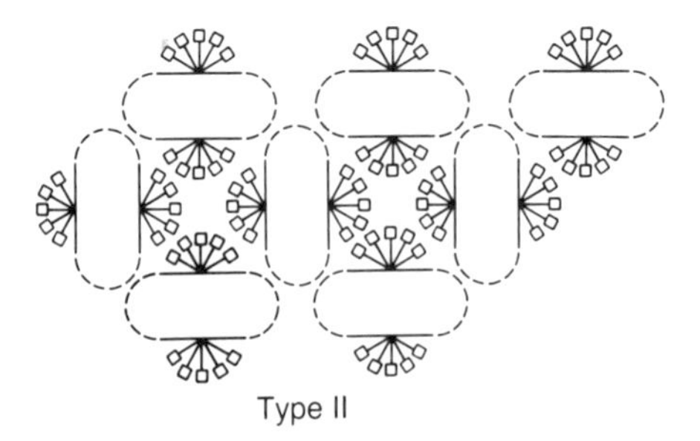

Figure 3.2. Two basic types of membrane proteins crystals. Type I consists of stacks of two-dimensional crystals which are ordered in the third dimension. Type II crystals have detergents bound to the hydrophobic surface. The dashed lines indicate hydrophilic domains of the protein. Adapted from ref. 979.

So far, the structure of only one class of membrane protein, the bacterial photoreaction center, has been solved to high resolution (see Section 3.51). Reportedly, a high-resolution structure of the matrix porin (OmpF protein) of the *E. coli* outer membrane is nearing solution (see Section 3.53).

Membrane proteins can be crystallized and, although the number of successful cases is small, several generalizations are justified concerning methodology.

i. The proteins are cocrystallized with detergent.
ii. The choice of detergent is important. Zwitterionic or nonionic detergents with a high CMC and small micelle size appear to be best (491).
iii. Small amphiphilic organic molecules can assist the crystallization, possibly

Table 3.5. Membrane proteins which have been crystallized.

Protein	References
1. Reaction center of *R. viridis*[l]	980, 319
2. Reaction center of *R. sphaeroides*[l]	19, 17, 18
3. Photosystem I reaction center of cyanobacterium *Phormidium laminosum*	448
4. OmpF (matrix porin) (*E. coli* outer membrane)[l]	490
5. OmpA (conjugin) (*E. coli* outer membrane)[l]	491
6. LamB (maltoporin) (*E. coli* outer membrane)	491
7. Bacteriorhodopsin (*H. halobium*)	978

[l]Suitable for X-ray diffraction analysis.

by influencing the detergent polar headgroup interactions (979). The *R. viridis* reaction center crystals were obtained with 1, 2, 3-heptanetriol.

iv. Both polyethylene glycol and ammonium sulfate, commonly used with soluble proteins, have been successfully used with membrane proteins to induce crystallization (1562).

Garavito et al. (491) have pointed out that crystallization conditions in some cases are near conditions where the detergent forms a separate phase. The role that detergent aggregation plays in the crystallization of protein–detergent comicelles is not known, but it is evident that the choice of detergent is critical to success.

3.42 Image Reconstruction and Two-Dimensional Crystals

Although three-dimensional crystals are difficult to obtain, a number of membrane proteins form two-dimensional ordered arrays. In some cases, the proteins exist naturally in such arrays, e.g., bacteriorhodopsin in the purple membrane. These proteins, as well as many others, can be induced to form "two-dimensional crystals" when purified and reconstituted with phospholipids under appropriate conditions. The two—dimensional ordered arrays can be used to obtain three-dimensional structural information by electron microscopy and image reconstruction techniques (see 1480).

Box 3.1 Image Reconstruction

The methodology for this was originally designed for studying virus particles (261), and the application to membrane proteins was pioneered by Henderson and Unwin (620) in their studies of bacteriorhodopsin. In principle, by this method one should be able to obtain sufficient structural detail to allow one to trace the polypeptide chain of a protein, though this has not yet been achieved in practice. The scattering of electrons is strong enough so that images of single molecules can be seen with an electron microscope. However, the beam intensity normally required to obtain an image is destructive to the biological sample, requiring much lower intensities to be used in order to obtain high resolution. In normal transmission electron microscopy one uses negative staining to enhance the contrast, but this is not useful for obtaining structural information about those portions of the protein buried in the bilayer, since these regions are not accessible to the stain. Thus, stains are not often used for image reconstruction, except to visualize aqueous channels.

In order to obtain sufficient data with the low electron beam intensity, it is necessary to sum the images of many molecules. It is for this reason that two-dimensional ordered arrays are useful. The images themselves represent two-dimensional projections of the electron density of the sample. By digitizing these images and carrying out a Fourier transform, one can, in essence, pick out the repetitive features of the image and eliminate the nonrepetitive noise. Another Fourier transform of the repetitive elements reconstructs the original image but without the random noise. This procedure basically is a trick allowing one to sum the images obtained from hundreds or thousands of molecules in the microscope field.

The electron density projection in two dimensions is insufficient to yield a three-dimensional structure. However, by tilting the sample, projections are obtained through the sample at different angles, and, together, a series of such views can be used to reconstruct the three-dimensional image of the object (see 1480). One obtains an electron density map at different levels in the membrane. Typically, electron density profiles every 15–25 Å are reported.

Table 3.6 lists some proteins which have been studied using this procedure. In all cases but bacteriorhodopsin, the resolution is only sufficient to indicate the general shape and dimensions of the molecules. However, even this can be very valuable. For example, many of these molecules exist as distinct multimers in the two-dimensional crystals. Porin and bacteriorhodopsin are clearly trimeric. In the case of porin, there is a clearly observed channel associated with each of the three individual polypeptides on the external surface of the cell, which merge to form a single channel on the periplasmic side of the *E. coli* outer membrane (384) (see Section 8.23). Similarly, connexin appears as a hexamer (1481) and the acetylcholine receptor as a symmetric pentamer (149) in these studies (see Section 8.2). The mitochondrial respiratory enzymes, ubiquinol-cytochrome *c*

Table 3.6. Membrane proteins for which structural information has been obtained by image reconstruction techniques using two-dimensional crystalline arrays.

Protein	Three-dimensional resolution	References
1. NADH: ubiquinone oxidoreductase (Mitochondrial)	13 Å	113
2. Cytochrome *c* oxidase (Mitochondrial)	20 Å	314, 466
3. Ubiquinol: cytochrome *c* oxidoreductase (Mitochondrial)	25 Å	843
4. Chlorophyll *a*/*b* light harvesting complex (chloroplast)	16 Å	795
5. Photoreceptor unit of *R. viridis*[1]	20 Å	1381
6. Connexin (gap junction protein)	18 Å	1481
7. Acetylcholine receptor	17 Å	149, 1633, 760, 119
8. OmpF (matrix porin)	23 Å	384
9. Na^+/K^+-ATPase	20 Å	1109, 1005
10. Bacteriorhodopsin	6.5 Å	620, 981
11. Rhodopsin (bovine)	20 Å	346

[1]This complex consists of the reaction center, whose structure has been solved to high resolution, plus three polypeptides making up the light harvesting complex.

oxidoreductase (843), and cytochrome oxidase (314, 466) are dimeric, though it is not clear if the dimeric units are relevant in vivo. In some cases, the extent to which these proteins protrude beyond the bilayer surface is remarkable. For example, portions of cytochrome *c* oxidase are 50 Å beyond the bilayer surface (314, 466) (Figure 3.3), and the same is true of the acetylcholine receptor (149) (see Figure 8.8). By contrast, porin (384) and bacteriorhodopsin (620) do not protrude measurably beyond the membrane surface. Figure 3.3 shows the low-resolution structural picture of cytochrome *c* oxidase.

3.5 Three Examples of Structural Studies on Membrane Proteins

The following three examples exemplify the wide range of techniques used to study intrinsic membrane proteins. The best known structures are the reaction centers of *R. viridis* and *R. sphaeroides*, where X-ray crystallography has been successful. The next best understood structure is bacteriorhodopsin of *H. halobium*, where the image reconstruction technique has been the most informative and where numerous other solution techniques have been employed to supplement these data. Studies on porin and related proteins from the *E. coli* outer membrane have relied more on solution methods combined with the use of genetics and molecular biology to identify functionally important regions.

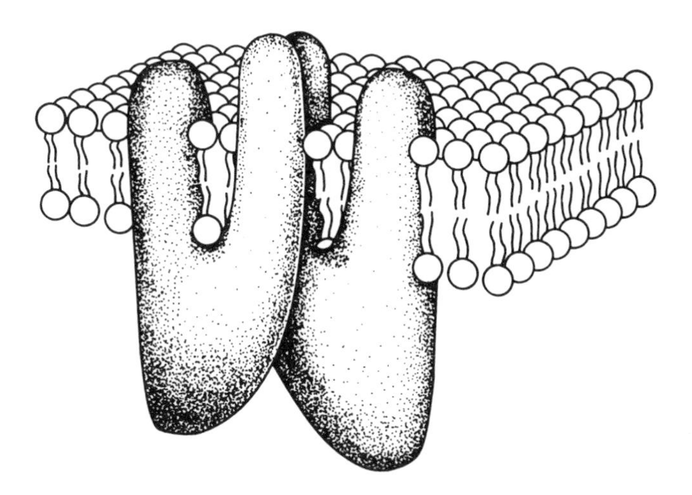

Figure 3.3. Model of the dimer of mitochondrial cytochrome *c* oxidase determined using image reconstruction techniques. Each Y-shaped monomer is composed of at least 12 polypeptide subunits. The upper surface represents the matrix side of the membrane, and the bulk of the protein extends into the mitochondrial intermembrane space (lower side). The depiction of lipid molecules in the crevice is fanciful and not demonstrated. Adapted from ref. 466. Drawing kindly provided by Dr. T. Frey.

3.51 Structures of the Photosynthetic Reaction Centers of *R. viridis* and *R. sphaeroides*

Photosynthetic reaction centers are protein-pigment complexes which carry out the primary charge separation in photosynthetic membranes (318). Those from purple nonsulfur bacteria are best characterized, generally consisting of three protein subunits, called H (heavy), M (medium), and L (light). In the case of the reaction center from *Rhodopseudomonas viridis*, there is also a fourth subunit which is a *c*-type cytochrome. The prosthetic groups of this complex are four heme groups, four bacteriochlorophyll *b*, two bacteriopheophytin, one nonheme iron (not an iron-sulfur center), one menaquinone (Q_A), and one ubiquinone (Q_B). Upon excitation with light, an electron is transferred from the primary electron donor, the "special pair" bacteriochlorophyll dimer, to bacteriopheophytin and then to the primary quinone acceptor (Q_A). Eventually, the electron ends up reducing a secondary quinone acceptor (Q_B) in a reaction in which protons are taken up from solution by the reduced quinone (see Figure 3.4). The Q_B is in equilibrium with the quinone pool in the bilayer. The oxidized primary electron donor, the "special pair," is reduced by the *c*-type cytochrome. Since the cytochrome and the quinone sites are on opposite sides of the photosynthetic membrane, the light–driven electron transfer is electrogenic and generates a voltage difference across the bilayer.

The reaction center from *R. viridis* (980, 319, 318, 982, 983) has a total molecular weight of about 150,000, with apparent subunit molecular weights by SDS-PAGE of 38,000 (cytochrome, 333 residues) 35,000 (H, 258 residues),

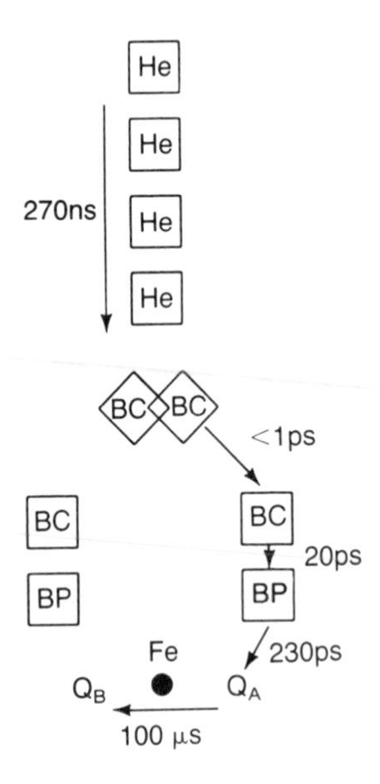

Figure 3.4. Schematic of the redox prosthetic groups in the bacterial photosynthetic reaction center of *R. viridis*. The four hemes (He) are in the cytochrome subunit located on the periplasmic side of the membrane. Electron flow, initiated by light, is indicated by the arrows and proceeds from the hemes to the special pair bacteriochlorophyll dimer (BC-BC) and from there to another bacteriochlorophyll (BC), bacteriopheophytin (BP), bound quinone (Q_A), non-heme iron (Fe), and, finally, to the terminal quinone (Q_B). The numbers are the characteristic half-times for each step in the reaction. Note that only one of the two nearly symmetric pathways is used. Adapted from ref. 318.

28,000 (M, 323 residues), and 24,000 (L, 273 residues). Note that the aberrant migration in SDS-PAGE gives incorrect molecular weights for the H (heavy), M (medium), and L (light) subunits, as indicated by the number of amino acid residues in each polypeptide, obtained by DNA sequencing (319). The purified complex was crystallized by Michel (979) using ammonium sulfate as the precipitating agent in the presence of the detergent *N,N*-dimethyldodecylamine *N*-oxide (LDAO, Table 3.1) and the organic amphiphile heptane 1,2,3-triol. The crystals were well ordered and were shown to be photochemically active. The structure of the four-subunit protein has been solved to about 3 Å resolution. The detergent is not ordered in the crystals, so the locations of the boundaries of the membrane-spanning regions are not known with absolute certainty. The complex has dimensions of 30 × 70 × 130 Å. The L and M subunits are substantially integrated into the membrane along with a short segment of the H subunit. The L and M subunits each contain five α-helical segments which are transmembranous, and the H subunit has one helical segment in the membrane. Hence, all the portions of this protein complex which extend into the bilayer have an α-helical configuration. Each helical segment is about 40 Å long, sufficient to completely traverse the membrane.

The overall structure of the protein complex is somewhat like a sandwich (Figure 3.5). The L and M subunits are folded quite similarly and are in the center of the sandwich. These subunits cross the bilayer and bind to all of the prosthetic groups except for the hemes. The segments of L and M which connect the transmembrane segments on either side of the membrane are involved in binding to the cytochrome and H subunits. The cytochrome forms a cap on the extracytoplasmic (periplasmic) surface of the bilayer, and the hydrophilic portion of the H subunit forms a similar cover on the cytoplasmic surface. The transmembrane α-helix at the amino terminus of the H subunit is in contact with the cytochrome on the opposite side of the membrane. The intermolecular contacts in the crystal are between portions of the H subunit and the cytochrome which are normally in contact with water (Type II crystals, Figure 3.2). These two proteins, which serve as hydrophilic caps, may be important in aiding the formation of well-ordered crystals. In general, the crystals are handled in the same way as are those of water-soluble proteins.

Several structural features of the protein are worth emphasizing.

1. Each of the eleven transmembrane segments is α-helical (Figures 3.5 and 3.6) and the amino acids within these stretches are largely nonpolar. Each of the transmembrane helices in the L and M subunits has a stretch of at least 19 residues without an acidic or basic amino acid (982).
2. The transmembrane helices (named A, B, C, D, E) within each subunit (L or M) generally run antiparallel to their neighbors, but helices C and E run parallel. The helices are tilted less than 25° away from the presumed normal to the membrane, with the exception of helix D (in L and M), which has a tilt of 38°. The helices vary in length from 24 to 30 residues each.

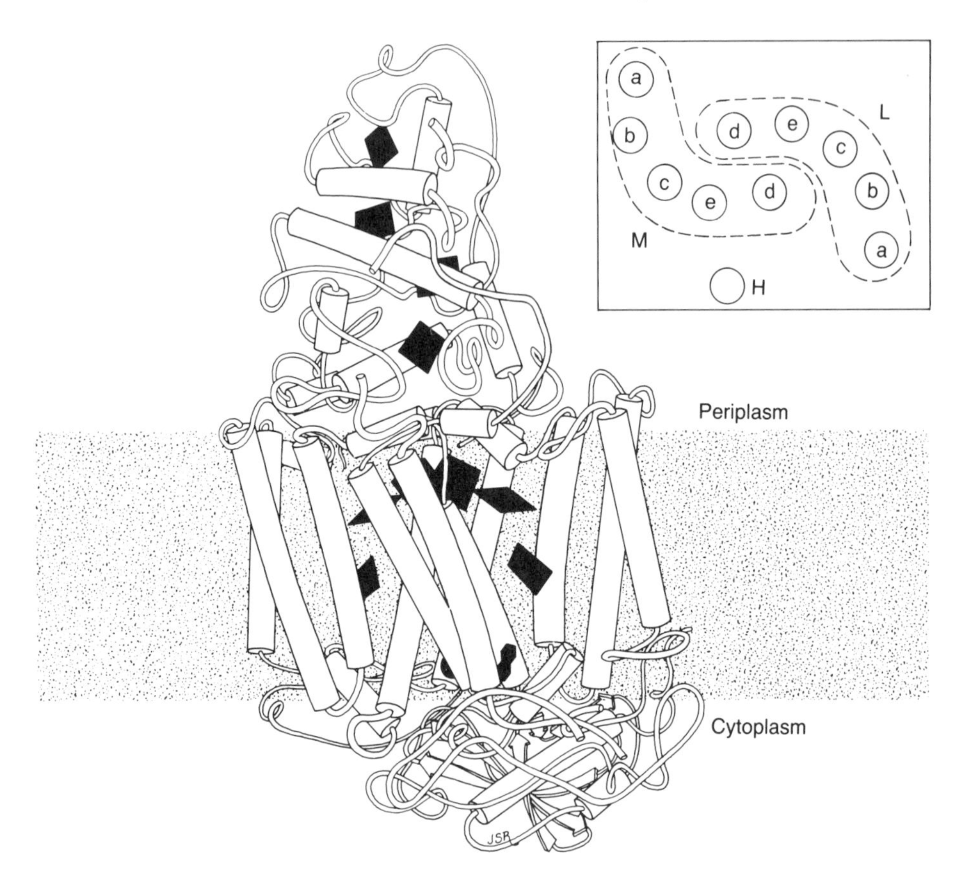

Figure 3.5. Schematic drawing showing the polypeptide backbone of the four subunits of the *R. viridis* reaction center. The eleven membrane-spanning α-helices are indicated by cylinders and the prosthetic groups are black. Note that the boundaries of the lipid bilayer are not known with certainty and the drawing is only meant to indicate the approximate location of the membrane. Drawing adapted from a version kindly provided by Dr. Jane Richardson. The inset shows the arrangement of the eleven membrane-spanning helices in cross section (618).

3. The portions of the L and M subunits connecting the transmembrane segments largely lie flat on each side of the membrane and form the contacts for the two hydrophilic subunits.
4. The charged amino acids in the L and M subunits are asymmetrically distributed, with the polar ends of the transmembrane helices and their respective helix connections being more negatively charged on the periplasmic side than on the cytoplasmic side of the membrane. This is energetically favorable, since the membrane potential is negative on the cytoplasmic side (see 982).
5. The most intimate contacts between the L and M subunits are at the cytoplasmic side at the H contact surface. In this region, helices from the two subunits

Figure 3.6. Ribbon representations of the three transmembrane polypeptides, L, M, and H, of the *R. viridis* photosynthetic reaction center. Kindly provided by Drs. H. Michel and R. Huber. From ref. 319. Reprinted by permission from Nature, vol. 318, pp. 620-621, Copyright © 1985 MacMillan Magazines Limited.

intercalate and the nonheme iron is bound between four helices, two each being contributed by the L and M subunits (see 983).

This structure also provides a wealth of information essential for understanding the photochemistry of this important complex (see 318, 983).

1. Although the redox centers appear to form two parallel pathways from the special pair bacteriochlorophyll, the two branches are not quite identical (983), and only one branch appears to be important in the electron transfer reaction across the membrane. The photosynthetic pigments are rigidly held in place by both hydrophobic interactions and hydrogen bonds to the protein (983), and they are unlikely to move during the reaction.
2. Electron transfer from the special pair to bacteriopheophytin is very rapid (less than 20 psec), consistent with the prosthetic groups involved being in van der Waals contact (see Figure 3.4). Electron transfer from bacteriopheophytin to Q_A has a half-time of 230 psec, and the two groups are separated by a center-to-center distance of about 14 Å. The isoprenoid sidechain of the quinone, however, is in direct contact with bacteriopheophytin.

3. The slow electron transfer from the cytochrome to reduce the special pair (270 nsec) is consistent with the closest heme having a center-to-center distance of 21 Å to the special pair. The larger distance, in part, results in a slow electron transfer rate.
4. The $Q_A \rightarrow Q_B$ electron transfer may be mediated by the nonheme iron, though it appears not to be necessary for this function (318). The Q_B quinone binds weakly and is lost during the preparation of the sample. The functional role of the H subunit may involve Q_B binding.

The structure of the three-subunit reaction center of *Rhodobacter sphaeroides* has also been determined by X-ray diffraction to a resolution of 2.8 Å and it is very similar to that of *R. viridis* (1619, 17, 18). Modeling studies (1619) suggest that the nonheme iron and quinones are located at the level of the phospholipid polar headgroups within the bilayer, though they are completely surrounded by protein. The special pair bacteriochlorophyll is estimated to be about 5 Å below the lipid polar headgroups. The eleven transmembrane helices are packed together as efficiently as are residues in the interior of water-soluble proteins. Those amino acid residues within the transmembrane region which are in contact with protein are generally hydrophobic, which is also the case for amino acids which are buried within soluble proteins. There are few polar interactions, such as hydrogen bonds or electrostatic salt bridges, between transmembrane helices. The three-subunit structure is stabilized by interactions between the solvent-exposed portions of the subunits, by favorable dipolar interactions between antiparallel α-helices (D and E) which are closely packed together, by van der Waals forces between closely packed helices, and by shared ligation of the single iron atom by four histidine residues in the D and E helices of subunits L and M. The lack of polar interactions between transmembrane helices is in contrast to most proposed models of bacteriorhodopsin, which is discussed in the next section.

Finally, the bacterial reaction centers exhibit some similarity to the photosystem II reaction center of higher organisms, which is the complex responsible for the oxidation of water and oxygen evolution (see 69, 982). Some aspects of the photosynthetic electron transport chains are discussed in Section 6.6.

3.52 The Structure of Bacteriorhodopsin

Bacteriorhodopsin is by far the most intensively studied membrane protein, and the structural models of this protein have had a large influence on how we generally conceive all intrinsic membrane proteins (see 1394, 1108 for reviews). Bacteriorhodopsin is found in a marine archaebacterium, *Halobacterium halobium*. The protein exists naturally in specialized patches of membrane within the bacterial cytoplasmic membrane called purple membranes, in a highly ordered two-dimensional array. The protein has a molecular weight of 27,000 (248 amino acid residues) and contains a single, covalently bound prosthetic group, retinal, attached via a Schiff base to lysine-216 (838). The function of this protein is to act as a photochemical proton pump and to create a proton electrochemical potential difference across the cellular membrane, which can then be used by the cell for solute transport and ATP synthesis. The absorption of a single photon by the retinal induces a series of identifiable changes, including a *trans* → *cis* isomerization at the C_{13}=C_{14} bond of retinal, and deprotonation of the Schiff base nitrogen, finally resulting in the electrogenic transfer of one or two protons from inside the cell to outside.

The natural occurrence of the two-dimensional lattice arrangement has made this protein amenable to the use of image reconstruction methods for obtaining three-dimensional structural information. The relatively small size, interesting and important function as a proton pump, and feasibility of monitoring retinal changes using optical methods have made the system the focus of considerable efforts to determine the details of its mechanism. Although much is known about the photochemistry of the retinal Schiff base, little is known about how this affects the protein and results in the observed proton translocation. The structural data have not provided solid clues as to how the protein might function, though models of a proton "bucket brigade" or a transmembrane hydrogen bond network have been proposed (1050) (see Section 8.52). Conversely, the functional mechanistic studies have not provided constraints or guides to interpreting the structural information.

The structural framework of bacteriorhodopsin has been obtained by image reconstruction techniques using the electron microscope, and represents the most successful application of this approach (1480).The resulting electron density maps resolve projected structures to better than 3.7 Å in the plane of the membrane (609) and to about 14 Å along the bilayer normal. Figure 3.7 shows an example of an electron density map and the three-dimensional model. The protein is associated in trimeric units and each polypeptide appears as seven columnar structures running through the membrane and approximately perpendicular to the plane of the membrane. These columns of electron density are usually interpreted as α-helical segments of the protein. They have been resolved to a length of about 45 Å, which is in good agreement with X-ray data on the total thickness of the purple membrane (about 49 Å). If the polypeptide is, indeed, in an α-helical configuration, this would require abouit 30 residues, since the α-helix has a 1.5Å displacement per residue along the helix axis. Hence, the seven putative helices account for better than 80% of the total amino acid residues. CD studies, in fact, are consistent with the large fraction of the protein being α-helical (1539, 1047, 1526), though the spectrum has also been interpreted as indicating substantial ß-structure (518).

The resolution obtained from the image reconstruction methods is not sufficient to trace the polypeptide and, for the most part, those portions of the protein connecting the 7 putative helices are not visible. Numerous models have been proposed since 1980 for the bacteriorhodopsin structure consistent with the electron diffraction data (386, 387, 1108, 731, 5, 1468), and many other techniques have been applied in attempts to fill in the missing information. By far the most important approaches have been (1) determination of the amino acid sequence (354, 745, 1107) and (2) use of proteolysis and of immunological methods to

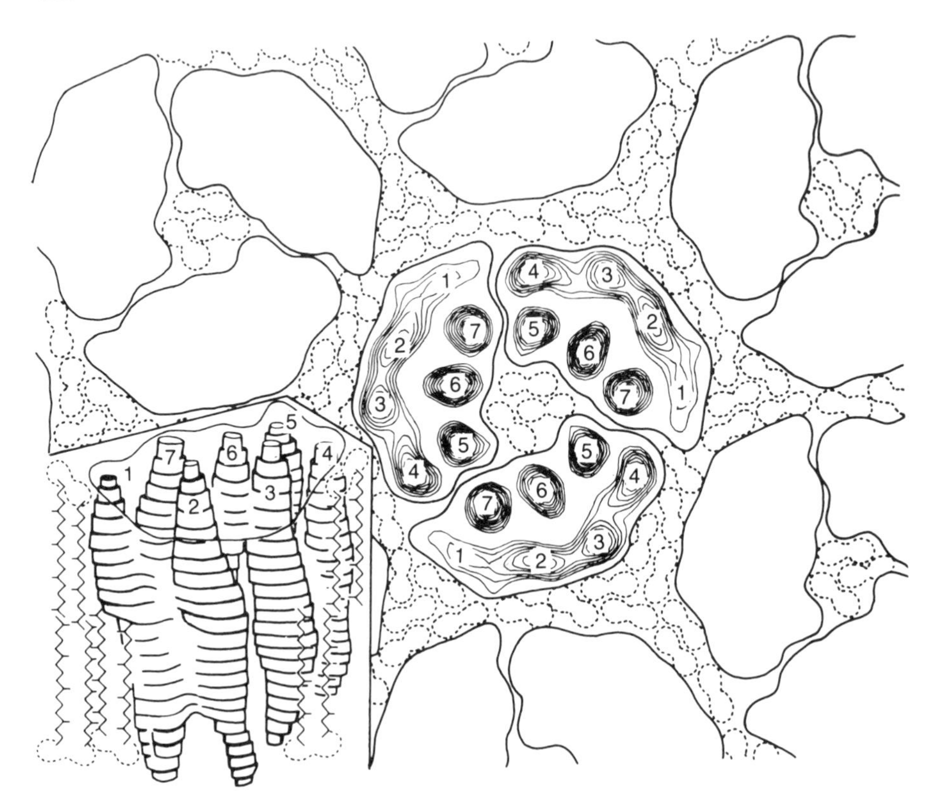

Figure 3.7. An electron density map and three-dimensional structural interpretation of bacteriorhodopsin derived by image reconstruction techniques. Adapted from ref. 428.

determine those portions of the polypeptide which are outside the bilayer and their orientation with respect to the cytoplasmic and extracytoplasmic sides of the membrane (506, 1536, 753, 1108, 1110).

The models proposed are all similar in the overall folding pattern and topology of the molecule. Figure 3.8 shows a model (355), similar to one initially proposed by Engelman et al. (385). It is estimated to take about 21 amino acid residues in an α-helix to cross the nonpolar portion of the bilayer, a distance of about 33 Å, and within the protein sequence there are seven stretches of approximately that length with a preponderance of nonpolar amino acids. These are presumed to form the seven putative α-helices observed on the image reconstruction maps. The presumed transmembrane α-helices in the sequence are usually labelled A, B, C, D, E, F, and G, going from the amino to the carboxyl terminus. How these match with the seven helices deduced from the electron microscopy studies (Figure 3.7) is not known and has been the focus of considerable attention and disagreement in the literature. There is little doubt that the overall topology illustrated in Figure 3.8 is correct (see 1468). The various models in the literature differ in some significant aspects, the dispute concerning the boundaries of several of the putative helices which are within the bilayer, which vary by up

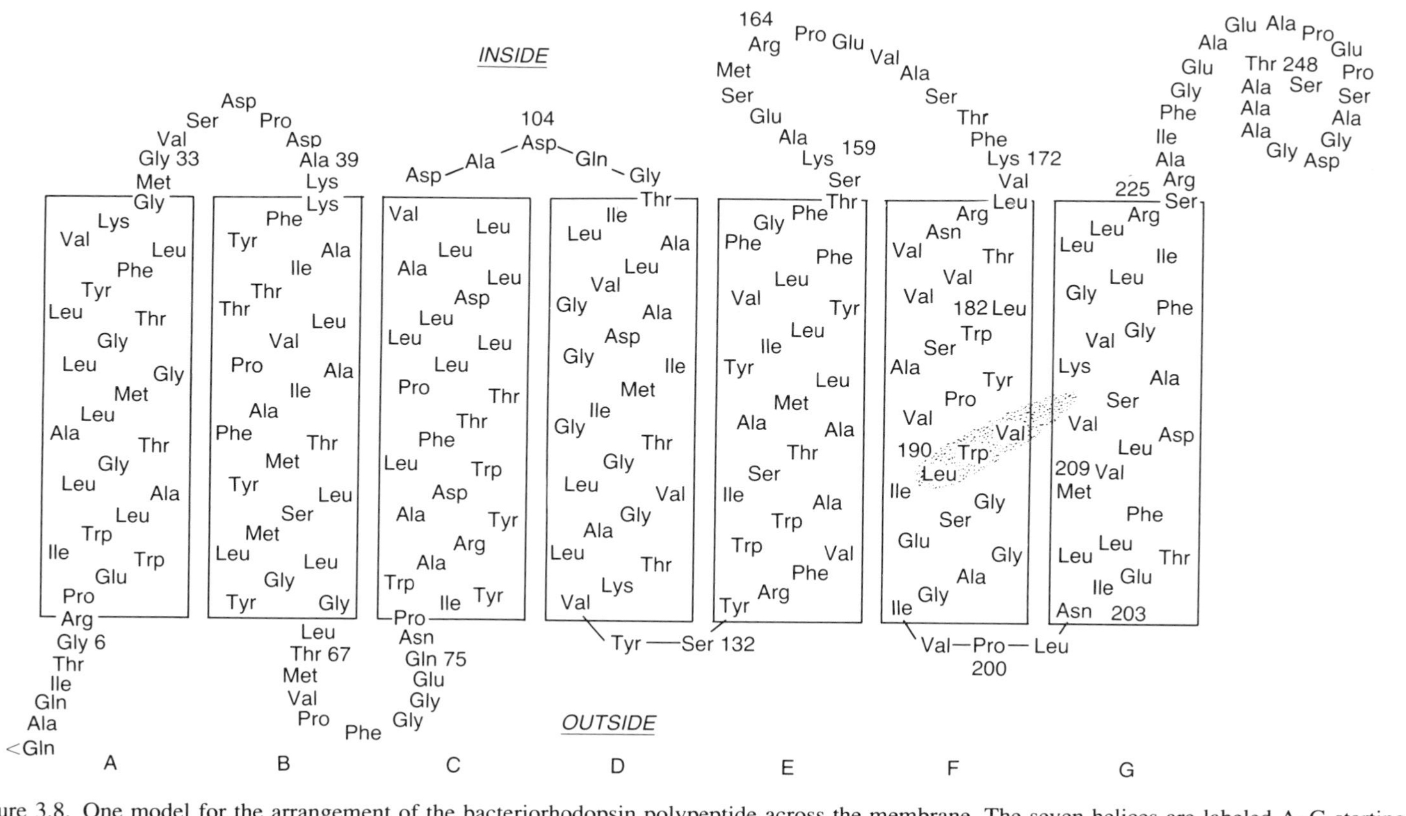

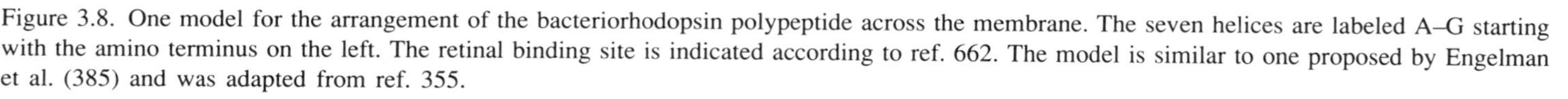

Figure 3.8. One model for the arrangement of the bacteriorhodopsin polypeptide across the membrane. The seven helices are labeled A–G starting with the amino terminus on the left. The retinal binding site is indicated according to ref. 662. The model is similar to one proposed by Engelman et al. (385) and was adapted from ref. 355.

to 10 residues in different models. Despite these disagreements, there are major features of this structure on which there is a consensus, though not unanimity.

1. The protein crosses the bilayer in the form of α-helices with overall nonpolar character. The helices seem to stretch across the entire membrane (about 45 Å) and not just the non-polar region of the bilayer (about 33 Å). The helices are perpendicular, or nearly so, to the membrane surface, and are packed together with a center-to-center distance of about 10 Å.
2. The charged and polar amino acid residues are largely, but by no means exclusively, located in the stretches connecting the transmembrane helices. There are as many as nine ionizable residues within the nonpolar portion of the bilayer, depending on where the boundaries are drawn. Presumably, these must be neutralized in some manner to stabilize the structure (see Section 3.63). This may be done by forming ion pairs or hydrogen bonds with residues elsewhere in the protein. Considerations of this requirement can be useful in guessing how the different helices pack together in the membrane (387). It is likely that the ionizable residues which are located within the membrane have structural and/or functional significance. This is certainly the case with Lys-216, which binds to retinal. Most of the charged amino acid residues in the solvent-exposed part of the protein are at or near the cytoplasmic surface.
3. There are three prolines clearly located near the middle portions of helices B, C, and F (Pro-50,91, 186). The helices, however, do not appear to be substantially bent.
4. There is an asymmetric distribution of the aromatic amino acids, with many tyrosines and nearly all the tryptophans being near the extracytoplasmic surface. The significance of this, if any, is unknown.
5. The amino terminus is extracytoplasmic, whereas the carboxyl terminus is located on the cytoplasmic side.
6. The carboxyl terminal portion is apparently a random coil by CD criteria and can be removed by proteolysis without destroying the function of the protein (853, 1541).
7. Lysine-216 is bound to the retinal prosthetic group and is located near the center of helix G (715, 756).

An Inside-Out Protein

An important point which has been made about bacteriorhodopsin and which may be generally applicable is that the polypeptide may be folded so that the polar or charged residues within the seven transmembrane segments face inward whereas the nonpolar residues may face the membrane lipids. The protein is essentially "inside-out" (386) compared to globular proteins, where the nonpolar residues are largely buried within the protein and the polar residues face outward. This is sensible and has some experimental support. The major supporting evidence comes from neutron scattering, taking advantage of the large scattering differential between hydrogen and deuterium. Deuterated valine and phenylalanine were biosynthetically incorporated into bacteriorhodopsin and difference Fourier techniques were used to assess the distribution of these amino acids in the projected structure of the purple membrane. Assuming a generally accepted assignment of the sequence of the presumed helical portions, it was observed by model building that, in general, looking down the helix axes, valine occurs on the opposite side of each helix (A-G) from charged and polar groups and phenylalanine residues tend to be on the opposite side from valine. The neutron scattering showed valine to be predominantly facing the lipid in the membrane, so it was deduced that the charged and polar residues are clustered on the interior of the protein. Note that this result does not depend on any particular assignment of the polypeptide segments on the electron microscope map.

Another approach has utilized a hydrophobic photoreactive probe, ^{125}I-TID (see Table 4.1). This reagent partitions into the hydrophobic membrane core and, upon photolysis, reacts non-specifically with most accessible amino acid side chains. This can be used to identify the portions of the protein exposed to the lipid and has in at least two cases, including bacteriorhodopsin, labeled specific polypeptide segments in a periodic manner (653, 157), with the label distributed every three or four residues. If this labeling pattern is reflecting only the degree of exposure of the amino acid side chains to the lipid, the results indicate which face of the presumed α-helical stretch is facing the lipid. Since there are 3.6 residues per turn in an α-helix, a periodicity similar to that observed is expected if only one side of the helix is accessible to the probe. The data on helix C (see Figure 3.9) in bacteriorhodopsin are consistent with the inside-out model for this protein and also support a helical configuration for this portion of the polypeptide. In summary, the concept of an inside-out principle for intrinsic membrane proteins

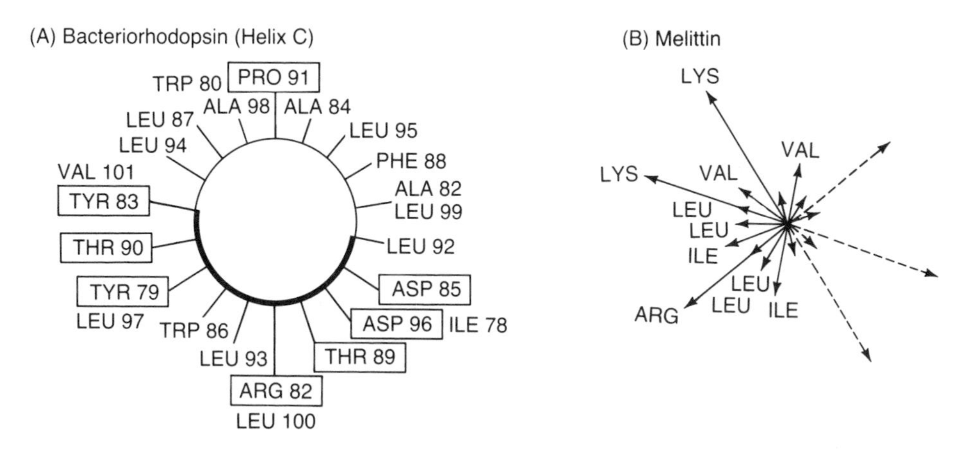

Figure 3.9. Examples of the graphical depictions of amphiphilic helices, viewed by looking down the helix axis. (A) Helix C of bacteriorhodopsin with hydrophilic residues indicated by boxes. This is a helical wheel representation. Adapted from ref. 47. (B) Vectorial contributions to the hydrophobic moment by each residue in the 5 to 22 region of melittin, which is highly amphiphilic. Each residue is represented by a vector whose magnitude is proportional to its hydrophobicity. The three charged residues (Lys and Arg) have negative values of hydrophobicity, as indicated by the dashed vectors. The hydrophobic moment is the vector sum. Adapted from ref. 375.

is attractive, but the experimental foundations are relatively weak so far. Recall that models of the photosynthetic reaction center place few, if any, ionizable residues within the core of the bilayer (Section 3.51).

Connectivity and Helix Assignment

Bacteriorhodopsin is an unusual case in that the intermediate resolution of the structure yields some structural detail but not enough to allow one to place the polypeptide in a unique configuration in the bilayer. The efforts that have been made to solve this problem are, so far, more instructive on the limitations of the methods than on the principles of membrane protein structure. The methods used include the following:

1. Theoretical considerations in placing the helices in a configuration to optimize charge neutralization by ion pair formation within the bilayer, and consideration of the lengths of the polypeptide connections between the helices (386).
2. Neutron scattering using deuterated amino acids and knowledge of the amino acid compositions within each helical stretch from the primary sequence (5).
3. Improved resolution in the plane of the membrane at the membrane surfaces using image reconstruction and electron diffraction approaches (731).
4. X-ray scattering differences before and after proteolytic cleavage of the carboxyl terminal peptide, to indicate where the carboxyl terminus resides in the electron density map (1542).
5. Neutron scattering using bacteriorhodopsin which has been reconstituted from proteolytic fragments in which one fragment has been deuterated (1468).
6. Neutron scattering using deuterated retinal incorporated either by reconstitution (756) or biosynthetically (715, 1324). The retinal is attached to Lys-216 in helix G. It is generally agreed that it is located within the bilayer, and linear dichroism studies show that it makes an angle of 75° with respect to the bilayer normal. It is nearly parallel to the membrane surface (628).
7. Photochemical crosslinking using a photoreactive retinal derivative substituted by reconstitution (662). The residues labeled should reveal proximity relationships in the folded structure.

Each of these approaches has its limitations and at this point there is no agreement. There are 5040 (7!) possible ways to place 7 helices (A–G) in 7 slots (1–7) and the techniques so far have yielded contradictory results as to which ones are most likely, let alone which one is correct.

3.53 Structure of Porin

Porins are a major class of proteins found in the outer membrane of enteric bacteria (see 1068, 579, 86 for reviews; also see Section 8.23). In *E. coli* and *Salmonella typhimurium* there are three porins which have been characterized, OmpF, OmpC, and PhoE. These proteins have molecular weights around 35,000 and have homologous amino acid sequences. Porins can be extracted by SDS from the outer membrane as stable trimers and can be reconstituted in phospholipid bilayers to form nonspecific pores, allowing the passage of small hydrophilic solutes (<600 daltons). Presumably, they confer this molecular sieve property to the outer membrane of the bacteria, permitting nutrients in and wastes out via the nonspecific channels (1053).

Considerable structural work has been reported on OmpF, also known as "matrix porin," and crystallographic studies leading to a high-resolution structure are in progress (490). However, already there is a sufficient amount known to make it clear that the structure is very different from either bacteriorhodopsin or the polypeptides in the photosynthetic reaction center. The primary sequence does not indicate any long hydrophobic stretches identifiable as transmembranous, and the amino acid composition is relatively polar (737, 1129, 1523). However, three-dimensional electron microscopy image reconstruction using crystalline sheets of reconstituted porin clearly show that the protein spans the bilayer with little protein protruding beyond the membrane boundary (384). Use of negative staining also shows the channels formed by each trimeric unit. Individual porin molecules form channels at the surface, and midway through the bilayer these three channels merge to form a single channel on the periplasmic side (384). Other evidence also indicates that porin, despite the lack of hydrophobic segments, is transmembranous (1129). The protein serves incidentally as a bacteriophage receptor, requiring exposed portions on the outer surface, and also displays an affinity for components of the cell wall on the periplasmic face. Also, purified porin can be reconstituted to form voltage-gated channels in phospholipid bilayers (see Section 8.23).

Infrared spectroscopy, circular dichroism, and high-angle diffuse X-ray diffraction are all consistent with porin being approximately two-thirds ß-sheet, with little α-helix (737, 1129, 1523). These studies, furthermore, indicate that the ß-strands are antiparallel, oriented perpendicular to the plane of the membrane, and have an average length of 10-12 residues, which is sufficient to cross the non-polar region of the bilayer. The way in which these ß-strands are arranged will have to await the analysis of the structure by X-ray diffraction. Models suggest that the ß-strands could be arranged to form a ß-barrel with polar and charged amino acid residues forming the lining of a water-filled channel (1129, 1523) (see Figure 8.7).

What is known so far about the structure of porin makes it clear that the hydrophobic α-helix is not required as a transmembrane element. Furthermore, it points out the limitations of the use of most predictive algorithms for modeling transmembrane proteins, since these are largely based on the assumption of hydrophobic stretches being required to cross the bilayer. The exact structure of porin is not yet known, nor is it clear whether the structure will be similar to any other membrane proteins. There are, however, other bacterial outer membrane proteins which have predominantly ß-structure. One is the OmpA protein (1068, 1019), which also serves as a phage receptor site but does not appear to exist as discrete trimers or to form pores as do the porins. Another outer membrane protein which contains considerable ß-structure is LamB (maltoporin), which is the lambda bacteriophage receptor, but whose function is to serve as a specific diffusional channel for maltodextrins (1292, 1068; see Section 8.23).

Although it is possible that the unusual structure of the porins is related to the unique structure and composition of the bacterial outer membrane (see Section 4.42), this is not likely. However, it is possible that the unusual structure reflects one way in which large aqueous channels can be formed across the bilayer (see Chapter 8).

3.6 Principles of Membrane Protein Structure and Predictive Algorithms for Transmembrane Proteins

As discussed in the previous section, there is only one class of membrane protein, the bacterial reaction center, whose structure is known to high resolution, and, in this example, the location of the lipid bilayer is not known with certainty. To extrapolate from this single example to a set of structural principles with predictive power for other membrane proteins is obviously somewhat risky. The risk is lessened by using thermodynamic principles as a guide and by the knowledge that a large body of experimental data from many membrane proteins is at least consistent with the presumption that the predominant transmembrane element in membrane proteins is α-helix. Thermodynamic considerations provide limits for the kinds of protein–lipid structures which would be expected to be stable.

3.61 Membrane Proteins Are Amphiphilic

Any membrane protein which interacts directly with the hydrophobic core of the lipid bilayer must be amphiphilic. Those portions of the polypeptide which are solvent-exposed can be expected to be enriched in amino acid residues with polar and ionizable side chains, whereas the residues in contact with the lipid hydrocarbon chains can be expected to be primarily nonpolar. This follows logically from energetic considerations similar to those described in Section 2.31. This does not exclude charged or polar amino acids from being located within the bilayer, but does set limits (see Section 3.62).

We can consider three levels of amphiphilic structures in membrane proteins: primary, secondary, and tertiary amphiphilicity.

(1) *Primary amphiphilic structures* are those in which there is a continuous stretch of predominantly nonpolar amino acid residues of sufficient length to cross the bilayer. This is experimentally observed in both the reaction center and in bacteriorhodopsin, in which all the membrane-spanning elements are α-helices. The α-helix is favored because this structure satisfies all the hydrogen bonding potential of the polypeptide backbone. An alternate structure which lacks a single hydrogen bond would be less stable by about 5 kcal/mol. For this reason, it is argued (see 389) that it is not likely that polypeptide turns are present within the

membrane (see 826 for example of such a model). The three to five amino acids in these turns could not form backbone hydrogen bonds with other residues in the structure, and this would be destabilizing by about 15 to 20 kcal/mol. In globular, water-soluble proteins, turns are located preferentially at the protein surface, where the amide groups can hydrogen bond with water, and it is likely that turns will also be located in solvent-exposed regions in membrane proteins (see 1129).

It is possible that ß-sheet could form a transmembrane element, possibly in the form of a ß-barrel as suggested for the porins (Section 3.53 and Figure 8.7). In these structures, the hydrogen bonding requirements of the polypeptide backbone within the membrane are also internally satisfied, though associations between separate ß-strands are required for this. This presents a problem of how such a structure might be inserted into the membrane, but the constraints imposed by the mechanisms for the assembly of membrane proteins are not at all clear (see Section 10.3).

(2) *Secondary amphiphilic structures*, in which hydrophobic residues occur with a periodicity along the sequence such that they form a continuous surface when the polypeptide is folded into a defined secondary structure. The repeat period for several secondary structural elements is indicated in Table 3.7. One example where secondary amphiphilic structure appears to be important is provided by the model for porins, in which amino acid residues are alternately polar and nonpolar within each of the proposed ß-strands (see Figure 8.7). The polar residues are all on one side of the pleated sheet, forming the lining of a water-filled pore in this model. Note that the model is not a proven structure.

An α-helix where hydrophobic residues occur every third or fourth residue would have a hydrophobic surface and a polar surface. This kind of structure is often visualized as a helical wheel in which the helical segment is viewed in projection from one end, and the locations of the side chains are indicated (1294). An example is shown in Figure 3.9. Secondary amphiphilic structures would be expected in several situations, as illustrated schematically in Figure 3.10.

i. Surface-active portions of a protein, where one part of a helix interacts with the hydrophobic portion of the lipid bilayer and the polar face is in contact

Table 3.7. Parameters of secondary structural elements[l].

Structure	Repeat period or residues per turn	Rise per residue	Radius or width
Untwisted ß-strand	2.0	3.2–3.4 Å	0.9–1.1 Å
Twisted ß-strand	2.3	3.3 Å	1.0 Å
3_{10}-helix	3.0	2.0 Å	1.9 Å
α-helix	3.6	1.5 Å	2.3 Å

[l]Data from ref. 1305.

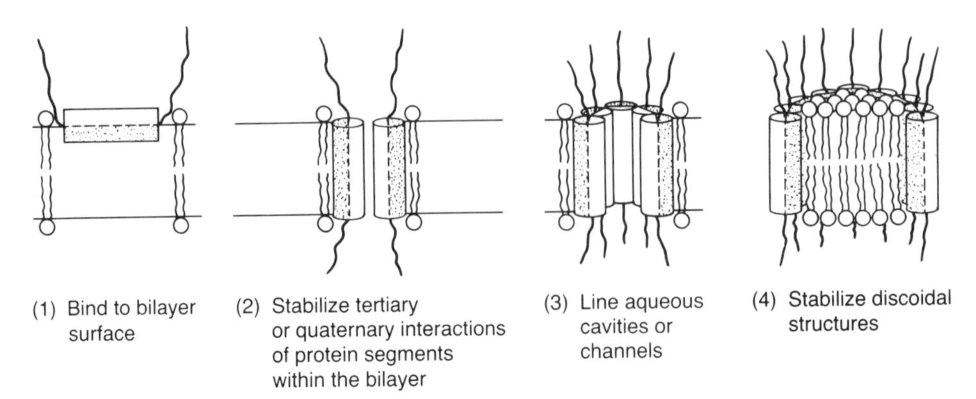

Figure 3.10. Some proposed structural roles for secondary amphiphilic helices interacting with the lipid bilayer.

with the aqueous phase and the polar region of the bilayer. A number of peptide hormones as well as membrane-disrupting peptides such as melittin can form amphiphilic a-helices (Section 3.7).

ii. Transmembrane elements where the nonpolar surface of the helix is facing the lipid and the polar half is part of the lining of an aqueous channel extending through the bilayer. This is a popular notion which has been mostly associated with studies on the nicotinic acetylcholine receptor, which functions as a chemically gated channel (see Section 8.24). However, the experimental claims that an amphiphilic helix is membrane-spanning in this protein (258, 1623) have been disputed (776, 1198). Amphiphilic ß-strands can also be modeled to form a water-filled channel, as in porin (Figure 8.7).

iii. Transmembrane elements in which the nonpolar surface is in contact with the lipid and the polar groups are in contact with the polar groups of other transmembrane elements. This is the basis of the "inside-out" structure postulated for bacteriorhodopsin (386). Polar interactions between amphiphilic helices could, in principle, stabilize subunit interactions in oligimeric proteins.

(3) *Tertiary amphiphilic structures* are not yet known with certainty, but they are likely to exist. In such proteins, a hydrophobic surface would be formed by tertiary folding, bringing together residues from disparate parts of the polypeptide chain. Proteins which bind to the bilayer but do not have identifiable hydrophobic domains defined by either of the above criteria may exhibit this kind of structure. A possible example might be α-lactalbumin (91).

3.62 Ionizable Amino Acid Residues in Transmembrane Segments (see 389, 646)

Many of the models of membrane proteins place ionizable residues in the bilayer interior, including the proposed structure of bacteriorhodopsin (Section 3.52).

These residues (Arg, Lys, His, Asp, and Glu) are likely to play important functional and/or structural roles in membrane proteins. There are several examples where this is already clear: (1) the lysine residues in bacteriorhodopsin (Figure 3.8) and rhodopsin (Figure 4.1) which form a Schiff base with the retinal prosthetic group required for light excitation and which can exist as a charged group within the bilayer (179), (2) the histidine residues in the bacterial reaction center polypeptides required for ligation to the photosynthetic pigments (Section 3.51), and (3) the charged residues in lactose permease from *E. coli* which are essential for the transport function of that protein and which might form a hydrogen-bonded network within the protein (Section 8.32).

It is energetically very costly to transfer an ion from water to the low dielectric medium inside the membrane (~25–40 kcal/mol) (645, 389), and these groups must be stabilized in some way for the proposed structures to be feasible. It is often assumed that ion pairing will suffice to provide the required stabilization, and Engelman and Zaccai (386) have used this principle as one criterion in building a three-dimensional model of bacteriorhodopsin. However, calculations (645) have indicated that the free energy of transfer of an ion pair from water to a low dielectric medium is still very unfavorable (10-15 kcal/mol). Additional polar interactions are required for further stabilization, and these can be provided by interactions with other polar groups, such as carbonyl residues (1552, 1553) or by hydrogen bonds (804, 645), as found in ion pairs inside globular, water-soluble proteins (1197).

In principle, even a single charged group within the membrane interior could be stabilized by interactions with polar groups and by hydrogen bond interactions which effectively delocalize the charge. There are several examples of isolated, desolvated ions which are stabilized by interactions with the polypeptide backbone in the interior of water-soluble proteins (1187). The same principles would presumably be operative for charged residues on the inside of membrane proteins.

However, it is more likely that ionizable amino acids will exist as neutral species within the membrane by virtue of protonation or deprotonation (see 389). The free energy costs of neutralizing the charged amino acids are easily computed to be in the range of 10-17 kcal/mol. Unless there is a specific set of polar interactions to stabilize a charged residue within the membrane, it is very likely that it will be present as a neutral species.

3.63 Charged Amino Acids in the Solvent-Exposed Segments (see 1529)

It has already been pointed out that the distribution of charged residues on the two sides of the bacterial reaction center polypeptides is asymmetric (Section 3.51; 982). This pattern is also observed in several other bacterial inner membrane proteins (1529). There is a four-fold bias for basic residues Lys and Arg to be found in the loops connecting membrane-spanning elements on the cytoplasmic side rather than the periplasmic side (outside) of the membrane. There is no similar bias for acidic residues Asp and Glu. It is speculated that the cause of this asymmetry may be related to the mechanism of membrane protein assembly, but this is not clear. Furthermore, it is not clear whether this observation can be generalized or if it has any predictive value.

3.64 A Special Role for Proline?

In globular, water-soluble proteins, prolines are rarely found in the middle of an α-helix. In one study of 58 proteins containing 331 α-helices, there were 30 examples of proline residues in the middle of an α-helix (1155). In half of these examples, the proline corresponds to a break in the helix, and, in the other half, it is in a region where there is a distortion or irregularity in the helix.

Surprisingly, the presumed structure of bacteriorhodopsin (Figure 3.8) has proline residues in the middle of three of the seven transmembrane helices. Similarly, rhodopsin (Figure 4.1) has proline in five of the seven presumed membrane-spanning helices. Other studies have also indicated a bias toward prolines in the presumed membrane-spanning segments of membrane proteins, especially those involved in transport functions (315, 133). The significance is not known. However, it should be noted that because of the cyclic side chain, proline cannot form a hydrogen bond with the residue in the preceding turn of an α-helix. This might favor structures where that hydrogen bond is provided by a specific interaction with a residue from another membrane-spanning segment. This kind of polar interaction within the bilayer could stabilize the three-dimensional structure of a membrane protein.

3.65 Algorithms for Identifying Primary Amphiphilic Structures

Although definitive structural data on membrane proteins are sparse, there is an abundance of amino acid sequence information, thanks to the advent of DNA sequencing. A number of algorithms have been proposed to analyze protein sequences to identify membrane-spanning α-helices, which are assumed to be about 20 residues long and to consist primarily of hydrophobic amino acids (for review see 389). Each of the algorithms is based on a scale in which each amino acid is numerically ranked with a parameter reflecting the likelihood of finding that residue in a transmembrane segment.

Two kinds of scales have been developed. One type of scale ranks the amino acids on their relative polarity or "hydrophobicity." These are thermodynamic scales and are meant to correlate to the standard state free energy change upon transferring the amino acid from an aqueous to a nonaqueous hydrocarbon environment. However, there are numerous ways in which amino acid hydrophobicity can be quantified and they are not all consistent (see 372, 389). Often, data relating to more than one physical characteristic are amalgamated (e.g., 47, 804). The most commonly used "hydropathy" scale of Kyte and Doolittle is of this type (804), taking into account hydrophobicity as measured by the hydration potential

(1602) and also the extent to which residues are found buried in globular, water-soluble proteins (217).

The scale of Goldman, Engelman, and Steitz (389) is based on a rational attempt to quantify the transfer free energy of α-helices from water into a membrane. Figure 3.11 compares the Kyte-Doolittle (KD) scale with that of Goldman, Engelman, and Steitz (GES).

Engelman et al. (389) estimate a favorable free energy change of about 30 kcal/mol to insert into the membrane an α-helix of polyalanine which is 20 residues long. The calculation is based on the surface area of the helix which is exposed to solvent, similar to that illustrated in Figure 2.16 for hydrocarbon chains. The hydrophobic free energy contribution of each amino acid side chain is computed based on its solvent-exposed surface area within a helix. The free energy cost of transferring polar groups into the bilayer is also taken into account. For example, it is assumed that glutamate will be protonated upon transfer into the bilayer, at a free energy cost of 10.8 kcal/mol. Similarly, the transfer of hydroxyls (Ser, Thr) is estimated to cost about 4.0 kcal/mol.

This approach makes it clear that the favorable hydrophobic component of the α-helical transfer free energy can be used to "drag" polar side chains into the bilayer. For example, a single arginine could be accommodated in the middle of an otherwise nonpolar transmembrane helix as long as it is deprotonated, which requires 16.7 kcal/mol at pH 7.0 (assuming 99% deprotonation). The net transfer free energy for the helix is still favorable. This would not be the case, however, if there were two arginine residues (deprotonated) or if the arginine were positively charged in the bilayer. It is recognized that polar residues can be stabilized within the bilayer by specific interactions, but this is not easy to build into a pre-

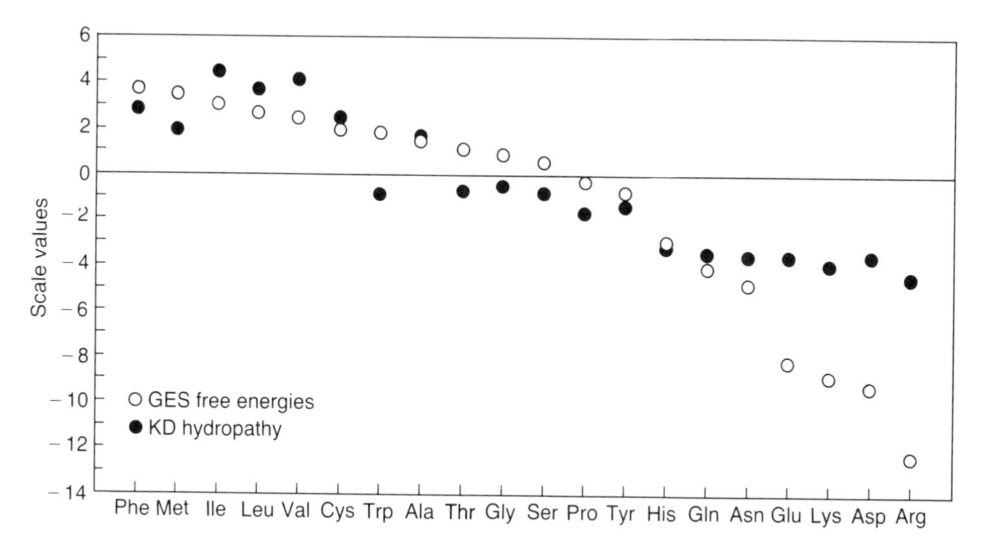

Figure 3.11. A plot comparing two different scales used to predict membrane-spanning helices: Kyte–Doolittle (KD) hydropathy, and the transfer free energy values used in the ranking of Goldman–Engelman–Steitz. Adapted from ref. 389.

dictive algorithm. For example, serine, cysteine, and threonine side chains can form hydrogen bonds to the polypeptide backbone (541). Acidic and basic residues can form ion pairs, and this can be predicted to occur if they are located four or five residues apart within an α-helix (389).

The second type of scale which is used to rank amino acids is based on the frequency with which amino acids are actually found in membrane-spanning segments. This will empirically take into account hydrophobicity as well as any other factors which are not quantified as hydrophobicity. The disadvantage of this semi-empirical approach is that there is not a large body of hard data defining the boundaries of transmembrane segments. Nevertheless, these scales can be shown to be as useful as the thermodynamics-based scales. Examples are the membrane propensity scale of Kuhn and Leigh (799) and the membrane buried helix parameter of Rao and Argos (1194). The four most hydrophobic residues on the scale of Goldman-Engelman-Steitz (Phe, Met, Ile, Leu) are also the four with the highest membrane buried helix parameter values (1194).

Figure 3.12 compares the profiles of three different membrane proteins using several different scales. These profiles are generated by taking the average value of the scale numbers assigned to each amino acid within a selected "window" and plotting the average as a function of residue position in the polypeptide. For example, if the window is 19 residues, the value assigned to position 40 will be an average of the scale numbers for all the amino acids from 31 to 49, inclusive. The value assigned to position 41 would be an average for residues 32 to 50, and so on. Peaks indicate regions which are hydrophobic or which are judged most likely to form a membrane-spanning helix. The window size is important and most of the curves in Figure 3.12 were generated with a window of 19.

Having generated these profiles, it is still necessary to interpret them. The scale of Goldman–Engelman–Steitz is designed so that peaks above zero represent predicted transmembrane helices. The Kyte–Doolittle scale is more of a problem, and the value of 1.25 is indicated in Figure 3.12A as being the lowest value of a known membrane-spanning helix in the L subunit of the reaction center of *R. viridis*. All three of the scales illustrated in Figure 3.12 give comparable profiles for the subunits of the reaction center.

Figure 3.12B shows two profiles of microsomal cytochrome P450. This has been selected because it was originally predicted on the basis of its primary structure to have at least eight transmembrane helices, but experimentally it has been shown to have only a single amino-terminal anchor in the membrane (306, 1274). The Kyte–Doolittle and Goldman–Engelman–Steitz profiles both pick out the amino-terminal span, but they also incorrectly predict one or more additional transmembrane segments. It is useful to emphasize the fact that the many models of membrane proteins which are in the literature and which are based purely on a consideration of the amino acid sequence are not necessarily correct.

Figure 3.12C compares three profiles of bacteriorhodopsin. Although the profiles are similar, one can see that there are differences in the delineation of the seven membrane-spanning segments. The version of the Goldman–Engelman–Steitz algorithm which was used does not take into account stabilization from ion pair formation from nearby charged residues within the

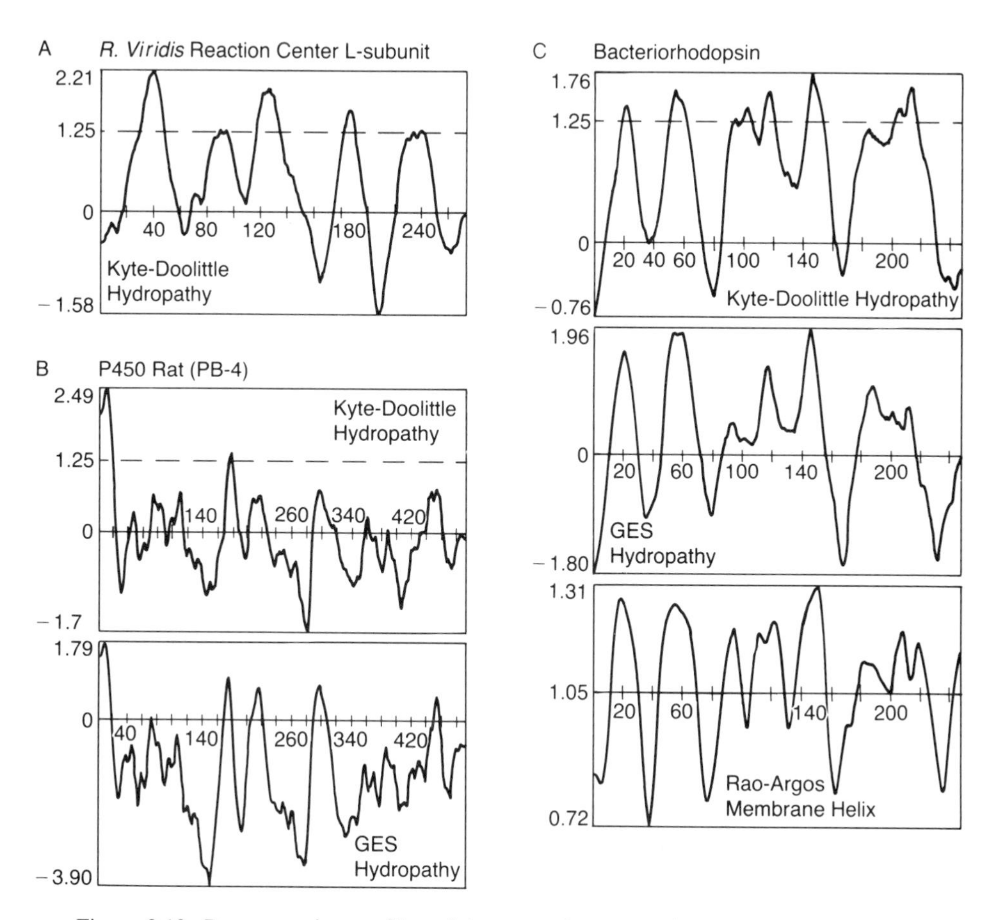

Figure 3.12. Representative profiles of three membrane proteins used to predict membrane-spanning helices. The amino acid scales of Kyte-Doolittle (804), Goldman–Engelman–Steitz (GES) (389), and Rao–Argos (1194) were used. A computer software package (SEQANAL) provided by Dr. A. Crofts (Univ. of Illinois) was used to generate these profiles. For comparative purposes, the Kyte–Doolittle and GES plots were obtained using a window of 19 residues and then smoothed using a second pass with a window of 7. The average value at each residue position is plotted as a function of residue number starting with the amino terminus on the left in each case. The values plotted for the Kyte–Doolittle and GES scales represent average hydropathy and transfer free energy per residue (kcal/mol). The Rao–Argos plot used a span of 7 residues and was smoothed used two additional passes with the same span of 7, as recommended by the authors. The scale values reflect the relative preference for being in a membrane-spanning helix. Note that the version of the GES algorithm which was used does not take into account possible ion pair formation. See text for details.

same helix. When this is included, the separation between the last two helices (F and G) becomes more clear (see 389).

One problem which is addressed by all the algorithms is to avoid picking hydrophobic segments in the sequences of known globular proteins that are clearly not membrane-spanning but are located internally within the protein. By and large, this is not a problem if one searches for a sufficiently long stretch (e.g., see 389, 1194).

Note that algorithms for predicting α-helical structures in soluble, globular proteins, e.g., Chou-Fasman (218), are not useful for picking out transmembrane elements in proteins (1540). These algorithms are not designed for use with non-globular proteins, including the portions of proteins within the bilayer.

The algorithms designed to locate membrane-spanning regions will not successfully locate segments which exhibit secondary amphiphilicity or which cross the membrane as ß-pleated sheet. In the former case, the polar residues would prevent the region from being selected, and in the latter case, the membrane-spanning element would be too short, since only 10-12 residues of ß-strand are required to traverse the bilayer. There are several examples where algorithms have been designed to search for ß-turns rather than the transmembrane elements themselves (826, 1129). Although this avoids some of the problems of searching for different classes of membrane-spanning elements, it is not clear how accurate these will be when generally applied.

3.66 Algorithms for Identifying Secondary Amphiphilic Structures

A number of algorithms have also been developed to detect secondary amphiphilicity or asymmetry in the distribution of hydrophobic residues in segments of a polypeptide chain. It is fairly common that α-helices and ß-pleated sheet in globular proteins have sequence periodicity of hydrophobicity (245). The helical wheel (e.g., Figure 3.9A) as a qualitative indicator can be misleading (441) and more quantitative measures are required. The basic approach is to pick out periodicities in the occurrence of hydrophobic residues by Fourier transform techniques. One variation of this is the hydrophobic moment.

(1) Hydrophobic moment: This parameter was introduced by Eisenberg et al. (372, 373, 375, 374) and is defined as

$$\mu_H = \{[\sum_{n=1}^{N} H_n \sin(\delta n)]^2 + [\sum_{n=1}^{N} H_n \cos(\delta n)]^2\}^{1/2} \qquad [3.4]$$

This is simply a way to vectorially add the hydrophobicities of individual residues in an N-element segment, where the hydrophobicity of the nth residue (H_n), is represented as a vector whose direction is along the angle ($n\delta$) made by the side chain as it emerges from the backbone as seen in the projection down the axis. For an α-helix, $\delta = 100°$. In Figure 3.9B the hydrophobicity "vectors" are observed in a helical wheel type of projection, and the hydrophobic moment is the vector sum of these. A hydrophilic residue is represented by a vector in the negative direction, that is, 180° away from the residue direction. A random sequence will have a value of μ_H which is small, since the hydrophobic residues will be evenly distributed. In contrast, the peptide melittin (Fig. 3.9B) has hydrophobic residues on one side and polar residues on the opposite side (Section 3.71). The hydrophobic moment has a numerical value which is assigned to the amino acid in the center of the segment being analyzed. Hence, one can "scan" a sequence and assign a value of μ_H as well as an average hydrophobicity to each residue position.

Eisenberg and co-workers have analyzed 11-residue segments from many proteins and peptides (373). They plotted the hydrophobic moment (μ_H) and average hydrophobicity, $\langle H \rangle$, of each segment and observed patterns. Polypeptide segments in globular proteins tend to have low values of both $\langle H \rangle$ and μ_H. Transmembrane elements of the type picked out on the basis of hydrophobicity have high values of $\langle H \rangle$ but low values of μ_H, being mostly nonpolar. Peptides and portions of proteins labeled "surface-active" have high values of μ_H because of the large asymmetry in the distribution of polar and nonpolar residues. Some segments of surface active proteins have also been identified using this algorithm, such as parts of diphtheria toxin (see 373) and a portion of *E. coli* pyruvate oxidase (1206).

The hydrophobic moment is designed to quantify the extent to which a region of polypeptide exhibits a particular periodicity, depending on the choice of δ. It is, essentially, one component of a Fourier transform of the hydrophobicity function. More general procedures, described below, are designed to scan all the Fourier components, and to pick out any periodicities which might exist.

(2) *Sequence periodicity*: A number of procedures have been described for identifying regions in proteins which exhibit sequence perodicity of hydrophobicity (e.g., 374, 998, 429). These algorithms all involve a Fourier transform of the function defined by the hydrophobicity values of the individual residues along the polypeptide, and a search for periodic variations. A peak at a period of 3.6 indicates the periodic appearance of a hydrophobic residue on the average every 3.6 residues within the segment of polypeptide being analyzed. This would suggest that the protein segment is an α-helix with a strip on one side containing predominantly hydrophobic residues. This technique has been used to identify amphiphilic segments in some channel and transport proteins, including the acetylcholine receptor (429, 258), the sodium channel (780), the glucose transporter (1027), the mitochondrial uncoupling protein (46), and Band 3 of the red blood cell, an anion transporter (774). There is little evidence that these putative amphiphilic helices are membrane-spanning (see Section 8.24).

These methods also have been useful in analyzing peptides which interact with the membrane surface (372) and apolipoproteins (784, 1172).

3.7 Peptide Models for Membrane Proteins

The use of peptides to study protein–lipid interactions goes back many years. The bulk of the work has been focused on naturally occurring membrane-active peptides, the three most important being gramicidin A, alamethicin, and melittin. Now, synthetic peptides are increasingly being used to test models of protein–lipid interactions. Two general remarks are applicable: (1) the importance of both primary and secondary amphiphilicity in membrane binding is evident; (2) the peptides frequently demonstrate polymorphisms, i.e., they can exhibit quite different conformations depending on their environment. Future studies may eventually allow for the systematic detailing of protein–lipid interactions by the use of synthetic peptides, but we are very far from this at present.

3.71 Naturally Occurring Peptides

(1) *Gramicidin A* is a hydrophobic peptide consisting of 15 amino acids with alternating L and D configurations. The sequence (1282) is

HCO-L-Val-Gly-L-Ala-D-Leu-L-Ala-D-Val-L-Val-D-Val-L-Trp-

D-Leu-L-Trp-D-Leu-L-Trp-D-Leu-L-Trp-$NHCH_2CH_2OH$

When incorporated into membranes, gramicidin A forms channels specific for monovalent cations (see 430; Section 8.15). Numerous studies using planar membranes and liposomes have shown that the channel is a dimer of the peptide. A number of models have been proposed, but the one which was first proposed by Urry (1484) appears to be correct (1564). In this model (Figure 3.13), Gramicidin A exists as a ß(L,D) helix [previously called a π(L,D) helix] in which two monomers are associated noncovalently in the middle of the bilayer via their amino termini. The total length is about 30 Å and the outer and inner diameters are ~15 Å and 5 Å respectively. The hydrophobic side chains are all on the outside of the helix and hydrophilic peptide backbone carbonyls line the pore. Crystals of gramicidin A have been obtained from organic solvent, and the structure obtained to 2.5 Å resolution by X-ray diffraction (1537). The structure, however, is not the same as that which is believed to be present in the membrane.

Due to the unusual D-amino acids present in this structure, gramicidin A cannot be used as a model for structures which may occur in membrane proteins. However, this peptide does illustrate one way to form a selective pore through

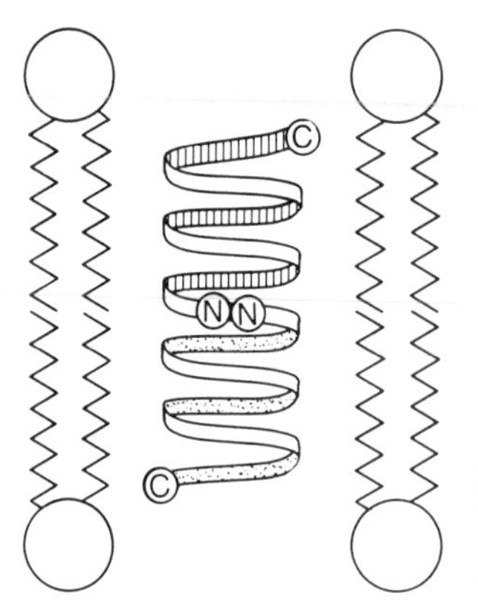

Figure 3.13. Illustration of the structure of the channel formed by a dimer of gramicidin A in the bilayer. Adapted from ref. 1564. Reprinted with permission from "Conformation of the Gramicidin A Channel in Phospholipid Vesicles," by S. Weinstein et al. Biochemistry, vol. 24, p. 4375. Copyright 1985, American Chemical Society.

the bilayer (see Chapter 8). Also, gramicidin A, because of its availability, has frequently been used as a "model membrane protein" in studying the perturbing influence of membrane proteins on lipids (831). At sufficiently high levels of incorporation (>5 mol %) gramicidin A aggregates to form tubular structures and induces the formation of hexagonal H_{II} phase in model membranes (137).

(2) *Alamethicin* is a 20-residue peptide antibiotic which can form voltage-gated channels in membranes (1030; Section 8.15). The alamethicin sequence, given below, includes unusual residues, α-aminobutyric acid (Aib) and L-phenylalaninol (Phl).

Ac-Aib-Pro-Aib-Ala-Aib-Ala-Gln-Aib-Val-Aib-

Gly-Leu-Aib-Pro-Val-Aib-Aib-Glu-Glu-Phl

The dependence of the conductance by alamethicin channels in planar membranes on the peptide concentration suggests that each channel contains at least 6-11 molecules. The crystal structure of alamethicin has been solved to 1.5 Å resolution (456), indicating that the peptide is mostly α-helical, with a bend in the middle at proline-14. This structure contains a strip of solvent-accessible polar atoms running the length of the molecule, making it an amphipathic structure. This polar strip includes backbone carbonyl groups, similar to the polar groups proposed to form the interior of the gramicidin channel. Fox and Richards (456) have proposed a model for the alamethicin channel based on this molecular configuration, by optimizing intermolecular interchain bonding and packing considerations. The model (Figure 3.14) has each channel formed by an oligomeric (≈8) cluster of molecules, hydrogen bonded in a stable structure with the polar atoms exposed

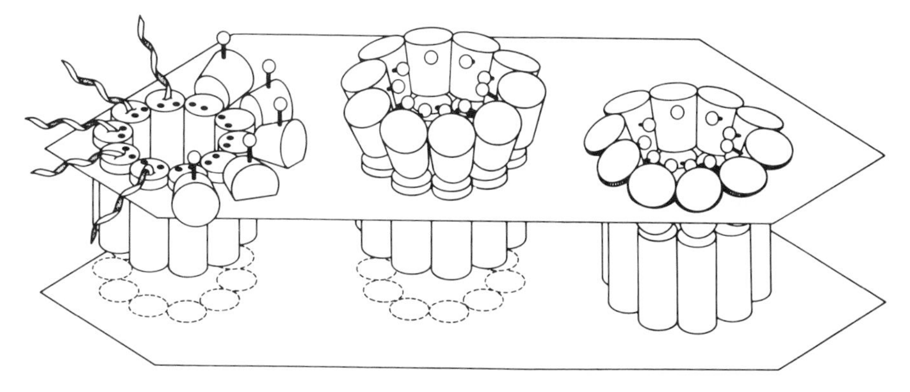

Figure 3.14. Proposed structures formed by an oligomer of alamethicin in a lipid bilayer. Three conformations are indicated. (1) The conformation in the absence of an applied transmembrane voltage (left). (2) The open channel conformation in the presence of an applied voltage (right). (3) An intermediate conformation (middle). Also see Section 8.15 for a discussion of the channel-forming properties of this peptide. From ref. 456. Kindly provided by Dr. F. Richards. Reproduced from the Biophysical Journal, 1982, vol. 37, p. 357 by copyright permission of the Biophysical Society.

to the solvent. A conformational change is postulated to result from the imposition of a voltage across the bilayer, stabilizing an "open" form of the channel and, thus, the voltage-gated conductance. The channel length would be 32 Å, sufficient to extend across the nonpolar parts of the bilayer.

(3) *Melittin* (see 372, 721), unlike gramicidin A or alamethicin, is a water-soluble peptide and is a component of bee venom. The sequence (below) of the 26-residue peptide includes many polar residues (566).

H_2N-Gly-Ile-Gly-Ala-Val-Leu-Lys-Val-Leu-Thr-Thr-Gly-Leu-

Pro-Ala-Leu-Ile-Ser-Trp-Ile-Lys-Arg-Lys-Arg-Gln-Gln-NH_2

Melittin integrates in membranes, causes cell lysis, and can also form voltage-gated channels in planar membranes (740). It is a "surface-active" peptide which forms monolayers at an air–water interface. The structure of melittin has been solved by X-ray crystallography, in two crystal forms (1445). In both cases, melittin is a bent α-helical rod, similar to alamethicin. The α-helix is distinctly amphiphilic (see Figure 3.15) due to the asymmetric distribution of the polar side chains. Note that the structures of gramicidin A and alamethicin are also amphiphilic, but the polar atoms are largely provided by backbone carbonyls and not by polar side chains. Melittin, like the other peptides, displays polymorphism. It exists as a monomer in aqueous solution which, by CD analysis, contains only 7% α-helix. It can also form tetramers in aqueous solution and can bind to membranes. In the latter cases, melittin is mostly (~70%) α-helix (1522). The way in which

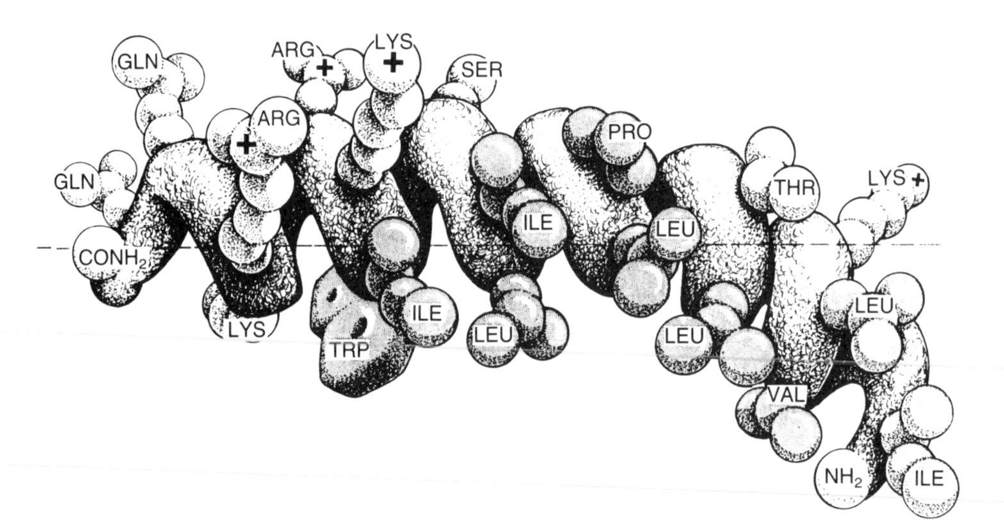

Figure 3.15. Proposed structure of melittin as it binds to the surface of a phosphatidylcholine bilayer. Hydrophobic residues are on the lower face and charged, hydrophilic residues are opposite. From ref. 1445. Kindly provided by Dr. D. Eisenberg. Reprinted by permission from Nature, vol. 300, p. 329, Copyright © 1982 Macmillan Magazines Limited.

melittin binds to membranes and the mechanism(s) by which it disrupts the bilayer are not known with certainty (1445, 302, 353). One model (1445) suggests that the disruptive effects are due to a "wedge" effect (299) in which the melittin binds to one side of the bilayer and eventually destabilizes the bilayer by increasing the surface area of the one leaflet in the same way as do detergents (614). It has also been shown that melittin can form voltage-gated channels in planar membranes (164). It is very likely that melittin can bind to the bilayer in a number of different ways depending on the experimental conditions (283, 353).

(4) *Peptide hormones* often contain potential amphiphilic α-helical segments similar to that in melittin (see 721). Notable examples are ß-endorphin (107), calcitonin (392), corticotropin (1506), and corticotropin-releasing factor (821). It is speculated that these hormones bind to the membrane via these amphiphilic helices and that the peptide secondary structure which is stabilized by the membrane facilitates binding to the specific hormone receptor. These are all water-soluble peptides but have an affinity for phospholipid bilayers. The α-helical segments can be at the amino terminus (corticotropin), at the carboxyl terminus (ß-endorphin), or internal (corticotropin-releasing factor). In the case of ß-endorphin, it was demonstrated that it is the amphiphilicity itself, not the amino acid composition of the helical segment, which is essential for biological activity (107). These types of structures indicate one way in which segments of proteins may bind to a membrane (e.g., pyruvate oxidase) (1206).

3.72 Synthetic Peptide Models (see 721)

Two classes have been examined: (1) peptides which should exhibit primary amphiphilicity, modelling a transbilayer hydrophobic α-helix; (2) peptides which have potential secondary amphiphilicity, either as α-helices or ß-sheet.

(1) *Primary amphiphilic peptides:* A few studies have been reported (e.g., 1021) in which amphiphilic peptides, such as Gly-Leu_N-Lys_2-Ala-amide, have been examined in complexes with synthetic phospholipids using NMR and differential scanning calorimetry. The temperature–composition phase diagrams of the peptide–lipid mixture have been determined and used to test thermodynamic models of the binary system. Conclusions include: (1) explanations of the thermodynamic behavior do not require postulation of an annulus of lipids bound to the peptide and sequestered from the bulk lipid and (2) mismatch of the thickness of the hydrophobic portion of the lipid bilayer and the length of the proposed α-helical peptide does not have any dramatic effects on the phase behavior of the system.

(2) *Secondary amphiphilic peptides:* Numerous studies have been performed with peptides designed to form amphiphilic α-helices. These have modeled potential α-helical segments in peptide hormones such as ß-endorphin (107), calcitonin (1001, 392), melittin (302), and apolipoproteins. Lipoproteins are globular complexes of proteins and lipids that are present in serum. The apolipoproteins purfied from the complexes have been extensively studied, and those in classes A, C, and E contain potential amphiphilic α-helical segments, presumably in-

volved in the lipid interactions. Some of the synthetic peptides designed to model the amphiphilic elements of the apolipoproteins demonstrate remarkable properties (214, 33, 470). They are able to stabilize bilayer disks (~100Å diameter) in which the protein is presumed to bind around the edge of the phospholipid disk. These discoidal complexes are similar to those observed with apolipoprotein complexes with phospholipids. Other amphiphilic peptides such as melittin (1328, 353) and glucagon (393) can also stabilize discoidal particles or flat sheets of lipid bilayer, presumably by a similar mechanism. The molecular explanation for why various amphiphilic peptides interact very differently with phospholipids (e.g., bind to the surface, stabilize discoidal complexes, form transbilayer channels, or disrupt the bilayer like a detergent) is not known.

Some peptides which form amphiphilic ß-sheet have also been synthesized as models of surface-active proteins with potential amphiphilic ß-sheet (1103). One point worth noting is that the structure of some peptides in aqueous solution may be entirely different from that predicted using the algorithm of Chou and Fasman (218), which is designed for globular proteins.

3.8 Membrane Proteins Covalently Bound to Lipids

(for reviews see 1322, 260, 875)

A large number of eukaryotic and prokaryotic membrane proteins have covalently bound lipids which are posttranslationally added to the polypeptide (Table 3.8). In some cases, the role of the lipid is to provide a hydrophobic "anchor" by which the protein is attached to the membrane. In other cases, the roles are less clear and may involve directing the protein to the appropriate subcellular location (sorting) or possibly, in the case of enveloped viral proteins, facilitating membrane fusion.

The best characterized example from a prokaryote is Braun's lipoprotein (591), the major lipoprotein from the outer membrane of *E. coli*. The mature form of this protein contains a fatty acylated glycerol linked by a thioether linkage to the amino-terminal cysteine. In addition, the amino-terminal amino group is bound to a fatty acid by an amide linkage. The membrane-bound form of the penicillinase from *Bacillus licheniformis* is attached to the cytoplasmic membrane by an amino-terminal fatty acylated glyceride similar to that in the outer membrane lipoproteins (811).

Eukaryotic membrane proteins are often found to have covalently attached lipids. There are three general classes: (1) proteins that are bound to myristic acid, (2) proteins that are bound to palmitic acid, and (3) proteins that are bound to glycosyl-phosphatidylinositol. Fatty acylated proteins appear to be primarily localized to the cytoplasmic surface of the plasma membrane (1583), whereas proteins attached to phosphatidylinositol appear to be located on the outer surface of the plasma membrane (184, 418).

Myristic acid is attached to proteins uniquely through an amide linkage to an amino-terminal glycine, and the myristylation appears to occur cotranslationally on soluble ribosomes. Not all myristylated proteins are associated with the plasma membrane (1583). Palmitic acid is most often attached to proteins by a thioester linkage to cysteine or by hydroxyester bonds to serine or threonine. These residues are usually located within the body of the polypeptide, and are near membrane-spanning segments, usually on the cytoplasmic side of the membrane. The addition of palmitic acid occurs posttranslationally.

A small class of eukaryotic plasma membrane proteins are anchored to the outer surface of the cell by a covalently attached glycophospholipid which is a derivative of phosphatidylinositol. This linkage is invariably located at the carboxyl-terminal amino acid. The phospholipid anchor is added posttranslationally after the proteolytic removal of 17 to 31 amino acids from the carboxyl-terminus of a precursor. In several cases, it is clear that the sole mode of membrane attachment is via the phospholipid anchor, since the addition of phospholipase C releases the protein from the membrane. The reason why a small group of proteins is attached to the plasma membrane by a phospholipid anchor is not known.

Table 3.8. Examples of membrane proteins covalently linked to lipids

I. *Prokaryotic*
   1. Bacterial outer membrane lipoproteins (*E. coli*) (591)
   2. Penicillinase (*B. licheniformis*) (811)
   3. Cytochrome subunit of reaction center (1569)

II. *Eukaryotic*
   (A) *Myristylated Proteins* (see 1322)
      1. $p60^{src}$
      2. Catalytic subunit of cAMP protein kinase
      3. NADH-cytochrome b_5 reductase
      4. α-Subunit of guanine nucleotide binding protein (172)
   (B) *Palmitylated Proteins* (see 1322)
      1. $p21^{ras}$
      2. G glycoprotein of vesicular stomatitis virus
      3. HA glycoprotein of influenza virus
      4. Transferrin receptor (mammalian)
      5. Rhodopsin
      6. Ankyrin
   (C) *Proteins with glycosyl-phosphatidylinositol anchor* (see 1322, 260, 875)
      1. Thy-1 glycoprotein
      2. Variant surface glycoprotein of trypanosomes
      3. Acetylcholinesterase
      4. 5'-Nucleotidase
      5. Alkaline phosphatase
      6. Neural cell adhesion molecule (N-CAM 120)

3.9 Membrane Proteins Covalently Bound to Carbohydrate (see 1245, 634)

The surface proteins of mammalian cells are almost always glycosylated, including most receptors and transport proteins (see Chapters 8 and 9). The reason for this is not clear, though it may be related, in some cases, to the biosynthetic sorting required to direct the proteins to the plasma membrane (see Chapter 10). Possibly, the sugar residues help protect against proteolysis or play specific recognition or adhesion roles (see Section 9.3). Whatever the functions may be, it is clear that the sugar residues on membrane glycoproteins are exclusively located on the extracytoplasmic side of the membrane. There are also examples of glycoproteins which are cytosolic (643).

There are two major types of membrane glycoprotein oligosaccharide structures: (1) *N*-glycosidic oligosaccharides, linked to the proteins through the amide of asparagine; (2) *O*-glycosidic oligosaccharides, linked through hydroxyl groups of serine or threonine (see Figure 3.16). The biosynthesis of the *N*-glycosidic linkages is best understood (see Chapter 10; 1245, 634). Within this class there are three subclasses:

1. Simple or mannose-rich, in which the oligosaccharide contains mannose and *N*-acetylglucosamine in a core heptasaccharide (Figure 3.16).
2. Normal complex, in which the mannose-rich core has additional peripheral side branches with other sugar residues, such as sialic acid.
3. Large complex, in which there is a large polymer containing repeats of Galß1-4GlcNAc. This is found in Band 3, the anion carrier, in the red blood cell membrane (see Section 8.33).

Most membrane glycoprotein oligosaccharides are of the mannose-rich or normal complex variety. Within each of these classes there are great similarities, so it is unlikely that the sugar residues serve specifically coded, distinct functions for each glycoprotein.

Figure 3.17 shows the structure of a representative glycoprotein, glycophorin A of the erythrocyte membrane. This polypeptide has a single hydrophobic transmembrane element, indicated in the figure, and is extensively glycosylated with a single *N*-glycosidic oligosaccharide (complex class) and 15 serine/threonine-linked oligosaccharides. All are located in the amino-terminal portion of the polypeptide, which is extracytoplasmic.

3.10 Chapter Summary

Membrane proteins are bound to the membrane by a variety of means. Some peripheral membrane proteins are associated with the surface of the membrane by a combination of electrostatic and hydrophobic noncovalent interactions. In other cases, proteins have been shown to be covalently linked to lipids which can

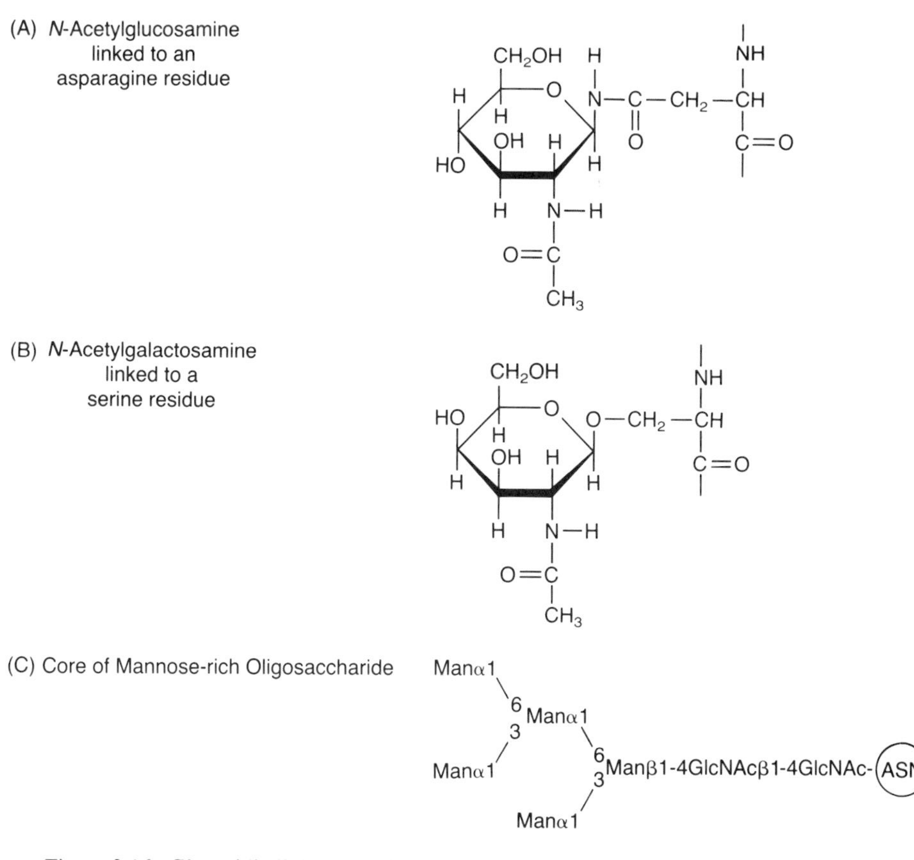

Figure 3.16. Glycosidic linkages which are commonly found in membrane glycoproteins. (A) An *N*-glycosidic bond. (B) An *O*-glycosidic bond. (C) The structure of the core of the mannose-rich oligosaccharide.

serve either a primary or secondary role in attachment to the membrane bilayer. Many membrane proteins have nonpolar domains which interact with the hydrophobic core of the bilayer. Experimental evidence from studies on bacterial photosynthetic reaction centers and on bacteriorhodopsin suggests that the predominant nonpolar element is a membrane-spanning α-helix. There are examples of integral membrane proteins proposed from amino acid sequence data to have as few as 1 or as many as 12 membrane-spanning α-helices. Predictive algorithms are useful for picking out candidates for these transmembrane helices. A note of caution is provided by studies on the porins in the outer membrane of gram-negative bacteria. These appear to have transmembrane ß-structure in the form of a ß-barrel. There is often considerable ambiguity in the interpretation and reliability of the predictive algorithms for identifying transmembrane segments, and there is a need for more and better experimental data on the folding patterns of membrane proteins.

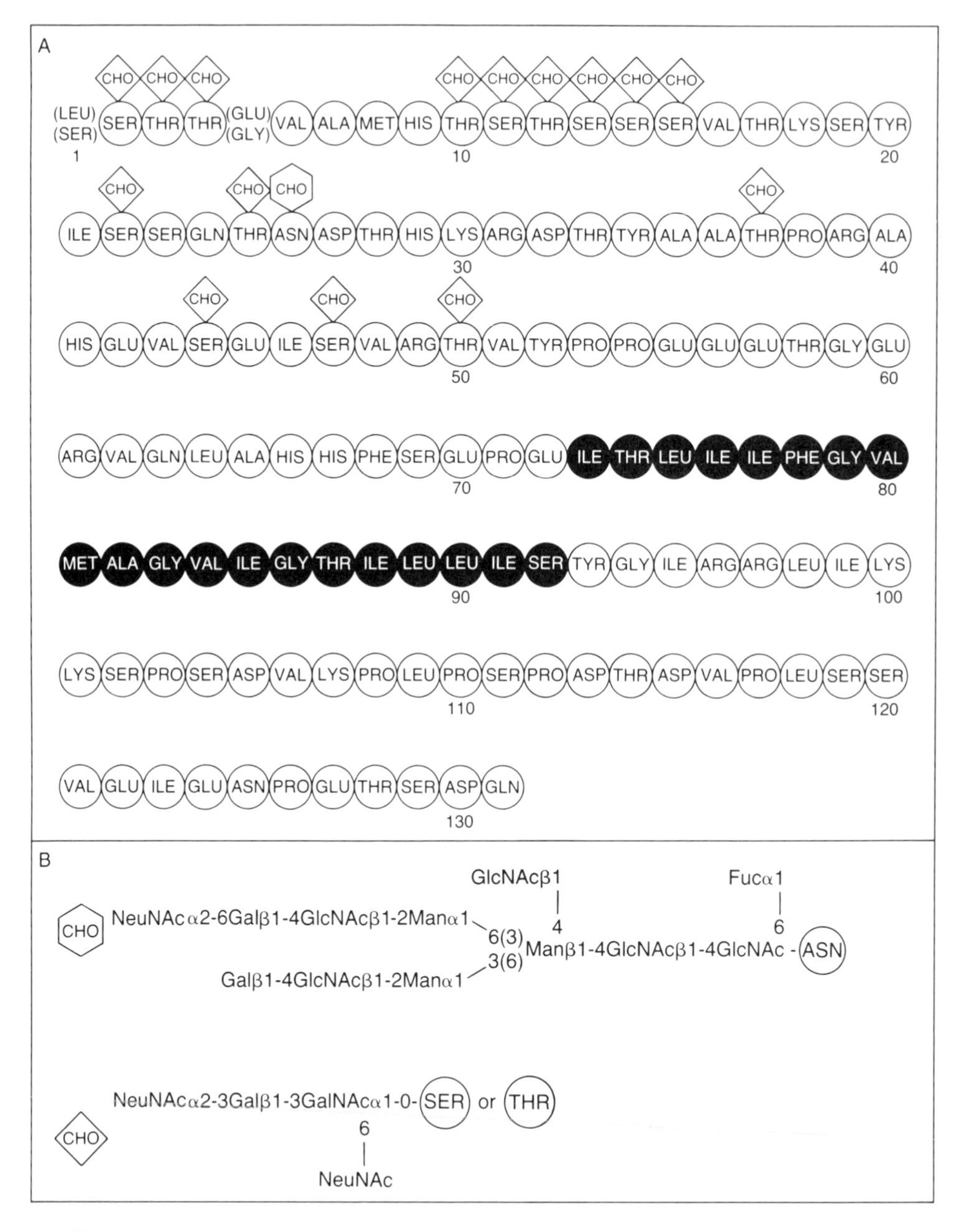

Figure 3.17. (A) Primary structure of human erythrocyte glycophorin A. The black circles denote the hydrophobic membrane-spanning portion of the polypeptide. The positions where carbohydrate (CHO) is covalently attached are indicated. Sequence differences between the M (top) and N (bottom) variants of glycophorin A are also indicated. Data from Dr. H. Furthmayr and references 1460 and 475. (B) Proposed structures of the N-linked (top) and O-linked (bottom) oligosaccharides attached to glycophorin A. The hexagon and diamond shaped codes are used in part A to designate the amino acids to which the sugars are attached. Data from refs. 475, 1460, and 469.

Information on the structure of membrane proteins is increasing due to the improvement in the art of purifying such proteins using a variety of commercially available detergents, and also due to the advent of DNA sequencing. In many cases, even in the absence of a pure protein, a DNA sequence can provide the primary sequence of the membrane protein, and a folding pattern with the bilayer can be proposed using the predictive algorithms for selecting transmembrane segments. The future promises more such models, and there is a need for improved methods for experimental confirmation.

Information on the secondary and quaternary structure of purified membrane proteins can be obtained using biochemical and spectroscopic methods. However, X-ray crystallography is required to provide structure at the resolution that is needed to address questions about the structure–function relationship of many membrane proteins. The spectacular success achieved by crystallographic analysis of the bacterial reaction centers provides hope that more structures will be forthcoming.

Chapter 4
Lateral and Transverse Asymmetry in Membranes

4.1 Overview

All biomembranes are asymmetric, and it is easy to rationalize why this should be so, since all membranes have two faces, exposed to different environments, e.g., cytoplasmic and extracytoplasmic. Therefore, transverse asymmetry, differentiating the two monolayer halves of the bilayer, is sensible. Membrane protein asymmetry is clearly a consequence of the way in which the proteins are originally inserted into the membrane. The rate of protein flip-flop across the bilayer is negligible. The term "flip-flop" is generally used to denote a rotation of 180 degrees about an axis in the plane of the membrane. Membrane lipids are also asymmetrically disposed, most convincingly demonstrated in the case of the red blood cell. How lipid transverse asymmetry originates and is maintained is not entirely clear at this time. Physical forces, such as those caused by extreme membrane curvature, may be important in some cases, but cytoskeletal interactions and possibly ATP-requiring enzyme "flipases" may be generally important.

In addition to transverse asymmetry, it is clear that biomembranes can have lateral inhomogeneities. Many eukaryotic cell surfaces are highly polarized and exhibit distinct macroscopic domains. Examples are the basolateral and apical plasma membranes of polarized epithelial cells (see Figure 1.2). These domains maintain different functions and compositions, and are physically separable. The thylakoid membranes in chloroplasts also have macroscopic domains called the appressed and nonappressed regions, containing different electron transport elements. These macroscopic lateral domains may be maintained by specific protein–protein interactions between membranes, as in the case of the thylakoids, or by specific structures within the membrane, e.g., "tight junctions," or by interactions with cytoskeletal elements, or by protein aggregation in the plane of the membrane. Little detail is known.

More controversial is the possibility of microscopic lipid domains within membranes, defined as small regions within the bilayer with distinct physical properties and composition. Model systems can be manipulated to manifest lateral phase separations (e.g., 607; Figure 2.25) and the possibility of similar phase separations in biomembranes under physiological conditions is intriguing. These might include nonbilayer elements or might be transient structures.

In this chapter, the asymmetric distribution of membrane proteins and lipids will be discussed. This can generally be termed as membrane topography. Included in this discussion will be a description of the cytoskeleton, since it is implicated as a significant factor in maintaining the organization of proteins and possibly also lipids in the eukaryotic plasma membrane. In the next chapter, the dynamics of the lipids and proteins within the membrane is discussed. Chapter 10 contains a discussion of how lipids and proteins are transported between membranes.

4.2 Membrane Protein Topography

Membrane proteins are inserted asymmetrically into the bilayer (see Chapter 10), and this asymmetry is kinetically maintained. The activation barrier to flip a protein, such that polar and charged residues which are normally on the surface are transiently within the hydrophobic bilayer, is very large. Hence, although some membrane proteins diffuse laterally and/or rotate about an axis normal to the bilayer (see Chapter 5), there is no evidence to suggest any spontaneous movement which would flip the orientation of a membrane protein with respect to the two sides of the bilayer.

This section reviews the major experimental approaches for determining how membrane proteins are oriented with respect to the bilayer. As discussed in Chapter 3, there are numerous algorithms to predict the location of membrane-spanning helices given the primary sequence of a protein. In some cases, the results are clear-cut and consistent with empirical studies (e.g., bacteriorhodopsin; Section 3.52), but in other cases this is not the case (e.g., acetycholine receptor; see Section 8.24). The computer-derived models serve as reasonable starting places, but are not definitive. It is quite clear that any detailed study of an intrinsic membrane protein must include an empirical topographic analysis. This is not experimentally simple, and often several approaches are best used to create a consensus model. Any particular technique is subject to artifacts. The details of some of the experimental methods have been reviewed (see 399, 593).

4.21 Methodology

Proteolysis

The use of proteases to define protein topography has been valuable in a number of cases (e.g., 593). An oriented membrane preparation (e.g., sealed vesicles of defined orientation) containing the protein of interest is exposed to a proteolytic enzyme such as trypsin or thermolysin, and the cleavage sites are identified and used to mark those portions of the polypeptide exposed on the outside of the membrane. To use this technique, it is usually necessary to be able to prepare highly oriented membranes (i.e., one side facing out) for in vitro study, in order to analyze the fragmentation of the protein of interest. Some membranes, such as those of the erythrocyte or enveloped viruses, can be examined directly without further isolation. In most other cases, membranes must be carefully prepared to yield an oriented system, either inside-out or right-side-out. This has been done for erythrocytes, eukaryotic plasma membranes, inner mitochondrial membranes, sarcoplasmic reticulum, and some bacterial membranes (see 1314, 853). In some cases, localization studies can be performed in highly oriented reconstituted proteoliposomes in which the protein being studied is the only one present. More often the protein is in a biomembrane along with other proteins. If it is a major protein, the proteolytic fragments can be isolated and characterized. If it is a minor protein component, indirect methods such as the use of monospecific antibodies or specific covalent labels must be used to identify fragments following SDS-PAGE. It is important to be certain that sites not normally exposed to proteolysis do not become susceptible due to damage done to the membranes by the initial proteolysis (e.g., 1314). It is also important to be able to inhibit the protease rapidly and irreversibly, since some proteases can remain active even after the addition of SDS prior to analysis. Note that negative results are not informative, since potential cleavage sites on the outer surface may not be accessible to proteolysis as a result of the tertiary structure of the membrane protein.

As is the case with many soluble proteins, proteolysis need not cause major changes in the tertiary or quaternary structure of a membrane protein in situ, though biological function can be totally eliminated. In fact, where one or two cleavage sites eliminate activity, proteolysis experiments can be used to identify and even isolate functional domains of the protein.

Some examples illustrating the use of proteolytic enzymes are Band 3 of the red blood cell (707), bacteriorhodopsin (853), the lactose permease of *E. coli* (1314), and subunit IV of cytochrome *c* oxidase (913).

Immunological Methods

Specific antibodies provide very useful techniques for determining membrane protein topography. In all cases, the technique requires measuring antibody binding to the protein in an oriented membrane preparation, such as *E. coli* membrane vesicles (e.g., 1314). Binding can be quantitated using standard immunological methods. Alternatively, the binding sites can be visualized using electron microscopy, by labeling antibody directly with colloidal gold or by using gold linked to a protein which binds to the specific antigen–antibody complex on the membrane (see 1198). Obviously, the methods used to prepare the membranes for this technique must not disrupt and alter the native topographic relationships. The more precise information one has about the antibody binding sites on the protein of interest, the more informative these experiments can be. Polyclonal antibodies directed against a purified polypeptide can be used to tell if the protein has portions exposed to either side of the membrane, if oriented membranes which are right-side-out and inside-out can be examined. However, such experiments do not identify which portions of the protein are exposed on the membrane surfaces. Two strategies can provide more detailed information. One is to use monoclonal antibodies, and the second is to use polyclonal antibodies directed against peptides corresponding to specific portions of the polypeptide.

A monoclonal antibody is directed against a very specific epitope or binding site on the polypeptide. The most useful monoclonals will bind to the protein in the membrane and also to denatured fragments of the protein and, thus, can be employed with proteins after SDS-PAGE analysis (i.e., Western blotting). The latter property is useful in order to map the location of the binding site on the polypeptide. In principle, the epitope can be localized to within a few amino acid residues. Examples where monoclonal antibodies have been used are rhodopsin (1110), bacteriorhodopsin (1108,1110; Section 3.52), acetylcholine receptor (1199; Section 8.24), and LamB, the phage λ receptor from the *E. coli* outer membrane (325).

The use of antibodies raised against synthetic peptides which correspond to part of a polypeptide allows one to specifically address the question of whether the designated part of the protein is available for antibody binding. If these sequence-directed antibodies bind to the native form of the protein in the membrane, they provide powerful tools to determine the protein topography. In this way, specific topographic models can be tested. Examples of the use of the techniques are work on the *E. coli* lactose permease (1314; Section 8.32), microsomal cytochrome P450 (306), and acetylcholine receptor (1199). Unfortunately, there is no guarantee that the anti-peptide antibodies will bind to the protein at all. The potential binding site in the native protein might be folded in a conformation or buried so as to be unrecognizable to the antibody or be inaccessible for binding. Indeed, in some cases, the anti-peptide antibodies may not bind to the protein even after it has been denatured in SDS (see 1314).

One clever variation which has been used is to use molecular genetics techniques to introduce a foreign epitope into the sequence of a membrane protein, and to use antibodies directed against the epitope to probe the topography of the polypeptide (422).

Chemical Labeling

This widely used approach has been used to identify proteins or portions of proteins exposed at the membrane surface or exposed to the hydrophobic membrane interior. The protein is covalently modified by a reagent that is presumably confined to one side of an oriented membrane or is localized within the lipid bilayer. To identify surface-accessible segments of proteins, the reagents must clearly be highly polar and must not adsorb onto or partition into the membrane. To locate portions of the protein in contact with the lipid, very nonpolar reagents which are partitioned into the hydrophobic core of the membrane must be used. The greatest source of artifacts in using these techniques is due to the label not being confined to the reactive space for which it has been designed. For example, a reagent designed to be surface-reactive may have enough nonpolar character to penetrate the membrane and gain access to the inner compartment. Also, the chemical modification of the membrane may be damaging, causing the membrane to become leaky to otherwise nonpenetrating reagents. Table 4.1 lists some of the chemical reagents used for the analysis of membrane-protein topography.

Surface labeling is often done using an enzyme, lactoperoxidase, to catalyze the iodination of the accessible surface tryosine or histidine residues. Since the reaction catalyzed between I^- and H_2O_2 occurs at the lactoperoxidase active site, this should be highly "vectorial," i.e., confined to a single surface. However, under some conditions I_2 can be formed, penetrate the membrane, and iodinate other residues on the inside (see 593). This is an example where stringent reaction conditions are important to avoid artifacts.

The other most frequently used reagents for surface labeling are diazonium salts, which react with lysine, cysteine, tyrosine, and histidine side chains, and photoreactive reagents such as NAP-taurine (see Table 4.1).

Table 4.1 also lists several nonpolar photoreactive reagents which are used to specifically label protein residues in contact with the bilayer interior. These reagents can be added to the membrane and allowed to partition into the bilayer before being photo-activated. The potential advantage of the resulting nitrenes and carbenes is that their reactions are much less specific than those of other reactive groups, so that specific amino acid side chains are not required to get covalent modification. However, nitrenes do show selective reactivity with nucleophiles (1382). Because they generally give higher reaction yields, carbenes are favored. One reagent, TID, has been used in several cases to deduce information about the secondary structure of transmembrane elements by determining which residues within a short sequence are lipid-exposed (652, 157, 653). For example, the helix C region of bacteriorhodopsin (Figures 3.8, 3.9) labeled along one face, consistent with it being a transmembrane helix with its nonpolar side facing the lipids (157).

Location of Specific Sites

Valuable information is sometimes available by knowing the location of specific sites on particular proteins. For example, sites of attachment of sugar residues in glycoproteins in the plasma membrane are always extracytoplasmic, so knowledge of their location in the polypeptide sequence yields topographic information. Specific location of sites of protein modification, such as phosphorylation, can be valuable if the location of the modifying enzyme is known (e.g., 1003). For example, the phosphorylated residues in plasma membrane proteins are on the cytoplasmic side. The location of specific binding sites can similarly be valuable. For example, the bacteriophage binding sites on the exterior portion of the *E. coli* outer membrane proteins OmpA (1019) and LamB (227) have been identified by analyses of mutations in those proteins.

Genetic Approaches

The ability to manipulate some membrane proteins genetically has made possible some novel approaches to topographic analysis. Although such methods are likely

Table 4.1. Some reagents used to probe the topography of membrane proteins.

Reagent	Target	Reference
(1) Diazotized sulfanilic acid (DSA, DABS) $N{\equiv}N^{+}{-}C_6H_4{-}SO_3^-$	Surface-accessible amino groups, tyrosine, histidine, cysteine.	1182
(2) Formyl-methionyl sulfone methyl-phosphate (FMMP) $HC(=O){-}NH{-}CH[(CH_2)_2{-}S^{+}(O^-)(=O){-}CH_3]{-}C(=O){-}O{-}P(=O)(O^-){-}O{-}CH_3$	Surface-accessible protein or phospholipid amino groups	145
(3) Trinitrobenzene sulfonic acid (TNBS) $(NO_2)_3C_6H_2{-}SO_3^-$	Surface-accessible protein or phospholipid amino groups	1332, 1251
(4) Isethionylacetimidate (IAI) $CH_3{-}C(={}^{+}NH_2){-}O{-}CH_2{-}CH_2{-}SO_3^-$	Surface-accessible protein or phospholipid amino groups	1059
(5) N-(4-Azido-2-nitrophenyl)-2-aminoethane sulfonate (NAP-Taurine) $N_3{-}C_6H_3(NO_2){-}NH{-}CH_2{-}CH_2{-}SO_3^-$	Non-specific label for surface-accessible groups. Photoactivated to reactive nitrene.	1182
(6) 1-azido-4-[^{125}I] iodobenzene (AIB) $N_3{-}C_6H_4{-}{}^{125}I$	Non-specific label for membrane-buried groups. Forms reactive nitrene.	296
(7) 3-(trifluoromethyl)-3-*m*-[^{125}I]iodophenyl diazirine (TID) $^{125}I{-}C_6H_4{-}C(CF_3)(N{=}N)$	Non-specific label for membrane-buried groups. Forms reactive carbene.	962, 157, 653, 1116
(8) 1-palmitoyl-2-[10-[4-[trifluoromethyl diazirinyl]phenyl]-[9-^{3}H] 8-oxaundecanoyl] -*sn*-glycero-3-phosphocholine (PTPC) $F_3C{-}C(N{=}N){-}C_6H_4{-}CH_2CH(^3H)\ O(CH_2)_6CO_2CH(CH_2OCO(CH_2)_{14}CH_3)(CH_2OP\bar{O}_2OCH_2CH_2\overset{+}{N}(CH_3)_3)$	Non-specific label for membrane-buried groups. Forms reactive carbene.	158 (also see similar labels, 1421)
(9) 1,2-dioleoylglycero-3-(4-[^{3}H]formylphenyl-phosphate) $CH_3{-}(CH_2)_7{-}CH{=}CH{-}(CH_2)_7{-}CO{-}CH_2$; $CH_3{-}(CH_2)_7{-}CH{=}CH{-}(CH_2)_7{-}CO{-}CH$; $H_2C{-}O{-}P(=O)(O^-){-}O{-}C_6H_4{-}C(=O){-}[^3H]$	Protein amino groups in contact with lipid polar headgroups.	957 see also 6, 170

Table 4.1 continued

Reagent	Target	Reference
(10) 1-palmitoyl-2-(2 azido-4-nitrobenzoyl)-*sn*-glycero-3-[^{3}H] phosphocholine. 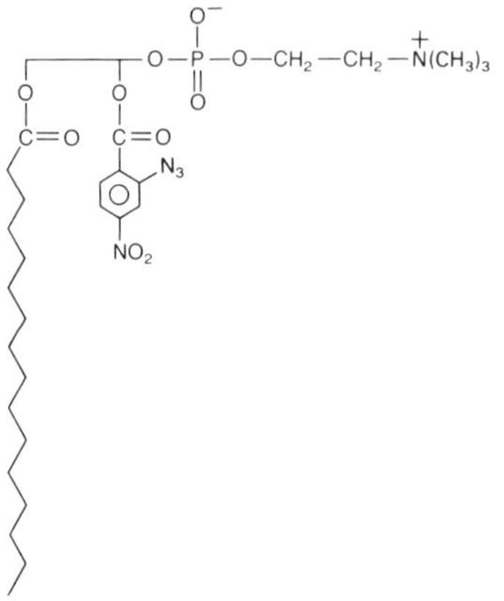	Non-specific label for membrane-buried groups near polar headgroup.	1182

to be applied with increasing frequency, so far the number of applications is too small to judge the reliability or general applicability of the methods. In one example, variants of the OmpA protein were made in which short inserts were made at specific points in the protein by genetic engineering (422). Susceptibility of the altered protein to proteolysis was used to judge whether the regions containing the inserted peptide were exposed externally. Phage binding sites and antibody epitopes have also been identified in outer membrane proteins by the analysis of mutational variants (325, 1019, 227).

Protein fusions are also proving to be valuable for the analysis of the topography of *E. coli* cytoplasmic membrane proteins (918, 11, 129). Hybrid proteins can be constructed so that the amino-terminal portion is from the membrane protein of interest and the carboxyl-terminal part is the catalytic portion of alkaline phosphatase. Alkaline phosphatase is normally located in the periplasm in *E. coli* and, apparently, must be transported to the periplasm in order to be enzymatically active. Since many membrane proteins appear to be assembled in the membrane in a linear fashion, starting at the amino terminus (see Chapter 10), if the fusion junction where the alkaline phosphatase sequence begins in the hybrid protein is located in a region which is normally on the periplasmic side, the alkaline phosphatase moiety in the hybrid will be transported to the periplasm and will be enzymatically active. If the junction site is on the cytoplasmic side, the alkaline phosphatase part of the hydrid protein will be located inside the cell and will exhibit low enzymatic activity. Hence, analysis of the junction sites of hybrid proteins that result in high alkaline phosphatase activity will designate external domains of the membrane protein.

4.22 Examples of Topographic Analysis of Membrane Proteins

One of the best examples of a membrane protein that has been topographically analyzed extensively is bacteriorhodopsin. Some of these studies were discussed in Chapter 3. In this case, it is clear from image reconstruction work that bacteriorhodopsin has seven transmembrane spanning elements, probably α-helices (see Figures 3.7, 3.8). Data collected using proteolysis, chemical labeling, and antibodies are consistent with this model, though there is no consensus on the precise locations of some of the boundaries defining the transmembrane elements.

A second example where there is general agreement on the protein topography is bovine rhodopsin (see 44, 1110), illustrated in Figure 4.1. Again, the primary

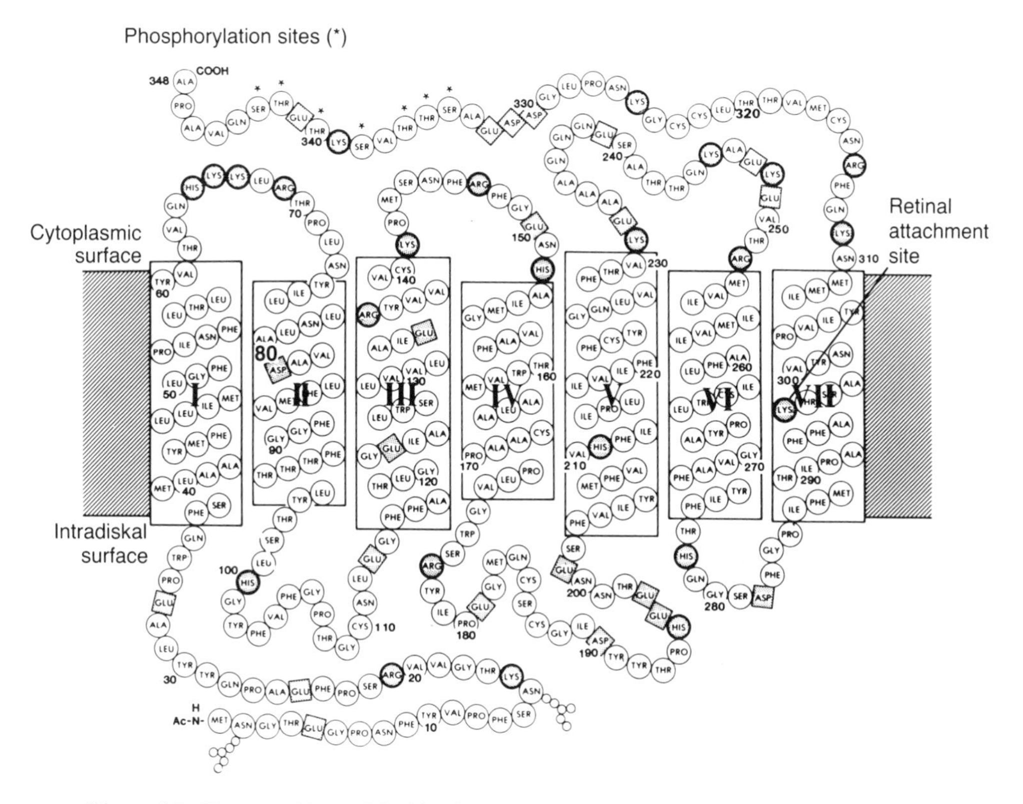

Figure 4.1. Topographic model of bovine rhodopsin indicating the seven transmembrane helices. Phosphorylation sites (asterisks) and N-linked glycosylation sites are indicated, as is the lysine in helix 7 which is the retinal attachment site. The aspartate in helix 2 is suspected to function as a counterion for the Schiff base formed by retinal. Cysteines 110 and 187 are suspected to form a disulfide bond. Note the presence of proline in five of the seven membrane-spanning helices. Adapted from ref. 593a. Kindly provided by Dr. P. Hargrave.

structure suggests seven transmembrane α-helices with connecting "loops." Studies with antibodies, proteases, and localization of sites of phosphorylation, carbohydrate attachment, and protein binding are all consistent with the indicated topography. Note that although bacteriorhodopsin and rhodopsin both bind the same prosthetic group, retinal, and both appear to fold similarly in the membrane, there is no sequence homology or functional relationship between these two proteins. Whereas bacteriorhodopsin is a bacterial, light-activated proton pump, rhodopsin is the visual pigment in rod cells in the retina. Rhodopsin undergoes a light-induced conformational change on the protein surface that results in an altered affinity for a GTP-binding protein, initiating a cascade of events culminating in the visual response (see Chapter 9).

The acetylcholine receptor subunits provide an additional example of a transmembrane protein which has been topographically analyzed (see 1198; Section 8.24). Several models of receptor topography have been proposed based on primary sequence analysis, and immunological methods have been used to test these models. At this point, the data are conflicting, pointing out, at the very least, that the application of these techniques is not necessarily straightforward.

A final illustration is provided by microsomal cytochrome b_5 (see 1421, 1420, 1267). This protein functions as an electron carrier and participates in several oxidation-reduction reactions in the endoplasmic reticulum (see Chapter 6). Cytochrome b_5 is an amphipathic protein in which a heme-binding, catalytically active domain (11,000 kDa) is connected by ten amino acid residues to a nonpolar membrane-binding domain (5,000 kDa) that serves as a membrane anchor. The two domains can be separated proteolytically and the structure of the heme-binding portion has been studied by X-ray crystallography. The structure of the membrane-anchoring peptide at the carboxyl end of the protein is not known. A simple topographic question has been addressed by a number of laboratories. Does the membrane anchor extend across the bilayer or does it go in about halfway and loop back out, leaving the amino and carboxyl termini on the same side? Despite considerable efforts, there is still no agreement on the correct answer. Most studies have used purified cytochrome b_5 reconstituted in phospholipid vesicles, and these have been complicated by the fact that different reconstitution procedures result in distinctly different molecular conformations. In "loose" binding cytochrome b_5, the carboxyl terminus is accessible to carboxypeptidase Y and is clearly located on the same side as the heme-binding domain. Different procedures, however, result in "tight" binding protein in which the carboxyl terminus is not accessible to proteolysis, and this probably corresponds to the topographic conformation in vivo. Khorana and his colleagues addressed the question by observing the labeling patterns in the two forms using photoactivatable phospholipid analogues (1421, 1420; Table 4.1). They concluded that in the "tight" binding form, the membrane anchor spans the bilayer. This conclusion, however, is not universally supported (1267, 48). Other approaches such as the use of fluorescence energy transfer, X-ray scattering, neutron scattering, or chemical labeling have either been unable to distinguish the two models of the "tight" binding mode or have resulted in conflicting data (see 1267, 48).

The main point of this and the other examples is that, in the absence of high-resolution structural information, it is difficult to determine the folding pattern of intrinsic membrane proteins and that the number of cases where this has been accomplished is actually very small.

4.3 The Cytoskeleton

In principle, both the tranverse and lateral distributions of membrane components could be strongly influenced by interactions with structures on either side of the membrane. This has been clearly demonstrated to be true in some instances, and there has been a justifiably increasing interest in membrane interactions with the cytoskeleton. The cytoskeleton is a complex network of different kinds of filaments and other structures found in eukaryotic cells (see 99, 574). No prokaryotic equivalent is known. In large part, the role of this system can be considered as mechanical. The cytoskeleton provides mechanical support of the plasma membrane, i.e., determines cell shape (380), and determines the positions and movements of cellular organelles, such as occur during mitosis. Internal membrane vesicle movements involved in endocytosis, exocytosis, and phagocytosis, as well as other plasma membrane movement such as seen in amoeboid cell migration, all probably involve the cytoskeleton. In short, the cytoskeleton acts as a dynamic scaffolding which is responsive to both internal and external stimuli. Part of the system is intimately associated with the plasma membrane, and by far the best characterized example of this is the cytoskeleton of the mammalian erythrocyte membrane (82). A second example which is not as well characterized biochemically is the cytoskeleton of the microvilli in the brush border membrane of intestinal epithelial cells (1015). More typically, the cytoskeleton consists of a three-dimensional network of filaments throughout the cell. Some attachment sites on the plasma membrane are known to be at regions involved in intercellular contacts or at focal contacts where the cells are attached to the substratum used in cell cultures.

Cytoskeletal interactions with the membrane have been implicated in stabilizing asymmetric transverse lipid distributions (Section 4.4), stabilizing protein lateral domains (Section 4.5), and directing the motion of membrane proteins (Chapter 5).

Three classes of filaments make up the cytoskeletal network: (1) microfilaments (~60 Å diameter), made up of actin and associated proteins; (2) intermediate filaments (80 Å-100 Å diameter), made up of keratins and related proteins; (3) microtubules (230 Å diameter), composed of tubulin. The biochemical details of the membrane associations are best defined for the actin microfilaments (597, 598) of which the actin–spectrin network in the erythrocyte (82) may be considered a specialized case.

4.31 Microfilaments

These are oligomeric or filamentous forms of actin. At least 20 proteins have been identified in various tissues as being actin associated (see 99, 598). These proteins are involved in such functions as bundling actin filaments together, binding the filament to the membrane, crosslinking actin, controlling filament length, contractibility, and providing Ca^{2+} responsiveness. Although, in a number of cases, the actin plays a role in contractility, this is not always the case. Microfilaments are disrupted by cytochalasins.

Microfilaments may be aligned parallel to the cytoplasmic membrane as in the contractile ring in a dividing cell, or may bind to the plasma membrane by an end-on association such as with adhesion sites where cell-cell or cell-substratum contacts are made. Several membrane proteins have been implicated in actin binding to membranes (e.g., 99, 598, 1236, 1455, 915). Based on physical proximity, biochemical associations of microfilaments, or disruptive effects of cytochalasin, microfilaments have been implicated in numerous membrane-related processes, included receptor-mediated endocytosis, patching and capping, cell motility, and cytokinesis. The best biochemically characterized actin-membrane association is that of the erythrocyte (Section 4.34).

4.32 Intermediate Filaments

These are polymers composed of one or two fibrous polypeptides which differ in different cell types (see 99, 1101, 458) and are encoded by a large differentially expressed multigene family. Examples are the keratins, from epithelial cells, and vimentin from mesenchymal cells. The functions of the intermediate filaments are not known, nor is much known biochemically about their interaction with membranes (e.g., see 506, 1138).

4.33 Microtubules

These are made of tubulin, which is a well-characterized αß heterodimeric protein. Microtubules form a cytoplasmic lattice that is thought to interconnect the plasma membrane and organelles, such as mitochondria (see 99). The translocation of endosomes and lysosomes appears to occur along microtubules (939). There is evidence for specific tubulin binding sites within membranes (1043, 93, 58), and one membrane protein, synapsin I, has been isolated which appears to interact with tubulin (58). During mitosis, the cytoplasmic microtubule network disassembles and reforms as the mitotic spindle. Microtubules are disrupted by the drug colchicine.

4.34 The Erythrocyte Membrane and Cytoskeleton

The membrane and cytoskeleton of the mammalian red blood cell have been extensively studied (see 82, 571). Table 4.2 lists the major proteins resolved by SDS-PAGE. The numbers assigned to each polypeptide derive from the relative

Table 4.2. Properties, associations, and functions of erythrocyte membrane proteins[l]

Peptide fraction	M_r	Copies per cell ($\times 10^{-5}$)	Associated state	Function	Association with
Band 1 (spectrin) Band 2 (spectrin)	240,000 220,000	2.2	Dimers of Bands 1 and 2 form tetramers and oligomers	Membrane skeleton	Band 3 (to Band 1); Ankyrin (to Band 2); Band 4.1
Band 2.1 (ankyrin) Band 2.2 (ankyrin) Band 2.3 (ankyrin)	210,000 183,000 165,000	1.1	Monomer	Connects spectrin with membrane through Band 3	Spectrin; Band 3
Band 3	95,000	12	Tetramer in equilibrium with dimers	Inorganic anion transport system; skeleton attachment site	Ankyrin; Band 4.1 (?); Band 4.2; glycolytic enzymes
Band 4.1a Band 4.1b	80,000	2.3	Dimer (?)	Connects skeleton with intrinsic domain; stabilizes spectrin–actin interaction	Spectrin; glycophorin
Band 4.2	72,000	2.3	Tetramer	Unknown	Band 3
Band 4.5/1 2 3 4 5 6	59,000 ↓ 52,000	1.3 1.4 0.7 1.0 1.3 1.3		Monosaccharide, possibly L-lactate and nucleoside transport systems	
Band 4.9	48,000	1		Actin binding?	Actin (?)
Band 5 (actin)	43,000	5.1	Oligomers of ~12 monomers	Membrane skeleton	Spectrin; Band 4.1
Band 6	35,000	4.1	Tetramer	Glyceral-dehyde-3-P–dehydrogenase	Band 3
Band 7	29,000		?	?	?
Glycophorin A Glycophorin B Glycophorin C	29,000 ↓ ~25,000	2 0.7 0.35	Dimer	Skeleton attachment site	Band 4.1

[l]Modified from ref. 571.

electrophoretic mobilities in the gel. Extraction of the membrane at low ionic strength removes the peripheral membrane proteins that are mainly components of the cytoskeleton. The major intrinsic membrane proteins are Band 3 and the glycophorins (A, B, C). Band 3 is the anion transporter (Section 8.33), and the glycophorins are major glycoproteins whose functions are not known (see Figure 3.17). The cytoskeleton can be visualized by electron microscopy as an ordered meshwork on the inner surface of the membrane (173). As seen in Table 4.2, the cytoskeletal proteins are major components, and this has facilitated the biochemical characterization of the system. In essence, the single-layered lattice (865) is composed of a spectrin–actin complex which is bound to the plasma membrane by interactions with both Band 3 and glycophorin, mediated through specific proteins, ankyrin and Band 4.1. Figure 4.2 illustrates the system. The main components have been purified to homogeneity and studied in vitro. The major interactions, such as between Band 3 and ankyrin or ankyrin and spectrin, are characterized by dissociation constants in the range of 10^{-7} *M*. Some of the interactions may be modulated physiologically. Ankyrin phosphorylation influences its affinity to spectrin (877), and the interaction between Band 4.1 and glycophorin appears to be modulated by phosphatidylinositols (37).

It is important to note that proteins closely related to the cytoskeletal com-

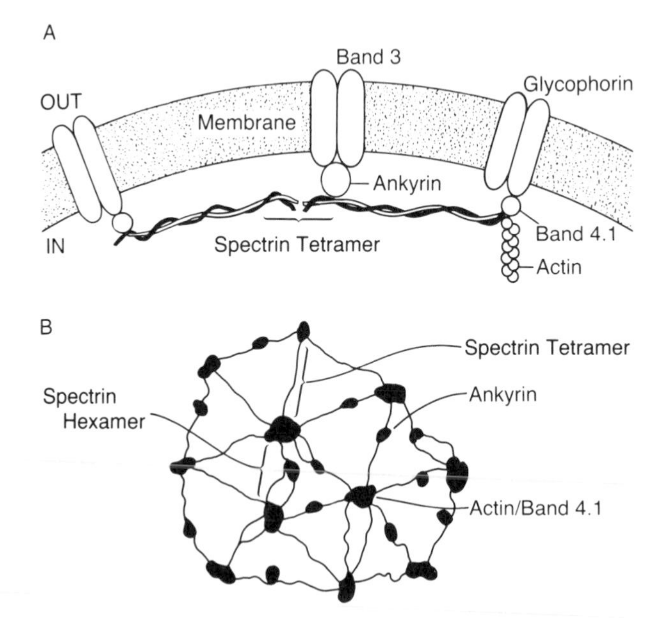

Figure 4.2. Diagrammatic views of the cytoskeleton of the erythrocyte. (A) Cross-sectional view showing the major protein components. Ankyrin and Band 4.1 mediate the interaction between the spectrin–actin complex and the transmembrane proteins Band 3 and glycophorin, respectively. The spectrin is probably folded and not fully extended, as indicated in the diagram (see 380). (B) Drawing of a spread membrane skeleton as visualized by negative-staining electron microscopy. Tentative assignments showing actin/Band 4.1, ankyrin, and a spectrin tetramer and hexamer are shown. Adapted from ref. 865.

ponents of the red blood cell have been found in a number of nonerythroid cells. Hence, interest in the erythrocyte cytoskeleton has been enhanced by the likelihood that the system is not unique to this cell type, but may be representative of the cortical cytoskeletons of other cell types as well. Some of the cytoskeletal proteins are described below.

(1) *Spectrin*. This is a tetramer of the $(\alpha\beta)_2$ type, in which two heterodimers $(\alpha\beta)$ are associated end-to-end. The molecule is very long, extending about 2000 Å for the tetrameric units. Spectrin (ß chain) binds to ankyrin and to Band 4.1 at distinct sites on opposite ends of the molecule. In addition, spectrin binds to actin, possibly in a complex with Band 4.1. Spectrin-like molecules such as fodrin (519) have been identified in a number of cell types.

(2) *Actin*. This is a globular protein which exists in this system as linear oligomers containing 12–18 molecules per unit. This appears in electron micrographs as a short rod, from which up to six spectrin tetramers can be attached (173).

(3) *Ankyrin*. This is the best characterized example of a soluble protein that mediates the interaction between an integral membrane protein and the cytoskeleton. This large polypeptide has distinct domains which can independently bind to spectrin and to the cytoplasmic domain of Band 3 (575). Ankyrin has also been detected on nonerythroid cells (292, 1058).

(4) *Band 4.1*. This is also in the class of proteins forming a link between the cytoskeleton and membrane. Band 4.1 binds to spectrin and actin (247), as well as to glycophorin (36). In addition, Band 4.1 can bind to Band 3 under certain conditions (1127). Synapsin I, found in synaptic vesicle membranes, appears to be related to Band 4.1 (57).

(5) *Band 3*. This is the principal anion transporter in the red blood cell (see Section 8.33). In addition, the Band 3 protein has a highly acidic amino-terminal region of its cytoplasmic domain which binds to several glycolytic enzymes (see Section 6.72) as well as hemoglobin. This cytoplasmic domain is not required for anion transport. The portion of the cytoplasmic domain closer to the membrane binds to ankyrin (Band 2.1) and to Band 4.2 (779). Based on the protein sequence, it has been proposed that Band 3 has 12 membrane-spanning segments (774), but empirical topographic studies are not sufficient to verify this model (e.g., 708).

(6) *Glycophorin A*. This is the major sialic acid-containing glycoprotein and, in contrast to Band 3, has a relatively small cytoplasmic domain with a single transmembrane segment (see Figure 3.17). It is implicated as a binding site for Band 4.1 (see 82). Other functions are not known.

4.4 Transverse Lipid Asymmetry

Membrane proteins, once in place within the bilayer, do not flip their orientation with respect to the two sides of the bilayer. They are put in the membrane with absolute asymmetry and remain kinetically trapped for their lifetime. Membrane lipids, in contrast, exhibit relatively rapid transbilayer motion, or flip-flop, in a number of biomembranes (see Section 4.43). Knowledge of the lipid transbilayer flip-flop rates not only is important in order to interpret the cause of lipid asymmetry but also is critical to evaluate the validity of the techniques used to measure the distribution of lipids between the two sides of the bilayer. In order to be valid, any measurement of, for example, phosphatidylserine on the outside of a membrane vesicle must be completed before the phosphatidylserine on the inside can flip to the outside. Since some of the techniques used to measure lipid asymmetry can perturb the system and enhance the rate of transbilayer motion, the resulting situation can be difficult to interpret. Critical reviews are available that evaluate and describe in detail the techniques for evaluating lipid asymmetry in membranes (1097, 1096).

4.41 Methodology for Measuring Lipid Distribution Across the Bilayer

Chemical Modification of Phospholipids

Only the aminophospholipids, i.e., phosphatidylserine and phosphatidylethanolamine, are relatively easy to label chemically. In these experiments, an oriented membrane preparation is exposed to a reagent which cannot penetrate the bilayer and which covalently reacts with the free amino groups of the externally facing aminophospholipid species. TNBS is the most commonly used probe (see Table 4.1). The percentage of phosphatidylethanolamine, for example, that reacts is a measure of how much is outside. Obviously, this will not be true if the reaction does not go to completion or if, during the course of the reaction, a significant amount of the phosphatidylethanolamine on the inside flips outside and becomes available for reaction. Both kinds of artifacts can occur.

A variation of this approach is to synthesize phospholipid analogues with reactive sulfhydryl groups and then utilize nonpermeating sulfhydryl-directed reagents (488). These lipid analogues must be incorporated into the membrane of interest prior to use, or studied in model systems.

Phospholipid Exchange

The spontaneous exchange of phospholipids between membranes or between vesicles and membranes is usually negligible. However, a class of proteins called phospholipid exchange proteins (PLEP) have been isolated which catalyze this exchange (see 616; Section 10.42). These are mostly isolated from mammalian tissue. The bovine liver phosphatidylcholine exchange protein is the most extensively studied and has an absolute specificity for catalyzing the exchange of phosphatidylcholine between membranes. Most of the other exchange proteins are much less specific as regards the lipid polar headgroups. These are soluble proteins which have high-affinity binding sites for phospholipid molecules. The mechanism of exchange is not known, but the proteins are useful in studying lipid asymmetry because they exchange only with lipids in the outer half of the bilayer with which the protein has contact. In a typical experiment (see 1494) the sealed membrane system is incubated with an excess of liposomes containing radiochemically labeled phospholipid in the presence of the exchange protein. The one-to-one phospholipid exchange that is catalyzed does not alter the composition of either the membrane, e.g., erythrocyte, or the liposome, but the specific radioactivity of the label in the membrane can be used to measure the size of the exchangeable phospholipid pool. If all the lipids in the membrane are available for exchange, the specific radioactivity of the lipid species in the liposomes and membrane will be the same at the end of the experiment. If the lipids on the inside are unavailable for exchange, the specific activity of the lipids in the membranes will be less than in the liposomes. If the lipids on the inside undergo transbilayer motion which is slow relative to the rate at which the exchange reaction goes to equilibrium, then the specific activity of the membranes will increase with time and this increase will monitor the rate of lipid flip-flop. In order to do these experiments it is necessary to separate the liposomes from the membranes being studied, and this is usually done by centrifugation.

The advantage of this technique is that it is nonperturbing, but the disadvantage is that equilibration is slow, on the order of an hour or more. Hence, it is useless when rapid lipid transbilayer movement occurs. This technique can be useful in conjunction with other methods. For example, radioactive lipids can be exchanged into the outer half of the bilayer and then phospholipases can be used to evaluate the rate at which these lipids translocate to the inside (see 986,1098; see Figure 4.5).

Phospholipases

Phospholipases are enzymes which degrade phospholipids. The bonds cleaved by the most commonly used phospholipases are shown in Figure 4.3. These enzymes are soluble proteins which will interact only on the outer surface of the bilayer and, hence, they provide a valuable tool for examining phospholipid asymmetry and the rate of transbilayer movement. Two significant problems in using phospholipases for this purpose are that (1) not all the lipids on the outer surface may be readily reactive, and (2) the products of the reaction, e.g., lysophospholipids, are usually destabilizing to the bilayer. Even if the integrity of the bilayer

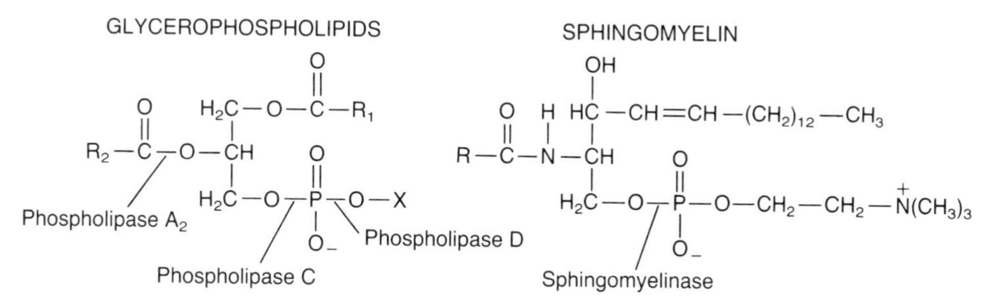

Figure 4.3. Action of phospholipases, showing the bonds cleaved by the several classes which have been well characterized. Adapted from ref. 1096. Phospholipase D can be used to catalyze headgroup exchange on phospholipids. Phospholipase A_1, which cleaves at the 1-position, also exists, but these enzymes are not as well characterized (see 1486 for review).

is not disrupted by the action of the lipase, the rate of lipid transbilayer movement could conceivably be enhanced (see 45). In short, the reactions should be carefully controlled and any conclusions derived from the results should be carefully examined.

Other Techniques

Several methods have been applied to measure specifically the transverse lipid distribution of particular lipids. Examples follow.

(i) *Cardiolipin* distribution has been measured by titrating the specific binding of adriamycin to this lipid (208).

(ii) *Glycolipid* distribution can be measured by oxidation either by galactose oxidase or by sodium periodate and then labeling with tritiated sodium borohydride (1410).

(iii) *Sterol* distribution has been measured in several ways (see 92). Sterols will spontaneously exchange between membranes without the need of specific exchange proteins (154). Hence, sterols in the outer half of the bilayer can be measured by direct exchange into or out of liposomes (226). The kinetics of specific complex formation with filipin has also been used to measure cholesterol distribution between the two halves of the bilayer (1309, 226). Finally, cholesterol oxidase has been utilized to quantitate cholesterol on the outer surface of the bilayer, but use of this enzyme has been criticized as leading to artifactual results (1452).

4.42 Examples of Lipid Asymmetry

Because of all the experimental problems in measuring lipid asymmetry, there are very few examples where the evidence is consistent and clear-cut. The human erythrocyte stands out as the one system which has been extensively examined and where all the techniques which have been utilized to determine lipid asymmetry yield compatible results. The outer membrane of gram-negative bacteria provides a second example of a highly asymmetric biomembrane, but this membrane is highly unusual in that lipopolysaccharide is a major component. There is, then, no doubt that biomembranes can have an asymmetric lipid distribution, but the convincing data are limited to relatively few cases.

Lipid Asymmetry in Phospholipid Vesicles

Small unilamellar vesicles (SUV) (see Section 2.55) composed of two different lipids have been shown in many cases to have asymmetric lipid distributions (see 1097, 802). For example, in vesicles containing mixtures with phosphatidylcholine, the outer surface is preferred by sphingomyelin or by phosphatidylglycerol. On the other hand, phosphatidylserine, phosphatidylethanolamine, phosphatidylinositol, and phosphatidic acid prefer the inner surface. It is generally agreed that the lipid asymmetry in these cases is largely due to packing differences on the two sides of the bilayer in the SUVs. Lipids with a more bulky polar headgroup will tend to favor the outer surface because of the larger available molecular surface area (see Chapter 2). However, biomembranes, with definite exceptions, do not have such extreme curvature, so these studies are not relevant to the general problem of lipid asymmetry in biomembranes. Large unilamellar vesicles can also be made which exhibit lipid asymmetry modulated by a transmembrane pH gradient (650).

Lipid Asymmetry in the Human Erythrocyte

The distribution of lipids in the red blood cell membrane has been clearly demonstrated to be highly asymmetric. As shown in Figure 4.4, phosphatidylcholine and sphingomyelin are largely found on the outer surface, while phosphatidylethanolamine and phosphatidylserine are largely found on the inner half of the bilayer.

There is some circumstantial evidence that the integrity of the cytoskeleton may be necessary to maintain the lipid asymmetry (see 347). The lipid asymmetry is decreased in cells which are deficient in either spectrin or Band 4.1 (1004). Treatments that perturb the cytoskeleton (347) also alter the lipid asymmetry,

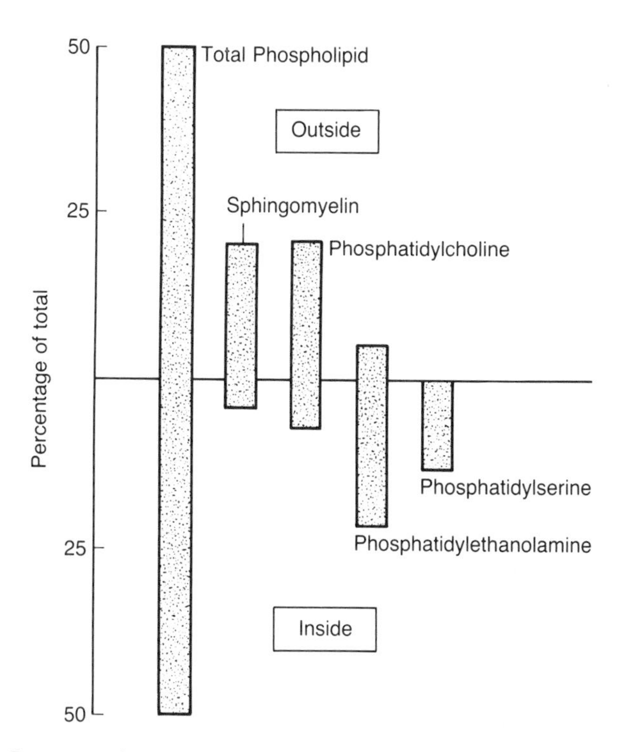

Figure 4.4. Representation of the relative distribution of the major lipids in the human erythrocyte membrane. Adapted from ref. 1510.

though it is not certain that the effects are a direct consequence of cytoskeletal damage (457). Finally, if small vesicles are made from the erythrocyte membrane, the distribution of lipids is much more random than in the membrane (802,1309). These data implicate the cytoskeleton as playing a role in the maintenance of the lipid asymmetry. The mechanism, however, is not clear, though one obvious possibility is a direct binding between the cytoskeletal proteins and the aminophospholipids (347, 1591).

Such direct stabilizing interactions, if they exist, are only part of the story. Exogenous phospholipids incorporated into the erythrocyte membrane using exchange proteins will redistribute after about 1 day to adopt the same asymmetric distribution as the endogenous lipids (985, 1455). Figure 4.5 shows the strategy used in these experiments. By measuring accessibility to phospholipases as a function of time, the lipid flip-flop rates can be determined (Figure 4.5). The flip rates for the lipids follow the order phosphatidylserine > phosphatidylethanolamine >> phosphatidylcholine > sphingomyelin. Similar data have been obtained with spin-labeled lipid analogues (1628) and with lysophospholipids (89). In short, those lipids which end up inside manifest relatively rapid transbilayer movement outside-to-inside. Those lipids which are preferentially found on the outside show less (or no) transbilayer movement.

A potentially very important observation is that the rapid transbilayer movement of the aminophospholipids appears to be ATP-dependent and is significantly slowed down in ATP-depleted cells (1455, 1628). This has resulted in speculation that enzymes ("translocases") are responsible for catalyzing the lipid-specific flip-flop (1628). There is a growing body of evidence supporting this conclusion (see Section 4.43), but much of the evidence is still circumstantial.

At this point, speculation on the maintenance of lipid asymmetry has taken two directions that are compatible and may both be important. One is static, with stabilization by specific binding interactions with the cytoskeleton, and the other is dynamic, with energy-dependent translocases used to move lipids preferentially across the bilayer to maintain a steady-state asymmetry (see 1591).

Lipid Asymmetry in the Bacterial Outer Membrane

Unlike the erythrocyte membrane, the outer membrane of gram-negative bacteria cannot be claimed as a model representative of other membranes. This is because a major structural component, lipopolysaccharide, is unique to this system. However, the outer membrane has been thoroughly studied (see 1068, 885), and it is clear that this membrane is an example of a highly asymmetric structure (see Figure 4.6). Lipopolysaccharide (Figure 4.6) is found uniquely on the outer half of the bilayer, and most if not all of the phospholipid is in the inner half, facing the periplasm. The lipopolysaccharide plays an important role as a permeation barrier, making the bacteria resistant to a number of antibiotics (806). One of the major components of the outer membrane is the Braun lipoprotein, which is attached to the membrane in part by covalently bound lipid (see Section 3.8) and is also attached to the peptidoglycan cell wall both covalently and noncovalently (see 1068, 885).

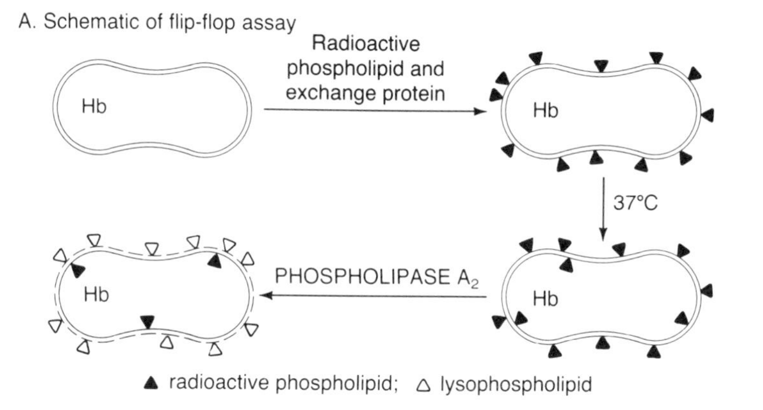

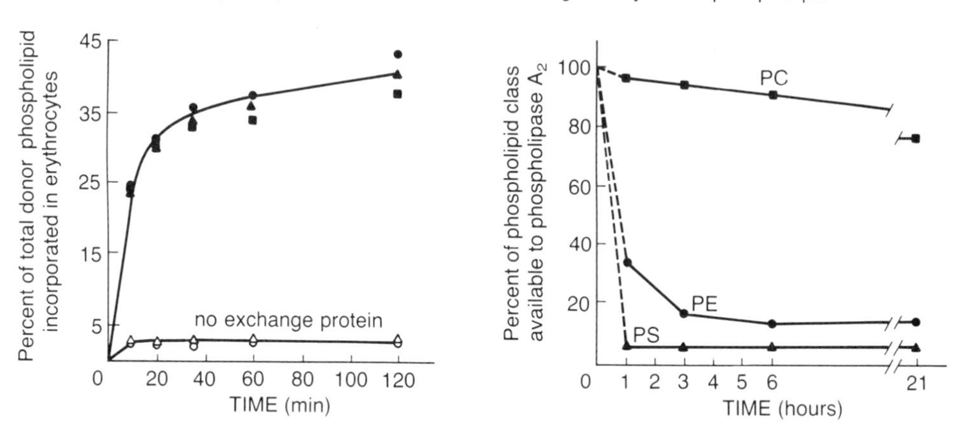

Figure 4.5. (A) An assay for determining the rate of phospholipid flip-flop in the erythrocyte. Radioactive phospholipids are incorporated into the outer half of the erythrocyte membrane using a phospholipid transfer protein. After isolation and washing, the cells are incubated to allow translocation of the radioactive lipid to the inside. Treatment with phospholipase A_2 is used to determine how much of the phospholipid originally present in the outer leaflet has become inaccessible to the phospholipase by transbilayer movement. Hb denotes hemoglobin. Adapted from ref. 1098. (B) Data showing the transfer of radioactive phospholipids from unilamellar vesicles to the erythrocyte in the presence and absence of the phospholipid transfer (exchange) protein. The data are similar for phosphatidylcholine (squares), phosphatidylethanolamine (circles), and phosphatidylserine (triangles). From ref. 1455. (C) Data showing transbilayer movement of exogenously added phospholipids as determined by the assay described in part A. PC, phosphatidylcholine; PE, phosphatidylethanolamine; PS, phosphatidylserine. From ref. 1455.

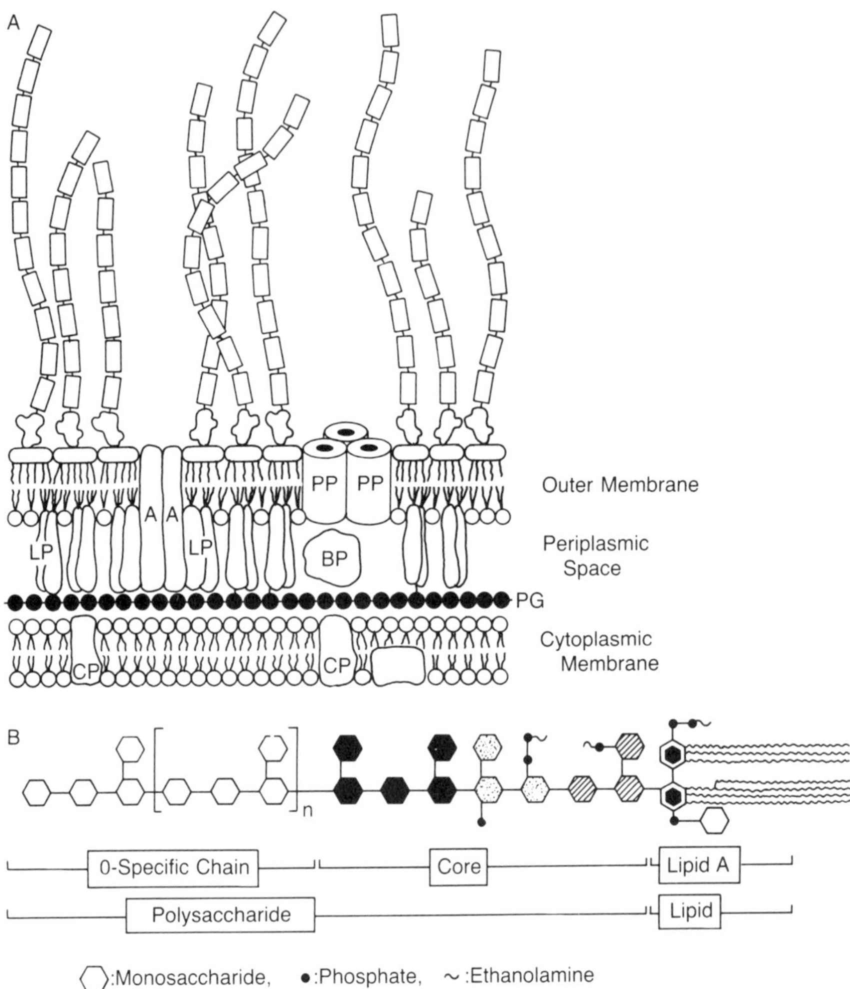

Figure 4.6. (A) Molecular organization of the outer membrane of gram-negative bacteria showing the cytoplasmic membrane, periplasmic space, and outer membrane. Note the asymmetric distribution of lipopolysaccharide and phospholipid in the outer membrane. Three outer membrane proteins are included: LamB (PP), OmpA (A), and the major lipoprotein (LP). A solute transport binding protein (BP) is shown in the periplasm, as is the thin peptidoglycan cell wall (PG). A cytoplasmic membrane protein (CP) is also represented. From ref. 885. (B) Schematic showing the structure of the lipopolysaccharide from *Salmonella typhimurium*. The number of hydroxylated fatty acids depicted is arbitrary. From ref. 882.

4.43 Transbilayer Motion of Lipids

Phospholipid biosynthesis and membrane assembly occur asymmetrically (see Section 10.4). The active sites of phospholipid biosynthesis are located on one side, not both, of the membrane. For example, phospholipids are made and inserted into the cytoplasmic face of the endoplasmic reticulum in rat liver (103) and on the inside of the bacterial cytoplasmic membrane. It is clear that these lipids must undergo a translocation to get to the other side of the bilayer. The methods previously summarized in this chapter have been used to measure the rates of lipid translocation in a number of systems.

Transmembrane movement of phospholipids in phospholipid vesicles is negligible, with half-times on the order of several days or longer (1097, 488). This can be speeded up by the presence of intrinsic membrane proteins such as glycophorin or by perturbing the bilayer such as by exposing the vesicles to phospholipase (see 1097). As expected, it is the polar headgroup which prevents translocation, as studies have shown that diacylglycerol derivatives translocate very rapidly across the bilayer ($t_{1/2} < 15$ sec) (103). Several biomembranes also exhibit very slow transbilayer movement of phospholipids, including the influenza virus membrane and the inner mitochondrial membrane (see 1097, 103).

Some membranes show extremely fast lipid translocation, with half-times on the order of a few minutes. This result was obtained with rat liver endoplasmic reticulum (670, 735) and with the cytoplasmic membrane of the gram-positive bacterium *B. megaterium* (1251). These are membranes where lipids are synthesized, and there may be a specific translocase enzyme present in these membranes that facilitates rapid translocation. This has been claimed to be the case in the endoplasmic reticulum (103), but biochemical characterization has not been reported.

The outside-to-inside lipid translocation rates measured in the erythrocyte membrane are intermediate, having half-lives on the order of hours depending on the lipid structure (985, 1455). The translocation rates of spin-labeled phospholipid analogues and of exogenous lysophospholipids have also been measured, with similar results (1004, 1628, 89). The rates have been shown to increase when the cytoskeleton is perturbed (1098, 1004) or in the presence of lipid perturbants such as gramicidin A (224). The cytoskeleton may play a role in reducing the rate of lipid translocation across the bilayer, possibly by binding to the aminophospholipids (1591). It is noteworthy that neither the endoplasmic reticulum nor the bacterial cytoplasmic membrane is associated with a cytoskeleton, whereas both demonstrate high rates of lipid transmembrane movement.

The observation that the rate of lipid transmembrane movement is ATP-dependent in the erythrocyte (1455, 1628) suggests the presence of an energy-driven translocase on this system. ATP-dependent transbilayer translocation of aminophospholipids that is likely to be protein-mediated has also been observed in the plasma membranes of fibroblasts (929) and lymphocytes (1627). However, no phospholipid translocase has been isolated to date. At this point, the existence of such enzymes, and their possible role in the maintenance of lipid asymmetry or in membrane biogenesis, appears likely but is still speculative.

4.5 Lateral Heterogeneity in Membranes

The unadorned fluid mosaic model (see Chapter 1) presumes a homogeneous distribution of protein and lipid components in the plane of the bilayer. It is clear, however, that there are domains or regions within some membranes which have distinct compositions and which may be separate from other portions of the membrane with respect to the diffusional exchange of components. There are several kinds of membrane domains which can be easily described within the constraints of the fluid mosaic model by postulating additional stabilizing interactions.

(1) *Macroscopic domains* are mostly large, morphologically distinct regions of the cell surface which are separated by barriers. Examples are the apical and basolateral domains of polarized epithelial cells (Figure 1.2). Also, in thylakoids, the appressed (stacked) and nonappressed (stroma-exposed) regions of the photosynthetic membranes have different compositions, apparently stabilized by intermembrane interactions in the stacks.

(2) *Protein aggregation* within the plane of the membrane can result in relatively large patches or domains that are enriched in a particular protein, accompanied by whatever other components are favored by that environment. Examples are the purple membrane patches in *H. halobium* containing bacteriorhodopsin (Section 3.52) or the gap junctions containing connexin (Section 8.21).

(3) *Cytoskeleton defined domains* can be formed, in principle, by organizing specific membrane proteins via interactions with proteins within the cell. The kinds of interactions observed with the erythrocyte cytoskeleton could be used to organize membrane proteins laterally. There is no clear-cut example of this, though it is likely that the phenomenon of patching and capping of cell surface antigens involves cytoskeletal direction and that the concentration of specific receptors in coated pits prior to endocytosis involves interactions with cytoskeletal elements or with clathrin (see Section 9.41).

(4) *Lipid microdomains* may be thermodynamically stable within the biomembrane, just as they have been demonstrated to be in simple lipid model systems (e.g., 607, 964). There is much suggestive evidence for this but no definitive proof.

These four categories are somewhat arbitrary and it should be noted that they are not mutually exclusive ways to stabilize lateral heterogeneity in membranes.

4.51 Macroscopic Domains and Plasma Membrane Barriers (see 560)

The plasma membrane of cells is often divided into distinct domains which can be isolated and characterized. These are typically divided by barriers which prevent proteins and possibly lipid components from crossing from one domain to another. Proteins and lipids are free to diffuse within the boundaries defined by the barriers. Several examples follow (Figure 4.7).

1. The apical and basolateral membranes of polarized epithelial cells have distinct compositions. It has been shown, for example, that gangliosides do not move across the barrier from the apical to basolateral membrane (1369). The barrier, in this case, corresponds to the tight junctions between cells (see 1495).
2. The sperm cell plasma membrane is highly polarized into separable membranes with distinct compositions (300). The boundary between domains prevents the diffusion of membrane proteins (255).
3. The cilium of the rod cell connects the outer and inner segments. Rhodopsin is inserted into the membrane of the inner segment, but is highly concentrated in the membranes of the outer segment. Presumably a diffusional barrier exists to maintain this concentration differential (1144).
4. Sodium and potassium channels are localized in different segments of the myelinated axon membrane (1560). Possibly, this organization is stabilized by the junctions with the glial or Schwann cells which provide the myelin sheath around the nerve fiber.

It should also be noted that prokaryotic membranes can also exhibit macroscopic lateral heterogeneity. For example, the membranes of gram-negative bacteria contain adhesion zones that appear to couple the inner and outer membranes (681). Also, the purple nonsulfur photosynthetic bacteria such as *R. sphaeroides* concentrate their photosynthetic apparatus in specialized membranes derived by invagination from the cytoplasmic membrane (348).

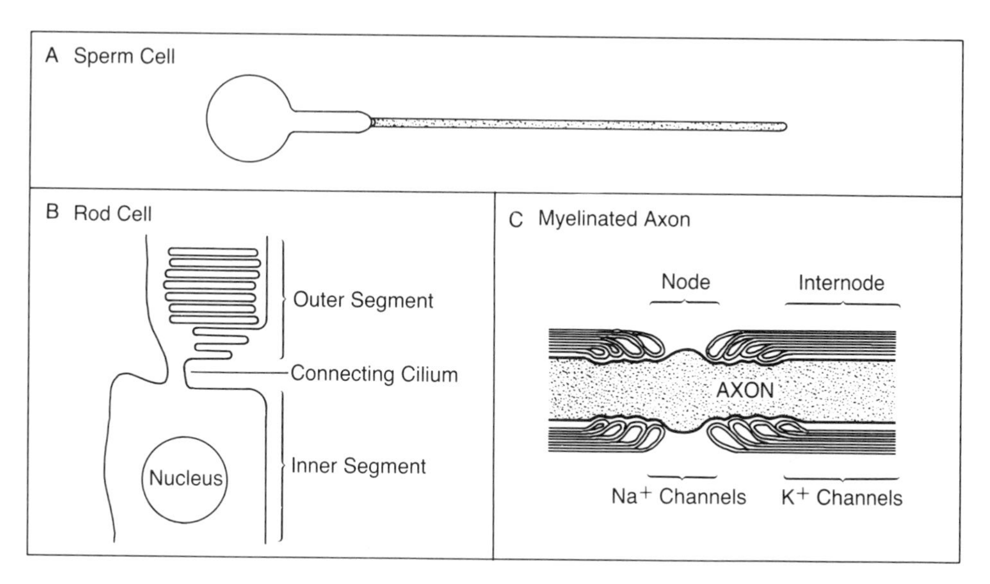

Figure 4.7. Plasma membrane domain boundaries in various cell types. Adapted from ref. 560.

The experimental problem is to define the nature of the barriers separating these domains. Very little is known at this time.

4.52 The Thylakoid Membranes (see 989, 1042, 1573)

The thylakoid membranes in the chloroplast of higher plants contain the apparatus for photosynthesis. These membranes are organized to form stacks, called grana (Figure 4.8). The appressed and nonappressed regions of the thylakoids are morphologically distinct and can be isolated individually. It is clear that the two regions have different compositions, presumably stabilized by the interactions between membranes. One stabilizing element in the stacking interaction is the direct binding of the light harvesting complexes to each other in adjacent

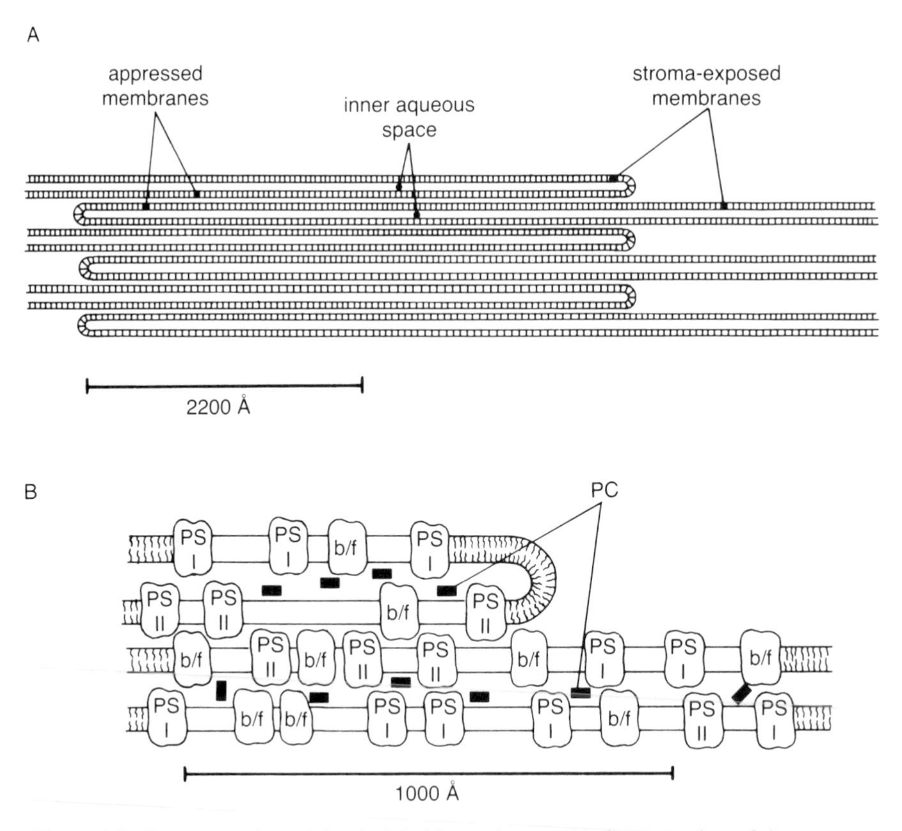

Figure 4.8. Representation of the thylakoid membrane. (A) Cross-section of the granum stack extending into the stroma-exposed portion of the membrane. (B) An enlarged view of the granum showing several protein components at approximately their calculated concentrations: cytochrome b/f complex (*b/f*), photosystem I (PSI), photosystem II (PSII), and plastocyanin (PC). Adapted from ref. 1573. Kindly provided by Dr. J. Whitmarsh. See also Figure 6.5.

membranes. Electrostatic interactions are also clearly important in stabilizing the stacks. Although the details of these interactions are far from clear, it is evident that these stacking interactions in some way result in the marked lateral separation of components, and this is functionally important. For example, the two photosystems, I and II, that are in a linear electron transport chain, are physically separated in different membrane domains (965) and are biochemically coupled by a diffusible plastoquinone (see Section 6.6). Also, the partitioning of the light harvesting complex between the two domains is regulated by phosphorylation of this protein.

4.53 Enveloped Viruses (see 1224, 1571)

Enveloped viruses consist of a nucleocapsid surrounded by a lipid bilayer. The membrane envelopes are derived from the preformed host membrane lipids by a budding process (Figure 4.9). Rous sarcoma virus and vesicular stomatitis virus are examples of viruses that bud from the plasma membrane of the host, whereas other viruses bud into internal compartments such as the Golgi or endoplasmic reticulum (1240). These viruses have been very useful models for studying

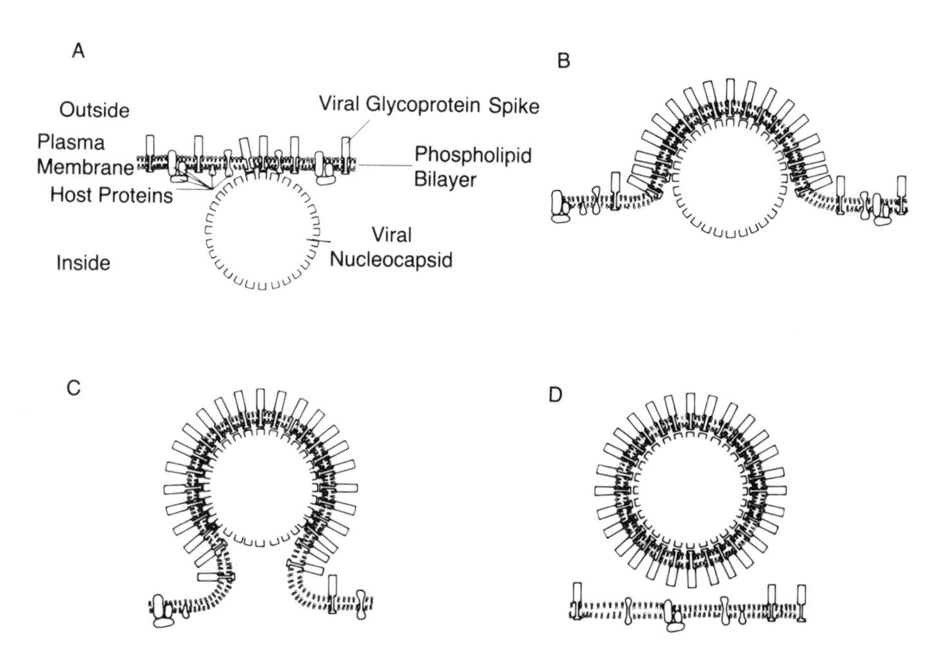

Figure 4.9. Postulated stages in the budding of an enveloped virus from the plasma membrane of a host cell. The interaction of the viral spike protein with the nucleocapsid is mediated by the matrix protein component of the virus. The matrix protein might also interact directly with the phospholipid bilayer. At the end of the procedure, the host cell proteins are excluded from the viral membrane. Adapted from ref. 1347.

membrane biogenesis and membrane traffic within the cell (see Chapter 9). In addition, they are useful as models for domain formation. As shown schematically in Figure 4.9, the budding process involves the interaction with the nucleocapsid and the transmembrane virus spike protein. The spike glycoproteins are inserted in the host plasma membrane, where they can interact via their cytoplasmic domains with the viral matrix protein, which binds to the viral nucleocapsid. During the budding process, the spike protein concentration increases and host plasma membrane proteins are totally excluded from the growing bud. The final virus envelope contains only the spike protein and no host proteins. Presumably, it is the interaction between the matrix protein and viral spike protein which drives the lateral separation that occurs in the plasma membrane. Interestingly, the lipid composition of the viral envelope is not identical to that of the host plasma membrane (1142). The most simple explanation is that this results in part from preferential interactions between the spike glycoprotein and particular lipid components. Clearly, this model system will be useful for further studies on lateral heterogeneities in membranes.

4.54 Lipid Microdomains (see 693, 728, 1270)

As indicated in Chapter 2, under appropriate conditions, lipids will undergo lateral phase separations yielding stable coexisting lamellar domains (see 693, 607, 702). These lateral phase separations can be induced by changes in temperature, pressure, or ionic strength (see 1270) or by the addition of divalent cations (607) or proteins (702). The question of whether such lipid microdomains observed in model lipid systems also exist in biomembranes has motivated considerable research efforts. The results are not entirely satisfying because lipid domains are not readily isolated and examined as unique species, in contrast to the examples of lateral heterogeneity discussed previously. The best that can be accomplished is to disrupt the membrane and demonstrate a heterogeneous range of compositions in the population of the resulting vesicles or fragments (e.g., 305). Electron microscopy has also been used to demonstrate lateral inhomogeneities in biomembranes (1329). Often, lateral heterogeneity has been inferred from biophysical techniques where the observed signal indicates multiple populations rather than the expected homogeneous population. An example of this is the measurement of the diffusion coefficient of a fluorescent lipid analogue in soybean protoplasts (976). Membrane microheterogeneity is sometimes inferred from the behavior of enzymes, where the enzymatic activity does not appear to be correlated with the average lipid physical state exhibited by the bulk of the membrane (e.g., 639). Often, these methods suggest regions with different lipid fluidity, as would be expected of coexisting gel and liquid crystalline phases within the biomembrane. This is also the conclusion of a series of studies where differential effects of perturbation caused by *cis* and *trans* fatty acids are interpreted in terms of preferential partitioning of the two into liquid crystalline and gel-like domains (see 728).

All these studies are consistent with the hypothesis of lipid microdomains within biomembranes, but for the most part the interpretations in terms of microdomains are not unique or totally compelling. Nevertheless, the concept of lipid microdomains is quite attractive. It provides for enzymes in the same membrane to be sequestered in unique environments, i.e., separate microdomains, where their activities may be optimized or modulated by specific lipid or protein interactions. In addition, the boundaries between domains represent regions where there should be packing "defects" in the lipid bilayer. Such boundaries, where the lipid packing undergoes a discontinuous change, are implicated in a number of events critical to bilayer function (see 1270). Studies with simple lipid model systems have demonstrated that at the phase transition, where gel and liquid-crystalline phases coexist, passive transport of both hydrophobic and hydrophilic species is enhanced severalfold over the rates observed in the fluid phase (see Chapter 7). Apparently, pores can develop in the regions of the bilayer where there are packing defects and where the lipid compressibility is high (1489). Phospholipid flip-flop also appears to be enhanced, as are the accessibility of lipids to phospholipases (Section 6.74) and the potential for lipid vesicles to fuse (Section 9.51) is enhanced due to boundary-associated defects in the bilayer.

The postulate of lipid microdomains within membranes may be correct, and it certainly provides a ready explanation for numerous observations. It is a very seductive concept awaiting definitive proof.

4.6 Chapter Summary

There is a large body of evidence showing that biological membranes can exhibit both lateral and transverse asymmetries. Transverse asymmetry implies that there are differences in the compositions of the halves of the membrane bilayer. It is absolutely certain that integral membrane proteins are inserted into the membrane in an asymmetric manner and that this asymmetry is not altered. Hence, the protein domains exposed on the cytoplasmic or extracytoplasmic faces of the membrane are distinct. There is also substantial evidence, particularly from studies on the red blood cell, that the phospholipid compositions of the two halves of the bilayer can be quite different. How this lipid asymmetry is created and maintained is not yet clear, but some data suggests the existence of ATP-driven phospholipid translocases that catalyze lipid flip-flop across the bilayer. Other factors in considering membrane asymmetry in animal cells are the interactions with the cytoskeletal elements on the cell interior and with the extracellular matrix components on the outside.

There is also evidence showing that there are lateral inhomogeneities in biological membranes. These can be large, specialized regions such as the apical and basolateral domains of the plasma membrane of polarized epithelial cells. On a smaller scale, the appressed and nonappressed regions of the thylakoid membrane provide a model demonstrating that adjacent portions of a single membrane can have distinct compositions and functions. These asymmetries represent a significant modification of the fluid mosaic model of Singer and Nicolson as it was originally proposed.

Chapter 5
Membrane Dynamics and Protein–Lipid Interactions

5.1 Overview

All biological structures are dynamic, and the rate and extent of motion are important in considering biological function. This is true for enzymes, for polynucleotides, and most certainly for membranes. The fluid mosaic model (Chapter 1) has helped focus attention on the mobility of membrane components by conceptualizing the membrane as a sea of lipid in which embedded globular proteins are freely floating. Over the past two decades an enormous literature has developed the quantitative and qualitative aspects of the dynamics of membrane components. Much of this utilizes the sophisticated applications of optical and magnetic resonance spectroscopies. From all this has emerged a useful physical picture of the membrane which encompasses the ways in which membrane proteins and lipids move and how they interact with each other. In addition, techniques have been developed to address specifically questions pertinent to membrane dynamics. There are considerable uncertainties remaining, but they point to areas which will be the focus of future research efforts.

The major motivation for studying membrane dynamics is its relevance to biological function. Some enzymatic functions require the membrane-bound components to be freely diffusing within the plane of the bilayer (see Chapter 6), whereas other processes clearly rely on constraints imposed on the mobility of membrane components. In the previous chapter, several examples were cited of plasma membranes separated into large domains with distinct functions and compositions, where barriers somehow prevent free lateral diffusion of components between the regions. This specialization into macroscopic domains is essential for the normal function of these cells. Also, intracellular membrane flow (Chapter 9) interconnects the rough endoplasmic reticulum, smooth endoplasmic reticulum, the Golgi stacks, and the plasma membrane, and yet these are all compositionally as well as functionally distinct. Again, selective connections or transfers between these membranes must be operative. To understand these and many other biological processes it is essential to first understand the fundamentals of membrane dynamics. The rate of transbilayer flip-flop of lipids is clearly related to the maintenance of transverse lipid asymmetry (Chapter 4) and may be related to the passive permeability of the bilayer (Chapter 7). The rate of exchange of membrane lipids between distinct membranes is relevant to membrane biogenesis (e.g., see 1044, 301) (Chapter 10). The rate of lateral diffusion of membrane components is pertinent to enzyme reactions involving multiple components which are membrane-bound, and the rate with which lipids adjacent to proteins exchange with bulk lipid is essential to characterize the nature of protein–lipid interactions.

It is important to realize the very wide range of motions which have been examined in membranes. These range from molecular vibrations occurring in about 10^{-14} sec to transbilayer flip-flop of lipids which can take many days to occur (see Chapter 4). Figure 5.1 summarizes some of these processes and also indicates the time ranges in which various biophysical techniques are sensitive to molecular motions. Considering that 20 orders of magnitude of time are represented in Figure 5.1, the terms "fast" and "slow" are clearly insufficient in themselves to characterize any motion. This figure also points out that some techniques yield a static picture of the membrane since the molecular motions are slow relative to the measurement, whereas other techniques only see a time-averaged picture of molecules which move between different environments very rapidly relative to the measurement time.

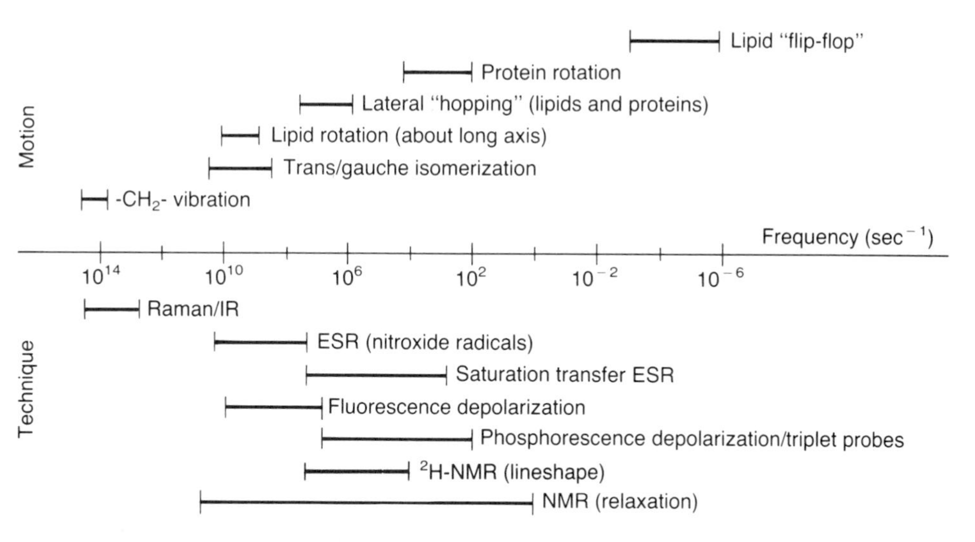

Figure 5.1. The characteristic frequencies of molecular motions of membrane proteins and lipids compared with the frequency ranges in which various spectroscopic techniques are sensitive to molecular motion. The characteristic times are obtained by taking the reciprocal of the indicated frequencies. Boundaries are very approximate.

Two broad classes of experiments are discussed. First, the use of probes within the membrane to measure membrane "fluidity" is described. Small-molecule ESR spin probes and fluorescent probes have been useful as indicators of the physical state of the membrane and have also been used to examine protein–lipid interactions.

A second class of experiments entails the direct measurement of lateral diffusion of membrane proteins or lipids and the rotational mobility of proteins within the bilayer. These types of experiments have also been used to monitor molecular interactions within the bilayer, since such interactions will influence the dynamics of the species being examined.

5.11 Some Simple Models for the Motion of Membrane Components

Figure 5.2 schematically illustrates some models for the rotational and translational motion of molecules within the membrane bilayer. Such models are necessary to interpret experimental results in terms of molecular motion.

(I) *Isotropic rotation* implies equivalent rotation in all directions with no

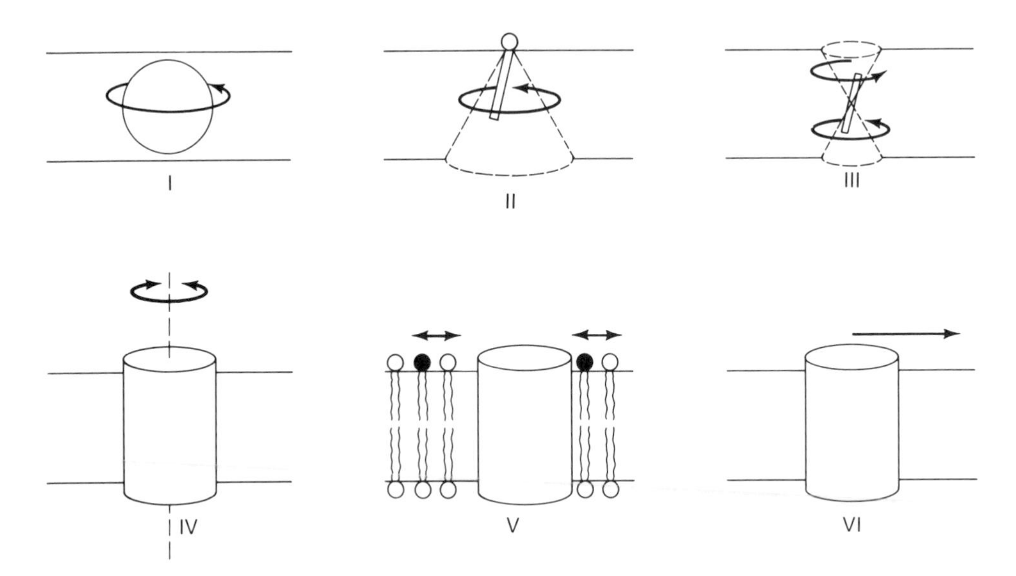

Figure 5.2. Some models used to analyze rotational and translational motion within membranes. (I) Isotropic rotation of a spherical molecule within the hydrophobic core of the membrane. (II) Wobbling-in-cone model for a "rod-like" molecule tethered at the surface and able to move rapidly within a cone-shaped region defined by an angle with respect to the bilayer normal. (III) Wobbling-in-cone model for rod-like molecule within the hydrophobic core. (IV) Rotation of a cylindrical transmembrane protein. (V) Lateral motion of lipids within the bulk (left side) and exchanging between bulk lipid and lipid adjacent to protein (right side). (VI) Lateral motion of a cylindrical transmembrane protein.

favored rotational axis. This would be expected for a sphere rotating in a continuum and has been applied to the interpretation of the rotational properties of small, hydrophobic probes such as TEMPO (Table 5.1) dissolved in the bilayer. Isotropic diffusion of a sphere in three dimensions is viewed as a random walk and quantified as a rotational diffusion coefficient, defined by the mean square angular displacement $<\theta^2>$ that occurs in time *t*.

$$D_{rot} = \frac{<\theta^2>}{4t} \quad [5.1]$$

Note that D_{rot} has units of sec^{-1}. Often the rotational diffusion of a molecule is reported in terms of either a rotational correlation time ($\tau_c = 1/[6D_{rot}]$) or a relaxation time ($ø = 1/[2D_{rot}]$).

(II) *Conical constraint* describes the motion of amphipathic probes such as fatty acid derivatives. These can be considered in simple models as rigid rods tethered at the membrane surface. The range of motion is constrained to a maximum deviation about the favored orientation, perpendicular to the plane of the membrane and, thus, describes a cone.

(III) *Conical constraint* is used by some (see 757) to describe the motion of DPH (see Table 5.1), a hydrophobic fluorescent probe, represented as a rigid rod. This model of the favored orientation and motion of DPH is certainly an oversimplification and alternatives have been proposed (see 29).

(IV) *Protein rotation* has, in simple models, been described as the motion of a cylinder embedded in the bilayer and able to rotate only about the axis perpendicular to the plane of the membrane (see 211 for review). This is anisotropic rotation. The relaxation time to characterize this motion is usually reported as $ø_{\parallel} = 1/[D_{rot}]$, denoting rotation parallel to the cylindrical axis.

(V and VI) *Translational motion of lipids and proteins* is modeled as a two-dimensional diffusion problem (see 366, 1504, 690, 691 for reviews). Of particular interest is the exchange rate between lipids adjacent to membrane proteins and those in the bulk lipid regions. Isotropic translational diffusion in two dimensions is quantified in terms of the mean square displacement $<l^2>$ which occurs in time *t*.

$$D_{trans} = \frac{<l^2>}{4t} \quad [5.2]$$

Note the units of D_{trans} are cm^2/sec. A particle with $D_{trans} = 10^{-8} cm^2/sec$ will diffuse about 2 μm in 1 sec.

5.2 Membrane Fluidity and the Application of Membrane Probes (see 328, 1333, 199 for reviews)

For an ordinary liquid, such as water, the term "fluidity" is defined as the inverse of viscosity, which is a well-defined and easily measured physical characteristic. Viscosity is essentially a measure of the frictional resistance encountered when

Table 5.1. Some probes used to study the dynamics of membranes.[l]

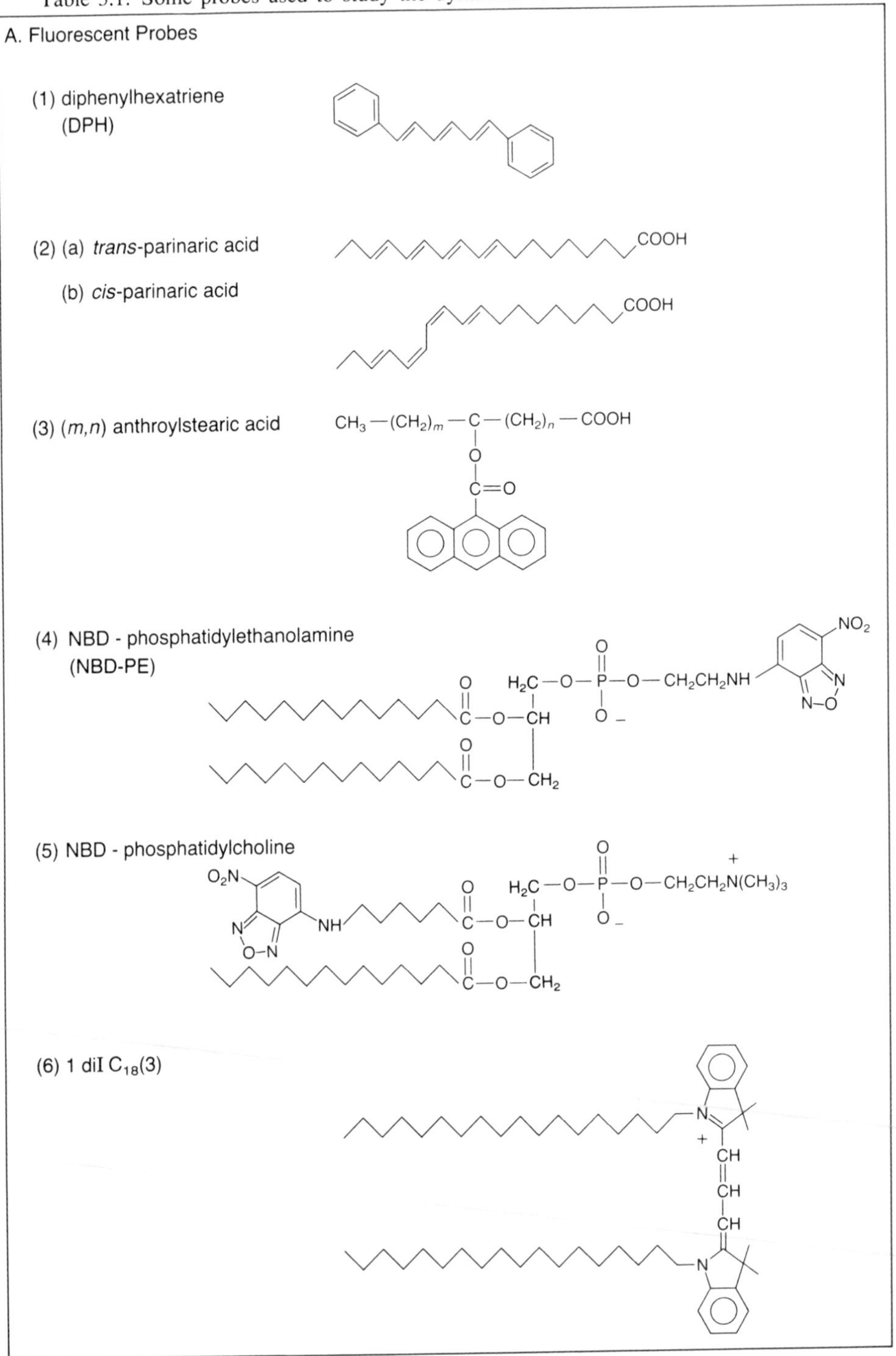

[l]Adapted from ref. 330.

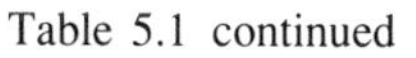
Table 5.1 continued

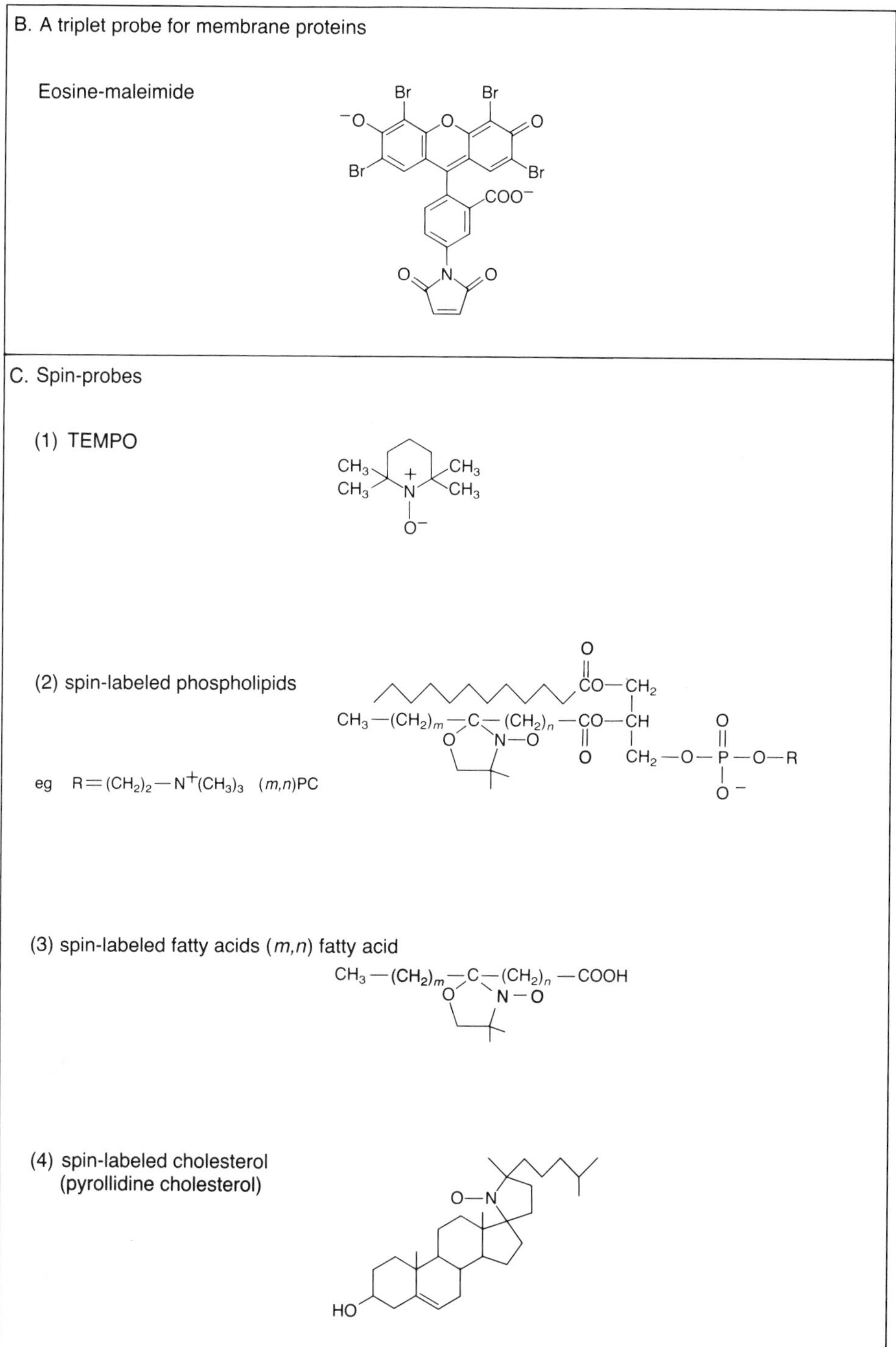

adjacent "layers" of fluid are moving with different velocities. Viscosity can be measured by simply observing the velocity with which a marble falls through the liquid. As applied to membranes, the term "fluidity" is usually used in a more qualitative sense, generally meant to be a measure of the resistance to movement of various types in the membrane. Generally, fluidity is measured by observing the motion of spin probes or fluorescent probes incorporated in the bilayer. These probes are usually small molecules, comparable to the size of the phospholipids in the membrane. Several are shown in Table 5.1. The conceptual basis of how the spectroscopic methods are used to measure molecular motions is described in subsequent sections. Since the measurements are sensitive to both the rate of motion and any constraints to that motion, information about dynamics and molecular order gets intermixed. Several considerations point out the problems in the quantitative interpretations of data relating to the motion of probes within biomembranes.

1. The lipid bilayer is not a simple three-dimensional homogeneous liquid, but a thin fluid which is quite different structurally and dynamically in the center as compared to the polar regions.
2. The probes utilized do not simply rotate isotropically, that is, like spheres with no preferred axis of rotation. Often the probes have preferred orientations within the bilayer and their range of motion is constrained (e.g., Figure 5.2). The interpretation of experimental data can depend on the model used to represent the molecular motion.
3. The probes may reside in different locations within the biomembrane. For example, the probe could be adjacent to a protein molecule, or trapped by protein aggregates, or within lipid domains which could conceivably be in different physical states.

Fluidity has been quantified by parameters derivable from the spectroscopic properties of small probes. These include (1) the rotational correlation time of spin probes or fluorescent probes; (2) the order parameter, also derived from spin probes or fluorescent probes; (3) the steady-state anisotropy (or polarization) of fluorescent probes; and (4) the partition coefficent describing the distributions of certain probes between the membrane and aqueous medium. It is clear that membrane fluidity, summarized by a single parameter, is not sufficient to characterize the physical state of the bilayer. Nevertheless, it has proved to be a useful measurement, especially in characterizing alterations in the physical state of the membrane due to changes, for example, in temperature, pressure, cholesterol content, phospholipid composition, or ionic composition (see 1333, 199), which we will not examine in any detail. Generally, fluidity as measured by spin probes and fluorescent probes is a strong function of lipid packing in the bilayer, at least in model systems. Perturbations that result in decreasing the area per lipid molecule, such as increased hydrostatic pressure, lower temperature, or the addition of cholesterol to phospholipids in the liquid crystalline state, all result in a decrease in fluidity (see 1333). This is consistent with the free volume theory (483, 1333) which quantitates the inverse relationship between fluidity and density. The probe motion will be more constrained in a more densely packed membrane. This approach works well in treating the fluidity of most non-associating liquids.

5.21 Physiological Relevance of Membrane Fluidity

Biomembranes are generally in the fluid liquid crystalline phase, and it appears that maintenance of membrane fluidity is critical to function. The fluidity of the bilayer drops by about two orders of magnitude during a phase transition from the liquid crystalline phase to the gel phase. The structural and dynamic properties of the gel phase bilayer are, simply, incompatible with the organization and proper functioning of the protein components in the membrane. There are some notable exceptions to this, including the semi-crystalline purple membrane patches containing bacteriorhodopsin in *H. halobium* (see Section 3.52).

Perhaps the most dramatic evidence that fluidity, as measured by spin probes or fluorescent probes, is physiologically relevant is from the studies showing adaptations of various organisms to environmental stress (see 1333, 252 for reviews; Section 10.5). This is most often observed with thermal stress in which microorganisms, plants, poikilotherms, or hibernating animals are acclimated to low temperatures. The organisms adapt by altering their membrane lipid compositions, usually by increasing the degree of lipid unsaturation or decreasing the average acyl chain length. The alterations tend to decrease the lipid packing density in the membrane and result in the maintenance of membrane fluidity. One or more membrane functions must be critically dependent on the maintenance of optimal membrane fluidity, though what this means on a molecular level is not known.

5.22 The Range and Rate of Motion Measured with ^{2}H-NMR, ESR, and Fluorescence Probes

It is important to have some appreciation of what is being measured with membrane probes, especially the relationship between the time scale of the motion being studied and the time scale of the technique used for the measurement. Consider a general case which could equally apply to an electron spin resonance (ESR), fluorescence, or ^{2}H-NMR measurement of a membrane probe. All three spectroscopic techniques are sensitive to the orientation of the molecules with respect to an axis system defined in the laboratory. For example, in ^{2}H-NMR, the absorption spectrum is sensitive to the orientation of the C-D bond in relation to the externally applied field. This is because the local field gradient must be vectorially added to the external field to obtain the resultant sensed by the deuterium. In ESR, the situation is similar, where the spectrum is sensitive to the orientation with respect to the applied magnetic field of the N-O (nitroxide) bond, present in most of the commonly used probes (Table 5.1). In fluorescence spectroscopy, the measured polarization of the emitted light is dependent on the orientation of the molecular transition dipole moment with respect to the direction defined by the polarizer used to make the measurements.

In NMR and ESR, if the molecules in the sample are randomly oriented and totally immobile, then each subpopulation with a particular molecular orientation is represented by a well-defined spectrum, and the experimental spectrum of the sample is simply obtained by adding up the spectra from each subpopulation in proportion to their concentrations. The result is called a powder spectrum or rigid glass spectrum, and it is typically broad. This is very different from the spectrum obtained if the molecules are randomly oriented but moving very rapidly relative to the time scale of the measurement. For the problems we are considering, the relevant frequencies are 10^5 sec^{-1} for ^{2}H-NMR and 10^8 sec^{-1} for ESR. In the case of fast isotropic motion, each molecule sees an average magnetic field, but all molecules see the same average, so the resulting spectrum is very sharp and defined. Examples of ESR spectra are shown in Figure 5.3. Because of time scale

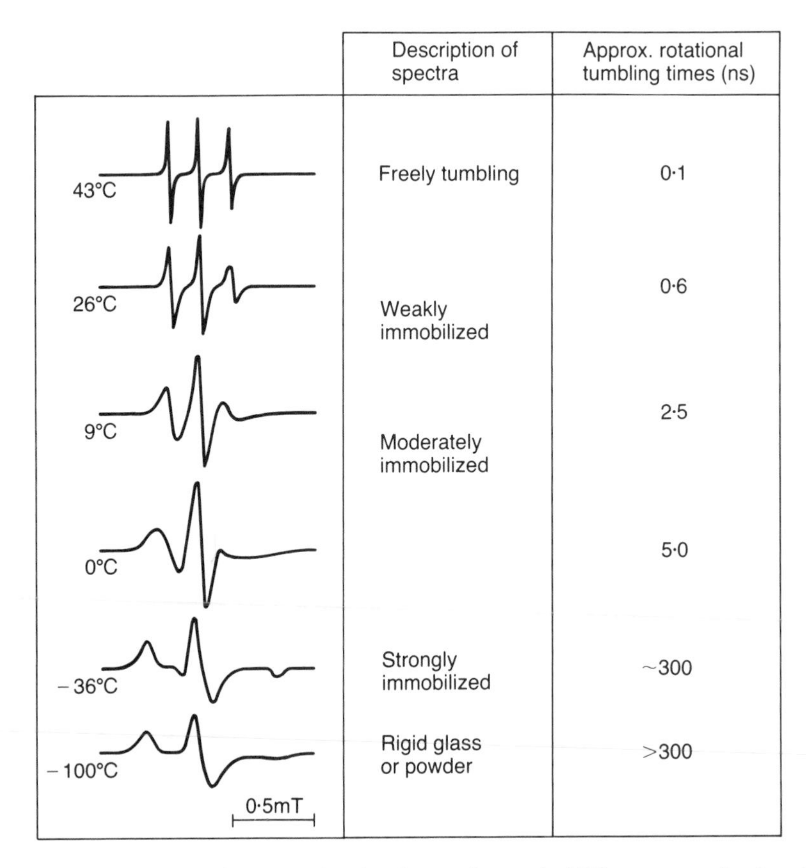

Figure 5.3. The effects of the rate of molecular rotation on the ESR spectrum of a nitroxide spin probe. These first derivative spectra were obtained by changing the temperature and, hence, the viscosity of the medium. Adapted from ref. 182.

differences, a molecule tumbling with a correlation time of 10^{-6} sec would appear to be static by ESR techniques (i.e., broad spectrum), but be moving very rapidly by ^{2}H-NMR criteria (i.e., sharp spectrum).

In fluorescence depolarization, one excites a subpopulation of the probe molecules with polarized light. After a few nanoseconds most of the excited molecules emit a photon. If the molecules are rigid during the time interval between the absorption and emission processes ($\approx 10^{-8}$ sec), the polarization of the emitted light will be determined largely by the polarization of the light used for the original excitation. If the molecules rotate isotropically and rapidly during this time, there will be no preferred orientation of the electric field vector of the emitted light, and the polarization will be zero. Polarization values between these two extremes could represent either slow motion where the molecules have not had enough time in 10^{-8} sec to randomize, or, alternatively, rapid motion which is constrained, thus preventing complete randomization (e.g., Fig. 5.2, models II, III).

All three spectroscopic techniques are used to obtain an order parameter, S, which is briefly discussed in Chapter 2.

$$S = 1/2\ (3\langle\cos^2\theta\rangle - 1) \qquad [5.3]$$

The term $\langle\cos^2\theta\rangle$ is a time-averaged orientation of the molecule (or of a transition dipole vector or bond in the molecule) with respect to the direction perpendicular to the plane of the membrane. The appropriate time over which the average is made depends on the technique: 10^{-5} sec for ^{2}H-NMR, 10^{-8} sec for ESR and for fluorescence depolarization (of a typical probe like DPH). The long time scale for ^{2}H-NMR makes this order parameter sensitive to slow motions in the range of 10^{-5} sec. In contrast, for either ESR or fluorescence depolarization, the order parameters measure the degree to which the probes can reorient in $\sim 10^{-8}$ sec, and information about slower motions is lost.

One can obtain dynamic information more directly from all three techniques. NMR can be used to obtain relaxation times which depend on molecular motions. The ESR spectrum can, under certain circumstances, be interpreted in terms of a rotational correlation time of the probe, but this relies on the assumption of isotropic rotation of the probe within the bilayer. Fluorescence depolarization can yield dynamic information apart from consideration of the constrained range of motion if the time dependence of the depolarization is measured (see Section 5.24).

5.23 ESR Probes for Measuring Membrane Fluidity (see 328, 1303, 182 for reviews)

ESR is extremely useful in the study of membranes. Typically, stable paramagnetic labels containing a nitroxide radical are used, such as those pictured in Table 5.1. The unpaired electron in the presence of a strong magnetic field ($\sim$0.3 T) can be induced to undergo energy level transitions by microwave radiation ($\sim 10^{10}$ Hz). The technique is very sensitive, typically requiring only about 10^{-6} M spins in a 50 μl sample. The spectrum is usually reported as a first derivative of the absorption spectrum.

The spectrum of the nitroxide radical has three peaks, due to the interaction between the spin of the unpaired electron and the spin of the nitrogen nucleus. Both the position of the spectrum (*g*-value) and the splitting due to the nuclear spin (*A*-value or hyperfine coupling) are dependent on the orientation of the molecule with respect to the external field. As described in the previous section, this results in the sensitivity of the spectrum to rotational motion. Figure 5.3 shows a series of spectra ranging from the limit of free rotation to a powder spectrum. A simple equation utilizing peak heights of the spectra can be used to directly obtain the value of τ_c, assuming the molecule is tumbling isotropically. This would best be applied to a probe such as TEMPO.

Spin probes which are derivatives of fatty acids or phospholipids surely do not rotate isotropically and, in this case, a second simple formula is used to yield directly the order parameter, S, from the splittings between the spectroscopic peaks and troughs (see 328, 1303). If one assumes the motion is modeled as a rigid rod, as pictured in Figure 5.2 (II), the order parameter can be interpreted to yield the maximal angular deviation from the direction perpendicular to the membrane surface.

The many assumptions involved in interpreting the ESR spectra in terms of either a rotational correlation time (τ_c) or an order parameter, S, make them less useful as quantitative measures of specific molecular characteristics, but they remain valuable qualitative indicators (see 328). The order parameter measured using spin probes attached at different depths in the membrane shows a gradient of increasing disorder from the surface to the middle, similar to that obtained by ^{2}H-NMR (see Figure 2.15), though not identical in detail (see 15). Numerous studies have compared membrane fluidity, quantified by either τ_c or S values from ESR probes, due to numerous kinds of membrane perturbations (see 328, 603).

5.24 Fluorescence Probes for Measuring Membrane Fluidity (see 1333, 182)

Fluorescence depolarization has long been utilized to measure molecular rotational diffusion. When applied to membranes, the rotational motions of some of the probes have often been compared to the measurements made in oils of known viscosity, and quantified in terms of "microviscosity." The units (cgs) are poise. The term "microviscosity" emphasizes the fact that the probe is sensing the resistance to movement only in its immediate surrounding, and it is not a true macroscopic parameter. The meaning of this measurement in terms of molecular motion is not clear, but as is the case with spin probes, the results have proved to be of qualitative value.

Some probes which have been used to measure membrane fluidity are shown in Table 5.1. DPH is the most frequently used probe because it partitions very favorably into membranes, has an intense fluorescence, does not appear to bind to proteins, and is sensitive to the membrane physical state. However, lifetime measurements indicate heterogeneity whose origin is not known, and there is no consensus on the orientation and nature of the motions of the probe on the bilayer (see 29).

Box 5.1 The Depolarization of the Emitted Light Is Used to Measure Molecular Rotation

Figure 5.4 illustrates the principles behind the use of both fluorescence and phosphorescence for measuring molecular rotation. Within about 10^{-11} sec following the absorption of a photon, the electronically excited molecule is in the first excited singlet state S_1. The term"singlet" indicates that there is no net electron spin because no change in electron spin has taken place during the absorption process. Typically it takes about 10^{-8} sec to return to the ground state, ($S_1 \rightarrow S_0$). This can occur by collision with neighbors and loss of the energy as heat, or it can occur by emission of a photon, which constitutes fluorescence. The fluorescence lifetime (τ_F) is the characteristic time for the population of excited molecules to return to the ground state following a flash of excitation light. In some molecules, there is a high probability that an electron spin will change while the molecule is in the excited state,which then yields the triplet excited state. Emission of a photon from this state ($T_1 \rightarrow S_0$) is called phosphorescence, and the lifetime of the triplet state is relatively long (>10^{-3} sec), because of the requirement that the spin state of the electron change along with the emission of the photon.

The key to the depolarization technique is that the probability of absorption and of emission is directional. Light polarized along the z-axis (Figure 5.4B) will preferentially excite molecules oriented with their transition dipole moment in the same direction. The probability varies with $\cos^2\theta$, where θ is the angle between the transition dipole moment and the electric field vector of the light. The same $\cos^2\theta$ dependence is obeyed for the emission of a photon, so a molecule oriented with its transition dipole moment along the z-axis will be likely to emit a photon with the same polarization.

The emitted light is monitored by using polarizers to quantify the intensity of the components parallel ($I_{\parallel}$) and perpendicular ($I_{\perp}$) to the original direction of polarization (see Figure 5.5). The values of $I_{\parallel}$ and $I_{\perp}$ are used to obtain the anisotropy, r:

$$r = \frac{I_{\parallel} - I_{\perp}}{I_{\parallel} + 2I_{\perp}}$$

If no molecular rotation occurs between the time of absorbance and the time of emission, $I_{\parallel}$ will be greater than $I_{\perp}$, and r has a maximal value of 0.4. If the molecules undergo appreciable isotropic rotation within the lifetime of the excited state ($\tau_c << \tau_F$), then the value of r will approach zero, since the molecules will be randomly oriented at the time of emission ($I_{\parallel} = I_{\perp}$). For $\tau_c \cong \tau_F$, the value of r is sensitive to molecular motions. Note that the rates of rotation and of fluorescence emission are given by τ_c^{-1} and τ_F^{-1}, respectively.

In order to separate dynamic information from effects due to constraints on the range of motion, it is necessary to measure the time dependence of the anisotropy, $r(t)$, following a flash of exciting light (Figure 5.5) (see 627). The value of r at any time is a measure of how much the molecules have randomized with respect to the polarization direction defined by the pulse of exciting light. The decay in $r(t)$ from its maximal value, r_0, is given approximately as

$$r(t) = r_\infty + r_0 - r_\infty\, e^{-t/\tau_c} \qquad [5.4]$$

The rate of decay defines a rotational correlation time (τ_c). The limiting value r_∞ will be zero if the molecules are able to become totally random after infinite time. However, if the molecules are constrained to move within a limited range, the molecules will not become random. In this case, r_∞ will be nonzero and will be a direct measure of the constraints imposed by the probe surroundings. The order parameter (627) is obtained directly from these values, and this has the same definition as for ESR or NMR experiments.

$$S = \left(\frac{r_\infty}{r_0}\right)^{1/2} \qquad [5.5]$$

Unfortunately, such time-dependent measurements of r_∞ and r_0 are rare, and most measurements are made in the steady state with continuous excitation and emission. The value of $\bar{r}$ obtained from this is an average value, and it can be shown (see 627) that

$$\bar{r} = r_\infty + (r_0 - r_\infty)\frac{\tau_c}{\tau_c + \tau_F} \qquad [5.6]$$

Consider three cases:

1. For very rapid rotation ($\tau_c << \tau_F$), the result is that $\bar{r} = r_\infty$, the limiting value.
2. For very little or no rotation ($\tau_c >> \tau_F$), the result is that $\bar{r} = r_0$, the maximal value.
3. For cases of interest, $\tau_c \approx \tau_F$ and $\bar{r}$ will depend on both the rate of motion (τ_c^{-1}) and limitations on the range of motion (r_∞).

Without suitable corrections (see 1333, 627) measurements of microviscosity based on values of $\bar{r}$ are not simple dynamic measurements.

Because of the effective viscosity of the bilayer (1-10 poise), the small fluorescent probes like DPH and the ESR probes like TEMPO have rotational correlation times in the range of 10^{-8} to 10^{-9} sec. In water, which has viscosity of 0.01 poise, molecules of this size would rotate at least 100 times faster. Membrane proteins are much larger than isolated probes and as a result rotate much more slowly. Probes attached to these proteins must have excited state lifetimes in the range of 10^{-3} sec in order to be sensitive to protein rotations.

5.3 Rotation of Membrane Proteins (see 211, 366 for reviews)

The rotational diffusion coefficient for rotation of a membrane protein in the plane of the bilayer, $D_{\parallel}$, can be estimated using a hydrodynamic model of a cylindrical

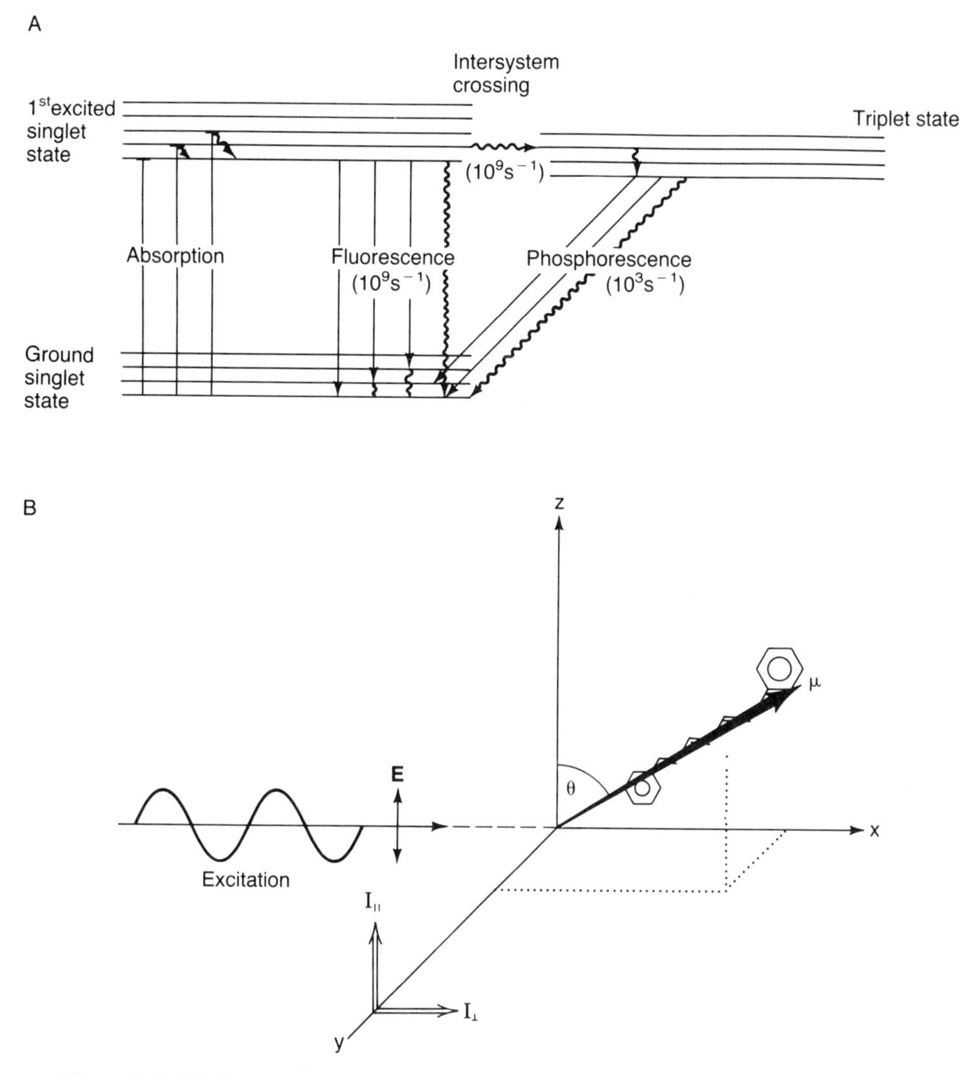

Figure 5.4. (A) Some pathways of relaxation for electronically excited states. The wavy arrows denote processes in which energy is lost without emission of a photon, and the straight arrows denote either photon absorption or emission. Approximate frequencies of the processes are also indicated. Adapted from ref. 182. (B) Geometry for measuring fluorescence or phosphorescence anisotropy. The excitation beam is polarized with the oscillating electric field vector in the z-direction. The probability of absorption depends on $\cos^2\theta$, where θ is the angle between the z-axis and the transition dipole moment vector in the molecule. Note that this angle (θ) is not the same as the angle defined in equation 5.3. Emitted light is viewed along the y-axis with polarizers set to measure the parallel and perpendicular components of the fluorescence or phosphorescence intensity. In the figure, diphenylhexatriene (DPH) is pictured with its transition dipole moment along the long axis of the molecule. Anisotropy measures any change in the angle θ due to molecular rotation occurring between the absorption and emission processes.

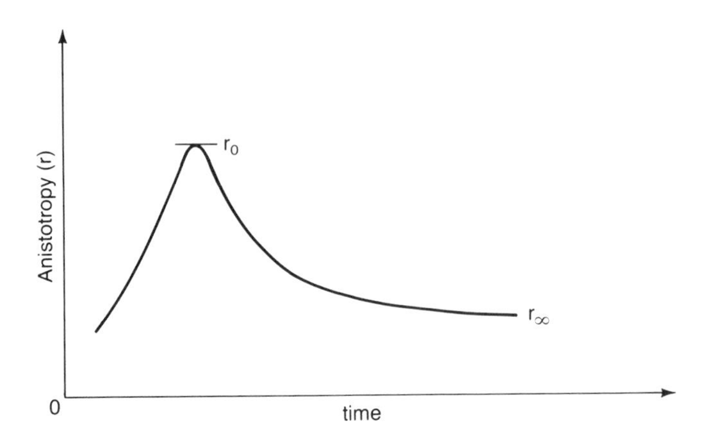

Figure 5.5. Schematic illustrating the time dependence of the anisotropy of fluorescence or phosphorescence. For an isotropically rotating molecule ($\tau_c << \tau_F$), r_∞ will go to zero.

protein rotating about a single axis as pictured in Figure 5.2. The model assumes the membrane has viscosity η and thickness h and the protein has radius a:

$$D_{\parallel} = \frac{kT}{4\pi a^2 h\eta} \quad [5.7]$$

A rotational relaxation time, $\phi_{\parallel} = 1/[D_{\parallel}]$, is frequently reported. For a protein of radius 25 Å, in a membrane with a thickness of 40 Å and viscosity of 5 poise, the predicted value of $\phi_{\parallel}$ is about 35 µsec. The quantitative aspects of this equation describing protein rotation in the bilayer have not been rigorously tested, but the sensitivity of the rotational relaxation time to the effective radius of the rotating species has been demonstrated. This has proved useful for detecting protein aggregation within the membrane (see 211, 1032, 561). The techniques employed to measure protein rotation in the bilayer must be sensitive in the time range of 10^{-5} to 10^{-3} sec. Ordinary fluorescence depolarization is not useful because the lifetime of the excited state is about 10^{-8} sec, during which time the proteins would appear to be static. Three techniques have been used successfully.

(1) To the protein of interest, one attaches a probe which has a long-lived triplet state (Figure 5.4A). If the probe is rigidly bound to the protein, the anisotropy of the phosphorescence can be used to monitor rotational motions of the protein. Eosin derivatives (Table 5.1) have proved useful for this purpose, since the phosphorescence lifetime of eosin is about 2 msec. The principles of the approach are the same as described in the previous section on the fluorescence depolarization technique, except that the molecular model for the type of motion causing the depolarization is different. The time-dependent value of the phosphorescence anisotropy is measured and yields a characteristic decay time. Problems may be encountered when there is heterogeneity in the system, and the resulting multiple exponential decay of the anisotropy is difficult to analyze quantitatively. Problems also result if the probe can rotate locally on the surface of the membrane protein, or if the protein exhibits any segmental flexibility.

Alternatively, the same probes can be used to measure protein rotation by measuring the time dependence of the absorption dichroism (see 496). In this case, changes in orientation of the transition dipole moment are monitored by the absorbance, measured with polarized light parallel and perpendicular to the polarization of the initial pulse of light used to create the excited state population.

(2) In a few cases, molecules naturally form long-lived states upon flash photolysis which can be monitored optically by absorption dichroism. Examples are rhodopsin and bacteriorhodopsin, where the excited states of bound retinal are exploited, and the excited state formed upon the photolysis of cytochrome-CO complexes, used with cytochrome *c* oxidase and cytochrome P450. The measurements can be done in situ, e.g., in the mitochondrial membrane, or with purified protein in phospholipid vesicles.

(3) Although ordinary ESR spectroscopy is insensitive to movements in the range relevant to the rotation of membrane proteins, a specialized technique called saturation transfer ESR (ST-ESR) is sensitive over a wide range of motions extending from 10^{-7} sec to 10^{-3} sec (see 1448 for review). This technique has been applied to measuring the rotation of several membrane proteins, using spin probes covalently attached to the protein (see 211, 366, 1057). A disadvantage of this technique is that the interpretation of spectra resulting from anisotropic molecular motion is difficult.

5.31 Examples of Protein Rotation

A wide range of rotational relaxation times have been obtained for different intrinsic membrane proteins (see 211, 366). Rhodopsin, for example, appears to be freely rotating in the rod outer segment disk membrane, with $\phi_{\parallel} = 20$ μsec. At the other extreme is bacteriorhodopsin, which exists in the purple membrane in an ordered crystalline array (see Chapter 3) and is rotationally immobile. Several purified proteins studied in reconstituted phospholipid vesicles show concentration dependence of $\phi_{\parallel}$, suggesting that they self-aggregate when the lipid/protein ratio decreases. This is true for bacteriorhodopsin (212), cytochrome *c* oxidase (736), Band 3 (1032), Ca^{2+}-ATPase (1057), and cytochrome P450 (561). Self-aggregation can be one of the causes of the heterogeneity which has been observed in several cases, e.g., Band 3 in erythrocyte ghosts and the Ca^{2+}-ATPase from sarcoplasmic reticulum (see 211, 366, 1057). Generally, membrane proteins appear to rotate in the plane of the membrane, and the rotational rate is consistent with the expectations of the simple hydrodynamic model.

Box 5.2 The Rotation of Band 3 and of Cytochrome P450 Can Be Used to Monitor Intermolecular Associations

Two particularly interesting examples are Band 3 (1032, 96; Section 8.33) and cytochrome P450 (561). The studies on Band 3 are of special interest because of the biochemical data suggesting the association of this anion transporter with the spectrin–actin cytoskeletal network. The finding of a substantial fraction of rapidly rotating protein suggested models where not all the Band 3 is immobilized or where the association with spectrin is transient. However, an additional complication appears to be that the lysis of the cells to create erythrocyte ghosts can disrupt the cytoskeletal association. In the intact cell, Band 3 exhibits only slow rotational motions in the range of 0.1 to 1 msec (96). This is more in keeping with the lack of translational motion of Band 3 (see next section), though the physical cause of the motional restraints has not been directly demonstrated.

Cytochrome P450 accepts electrons from cytochrome P450 reductase in the microsomal electron transport chain (see Chapter 6). A question addressed by several laboratories is whether these molecules form a long-lived complex in the membrane or whether they interact by collisions as freely diffusing species. When reconstituted by itself at low lipid/protein ratios, cytochrome P450 self-aggregates (561), which slows the protein rotation. However, reconstitution with a stoichiometric amount of the reductase under similar conditions results in a fully mobile cytochrome P450 with $\phi_{\parallel} = 40$ µsec. This strongly suggests an interaction of these two proteins to form a 1:1 complex, which prevents the formation of the cytochrome P450 aggregates.

5.4 Lateral Diffusion of Lipids and Proteins in Membranes (see 211, 691, 366, 690, 1503 for reviews)

The ability of membrane components to diffuse laterally in the plane of the bilayer is explicit in the fluid mosaic model. Several techniques have been developed to measure lateral diffusion coefficients (see 211, 366). There are three approaches.

(1) Cells which have different cell surface markers are fused to form a heterokaryon. The spreading rates of the surface markers can be followed after the fusion event using fluorescence microscopy with antibodies labeled with different fluorescence probes directed against specific surface antigens.

(2) Membrane proteins can be redistributed by electrophoresis in situ. The proteins are concentrated in one part of the membrane by placing the cells in an electric field, and the lateral diffusion coefficient is determined by following their randomization rate after the electric field is removed. The protein distribution can be monitored by electron microscopy or by fluorescence techniques (see 1363, 517 for examples).

(3) Fluorescence recovery after photobleaching (FRAP or FPR) is by far the most widely used technique. The technique is illustrated in Figure 5.6. A uniformly labeled membrane is bleached in one spot (diameter ~1 µm) using an intense laser beam. The increasing fluorescence from that spot is monitored as a function of time after the bleaching, and the rate of recovery is a direct measure of the lateral diffusion of the fluorescent species from the surrounding medium into the spot. Amphiphilic fluorescent probes, labeled phospholipids, and proteins carrying fluorescent labels can all be studied using this technique. Some of the commonly used probes are illustrated in Table 5.1. The application of video techniques to study the redistribution of fluorescent molecules in the membrane

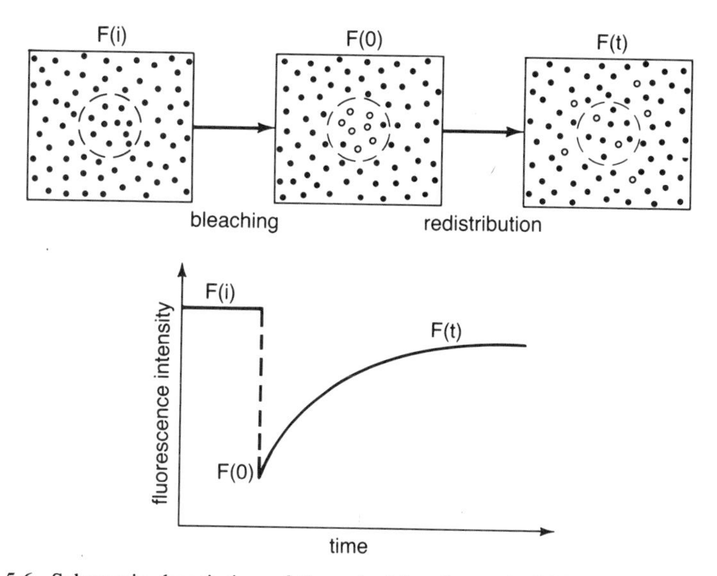

Figure 5.6. Schematic description of the principle of a typical FRAP experiment. The initial fluorescence intensity, F(i), arising from fluorophores in a small area of a membrane surface is measured (upper left) with a nonbleaching intensity of exciting laser light. The next step involves an irreversible photobleaching of a substantial fraction of fluorophores in the same area by using an intensified laser beam. This is followed immediately by monitoring the postbleach fluorescence intensity as a function of time, $F(t)$. The temporal behavior of the fluorescence intensity from the observation area is shown in the trace in the lower part of the figure. Adapted from ref. 1504.

can be expected to broaden the utility of this technique (726, 691). Whereas the first two techniques cause major perturbations to the cell or membrane, the FRAP technique is much less damaging. Although serious questions have been raised about the perturbations or damage caused by the laser beam, these appear to have been addressed convincingly (see 366, 690). Furthermore, this technique can also be applied to model membrane bilayer (see 1504) and monolayer (1403) systems, as well as intact cells or biomembrane fragments. Also, by focusing the laser beam (~1 μm diameter) in different places, the lateral diffusion of components in different regions of the membrane can be monitored.

The translational diffusion coefficients that can be measured using the techniques range from about 10^{-7} cm^2/sec to 10^{-12} cm^2/sec. A diffusion coefficient of 10^{-12} cm^2/sec is considered effectively immobile in these experiments. A value of D of 10^{-8} cm^2/sec, typically measured for lipids in biomembranes, corresponds to a net distance traversed of about 2 μm in 1 sec.

5.41 Theoretical Models of Lateral Diffusion (see 1503 for review)

A model often applied to the translational diffusion of membrane proteins is the hydrodynamic model of Saffman and Delbrück (1273). This considers the protein to be diffusing in a thin viscous sheet, where the solvent is treated as a continuum; i.e., solvent molecules are small compared to the diffusing species. These considerations yield the following relationship.

$$D_{trans} = \frac{kT}{4\pi\eta_m h} \left[\ln \frac{\eta_m h}{\eta_w a} - 0.5772\right] \qquad [5.8]$$

where η_m and η_w are the viscosities of the membrane and aqueous phases respectively, a is the radius of the cylindrical protein, and h is the membrane thickness. This model predicts that the rate of translational diffusion will be very insensitive to molecular size, which is largely verified by the experimental results so far (see Table 5.2).

This hydrodynamic theory does not appear to apply to the diffusion of lipid probes, since these molecules are certainly not properly considered to be moving in a continuum. Variants of the free volume model (e.g., 483) appear to explain the lipid data (658, 1503). The free volume model basically views diffusion as a process whereby molecules move into spaces caused by spontaneous fluctuations, and the critical parameter for membranes is the ratio of the area per lipid compared to the minimal area per lipid under close-packed conditions, i.e., lipid packing density. It was already pointed out that rotational properties of fluorescent probes are correlated to lipid density in the membrane (1333). The reason why

Table 5.2. Lateral diffusion coefficients of some membrane lipids and proteins.[l]

	$D_{trans} \times 10^8$	% mobile
(A) *In Phospholipid Vesicles*		
Lipid probes (diI, NBD-PE) in phospholipid vesicles (see Table 5.1)	1–5	100
NBD-gramicidin	2	100
Glycophorin	2	100
Rhodopsin	1–3	100
Acetylcholine receptor	1–2	100
Band 3	1–2	100
(B) *In Natural Biomembranes*		
Lipid probes	(0.2–2)	100
Nerve growth factor receptor (neurites)	0.06–0.08	22–64
IgE receptor (rat mast cell)	0.03	50–80
Rhodopsin (rod cell disk)	0.4	100
Acetylcholine receptor (myoblast patches)	<0.0003	—
Band 3 (erythrocyte ghost)	0.004	10
(C) *In Cytoskeletal-Deficient Membrane*		
Acetylcholine receptor (membrane "blebs" from motor end plate)	0.3	100
Band 3 (spectrin-deficient erythrocyte)	0.2	—

[l]Data from refs. 366 and 690.

this model should apply at all to either lipid rotational or translational diffusion is not clear, since it was originally formulated for nonassociating solvents, which would not seem to apply to all membrane lipids. In any event, the inverse relation between lipid translational diffusion and the lipid packing density in the bilayer appears to be generally valid (see 1503).

Finally, several approaches have been made toward a theoretical explanation of the influence of membrane proteins on the lateral diffusion of both proteins and lipids in the membrane (see 691 for review). Essentially, the proteins get in the way of diffusing molecules and cause the motion to be directed along channels in between the proteins. This is the archipelago effect (see 1285, 1162, 378). Generally, it appears to be able to account for translational diffusion coefficients being about 10-fold smaller (slower diffusion) in biomembranes than in pure lipid bilayers (see 1503, 1363, 522, 378).

5.42 Examples of Lateral Diffusion of Membrane Components (see 366, 1504, 690, 1503, 691)

The translational diffusion coefficients of some examples of membrane lipids and proteins are listed in Table 5.2. Phospholipid probes and lipid analogues (Table 5.1) have diffusion coefficients of about 10^{-8} cm^2/sec in pure phospholipid multibilayer or large unilamellar vesicles. When the same probes are studied in a wide variety of natural membranes, the diffusion is usually smaller by about 10-fold. This slower diffusion in biomembranes is usually ascribed to the presence of proteins, which can be viewed essentially as getting in the way (see 1503, 1285, 1162, 522, 378). There is no significant difference in the lateral diffusion of phospholipids with different polar headgroups, but gangliosides appear to diffuse more slowly in reconstituted vesicles (521). Of course, when the bilayer is in the gel phase, the rate of lateral diffusion drops dramatically by more than two orders of magnitude ($D < 10^{-10}$ cm^2/sec). Generally, lipids diffuse freely in the fluid bilayer of model membranes and in biomembranes, though there are rare exceptions (1599).

The lateral diffusion of membrane proteins presents quite a different story. Several proteins have been examined in reconstituted vesicles and, although it is premature to generalize, the data (Table 5.2) indicate that the translational diffusion coefficients are very similar regardless of the size of the protein, and the proteins diffuse only slightly more slowly than do the lipids (see 1503, 691). The proteins represented differ widely in the number of transmembrane segments and state of association, yet they diffuse at the same rate. The very weak dependence on molecular size is predicted by the hydrodynamic model of Saffman and Delbrück (1273). In reconstituted vesicles with high protein content, protein aggregation and lateral phase separations can occur which result in effective immobilization of a portion of the protein (see 725).

In biomembranes, the lateral diffusion coefficients of proteins are usually about 100- to 1,000-fold smaller than in the model systems, where the protein concentration is low (i.e., dilution limit). Rhodopsin is an example of a protein which appears to be freely diffusing in the biomembrane, $D \approx 4 \times 10^{-9}$ cm^2/sec. Since the disk membrane in which rhodopsin resides has a high protein/lipid ratio (see Table 1.4), it is clear that the presence of proteins in the bilayer cannot alone account for the generally observed slow diffusion of other membrane proteins. Another frequent observation is that the recovery of fluorescence in the FRAP experiments is not complete and is often less than 75%. This implies that part of the population of the protein being observed is effectively immobile ($D < 10^{-12}$ cm^2/sec).

A question of central importance is what restrains the lateral diffusion of membrane proteins (see 366, 691). It is not likely the aggregation of membrane protein is responsible, since quite large aggregates would be required to explain most of the observations. One possible explanation is that membrane proteins are restrained due to interactions with either the extracellular matrix or the intracellular cytoskeleton. It has been demonstrated that interactions with the extracellular substrate can influence the lateral diffusion of membrane proteins (1052), but other data tend to minimize this as an important factor (see 366, 690, 1073).

Several lines of evidence suggest that interactions between membrane proteins and cytoskeletal elements are responsible for some of the observed restraints on protein lateral diffusion (see 1503, 367, 1331, 1436). Band 3 in normal erythrocyte membranes is mostly immobilized, but in cells which are spectrin-deficient, the lateral mobility increases by at least 40-fold (1331) (Table 5.2). Perturbation of the cytoskeletal network in erythrocyte ghosts also increases the lateral diffusion of Band 3 (1472). Similarly, acetylcholine receptors can be examined in membrane "blebs" which appear to lack actin and the cytoskeletal connections. The mobility of the protein is substantially increased in this system compared to cells with the intact cytoskeleton (1436) (Table 5.2). Despite these experiments, the case for specific cytoskeletal interactions resulting in restrained lateral diffusion of membrane proteins has yet to be fully demonstrated.

In at least one case, it seems that cytoskeletal interactions are not responsible for hindered lateral diffusion. The G protein of vesicular stomatitis virus resides in the plasma membrane of infected animal cells and has a single membrane-spanning segment. Mutational variants of this protein have been expressed in which the cytoplasmic domain has been eliminated (1313). None of the mutations resulted in the rapid lateral diffusion which is characteristic of membrane proteins which have been reconstituted in artificial bilayers. The results rule out direct interactions between this protein and cytoplasmic proteins as the cause of the hindered lateral motion.

Finally, it should be noted that the FRAP technique, which is used for most of these experiments, measures diffusion over a range of several hundred angstroms to 1 μm. A protein may be freely diffusing within a small domain but trapped by the presence of other integral or peripheral proteins. This would still be measured as immobilized by the FRAP method. Also, if the interaction between a membrane protein and cytoskeletal protein were relatively weak and the on-off rates were rapid, then the protein could effectively hop from one binding site to another. The net rate of lateral diffusion would be slowed down since the protein would spend most of the time bound to the cytoskeleton, but rapid lateral diffusion through the bilayer would still occur when the protein moved between binding sites (see 692). Clustering of membrane proteins due to attractive interactions can also result in reduced lateral diffusion (1164).

5.5 Protein-Lipid Interactions (see 328, 1360, 925, 828, 330 for reviews)

Many of the same techniques used to study membrane order and dynamics have been used to study protein–lipid interactions. This has largely been directed at measuring the influence of membrane proteins on the lipid physical state. Consider a typical membrane with a weight ratio of lipid to protein of 1:1 (see Table 1.4). Using an average protein molecular weight of 50,000, this would mean a molar lipid/protein ratio of about 60:1, if only phospholipids were present. By comparison, estimated ratios for the rod outer segment disk membrane and the sarcoplasmic reticulum are 75:1 and 110:1. If the protein is represented as a cylinder extending about 10 Å beyond the bilayer on both sides, its radius would be about 18 Å. A phospholipid in a fluid membrane occupies an area on the order of 60 $Å^2$, corresponding to a headgroup radius of about 4.4 Å, and it would take about 16 of these lipids on each side of the bilayer to completely surround the protein. Hence, about 50% of the lipids at any time would be adjacent to a protein molecule in this model. About 35% of the surface area of this membrane is occupied by protein. Even considering the crudeness of this model, it is clear that physical probes of the lipid physical state should be sensitive to the influences of proteins in biomembranes.

The critical questions which are relevant to understanding of the structure and function of the membrane are: (1) Do intrinsic membrane proteins bind tightly to lipids, and what is the nature of the layer of lipids adjacent to the protein? (2) Do membrane proteins have long-range effects on the order or dynamics of membrane lipids? (3) How do the lipids influence the structure or function of intrinsic membrane proteins? (4) How do peripheral membrane proteins which bind to the bilayer surface interact with the lipids and influence their behavior?

5.51 The Binding of Lipids to Intrinsic Membrane Proteins in the Bilayer

This subject has been the objective of a large research effort (e.g., see 328, 330, 198, 925 for reviews) involving many approaches. In essence, one wants to know whether proteins have binding sites which are specific for particular lipids and whether the protein–lipid complexes can be considered to be long-lived, on the order of the time required for the turnover a typical enzyme, $\approx 10^{-3}$ sec. This is now known in several examples, primarily through the use of ^{2}H-NMR, ESR, and fluorescence methods. In order to appreciate fully the results it is useful to consider the thermodynamics of a simple exchange reaction where a lipid of one type (L_1) is displaced from the binding site on a protein (P) by another lipid (L_2).

$$PL_1 + L_2 \underset{k_{-1}}{\overset{k_1}{\rightleftarrows}} PL_2 + L_1$$

The equilibrium constant is given as

$$K = \frac{[PL_2][L_1]}{[PL_1][L_2]}$$

where K is the relative binding constant for lipids L_1 and L_2 binding to the protein site(s),

$$\text{and } \Delta G^\circ = -RT \ln K$$

If the two kinds of lipids are present at the same concentration in the membrane ($[L_1] = [L_2]$), and if the affinity of the two lipids for the protein is the same, then $K = 1$. Consider a hypothetical case where L_2 is a minor lipid component, comprising only 5% of the total lipid. In this case,

$$[L_1]/[L_2] = 19/1$$

Suppose that L_2 binds preferentially to a protein binding site so that 90% of the binding site is occupied by L_2 at equilibrium; then

$$[PL_2]/[PL_1] = 9/1$$

Hence, $K = 171$, which corresponds to $\Delta G^\circ = -3.1$ kcal/mol. Only a small difference in binding free energy is sufficient to make a large difference in the equilibrium distribution of lipids binding to the protein. This is because the effective concentration ratios between competing lipids within the membrane are relatively small, within a factor of 100 and in most cases much smaller. In other words, even minor lipids are present in the membrane at concentrations that are no less than 1% of the concentration of the major lipid species with which they are competing for binding sites on proteins.

There are two approaches used to measure the relative affinity of lipids binding to specific membrane proteins (see 828 for critical review). These involve the use of lipid analogues in reconstituted phospholipid vesicles containing the protein of interest.

(1) Spin-labeled phospholipids are motionally restricted when they are adjacent to membrane proteins. The result is a component in the ESR spectrum which is broadened, similar to those shown in Figure 5.3. For probe molecules adjacent to proteins, the motions characterized by frequencies $\geq 10^8$ sec^{-1} are substantially reduced. The experimental spectrum can be analyzed as the sum of two components, a rapidly tumbling species in the "bulk" lipid phase with a sharp spectrum, and a motionally restricted component adjacent to the protein. In order for a substantial portion of the experimental spectrum to originate from the "bound" species, it is necessary that the protein/lipid ratio be sufficiently high. The potential

for lipids to be "trapped" in protein aggregates in such systems is a complicating factor. Trapped lipid probes would also be motionally restricted and can represent a third component of the ESR spectrum. Lipid spin probes covalently attached to the protein surface can be used to detect and distinguish such trapped species at high protein concentrations (see 328, 330, 297, 550).

(2) Spin-labeled and brominated lipid derivatives have been shown to be able to quench the intrinsic tryptophan fluorescence from membrane proteins. The efficiency of quenching depends on the distance between the lipid derivative and the fluorescent tryptophans (direct contact may be required), so the presence of the lipids in the layer immediately adjacent to the protein will result in quenching of the fluorescence intensity from the protein. The relative ability of these lipids to bind to a protein in the presence of different competing lipids can be monitored by measuring the protein intrinsic fluorescence.

Analysis of Experimental Results (see 328, 330, 766, 151, 925, 828)

The use of spin-labeled lipids to quantify the relative affinities of different lipids to membrane proteins has been analyzed using relatively simple multiple equilibrium models (766, 151; see 328, 828 for reviews). The methods rely on measurements of the fraction of the spin probe which is bound as a function of the total lipid per protein. In principle, the technique could be used to quantify the binding of different classes of bound or trapped probe (e.g., strong and weak binding), but this is difficult in practice. Generally, it is assumed that all the binding sites are identical. One way to do these experiments is to study the binding of a small amount of a spin-labeled lipid of one type in the presence of a large amount of lipid of another type. Spectra are taken from samples with different lipid/protein ratios. Assuming N identical binding sites, one can show (see 44):

$$\frac{[\text{Probe}]_{\text{free}}}{[\text{Probe}]_{\text{bound}}} = \frac{[\text{Total lipid}]/[\text{Protein}]}{NK} - \frac{1}{K} \qquad [5.9]$$

The spectra yield the ratio on the left as a function of the ratio of [total lipid]/[protein], and if the two ratios are plotted, the predicted straight line will yield N and K, the total number of binding sites and the relative association constant of the probe compared to the unlabeled lipid. If the probe is a spin-labeled derivative of the lipid which is serving as the unlabeled solvent, K should equal 1, which is what is found experimentally.

The use of fluorescence quenching can also be analyzed, using similar assumptions, to yield the value of K (363, 872, 925). The potential for heterogeneity of the binding sites, both in their intrinsic affinities for lipids and in the ability of the bound lipids to quench the tryptophan fluorescence, can be a complication in interpreting the results. The fluorescence technique has the advantage that high lipid/protein ratios can be used so that protein aggregation is less of a problem than with the ESR technique.

It is important to realize that neither of the methods used to quantify the relative binding constants of lipids to membrane proteins will be likely to detect a small number of sites with high affinity, so the fact that such sites are not found experimentally by these methods does not mean that they do not exist.

Box 5.3 Some Proteins Exhibit Selectivity in Binding to Phospholipids with Different Headgroups (see 828 for review)

The lipid binding selectivity of only a small number of proteins has been examined, including mitochondrial cytochrome *c* oxidase, Na^+/K^+-ATPase, Ca^{2+}-ATPase from the sarcoplasmic reticulum, and vertebrate rhodopsin. Generally, the data reveal only weak selectivity between lipids, with the largest value of K in the range of 5. This is still sufficient to make a substantial difference in which lipids are bound to particular proteins, depending on the concentrations of the different lipid species present in the membrane.

(1) Rhodopsin (see Figure 4.1) binds to about 24 lipid molecules per molecule of protein (i.e., $N = 24$), with very little preference between cardiolipin and phosphatidylcholine (926).

(2) Na^+/K^+-ATPase from *Squalus acanthus* (see 151, 395; Sections 6.5 and 8.41) binds about 60 phospholipid molecules, with a preference for negatively charged lipids compared to phosphatidylcholine: cardiolipin ($K = 3.8$), phosphatidylserine ($K = 1.7$), and phosphatidic acid ($K = 1.5$). Phosphatidylethanolamine and phosphatidylglycerol bind with the same affinity. The preference for acidic lipids cannot simply be due to electrostatic effects (396). The biochemical significance of the preferential lipid interaction is not known (see 395).

(3) Cytochrome *c* oxidase (see Section 8.51) binds between 40 and 55 lipids (550, 766, 926, 1171). This enzyme has a definite preference for binding cardiolipin compared to phosphatidylcholine ($K \approx 5$). In most, but not all preparations, cardiolipin is reported to copurify with this enzyme when isolated from mitochondria (see 1317). The suggestion that the enzyme has a small number of high-affinity sites for this lipid and that it is specifically required for catalytic activity has been disputed (1171). Additional studies appear necessary to clarify the significance of this binding preference. Note that since cardiolipin is a dimeric phospholipid, there are problems in the analysis of competitive binding experiments (828).

(4) Ca^{2+}-ATPase (see Sections 6.5 and 8.41) has received considerable attention because of the suggestion that the composition of the so-called "lipid annulus" or "boundary lipid" which is immediately adjacent to the protein is what determines the enzyme activity (see Section 6.5). On the basis of the insensitivity of the enzyme activity to the cholesterol content in reconstituted vesicles, and the fact that pure cholesterol will not support activity, it was postulated that cholesterol is excluded from this layer of lipid (1551). Binding studies, however, show that this is not the case, but that the relative association constant of cholesterol to the protein is about half of that for phosphatidylcholine (1344, 1342). There is little cholesterol present in the natural membrane in which this enzyme is located. The enzyme shows little variation in relative binding affinities for phospholipids with different polar headgroups (363) or with a wide variety of acyl chains (229), although there is a wide variation in enzyme activity when the enzyme is reconstituted with different lipids (229, 364).

Exchange Rate of Boundary Lipid with Free Lipid

The term "boundary lipid" historically has been associated with the idea of a static layer of lipid bound to the surface of a protein such as cytochrome oxidase. Static implies nonexchanging on the time scale of enzyme turnover, $\sim 10^{-3}$ sec, so the catalytic species would be viewed as a long-lived complex. In all cases examined so far, this is most certainly not the case, but the term boundary lipid remains a useful descriptive term for the nearest lipid neighbors to the protein whose structure and dynamics would be most likely to be perturbed by the interaction with the protein.

The application of ^{2}H-NMR has clearly shown that this boundary lipid is rapidly exchanging with other lipids in the bilayer. Whereas ESR spectra of labeled phospholipids are interpretable in terms of a bound and free components, the analogous experiments with deuterated lipid probes show very little perturbation of either the lipid order or dynamics due to the presence of proteins such as cytochrome oxidase (see 328, 1360). The explanation that is largely accepted for the dramatic difference in results is that it is due to the time scale difference in the sensitivity of the two techniques, as described in Section 5.1. The boundary lipid is apparently exchanging on a time scale of 10^{-6} to 10^{-7}sec, so ^{2}H-NMR sees only a single species, whereas in ESR experiments two distinct populations can be discerned.

Given the lateral diffusion coefficient of phospholipids in the bilayer (Table 5.2), the average time for a molecule to hop from one lipid position to a neighboring position in the bilayer (one molecular diameter, ~10 Å) is about 10^{-7} sec. The rapid exchange observed between the boundary lipid and free lipid ($\sim 10^{-7}$ sec) (1265) suggests that, in the cases examined, there is little preference for the lipid to be adjacent to the protein or surrounded by other lipids. It should be stressed that this is only a crude estimate and that only a relatively small number of proteins have been examined, primarily with derivatives of phosphatidylcholine.

5.52 Perturbations of the Lipid Bilayer Due to the Presence of Integral Membrane Proteins. (see 328, 198 for reviews)

Most ^{2}H-NMR experiments with deuterated phospholipids demonstrate that the presence of protein has little effect on either the order parameter of the lipids in the bilayer or the lipid dynamics, as measured by relaxation times. Consistent data have been obtained with ^{1}H-, ^{19}F-, and ^{31}P-NMR (see 328, 381, 893). There are reports that boundary lipid can be detected in some natural membranes by ^{31}P-NMR, suggesting immobilization of the polar headgroup by long-lived (10^{-3} sec) protein-lipid complexes (see 13), but this has been disputed (see 381) and is not generally found. The overall view resulting from NMR experiments is (1) that the exchange rate between boundary and free lipids is rapid (10^7 sec^{-1}), (2) that the order parameters of the bound lipid (i.e., *trans*/*gauche* configurations of the acyl chains) are barely affected by being adjacent to proteins, (3) that the dynamics of the acyl chain reorientations are slowed only slightly (~20%) in the frequency range of 10^9 sec^{-1} (but see ref. 960), and (4) that the orientation and dynamics of the polar headgroups are similarly unaffected in any substantial manner by being adjacent to transmembrane proteins (see 328). There are subtle changes in lipid order and dynamics which are seen by NMR, ascribed to the "roughness" or "rigidity" of the protein surface in contact with the lipid, but these are relatively small effects.

A number of other techniques have been applied to the question of how integral membrane proteins influence the lipid bilayer. In view of the rather benign picture derived from NMR techniques, it is useful to review briefly some of these results.

(1) Spin-labeled lipids are clearly restricted in their motion when adjacent to proteins and when trapped in lipid–protein aggregates (e.g., 1128). Most NMR relaxation studies do not show similar dramatic effects in the same frequency range (>10^8 sec), suggesting that the effects may be due to the presence of the bulky nitroxide spin label. However, in at least one case both NMR and ESR experiments showed a decrease in acyl chain motion in this time range due to protein–lipid interactions (960). There is no consensus as to whether the effects of protein on the motion of the spin probes extend beyond the boundary layer. Spectroscopic data have been used as evidence for the existence of a layer of lipid beyond the boundary layer which senses the presence of the protein (e.g., 766) and have been analyzed simply in terms of a single layer of perturbed lipid (e.g., 1166). In any event, it is important to keep in mind that large motional restrictions are not usually observed by NMR techniques.

(2) Fluorescence depolarization of membrane probes such as DPH (Table 5.1) is sensitive to the presence of integral membrane proteins. The apparent microviscosity increases progressively as more protein is incorporated in the bilayer, using steady-state polarization techniques. Simple modeling has been used to fit the experimental data with the assumption that only those probes which are adjacent to the protein are motionally restricted, i.e., have a high anisotropy (e.g., 1166). However, a detailed analysis including time-resolved anisotropy measurements has indicated in one case that (a) the effect of protein is largely due to an increase in the order parameter, i.e., restricted range of motion, and (b) the effect extends beyond the boundary layer (1208). Again, effects observed with these probes do not necessarily reflect similar alterations in the dynamics or order of the phospholipids.

(3) Fourier transform infrared spectroscopy can be used to measure the acyl chain configurations (see Chapter 2) and has been used to examine the effects of proteins on lipids. The results are consistent with the ^{2}H-NMR data, indicating only small changes in the acyl chain order (*trans/gauche* configurations) due to the presence of protein (e.g., see 34). The nature of the effects which are observed is dependent on the identity of the acyl chain, so general models may not be easily applicable (34). The use of deuterated lipids allows different lipids to be selectively studied in binary lipid mixtures and has demonstrated a preferential interaction between glycophorin and phosphatidylserine (968).

(4) Differential scanning calorimetry (see Section 2.41) has been used to monitor the thermotropic phase transition of lipids and the influence of membrane proteins on this transition (see 949 for review). Generally, the presence of an integral membrane protein results in (a) a slight decrease or no effect on the transition midpoint temperature, (b) an increase in the breadth of the transition, and (c) a decrease in the molar enthalpy change required to complete the transition (see 840). In the presence of protein (e.g., 840), when only one transition is observed, this is ascribed to lipid which is unperturbed by the protein. The presence of protein sequesters lipid and essentially removes it from participation in the phase transition. This results in the decrease in the molar enthalpy change (ΔH°) for the transition due to the presence of protein. The decrease in ΔH° as a function of protein content is usually linear, easily yielding a value for the number of lipids "bound" per protein molecule. These values range from about 20 for bacteriorhodopsin (24) to 685 for Band 3 (213). These numbers do not simply reflect boundary lipid, but include lipid which is trapped in protein aggregates, sequestered in protein-rich domains, or possibly perturbed due to interactions with glycoprotein carbohydrate chains (e.g., 213, 891, 1165). The complications due to lateral phase separations and protein aggregation make it difficult to use this approach to obtain detailed information about protein–lipid interactions. However, the effects of proteins on the thermotropic transitions of binary lipid mixtures have been used to demonstrate preferential interactions between glycophorin and phosphatidylserine (968) and between cytochrome *c* oxidase and cardiolipin (1326).

In some cases, additional phase transitions can be identified which are ascribed to lipids under the influence of the protein. These may be trapped in a protein-rich domain, but could conceivably be due to an effect by the isolated protein molecules in a region of lipids beyond the boundary layer (see 840).

Box 5.4 Phospholipid Polymorphism Can Be Influenced by Proteins

It should be noted that some proteins have been shown to have a strong influence on phospholipid polymorphism, i.e., stabilizing lamellar or hexagonal forms of the lipid. This is of interest in view of the hypothesis that inverted hexagonal cylinders (Chapter 2) or lipidic particles play a role in transbilayer lipid movement and membrane fusion (see 265). The hydrophobic polypeptide gramicidin A will convert dioleyl phosphatidylcholine from a bilayer to hexagonal H_{II} form when the lipid/protein molar ratio is 10:1 (749; Section 3.71). The mechanism apparently involves aggregation of gramicidin and subsequent dehydration of the lipid which stabilizes the nonbilayer form. In contrast, glycophorin stabilizes the bilayer form of dioleyl phosphatidylethanolamine, which is normally in the H_{II} hexagonal form (1439). Positively charged proteins such as cardiotoxin, cytochrome *c*, and even poly-L-lysine will influence the polymorphic form of cardiolipin (see 74). Cardiotoxin interacts by both electrostatic and hydrophobic interactions, which appears to be the case with other membrane-surface-binding proteins (See section 5.53).

5.521 Possible Role of Elastic Distortions in the Bilayer Due to Proteins (see 1269 for review)

The perturbations described in the previous section can largely be explained as the result of the lipid acyl chains conforming to the shape of the protein surface and being somewhat constrained in certain rapid motions as a result of being near the protein. These effects would be expected to be short range, not extending far beyond the boundary layer in the bilayer. It is interesting to consider the consequences of more complex interactions in which the proteins inserted in the bilayer require more dramatic adjustments in the surrounding lipids. Some are pictorially shown in Figure 5.7. In principle, strains introduced into the bilayer can have long-range influences in the bilayer, resulting in reorganization of lipid and/or protein components for maximum stability. Several theoretical approaches have been described (see 1269, 1131, 1025), but there are few experimental tests (see 1141).

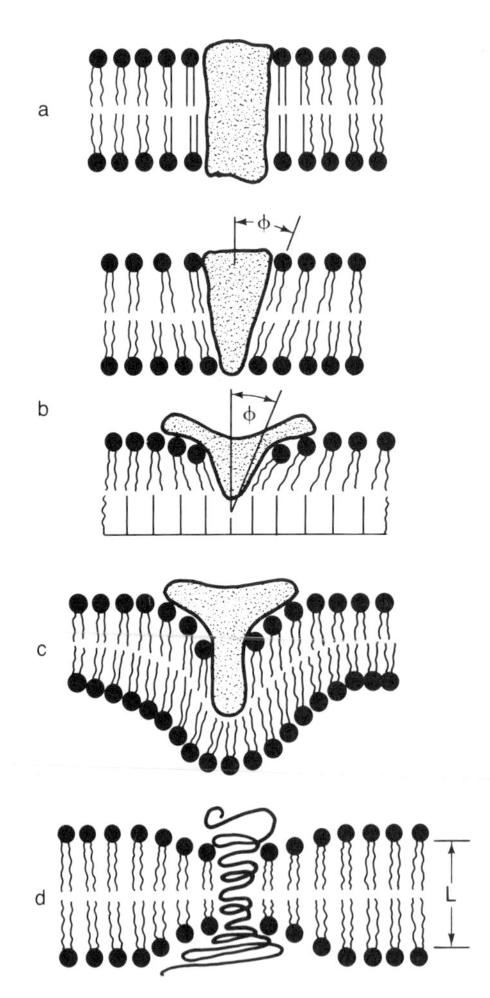

Figure 5.7. Some possible protein-induced perturbations of the lipid bilayer. (a) Local change in order parameter. (b) Elastic distortion by wedge-shaped protein or by protein which partially penetrates the membrane. (c) Change in local curvature of the bilayer. (d) Compression (or dilation) of the bilayer due to mismatch in the length of the hydrophobic regions of the lipids and protein. Adapted from ref. 1271.

(1) A wedge-shaped protein or one which penetrates only one monolayer (Figure 5.7b) could cause the lipids to tilt in one or both leaflets of the bilayer, depending on whether the protein extends all the way across the membrane. This tilting could be propagated well beyond the protein surface and influence the lipid interactions with other membrane proteins in this region. Reorganization of the lipids could help relax this strain. For example, lipids with a relatively small polar headgroup (cone shape, Figure 2.18) might pack favorably around a protein, which causes the polar headgroup region of the bilayer to be expanded more than the interior of the bilayer. Part of the advantage of having membranes composed of many lipid structures may be to optimize the lipid packing around individual membrane proteins, in order to relieve potential deformations. This would minimize packing defects which could form at the protein–lipid boundaries.

(2) Induced lateral curvature is another possible deformation which could result from a protein as pictured in Figure 5.7c. The loose binding form of cytochrome b_5 (see Section 4.22) might be an example of a protein which does this.

(3) A mismatch between the size of the hydrophobic region of the bilayer and the membrane thickness would be another source of bilayer deformations. Presumably either the protein or lipids would adjust to prevent the exposure of any hydrophobic surface to water. If the protein is not deformed to make the adjustment, several alternatives are possible. (a) Either the lipid acyl chains or protein can tilt with respect to the normal to the bilayer, depending on whether the bilayer is too thin or too thick. This has been experimentally tested in one case, but not observed (1021). (b) The lipid acyl chains could become deformed. (c) In a heterogenous lipid mixture, the lipids could rearrange so that lipids which are the correct length could interact with the protein, as indicated in Figure 5.7d.

Note that the elastic deformations can, in principle, cause specific lipids to bind to particular proteins to minimize deformational strain in the bilayer, by shape or size matching and not due to specific chemical interactions. There is no evidence that this actually occurs, however. Also, in principle, the long-range deformations in the bilayer could affect protein–protein associations. This has been examined experimentally by looking at the effects of bilayer thickness on the distribution of bacteriorhodopsin and rhodopsin in reconstituted phospholipid vesicles using electron microscopy (see 1131, 1132, 851). Dark-adapted rhodopsin does exhibit a lipid-dependent self-association which may be the result of lipid tilt induced by incorporating the enzyme in a bilayer which is too thick for ideal packing around the protein (1132).

5.53 Backbone and Side Chain Dynamics of Membrane Proteins

Solid-state NMR techniques have the potential to yield detailed information about the dynamics of individual amino acid residues within a membrane protein. However, the techniques can require large amounts (100 to 200 mg) of isotopically labeled protein and are most informative when applied to small proteins where spectroscopic assignments are possible. Studies on the coat proteins of filamentous bacteriophage (e.g., fd, M13) illustrate the power of these methods (842, 621). The proteins are inserted into the *E. coli* cytoplasmic membrane during infection (see Figure 10.9) and contain a single membrane-spanning helix. Upon phage assembly, lipids and host proteins are excluded from the phage coat. Small size, abundance, and ease of preparation make these proteins a choice for NMR as well as other studies.

^{2}H- and ^{15}N-NMR have been used to examine the dynamics of the residues in the 50-residue fd coat protein in reconstituted phospholipid bilayers. The results showed that the polypeptide backbone is relatively rigid for residues 5 to 43, including but not bounded by the region thought to be within the lipid bilayer. Several residues at either end of the polypeptide are free to undergo large-amplitude motions. Most of the side chains are free to undergo large-amplitude rapid motions, such as 180° ring flips for Phe and Tyr residues, even when they are within the bilayer.

The number of studies using these techniques has been too few to yield any general insights into how lipid–protein interactions influence the internal dynamics of membrane proteins.

5.54 The Binding of Peripheral Membrane Proteins to the Lipid Bilayer (see 328, 116, 330 for reviews)

Although much of the research effort in the area of protein–lipid interactions has been directed at transmembrane proteins, there is considerable and growing interest in the details of how peripheral membrane proteins bind to and influence the bilayer. Many peripheral membrane proteins bind to the membrane primarily through interactions with integral membrane proteins. But there is a diverse group of proteins which interact directly with the surface of the lipid bilayer. Some, such as myelin basic protein (1351, 1413), spectrin (912), and the matrix protein (1581) of vesicular stomatitis virus, appear to have mainly structural roles. A number of soluble proteins can bind to the bilayer surface transiently or under specific conditions. There are several examples where the membrane binding is necessary for enzymatic activity, notably protein kinase C (586), several blood clotting factors (818, 945, 773, 1180), and pyruvate oxidase (536) (see Chapter 6). Still other examples of important protein interactions with the surface of the bilayer are provided by the amphipathic peptide hormones (see Section 3.71) and possibly the signal sequences (e.g., 394, 147) required during biosynthesis to direct secreted or membrane proteins to their proper destination (Section 10.32).

The lipid binding appears to be of two general kinds, which need not be mutually exclusive. (1) Binding may be mediated by an amphipathic secondary structural domain in the protein, usually an α-helix. The secondary structure may be induced and stabilized by the lipid interaction. (2) Binding may be primarily electrostatic, involving a positively charged region of the protein and acidic phospholipids. This kind of interaction is also likely to involve a substantial hydrophobic component, depending on how much the protein penetrates the surface of the bilayer. In a number of cases, Ca^{2+} is required to bind to acidic phospholipids, but the exact role of the divalent cation, e.g., cross–bridging to acidic protein residues, is not exactly defined (e.g., see Section 6.73).

A number of techniques have been applied to studying peripheral membrane proteins interacting with phospholipids. The binding to vesicles can be monitored by light scattering (e.g., 818) or by changes in protein fluorescence (e.g., 955) and can be quantitatively analyzed to yield a dissociation constant (e.g., 955). The perturbation of the bilayer due to protein can be demonstrated by changes in vesicle permeability (e.g., 201, 198) or by changes in the thermotropic phase transition of the lipids (e.g., 1180, 188), though it is difficult to interpret these results in terms of molecular detail. Monolayers have also been useful (e.g., 945, 773), and the degree to which the protein penetrates the monolayer can be estimated by the changes in monolayer surface area resulting from the protein–lipid interaction. For detailed information at a molecular level, one of the most valuable approaches has been NMR.

The effects on the lipid bilayer caused by the interaction with the myelin basic protein and cytochrome *c* have been examined in detail by NMR (see 1351, 329). Both of these proteins interact with acidic lipids primarily by electrostatic interactions, though the physiological relevance of the interaction between cytochrome *c* and lipids is not clear. In contrast to the consistency of the results obtained with different transmembrane proteins, the details elucidated so far with these two peripheral proteins indicate that there are substantial differences in how specific peripheral membrane proteins and phospholipids interact. The myelin basic protein specifically interacts with dimyristoyl phosphatidylglycerol when studied with vesicles also containing phosphatidylcholine (1351). Fourier transform infrared spectroscopy shows that upon binding to phosphatidylglycerol, the protein adopts a highly ordered secondary structure, largely ß-sheet, which is not present in the absence of the lipid (1413). The protein–lipid interaction results in substantial changes in the headgroup of the acidic phospholipid as monitored by NMR (1351). Similar studies with cytochrome *c* show that binding causes little perturbation of the headgroup of phosphatidylserine, as monitored by ^{2}H-NMR, using vesicles also containing phosphatidylcholine. In both systems, no lateral phase separation of the lipids was observed and there was rapid exchange ($>10^5$ sec^{-1}) between protein-bound and free lipids. Cytochrome *c* interacts, however, very differently with other acidic phospholipids (see 329). For example, it causes lateral phase separations in vesicles containing both cardiolipin and phosphatidylcholine and stabilizes nonbilayer lipid forms in the vesicles containing cardiolipin and phosphatidylethanolamine.

In summary, there seems to be a rich diversity of interactions between peripheral membrane proteins and the phospholipid bilayer.

5.6 Chapter Summary

The function of biological membranes cannot be fully appreciated without considering the dynamics of the individual components. Spectroscopic techniques have been developed to measure the rates at which the lipid and protein components rotate within the membrane and the rates at which they diffuse laterally in the plane of the membrane. These methods largely rely on the use of spin probes or fluorescence probes incorporated in the membrane or attached to particular proteins. Generally, membrane lipids can diffuse freely within the plane of the membrane at rates comparable to those measured in model membranes. Integral membrane proteins in biological membranes, in contrast, are often hindered in their lateral motion. The hindered lateral diffusion is likely to be due to associations with other membrane proteins, or with elements of the cytoskeleton or extracellular matrix, depending on the particular protein being examined. Many proteins have been shown to be able to rotate freely in the plane of the membrane, but this motion can also be hindered by protein–protein associations.

At any instant in time, a substantial fraction of the lipids in a biological membrane are adjacent to protein. The layer of lipids immediately surrounding a protein is called the boundary layer or the lipid annulus. However, these lipids exchange rapidly with the bulk lipid in the bilayer (10^7 sec^{-1}) and, generally, there is little preference for a lipid to be in the bulk or adjacent to protein. Individual proteins do bind to lipids with some selectivity, but among the cases examined so far the maximum differential binding comparing the affinities of various lipids to the boundary layer is about 5. Though this is modest selectivity, it does demonstrate that the lipids surrounding a particular protein will not necessarily have the same composition as the bulk membrane.

Chapter 6
Membrane Enzymology

6.1 Overview

In the previous chapter, the membrane is described as a physically dynamic structure. In this chapter, the biological activity of membranes is emphasized. The biomembrane is not simply a passive structure whose function is to delineate and enclose aqueous compartments. A brief consideration of the kinds of enzymes which are membrane-bound will illustrate the extraordinary diversity of catalytic activities associated with membranes.

(1) *Transmembrane enzymes involved in coupled reactions on opposite sides of the membrane*: Typical enzymes in this class have multiple active centers. Prominent examples are redox enzymes such as the plant and bacterial photosynthetic reaction centers or mitochondrial cytochrome *c* oxidase, where catalytic active centers on opposite sides of the membrane are coupled by electron flow generating a voltage across the membrane. Many receptors can also be considered in this class of membrane enzymes, in the sense that the binding of a ligand (a hormone, for example) to an extracytoplasmic domain of the receptor results in changes in the cytoplasmic domain which, in turn, initiate the cellular response. In this case, information is being transferred across the membrane and not charge or solute molecules. Several receptors have been shown to be protein tyrosine kinases (see Section 9.7) and are, therefore, membrane enzymes with catalytic activity. Many membrane receptors do not catalyze chemical reactions per se, and should not be considered enzymes. Receptors are discussed in Chapter 9.

(2) *Transmembrane enzymes involved in solute transport*: Many membrane proteins are involved in the transport of molecules across the bilayer. Active transport may be linked to the hydrolysis of ATP, as in the Ca^{2+}-ATPase of the sarcoplasmic reticulum (Sections 6.5, 8.41). Alternatively, active transport of a solute can be driven by ion gradients, as in the case of the lactose permease of

E. coli, where lactose uptake is coupled to that of protons and is driven by a gradient in the proton electrochemical potential (Section 8.32). Proteins involved in solute transport are discussed in Chapter 8 in more detail.

(3) *Enzymes participating in electron transport chains*: The most prominent examples in this category are the mitochondrial respiratory chain terminating in cytochrome *c* oxidase, the microsomal electron transport systems, involving cytochrome P450 and cytochrome b_5, and the photosynthetic electron transport chain in thylakoids. The localization of the components of the electron transport chains on the membrane concentrates them, allowing for more rapid intermolecular electron transfer rates. A major question is the extent to which the components of the respective electron transfer chains exist as long-lived complexes, as opposed to freely diffusing species in the plane of the membrane (see Section 6.6).

(4) *Enzymes that utilize membrane-bound substrates*: This class would include those enzymes involved in the metabolism of membrane components, including phospholipids, glycolipids, polyisoprenoid compounds, and sterols, as well as enzymes involved in the processing of membrane proteins and secreted proteins. In most cases, these enzymes are integral membrane proteins, but some, such as phospholipases, are soluble proteins which are bound to the membrane only transiently (see Section 6.7; Chapter 10). Interesting examples are the leader peptidase of *E. coli* (see Section 10.33) and the phospholipase C from *Trypanosoma brucei* which specifically releases proteins bound to the membrane by a glycosylphosphatidylinositol anchor (623).

(5) *Enzymes that utilize water-soluble substrates*: Many membrane-bound enzymes use soluble substrates. In some cases, the membrane localizes the enzyme in a region where the substrate is highly concentrated. An example of this is the acetylcholinesterase, which is probably anchored to the post-synaptic membrane by a covalent linkage to a phosphatidylinositol glycolipid (1104; Section 3.8). This enzyme is required for the rapid hydrolysis of acetylcholine. The intestinal microvillar membrane contains numerous enzymes which serve to digest starch and proteins (see 1080), and which are anchored to the membrane by a domain in the amino-terminal part of the polypeptides. Presumably, the presence of these digestive enzymes on the membrane creates a locally high concentration of solute molecules which can be subsequently absorbed. Two enzymes in this category are sucrase–isomaltase (666) and maltase–glucoamylase (1080).

(6) *Enzymes localized in membrane-bound complexes to facilitate substrate channeling*: The membrane can serve as an organizational framework to which peripheral enzyme components can bind to form a multienzyme complex. There is circumstantial evidence that the mitochondrial matrix enzymes involved in the Krebs cycle may be bound to the membrane in such a way that the product of one enzyme is "channeled" to become the substrate of another enzyme which is within the multienzyme complex. This would presumably facilitate the metabolic conversions. Because the complexes are weak, it has been difficult to demonstrate definitively. This is discussed in Section 6.72.

(7) *Enzymes that shuttle between cytosol and the membrane and whose activities are modulated by membrane binding*: This is a newly recognized group of membrane enzymes. These enzymes may bind directly to the phospholipid bilayer

surface, or may have specific receptor proteins to which they bind. In most examples, the enzymes are activated when bound to the membrane, but inactivation can also occur. Prominent examples of enzymes which are activated are *E. coli* pyruvate oxidase, protein kinase C, and several enzymes involved in the blood coagulation cascade (see Section 6.7).

The purpose of this chapter is not to provide a survey of reaction mechanisms for various membrane enzymes, but rather to point out special considerations which must be taken into account in understanding and in dealing with membrane enzymes. These include, foremost, the crucial role of the lipid environment in the activity of membrane enzymes. This factor can severely complicate the interpretation of kinetic data and has led to considerable efforts to learn how to reconstitute purified integral membrane enzymes with phospholipids so as to mimic the in vivo environment or at least place the enzyme in a defined environment.

6.2 Some Special Considerations Relevant to the Activity of Membrane Enzymes

By definition, the study of membrane enzymes is a study of heterogeneous catalysis. The enzymes are not present in a continuous, isotropic medium, but are localized in a biomembrane, micelle, vesicle, or other membrane model. The enzymes are sensitive to their local environment, which may be very different from the average bulk environment in the solution. Furthermore, in order to interact, the enzyme and a membrane-bound substrate must be present in the same membrane or vesicle, and there are often partitioning problems to consider in analyzing the kinetic properties of membrane enzymes. Several considerations relevant to the kinetics of membrane enzymes both in situ and in model systems are indicated below.

(1) *Enzyme and substrate partitioning*: The enzyme and substrate must be able to interact. Consider a purified integral membrane enzyme studied in a detergent solution with a nonpolar lipophilic substrate [e.g., ubiquinol:cytochrome *c* oxidoreductases (1566)]. Both the enzyme and substrate are solubilized in detergent micelles, but they must be present in the same micelle in order to observe catalysis. Excess detergent could ensure that the substrate and enzyme are not present in the same micelle, and the rate of transfer of the substrate into the enzyme–detergent micelles could conceivably become rate limiting. The enzyme velocity depends on the surface concentration of the substrate in the micelle and not on the bulk concentration (see Figure 6.7, for example). Another problem frequently encountered with very hydrophobic substrates is that they may not be totally solubilized in the micelles or membrane vesicles, but a fraction may form clumps or microscopic crystals which are not accessible to the enzyme. A few enzymes can be assayed in inverted micelles, which are water-containing structures dispersed in organic media (631), but this is unusual.

Other problems in assaying membrane enzymes arise when either the enzyme or the substrate exists in equilibrium between membrane-bound and soluble forms. Examples are interfacial enzymes such as lipases or the blood coagulation factors (see Sections 6.73, 6.74). To understand the kinetics of these systems, it is necessary to know which enzyme forms exist under assay conditions and the extent to which each form is catalytically active. In all these cases, the meaning of V_{max} and K_m values may be quite different than for enzymes assayed in a homogeneous medium, and the interpretation of these parameters may be difficult (see 164).

(2) *Hysteresis and heterogeneity*: These are very common complications which make it difficult to interpret steady-state enzyme kinetics of membrane enzymes once they are solubilized. Membrane enzymes often show a range of catalytic activity depending on the detergent and phospholipid present. It is usual for membrane enzymes to be assayed in mixtures containing detergent plus exogenously added phospholipid, as well as whatever endogenous lipids have copurified with the enzyme. The physical state of the enzyme, in particular the state of aggregation, under these circumstances is generally ill-defined and very likely heterogeneous. Often one can prepare assay mixtures with exactly the same composition, but mixed in different ways, which have very different levels of enzyme activity. This dependence on the history of the assay sample is an example of hysteresis and is very common with membrane enzymes. In essence, the enzyme gets trapped in a metastable state, unable to reach the most stable configuration. For example, simply mixing together a solubilized membrane protein with phospholipid vesicles will probably not result in insertion of the protein into the vesicle. Procedures have been specifically developed to assure proper reconstitution of membrane proteins in phospholipid vesicles by avoiding kinetic trapping in undesirable metastable states (see Section 6.3). One example of an enzyme which can be trapped in a highly aggregated state is bactoprenol kinase, a very hydrophobic enzyme from *Staphylococcus aureus* (504). Its activity is not affected by the state of aggregation, but this is not always the case with other enzymes (e.g., see 379).

Because hysteretic effects may be highly dependent on the phospholipid, reports of lipid specificity in stimulating the activity of particular membrane enzymes are often highly unreliable (see 503).

(3) *Enzymes in vesicles*: Membrane enzymes are frequently assayed when they are incorporated in the bilayer of sealed vesicles. This is usually the case for enzymes assayed in situ, in isolated natural membranes. Also, many purified enzymes are assayed after being reconstituted in liposomes. There are particular problems which arise in such experiments.

The most obvious problem is that the active site of the enzyme may be located on the inside of the vesicle and, therefore, inaccessible to a water-soluble substrate. This orientation problem is responsible for cryptic or latent enzyme activity, which is manifested only when the vesicles are permeabilized or disrupted in some way. This phenomenon can often be used to advantage to assess the orientation of a particular membrane enzyme in the vesicles. The fraction of cryptic activity can be a direct measure of the fraction of enzyme whose active site is located inside the vesicle. This can be done only with a substrate which does not penetrate the membrane, e.g., cytochrome *c* used to assay cytochrome *c* oxidase.

Another kind of problem arises in assaying transmembrane enzymes which catalyze reactions involving solute or charge transport across the bilayer. Examples are cytochrome *c* oxidase (Section 8.51), which catalyzes both electron transfer across the membrane and proton transport in the opposite direction, and the various ATP-driven ion pumps such as the Ca^{2+}-ATPase (see Section 8.41). When these enzymes are present in vesicles and are oriented largely in the same direction with respect to the inside and outside of the vesicles, they will generate solute or voltage gradients across the membrane. This is their physiological function. However, in vesicles with a small internal volume, the size of the gradient can build up rapidly and actually reduce the rate of enzyme turnover unless precautions are taken. This is because the amount of chemical work required to move a molecule, ion, or electron against an existing gradient increases as the size of the gradient increases. Beyond a certain point, the enzyme will cease to function. A system which exhibits this reduced activity is termed "coupled," and this is a measure of how well sealed the vesicles are to prevent the ion or molecule from leaking down the gradient formed by the enzyme. The extent of coupling can be measured by assaying the enzyme under conditions where the gradient is not allowed to build up. For example, the voltage generated across the membrane by cytochrome *c* oxidase in vesicles can be prevented by including an ionophore (Section 7.52) in the assay which increases the ion (usually H^+ or K^+) permeability of the bilayer (see 191,903). It is also necessary in this case to use a highly buffered solution inside the vesicles because the utilization of protons on the inside of the vesicle to form water would otherwise result in rapid and extreme alkalinization of the inside of the vesicle, and this can influence the enzyme activity.

A number of ion channels (see Chapter 8) and some enzymes (e.g., 472) are directly regulated by a transmembrane potential. It has also been reported that a voltage difference across the bilayer can significantly increase the microviscosity of the bilayer, as measured by fluorescent and ESR spin probes (1087). This could also affect enzyme activity. Finally, there are claims that a transmembrane voltage difference can alter the state of aggregation of certain membrane proteins (524,361), but the physiological significance of this is not known. All of these effects are dependent on having sealed vesicles.

(4) *Influence of the surface potential*: Most biomembranes contain a significant proportion of negatively charged phospholipids and, consequently, carry a net negative charge. This negative charge density at the membrane surface results in a surface electric potential (see Section 7.33). This reduces the concentration of negative ions in solution near the membrane surface to a value below the average bulk concentration, and increases the local concentration of positive ions near the membrane surface. The surface potential can have a substantial influence on the behavior of enzymes whose active sites are located near the membrane surface, because it will affect the local concentrations of charged substrates, as

well as protons. At physiological ionic strength, this effect will be largely restricted to the immediate vicinity of the charged membrane surface, but it can be substantial. This is manifested by changes in the measured K_m for charged substrates and alterations in the pH dependence of the enzyme because the local concentration of all charged species can be higher or lower than the measured bulk values. Thus, the kinetic characteristics of a membrane enzyme may differ when it is incorporated in various phospholipid vesicles, depending on the surface charge density of the vesicles, and may be different from those of the enzyme assayed in a neutral detergent micelle.

These effects have been documented for several mitochondrial and microsomal enzymes both in situ and in reconstituted phospholipid vesicles. Examples are arylsulfatase C (932, 1056), which has an anionic substrate, and monoamine oxidase, which has a cationic substrate (1056). Changes in the lipid environment of ß-hydroxybutyrate dehydrogenase also affect the K_m values of NADH, attributable to changes in surface charge density (221, 222).

Finally, it should be noted that the surface charge of thylakoid membranes has been extensively discussed (68) as a factor in regulating the lateral distribution of the membrane proteins and the interaction between membranes (see Section 4.52). In this case, the charges on the membrane surface are mostly due to the protein and not lipid constituents. Nevertheless, the main point is that an appreciation of the role played by surface potential is necessary to understand how many membrane proteins function in vivo and can be an important consideration when trying to mimic the natural system by reconstitution with purified components.

6.3 Reconstitution of Membrane Enzymes (for reviews, see 405, 282, 847, 191, 694)

Once a membrane enzyme has been purified, it is advantageous, and often necessary, to reconstitute it with phospholipids to study its catalytic activity. There are many enzymes whose kinetic characteristics after purification and reconstitution are similar to those observed in situ in the natural membrane. There are many reasons to justify the study of purified, reconstituted enzymes which offset the risk that the enzyme properties might be altered in the artificial system. Competing reactions and other complicating background problems in the biomembrane systems are, of course, eliminated. The use of reconstituted components allows one not only to characterize the isolated system but also to determine the minimal number of required components for particular biochemical activities. For example, lactose uptake in *E. coli* was clearly demonstrated to require only one protein, the lactose permease (Section 8.32), by reconstitution into phospholipid vesicles (938). Similarly, the respiratory chain of *E. coli* (770) and the adenylate cyclase hormonal response system (942; see Chapter 9) have been clearly defined in reconstituted systems in terms of a small number of purified protein components.

Many procedures have been developed to reconstitute detergent-solubilized purified membrane components into model membrane systems. Most frequently, the enzymes are reconstituted into unilamellar phospholipid vesicles. However, proteins that either actively or passively facilitate ion conductance across membranes are often characterized in planar membrane systems (282, 1010; see Sections 2.53 and 8.14). The advantage of the planar membranes is that electrical measurements are easily made to quantify the ion conductance, as well as any characteristics that can be monitored by conductance, such as the voltage dependence of channel opening. However, it is difficult to characterize the protein biochemically in these systems, and catalytic activity, other than that coupled to ion conductance across the bilayer, is difficult or impossible to measure.

Box 6.1 The Reconstitution of Membrane Enzymes into Lipid Vesicles

To reconstitute a membrane protein into lipid vesicles, it is necessary to remove most of the detergent present in the enzyme preparation because substantial amounts of detergent destabilize the phospholipid bilayer. The detergent is often removed from a mixture containing both the protein and phospholipids, but in some cases the detergent is removed prior to the reconstitution procedure. Methods for detergent removal include gel filtration, dialysis, and the use of polystyrene beads (BioBeads SM-2), which are used primarily to remove Triton X-100. All these procedures can be effective, but it should realized that residual detergent usually remains even after intensive efforts to remove it (e.g., see 20).

Methods of reconstitution (see 191) fall into two general categories:

I. Procedures in which proteins, which have previously had most of the detergent removed, are induced to incorporate into phospholipids.
II. Procedures in which proteins and phospholipids are mixed together in the presence of detergent, which is then removed to yield proteoliposomes. The only real limitation on the choice of phospholipid is that it should readily form a stable bilayer, e.g., not unsaturated phosphatidylethanolamines. Often, soybean phospholipids are used since they are inexpensive, readily available, and have been successfully employed in many cases (see 191).

I. *Reconstitution Without Excess Detergent* (see 694 for review)

The conditions which favor the incorporation of membrane proteins into preformed phospholipid vesicles often favor phospholipid vesicle fusion (see Section 9.5). Both processes may rely on some perturbation in the bilayer structure or the generation of "defects" which could facilitate both protein incorporation and vesicle fusion. The mechanisms are poorly understood. Protein aggregation is a potential problem in applying these methods.

(1) *Co-incubation of the protein and preformed vesicles*: This procedure is not generally applicable, but has been used with proteins with limited hydrophobic surface such as cytochrome b_5 and ß-hydroxybutyrate dehydrogenase (Section 6.5), which need not penetrate across the bilayer. In the case of cytochrome b_5, recon-

stitution by this method yields a form of the protein which is experimentally distinguishable from the native structure (see Section 4.22).

(2) *Inclusion of amphiphilic catalysts*: Several reports have indicated that low levels of amphiphiles present in the protein-phospholipid mixture will facilitate the incorporation of membrane enzymes, such as bacteriorhodopsin or cytochrome *c* oxidase, into vesicles. Cholesterol (1311), short-chain phosphatidylcholines (323), and fatty acids (1310) have been used. These procedures have the advantage that harsh techniques, as well as excess detergents, are avoided, but they have not yet been widely utilized.

(3) *Freeze-thaw/sonication*: The use of a freeze–thaw cycle and sonication has been successful in some cases (see 405, 191), but it is not widely used because of the potential of denaturing the protein. Sonication has also been used alone to facilitate reconstitution. Most likely, the protein initially inserts into small vesicles because of the high degree of curvature of these vesicles. The freeze–thaw cycle may serve to fuse the vesicles to form larger, more nearly uniform species.

II. *Reconstitution using Detergents*

By far the most frequently used reconstitution procedures involve the use of detergents to cosolubilize the protein and phospholipid, followed by removal of the detergent. The remaining protein and phospholipid spontaneously form unilamellar vesicles, which usually have excellent properties for enzymological studies. Usually, a detergent with a high critical micelle concentration and small micelle size is chosen so that it can be readily removed by dialysis or by gel filtration. Sodium cholate and octylglucoside are the most common choices. The ratio of detergent to phospholipid and the method and rate of detergent removal are important considerations in the size distribution of the resulting vesicles (see 23, 1475). Dialysis is slow and is usually the method of choice. Examples include cytochrome *c* oxidase (191) and the Na^+/K^+-ATPase (39).

Lipids other than phospholipids, such as cholesterol or ubiquinone, can also be incorporated into vesicles using the procedure. In some cases it is advantageous to eliminate the detergent more rapidly, especially if the protein is unstable in the presence of the excess detergent. Gel filtration chromatography has been successful, and will effectively separate the protein–lipid vesicles from the detergent. Examples include glycerol-3-phosphate acyltransferase (544) and phosphatidylinositol synthase (431). Figure 6.1 illustrates the elution profile of such a column.

Even more rapid is the dilution procedure, in which the protein–lipid–detergent mixture is diluted far below the critical micelle concentration of the detergent. Protein–phospholipid vesicles spontaneously form, and can be separated from the detergent by pelleting in a centrifuge (e.g., see 938).

Triton X-100 is commonly used for the purification of membrane enzymes, but it is not ideal for performing reconstitution experiments because it is difficult to remove. Polystyrene beads have been useful for removal of this detergent, leaving protein and phospholipid behind. One problem is the loss of protein on the resin (see 20). The sodium channel is an example of a protein which has been reconstituted using this method (416).

Finally, reverse phase evaporation, in which diethyl ether is used to disperse a cholesterol–phospholipid–protein mixture, has also been used for reconstitution (240). It is unlikely that many proteins will tolerate this procedure.

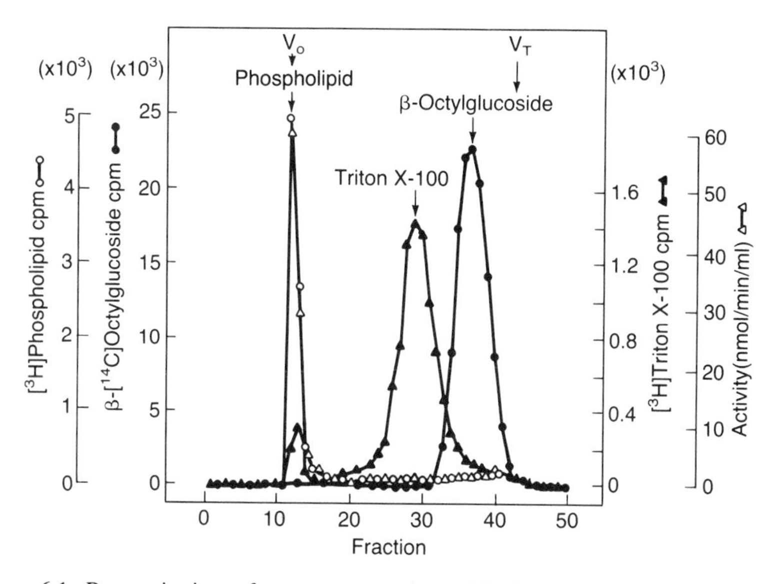

Figure 6.1. Reconstitution of an enzyme, glycerol-3-phosphate acyltransferase, into phospholipid vesicles by gel filtration chromatography with Sepharose G-50. The sample contained a mixture of phospholipids, ß-octylglucoside, Triton X-100, and the purified protein. The column is able to resolve the protein–phospholipid vesicles from micelles of Triton X-100 and ß-octylglucoside, which elute separately. Adapted from ref. 544.

6.31 Some Characteristics of Reconstituted Protein-Phospholipid Vesicles (Proteoliposomes)

The procedures which have been developed for reconstitution of membrane proteins are all designed to yield unilamellar vesicles. In a number of case studies, the characteristics of these vesicles have been examined in detail, including vesicles containing cytochrome *c* oxidase (191, 903), the Na^+/K^+-ATPase (39), cytochrome b_5 (219, 547), and glycophorin (993). Properties of particular interest are (1) average vesicle size and the size distribution, (2) distribution of the protein in the vesicle population, (3) orientation of the protein with respect to the plane of the bilayer, and (4) permeability properties of the vesicles. These properties are of particular importance when the enzyme catalyzes transmembrane flow of a solute or ion. Quantifying the results of assays in these cases requires knowledge of the vesicle internal volume, vesicle permeability, and the protein distribution. For enzymes engaged in vectorial reactions, such as cytochrome *c* oxidase, which pumps protons in one direction, protein molecules with opposite orientations within the same vesicle can cancel each other out in terms of the resulting ion gradient generated.

(1) *Vesicle size*: This depends very much on the reconstitution procedure, and is best characterized by electron microscopy or gel filtration chromatography. Detergent dialysis procedures yield proteoliposomes with diameters ranging from

about 500 Å to 2,500 Å, depending on the protein and method used. There is always a wide range represented in any given reconstitution and the vesicles can be subsequently sized by gel filtration chromatography (e.g., 993, 903).

(2) *Protein distribution*: Using the detergent dialysis procedures with excess lipid, the distribution of proteins among the vesicles appears to follow a Poisson distribution (e.g., see 39), but this may depend on the protein, and the details of the procedure utilized. It should be noted that the spontaneous incorporation of some proteins into preformed vesicles shows a very biased tendency toward small vesicles. Cytochrome b_5, for example, favors small vesicles (200 Å diameter) over large ones (1,000 Å diameter) by 200-fold (547).

(3) *Protein orientation*: This is particularly important for any enzymological studies, since proteins with their active sites facing inward may not have access to substrate. It is surprising that a large number of membrane enzymes can spontaneously insert in a highly oriented manner, usually with 75% to 95% of their active sites facing outward. Examples include cytochrome *c* oxidase (903, 191), Na^+/K^+-ATPase (39), glycerol-3-phosphate acyltransferase (544), and bacteriorhodopsin (323). Vesicles containing cytochrome *c* oxidase have been resolved by DEAE chromatography into two populations with the enzyme facing inward and outward (1629), and similar methods could, in principle, be used with other proteins.

The reason for the biased orientation is not known. In some, but not all, cases, the outward-facing part of the enzyme has a larger size which may disfavor reconstituted forms with the bulk facing inward. The fact that the orientation is observed with large vesicles where the curvature is not extreme suggests that vesicle curvature is not the critical factor, as it is in the case of lipid asymmetry of small vesicles (Section 4.42). Kinetic factors are probably important, but this is difficult to evaluate since the nature of the intermediates is not known.

It should also be noted that some proteins have been shown to reconstitute into vesicles improperly, i.e., in ways differing from that found in the natural membrane. The best example is cytochrome b_5, in which two forms can be distinguished depending on the reconstitution procedure (219; Section 4.42). Other examples are the H-2K component of the mouse major histocompatibility complex (185), and possibly the M13 bacteriophage coat protein (1574). These proteins all have one transmembrane hydrophobic domain and, under some circumstances, they might insert into the bilayer in a "U" configuration with both amino terminus and carboxyl terminus facing outward.

(4) *Permeability*: This is of obvious importance for any enzyme which catalyzes solute or ion flux across the bilayer. Many transport systems and ion pumps have been studied in proteoliposomes (see Chapter 8), so it is evident that proteoliposomes can be prepared with sufficiently low permeability properties to allow these studies. The presence of protein in the vesicles will, however, tend to increase the permeability of the vesicles, but this will depend very much on the choice of lipid, as well as the amount of protein per vesicle (e.g., see 993, 185). It has been speculated that particular lipids, because of their shape, will pack better around proteins inserted in the bilayer, thus minimizing solute permeability

due to packing defects in this region (185). This has not yet been directly demonstrated, however.

6.4 The Role of Lipids in the Activity of Membrane-Bound Enzymes (see 463, 1280, 1278, 1279, 503)

Lipids are essential for the catalytic activity of many membrane-bound enzymes. The lipids play two distinct functions: (1) to provide a medium in which the enzyme resides, and (2) to act as allosteric effectors in modulating enzyme activity. In providing a medium for the enzyme, the lipids not only serve to stabilize the protein against denaturation, but they also facilitate the interactions between the enzyme and other membrane-bound components, such as a lipophilic substrate. As allosteric effectors, the lipids usually activate the enzyme, presumably by stabilizing a particular conformation of the protein. Although these two roles of lipids are quite distinct theoretically, they are usually very difficult to distinguish experimentally. In an ideal experimental system, a specific lipid would serve as the allosteric effector and the bulk of the lipids in the bilayer would serve to provide the working environment for the enzyme. Unfortunately, there is only one well-documented example where an enzyme exhibits an absolute specificity for a particular lipid for activation. ß-Hydroxybutyrate dehydrogenase specifically requires phosphatidylcholine for catalytic activity (1277). In most other cases, many different lipids can effectively activate the enzyme, making it difficult to distinguish bulk effects from specific allosteric effects, since any given lipid can fulfill both functions to some degree. The physical state of the bilayer, including surface charge density and fluidity, can influence enzyme activity in a number of ways. Systematic changes in lipid structure and physical state can be correlated to effects on enzyme velocity, but this kind of study alone cannot unravel the two roles played by the lipids. Changes in the physical state of the bilayer could influence the interaction between the enzyme and whatever lipids may be acting as allosteric effectors. Studies of enzymes which require lipophilic substrates with reconstituted vesicles present special problems, since the substrate must be incorporated into the bilayer either before or after formation of the vesicles. Large amounts of a substrate incorporated into the bilayer would undoubtedly influence the physical state of the model membrane.

One strategy for studying the influence of lipids on membrane enzymes is to use a mixed micelle assay system, in which the enzyme is dispersed in a detergent which does not activate the enzyme and lipids are added to this mixture (e.g., see 1276, 1545). Many membrane enzymes are active to some degree in detergents, and this is strongly dependent not only on the particular enzyme, but also on the detergent and, possibly, on the presence of endogenous lipids in the preparation of the purified enzyme. Delipidation of the enzyme preparation is usually advisable, though this can sometimes require somewhat harsh treatments which can denature the enzyme. Methods of delipidation usually involve passing the enzyme through a large excess of detergent either by gel filtration chromatography or sucrose density gradient centrifugation or by binding the enzyme to a solid support such as DEAE-Sepharose and washing with excess detergent (e.g., see 1105, 1551).

The mixed micelle system has the advantage that lipophilic substrate can be added in micelles of the same detergent, though partitioning problems, discussed in Section 6.2, can sometimes be a problem (see 1545). Presumably, lipid effects observed in this kind of system are allosteric in nature, since there is no bilayer and the lipids are presumably binding to the protein within a globular micelle with a large amount of detergent present. However, the physical state of the enzyme–detergent–phospholipid complex is ill-defined, which is a disadvantage of this approach.

Any effects ascribed to a particular lipid in such a mixed micelle system should be observable using different nonactivating detergents to disperse the protein and lipids. This would demonstrate that the detergent is playing only a neutral role, but such experiments are rarely performed.

There are several generalizations which summarize what has been learned from many systems.

(1) Enzyme activation very rarely requires a particular lipid structure. Of course, for any given enzyme, some lipids will activate to a greater extent than others. However, this preference can be assay-dependent, and, in any event, the physiological relevance is unclear. It is not unusual to find that a lipid which is highly effective as an activator is not present in the membrane from which the enzyme was purified. Claims of lipid specificity should be viewed with caution (see 1278).

(2) The dependence of enzyme velocity on the concentration of lipid in the mixed micelle type of assay is usually indicative of a highly cooperative process (1279). One interpretation of this is that lipids bind to a number of equivalent sites on the enzyme noncooperatively, but that activity is manifested only in those enzyme species which have most of the lipid binding sites occupied (1279, 1277). Often there is an optimum lipid concentration for enzyme activation, probably related to the lipid–detergent ratio and the kinds of structures formed in these mixtures.

(3) It is evident from the many enzymes which are active in detergent or lipid–detergent micelles that the bilayer structure per se is not a requirement for enzyme activity. There are examples where lipids which do not form a stable bilayer will effectively activate membrane enzymes (e.g., 709).

(4) The physical state of the bilayer is definitely an important parameter in determining the activity of most membrane enzymes. There is no solid indication, however, that enzyme activity in vivo is regulated physiologically by alterations in the membrane physical state. Most enzymes become inactive or at least show a dramatic loss of activity when the surrounding lipids are in the gel phase. There are many examples where changes in enzyme velocity with temperature correlate with changes in the physical state of the lipids (e.g., see 463). Generally, these

studies focus on correlations between the temperatures where there are discontinuities in the thermal dependence of enzyme activity and the thermal dependence of some measure of the lipid physical state (e.g., fluorescent probe anisotropy). Efforts to demonstrate a dependence of enzyme activity on membrane fluidity by observing thermal effects are difficult to interpret because temperature influences many parameters simultaneously. It is useful to keep in mind that membrane fluidity is largely correlated with the lipid packing density, and changes in this account for most of the thermally induced changes in fluidity. There are several examples where there is clearly a linear correlation between enzyme activity and membrane fluidity (504, 216). However, the interpretation in terms of enzyme mechanism is unclear. In only a few examples is there even a hint of what step or steps in the enzyme mechanism are rate limiting and are being affected by lipid binding (see 1278). It should be noted that lipids, in addition to affecting the maximal velocity, can also have a dramatic influence on substrate and cofactor binding properties (e.g., 640, 1083).

(5) Generally, where large numbers of lipids are studied with a particular enzyme, there is no correlation between the resulting enzyme activity and any single parameter (e.g., fluidity) (e.g., 117, 189, 364). As long as the lipid is in the liquid crystalline phase, the identity of the headgroup is often more important than the bilayer fluidity in determining enzyme activity. In other words, the chemical structure of the individual lipids is important, but the reasons are not known.

6.5 Some Examples of Lipid-Requiring Enzymes

These five examples have been selected because they are particularly well-studied and characterized, because they represent several kinds of membrane enzymes, and because they illustrate several of the major themes of lipid-activation studies.

6.51 ß-Hydroxybutyrate Dehydrogenase (BDH) (941, 379, 1277, 249)

This is one of the best characterized lipid-activated enzymes and is the only good example of an enzyme which exhibits an absolute specificity for lipid activation. Only phosphatidylcholines will activate the enzyme. BDH is located on the inner surface of the inner mitochondrial membrane and catalyzes the oxidation of D-ß-hydroxybutyrate by NAD^+, both water-soluble substrates. The enzyme has a subunit molecular weight of 31,000, but the complete sequence is not yet known. It does not appear to be a transmembrane protein. The purified enzyme is obtained free of phospholipid in a detergent-free form which will spontaneously insert into preformed phospholipid vesicles. Although BDH binds to vesicles which do not contain phosphatidylcholine, catalytic activity is manifest only when phosphatidylcholine is present. Binding appears to involve more than an interaction at the surface of the bilayer, and may involve substantial penetration of the bilayer as well as changes in protein conformation (941, 379). The kinetic properties of the reconstituted enzyme and its tetrameric state are the same in the mitochondrial membrane and when reconstituted with mitochondrial lipids.

Figure 6.2 shows the activation of BDH by phospholipid vesicles containing various amounts of phosphatidylcholine. In all cases, the enzyme is bound to the vesicles. The increase in enzyme velocity as a function of the amount of phosphatidylcholine in the bilayer indicates a highly cooperative process, consistent with two lipid binding sites required for activation (1277, 249). Of particular interest, phosphatidylcholine proximity to BDH within the bilayer was also measured by monitoring the tryptophan fluorescence as it was quenched by a derivative of phosphatidylcholine (1277). To the extent that fluorescence quenching is actually measuring lipid binding, the results show that phosphatidylcholine does not bind to the enzyme in a cooperative fashion, but binds noncooperatively to about 12 sites. Noncooperative binding of lipids to a fairly large number of sites is generally found for membrane proteins within the bilayer (see Section 5.5). In addition, most of the fluorescence quenching occurs at a lower concentration of phosphatidylcholine than is required for activation. One possible explanation of the discrepancy in the lipid binding and activation profiles is that only a small subclass of sites is involved in activation and that these are not observed in the quenching experiments. This experiment is of interest because it compares the physical interaction of a particular lipid with an enzyme and the biochemical response (activation).

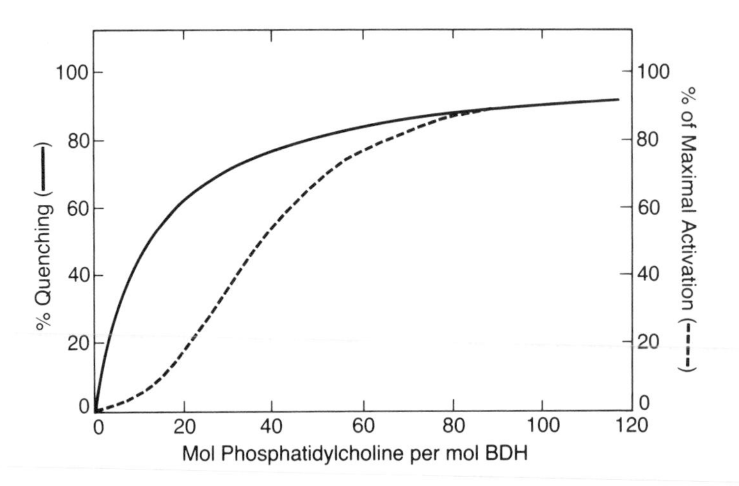

Figure 6.2. Comparison of the activation of D-ß-hydroxybutyrate dehydrogenase (BDH) and the quenching of the tryptophan fluorescence from the enzyme. Each sample contained the same amount of nonactivating lipids (112 mol per mol enzyme) plus the indicated amount of phosphatidylcholine. The phosphatidylcholine added to the samples was a 1:1 mixture of dioleoyl phosphatidylcholine and 2-pyrenyl phosphatidylcholine, with the latter responsible for the fluorescence quenching. Note that the biochemical response (activation) and physical interaction (quenching) do not occur in parallel. Adapted from ref. 1277.

One potential complicating factor in these activation experiments is that the state of aggregation of the enzyme may strongly influence the enzyme activity, and the enzyme appears to be activated to the same extent as it is disaggregated in different lipids (1173). BDH is also remarkable in that it is one of the few enzymes which is activated by short-chain phosphatidylcholines which bind and activate below their critical micelle concentrations (248). These studies also suggest that lipid activation involves only a small number of binding sites on the protein.

6.52 Pyruvate Oxidase

This is a flavoprotein dehydrogenase from *E. coli* which oxidizes pyruvate to acetic acid plus CO_2 and reduces ubiquinone in the cytoplasmic membrane. The enzyme feeds electrons into the aerobic respiratory chain (770, 40). The enzyme has several remarkable features. It is a water-soluble protein which shows none of the expected characteristics of a membrane protein. However, in the presence of substrate and cofactor (thiamin pyrophosphate), the enzyme undergoes a conformational change exposing a membrane binding site. Under these conditions, the affinity of the enzyme for detergents and phospholipid vesicles increases dramatically, and the protein manifests the solution properties expected of a true membrane enzyme. This may be an example of an enzyme which is cytoplasmic until recruited to the membrane when the substrate levels are sufficiently high. Other examples similar to this are described in Section 6.7.

Studies on this enzyme have focused on its dramatic activation by lipids. The enzyme can be assayed using a water-soluble artificial electron acceptor such as ferricyanide, in place of the natural electron acceptor which is membrane-bound. The catalytic activity is stimulated up to 50-fold in the presence of lipids. Unlike ß-hydroxybutyrate dehydrogenase, lipid activation is observed with an extraordinarily wide variety of phospholipids and detergents (106). Both anionic and cationic detergents fully activate this enzyme below their critical micelle concentrations, and detergent binding studies suggest that a small number of binding sites are involved in the activation (1304). Binding to phosphatidylcholine vesicles or detergents is not observed unless the substrate and cofactor are present. Under these conditions, the enzyme binds to and is activated by a number of different phospholipid vesicles. Binding is not primarily electrostatic but hydrophobic (106, 1304). For pyruvate oxidase, lipid binding and activation appear to be inseparable, though the extent of activation is not identical with all phospholipids and amphiphiles. Activation by lipids results in changes in the absorption spectrum of the bound flavin, apparently making it more accessible to electron transfer reactions (936).

The conformational change in pyruvate oxidase that results in exposure of the lipid binding domain also results in exposure of a protease-sensitive region near the carboxyl terminus. Proteolysis destroys the lipid binding properties of the enzyme but, surprisingly, activates the enzyme when using a water-soluble electron acceptor (1263). The protease-activated enzyme will not, however, reduce a membrane-bound electron acceptor because it has lost the ability to bind to the membrane.

The gene encoding pyruvate oxidase has been cloned and sequenced. The inferred protein sequence shows no long hydrophobic stretches. Genetics studies (197) along with the proteolysis studies (1263) suggest that lipid binding involves portions of the polypeptide near the carboxyl terminus. There is a short potential amphipathic α-helix (see Section 3.6) in this region which may be involved in binding to the bilayer surface (1206). A mutational variant of pyruvate oxidase has been made which lacks the last 24 amino acid residues (536). This enzyme variant is inactive in vivo but is active in vitro with a water-soluble electron acceptor. Presumably, this mutational variant of pyruvate oxidase cannot bind to the membrane even in the presence of a high concentration of substrates and is, therefore, unable to function, since it must reduce ubiquinone in order to turn over. This demonstrates that the lipid binding and activation data on the enzyme in vitro are relevant to the function inside the cell.

Both pyruvate oxidase and ß-hydroxybutyrate dehydrogenase can be activated either by (1) binding to a small number of monomeric amphiphiles or (2) complexing with the surface of the bilayer, with some portions of the protein inserting into the membrane. Both of these enzymes clearly demonstrate the allosteric role played by phospholipids.

6.53 Ca^{2+}-ATPase (also see Section 8.41; 629 for review)

Whereas the previous two examples illustrate the role of lipids as allosteric effectors of membrane enzymes, this example and the next two illustrate the influence of lipid structure and the physical state of the bilayer on enzyme activity. The Ca^{2+}-ATPase is a single polypeptide of molecular weight 115,000 isolated from the sarcoplasmic reticulum of skeletal muscle. This enzyme actively transports Ca^{2+} into the sarcosplasmic reticulum and, thus, lowers the cytoplasmic concentration of Ca^{2+} during muscle relaxation. Two Ca^{2+} ions are transported across the membrane for each ATP hydrolyzed, and the reaction is electrogenic. That is, the movement of charge across the bilayer generates a transmembrane potential. The sequence of the protein has been deduced from the DNA sequence of its gene, and it has been proposed that the protein has 10 transmembrane helices (134; Figure 8.13). The structure and proposed mechanism for ion transport are discussed in Section 8.41.

The enzyme appears to be a dimer in the sarcoplasmic reticulum membrane (671, 1057a), but the monomeric form in detergent micelles retains most of the kinetic properties (928), although ion translocation per se cannot be measured under these circumstances. The purified Ca^{2+}-ATPase can be successfully reconstituted, using a deoxycholate-dialysis procedure (e.g., 209). Some reconstituted enzyme preparations have a tendency to aggregate beyond the dimeric form, although the physiological significance of this is not known (e.g., 406).

The purified enzyme has been completely dilipidated and reconstituted with many lipids and lipid mixtures by several laboratories. Figure 6.3 shows the

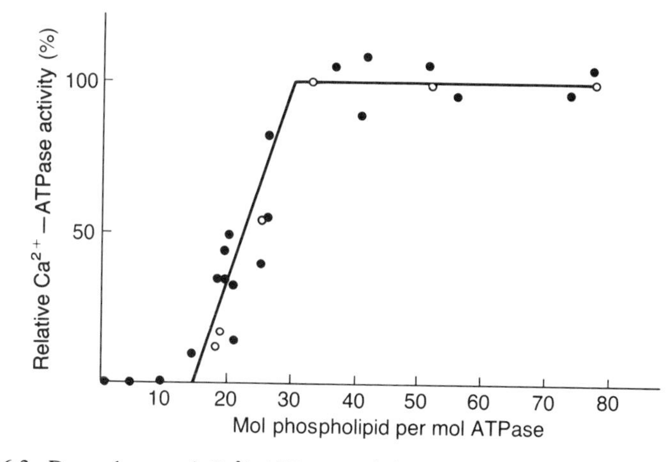

Figure 6.3. Dependence of Ca^{2+}-ATPase activity in the sarcoplasmic reticulum on phospholipid content. Sarcoplasmic reticulum vesicles were treated with cholate to remove phospholipid and to prepare samples with different lipid/protein ratios (open circles). The replacement of phospholipids by cholesterol (filled circles) gave essentially the same result as removing phospholipid. The data show that cholesterol does not support enzyme activity and suggest that cholesterol does not bind to the lipid annulus since the activation by phospholipids is the same regardless of the presence of cholesterol. Adapted from ref. 1551. Reprinted by permission from *Nature*, vol. 255, p. 685. Copyright © 1975 Macmillan Magazines Limited.

dependence of enzyme ATPase activity as a function of the amount of lipid complexed to the enzyme in the absence of detergent (1551). The data suggest that 30 phospholipid molecules are required per molecule of ATPase, which is estimated to be about enough to cover each protein molecule with a single layer of phospholipid. It is probably not correct to interpret the meaning of this value strictly in this way, since the physical state of the mixture is not well defined. The concept of a lipid "annulus" or boundary lipid, meaning the layer of lipids adjacent to the protein, derives from this work (see 98). The lipid composition of this annulus should be a critical factor in determining enzyme activity. The lack of influence of cholesterol when included in protein–phospholipid complexes suggested that cholesterol was excluded from this annulus (1551), but subsequent binding studies showed this was incorrect (see Chapter 5.5). It is difficult to utilize kinetic data to obtain unique models of lipid binding.

Several studies have clearly shown that above the lipid phase transition there is no correlation between bilayer fluidity or order parameter and enzyme activity (364, 209, 174). This is true whether one measures Ca^{2+}-stimulated ATPase activity or Ca^{2+} transport into vesicles in a well-coupled system. Certainly, individual lipid structures play a more important role in regulating enzyme activity than any single parameter reflecting the physical state of the bilayer. The reason why some lipids are better than others at supporting enzyme activity in reconstituted vesicles is not known.

6.54 Na^+/K^+-ATPase (also see Section 8.41; 39, 714 for reviews)

This enzyme catalyzes the ATP-driven transport of Na^+ and K^+ across the plasma membrane of animal cells. For every ATP hydrolyzed, three Na^+ ions are pumped out and two K^+ ions are pumped into the cell. The enzyme is an electrogenic ion pump and generates a voltage across the bilayer. The Na^+/K^+-ATPase has been purified from several sources and is always composed of two subunits, a large catalytic subunit (α, molecular weight ≈ 115,000) and a small glycoprotein (ß, molecular weight ≈ 55,000) of unknown function. The complete sequences are known for both subunits from sheep kidney (1337, 1338) and from the electric organ of *Torpedo californica* (1077, 733). The catalytic subunit is homologous to the Ca^{2+}-ATPase, with numerous putative transmembrane helices. The ß-subunit also appears to cross the membrane, probably with one transmembrane helical segment. See Section 8.41 for additional discussion of structure and mechanism.

The enzyme appears to function either as an αß heterodimer or an $(\alpha\beta)_2$ tetramer, and there is no consensus on which is the minimal unit required for ion transport. The purified enzyme has been reconstituted with a number of phospholipids (e.g., 752), and phosphatidylserine and phosphatidylglycerol were found to be substantially more effective at restoring enzyme activity. The reason for this is not known, but it has also been found that this enzyme has a slight preference for binding acidic phospholipids in the bilayer (see Section 5.5). In addition to influencing the catalytic activity, the lipid environment also modulates the sensitivity of the enzyme to an inhibitor, oubain (1).

Although the identity of the phospholipid headgroup appears to be of primary importance in reconstitution experiments, this enzyme provides an example where bilayer fluidity is also probably important. This was demonstrated in experiments with a preparation of the enzyme in its natural membrane, in which the fluidity was altered by increasing the hydrostatic pressure (216). High pressure stabilizes the system in configurations of minimal volume. Hence, the free volume and fluidity of the bilayer are decreased (i.e., higher packing density) as the hydrostatic pressure is increased. Figure 6.4 shows that there is an excellent correlation between the membrane fluidity, as measured by fluorescent probe anisotropy, and enzyme activity. It is speculated that the bilayer properties at high pressure stabilize particular enzyme conformations relative to others in the catalytic cycle and, thus, influence one or more rate-limiting steps in the mechanism (see Section 8.41). However, these data do not rule out the distinct possibility that the pressure effects are directly on the protein and are not mediated through the lipids. This experiment is significant, however, because it illustrates the usefulness of hydrostatic pressure changes to vary the membrane physical state.

6.55 Glucose Transporter (also see Section 8.31; 189 for review)

Like the previous two examples, this protein is involved in transport, but facilitated diffusion rather than active transport. This protein facilitates the flux of

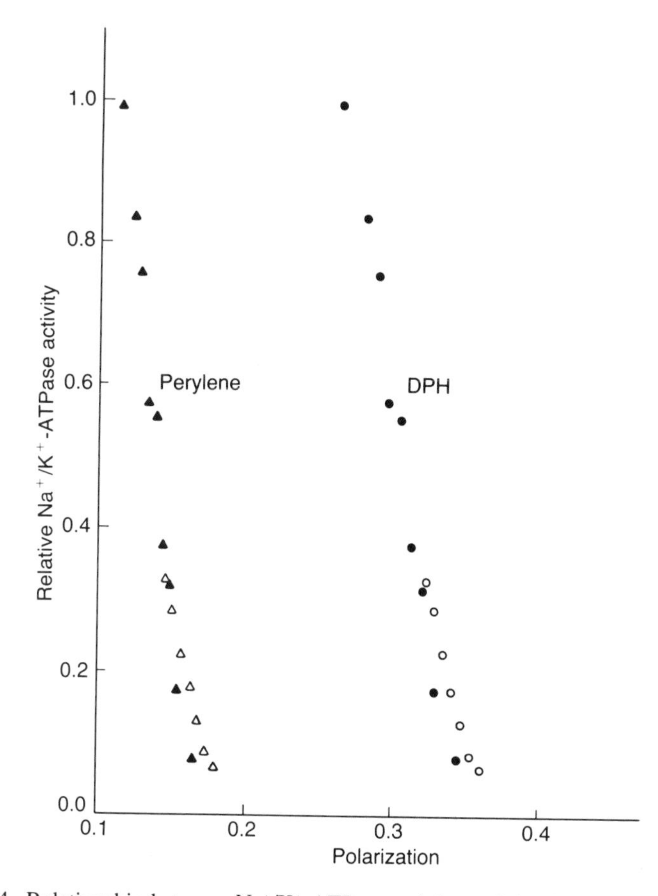

Figure 6.4. Relationship between Na^+/K^+-ATPase activity and the fluorescence polarization of lipid probes (perylene and diphenylhexatriene, DPH) in the same membrane vesicles. The open symbols and filled symbols represent experiments performed at 22°C and 36°C, respectively. At each temperature, the enzyme activity and polarization were measured at different pressures. For each probe, combinations of different temperatures and pressure which result in the same polarization also have the same enzyme activity. The simplest interpretation is that lipid order or fluidity influences the enzyme turnover. Adapted from ref. 216.

sugars down a concentration gradient across the erythrocyte membrane. It consists of a single glycoprotein of molecular weight 55,000 whose primary structure suggests 12 transmembrane helices. Reconstitution experiments have been performed with many different lipids and lipid–cholesterol mixtures. The general conclusion is that the phospholipid polar headgroup is the most important determinant of the protein turnover, and that bilayer fluidity or order is among the least important. Both the maximal velocity and affinity for sugar (K_m) are influenced by lipids. Particularly dramatic is the fact that the maximal velocity of this transporter is not perturbed by a liquid crystalline to gel thermal phase transition when reconstituted in phosphatidylglycerol, but it is markedly affected in phosphatidylcholine.

6.6 Membrane-Bound Electron Transfer Chains

A substantial fraction of work on membrane enzymes concerns components of the various membrane-bound electron transfer systems. These systems carry out critical biochemical functions, and their components contain flavins or hemes or nonheme iron prosthetic groups which permit their kinetics and conformational properties to be monitored spectroscopically, both in the natural membrane and after purification. Attention has been focused on four major electron transfer systems, illustrated in Figure 6.5. These are (1) the mitochondrial cytochrome P450 system of the adrenal gland, involved in steroid biosynthesis, (2) the microsomal system including numerous cytochromes P450 as well as cytochrome b_5, (3) the mitochondrial respiratory system, and (4) the photosynthetic system of green plant thylakoids. These four systems are major components of the membranes in which they are located, and they illustrate different levels of complexity in membrane enzymology. The first two systems carry out anabolic and catabolic reactions, requiring molecular oxygen and a substrate that is usually lipophilic and membrane-bound. The terminal enzymes in these electron chains, P450 cytochromes and the fatty acid desaturase, have very low turnover numbers (<1 sec^{-1}). Systems (3) and (4) are bioenergetic assemblies whose primary function is the net translocation of protons across the membrane. The proton chemiosmotic gradient which is generated (the proton motive "force") is used to drive the synthesis of ATP. The respiratory and photosynthetic chains can operate at relatively high turnover, 200-300 sec^{-1}, and clearly carry out transmembrane reactions, in contrast to the microsomal and mitochondrial P450 systems. The following brief descriptions are intended to focus attention on particular aspects of enzymological interest, indicating some common features of the electron transfer systems as well as points of special interest or controversy.

One question which arises in considering each of these systems is the extent to which the various membrane-bound components of these electron transfer chains interact with each other to form long-lived complexes. Theoretical considerations (540) of the association of membrane proteins point out the importance of several factors: (1) high concentration of the reactive proteins by virtue of being confined to the membrane, (2) pre-orientation of the proteins, since they cannot tumble freely, though they can rotate, (3) excluded volume due to

→

Figure 6.5. Schematic in which the four most well studied electron transport chains are compared. Electron transfer reactions (heavy arrows) and proton fluxes are indicated. The dashed arrows in part II indicate evidence for physical interaction between cytochrome b_5 and the two enzymes shown (see text). LHC denotes light-harvesting complexes. The approximate stoichiometries and turnover numbers are also indicated. The sizes and shapes of the enzymes are not meant to be accurate depictions.

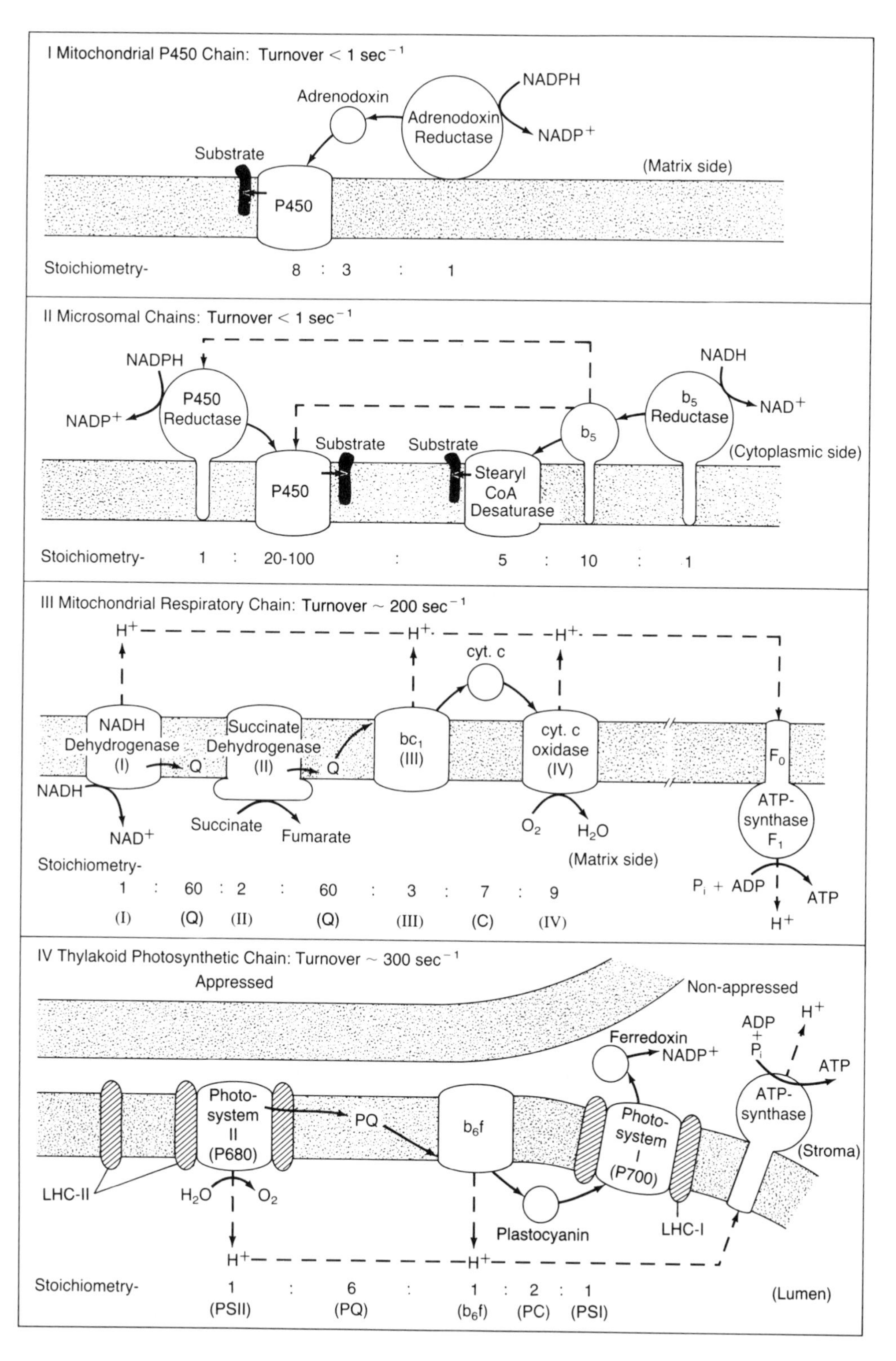

I Mitochondrial P450 Chain: Turnover < 1 sec⁻¹
NADPH
Adrenodoxin
Adrenodoxin Reductase
NADP+
Substrate
(Matrix side)
P450
Stoichiometry- 8 : 3 : 1
II Microsomal Chains: Turnover < 1 sec⁻¹
NADPH
NADH
P450 Reductase
b5 Reductase
NADP+
NAD+
b5
Substrate
Substrate
(Cytoplasmic side)
P450
Stearyl CoA Desaturase
Stoichiometry- 1 : 20-100 : 5 : 10 : 1
III Mitochondrial Respiratory Chain: Turnover ~ 200 sec⁻¹
H+
H+
H+
cyt. c
NADH Dehydrogenase (I)
Succinate Dehydrogenase (II)
Q
Q
bc1 (III)
cyt. c oxidase (IV)
F0
NADH
NAD+
Succinate
Fumarate
O2
H2O
ATP-synthase F1
(Matrix side)
Stoichiometry-
1 : 60 : 2 : 60 : 3 : 7 : 9
(I) (Q) (II) (Q) (III) (C) (IV)
Pi + ADP
ATP
H+
IV Thylakoid Photosynthetic Chain: Turnover ~ 300 sec⁻¹
Appressed
Non-appressed
ADP + Pi
H+
ATP
Ferredoxin
NADP+
Photo-system II (P680)
PQ
b6f
Photo-system I (P700)
ATP-synthase
(Stroma)
LHC-II
H2O
O2
Plastocyanin
LHC-I
H+
H+
Stoichiometry- 1 : 6 : 1 : 2 : 1
(PSII) (PQ) (b6f) (PC) (PSI)
(Lumen)

the presence of other proteins in the membrane, which further increases the effective concentration of the reactive species.

6.61 The Mitochondrial Steroidogenic System (see 1423, 592)

Steroid hormones, such as cortisol, which are synthesized in the adrenal gland are made from cholesterol by a series of reactions catalyzed by enzymes located in the mitochondria and the endoplasmic reticulum. Metabolic intermediates must be shuttled back and forth between these two organelles during the process. The reactions are catalyzed by membrane-bound heme proteins called P450 cytochromes. The name of this large family of enzymes derives from an absorption peak they have at 450 nm. They are mixed function oxidases, in which molecular oxygen is split, one atom of oxygen forming water and one atom of oxygen incorporated into the substrate. There are four P450 enzymes involved in steroid biosynthesis, two located in the mitochondria and two in the endoplasmic reticulum of steroidogenic tissue. These reactions require electrons, which are provided by relatively simple electron transfer chains.

In the adrenal mitochondria, the electrons are provided by NADPH, which reduces a flavoprotein, adrenodoxin reductase, and a ferredoxin, adrenodoxin. These proteins are located on the matrix side of the inner mitochondrial membrane. The two P450 cytochromes, $P450_{scc}$ (side chain cleavage) and $P450_{11\beta}$, are integral membrane enzymes, suitable for interacting with membrane-bound substrates. Adrenodoxin reductase may be partly embedded in the membrane, but it is readily removed from the membrane. Adrenodoxin is a small soluble protein (12,000 Da) which shuttles electrons from the flavoprotein to the P450 cytochromes (815, 816). Adrenodoxin binds to the reductase and to cytochrome P450 primarily through ionic interactions. Apparently there is a common binding domain on adrenodoxin for both reaction partners. The theme of having a small soluble protein shuttle electrons between membrane-bound proteins in an electron transfer chain is repeated in the respiratory (cytochrome *c*) and photosynthetic (plastocyanin) systems. The binding of steroid substrates to the P450 cytochromes can be monitored by spectroscopic changes induced in the heme protein, allowing the binding site properties to be evaluated. The binding site for cholesterol appears to be facing the bilayer lipids (see 772). It is also evident that the binding of cholesterol increases the affinity of cytochrome P450 for adrenodoxin by over 10-fold (814).

The slow turnover of these enzymes means that one adrenodoxin reductase molecule can turn over rapidly enough to provide electrons to many cytochrome P450 molecules. The stoichiometry of components, indicated in Figure 6.5, shows a cascade-like effect, where the ratio of reductase:adrenodoxin:P450 is 1:3:8, with the rate-limiting step being the turnover of the P450 enzymes (see 592). The three components of this simple chain obviously are not designed to form a single ternary complex at any time, and, in fact, this would not benefit the electron transfer efficiency of this chain.

6.62 The Microsomal Electron Transfer Chains

The electron transfer chains present on the cytoplasmic surface of the endoplasmic reticulum also carry out metabolic reactions on lipophilic substrates. As illustrated in Figure 6.5, there are two systems, but they are now known to be interactive. In one system, NADH is oxidized by a flavoprotein, cytochrome b_5 reductase, which reduces cytochrome b_5, which, in turn, reduces stearyl-CoA desaturase. The desaturase is an induced enzyme in liver microsomes. The second system involves the oxidation of NADPH by another flavoprotein, cytochrome P450 reductase, which then provides electrons to a family of cytochromes P450. Some of these cytochromes P450 are also inducible and some are constitutive. Note that in both of these systems, the number of terminal enzymes (desaturase or P450 species) is far in excess over the number of reductases at the front of each chain (see Figure 6.5). As in the mitochondrial P450 system, the turnover of the terminal enzymes in the microsomal system is so slow that many enzyme molecules can be provided with reducing equivalents by a small number of reductases. The microsomal system has no soluble protein equivalent to adrenodoxin, and all the protein components are integral membrane proteins which have been purified and studied in reconstituted form.

In a series of influential papers, Stritmatter and his colleagues argued strongly that cytochrome b_5 reductase and cytochrome b_5 were randomly distributed and were diffusion-coupled (see 1238, 1237). That is, the two enzymes diffuse laterally on the membrane surface, and electron transfer occurs upon a chance collision. This was shown both in reconstituted vesicles (1238) and in the microsomal membrane (1237). The diffusion-dependent step, however, is not rate limiting. There is normally a 10-fold excess of cytochrome b_5 over the reductase. An additional 10-fold excess of exogenous cytochrome b_5 bound to the microsomal membranes was kinetically identical to the endogenous pool of cytochrome b_5 (1237). Both cytochrome b_5 (see Section 4.22) and cytochrome b_5 reductase (743) are two-domain proteins in which a globular soluble domain binds the prosthetic group (heme or flavin) and a single hydrophobic tail anchors each protein to the membrane. The heme-binding and flavin-binding domains are designed to interact to optimize the rate of electron transfer from the flavin to the heme (569). Reconstitution studies have also been extended to include the desaturase (391).

More recent studies have strongly indicated, however, that a portion of the cytochrome b_5 in the microsomes is not freely diffusing, but is, at least transiently, involved in stoichiometric complexes with cytochrome P450 species (1427, 125, 1426) and, probably, with the NADPH-cytochrome P450 reductase (1072, 1428). It is quite clear that cytochrome b_5 plays an important role in the electron transfer chain involving at least some of the liver microsomal P450 isozymes (1427).

The various cytochromes P450 in liver microsomes are involved in the oxidative reactions of a variety of endogenous lipophilic substrates as well as foreign molecules (xenobiotics). The primary structure of P450 isozymes suggested that these proteins probably have numerous transmembrane segments (e.g., 1114; Figure 3.12), but experimental studies indicate that they are anchored to the microsomal membrane by a single transmembrane helix at the amino terminus (306). The NADPH-cytochrome P450 reductase has a similar structure, with a large hydrophilic domain anchored to the membrane by a short hydrophobic tail (see 1072). Two types of mechanisms have been proposed for how a single reductase molecule services an excess of P450 molecules. One model postulates a stable cluster, in which the reductase sits in the center surrounded by cytochrome P450 molecules, and the second model postulates that electron transfer requires free diffusion and collision of reactive species (1435). Some evidence suggests that cytochrome P450 and its reductase can form equimolar complexes in reconstituted phospholipid vesicles (e.g., 561, 771, 465). Presumably, if these form in the microsomal membrane (see 1428) they are transient species, rapidly changing partners to allow many cytochromes to be reduced by a single reductase molecule. The final picture has yet to emerge, but it seems evident that the components of the microsomal chain should not be viewed as isolated species, and that a complicated set of protein–protein interactions are probably important to the proper function of the system.

6.63 The Mitochondrial Respiratory System (see 568, 837, 1190, 1532)

The inner mitochondrial membrane is designed for oxidative phosphorylation. The role of respiratory electron transfer is to couple the flow of electrons, derived from the oxidation of organic substrates and going to oxygen, to the translocation of protons out of the mitochondrial matrix, across the membrane, and into the intermembrane space. As with the previous two electron transfer chains, the components are not present in equimolar amounts, indicating that they cannot function as a long-lived multicomponent complex. However, in this case the terminal enzyme, cytochrome *c* oxidase, can turn over rapidly, and the rate-limiting step(s) is not known with certainty (e.g., 568, 837). The chain contains four multisubunit complexes which are transmembranous, plus a soluble cytochrome *c* and ubiquinone-10 (see Figures 6.5, 6.6). Electron transfer through

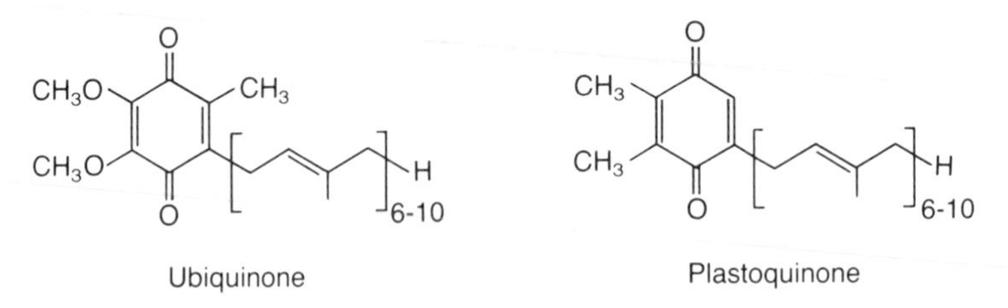

Figure 6.6. Structures of the lipophilic hydrogen carriers used in the respiratory electron transport chain (ubiquinone) and in the photosynthetic electron transport chain (plastoquinone).

complex II (succinate dehydrogenase) does not result in proton translocation. However, the reactions of complex I (NADH dehydrogenase), complex III (ubiquinol:cytochrome *c* oxidoreductase or bc_1 complex), and complex IV (cytochrome *c* oxidase) each result in the movement of protons across the membrane. These reactions are electrogenic and result in a voltage difference across the membrane. These are called "coupling sites" in the chain. There is no consensus as to the mechanisms of proton translocation, though there are detailed models, in particular for complex III (see 1532) and complex IV (502). The resulting difference in proton electrochemical potential across the membrane is dissipated by protons moving back into the mitochondrial matrix through the proton channel of the ATP synthase (or H^+-ATPase), and the energetically downhill proton flux is used to drive ATP synthesis by this enzyme (see Chapter 8).

Although the components of the respiratory chains are present in high amounts in the mitochondrial membrane, there is little evidence to suggest the formation of equimolar complexes between the multisubunit enzyme units (568). Models in which transient long-lived protein aggregates are involved in electron transport have been proposed, but such aggregates may not be necessary to explain the kinetic data (568). Cytochrome *c* serves to shuttle rapidly electrons from complex III to complex IV, in a manner very similar to that described for adrenodoxin in the mitochondrial P450 system. At physiological ionic strength, cytochrome *c* can diffuse not only along the bilayer surface but also through solution, and this increases its capacity for rapid electron transfer (see 568). This system illustrates several features not present in the previous examples of electron transfer chains.

(i) The major components of the electron transfer chain are organized into non-dissociating complexes. Complex III, for example, contains numerous (eight to eleven) subunits bound to a total of three hemes, and an iron–sulfur center. Electron transfer between the prosthetic groups within each of these complexes is rapid and not dependent on chance collisions.

(ii) A lipid-soluble hydrogen carrier is used to shuttle reducing equivalents not only between reactive species (complexes I and II to complex III) but also from one side of the bilayer to the other. Ubiquinone has been shown to be able to rapidly diffuse laterally in the membrane, though the magnitude of the lateral diffusion coefficient is currently in dispute (see 568, 837). This molecule has also been shown to carry reducing equivalents across the bilayer in model systems (see 837). Although confined to the membrane, ubiquinone can react so as to pick up or release protons from the lipid–water interface on either side of the membrane. The kinetics of the isolated enzymes which use ubiquinone (or ubiquinol, the reduced form) as a substrate can be successfully analyzed using Michaelis–Menten kinetics with K_m and V_{max} values if the actual quinone concentration in the bilayer is known (837). However, the steady-state kinetics of a system where one enzyme population (complexes I, II) is coupled to another population (complex III) by a freely diffusing pool of quinone as intermediary must be analyzed using specific equations for describing such "pool" kinetics (see 114). There is little doubt that components of the respiratory chain are diffusion-coupled via the quinone pool, but there is no consensus about whether quinone diffusion is a rate-limiting step in the electron transport chain (see 568, 837).

(iii) The protons driven across the membrane by the respiratory chain return back through the ATP synthase, completing a proton circuit. The mechanism by which the protons flow from the electron transfer chain components to the ATP synthase is another point of debate, unlikely to be settled in the near future (see 419 for review). The majority view for some time was that the protons driven across the bilayer rapidly equilibrate with the bulk aqueous phase. However, other views are that various forms of localized proton flux carry protons either along the bilayer surface (1174), within the bilayer, or directly to the ATP synthase (1256). Most of the arguments are somewhat indirect (see 419), but there is experimental evidence for rapid proton conductance under nonphysiological conditions, along the surface of a phospholipid monolayer, suggesting that localized proton flux could possibly be significant in some circumstances (1174; Section 7.51).

In summary, this electron transfer system is structured as a small number of highly organized complexes, coupled by small mobile carriers, both lipophilic (quinone) and water-soluble (cytochrome *c*). The kinetics can be explained in terms of freely diffusing species, where the diffusion can occur over distances greater than 100 Å.

6.64 The Photosynthetic Electron Transport System of Thylakoids (see 534, 1573, 1042)

The photosynthetic electron transfer chain has many characteristics in common with the mitochondrial respiratory system, but with an additional level of complexity. As shown in Figure 6.5, the photosynthetic electron transfer system contains three major multisubunit, transmembrane complexes, photosystem I, photosystem II, and the cytochrome b_6f complex. In addition, there are small water-soluble proteins, plastocyanin and a ferredoxin, as well as ferredoxin-NADP$^+$ oxidoreductase, a peripheral protein, and a membrane-bound hydrogen carrier, plastoquinone (Figure 6.6). Light-harvesting complexes also are present to collect light and funnel the excitation energy to the two reaction centers. The trapping of the excitation energy by the reaction centers results in a charge separation, creating a strong oxidant and strong reductant on opposite sides of the membrane. This results in the oxidation of water to oxygen by photosystem II and produces the driving force for the electron transfer chain, whose function is to move protons across the membrane. The proton electrochemical potential difference across the bilayer is used by the ATP synthase, called the coupling factor, to synthesize ATP. The structure of the reaction center of photosystem II may be similar to that of the bacterial reaction center, which has been determined by X-ray diffraction (see Chapter 3). The cytochrome b_6f complex is very similar to complex III (cytochrome bc_1 complex) of the mitochondrial respiratory chain, and the coupling factor is very similar to the mitochondrial ATP synthase.

The role of plastocyanin is similar to that of cytochrome *c* in the respiratory system, serving as a soluble electron shuttle. Plastoquinone plays an analogous role to that of ubiquinone in the mitochondrial electron transfer system.

Unlike the respiratory system, the major membrane complexes in the photosynthetic system are present in approximately equimolar amounts (1573) (see Figure 6.5). However, there is no strong evidence to suggest they form complexes. Quite to the contrary, this system displays dramatic lateral heterogeneity that results in photosystem II being localized within the appressed regions, and photosystem I and the coupling factor localized within the nonappressed regions (see Section 4.52). The plastoquinone and cytochrome b_6f complex appear to be evenly distributed in both regions. This lateral separation requires diffusion over a distance of at least 1,000 Å to couple the two photosystems. Most likely, plastoquinone is the main carrier of reducing equivalents over this distance; however, its rate of lateral diffusion is not known with certainty. The rate-limiting step under optimal conditions appears to be the oxidation of plastoquinol by the cytochrome b_6f complex, but whether this is limited by the rate of lateral diffusion is not known.

The lateral separation in the thylakoids is obviously not designed to optimize electron transfer between components. It may be that the lateral separation is related to the efficient distribution of light excitation between the two photosystems (see 534). The phosphorylation of light-harvesting complex II results in its redistribution within the appressed and nonappressed membranes favoring interactions with photosystem I in the nonappressed regions, making more of the energy available to this complex under certain conditions.

6.7 The Interactions Between Membranes and Soluble Enzymes

Biomembranes play important roles in the function of a number of soluble enzymes. Upon disruption of the cell, many enzymes are located, in part, in the soluble fraction and, in part, bound to membranes. The categorization of an enzyme as a peripheral membrane protein depends on the strength of the protein–membrane interaction and on the manner of membrane isolation. There are also some soluble enzymes which are recruited to the membrane under specific conditions and will be localized to the membrane or cytosol depending on the physiological state of the cell. There is an additional group of soluble enzymes whose substrates are membrane-bound and which, therefore, must bind to the membrane, at least transiently, in order to function.

The membrane serves several purposes in these cases: (1) localization or compartmentalization of an enzyme or assembly of enzymes; (2) allosteric activation or inactivation of enzymes under specific conditions and at specific locations in the cell; (3) providing a medium where lipophilic substrates can be converted to products.

6.71 Soluble Enzymes Which Are Recruited to the Membrane when Needed

There is a small but significant list of enzymes in this category, and it is likely that more examples will be found in the near future. The most prominent example is protein kinase C, but others are also well characterized.

Protein Kinase C (see 1071, 80, 1121 for reviews)

This enzyme is at the heart of the signal transduction system which is initiated by the breakdown of phosphatidylinositols in the plasma membrane (see Section 9.73). Extracellular agents, such as neurotransmitters, hormones, and growth factors, bind to specific cell surface receptors, resulting in the generation of second messengers through the activity of a phospholipase C, which causes the hydrolysis of phosphatidylinositols. One product of the phospholipid degradation is inositol 1, 4, 5-triphosphate, which mobilizes internal calcium. The second product, *sn*-1, 2-diacylglycerol, activates protein kinase C, which in turn specifically phosphorylates targeted proteins, many of which are membrane-bound, such as the epidermal growth factor receptor. This system is now implicated in numerous cellular functions including mitogenesis, cellular differentiation, and exocytosis.

Prior to treatment with the external signalling agent, protein kinase C is a cytosolic, inactive enzyme. However, after stimulation of the cells, the enzyme rapidly partitions into the membrane fractions and is activated. Studies in vitro have demonstrated that this shift to the membrane and activation require acidic phospholipids, with phosphatidylserine being particularly effective, as well as Ca^{2+}, and diacylglycerol. The phospholipid specificity of activation depends to some extent on the substrate (76). Short-chain diacylglycerols which are cell-permeable, such as dioctanoylglycerol, can be added to cells and activate the phosphotransferase activity of the enzyme (80). The natural second messengers, long-chain diacylglycerols, are insoluble and remain bound to the membrane. Phorbol esters (Figure 9.17), which are tumor promoters, appear to act as analogues of diacylglycerols, and have been shown to bind tightly to the enzyme and to act in the same way as the endogenous signal. Upon binding of phorbol esters or diacylglycerols, the affinity of the enzyme for Ca^{2+} and for phosphatidylserine is enhanced. Sphingosine has been shown to be a competitive inhibitor of the enzyme (585).

Protein kinase C actually represents several distinct but closely related polypeptides with a molecular weight of about 80,000. Three forms have been purified from rabbit brain, for example (699). Protein kinase C has a two-domain structure (1121). One domain contains the catalytic sites for ATP and the protein substrate and functions as a serine and threonine phosphotransferase. The second domain appears to be involved in binding to phosphatidylserine, diacylglycerol, and Ca^{2+}. The two domains can be separated by mild proteolysis, resulting in a fully active catalytic fragment with molecular weight about 50,000. Hence, the catalytic functions are activated by two mutually exclusive pathways, proteolysis and membrane binding. This is also observed for some other lipid-requiring enzymes such as pyruvate oxidase (see Section 6.5). The physiological relevance of the proteolytic activation is not clear.

The pure enzyme has been examined in phospholipid vesicles (1361) and in phospholipid–Triton X-100 mixed micelles (588). Both approaches have shown that lipid binding per se is necessary but not sufficient for enzymatic activation. For example, 2% phosphatidylserine is sufficient for binding of the enzyme to phospholipid vesicles, but 50% phosphatidylserine is required for optimal phosphorylation (1361). The normal content of phosphatidylserine in plasma membrane is 8–10% (588). The fact that the enzyme is activated by mixed micelles containing about 20 mol % phosphatidylserine in Triton X-100 shows that a bilayer structure is not required, which is generally the rule for lipid-requiring enzymes. The activity observed as a function of the concentration of phosphatidylserine indicates cooperative interactions between the enzyme and lipid, a result which is also similar to findings with other enzymes (see Section 6.4). It is estimated (80, 588) that the enzyme functions as a complex containing one enzyme monomer, one molecule of diacylglycerol (or phorbol ester), one or more Ca^{2+}, and at least four molecules of phosphatidylserine. Binding is certainly to the membrane surface, but the role of Ca^{2+} is not known. It may chelate both groups on the enzyme and the phosphatidylserine carboxyl groups and/or may play an allosteric role upon binding to the protein. The fact that the enzyme can be labeled with iodonaphthalene l-azide (see Table 4.1) suggests that there is some penetration of the hydrophobic core of the bilayer, though this is not certain (1361).

Box 6.2 Surface Dilution Effects in Mixed Micelles

Figure 6.7 illustrates an interesting aspect of protein kinase C and other membrane enzymes which are assayed in mixed micelle systems. In this experiment, the ability of the enzyme to bind to phorbol ester is measured as a function of the concentration of Triton X-100. If the mol percent of phosphatidylserine is maintained as the Triton X-100 concentration is increased, the enzyme binds to phorbol ester, even at high detergent concentration. However, if the phospholipid concentration is kept constant as the amount of detergent is increased, the enzyme does not bind to phorbol ester at high detergent concentrations. It is not the bulk concentration of lipid which is important in the activation of the enzyme, but rather the surface concentration or amount of lipid per micelle, which is critical. As this value drops, the ability of the enzyme to bind to these mixed micelles and be activated also diminishes. This is called surface dilution.

Other Examples

The rate-limiting step in the biosynthesis of phosphatidylcholine is the synthesis of the intermediate CDP-choline. The enzyme which catalyzes the reaction is CTP:phosphocholine cytidyltransferase. This is a cytosolic enzyme, but it has

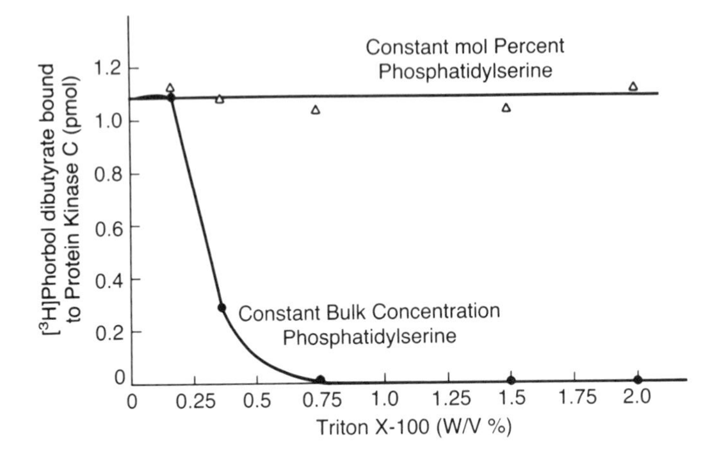

Figure 6.7. The effect on a membrane enzyme of surface dilution of a required lipid component. The binding of a phorbol ester to purified protein kinase C was measured in mixed micelles. Phosphatidylserine is required for this interaction. In one experiment, the ratio of phosphatidylserine to detergent (Triton X-100) was kept constant at 16 mol %. An increase in Triton X-100 did not reduce phorbol ester binding. The mixed micelles of phospholipid and detergent have the same composition regardless of the total amount of detergent added. In a second experiment, the amount of phosphatidylserine is kept constant at an amount required to be 16 mol % for 0.2 % Triton X-100. Increasing the detergent results in dilution of the phospholipid among the increased number of micelles. The surface dilution effect corresponds to loss of phorbol binding by the enzyme. Adapted from ref. 588.

recently been recognized that this enzyme binds to the endoplasmic reticulum where it is activated (see 1499, 1609, 244). This is the site where phosphatidylcholine biosynthesis occurs (see Section 10.4). Note that both substrates for this enzyme are water-soluble and not bound to the membrane. The signal that causes the enzyme to be recruited to the membrane is not yet clear. Possibly, the presence of diacylglycerols in the membrane provides the signal. Other possibilities include (1) an increase in long-chain fatty acids or acyl CoA derivatives (see 244), (2) the depletion of phosphatidylcholine in the microsomal membrane, and (3) the dephosphorylation of the enzyme itself, converting it to a membrane-binding, active conformation. More work is required to clarify the situation.

Diacylglycerol kinase has been shown to translocate from the cytosol to membranes dependent on the content of membrane-bound diacylglycerol (95). This enzyme converts diacylglycerol to phosphatidic acid and is at least partially responsible for the half-life of diacylglycerol in the membrane. Diacylglycerol and fatty acids have also been implicated in the binding of α-actinin to membranes (169). This protein has been speculated to play a role in the anchoring of microfilament bundles to the plasma membrane. The stimulation of the α-actinin/actin complexes by diacylglycerol and fatty acids may be important in the physiological activation of platelets.

Two additional examples of enzymes that bind conditionally to biomembranes are pyruvate oxidase and phosphatidylserine synthase from *E. coli*. In both cases, the presence of the substrate stimulates recruitment of the enzyme from the cytosol to the cytoplasmic membrane. Pyruvate oxidase has already been described as an example of a lipid-requiring enzyme (Section 6.5). It takes on the solubility characteristics of a membrane protein in the presence of pyruvate, which reduces the protein-bound flavin cofactor. This enzyme reduces ubiquinone within the membrane, and so must be membrane-bound in order to function. Phosphatidylserine synthase is normally purified associated with ribosomes, but the presence of either the substrate, CDP-diacylglycerol, or the product, phosphatidylserine, results in membrane binding (874).

6.72 Soluble Enzymes and Enzyme Assemblies Which May Be Membrane-Bound In Vivo (see 934, 1374 for reviews)

There is a growing literature on the possible association of a number of soluble enzymes with membranes. Most of the work concerns enzymes involved in the citric acid cycle (144) and the ß-oxidation of fatty acids (1407) binding to the inner mitochondrial membrane, as well as the binding of glycolytic enzymes to the membrane of the red blood cell. There is some evidence that the enzymes in the mitochondrial matrix may be organized in multienzyme complexes bound to the surface of the membrane (see 934, 1374), though the evidence is generally not compelling. In some cases, specific binding sites have been implicated, such as NAD^+-linked dehydrogenases (e.g., pyruvate dehydrogenase complex) binding to the NADH:ubiquinone oxidoreductase (complex I) (1406, 1170). Creatine phosphokinase reportedly binds specifically to cardiolipin in the inner mitochondrial membrane (1035), and hexokinase also binds to mitochondria, possibly to the outer membrane (1517, 1055). The distribution of hexokinase between soluble and mitochondrial-bound forms appears to be modulated by hormones or metabolites (see 1517). In all these cases, the rationale for the association of these enzymes to the membrane, and possibly to other soluble enzymes, involves substrate channeling (see 1375 for review). For example, ATP transported across the inner mitochondrial membrane could be more efficiently utilized by hexokinase located near the membrane. Evidence that this is correct, however, is largely circumstantial.

A number of studies have characterized the binding of glycolytic enzymes to the acidic cytoplasmic domain of the Band 3 anion channel in the erythrocyte membrane (see 876). Examples include aldolase (705), phosphofructokinase (705, 706), and glyceraldehyde-3-phosphate dehydrogenase (1215). It is suggested that, in some cases, the enzyme may be stored in an inactive form on the membrane and readily mobilized to the active cytoplasmic form in the presence of appropriate metabolites (705). There is no consensus, however, on whether these associations occur in vivo or are an in vitro artifact. The associations are relatively weak and dependent on ionic strength. The same is true for some of the putative mitochondrial membrane associations (e.g., 144). However, the coupling of glycolytic functions to membrane activities, and compartmentalization of these enzymes, is a generally attractive concept and evidence exists for this in other systems (e.g., 1156).

Convincing evidence for the physiological relevance of the membrane interactions of these enzymes and enzyme assemblies has been hard to produce, but this area will likely receive additional attention in the near future.

One additional class of soluble proteins that interact with membranes are cytotoxins. These are discussed in the context of pore-forming agents in Section 8.61.

6.73 Blood Coagulation Factors—Extracellular Enzymes Activated by Membrane Binding (see 688, 917 for reviews)

Blood coagulation occurs as a result of a complex cascade involving many serum factors. The cascade consists in large part in the sequential proteolytic activation of a series of zymogens which, once activated, proceed to activate one or more other zymogens. The final result is the conversion of fibrinogen to fibrin. Many of the steps in this cascade require the activated protease and/or the substrate zymogen to be membrane-bound to platelets, endothelial cells, or other types of cells. Specific questions of interest are (1) How are the proteins bound to the membranes? and (2) What role does the membrane play in activating the enzymes involved? The answers are not clear-cut.

All of the blood coagulation factors which bind to membranes require acidic phospholipids. Several of these, including Factors VII, IX, X, and prothrombin, also require Ca^{2+}. These four proteins are modified in a vitamin K-dependent reaction to contain γ-carboxyglutamic acid residues, which are located in homologous amino-terminal portions of each polypeptide. These provide most of the high-affinity Ca^{2+} binding sites which are necessary for the proper interaction with the acidic phospholipids (e.g., phosphatidylserine) in the membrane (1018). However, little is known about the specifics of how Ca^{2+} facilitates the membrane–protein interactions, though crossbridging between protein and lipid groups is one possibility. Not all of the blood coagulation factors require calcium. Factor V can interact directly with the negatively charged phospholipids (684), and another component, "tissue factor," is an integral membrane protein (see 449). For the most part, however, the factors interact primarily with the bilayer surface via electrostatic interactions, though some penetration of the membrane may occur (e.g., 827, 787).

The presence of phospholipids greatly alters the steady-state kinetics of the proteolytic activation reactions (see 577). The K_m for the protein substrate is drastically reduced, for example, from 181 μM to 0.058 μM for the activation of Factor X by Factor IXa (the "a" designates the activated form). The addition of another protein, Factor VIIIa, increases the V_{max} by more than 200,000-fold. Since both the enzyme carrying out the reaction and the substrate exist in membrane-bound and free forms during these assays, it can be difficult to distinguish the actual cause of the effects of lipids in these reactions. For example, it is reported that proteolytic activation of Factor X occurs when Factor X is membrane-bound, but that the free and bound forms of the protease, Factor IX, are equally effective catalytically (974). Factor X can also be activated by a complex of Factor VIIa and tissue factor. In this case, the proteolytic activation of Factor X occurs when Factor X is free in solution and not bound to the membrane (449). Another example is provided by the activation of prothrombin by Factor Xa (1496, 699). A low K_m observed in this case correlates with the concentration of the prothrombin substrate on the phospholipid vesicle surface. However, when a cofactor, Factor Va, is added, forming a complex with Factor Xa, there is no longer any dependence on the surface concentration of prothrombin on the phospholipid vesicles (1496). The main point is that these systems are quite complex, and that the role of the lipids goes beyond simply providing a convenient surface on which components are concentrated.

The blood coagulation factors are part of a group of Ca^{2+}-dependent lipid binding proteins (1561). In some cases, the functions of these proteins are not known (1561), though some are associated with the cytoskeleton (369). Phospholipases, discussed below, are also Ca^{2+}-dependent.

6.74 Phospholipases—Soluble Enzymes Which Act on Membrane-Bound Substrates (see 324, 1486 for reviews)

Many soluble enzymes utilize phospholipids as substrates, including a number of phospholipases. By far the most intensively studied class of enzymes is phospholipase A_2, which hydrolyzes phospholipids at the *sn*-2 position to form fatty acid and lysophospholipid products (see Figure 4.3). These enzymes have been isolated primarily from cobra and rattlesnake toxins and from bovine and porcine pancreas. These are all small proteins with molecular weights around 14,000 and with substantial sequence homology. The three-dimensional structure is known to high resolution for several of the enzymes, and they also reveal great similarity (see 1210). The pancreatic enzymes are made as inactive zymogens which are activated proteolytically by removal of seven residues from the carboxyl-terminal end.

Phospholipase A_2 is particularly interesting from the viewpoint of membrane enzymology because the enzyme is activated in some way when it interacts with aggregated forms of the substrate, such as in micelles or in bilayers. Figure 6.8 shows the rate of hydrolysis of a short-chain phosphatidylcholine catalyzed by the porcine pancreatic phospholipase A_2 and by the zymogen form. This substrate is monomeric below about 1.5 m*M*, but forms micelles above this concentration. Both the zymogen and active enzyme show similar low activity with the monomeric substrate, but phospholipase A_2 is substantially more active when the substrate is in a micelle.

The preference shown for aggregated substrates has been the focus of much study. The enzyme kinetics have been examined using substrates presented as monomers, in pure lipid micelles, in mixed micelles with Triton X-100, in

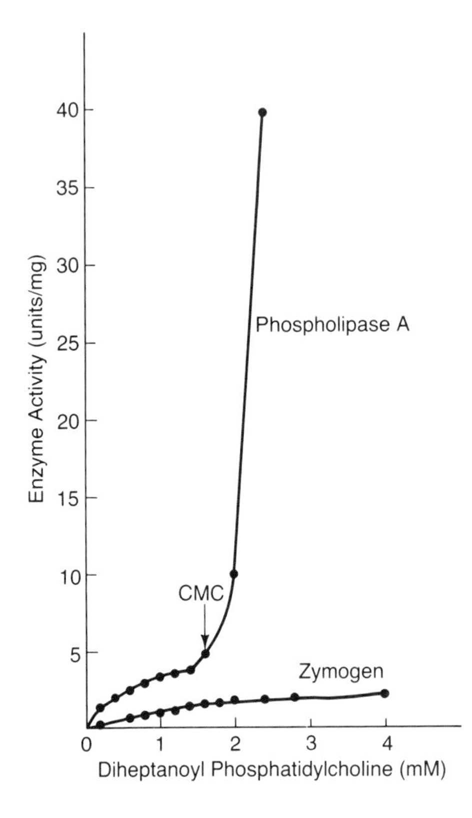

Figure 6.8. Activity of phospholipase A_2 and its zymogen as a function of concentration of a short-chain phosphatidylcholine. Note the increase in specific activity at concentrations of lipid substrate above the critical micelle concentration (CMC). Adapted from ref. 1157. Reprinted with permission from "Zymogen-Catalyzed Hydrolysis of Monomeric Substrates and the Presence of a Recognition Site for Lipid-Water Interfaces in Phospholipase A_2," by W. A. Pieterson et al., *Biochemistry*, vol. 13, p. 1456. Copyright 1974 American Chemical Society.

monolayers at an air–water interface, and in phospholipid vesicles. The enzyme requires Ca^{2+} for activity in all cases, and the binding site for a single Ca^{2+} is observed in the X-ray studies. Unlike the blood coagulation factors, this membrane-binding protein does not contain γ-carboxyglutamic acid and does not require acidic phospholipids.

Box 6.3 Several Theories Have Been Proposed to Explain the Activation of Phospholipases

Several studies have shown that binding to micelles or to bilayers precedes and is experimentally distinct from the activation step, in which enzyme turnover is dramatically enhanced (see 967). This is similar to other lipid-activated enzymes described in this chapter. Despite the large body of kinetic and binding data and the structural information which is available, there is no consensus as to what occurs to activate phospholipase A_2 in the presence of lipid bilayers or micelles. Several theories have been postulated.

1. The enzyme binds to the bilayer by interaction with an interface recognition region, which is distinct from the active site, requires Ca^{2+} to be in the proper

configuration, and is postulated to penetrate beyond the surface of the membrane. This model is based, in part, on chemical modifications in the amino-terminal portion of the polypeptide which specifically influence the interaction with aggregated substrates (e.g., 335, 697). Following interaction with the interface recognition region, activation is presumed to involve conformational changes induced in the protein. It should be noted that the bovine and porcine enzymes are crystallized as monomers, whereas phospholipase A_2 from rattlesnake venom is a dimer (1210). The proposed interface recognition region observed in the monomeric phospholipases is largely inaccessible and is in the intersubunit contact surface of the dimeric enzyme (1210).

2. The dual phospholipid model postulates that the enzyme has two or more binding sites for phospholipids, and is based largely on kinetics studies of phospholipids in mixed micelles. This model can accommodate a role for enzyme aggregation to form dimers or higher oligomers as an essential part of the activation scheme (see 967), as well as a possible conformational change resulting in enhanced catalytic activity.
3. The phospholipid substrate has been postulated to be in a different conformation when in an aggregated state compared to the monomeric form, and this has been used to explain greater catalytic activity with aggregated forms.
4. The facilitation of product removal in micelles or bilayers has been postulated to be responsible for enhanced catalysis. Also, the accumulation of products per se can accelerate the activity of phospholipase A_2, though the mechanism is unclear.

One of the difficulties in analyzing the activation process is to separate the phenomena of lipid binding and lipid activation. It is clear from studies with unilamellar phospholipid vesicles that the physical state of the bilayer is a critical parameter in both steps. For example, phospholipase A_2 has been shown to bind optimally to dipalmitoyl phosphatidylcholine in the gel phase (967, 854), and the initial binding does not require Ca^{2+}. The activation of the enzyme in this system appears to be the step requiring Ca^{2+} and, in studies with vesicles, also appears to require that the phosphatidylcholine bilayer be in a state where there are packing defects or structural fluctuations such as those present during a thermal phase transition (see 967, 854). Protein interactions, such as dimerization, may be involved in either the binding or activation step. Under some circumstances, once activated, the enzyme can remain activated for at least 30 minutes (967).

Phospholipids with short acyl chains and with small headgroup substituents on the phosphate are the best substrates (697). Although acidic phospholipids are not required, anionic charge at the interface results in an increased affinity (695). The enzyme, once bound to the interface, can "scoot" along the bilayer surface and hydrolyze several thousand phospholipid molecules per minute, before dissociating from the bilayer (695, 696). The residence time of the protein at the bilayer interface is highly dependent on the lipid and solution conditions.

The details of the conformational changes resulting in activation and of the mode of binding to the bilayer are not known. The role of bilayer defects in the kinetic acceleration, which has been elucidated with model bilayer systems (1489), is intriguing, though its meaning for the enzyme in vivo is not clear.

Finally, it should be noted that phospholipase A_2 is responsible for the release of arachidonic acid from membranes, and that subsequent conversion of this fatty acid to leukotrienes and prostaglandins is part of the inflammatory response (661). Anti-inflammatory steroids induce proteins called lipocortins, which specifically inhibit phospholipase A_2 and which mimic the anti-inflammatory effects of the steroids. The lipocortins are also substrates for protein kinase C and for protein tyrosine kinases, suggesting that their activity may be regulated by this mechanism (661, 155). Lipocortins do not appear to inhibit by forming a strong complex with phospholipase A_2, but they appear to interact directly with membranes (155, 286) and may be members of a family of Ca^{2+}-regulated membrane binding proteins (499, 920).

6.8 Chapter Summary

Many cellular processes are catalyzed by membrane-bound enzymes. The membrane can serve several functions. By being bound to a particular membrane or region of a membrane, the site of catalysis can be localized to a portion in the cell. There are numerous examples in which several enzymes which act sequentially are localized together in this manner, thus enhancing the overall reaction rate. Many membrane enzymes are membrane-spanning proteins which are involved in conveying solute molecules across the bilayer or in transmitting information across the bilayer in the form of transmembrane allosteric signals. Other membrane enzymes are peripheral membrane proteins which, in some cases, are bound to the membrane only in response to a physiological signal.

In many examples, the lipid bilayer is a relatively passive participant in the activity of membrane-bound enzymes, providing a medium for enzymes and substrates to interact. In other cases, the membrane enzymes function optimally only in the presence of particular lipids, though there are few cases where there is an absolute specificity of a particular lipid for activating an enzyme. Studies on the effects of lipids on isolated membrane enzymes are often not easily interpreted. Indeed, the in vitro examination of purified membrane enzymes presents numerous problems that are not considerations for soluble enzymes.

Chapter 7
Interactions of Small Molecules with Membranes: Partitioning, Permeability, and Electrical Effects

7.1 Overview

At the beginning of the previous chapter, the point was made that biomembranes are not merely passive barriers defining the boundaries of cells or organelles. In this chapter, we recognize that, in fact, being a barrier to solute flux between aqueous compartments is an essential part of membrane function. The major focus of this chapter is the interaction between the phospholipid bilayer and small solute molecules, both ions and nonelectrolytes. We will first discuss the binding of small molecules to the membrane, either adsorbed at the surface or partitioning into the interior of the bilayer. Any solute that can partition into the membrane can diffuse across the bilayer and exit from the opposite side. Membrane permeability to nonelectrolytes is discussed in this context. In order to understand the way in which ions interact with the membrane, the total electric potential energy profile of the membrane must be considered. The electrical component of the free energy of an ion near or within the membrane is critical not only for understanding how divalent and monovalent metal ions bind to the membrane surface, but also for estimating the local pH at the membrane surface, and for modeling how some enzymes and ion channels might be regulated by a voltage across the bilayer.

Whereas in this chapter our major concern is to describe the properties of the lipid bilayer, in Chapter 8 the focus is on solute transport in biomembranes which is protein-mediated.

7.11 Analysis of the Adsorption of Ligands to the Bilayer

Before proceeding, it is important to discuss how to analyze the adsorption of ligands to the bilayer. This problem arises in numerous experimental situations and there are several formalisms which have been employed. We are particularly interested in the adsorption of ions and amphiphilic molecules to the bilayer, but the same formalism can also apply to any molecule which partitions into the membrane or binds to the bilayer surface. Studies of the binding of small molecules to proteins generally involve specific, well-defined binding sites, such as the ATP-binding sites of an ATPase. In such studies, one can obtain (1) the number of binding sites per protein molecule, (2) the strength of the interaction, and (3) the degree to which there is interaction between binding sites, i.e., cooperativity. Experimentally, this is obtained by analysis of the binding curve (or isotherm) in which the amount of bound ligand (e.g., ATP) is measured as a function of the concentration of free (or unbound) ligand. The analysis is relatively straightforward because there is a fixed total number of binding sites, which are either occupied or not, and this is experimentally determined. When a ligand binds to a defined binding site in a membrane, such as the binding of a hormone to a hormone receptor, the same formalism applies.

However, the analysis of the adsorption of a molecule to the lipid bilayer surface is more complicated because the definition of the binding site is not so clear. The formalism which is used and the information which can be obtained depend on the experimental situation. Several approaches are indicated below.

Partition Equilibrium

This model treats the membrane as a separate phase. Small molecules partition between the aqueous phase and the membrane according to a partition coefficient, K_p:

$$K_p = C_b/C_f \qquad [7.1]$$

where C_b and C_f are the concentrations of ligand bound to the membrane and free in solution. In this model, there is no saturation and, in principle, the concentration of bound ligand will increase indefinitely as the concentration in the aqueous phase in increased. Certainly, this is not realistic, and this model can be used only when a relatively small amount of ligand is bound. Note that this model does not explicitly have any information about binding sites built into it. At low occupancy, i.e., a small fraction of the potential binding sites filled, the binding equations associated with virtually any model will simplify to the form of a partition constant (equation 7.1).

The partition coefficient can be expressed in several ways. A common method is to express the concentration of bound ligand as a surface concentration, N_b, with units of mol/cm^2. The partition coefficient, ß, then has units of length (cm):

$$ß = \frac{N_b\,(\text{mol/cm}^2)}{C_f\,(\text{mol/cm}^3)} \qquad [7.2]$$

If N_b is divided by the membrane thickness, δ, the concentration in the membrane is converted to units of (mol/volume), where the volume refers to the volume of the bilayer, or that portion to which the solute is restricted.

$$C_b = (N_b/\delta), \text{ so } K_p = ß/\delta$$

Both types of partition constant are used (e.g., see 439).

Langmuir Adsorption Isotherm

In this model, the membrane is treated as a regular lattice of potential binding sites. If the maximum concentration of ligand which can be bound is C_b^{max}, one can define θ as the fraction of sites occupied by ligand as

$$\theta = C_b/C_b^{max} \quad [7.3]$$

The fraction of unoccupied sites is $(1 - \theta)$, and a simple association constant can be defined as

$$K_L = \frac{\theta}{(1-\theta)C_f} \equiv \frac{C_b}{[\text{Sites}]_{free} \cdot C_f} \quad [7.4]$$

This model assumes a fixed concentration of independent, equivalent, and immobile binding sites. It is a simple model, but in some cases has been successfully used to analyze adsorption data (see 26). A serious problem with this approach is that in many cases the binding sites are not discrete entities which can be represented by a single fixed lattice position, and a calculation of the number of available binding sites is not as simple as in this model.

For example, if a ligand binds to a maximum of 1 per n phospholipid molecules, a binding site can be defined in terms of n lipids binding in some way to the ligand. The simple Langmuir model would define the number of binding sites; e.g., for $n = 2$ and 10,000 lipid molecules there would be 5,000 binding sites. This is not correct, however, because the potential binding sites are overlapping (see Figure 7.1). For 10,000 lipid molecules in a square lattice, there are 20,000 possible pairs of neighboring sites, ignoring edge effects. The correct computation of the number of available binding sites is at the heart of the problem of obtaining a correct expression for the binding isotherm.

Complexation to *n* Lipids

Part of the problem of correctly accounting for overlapping binding sites can be solved by simply expressing the equilibrium in terms of the n phospholipid molecules which are involved in ligand binding. One then has an expression for complex formation between the ligand, L, and n phospholipids, P.

$$L + nP \rightleftarrows LP_n$$

$$K_a = \frac{[LP_n]}{[L][P]^n} \quad [7.5]$$

This has been used successfully to analyze Ca^{2+} binding to the phosphatidylcholine bilayer, for example, with $n = 2$, where the Langmuir isotherm does not work well (27). The equations derived from this model predict a nonlinear Scatchard plot with concave curvature (see 1379, 1378).

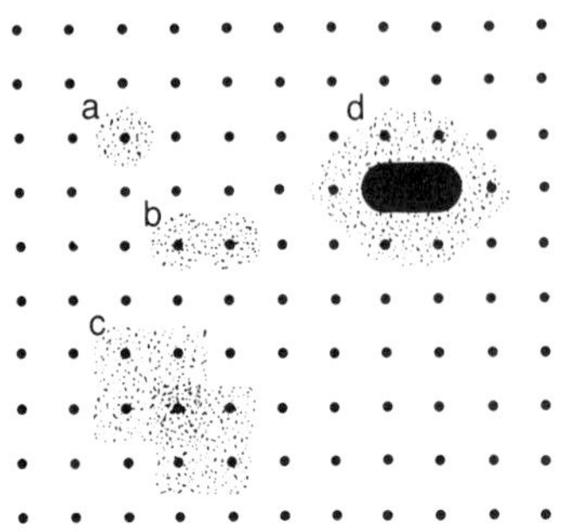

Figure 7.1. Models of ligands adsorbed by a lipid bilayer. The bilayer is viewed as a square lattice where each lattice point represents a lipid. (a) A ligand which binds to a single lipid at the surface; (b) a ligand which can bind to two lipids; (c) a large ligand which sits over four lipids, shown in two overlapping potential binding positions; (d) a ligand such as an amphiphile or hydrophobic ion which inserts into the membrane but is constrained to the surface, effectively binding to the neighboring lipid molecules. This molecule occupies two lattice positions and binds to six lipids in this model.

Taking Ligand Shape into Account

Even the above analysis does not fully solve the analytical problem, because it does not take into account the shape of the ligand. Viewing the membrane as a two-dimensional lattice, a complete analysis must account for large ligands of particular shapes which bind to several lattice positions in a particular geometric arrangement. For example, a rod-shaped molecule will bind to lattice sites arranged linearly. Every molecule that binds to the lattice blocks a certain number of additional binding sites, and this will be shape-dependent. A complete analysis, taking into account overlapping binding sites as well as the shape of the ligand, is quite complicated, but has been carried out (see 1378). This analysis can also be extended to include ligands which insert into the membrane and occupy additional lattice positions (1378).

The main point is that the formal thermodynamic analysis of the binding of even small molecules to the membrane can be quite complicated and often requires more than the simple application of a standard equation. In addition to the problem of overlapping binding sites restricted by steric interactions, we will see that electrostatic interactions must also be taken into account in order to completely understand the binding of charged ligands to the membrane surface.

7.12 Classes of Ligands Which Interact with the Lipid Bilayer

There is an enormous literature describing the interaction of small molecules with membranes and model membranes. For convenience, these molecules can be crudely classified according to their polarity characteristics: nonpolar, amphipathic, and ionic. It is useful at this point to review briefly some of these studies to see the motivation for much of this work.

Class I: Nonpolar Solutes

Since the lipid bilayer is generally thought of as a two-dimensional fluid, its ability to actually serve as a solvent for small nonpolar molecules is of interest.

These data may be valuable for generally understanding how nonpolar portions of proteins interact with the bilayer. One example is the study of hexane dissolved in dioleoyl phosphatidylcholine (755). Generally, nonpolar molecules like hexane appear to be localized in the center of the bilayer. This has been confirmed for hexane by neutron diffraction studies, though there appears to be a more complicated interaction at high levels of hexane.

Class II: Amphipathic Molecules

This is by far the largest group of molecules which has been examined. This group includes many anesthetics, drugs, and tranquilizers whose pharmacological activities are dependent on their ability to interact with membranes. A number of antibiotics and other natural products, such as bile salts and fatty acids, are also in this category. In addition, many membrane-directed fluorescent and EPR spin probes are amphipathic molecules. All of these molecules have distinct polar and nonpolar moieties and tend to interact with the membrane surface. At sufficiently high concentrations, some of these amphipathic molecules will exhibit detergent-like action and disrupt the bilayer, and it has been reported that membrane damage also results when biomembranes interact with some amphipathic species even at moderate concentrations (904).

Anesthetics

There are many classes of anesthetics, whose structures range from atomic xenon to complex organic heterocycles. In general, there is a good correlation between the pharmacological activity of a compound and its oil/water partition coefficient. This suggests that the mechanism may involve nonspecific interactions with membrane lipids in the nerve cell membranes. A number of studies have been directed at examining the nature of the interaction between anesthetics and biomembranes or model membranes. Although these studies have shown that these compounds can perturb the lipids in the bilayer, the mechanism(s) of action is still not known in any case and may very well involve direct or indirect effects on specific proteins in the neural membranes (545, 462, 382). The effects of anesthetics on membranes have been examined by NMR methods. Studies include the general anesthetic chloral hydrate (450), local anesthetics such as procaine (739), tetracaine (739), and dibucaine (944), and anesthetic steroids (911). Anesthetic steroids do not partition into the membrane differently than non-anesthetic steroids, and specific structural features of these molecules appear to result in particular lipid perturbations (911).

Drugs

By far the most studied drug is chlorpromazine, whose partitioning into the lipid bilayer vesicles (886) and into red blood cell membranes (858) has been studied. The partition coefficients are similar for both cases (886). Substantial bilayer rearrangement and damage is caused by high amounts of incorporation of chlorpromazine into the red blood cell (858). The maximum amount taken up corresponds to a volume greater than the bilayer itself, and apparently large amounts of presumably the uncharged form of the drug aggregate in the center of the bilayer, causing a large increase in membrane thickness. More modest amounts of the drug cause lysis by forming holes with a size equivalent to a 14 Å diameter. The nature of these holes is not known.

Antibiotics

Several antibiotics are of particular interest in that their toxicity appears to be due to membrane interactions.

(1) *Polymyxin B* (792) is a cyclic polypeptide which has five cationic residues and a hydrophobic acyl side chain. This antibiotic is specific for gram-negative bacteria (e.g., *E. coli*) and is targeted primarily at the negatively charged lipids in the cell outer membrane and cytoplasmic membrane. Polymyxin has a high affinity for negatively charged lipids, which is apparent in studies with phospholipid vesicles, where lateral phase separation of lipid species can be induced by the antibiotic. Interaction with *E. coli* results in increased permeability and loss of cytoplasmic material, evidently caused by membrane damage.

(2) *Polyene antibiotics—amphotericin B (225, 351, 1597), nystatin (1085), and filipin (351)* are membrane-active antifungal agents. They can induce lysis in yeasts, erythrocytes, and various mammalian cells. Their lytic action has been suggested to arise from a specific association with sterols in the membrane to form a pore (see Section 8.15). This complexation has been observed by NMR (e.g., 351), and the binding constants to different sterols incorporated into phospholipid vesicles have been examined (e.g., 1597). The fluorescence of filipin complexed to sterols has been used to identify sterol-rich membranes (see Section 4.41).

(3) *Adriamycin* (530, 873, 138) is an antibiotic in the anthracycline glycoside family. It is an anti-cancer agent whose use is limited by cardiotoxicity. Adriamycin appears to have a specific affinity for cardiolipin in the membrane, and this may be related to its toxic effects. Molecular modeling and energy minimization have been used to suggest a conformation of adriamycin on the bilayer surface, stabilized through electrostatic interactions with the cardiolipin phosphate groups (138).

Detergents

The interaction between detergents and membranes has already been discussed in some detail (see Section 3.2). One particularly interesting example is digitonin and related saponins, which have an affinity for membrane-bound sterols and apparently bind to cholesterol in the bilayer with 1:1 stoichiometry (1070, 991). The polyene antibiotics, adriamycin, and digitonin are unusual examples of amphiphilic molecules with an affinity for particular membrane lipids.

Membrane Probes

Fluorescent and spin-labeled EPR probes are often amphipathic molecules. One example is the carbocyanine dye used for fluorescence recovery after photobleaching (see Table 5.1 and Figure 5.6). This probe binds to one side of the membrane and does not undergo flip-flop (1598), which could result in its binding to internal membranous organelles. This sidedness has been demonstrated (1598). Other examples are the EPR spin probes (177) and fluorescent probes (1511) which bind to the membrane and are used to monitor the electrical properties of the bilayer (see Figure 7.9). See Section 7.41 for further discussion of membrane probes.

Class III: Hydrophobic Ions

These are molecules where the charge is surrounded by nonpolar groups, in contrast to simple amphiphiles, where the polar and nonpolar parts of the molecule are separate entities. When an amphipathic molecule binds to the membrane, the polar group is located in or near the aqueous phase with the nonpolar group embedded in the bilayer. The charged portion of a hydrophobic ion, however, becomes submerged in the membrane, at least several angstroms beneath the surface (see 439). Examples are the tetraphenylboron and tetraphenylphosphonium ions (see Figure 7.9). Some ionophores (see Section 7.52) are also hydrophobic ions. The properties of hydrophobic ions will be discussed in more detail later in this chapter, in particular the relative ease with which they can cross the bilayer.

Class IV: Ions

Monovalent and divalent cations are of particular interest as counterions to the net negative charge on most membranes. Also, the distribution of protons is of special interest because of the pH dependence of many membrane processes catalyzed by membrane-bound enzymes. This topic will be discussed in Section 7.33.

7.2 Permeability of Lipid Bilayer Membranes to Nonelectrolytes (see 1546, 856 for reviews)

The ability of a small solute molecule to cross the membrane is quantified by its permeability coefficient. The study of membrane permeability is clearly of physiological relevance. Unless there is a protein-mediated transport process, small solute molecules must cross the lipid bilayer in order to enter cells. The study of membrane permeability is also of interest in that it reveals structural and dynamic aspects of the lipid bilayer. Whereas sophisticated techniques have been developed to measure the rates of rotational and lateral diffusion of molecules in membranes, permeability studies can yield the transverse diffusion coefficient for molecules crossing the bilayer (see 856). The mechanism by which this occurs is also of interest and may be dependent on transient packing defects.

For a molecule to cross the bilayer it must (1) enter the membrane, overcoming any interfacial resistance or free energy barrier to do this, (2) diffuse across the bilayer, and (3) exit the membrane on the opposite side, again overcoming any possible interfacial resistance. Any of these steps could, in principle, be rate limiting. The permeability of most nonelectrolytes through lipid bilayer membranes can be successfully analyzed by the *solubility-diffusion model*. This makes the assumption that the rate-limiting step is the diffusion within the lipid bilayer and that interfacial barriers for membrane entry and exit are negligible. This allows one to assume a rapid partition equilibrium between the aqueous and membrane phases (steps 1 and 3 in Figure 7.2). The partition coefficient, K_p, is defined by

$$K_p = \frac{[C^m{}_1]}{[C^{aq}{}_1]} = \frac{[C^m{}_2]}{[C^{aq}{}_2]} \qquad [7.6]$$

Let us now define the permeability coefficient, P, in terms of the net flux, from side 1 to side 2, of solute across a membrane of thickness d, where there is a difference of concentration on the two sides of the bilayer:

$$\text{flux} = \frac{\text{mol solute}}{\text{sec} \cdot \text{cm}^2} = P \cdot (C^{aq}{}_1 - C^{aq}{}_2) \qquad [7.7]$$

This same flux can be expressed in terms of the diffusion coefficient describing the flux of solute within the membrane, D_m, using Fick's first law. The diffusion

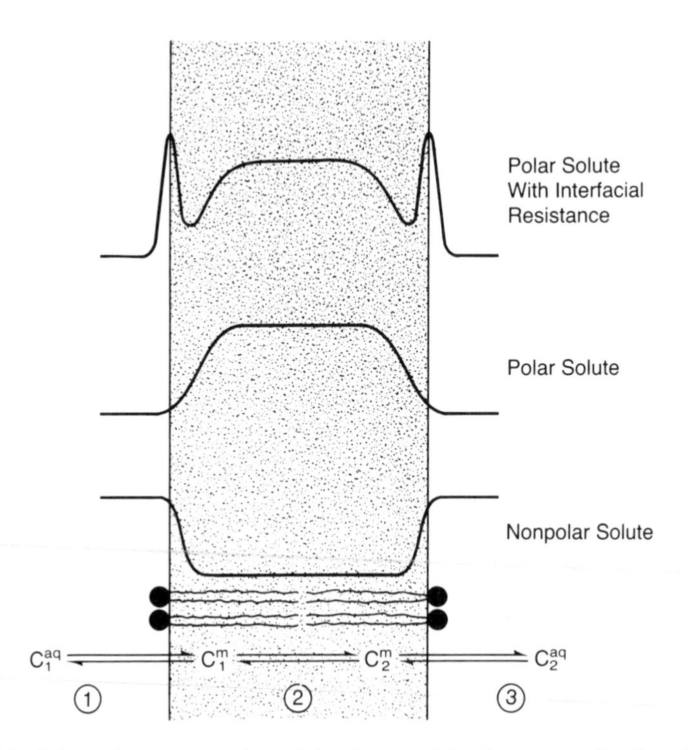

Figure 7.2. Schematic representation of the shapes of the free energy barriers for polar and nonpolar solutes crossing a lipid bilayer membrane. The solute concentrations at locations in the membrane (m) and aqueous (aq) phases are also indicated. The numbers at the bottom refer to the steps required for a solute to cross the membrane from left to right.

coefficient is defined as the factor relating flux to the concentration gradient within the membrane.

$$\text{flux} = -D_m \frac{dC}{dx} \approx -D_m \frac{(C^m_2 - C^m_1)}{d} \qquad [7.8]$$

where the concentration gradient is assumed to be linear across the bilayer and, therefore, can be approximated as above.

Substituting from equation 7.6 into equation 7.8 yields

$$\text{flux} = D_m \frac{(C^{aq}_1 - C^{aq}_2)}{d} K_p \qquad [7.9]$$

Comparing equations 7.7 and 7.9 gives

$$P = \frac{K_p D_m}{d} \qquad [7.10]$$

The permeability coefficient is a product of the partition coefficient and diffusion coefficient within the membrane, divided by the width of the membrane. It has units of cm/sec or velocity.

Alternatively, flux across the membrane can be defined in terms of a first-order rate constant, k, with units of sec^{-1}.

$$\text{flux} = k\Delta N_b \qquad [7.11]$$

where ΔN_b is the difference in solute concentration on the two sides of the membrane in units of mol/cm^2. The rate constant is also related to the permeability coefficient by the following relationship, which is derived in a manner similar to equation 7.10.

$$P = k\beta \qquad [7.12]$$

where β is the partition coefficient using surface density units (equation 7.2). The rate constant formalism is useful in that the reciprocal of k is a measure of the transit time across the membrane and k can be used to calculate the free energy barrier for passive transport.

Note that the equation 7.7 used to define permeability has a form similar to Ohm's Law for an electrical circuit. The physical meaning of permeability becomes more clear by this analogy. The concentration difference (ΔC) is a driving force, analogous to a voltage, and solute flux is equivalent to a current. The inverse of the permeability ($1/P$) is equivalent to a resistance to flux, and the permeability coefficient is equivalent to conductance.

$$\text{flux} = \frac{\text{driving force}}{\text{resistance}} \equiv \text{driving force} \times \text{conductance}$$

Just as one can have resistors in series which retard electrical flow in a circuit, there is a resistance associated with each of the steps shown in Figure 7.2 for solute flux across a membrane, with an interfacial resistance for entering or exiting the membrane and a resistance to flux within the membrane. Equation 7.10 is derived with the assumption that the interfacial resistance is negligible, which may not always be true (see 331).

Experimentally, the partition coefficient from the aqueous phase into the membrane is related to the partition coefficient of the solute into a nonpolar solvent (see 331, 1386). Figure 7.3 shows that the permeability of nonelectrolytes across a phospholipid bilayer is clearly a function of the partition coefficient into hexadecane over a million-fold range of permeability values. This correlation verifies Overton's rule, which states that permeability coefficients correlate with the oil/water partition coefficients. Observations contributing to Overton's rule go back to the last century and were early indications of the existence of a membrane barrier around cells (see Chapter 1). The linear correlation shown in Figure 7.3 should not be interpreted to mean that the diffusion coefficient in equation 7.10 is identical for all the compounds. However, these data do indicate that D_m for the set of compounds varies systematically with the partition coefficient so that it does not appear as an independent variable. This is not always the case,

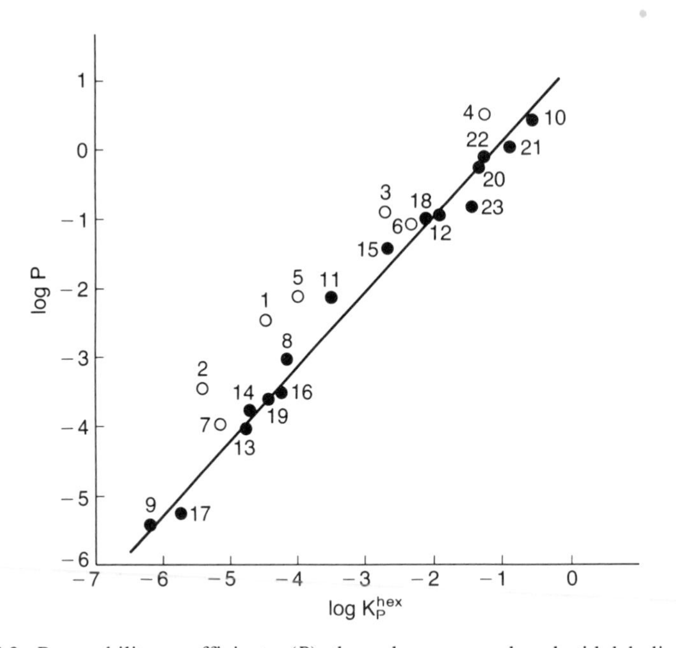

Figure 7.3. Permeability coefficients (*P*) through an egg phosphatidylcholine planar membrane plotted against the partition coefficients (K_p^{hex}) into hexadecane. The seven smallest solutes, indicated by open circles, were not included in the linear regression analysis. The solutes used were 1. water, 2. hydrofluoric acid, 3. ammonia, 4. hydrochloric acid, 5. formic acid, 6. methylamine, 7. formamide, 8. nitric acid, 9. urea, 10. thiocyanic acid, 11. acetic acid, 12. ethylamine, 13. ethanediol, 14. acetamide, 15. propionic acid, 16. 12-propanediol, 17. glycerol, 18. butyric acid, 19. 14-butanediol, 20. benzoic acid, 21. hexanoic acid, 22. salicylic acid, 23. codeine, 24. lactic acid. Adapted from ref. 1546. Kindly provided by Dr. J. Gutknecht.

and Stein and his coworkers (856, 1386) have shown that for some biomembranes, deviations from Overton's rule can be explained by the variation in the value of D_m for various solutes.

The line in Figure 7.3 is the best fit for the solutes with molecular weights between 50 and 300. The open circles represent very small solutes, including water, which have values of P which are 2- to 15-fold higher than expected based on data from the larger solutes (see 1546). These deviations from Overton's rule, and others which have been noted in permeability studies with red blood cell membranes (856), have been used to support a model for the diffusion process. The deviations are thought to reflect a steep size dependence of the membrane diffusion coefficient, D_m, for the solutes within the bilayer, which is a property previously observed in diffusion measurements in some polymers (1546, 856, 1386). Diffusion of molecules in a polymer depends quite differently on solute size and shape than diffusion through a liquid solvent. The Stokes–Einstein equation predicts a dependence of the translational diffusion coefficient on the inverse of molecular radius for a spherical particle. Translational diffusion through a polymer can be viewed as a series of steps into adjacent holes or pockets of free volume which spontaneously form in a fluctuating manner. The rate of diffusion will depend on how many holes of the correct size form and on how fast they form. The result is that the dependence of the diffusion coefficient in a polymer on solute size and shape is very different than in a liquid, where the hydrodynamic analysis is more appropriate. Biomembranes (856) as well as pure phospholipid bilayers (1546) appear to behave more like a soft polymer in terms of transverse diffusion coefficients of small nonelectrolytes.

Table 7.1 lists the permeability coefficients for several kinds of small molecules for biomembranes and model membranes. A detailed discussion of how these measurements are made is beyond the scope of this book and can be found in the book by Stein (1386). Most of the studies involve following radioactive tracers either into or out of cells or multilamellar liposomes or across a planar membrane. The planar membranes are best suited for measuring ion flux by electrical measurements, but have been widely used for studies with nonelectrolytes as well. Liposomes provide an enormous surface area and are very stable, allowing one to study molecules with very low permeability coefficients. It has been estimated that the surface area of a suspension of liposomes at 2 mg/ml is 10,000 $Å^2$ per ml (311). The method most widely used for monitoring permeability in liposomes is to follow the change in turbidity or light scattering. As can be seen in Table 7.1, water can relatively easily penetrate the membrane bilayer and will flow into or out of the liposome in response to osmotic changes induced by the solute flux (see Section 7.21). The shrinking and swelling of the multilamellar liposomes causes the change in turbidity, and this can be used to measure the solute permeability. One technical problem encountered in doing permeability studies is the presence of an unstirred layer near the membrane surface which may extend beyond the membrane surface by 50 μm or more. Local solute concentrations may be very different in this layer than in the bulk solution due to poor mixing (e.g., see 1546).

Table 7.1. Permeability and partition data on selected solutes.

Compound	Membrane[1]	Permeability coefficient (cm/sec)	Rate constant (sec^{-1})	Partition coefficient into hexadecane[2]	Ref.
1. Water	Egg phosphatidylcholine (planar)	3.4×10^{-3}	$6.0 \times 10^{+6}$	4.2×10^{-5}	(1546)
	Egg phosphatidylcholine (liposome)	2.0×10^{-4}			(311)
	Red blood cell	1.2×10^{-3}			(856)
2. Urea	Egg phosphatidylcholine (planar)	4.0×10^{-6}	$3.6 \times 10^{+7}$	2.8×10^{-7}	(1546)
	Red blood cell	7.7×10^{-7}			(856)
3. Glycerol	Egg phosphatidylcholine (planar)	5.4×10^{-6}	$2.5 \times 10^{+7}$	2.0×10^{-6}	(1546)
	Red blood cell	1.6×10^{-7}			(856)
4. Tetraphenyl-phosphonium (TPP^+)	Egg phosphatidylcholine (planar)	10^{-7}	$\sim 10^{-2}$	$\sim 10^{+2}$	(439)
5. Tetraphenyl-boron (TPB^-)	Egg phosphatidylcholine (liposome)	10^{-1}	$\sim 10^{+1}$	$\sim 10^{+5}$	(439)
6. Na^+	Egg phosphatidylcholine (liposome)	$\sim 10^{-14}$	—	—	(601)
7. Cl^-	Egg phosphatidylcholine (liposome)	$\sim 10^{-11}$	—	—	(601)
8. H^+/OH^-	Egg phosphatidylcholine	10^{-4}–10^{-8}	—	—	(1139, 563)

[1]Measurements on phospholipid bilayers were done either with planar membranes separating two aqueous compartments or using liposomes (see 311, 1386).
[2]The partition coefficients for TPP^+ and TPB^- are for the lipid bilayer, not hexadecane.

One major conclusion from the studies on nonelectrolyte permeability is that the permeability barrier of biomembranes to the passage of small molecules is provided by the phospholipid bilayer. In the absence of specific channels for solute flux, the behavior of model bilayer membranes and biomembranes is quite similar.

7.21 Water Permeability (see 311, 81)

It may seem surprising at first to learn that water can so readily penetrate the phospholipid bilayer. Studies summarized in Section 2.21 clearly show that there is no substantial water to be found inside the membrane beneath the carbonyl groups. The partition coefficient for water into hexadecane does not contradict this, because it indicates that the water concentration inside the membrane would be in the millimolar range. This is only about 1 water molecule per 10^3 phospholipids. Yet the permeability coefficient suggests that water can penetrate at the rate of about 10-100 molecules per second per phospholipid molecule if there is a driving force equivalent to a 0.1 M concentration difference across the membrane. The data in Figure 7.3 show that for a phospholipid bilayer, water is more permeable than expected in comparison to other nonelectrolytes, but not by very much. The rate constant for water flux is about 10^6 sec^{-1}, so it should only take on the order of 1 μsec for water to diffuse the width of the membrane. Hence, a very small steady-state concentration of solute in the membrane core is not inconsistent with a relatively large net flux across the bilayer.

It is also possible that the concentration of water in the hydrophobic core of the bilayer is higher than that estimated from the partition coefficient. In addition, transient hydrogen-bonded chains of water extending across the membrane have been postulated to exist (1049, 310). These structures are speculated to play a role in mediating proton flux across the bilayer (Section 7.51).

The water flux across the lipid bilayer can also be observed with unilamellar vesicles (1408). These vesicles act as perfect osmometers. If the salt concentration inside is higher than that outside, water will enter, causing an osmotic pressure and consequent swelling. The vesicle membrane can swell as much as a 5% increase in radius and remain stable, but the response is very much dependent on vesicle size.

Water permeability through biomembranes is generally within 10-fold of that measured with model membranes. By far the best studied system is the red blood cell (e.g., 81, 1463). The data from numerous laboratories show that treatment of the red blood cell membrane with a sulfhydryl-directed reagent, *p*-chloromercuribenzene sulfonate, results in inhibition of water permeability, reducing it to a level expected for the lipid bilayer. Apparently, this reagent acts by binding to the Band 3 anion channel (see Section 8.33), and possibly to Band 4.5 as well. It seems that water can pass through the membrane via the channel provided for protein-mediated anion transport. Similar observations have been made with renal membrane preparations, suggesting a water channel in this system as well (1487).

7.3 Electrical Properties of Membranes

Before discussing the permeability of ionic solutes through the lipid bilayer (Section 7.5), it is necessary to describe the electrical properties of membranes. We want to define the total electric potential profile across a membrane. This tells us how much electrical work is required to move a charge from one place to another. Obviously, this is essential if we are to understand the movement of ionic solutes across the membrane, and it is also essential for understanding how charges are distributed within or near the membrane. This would apply to ions in the aqueous phase at or near the membrane surface, as well as to the conformation of proteins with charged amino acid residues. We have already seen examples of enzymes whose function depends on the surface charge density on the membrane (Section 6.2). We have also seen how a peptide, alamethicin, forms channels across the membrane in response to a voltage across the bilayer (Section 3.71), and more examples of voltage-gated channels are described in the next chapter.

It will be useful to recall a few definitions (see 633).

(a) *Faraday constant*, F: This is the amount of charge in a mole of monovalent ion.

$$\begin{aligned} \mathrm{F} &= \text{Avogadro's number} \times \text{elementary charge} \\ &= (6.022 \times 10^{23}) \times 1.602 \times 10^{-19} \text{ coulomb} \\ &= 10^{5} \text{ coulomb/mol} \end{aligned}$$

(b) *Voltage*, V or ψ: The terms "voltage," "potential," "potential difference," and "voltage difference" are used interchangeably in reference to membranes. The voltage is a measure of how much electrical work is required to move a charge in a frictionless manner from one point to another. One volt is defined as the potential difference requiring one joule of work to move a coulomb of charge. One always needs a reference point for voltage measurements and for membranes this is usually a point very far from the membrane surface, defined as having zero potential.

(c) *Current*, I: An ampere is a measure of the flow of charge, with one ampere defined as one coulomb per second.

(d) *Conductance* (G) or *Resistance* (R): These measure the ease or difficulty of current flow between two points. They are defined by Ohm's law, relating current (flux) to potential difference (driving force).

$$I = GV \text{ or } I = (1/R)V \qquad [7.13]$$

Conductance is measured in units of siemens (S or mho), and its reciprocal, resistance, is measured in ohms (Ω). A conductance of one siemen corresponds to a current response of one coulomb per second per volt.

(e) *Resistivity* (ρ): This is the resistance to current flow used for a homogeneous material, such as salt solution, and can be used to estimate the ion flux through water-filled channels through the membrane. Resistivity is the resistance to current flow between two 1-cm^2 electrodes separated by 1 cm. It has units of (Ω-cm). Resistance between two electrodes of area A and separated by distance X can be computed from the resistivity of the material in between the electrodes.

$$R = \frac{\rho X}{A} \tag{7.14}$$

(f) *Capacitance* (C): The separation of charge is sufficient to create a potential difference between two points. For membranes, such charge separation can occur across the bilayer. The capacitance is a measure of how much charge separation is required to create a given voltage difference.

$$C = \frac{Q}{V} \tag{7.15}$$

where Q is the amount of charge on each side of the membrane, positive on one side, negative on the other, and V is the voltage difference created. Capacitance is measured in units of farads. This is an important electrical property of membranes because it relates how much charge must be transferred from one side to the other in order to create a given voltage difference. Specific capacitance is the capacitance per unit area, and depends on the charge separation per unit area of membrane.

In the simplest view, a bilayer membrane can be modeled electrically as a thin slab of nonconducting material separating two aqueous solutions. The membrane acts, thus, as a simple parallel-plate capacitor, where charge can be built up at either interface. The capacitance depends only on the distance between the two charged surfaces (d) and on the dielectric constant (ε) of the material (hydrocarbon) in between. The dielectric constant is a measure of the polarizability of the material and the degree to which any permanent electric dipoles which may be present in the material respond to an electric field (voltage difference). A high dielectric constant would essentially negate a portion of the charges at the interface, so more charge separation would be needed for a given voltage difference.

$$C = \frac{\varepsilon \varepsilon_0 A}{d} \tag{7.16}$$

where ε_0 is called the permittivity of free space and is a constant, 8.85×10^{-12} coulomb V^{-1} m^{-1}. For phospholipid bilayers and for biomembranes, the measured specific capacitance values are similar, around 1 μfarad/cm^2, which corresponds to a dielectric constant of 2 and distance of about 25 Å. Whereas the capacitance of biomembranes is clearly determined by the properties of the lipid bilayer, the resistance to current flow is dependent on the type and number of ion channels which are present, and this varies widely in different membranes (see Chapter 8).

We are now in a position to discuss the total electric potential energy profile of a membrane. There are a number of factors that contribute to the amount of work required to move a charge from a large distance away from the membrane up to the membrane and then into the bilayer core. These are:

1. Work associated with moving between media of different dielectric constants, e.g., from an aqueous phase to the membrane interior. This is due to a difference in dipolar polarization of the media and is the major unfavorable component destabilizing charges within the bilayer interior.
2. *Dipole potential*, probably mostly due to the oriented ester carbonyls on the phospholipid molecules. Oriented water is another possible contributor.
3. *Surface potential*, due to charged groups on the membrane surface.
4. *Transmembrane potential*, due to either protein-mediated or non-mediated ion permeability creating a charge separation across the bilayer either in an equilibrium situation or in a steady state.

In the subsequent sections, each of the components will be briefly discussed, especially as they pertain to important properties of membranes.

7.31 Work Required to Place an Ion Inside a Bilayer Membrane (see 646)

An ion in water is stabilized by the favorable interactions of the water dipoles, the hydration energy. To remove an ion from water and place it in the middle of a membrane is unfavorable because of the loss of this hydration energy. The most successful model for quantifying this is the Born model, which yields an expression for the work required to transfer an ion with a charge of q and radius r from a medium of dielectric constant ε_2 to one of dielectric constant ε_1.

$$W_B = \frac{q^2}{2r}\left(\frac{1}{\varepsilon_1} - \frac{1}{\varepsilon_2}\right) \qquad [7.17]$$

The dielectric constant for water (ε_2) is 80, and the value usually used for the interior of the membrane is $\varepsilon_1 = 2$, the value for hydrocarbons. For an ion of valence Z, this expression simplifies to $W_B = (81Z^2/r)$ kcal/mol. So, for a monovalent ion with a 2 Å radius, it takes about 40 kcal/mol to transfer it into the membrane core. This is obviously very unfavorable, and results in the fact that the lipid bilayer is a very effective barrier against the passage of ions. Note, however, that an ion with a large radius will not be as unstable in the membrane interior. This is important in understanding how some ions such as I_3^-, SCN^-, and charged ionophore–ion complexes can penetrate membranes.

In addition to the Born energy, there is a second component due to the polarization forces arising at the dielectric interface. The presence of a charge on one side of the interface causes the dipoles in the medium on the other side to reorient. This results in the "image" energy, so-called because of the mathematical method used to calculate its magnitude. For membranes it amounts to only a 10%-15% reduction of the Born energy value.

These models view the membrane as a simple layer of low-dielectric material. However, the polar headgroup region of the bilayer, which is about 10 Å wide on either side of the hydrophobic core, is obviously different. The dielectric constant in this region obviously must vary from 2 on one side to 80 on the other, and average values between 10 and 30 have been reported. A theoretical model suggests that the rotation of the polar headgroups contributes to this (1201). This region is where many membrane probes and amphiphilic molecules can be expected to bind and is, therefore, of interest.

7.32 Internal Dipole Potential (see 646, 439, 438)

There are experimental as well as theoretical reasons to believe that dipoles oriented at the membrane surface result in a potential in the center of the phosphatidylcholine bilayer which is about 240 mV, positive inside. This is equivalent to a free energy contribution of about 5.5 kcal/mol, favoring anions and disfavoring cations. The phosphocholine headgroup is oriented parallel to the membrane surface (Section 2.24) and would not be expected under any circumstances to create a dipole which is positive inside, since this would require the choline (positive) moiety to be closer to the center of the bilayer than the phosphate. The best explanation appears to be that this internal dipole arises from the orientation of the fatty ester carbonyls of each phospholipid molecule. The potential energy profile which results from this appears to be limited to the membrane interior and is pictured in Figure 7.4.

For small ions like Na^+ and Cl^-, the electric potential profile shows that this results in a slightly lower barrier for anions to cross the bilayer compared to cations. Nevertheless, the permeability barrier for small anions and cations is still very large and the resulting permeability coefficients are very small (Table 7.1).

This is not the case for hydrophobic ions, such as TPP^+ and TPB^- (Figure 7.9). These ions are surrounded by nonpolar structures, so there is a hydrophobic component to the total free energy change which occurs when they partition into the membrane. This counterbalances a good part of the unfavorable electrostatic energy loss (Born energy). In addition, the estimated size of these ions (4 Å) reduces the Born energy contribution. The free energy profile for hydrophobic ions is also shown in Figure 7.4. The data suggest that the hydrophobic component of the free energy for partitioning into the bilayer is about –7 kcal/mol for each and that the interaction with the internal dipole potential preferentially stabilizes the hydrophobic anion. Note the potential minima for the binding of the hydrophobic anion, located in the vicinity of the ester groups in the membrane (438).

The result is that the hydrophobic anion binds more strongly and is much more permeable than the hydrophobic cation, which has a very similar structure. The permeability coefficients differ by a factor of 10^6 due to this effect (see Table 7.1), and the binding constants to the bilayer also differ by 4 or 5 orders of magnitude.

It has been pointed out (439) that several hydrophobic ions, both anionic and cationic, saturate in their binding to phosphatidylcholine bilayers at about 1 per 100 lipids. This is due to electrostatic repulsion from the surface potential generated by the binding of ions to the bilayer surface. These effects are discussed in the following section.

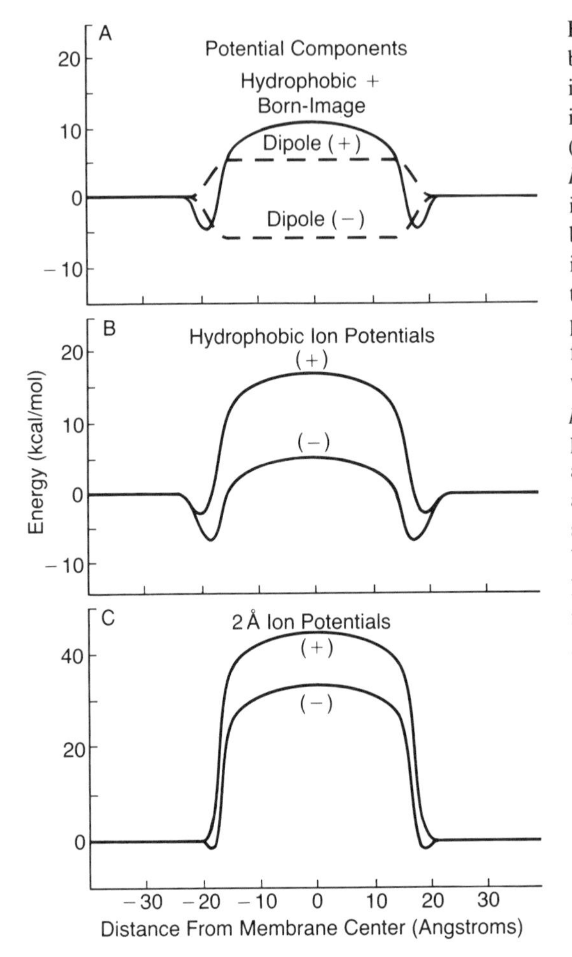

Figure 7.4. Theoretical membrane free energy profiles showing contributions from the Born, image, dipole, and hydrophobic (neutral) components. (A) *Hydrophobic ions*. The Born, image, and hydrophobic contributions (——) are the same for ions that are identical except for the sign of the charge. The dipole potentials (– – – –) are very different and this stabilizes anions within the bilayer. (B) *Hydrophobic ions*. Total free energy profiles for an anion and cation, assuming a 4 Å radius. Note the adsorption sites near the bilayer surface and the relatively low barrier for translocation of the hydrophobic anion across the membrane. (C) *Ions with 2 Å radii*. The total potential for ordinary inorganic ions presents a large barrier against translocation which is slightly less for anions due to the dipole potential. Adapted from ref. 646. Reproduced, with permission, from the *Annual Review of Biophysics and Biophysical Chemistry*, vol. 15. © 1986 by Annual Reviews Inc.

7.33 Membrane Surface Potential (see 952, 68)

The surface of most biomembranes is negatively charged, usually due to the presence of acidic phospholipids. Typically, 10–20% of the membrane lipids are anionic. Negative charges are also contributed by other membrane components, such as gangliosides or proteins. The negative charge on thylakoid membranes, for example, is primarily due to proteins (68). The anionic groups such as phosphates and carboxylates are fixed on the membrane surface and are, of course, electrically neutralized by counterions. However, the counterions are mobile and are not fixed to the membrane surface. They are distributed in the aqueous phase in a balance between their entropic drive to randomize and the favorable electrical attraction to the surface charges. The result is that, on the average, the counterions are located not at the surface, but at a distance away from the membrane surface, creating what is called a diffuse double layer. The physical situation is a membrane surface with a fixed charged density and a diffuse cloud of counterions. The electric potential in the solution near the membrane surface is a function of the charge density on the surface and the concentration and valence of ions in solution. There are substantial consequences due to this surface potential effect, which are very well explained using relatively simple theories and which are experimentally verified. In this section, we will not derive the theoretical relationships (see 952, 68) but, rather, focus on the consequences of having a charged membrane surface.

The theoretical challenge is to determine the value of the electric potential energy as a function of the distance from the membrane surface. Once this is known, one can calculate the local concentration of any ionic species and quantify other phenomena dependent on this, such as vesicle electrophoretic mobility or the electrostatic interaction forces between adjacent membranes. The theoretical model for solving this problem was conceptualized by Gouy and Chapman at the beginning of this century, and later modified by Stern in the 1920s. The resulting Gouy–Chapman and Gouy–Chapman–Stern theories have been remarkably accurate in describing the electrostatic phenomena associated with charged membranes.

There are four major assumptions in Gouy–Chapman theory: (1) the charges are considered to be smeared uniformly over the surface rather than treated as individual points; (2) the ions in solution are treated as simple point charges, ignoring their size; (3) the so-called image effects, the repulsion of mobile ions as they approach the dielectric interface, are neglected; (4) the dielectric constant in the aqueous phase is considered a constant and the same at the membrane interface as in the bulk solution. Each of these assumptions has been examined experimentally and demonstrated to be a reasonable approximation (951, 1596, 597). The Stern modification of the basic theory is to consider the size of the counterions binding at the surface, so that there is an upper bound to the number which can physically bind.

The theoretical predictions are best illustrated graphically. Figure 7.5 shows the electric potential profile due to different charge densities in a medium containing 0.1 *M* monovalent salt (e.g., NaCl). One charge per 300 $Å^2$ corresponds to about 20 mol % anionic phospholipid, and one charge per 60 $Å^2$ would apply to pure anionic lipids in a bilayer. The main point to notice is that the electrostatic effects extend far into the solution away from the membrane surface. The potential at the membrane surface ($x = 0$) is called the surface potential, ψ_0, and this is obviously a function of the surface charge density. Figure 7.5 also illustrates the effect of increasing the concentration of monovalent salt in solution. At high salt concentration, the electrostatic effects are effectively screened out. Both the magnitude of the surface potential and the extension of the electrical effects away from the surface are reduced at high salt concentration. This screening effect is much more potent with divalent ions, as shown in Figure 7.6. This is expected from electrostatic theory. Divalent cations, such as Ca^{2+}, can have substantial effects on the membrane surface potential at concentrations far lower than required for monovalent ions. Salt effects shown in Figure 7.6 have nothing

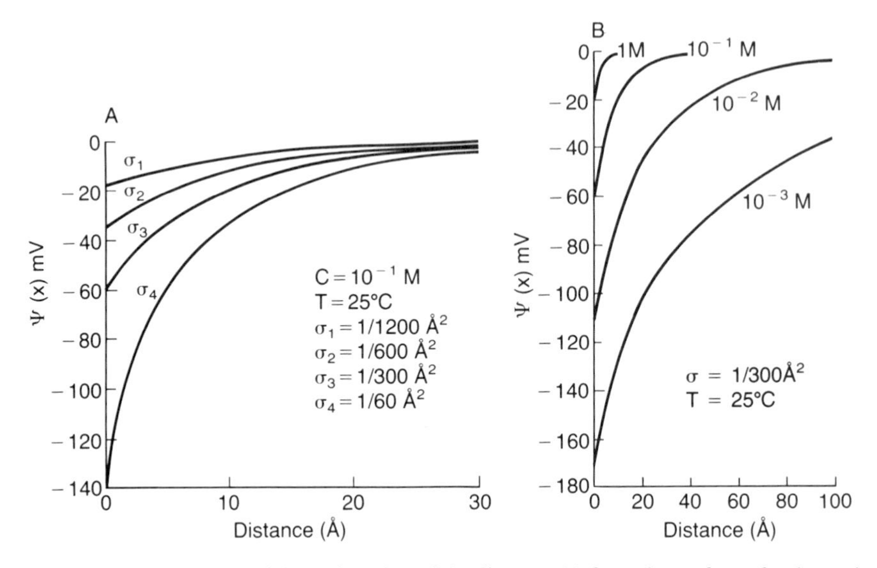

Figure 7.5. Electric potential as a function of the distance (x) from the surface of a charged membrane predicted by Gouy–Chapman theory. (A) Potential profiles at different surface charge densities (σ). One charge per 60 Å² corresponds to every lipid having one negative charge. (B) Potential profiles at concentrations of a monovalent electrolyte (e.g., NaCl) in the bulk aqueous solution. Note that both the surface potential and the length to which the electric potential extends into the solution are highly dependent on ionic strength. The calculations assume that about 20% of the lipids are anionic. Adapted from ref. 952.

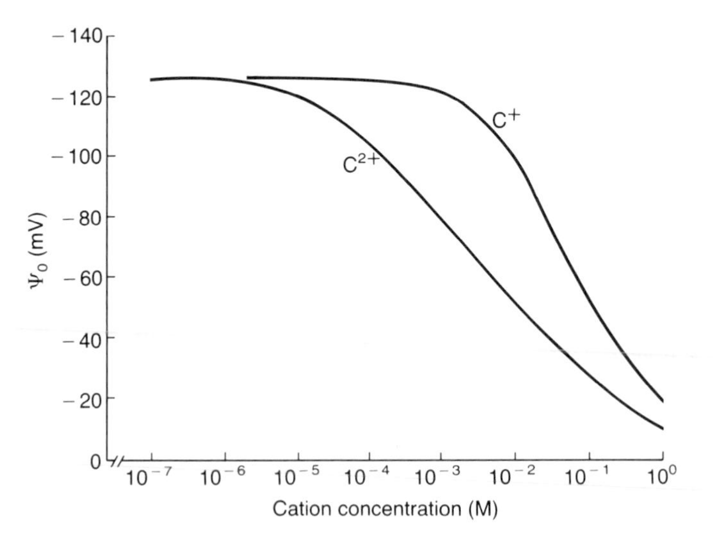

Figure 7.6. Effectiveness of divalent and monovalent cations in reducing the surface potential of a membrane with a negative charge density of about 1 charge per 10 lipids (–0.025 coulomb/m²). The calculations assumed 5 m*M* of a monovalent electrolyte (NaCl). The effects are not due to binding of the cations to the charge groups on the membrane but are due purely to electrostatic screening. Adapted from ref. 68.

to do with ion binding to the membrane surface, but are due to electrostatic screening. In addition, divalent cations often bind to the charged groups on the membrane surface, further reducing the surface potential (see Section 7.31).

Figure 7.7 shows the effects of this electric potential on the anion and cation distribution in solution. The local concentration of any ion can be computed easily by using the Boltzmann equation provided the electric potential is known:

$$C(X) = C(\infty)e^{-ZF\psi(X)/RT} \qquad [7.18]$$

where $C(X)$ and $\psi(X)$ are the ion concentration and electric potential at a distance X from the membrane surface, $C(\infty)$ is the ion concentration at an infinite distance from the surface (bulk concentration), Z is the valence of the ion (+2, +1, −1, etc.), and F is Faraday's constant. A value of 60 mV corresponds to about a 10-fold change in ion concentration at 25°C. Note again that multiple valency (e.g., $Z = +2$) has a substantial effect.

The effects of the surface potential are manifest in a number of different experimental situations. A large number of experiments have confirmed that the potential at the membrane surface, ψ_0, influences ion concentrations in a manner predicted by the Gouy–Chapman theory. This relates to protons, metal ions, and organic ions.

Local pH at the Membrane Surface

The proton concentration at a negatively charged membrane surface is higher than the value in bulk solution which is measured with a pH meter. This alters the apparent pK_a of any groups which are located at the surface. Examples include the titratable groups on phospholipids themselves (1471). Figure 7.8

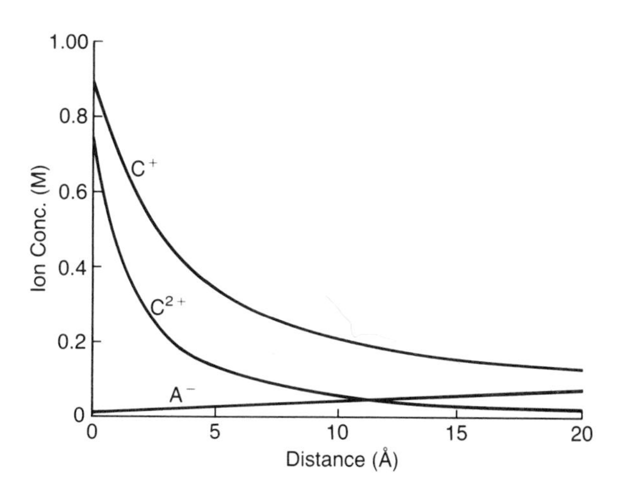

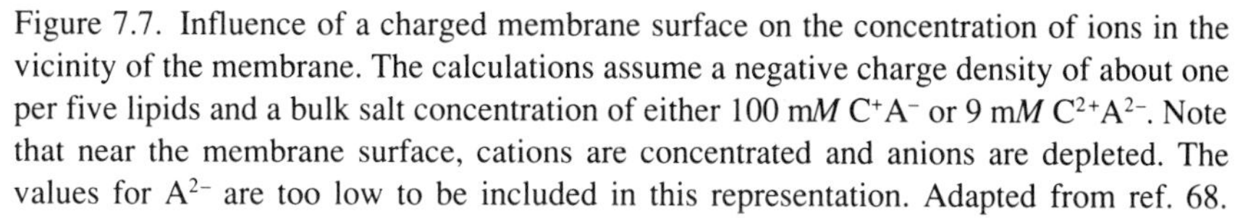
Figure 7.7. Influence of a charged membrane surface on the concentration of ions in the vicinity of the membrane. The calculations assume a negative charge density of about one per five lipids and a bulk salt concentration of either 100 m*M* C^+A^- or 9 m*M* $C^{2+}A^{2-}$. Note that near the membrane surface, cations are concentrated and anions are depleted. The values for A^{2-} are too low to be included in this representation. Adapted from ref. 68.

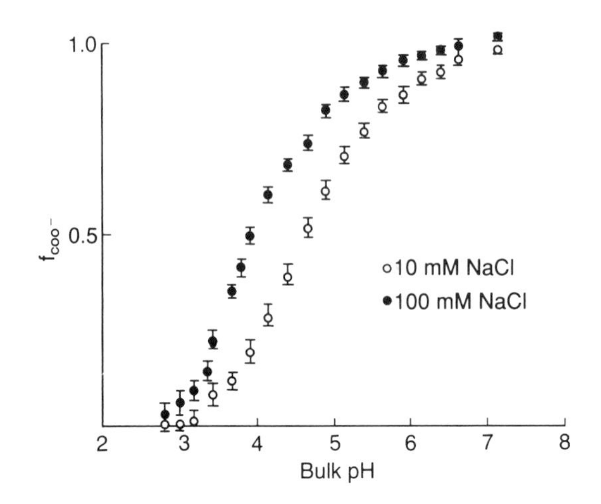

Figure 7.8. Experimental data showing the fraction of deprotonated carboxyl groups in phosphatidylserine (f_{coo^-}) as a function of bulk pH. The data were collected using vesicles containing 14 mol % phosphatidylserine and 86 mol % phosphatidylcholine. The apparent pK_a values for the vesicles in 10 m*M* and 100 m*M* NaCl are 4.7 and 3.9, respectively. The intrinsic pK_a is 3.6, and the salt effect is due entirely to the increase in the local proton concentration at the membrane surface at low ionic strength. Adapted from ref. 1471. Reproduced from the *Biophysical Journal,* 1986, vol. 49, p. 465, by copyright permission of the Biophysical Society.

shows the titration of carboxylate groups on phosphatidylserine which is incorporated into phosphatidylcholine vesicles. The apparent pK_a is 4.7 at 10 m*M* NaCl and 3.9 at 100 m*M* NaCl. This is due not to any change in the intrinsic pK_a of this residue but to the increased H^+ concentration at the membrane surface at low salt in response to the larger magnitude of ψ_0. The intrinsic pK_a is estimated to be 3.6, and the behavior is described very well by the Gouy–Chapman theory. The local pH is computed using equation 7.18, where $X = 0$. Similar results have been obtained with pH indicators bound to the surface of the membrane (e.g., 1502), where the local pH was shown to be different than the bulk pH by an amount predicted by the Gouy–Chapman theory.

Box 7.1 Intrinsic pK_a and Apparent pK_a of a Group on the Surface of a Membrane

The dissociation constant, K_a, for the protonation of a group (A^-) at the membrane surface is defined as follows.

$$K = \frac{[A^-][H^+]_{\text{surface}}}{[HA]} = \frac{[A^-][H^+]_{\text{bulk}}}{[HA]} e^{-F\psi_0/RT} \tag{7.19}$$

$$-\log K_a = -\log \frac{[A^-][H^+]_{\text{bulk}}}{[HA]} + \frac{F\psi_0}{2.303\, RT} \tag{7.20}$$

$$\mathrm{p}K_a \text{ (intrinsic)} = \mathrm{p}K_a \text{ (apparent)} + \frac{F\psi_0}{2.303\,RT} \quad [7.21]$$

$$\text{where } K_a \text{ (apparent)} = \frac{[A^-][H^+]_{bulk}}{[HA]} \quad [7.22]$$

Metal Ion Binding (see 953)

The interaction between phospholipids and metal ions, particularly Ca^{2+}, has received considerable attention and is clearly of physiological importance. Numerous experimental methods have been brought to bear on the question of how metal ions bind to the lipid bilayer and perturb the membrane structure, including calorimetry, Raman and infrared spectroscopies, neutron and X-ray diffraction, ^{31}P-NMR, and ^{2}H-NMR. Table 7.2 lists the intrinsic dissociation constants for some metal cations binding the acidic and zwitterionic lipids. Some of these values were obtained with phospholipid vesicles by monitoring the change in the surface potential due to metal binding. This will tend to decrease the negative charge density on negatively charged residues or to increase the positive charge density of zwitterionic vesicles. In all of these studies, the Gouy–Chapman theory has been successful in analyzing the data. The intrinsic binding constants must be evaluated using the free ion concentration in the vicinity of the membrane surface, and this will depend on the surface potential. An equation analogous to 7.19 is used and, without this correction, the binding data cannot be properly analyzed. There are several points worth emphasizing.

(i) Monovalent ions, such as sodium, do bind to acidic phospholipids, but the affinity is quite weak.

(ii) Ca^{2+} binds to the surface of zwitterionic phospholipid vesicles with an intrinsic affinity which is similar to or only slightly weaker than that for acidic phospholipids (27). Increased binding of Ca^{2+} to acidic phospholipids is due not to a larger intrinsic affinity of the anionic lipids for Ca^{2+} but rather to the

Table 7.2. Intrinsic dissociation constants for metal cation binding to some phospholipids.[l]

Phospholipid	Cation		
	Ca^{2+}	Mg^{2+}	Na^+
Phosphatidylcholine	72–333 m*M*	500 m*M*	—
Phosphatidylethanolamine	333 m*M*	500 m*M*	—
Phosphatidylglycerol	50–118 m*M*	167 m*M*	1.4 *M*
Phosphatidylserine	83 m*M*	130 m*M*	1.6 *M*
Cardiolipin	65 m*M*	—	1.3 *M*

[l]Data taken from refs. 953, 27, 376, 894, 895. The conditions are not identical for all of these measurements.

electrostatic effect of an increased negative charge density leading to a high local concentration of Ca^{2+} at the surface (894, 895).

(iii) The binding of either monovalent (269) or divalent (122, 12) cations appears to have little effect on the conformation of the lipid polar headgroup, at least for phosphatidylcholine and phosphatidylglycerol. Ca^{2+} binding to phosphatidylserine results in immobilization of the carboxylate group (340).

(iv) In most cases, Ca^{2+} binding is probably 1:1 with the phospholipid (47), though it is likely that one Ca^{2+} binds to two molecules of phosphatidylcholine (27). A 1:2 complex is also probably formed when metal binding causes membrane aggregation, and the divalent cation can bridge between two closely apposed membrane surfaces (953).

(v) Ca^{2+} binding can result in changes in the lipid physical state (e.g., 1354). For example, Ca^{2+} binding to phosphatidylserine or phosphatidic acid can result in formation of a gel-phase bilayer (602, 175). Interaction with cardiolipin stabilizes the hexagonal phase. The interaction with Ca^{2+} can also cause lateral phase separation to occur when acidic phospholipids are mixed with zwitterionic lipids (538). This can lead to the formation of large domains in unilamellar vesicles (607).

(vi) The interaction between Mn^{2+} and phospholipids can be used to advantage, since this paramagnetic ion perturbs the ^{31}P-NMR signal. The ^{31}P-NMR spectral perturbation is dependent on the local Mn^{2+} concentration and, hence, can be used to measure the surface potential (e.g., 948).

(vii) Polycations such as poly(L-lysine) (304) or gentamicin (220) also bind strongly to the surface of acidic bilayers.

Zeta Potential and Electrokinetic Effects

When a charged vesicle is placed in an electric field, it will migrate toward the electrode of opposite sign. The electrophoretic mobility of vesicles is related to what is called the zeta potential (see 952). This is the electric potential at the shear plane, which is the plane defining what migrates in the electric field, and is about 2 Å beyond the charged vesicle surface. The magnitude of the zeta potential is, thus, less than the surface potential (e.g., see Figure 7.5), and it can be related to the surface potential by the Gouy–Chapman theory. Measurement of the zeta potential is one of the standard methods used to estimate the surface potential and can be used to monitor ion binding to the surface of phospholipid vesicles (e.g., 954). The presence of charges on membrane components which stick out beyond the bilayer surface, such as proteins or gangliosides, will also influence the electrophoretic mobility (e.g., 948), and this method can be informative about the charge distribution near the vesicle surface.

Binding of Hydrophobic Ions and Membrane Probes

The surface potential will modulate the binding to the membrane surface of hydrophobic ions and amphiphilic membrane probes. This is the basis on which several probes function to monitor surface potential. Examples of such probes are shown in Figure 7.9. In all these cases, the intensity of the spectroscopic

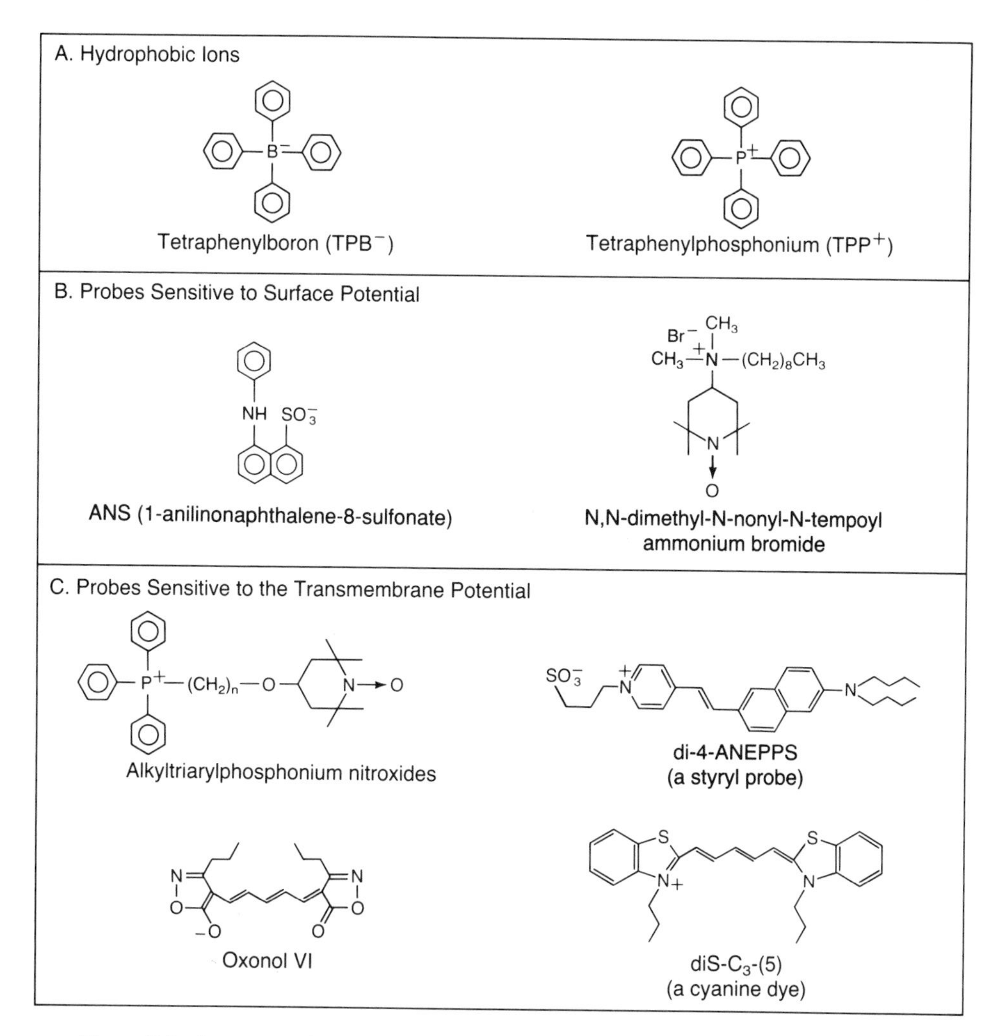

Figure 7.9. Structures of some probes used to monitor electrical effects in membranes (see text for details and references).

signal can be interpreted in terms of the amount of bound probe and, thus, reports the surface potential. There are numerous examples where such probes have been used with charged phospholipid vesicles (e.g., 948, 1409, 176). Studies with biomembranes are more problematic, because the probes often respond to more than the surface potential, and changes in pH or the transmembrane potential can make the interpretation of any observed changes in probe signal difficult. Examples of this include ANS (Figure 7.9) binding to mitochondria (1228) and neutral red responses in chloroplast particles (309).

Figure 7.10A shows the electric potential profile for a membrane with negative charges on both sides. Hydrophobic cations such as TPP^+ (Figure 7.9) or the K^+-valinomycin complex (Figure 7.12) will partition more favorably into this membrane

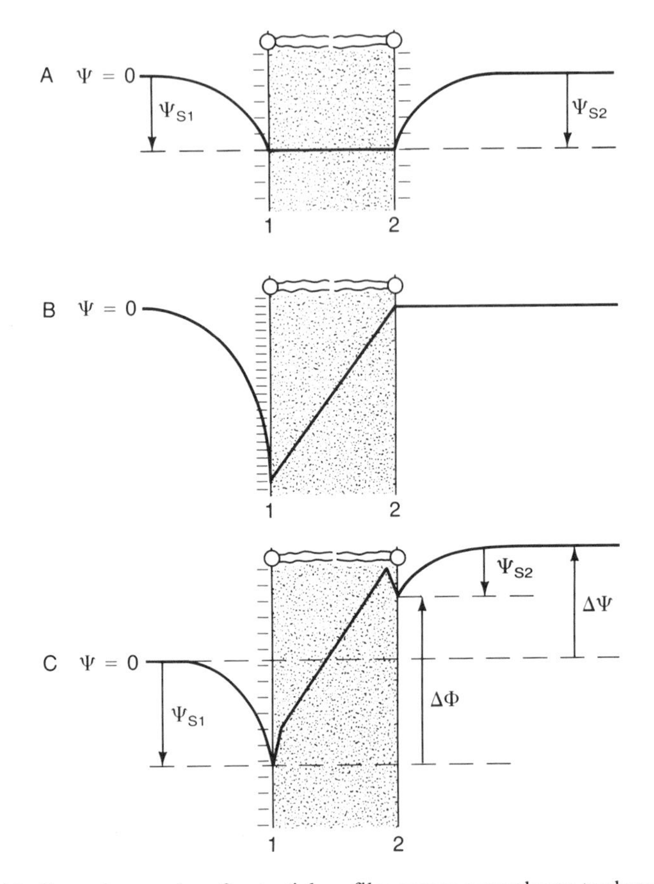

Figure 7.10. Several examples of potential profiles across a membrane to show the effects of surface charge on the transmembrane potential. The dipole potential is ignored in parts A and B. (A) A membrane with the same negative charge density on both sides results in identical surface potentials on the two sides (ψ_{S1}, ψ_{S2}) but no transmembrane potential. (B) A membrane with a large surface charge density on one side but no charge on the opposite side. There is a potential difference between the two membrane surfaces, but no transmembrane potential difference between the bulk solutions on the two sides. (C) A transmembrane potential difference imposed on a membrane with negative charge on one surface and none on the other side. In this case the potential profile also includes the dipole potential within the membrane, making the potential in the membrane interior more positive and creating two local minima for anions on either side. The figure shows the transmembrane potential ($\Delta\psi$), the two surface potentials (ψ_{S1}, ψ_{S2}), and the potential difference between the membrane surfaces ($\Delta\Phi$).

in response to the surface potential. This is because the local concentration of the ionic species adjacent to the membrane surface is increased over the bulk concentration. Recalling that the permeability coefficient can be expressed as the product of the partition coefficient (ß) and rate constant for crossing the membrane (k) (equation 7.12), it is clear that the permeability of a negatively charged membrane to hydrophobic cations will be greater than that of uncharged membranes due to the increase in ß. The dependence of membrane conductance on surface charge is described satisfactorily using the Gouy–Chapman expression for the surface potential (see 952).

Note that the influence of the internal dipole potential on hydrophobic ion permeability can result from an influence on binding or on the rate constant for crossing the membrane (Section 7.32, Table 7.1), whereas the symmetric surface potential influences the permeability entirely by its effect on the partition coefficient. The situation represented by an asymmetric surface charge distribution is more complicated (Figure 7.10B). We have seen in Chapter 4 that such an asymmetric lipid distribution may not be unusual. The voltage gradient across the bilayer in this case will make it harder for cations to cross from side 1 to side 2 than in the reverse direction. We will see in the next section how the transmembrane potential difference contributes to the electric potential energy profile.

7.4 The Transmembrane Potential

The transmembrane potential is defined as the difference in the electric potentials of the two bulk aqueous phases separated by the membrane. The relationship between the transmembrane potential ($\Delta\psi$) and the surface potentials (ψ_{S1},ψ_{S2}) is illustrated in Figure 7.10C. This schematic also shows that the potential difference between the two membrane surfaces, $\Delta\phi$, can be different from $\Delta\psi$, because of an asymmetric charge distribution on the two membrane surfaces. Any charged group within the membrane will move in response to $\Delta\phi$. $\Delta\psi$ is also called the resting potential, and this is the potential difference which is measured by a pair of electrodes when they can be used.

There are several ways in which a transmembrane potential can be generated. These are shown schematically in Figure 7.11.

(1) *Equilibrium condition*: If a membrane is permeable to a particular ion, e.g., Na^+, and impermeable to others, then a *diffusion potential* will developed in proportion to the logarithm of the concentration difference of the permeable ion on the two sides of the membrane. The ion diffuses across the membrane but, in doing so, creates a charge separation. The potential difference which is generated opposes the tendency of the ion to diffuse across the membrane. The amount of charge separation required to generate a given $\Delta\psi$ can be calculated from the membrane capacitance (equation 7.15). A $\Delta\psi$ of 100 mV requires an equivalent charge density of about 1 per 250 phospholipid molecules, so it is clear that there is little effect on the surface charge density.

At equilibrium, $\Delta\psi$ is determined by the Nernst equation.

$$(\psi_1 - \psi_2) = \Delta\psi = \frac{-RT}{2.303FZ} \log\left(\frac{[Na^+]_1}{[Na^+]_2}\right) \qquad [7.23]$$

This same equation will hold for any permeable ion of valence Z ($Z = 1$ for Na^+). In biomembranes, the ion permeability is determined by opening specific ion channels (see next chapter). Experimentally, ion permeability can be increased

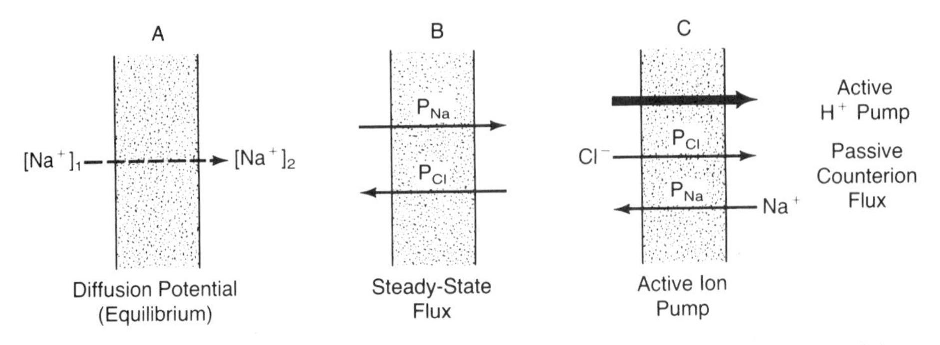

Figure 7.11. Schematic showing three situations resulting in a transmembrane potential. In vivo the diffusion of ions across biomembranes must be protein-mediated. (A) A membrane permeable to a single ion manifests a diffusion potential which can be calculated using the Nernst equation. (B) The steady-state flux of ions can also result in a transmembrane potential difference, which can be calculated using the Goldman–Hodgkin–Katz equation. (C) An energy-driven ion pump can move ions across the membrane, but other ions must diffuse in response to maintain bulk electroneutrality. If the counterion movement is sufficiently facile (high permeability) then the net charge separation and transmembrane potential difference generated by the ion pump will be nil.

by using specific ion carriers or ionophores such as valinomycin (K^+) (Section 7.52).

(2) *Steady-state diffusive ion flux* (see 633): When a membrane is permeable to several ions, the ions will flow across the membrane. In the steady state, a charge separation can develop across the membrane because of differences in the permeability coefficients. In essence, some ions diffuse rapidly across the membrane and the others will lag behind. The equation which describes this is called the Goldman–Hodgkin–Katz equation, shown below for Na^+ and Cl^- as the two permeable species.

$$\Delta\psi = \frac{-RT}{2.303F} \log\left(\frac{P_{Na}[Na^+]_1 + P_{Cl}[Cl^-]_2}{P_{Na}[Na^+]_2 + P_{Cl}[Cl^-]_1}\right) \qquad [7.24]$$

The ion flux will continue until equilibrium is reached.

(3) *Active ion translocation*: Charge separation across the membrane can also be created by active processes. Many enzymes catalyze reactions resulting in charge translocation across the bilayer. Examples are the various ATP-driven ion "pumps" such as the Ca^{2+}-ATPase (Section 8.41) and cytochrome *c* oxidase (Section 8.51), which is a proton pump. The enzymology of these systems is briefly described in the next chapter. It should be noted that these reactions are electrogenic, i.e., they result in charge movement across the bilayer, but there must be some neutralizing ion flux elsewhere across the membrane. This is illustrated in Figure 7.11C, which shows Cl^- counterion flux in response to a proton pump. Like the situation described for passive ion flux, the counterion flux will tend to lag behind the active process, resulting in a net charge separation across the bilayer and a $\Delta\psi$. If the permeability of the neutralizing ion is increased

sufficiently, the charge separation will not be developed. This is the principle of using ionophores to collapse the transmembrane potential across biomembranes or model membranes.

7.41 Measuring the Transmembrane Potential (see 54 for review)

The best way to measure the transmembrane potential is to use an electrode on each side of the membrane. However, this is feasible only for planar model membrane systems and for a few large cell types (e.g.,140). Usually, one is studying either phospholipid vesicles, biomembrane vesicles, or suspensions of cells or organelles, such as mitochondria or chloroplasts. Several methods have been developed to measure $\Delta\psi$ in these situations.

(1) *Partitioning of an ion according to the Nernst equation* (7.23): An ion which can penetrate the membrane is added to the system, and it will partition between the inside and outside in a manner described by the Nernst equation. Hydrophobic ions such as TPP^+ (e.g., 1255) or ^{86}Rb plus valinomycin are examples of molecular probes used in this way. It is necessary to know the concentration of ion inside the vesicle, organelle, or cell and this can be a problem. Artifacts can arise, for example, if the probe is binding excessively to cell membranes or if the internal volume is not correctly measured.

(2) *Spin-labeled EPR probes*: Several probes have been developed for this purpose (see 176,177) (see Figure 7.9). These are hydrophobic ions with a paramagnetic nitroxide group covalently attached. The fraction of the probe which is membrane bound can easily be determined from the EPR spectrum, and this changes in a predictable manner as the dye redistributes in response to a transmembrane potential. The amount of membrane-bound probe changes because the surface-to-volume ratio is much larger for the vesicle interior space than the external solution (see 177).

(3) *Optical molecular probes*: There are many optical probes which respond to a transmembrane potential. Most commonly used are fluorescent derivatives of merocyanine, oxonol, and cyanine dyes (see 1534, 230, 54 for reviews). All these probes bind to the membrane, and a number of different mechanisms appear to be involved in the probe response to transmembrane potential changes (e.g., 1511). The most common mechanism involves a change in the probe orientation in the bilayer through an interaction with an electric dipole in the probe molecule. The state of aggregation of some of the probes in the bilayer causes changes in the fluorescent quantum yield. Most of the probes are useful for transmembrane potentials which are negative inside, and others, such as the oxonols, are limited to use when the potential is positive inside (782).

Another class of probes are the "styryl-type" probes, which are highly conjugated molecules whose absorption spectra are sensitive to the transmembrane potential (443). The mechanism is called electrochromism (722). The molecules undergo an electronic redistribution in going from the ground to the excited state upon photon absorption. The energy of the electronic transition is sensitive to a potential gradient parallel to the direction of this charge shift. This response is also seen with some naturally occurring carotenoids in photosynthetic membranes (722). The advantage of these types of probes is that the response is not dependent on probe aggregation or partitioning and is very rapid. This makes this kind of probe particularly useful when rapid kinetic studies are required.

7.42 The Concept of an Energized Membrane (see 1065)

The term "energized membrane" is one which is rather loosely defined, but means that ion flux across the bilayer can be used to accomplish work. The ion most often utilized for this purpose is the proton, for example, in the mitochondrion and chloroplast, and the difference in the proton electrochemical potentials ($\tilde{\mu}_{H+}$) on the two sides of the bilayer is called the proton motive force.

$$\text{side 1:} \quad \tilde{\mu}_{H^+,1} = 2.303\ RT \log [H^+]_1 + F\psi_1$$

$$\text{side 2:} \quad \tilde{\mu}_{H^+,2} = 2.303\ RT \log [H^+]_2 + F\psi_2$$

$$\text{difference (1 minus 2):} \quad \Delta\tilde{\mu}_{H^+} = 2.303\ RT \log \frac{[H^+]_1}{[H^+]_2} + F\Delta\psi$$

$$\text{proton motive force (mV):} \quad \frac{\Delta\tilde{\mu}_{H^+}}{F} = \Delta\psi - \frac{2.303\ RT}{F}\Delta\text{pH} \qquad [7.25]$$

where $\Delta\tilde{\mu}_{H^+} = (\tilde{\mu}_{H^+,1} - \tilde{\mu}_{H^+,2})$ is the difference in the proton electrochemical potential expressed in joules/mol, and division by Faraday's constant converts this to mV units

$$\Delta\psi = (\psi_1 - \psi_2), \text{ referring to the bulk solution}$$

$$\Delta\text{pH} = (\text{pH}_1 - \text{pH}_2), \text{ referring to the bulk solution}$$

At 30°C

$$\Delta\tilde{\mu}_{H^+}(\text{mV}) = \Delta\psi - 60\ \Delta\text{pH} \qquad [7.26]$$

This is a measure of the free energy change for the transfer of protons from one side of the membrane to the other. In mitochondria and photosynthetic systems, the proton motive force is used to drive ATP synthesis, but it can also be coupled to the solute transport system (see Chapter 8). This is described by the chemiosmotic theory (see 1065).

To quantify the magnitude of $\Delta\tilde{\mu}_{H^+}$, it is necessary to know both $\Delta\psi$ and ΔpH. We have already discussed methods used to quantify $\Delta\psi$. ΔpH is usually measured by monitoring the distribution of weak acids or bases across the membrane (see 54 for review). The neutral form can permeate the lipid bilayer and the charged forms will accumulate on either side, depending on the pH values. Radioactive species can be used, but optical probes are also utilized (e.g., 131). One frequently used probe for these purposes is 9-aminoacridine, which accumulates inside acidic vesicles, resulting in quenching of its fluorescence signal.

7.5 The Permeability of Lipid Bilayer Membranes to Ions (see 311)

It is evident from the previous discussion that there is a very considerable energy barrier for metal ions to cross the bilayer. The permeability coefficient of Na^+ for unilamellar vesicles is in the range of 10^{-12} to 10^{-14} cm/sec (see Table 7.1). Even this slow rate is orders of magnitude faster than predicted from the solubility-diffusion model, using the Born equation to estimate the energy required to transfer the ion from the aqueous phase into the membrane core. There is no consensus on the mechanism by which ions cross the lipid bilayer, but there are several competing theories, focusing on defects in the packing order of the bilayer. These defects are postulated to form spontaneously, like fluctuating holes or lipid kinks, or possibly along boundaries between coexisting lipid gel and liquid crystalline phases. Generally, ion permeability is maximal at the transition temperature between the gel and liquid crystalline phases, but this is not the case for water or for protons (below). It should be remembered that steady-state ion flux across the bilayer must be electroneutral, so the flux of one ion must be linked to the flux of other ions to maintain neutrality.

In any event, it is clear that the bilayer is a remarkably good barrier against simple anions and cations. As we have seen, however, ions which have a larger radius can penetrate the membrane because the Born energy is less. Hydrophobic ions are also capable of crossing the membrane.

7.51 Permeability to Protons (see 310, 562 for reviews)

Measurements on model membranes have demonstrated that the lipid bilayer is remarkably permeable to protons. Experimentally, it is not possible to distinguish proton permeability from hydroxide permeability, so this is usually indicated as (H^+/OH^-) as in Table 7.1. We will refer to this simply as proton permeability. There is a wide spread of values reported for the permeability coefficient of protons, generally in the range 10^{-4} to 10^{-8} cm/sec. This variability has been explained by experimental differences in vesicle size, the size of the pH gradient imposed across the membrane, and lipid unsaturation (1139). Nevertheless, it is clear that proton permeability is at least 10^6 greater than for other simple ions, and this is true for biomembranes as well as the model membranes (563).

These data strongly suggest some special mechanism for proton permeability, and this is supported by data, for example, which demonstrate that the rate is not limited by simple electrostatic barriers in the membrane (1139). The nature of this mechanism is not known. One model which has been proposed is that there are transient hydrogen-bonded chains of water which extend across the bilayer and that protons are rapidly transferred across the bilayer through these water chains (see 1049, 310). There is no direct evidence for such chains of water, however. It has also been claimed, with experimental support, that the anomalously high proton permeability across phospholipid bilayers is largely due to the presence of weakly acidic contaminants such as fatty acids which act as proton carriers at physiological pH (562). However, this does not account for all of the anomalous proton flux in liposome systems (310).

It has also been shown that protons can move rapidly across the membrane-solution interface and protonate the anionic forms of weak acids adsorbed at the membrane surface (730). There is no interfacial barrier for protons in the bulk solution to rapidly equilibrate with groups at the membrane surface.

Since proton flux across the bilayer is a critical part of most bioenergetic systems (see 1065), the question of the mechanism of diffusive proton permeability is of interest (e.g., 789). Experimentally, proton permeability is of particular interest when proton-translocating proteins are examined in phospholipid vesicle systems. Studies have shown that the incorporation of proteins in these systems makes little difference to the proton permeability (1086, 1325) but that the counterions (1325) and magnitude of the transmembrane potential (492, 789) can be important factors.

A related question of significance in bioenergetics is how protons diffuse from one place to another at the membrane surface. The proton circuits pictured in Figure 6.5, for example, require the diffusion of protons from the proton-pumping enzymes to the ATP synthases. A point of controversy is whether these protons equilibrate with the bulk solution or whether there are localized pathways for proton transfer on or within the membrane. As stated above, it has been shown experimentally that there is no interfacial barrier preventing rapid proton equilibration between groups on the membrane surface and the bulk solution (730). However, a series of interesting studies has shown that the lateral diffusion of protons along the surface of a phospholipid monolayer can be 20 times faster than diffusion through the bulk solution (1175). This is postulated to occur through a two-dimensional hydrogen bond network formed from the lipid polar headgroups and water molecules at the membrane surface. The biological relevance of this observation, however, is yet to be demonstrated, and the conclusions are not universally accepted (730).

7.52 Ionophores (see 1176, 1065)

Ionophores are a diverse group of compounds which increase the permeability of membranes to ions. Some, like gramicidin A and alamethicin, form channels across the bilayer (Section 8.15). Other ionophores form stoichiometric complexes with cations and act as carriers to facilitate ion transport across the lipid bilayer. These are particularly useful tools for membrane research, especially in bioenergetic systems or in other cases where ion gradients are important. Since these ionophore–cation complexes can be relatively specific for particular ions, they can be used to manipulate the ion and voltage gradients across membranes. Some ionophore–cation complexes are uncharged, and the cation is transported in this charge-neutralized form. Other ionophore–cation complexes are charged, and they cross the bilayer as hydrophobic ions, as previously discussed. Some of the most frequently used ionophores are indicated below (see Figure 7.12).

CCCP (Carbonyl cyanide *m*-chlorophenylhydrazone)

This is closely related to FCCP (carbonyl cyanide *p*-trifluoromethyoxyphenylhydrazone), and both are weak acids. The protonated form is electrically neutral and it has been shown that this form easily crosses the membrane ($P \approx 17$ cm/sec) and that the deprotonated (anionic) form can cross back at about 1% of this rate (729). The anionic forms of this and other protonophores are soluble in the interior of the bilayer for several reasons (730). The negative charge is delocalized and the larger ionic radius reduces the unfavorable Born energy (equation 7.17). Also, the dipole potential, which is positive inside the bilayer, stabilizes the anion partitioned in the bilayer (Figure 7.4). Finally, the hydrophobic groups favor partitioning in the membrane.

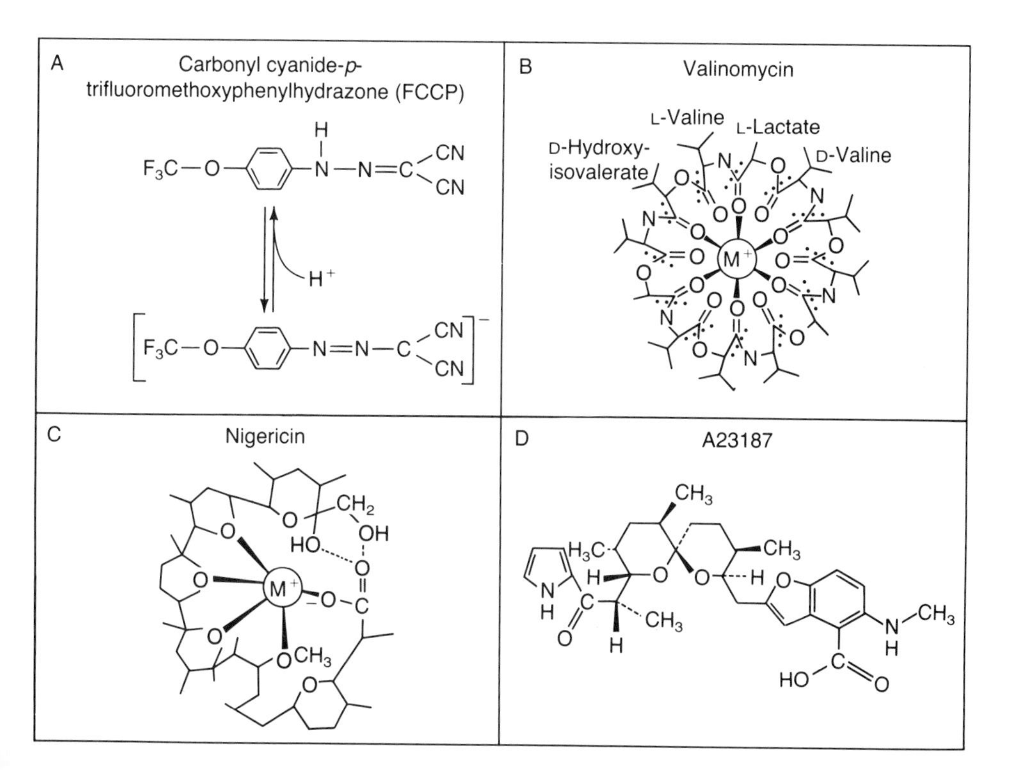

Figure 7.12. Structures of four ionophores. See text for details and references. Metal ion interactions indicated by X-ray crystallography are indicated by dark lines in parts B and C, which are adapted from ref. 1176.

CCCP, FCCP, and other weak acids effectively increase the membrane permeability to protons and allow them to reach electrochemical equilibrium across the membrane. Often, this results in the collapse of the transmembrane potential and pH differences across the bilayer.

Valinomycin

This is a cyclodepsipeptide which can form a 1:1 complex with monovalent cations with a preference for K^+. The structure of the complex somewhat resembles a cage with the potassium in the center, stabilized by interactions with ester carbonyls (Figure 7.12). The valinomycin–K^+ complex acts as a hydrophobic ion, and this easily crosses the bilayer. Use of this reagent allows a K^+ diffusion potential to be created in vesicles where the internal and external K^+ concentrations are different. In a system where a transmembrane potential is being generated by an active process, valinomycin–K^+ will collapse $\Delta\psi$ but will not directly alter the ΔpH value.

Nigericin and Monensin

These are monovalent polyethers which have carboxyl groups (see 1464). They also form 1:1 complexes with monovalent cations, but the complexes are electrically neutral. Nigericin (Figure 7.12) is selective for K^+ and monensin is selective for Na^+. These ionophores can also cross the bilayer in the neutral protonated form. They are used to promote the exchange of H^+ for Na^+ (monensin) or H^+ for K^+ (nigericin) across membranes. These reagents cause the H^+ and Na^+ (or K^+) gradients to equalize. They are most frequently used to cause the ΔpH to be dissipated across the membrane. Because the net exchange is electroneutral, the $\Delta\psi$ is not directly perturbed.

A23187

This is also a carboxylic cation carrier (Figure 7.12) which is particularly useful because it has a strong preference for binding divalent cations (see 337). It is usually used as a Ca^{2+} ionophore. It is likely that it functions by forming a neutral complex within the bilayer containing two molecules of A23187 and one molecule of Ca^{2+}, and such 2:1 complexes have been characterized (see 337).

7.6 Chapter Summary

A primary function of any biological membrane is to provide a selective permeability barrier between the aqueous compartments on either side. The thin hydrophobic core of the bilayer is a very effective barrier against the passage of inorganic ions but allows nonpolar solutes to pass across to varying degrees. The rate at which a nonelectrolyte passes across the bilayer is quantitatively related to the solubility of the solute in the bilayer as measured by a partition coefficient between water and organic solvents. Water can cross model membranes at a rate compatible with its solubility in organic solvents, but in some biological membranes water can also pass through protein channels.

Although inorganic ions cannot penetrate the lipid bilayer to a significant extent, the permeability of model membranes and biological membranes to protons is unusually high for reasons that are not entirely clear. Also, organic ions or chelates of organic molecules with inorganic ions can partition into the bilayer core as a result of their hydrophobic character and the larger size of these ions. Ionophores are molecules of this type.

The electrical properties of biological membranes are reasonably well understood. The permeability to inorganic ions is entirely due to proteins which function as ion channels across the bilayer. The transmembrane potential that results when there is charge separation across the bilayer is a result of the membrane capacitance, which is similar for both model lipid bilayers and biological membranes. Charges on the membrane surface create an additional surface potential which can substantially alter the concentration of any charged species in the immediate vicinity of the membrane. This can influence the catalytic behavior of membrane-bound enzymes.

Chapter 8
Pores, Channels and Transporters

8.1 Overview

It is clear from the data presented in Chapter 7 that the phospholipid bilayer is a very effective barrier to most small molecule solutes. Yet, there is a constant flux of polar and ionic substances across the cell plasma membrane, as well as across the membranes defining various organelles such as the mitochondria. This mass transport is all protein-mediated, and a wide variety of mechanisms have been elucidated by which solute flux across membranes is achieved. It is the primary purpose of this chapter to present a survey in which the general question addressed is how matter gets across a biomembrane. The next chapter addresses the question of how information is transmitted across the membrane and covers topics related to cellular communication, surface receptors, and signal transduction. The material in these two chapters is interrelated. For example, one response to certain signals (e.g., a molecule binding to a cell surface receptor) is to alter rapidly the membrane permeability to specific ions, resulting in an alteration in the transmembrane potential and/or changes in the intracellular concentration of the ions. Before discussing this and other mechanisms of signal transduction, however, it is important to understand how the cell regulates which molecules exit and enter various cellular compartments.

At this point, it is useful to define a few terms used to characterize proteins or structures involved in transmembrane transport. Table 8.1 contains a classification scheme for transport proteins. A major subdivision distinguishes *channels* (or *pores*) from *transporters*. Pores and channels are often pictured as tunnels across the membrane in which binding sites for the solutes being transported are accessible from either side of the membrane at the same time. Channel proteins do not require any conformational alteration in order for the solute entering from one side of the membrane to exit on the other side. In contrast, transporters all require a protein conformational change during the process of solute translocation

Table 8.1. Classification of some transport proteins based on mechanism and energetics.

- I. Channels
  - (A) Voltage-regulated channels (*example*: Na^+ channel)
  - (B) Chemically regulated channels (*example*: nicotinic acetylcholine receptor)
  - (C) Other (unregulated, pressure-sensitive, etc.)
- II. Transporters
  - (A) Passive uniporters (*example*: erythrocyte glucose transporter)
  - (B) Active transporters
    - (i) Primary active transporters
      - (a) Redox coupled (*example*: cytochrome *c* oxidase)
      - (b) Light coupled (*example*: bacteriorhodopsin)
      - (c) ATPases (*example*: Na^+/K^+-ATPase)
    - (ii) Secondary active transporters
      - (a) Symporters (*example*: lactose permease)
      - (b) Antiporters (*example*: Band 3)

across the membrane. The solute binds from one side and a conformational change is necessary in order for it to exit on the opposite side. The solute binding site of a transporter is accessible to only one side of the membrane at any time.

Channels and pores also undergo conformational changes, but these regulate whether they are open or closed to solute traffic and play no role in the mechanism of translocation per se. Two major groups of channels indicated in Table 8.1 distinguish channels which are *voltage-regulated* from those which are *chemically regulated*. Voltage-regulated (or voltage-gated) channels are those which open or close in response to a change in the transmembrane potential, and the best studied examples are those from electrically excitable cells such as nerve or muscle (see 1425). Chemically regulated channels respond to specific chemical agents, and the best characterized examples are those which bind to neurotransmitters such as acetylcholine. The response of the nicotinic acetylcholine receptor, for example, on binding to the neurotransmitter is to switch to an open conformation and allow the passage of monovalent cations across the membrane.

The terms *pore* and *channel* are generally used interchangeably, but *pore* is used most frequently to describe somewhat nonselective structures that discriminate between solutes primarily on the basis of size, allowing the passage of molecules that are sufficiently small to fit. The term *channel* is mostly reserved to describe ion channels, which are now known to be widely distributed in many cell types other than nerve and muscle [e.g., epithelial cells (1592)].

Transporters can be subdivided into those which are *passive* and those which are *active*. We shall reserve the use of the term *passive transporter* to those which allow the passage of a single kind of solute across the membrane. These are uniporters, and they can only facilitate net solute flux downhill energetically, i.e., down a gradient in the solute electrochemical potential. This is called *facilitated diffusion*. The best characterized example of a passive transporter is the glucose transporter in the erythrocyte.

Active transporters can be used to move solutes energetically uphill across the membrane, resulting in accumulation of the solute on one side of the membrane.

This requires that the solute transport be coupled to another process which is free energy-yielding and which can drive solute uptake. *Primary active transporters* are mostly, but not all, *ion pumps*, where ion translocation is mechanistically coupled to an energy-yielding chemical or photochemical reaction. An example is bacteriorhodopsin, which utilizes the energy derived from the absorption of a photon of visible light to translocate protons across the membrane. In most cases, the ion pumps are *electrogenic*, in that there is a net movement of charge associated directly with the primary pump, so the active pump generates a charge separation and voltage across the membrane.

Primary active transporters are used to generate voltage and ion gradients across the membrane. *Secondary active transporters* are designed to utilize such gradients to drive solute transport. The best characterized example is the lactose permease from *Escherichia coli*. This transporter utilizes the proton electrochemical gradient generated by the respiratory electron transport chain to drive the uptake of lactose into the cell. This is an example of a *symporter* in which two different solutes (e.g., protons and lactose) are simultaneously translocated across the membrane. *Antiporters* couple the transport of solutes in opposite directions, an example being Band 3 from the erythrocyte, which couples the transport of Cl^- and HCO_3^- in opposite directions across the red blood cell membrane.

The terms *permease*, *translocase*, and *carrier* are often used to refer to transporters other than primary active transporters. Historically, the term permease has been used in reference to bacterial transporters. It is probably best to use carrier to refer to ionophores or similar substances (Section 7.52) which bind to ions and literally carry them across the bilayer as a complex. It is quite clear that this is not the mechanism used by any of the transport proteins in biological membranes.

The classification system shown in Table 8.1 is largely based on the energetics and mechanisms of solute transport. The increasing availability of amino acid sequences of a wide variety of transport proteins allows these proteins to be grouped according to structural similarities. Table 8.2 shows several structurally related families of channels or transporters. These are *superfamilies* in that the structurally related family members carry out different functions. Consider the relationship between the mammalian glucose transporter and the bacterial H^+-arabinose transporter. The former is a uniporter which can catalyze only facilitated diffusion of glucose, whereas the latter can couple a proton electrochemical gradient to the active transport of another sugar, arabinose. It is clear that nature can adapt the same structural framework to a variety of ends. We will see this same theme again in the next chapter in considering superfamilies of cell surface receptors.

8.11 Channels Versus Transporters: A Range of Functions and Rates

The functional roles played by channels and transporters are quite diverse and can be illustrated with a few comparative examples. Gated ion channels are frequently involved in signal transduction, rapidly altering the membrane permea-

Table 8.2. Some superfamilies of structurally related channels and transporters.

(1) Voltage-regulated ion channels:
Na^+ channel
K^+ channel
Ca^{2+} channel (dihydropyridine-sensitive)
(2) Neurotransmitter-regulated channels:
Nicotinic acetylcholine receptor
γ-Aminobutyric acid (GABA) receptor
Glycine receptor
(3) Mitochondrial transporters:
ADP/ATP transporter
H^+-Phosphate transporter
Uncoupling protein (H^+/OH^- transporter)
(4) Sugar transporters:
Mammalian glucose transporter
E. coli H^+-arabinose transporter
E. coli H^+-xylose transporter
(5) E_1E_2-type ion-motive ATPases (also see Table 8.5)
H^+/K^+-ATPase (mammalian gastric mucosa)
Na^+/K^+-ATPase (plasma membrane)
Ca^{2+}-ATPase (sarcoplasmic reticulum)
H^+-ATPase (plasma membrane)
K^+-ATPase (*S. faecalis*)

bility to a specific ion in response to some external signal and, thus, changing the transmembrane voltage (see Chapter 9). The fundamental requirements are the ability to switch between the open and closed states rapidly in response to the signal and the ability to allow the voltage to attain quickly the new equilibrium or steady state. Speed is essential. An ion channel in its open configuration typically can allow the passage of 10^6-10^8 ions/sec across the bilayer. Such a large flux is the experimental criterion distinguishing channels from transporters (see Table 8.3). The large current passed by such ion channels means that opening a relatively small number of channels can cause a large, rapid change in the electrical properties of the membrane. It will be useful to consider some realistic numbers at this time and to see how to estimate the response time of the membrane.

As an example, suppose we have a membrane with a capacitance of 1 μfarad/cm^2 which contains a low density of K^+ channels, each with a conductance of 20 pS when open (see Section 7.3 for definitions of units). A conductance of 1 pS is equivalent to 6×10^6 ions/sec/volt, so each channel can pass 1.2×10^8 ions/sec/volt. Let us say that there is a K^+ gradient across the membrane, with $[K^+]_{in}/[K^+]_{out} = 52$. When the K^+ channels are closed, the membrane permeability to K^+ is effectively zero, so K^+ plays no role in determining the transmembrane potential. If the channels are open, however, K^+ will flow out until a new steady-state distribution is attained.

Table 8.3. Comparison of transport rates for selected systems[1]

System	Transport rate
Channels/Pores	
Sodium channel	$\sim 10^7$ sec^{-1}
Gramicidin A	$\sim 10^7$ sec^{-1}
Acetylcholine receptor channel	$\sim 10^7$ sec^{-1}
Permeases	
H^+-Lactose permease (*E. coli*)	30 sec^{-1}
Glucose transporter (erythrocyte)	300 sec^{-1}
Band 3 anion transporter[2] (erythrocyte)	100,000 sec^{-1}
Active transporters	
Bacteriorhodopsin	50 sec^{-1}
Na^+/K^+-ATPase[3]	450 sec^{-1}
Cytochrome *c* oxidase	1,000 sec^{-1}

[1]See text for references. Values are only approximate since they vary considerably depending on experimental conditions. The values for channels are estimated for 0.1 *M* NaCl, 100 mV, from ref. 1386.
[2]Exchange rate. See text.
[3]Na^+ transport rate (three per ATP hydrolyzed).

The equilibrium value of the transmembrane potential attained after opening the channels is obtained from the Nernst equation (see Section 7.4) and is easily computed to be 100 mV in this case, due to the charge separation created by K^+ diffusing across the membrane. The capacitance (C) defines the amount of charge separation required to maintain this voltage (equation 7.15) and is only about 10^{-12} mol/cm^2 (about 1 charge per 250 lipid molecules). This is a minuscule number of ions which are moved in order to give a large electrical signal. There is no detectable change in the K^+ concentration on either side of the membrane.

The time required to attain the new steady state once the channels are opened is a critical parameter and is determined by both the capacitance and the specific membrane resistance. If we assume an average of 50 open channels per square micrometer of membrane, the response time is 0.1 msec.

Box 8.1 The Electrical Response Time of a Membrane

The electrical response time (τ) is a measure of how rapidly the transmembrane potential achieves a new equilibrium position after specific ion channels have been opened. In excitable membranes, such as nerve and muscle, this is a critical parameter and is determined by the membrane capacitance (C) and specific membrane resistance (R). These relationships are easily derived, below.

Start by taking the derivative with respect to time of the equation defining capacitance (eq. 7.15):

$$C \cdot \left(\frac{dV}{dt}\right) = \left(\frac{dQ}{dt}\right) \qquad [8.1]$$

dQ/dt is equal to current, I, and is equal to zero when the voltage reaches its final value, V_K (= 100 mV). Empirically (see 633), the current through an ion channel is proportional to the difference between the actual voltage and the voltage determined by the Nernst equation defining the equilibrium situation where there is zero current. So $I = (V_K - V)/R$ by a variation of Ohm's law (eq. 7.13). Then

$$\frac{dV}{dt} = \frac{(V_K - V)}{RC} \quad [8.2]$$

which has the solution:

$$V = V_K\,(1 - e^{-t/RC}) \quad [8.3]$$

or

$$V = V_K\,(1 - e^{-t/\tau}) \quad [8.4]$$

where

$$\tau = RC$$

The voltage across the membrane rises exponentially from zero to its final value, 100 mV, with a time constant, τ, equal to RC. Hence, the product of the specific membrane resistance (R) and the capacitance (C) has the units of time and is a measure of the electrical response time of the membrane. Let us postulate that there will be an average of 50 channels open per square micrometer. This means that the specific membrane conductance is 10 mS/cm^2 (current per volt per cm^2) and the specific membrane resistance (R) is the reciprocal of this, 100 Ωcm^2. In this case, RC = 0.1 msec, which is within the range for excitable biomembranes (10 µsec to 1 sec) (633). This value will depend primarily on the specific resistance (R) of the membrane, since the capacitance (C) is determined primarily by the lipid bilayer (see Chapter 7).

Note that the specific resistance of a membrane depends on the number of channels, the fraction of time they are open, and the individual (single-channel) conductance of each channel in the open configuration. The use of single-channel recording methods, such as patch clamping, has allowed the direct measurement of these parameters (see Section 8.14).

Table 8.3 shows the turnover numbers of several ion channels and transporters. Note that the ion channels and pores are characterized by very high turnover values. In contrast, the lactose permease from *E. coli* has a maximal turnover of only about 30 sec^{-1}. If the ion channel in the previous example were to operate at this rate, it would require a 10 million-fold increase in the channel density to achieve the same specific membrane conductance. This would be physically impossible.

The physiological function of the lactose permease obviously does not require the enormous turnover exhibited by channels. Its function is to transport the sugar lactose for metabolic use by the cell. Similarly, ion "pumps" driven by electron transport or by ATP hydrolysis operate at maximal turnover numbers in the range of $10^2 - 10^3$ sec^{-1}, far slower than channels or pores but rather typical for enzymes.

Not all transporters are so slow. The anion transporter, Band 3, from the erythrocyte membrane plays a crucial physiological role in facilitating the rapid exchange of Cl^- for HCO_3^- (bicarbonate) across the membrane (see Section 8.33).

One of the roles of the erythrocyte is to facilitate the transport of CO_2 from the venous capillaries to the lungs, where it is excreted. CO_2 itself is sparsely soluble in water. In the venous capillaries, CO_2 diffuses rapidly across the erythrocyte membrane, where its conversion to H_2CO_3 is accelerated by carbonic anhydrase. H_2CO_3 rapidly equilibrates to $H^+ + HCO_3^-$, and the bicarbonate anion is transported across the membrane to the extracellular plasma via Band 3. As a result, the concentration of HCO_3^- in the plasma is increased as the cell is in transit through the capillaries, which takes less than 1 second. When the blood reaches the lungs, the CO_2 diffuses out to the atmosphere, and, by mass action, the H_2CO_3 is converted back to CO_2 and H_2O inside the erythrocyte by carbonic anhydrase. Mass action drives the uptake of HCO_3^- by the cell when the erythrocyte is in the lungs as it is rapidly converted to CO_2 and H_2O.

The transport system must be rapid, but, unlike the ion channels in axons, there is no need for an electrogenic reaction, which could only serve to slow down the rapid mass transport. The cotransport of a cation such as Na^+ along with HCO_3^- would not be desirable, since the changing salt concentration within the red cell would result in an osmotic imbalance. The solution is a transporter which acts as a strict antiport system, in which for every HCO_3^- transported a Cl^- is returned from the opposite side. This shuttle system works very rapidly, and is able to turn over at about 10^5 sec^{-1}, not much slower than a true channel.

8.12 Channels and Transporters Viewed as Enzymes: Application of Rate Theory

The Eyring transition state kinetic formalism used by enzymologists, and most familiar to biochemists, has been successfully applied to a variety of transport systems (see 824, 825, 632, 791). A crucial element of this approach is that the system is viewed as existing in a small number of discrete states, each of which can be assigned a standard state electrochemical potential. Interconversions between these states require that the system pass through transition states of higher free energy, and the rate constants are related to the heights of the free energy barriers. Wells in the free energy profiles (e.g., Figure 8.1) represent binding sites for the solute being transported. One can postulate one or a number of binding sites for the solute within the channel or transporter. The existence of solute binding sites necessitates that at sufficiently high solute concentration the site(s) can be saturated and the transport rate will reach a maximum value, equivalent to V_{max}, the maximal velocity of an enzyme. This is experimentally observed for all transporters and also for many channels.

Application to Channels

To see how this formalism can be applied to channels, consider a simple case of an ion channel with ions initially present on one side. We will assume, for simplicity, that there are barriers to entering or leaving the channel and one binding site present inside the channel. There may be numerous binding sites

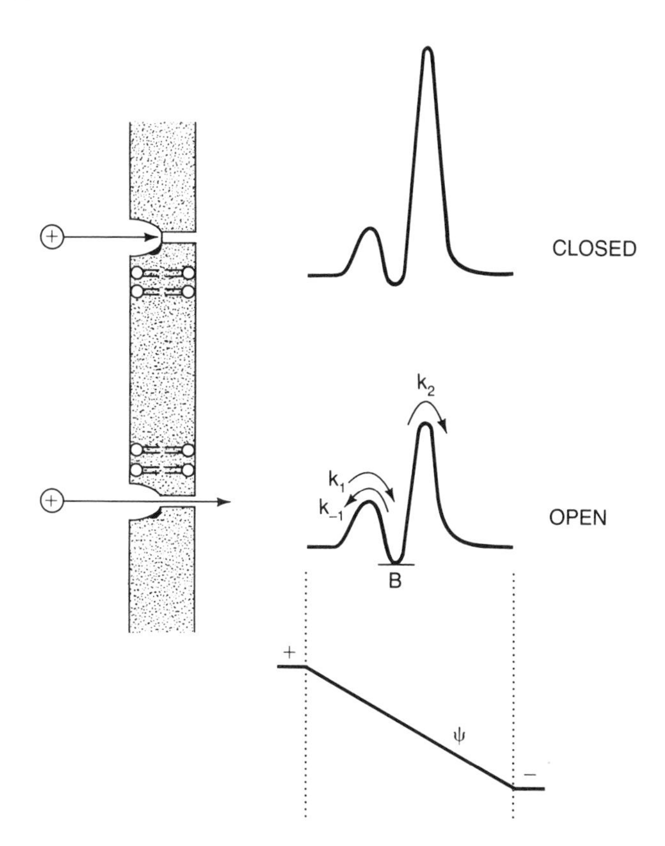

Figure 8.1. Schematic of a simple ion channel (left), pictured as having a wide mouth leading to a narrow portion through which the ion must pass. The free energy profiles on the right correspond to the case where there is a single binding site within the channel, possibly located somewhere in the mouth of the channel. The depth of the well ("B") is drawn to imply weak binding. In the presence of a transmembrane potential difference, the electrical potential profile (ψ) must be added to the free energy profiles. In this case, the barrier heights corresponding to k_1 and k_2 will be reduced, accelerating the passage of cations through the channel.

within the channel along which the ion might be relayed, but the velocity (ion flux) will still be described by the Michaelis–Menten equation. For the moment, we will assume there is no voltage across the membrane. The "reaction" is then written as follows

$$\mathrm{E} + S_i \underset{k_{-1}}{\overset{k_1}{\rightleftarrows}} \mathrm{ES} \overset{k_2}{\rightarrow} \mathrm{E} + S_o$$

where E is the channel protein (or enzyme) in an open configuration, S_i and S_o are the internal and external solute concentrations, and ES is the complex of the solute with the binding site inside the channel. Figure 8.1 shows a free energy profile for such a model. Also shown in Figure 8.1 is the profile for the channel in a closed configuration. This is easily obtained by raising the height of one of the energy barriers to prevent the ion from crossing the channel. No one knows what this actually corresponds to for any gated channel with the possible exception of the gap junction channel (1482). It could represent the movement of an α-helix or even the subtle movement of a side chain to a position effectively blocking transport through the channel.

Applying the steady-state criterion to the species ES yields the familiar Michaelis–Menten equation for the velocity (ions transported/sec):

$$V = \frac{[E]_o[S]k_2}{K_m + [S]} \qquad [8.5]$$

where $K_m = \dfrac{k_{-1} + k_2}{k_1}$ and $[E]_o$ is the total concentration of transporter.

Consider two situations: (1) saturating solute, $[S] >> K_m$; (2) low solute concentration, $[S] << K_m$.

(1) *Saturating solute*: Under these circumstances, the channel binding sites are filled ($[E]_o \approx [ES]$) and the rate reaches its maximal value, determined by the energy barrier to exit the channel. This is easily seen by simplifying the Michaelis–Menten equation [8.5] for the case $[S] >> K_m$.

$$V = V_{max} = k_2[E]_o \qquad [8.6]$$

(2) *Low substrate*: For $[S] << K_m$, we are essentially on the linear portion of the velocity–substrate curve and the kinetics appear to be second-order. The apparent second order rate constant is obtained by rewriting the Michaelis–Menten equation in the limit of small [S].

$$V = \left(\frac{k_2}{K_m}\right)[E]_o[S] = \left(\frac{V_{max}}{K_m}\right)[S] \qquad [8.7]$$

This equation is important because most, if not all, ion channels operate near this limit. The apparent second-order rate constant for this reaction is (k_2/K_m). This is not a true microscopic rate constant but is made up of a combination of rate constants even for this simple case.

$$\text{Transport rate constant} = \frac{k_2}{K_m} = \frac{k_1 k_2}{k_{-1} + k_2} \qquad [8.8]$$

This simple model illustrates several points:

(1) The measured rate constant which defines the transport velocity of the open ion channel contains information about the apparent affinity of the ion to the binding site (K_m) and the turnover number (k_2). In the absence of a voltage across the membrane, this rate constant is directly related to the permeability due to the single channel. If the internal barrier is small ($k_2 >> k_{-1}$), the transport rate constant will equal k_1, the second-order rate constant for the ion to enter the channel. This could be a diffusion-limited reaction, in which case the value for k_1 would be

quite large, at least 10^8 to 10^9 M^{-1} sec^{-1}. This is observed in some cases. If the internal barrier is large ($k_2 << k_{-1}$), the transport rate constant equals the product $(k_1/k_{-1}) \cdot k_2$.

(2) The voltage dependence of the transport velocity (i.e., ion flux or current) is quantified by the conductance. This is easily built into the rate model by evaluating how a voltage drop across the membrane affects the individual microscopic rate constants (k_1, k_{-1}, k_2 in this model). Superimposing a linear voltage drop onto the potential profile shown in Figure 8.1 results in a decrease in the barrier heights for k_2 and k_1 and an increase for k_{-1}, effectively increasing the second-order rate constant for ion transport, as expected. Simple models like this do not actually predict the linear current-voltage (i.e., ohmic) relationship which is usually experimentally observed, but the introduction of additional barriers in such models makes the current-voltage curve very close to linear (633, 846).

(3) Ion selectivity can be explained in several ways. Selectivity is the preferential ability of a channel to pass some ions better than others and can be quantified by the ratio of permeability or conductance values for the ions being compared. Since the transport rate constant (eq. 8.8) contains both k_1 and k_2, in principle, selective passage of ions can result from either a lower free energy barrier to enter the channel (k_1) or a lower free energy barrier for translocation within the channel (k_2). For example, the presence of negative charges or an arrangement of dipoles at the mouth of the channel could reduce the rate of anions entering the channel, resulting in a cation-selective channel. This appears to be the case for the several receptor channel proteins (713) and gramicidin A (1411). Simple considerations of solute size also restrict entry into channels. Obviously, a solute which is larger than the channel diameter cannot get into the channel. This is referred to as molecular sieving. Finally, differences in the translocation rates for different ions within the channel can cause selectivity. The Na^+-selective channel is an example where a barrier to internal translocation is substantially lower for Na^+ than for other ions, such as K^+, resulting in a higher conductance for Na^+. The molecular rationale for this "selectivity filter" effect is discussed further in Section 8.25. Note that the selectivity of a channel is unlikely to result from increasing the affinity of the ion for the channel. This corresponds to lowering the depth of the well (labeled "B") in the potential profile in Figure 8.1, which will result in reducing the escape rate out of the channel, i.e., lowering the maximal channel flux, as well as reducing the concentration of solute required to saturate the channel, i.e, lower K_m. The K_m values for those channels which primarily pass Na^+ or K^+ are quite high, in the range 200 mM to 300 mM, which is above the physiological ion concentrations normally encountered.

Application to Transporters (see 825, 632, 1386, 1286)

Let us now turn our attention to transporters. Consider a simple transporter with one binding site shuttling a solute molecule across a membrane. Figure 8.2 illustrates the essential features for both a primary active transporter and a permease. Minimally, there are four states of the protein to consider: (1) inward facing/bound

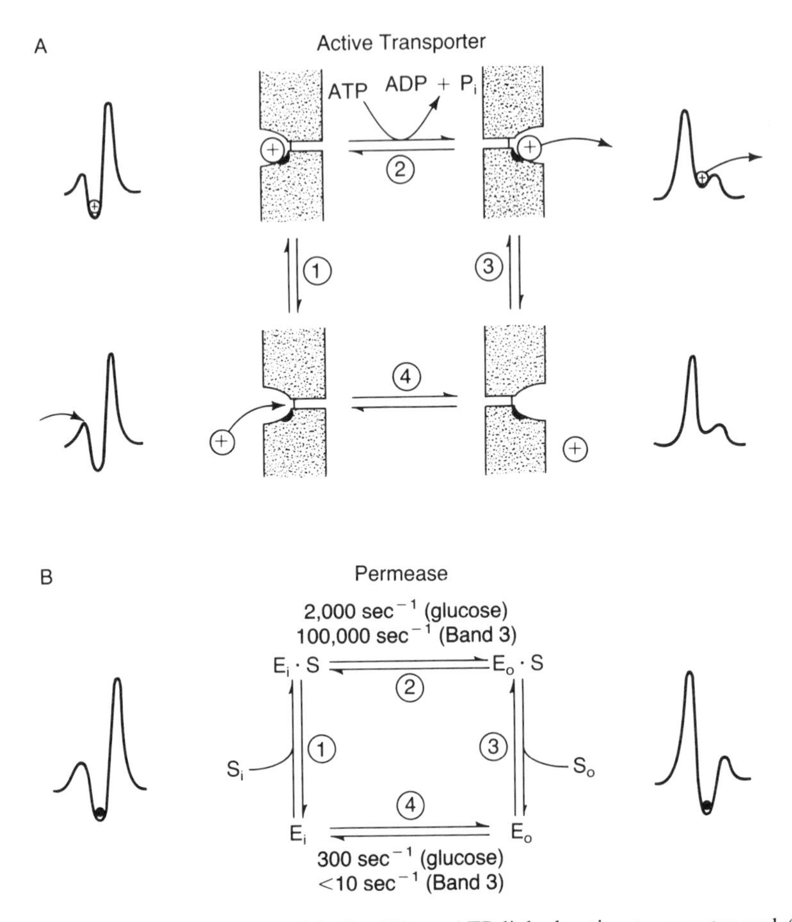

Figure 8.2. Simple four-state models for (A) an ATP-linked active transporter and (B) a permease catalyzing solute transport or exchange. The free energy profiles are indicated on either side, showing a single binding site and alternating barrier. In model (A), ATP hydrolysis is linked to a conformational change which alters both the access and binding affinity of the solute binding site. The permease, in part (B), spontaneously fluctuates between the inward-facing and outward-facing forms, with no change in binding affinity. Values are the rates for interconversions measured for the erythrocyte glucose transporter (23°C) and the erythrocyte anion exchanger (Band 3) (37°C). Note that the substrate in both cases enhances the rate of conformational change. Step 4 for Band 3 is exceedingly slow, so this protein catalyzes only anion exchange.

to solute, (2) inward facing/unbound, (3) outward facing/bound to solute, and (4) outward facing/unbound. One possible set of elemental, reversible, transport steps is as follows:

1. The solute binds to a site facing one side (defined as the *cis* side).
2. There is a conformational change, essentially lowering the kinetic barrier for forward motion of the ion and increasing the barrier for moving backward. This conformational change could be spontaneous, or might require energy

input such as from ATP hydrolysis. The site and bound solute are now facing the opposite side of the membrane (defined as the *trans* side).

3. The solute dissociates from the site and exits on the opposite side. For active transporters, the affinity for the solute is lower when the site is facing the *trans* side of the membrane.
4. There is a conformational change, returning to the original conformation with the site facing *cis* once again.

A key element of the function of all transporters is that there is a large central free energy barrier which requires alteration by a protein conformational change as a necessary part of the transport scheme. If energy input is required for this, then the system can act as an active transporter, such as the ATP-driven Ca^{2+}-pump (825). If the conformational change requires a solute molecule bound to the protein, i.e., step 4 does not occur, then the protein will only catalyze solute exchange across the bilayer, since it cannot isomerize in the "unloaded" form [e.g., Band 3 (791, 467)].

There are many models and schemes for viewing ion channels (see 633) and the various kinds of transporters (825, 1286). Although absolute rate theory has definite limitations, especially as regards the kinetic properties of channels (846), the application of rate theory has intuitive appeal, simplifies, and allows one to view a wide variety of transport mechanisms in a unified manner.

8.13 Steady State Assays (see 791, 1386, 41)

A variety of different steady-state assays can be used to kinetically characterize transport systems which catalyze facilitated diffusion or active transport. All these assays measure the rate of translocation of the solute across the membrane, but under different conditions. As examples, we will consider experiments on permeases in membrane vesicles. Usually radioactive tracers are used for these measurements. One key point is that permeases need not be symmetrical with regard to the solute interactions on the two sides of the membrane. Furthermore, the permease must not only carry the solute across the bilayer (*cis* → *trans*) but must also return (*trans* → *cis*). The net rate can depend on either of these steps. Some assay designs are briefly described below.

1. *Zero trans*: Solute is present on one side only (*cis*). The initial velocity of transport is measured for the unidirectional flux. Note that the transporter (permease) must return unloaded to obtain net flux. One measures flux as a function of $[S]_{cis}$. One can measure this in either direction, for example, the influx or efflux of solute into or out of vesicles.
2. *Equilibrium exchange*: The same concentration of solute is present on both sides of the membrane, but the radioactive label is only on one side (*cis*). In this case, the permease can return with unlabeled solute. One measures flux as a function of [S].
3. *Infinite trans*: The radioactive solute is present on the *cis* side and a saturating concentration of solute is present on the opposite (*trans*) side. One measures flux as a function of $[S]_{cis}$. As before, this measures unidirectional flow (*cis*

→ *trans*). This is also called *counterflow* because the labeled solute is transported against its chemical gradient.

4. *Infinite cis*: The radioactive specific activity of the solute on both sides of the membrane is the same in this assay. A saturating concentration of solute is present on the *trans* side, and $[S]_{cis}$ is varied. In this case, one measures the total net mass transport, since radioactive solute will cross in both directions. This measures the approach to equilibrium.

In all these cases, one measures a V_{max} and K_m, which need not be the same for these assays. Steady-state kinetic expressions can be derived for particular models and tested (e.g., 791). As in classical enzymological studies, the use of inhibitors can be very revealing (e.g., 791, 410, 409). Of course, the inhibitors can be added to either side of the membrane, yielding additional information.

These kinds of assays can be performed in cells or natural membrane vesicles or artificial reconstituted vesicles. Ion channels are usually assayed using electrical methods, which offer tremendous advantages, as described in the next section.

Box 8.2 Steady-State Transport Assays with Mutational Variants of the Lactose Permease of *E. coli*

As a qualitative example of how these various assays can be used, consider the analysis of one of the mutational variants of the lactose permease in which Glu-325 has been replaced by Ala-325 (717; Section 8.32). This residue is indicated in Figure 8.11 in helix 10. The transport assays are performed with radioactive lactose either with intact cells (influx) or with right-side-out cytoplasmic membrane vesicles (efflux, counterflow, and exchange). The wild-type permease catalyzes the symport of ß-galactosides (e.g., lactose) and H^+. In the presence of a H^+ electrochemical gradient, the permease utilizes the free energy from the downhill translocation of H^+ to drive uphill accumulation of ß-galactosides against a concentration gradient.

The Ala-325 replacement abolishes lactose active transport. The influx assay (zero *trans*) is performed by using intact cells in which the mutant permease has replaced the wild type. [1-^{14}C]Lactose is added to a suspension of respiring cells and the amount of lactose entering the cells is measured as a function of time. After about 3 minutes the uptake reaches a maximal value in the control cells (~50–100 nmol lactose per mg of protein), but uptake is negligible in the cells with the mutant permease. Efflux is measured by allowing membrane vesicles to incubate for several hours in the presence of a high concentration of radioactive lactose and by then diluting aliquots into medium without lactose. Rapid filtration of the vesicles followed by scintillation counting measures the rate of loss of lactose from the vesicles. The half-time for control vesicles was 10 sec, but it was 540 sec for vesicles containing the Ala-325 mutation. The mutant is defective in its ability to transport lactose in either direction across the membrane.

An equilibrium exchange assay, however, yields strikingly different results. This assay is performed identically to the efflux assay, except that the lactose-loaded vesicles are diluted into a medium containing the same concentration of unlabeled lactose. In this assay, the rates of loss of [1-^{14}C]lactose from the vesicles containing the wild-type and mutant permeases were identical. This assay shows that although the mutant permease cannot complete a full cycle to catalyze net flux of lactose, it can isomerize normally in the loaded form (bound to lactose and H^+) to allow lactose to exchange across the membrane (see Figure 8.11). Similarly, the mutant permease functions normally in a counterflow assay (infinite *trans*). In this assay, the vesicles are loaded with unlabeled lactose (10 m*M*) and diluted into medium containing a low concentration of [1-^{14}C]lactose (1.6 m*M*). Flux of the labeled lactose into the vesicles is measured. This assay also indicated that the form of the permease bound to lactose can isomerize normally so as to allow lactose to exchange across the membrane.

These data have been interpreted to conclude that the Ala-325 mutant cannot become deprotonated. Thus, the permease is defective in all coupled transport processes in which an H^+ and a lactose are cotransported in a cycle which requires the deprotonated form of the transporter to isomerize ($E_i \leftrightarrow E_o$ in Figure 8.11).

For more quantitative examples using these types of assays, the excellent book by Stein (1386) is recommended.

8.14 Single-Channel Recordings: Reconstitution in Planar Membranes (see 582, 1010) and Patch Clamping (see 24)

The rapid ion flux through a channel can be measured as current very easily. For example, the reconstituted nicotinic acetylcholine receptor, a chemically gated cation-selective channel found in postsynaptic membranes, has a conductance in the open state of about 45 pS with 0.5 *M* NaCl on one side. This is in the normal range of conductance values for ion channels. This means that each channel conducts a current of 4.5 pA or 2.7×10^7 ions/sec if the voltage across the membrane is 100 mV. This is easily detected with appropriate electronic circuitry. Hence, the current through a single molecular channel can be directly measured if the background leakage across the membrane is sufficiently small. In order to examine the properties of individual channel molecules, it is necessary to examine the current across a portion of membrane under conditions where it is unlikely that more than one channel will be open at the same time. This can be artificially created by making planar membranes reconstituted with an appropriate density of channel proteins (1012, 246, 582). Alternatively, one can use a patch clamp (243). This derives from a discovery in 1979 that when a clean glass micropipet is placed against a cell membrane and suction applied, the cell membrane forms a very tight seal which is mechanically stable. This allows the voltage-current behavior of any channels within this small portion of the membrane to be studied. The background current is usually considerably less than 1 pA, so individual channel current can easily be recorded. The application of this technique has revolutionized the study of ion channels because it has enabled a wide variety of channels to be examined in detail from a variety of membrane sources. Figure 8.3 schematically shows the four modes in which the tight patch-clamp membrane seal can used in studying biomembranes. The ability to vary the solution composition on one or both sides of a small patch of the membrane, and to examine the

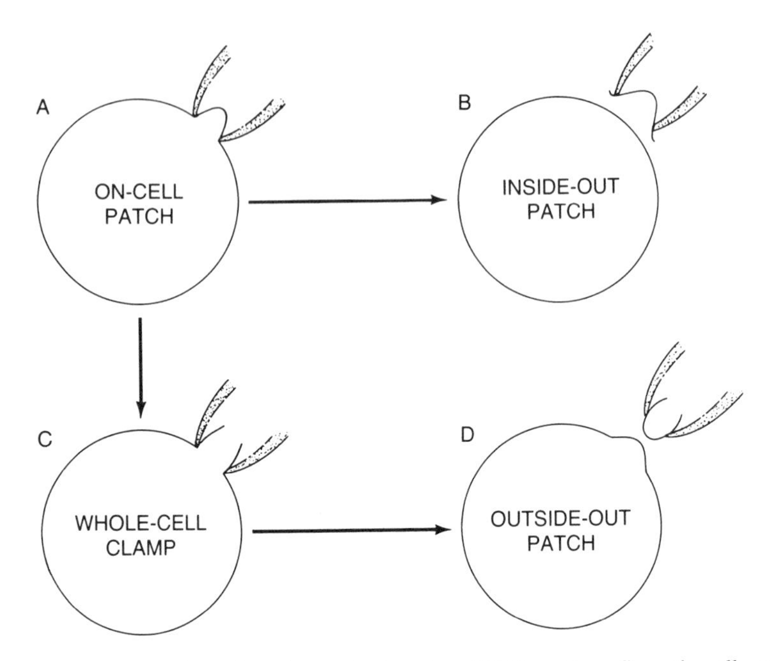

Figure 8.3. Four configurations for patch clamping. (A) On-cell configuration allows the intact cell to be studied. (B) Withdrawing the pipet breaks off a patch of membrane sealed tightly to the pipet tip. The side of the membrane which faces the cytoplasm is now facing the external bath, so this is called an "inside-out" patch. (C) Suction applied to the pipet ruptures the membrane, allowing the pipet to be used for imposing a voltage across the whole cell membrane. (D) Slow withdrawal results in resealing of a patch of membrane across the pipet tip, with an inverted orientation. Adapted from ref. 243.

conductance properties of single channels provides an extraordinarily powerful set of techniques. Patch clamping allows one to study the enzymology of single molecules.

To appreciate the power of these techniques, examine the single-channel recording in Figure 8.4 for the acetylcholine receptor channel reconstituted in bilayers across the tip of a patch pipet (1012) (see Figure 2.26). This is a gated channel, and individual channels can be seen to open and close spontaneously. Each downward deflection is a channel opening. Note that the current of the open channel is quantized and easily measured to yield the single-channel conductance (current/volt). The length of time each channel remains open can also easily be measured. One gets a distribution of "dwell times" in the open state, and this is related directly to the rate constant for channel closing. For the simplest model with a single open (O) or closed (C) state

$$\mathrm{O} \underset{k_{-1}}{\overset{k_1}{\rightleftarrows}} \mathrm{C}$$

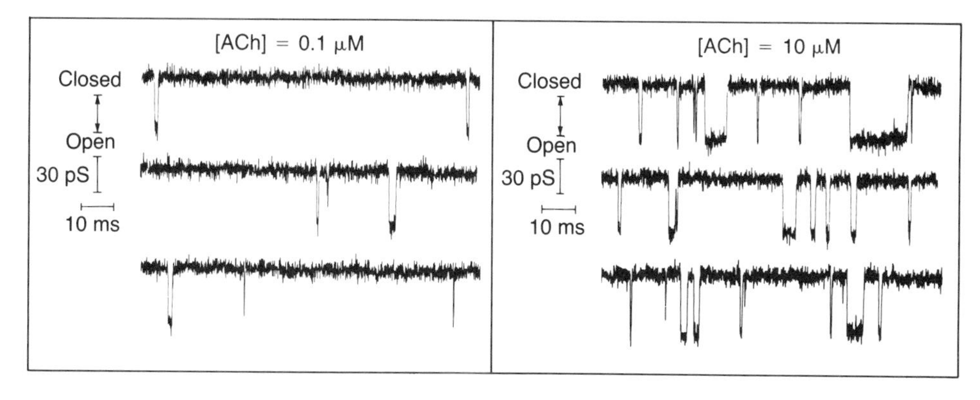

Figure 8.4. Single acetylcholine receptor channel currents activated by acetylcholine (ACh) (0.1 or 10 μ*M*) at an applied voltage of 100 mV. The channels were reconstituted in planar membranes across the tip of a patch pipet (see Figure 2.26). The solution contained 0.5 M NaCl on both sides of the membrane. From ref. 805. Reproduced with permission from the Society for Neuroscience. Kindly provided by Dr. M. Montal.

the distribution of dwell times is an exponential function, with the exponential decay time $\tau_{-1} = (1/k_{-1})$, the reciprocal of the first-order rate constant for channel closing. When a simple exponential distribution function of the open channel lifetimes is not obtained, as is the case for the acetylcholine receptor channel (1012), one needs more complicated models (e.g., more than one open state) to explain the data. One can study the effects of various chemical agents interacting with ion channels and differentiate effects on channel opening-closing kinetics from effects on single-channel conductance values.

Box 8.3 Determination of the Binding Constant of a Channel Inhibitor by Single-Channel Conductance Studies

An elegant example of the use of single-channel conductance is provided by studies on the interaction between charybdotoxin and the Ca^{2+}-activated K^+-specific channel (1358). Charybdotoxin, a small peptide isolated from scorpions, is a high-affinity inhibitor of this channel. Although these channels have not been biochemically characterized, they have been extensively examined using single-channel recording techniques. The method used in this case (1358) was to prepare membrane vesicles from rat skeletal muscle transverse tubules and to induce them to fuse with planar lipid bilayers (see Chapter 2). The recordings shown in Figure 8.5 illustrate the current from one individual channel molecule which is rapidly fluctuating between the open and closed states. If these data are plotted on a msec time scale, the individual opening and closing events can be seen as in Figure 8.4.

When charybdotoxin is added, long periods appear when the channel remains nonconducting. Toxin binding evidently blocks the channel in some manner. The frequency of appearance of these closed periods gives the on-rate for charybdotoxin binding to the channel, and the time that the channels remain closed is related to the rate constant for toxin dissociation.

$$\text{active channel} + \text{toxin} \underset{k_{off}}{\overset{k_{on}\,[\text{toxin}]}{\rightleftharpoons}} \text{inactive channel-toxin}$$

The average dwell time in the active state, τ_1, is inversely related to the rate of toxin binding, and the dwell time in the closed state, τ_{-1}, is inversely related to the rate of dissociation.

$$(1/\tau_1) = k_{on}\,[\text{toxin}]$$

$$(1/\tau_{-1}) = k_{off}$$

Data plotted in Figure 8.5 show, as expected, that the rate of channel blocking is dependent on toxin concentration, whereas the rate of dissociation is not dependent on the free toxin concentration.

From these data one can easily obtain both k_{off} and k_{on} for the toxin, and that ratio gives the dissociation constant (3.5 nM = K_d). What is remarkable is that these kinetic and thermodynamic values are obtained by examining events taking place on an individual molecule.

8.15 Small Molecules Which Serve as Models for Channels and Pores

The kinetic methods outlined in the previous section have been used to characterize numerous channels and transporters found in biomembranes. Methods of molecular biology have facilitated obtaining the amino acid sequences of many of these proteins. However, as pointed out in Chapter 3, there is no high-resolution crystal structure of any mass transport protein, and one must resort to model building to obtain speculative structures and mechanisms. Important clues are provided by several small molecules which are able to form pores across the bilayer. Three examples of particular interest are (1) nystatin, a polyene antibiotic which forms a complex with cholesterol, (2) gramicidin A, a peptide which forms a cation-selective channel which is the best characterized of all channels, and (3) alamethicin, a peptide which forms voltage-gated channels. The structures of the latter two have been described in Section 3.71.

Nystatin and Amphotericin B (see 762, 118; Section 7.12)

These are related polyene structures shown in Figure 8.6. When nystatin is added to sterol-containing planar membranes from one side, the permeability increases for water, univalent ions, and small non-electrolytes. Non-electrolytes larger than glucose cannot pass through the channel formed by these polyenes, suggesting a pore with a diameter of 8 Å. A plausible model shown in Figure 8.6 is that 8 to 10 of these molecules form a "barrel stave" structure with the hydroxylated portion of each molecule facing the inside, forming a water-filled pore extending

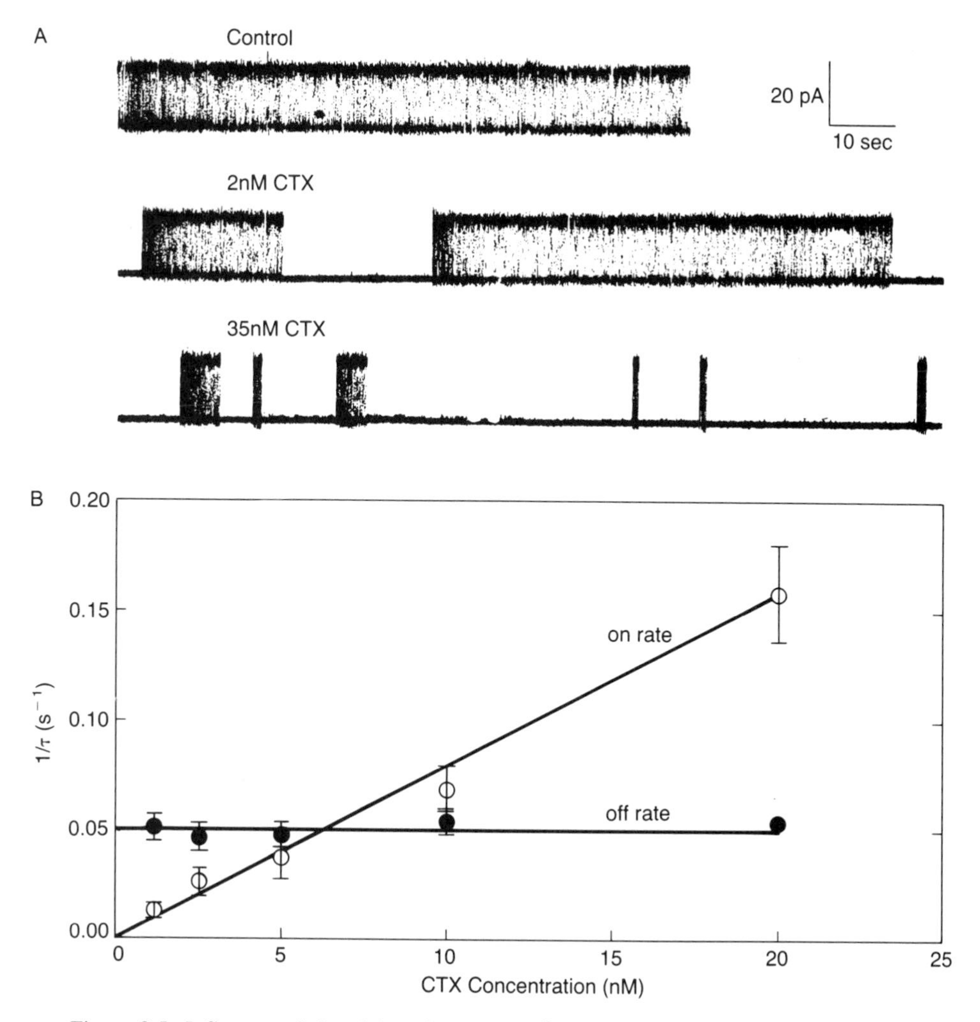

Figure 8.5. Influence of charybdotoxin on the Ca^{2+}-activated K^+-specific channels from rat skeletal muscle membranes. (A) Single-channel recordings from an individual channel molecule are shown in the upper panel. Note that when plotted on this time scale, individual opening/closing events cannot be observed as in Figure 8.4. Addition of charybdotoxin (CTX) results in channel inactivation during long periods. (B) The inverse of the average dwell times for the channel in the active or inactive states are plotted as a function of toxin concentration. Figure kindly provided by C. Miller.

across the bilayer. The exterior of this structure would be nonpolar, with room for an intercalated sterol molecule between each pair of nystatin molecules. The conductance of this channel is relatively small, 5 pS in 2 *M* KCl. When nystatin is added to the planar membrane from both sides, it appears that two such barrel-like structures join to form a channel which is twice as long, presumably distorting the bilayer as a result.

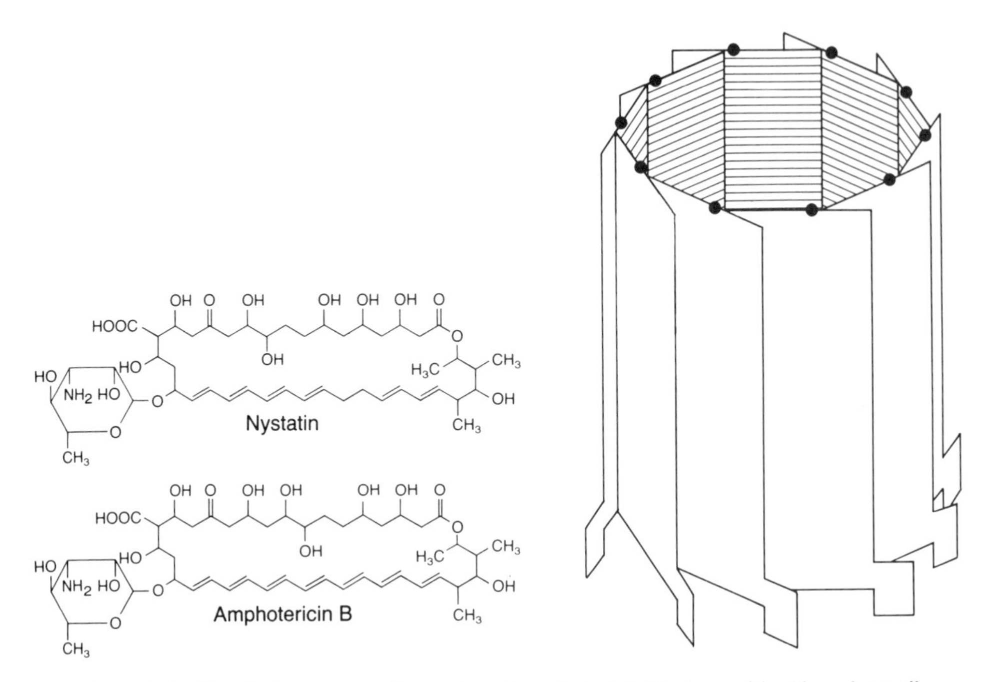

Figure 8.6. Chemical structures of nystatin and amphotericin B, along with a "barrel stave" model of the channel which forms when these polyene antibiotics are in membranes. The protuberance on the bottom represents the amino sugar and the shaded interior represents the hydrophilic polyhydroxyl portion of the molecule. The exterior surface of the channel is completely nonpolar. Adapted from refs. 762 and 959a.

One point to note is that this pore-forming molecule is distinctly amphipathic with polar and nonpolar surfaces. The polar portion in this case is provided by a series of hydroxyl and carbonyl groups. A second point is that the pore is apparently formed by an aggregate which is stable within the bilayer. Both of these structural features are themes which are repeated in considerations of numerous channel-forming proteins. Note, however, that the nystatin pore structure is purely conjecture, though based on sound reasoning and consistent with known data.

Gramicidin A (see 633, 751, 1411; Section 3.71)

This is, by far, the best characterized channel, made from a hydrophobic peptide with alternating L- and D-configuration. Experimental evidence supports the model for the channel shown in Figure 3.12, with two gramicidin A molecules forming a head-to-head dimer in a ß(L, D) helical configuration. The alternating L-, D-configuration results in a helix where all the side chains face out and the backbone carbonyls all face the inside of the channel. Because of the unusual alternating stereochemistry of the amino acids in gramicidin A, this type of helix is not found in proteins. From this perspective, gramicidin A is a poor structural model for how proteins form transmembrane pores. However, in many other respects, the gramicidin A channel has many features in common with other channels and pores. Points to note follow.

1. Gramicidin A is extremely hydrophobic with no polar amino acid side chains.
2. The single-channel conductance values of the gramicidin channel are quite respectable. For example, in 100 mM RbCl, the cation conductance is 30 pS per channel. This is, however, far below the theoretical limit for a diffusion-limited pore of this size (300 pS) (see 633).
3. The gramicidin channel is cation selective. Whereas small inorganic and organic cations pass through the channel, Cl^- is totally excluded. It neither passes through nor blocks the channel, though it is sufficiently small to easily pass through a pore of this size. This is quite remarkable, considering that there are no polar or charged amino acid side chains in gramicidin. There is a repulsive electrostatic interaction with anions due entirely to dipoles associated with carbonyl groups situated at the mouth of the channel (1411).
4. The channel is permeable to monovalent cations in the order $Cs^+ > Rb^+ > K^+ > Na^+ > Li^+$, which is also the order of the free diffusion of these ions in bulk water. The selectivity exhibited by this channel for these monovalent cations is very modest, spanning only about a 5-fold range of permeability coefficients. The activation energy of the conductance is small, about 5 kcal/mol, similar to that of ions diffusing in bulk water. In this sense, the channel acts as if it is water-filled. The free diffusion properties of cations in solution are largely a function of the strength of the electrostatic interactions with water. Small ions (e.g., Li^+) interact with water molecules more strongly than do large ions, and the electrostatic polarization tends to cause more drag, i.e., slower diffusion. The same kinds of forces apparently are involved in the rate-limiting step(s) as the cations proceed through the long, thin gramicidin channel.
5. The channel can pass molecules through only in single file since it is only 4 Å in diameter. Hence, an ion passing through must be partially dehydrated as it enters the channel. The channel can be filled with 5 to 7 water molecules in addition to the cation. In the channel, the ion will be in direct contact with a water molecule in front and behind. The channel must provide groups to interact favorably with the ion in place of the lost waters of hydration. Since dehydration is very costly energetically (see 633), it is this step which is likely to be rate limiting.
6. The gramicidin A channel exhibits saturation at sufficiently high salt concentrations, indicative of binding sites for cations. For example, the half-saturation concentration for Na^+ conductance is 0.31 *M*. Models consistent with the flux kinetics suggest that there are two binding sites for cations, one located near each channel entrance. The cation binding is the result of favorable dipolar interactions, in particular with the carbonyl of residue 11 (1411).
7. Water passes through at a rate of about 10^8 water molecules/sec at low ionic strength. Whenever a cation passes through the channel under the influence of an electric field, the 5 to 7 water molecules in the channel are also caused to flow, resulting in electro-osmosis.

8. The gramicidin A channel opens and closes corresponding to the association of monomers to form the dimer (open) and the dissociation of the dimer (closed). The lifetime of the dimer is on the order of 10 msec to 1 sec depending on lipid composition and other factors.
9. When gramicidin A is in a negatively charged lipid bilayer, the electrostatic effect of the surface charge density can result in an increase in the local cation concentration near the mouth of the gramicidin pore. This dramatically increases the conductance values obtained at low salt concentrations.

The above list of characteristics clearly shows why gramicidin A has been such a valuable experimental system. Many of its channel properties are similar to those exhibited by physiologically important channels such as the acetylcholine receptor channel.

Alamethicin (see 969, 1223, 397)

This is one of a family of naturally occurring peptides which form voltage-gated channels across membranes and, thus, provides a good model for how channel opening and closing can be controlled by the transmembrane potential. Alamethicin contains 20 amino acids, including a number of α-aminoisobutyric acid residues (see Section 3.21). The key structural feature of alamethicin and its homologues is that they readily form α-helices. Unlike the ß(L, D) helix formed by gramicidin A, the center of the α-helix is too small to allow an ion to pass through. Rather, the data strongly support a barrel stave model in which upwards of 12 alamethicin molecules aggregate to form a pore (see Figure 3.13). As pointed out in Chapter 3, when alamethicin is in such a configuration, polar atoms line the interior of such a structure. Note that polar amino acid side chains are not required to provide a polar lining for a water-filled channel.

At low voltage (<100 mV), no current will be conducted across planar membranes containing alamethicin. As the voltage is increased, there is a critical "switch-on" voltage beyond which current is conducted by alamethicin channels. The voltage can be in either direction across the membrane. The single-channel conductance of these channels is very large, 5,000 pS at 1 *M* KCl, indicating a very large pore. A model with 12 aggregated monomers would have a diameter of 20 Å, consistent with the large single-channel conductance.

There are several models for how alamethicin is voltage-gated. The models all have as a critical feature the recognition that in an α-helix, the dipoles associated with each peptide bond add up to create a net dipole equivalent to one-half of a positive charge on the amino-terminal and one-half of a negative charge on the carboxyl-terminal end. The transmembrane potential is sufficiently large to interact with this permanent dipole and to cause the helical peptide to substantially alter its orientation in the membrane (969) or to increase its partition coefficient into the membrane (1223).

The simplest model for voltage gating is that the electric field causes more alamethicin to partition into the membrane and that, once bound, there is a highly cooperative tendency to aggregate to form a barrel-type channel (1223). The more widely held opinion is that alamethicin is already bound to the membrane in a nonconducting form in the absence of a voltage and can only aggregate in an open-channel configuration once the field is applied. The electric field in these models causes a dipole "flip-flop" of the α-helical peptide, stabilizing an aggregate in which the α-helices are parallel to each other in the barrel-type channel (969).

Alamethicin provides a good model for how associated α-helices can form a pore, and also demonstrates, in a simple system, that the transmembrane potential can stabilize particular protein conformations, and drastically alter channel conductance. These properties are not unique to the alamethicin family of peptides. Synthetic peptides (723) and melittin (1524) provide additional examples.

8.2 Several Examples of Pores and Channels (see 377 for review)

Progress is accelerating in our understanding of the molecular structures of pores and channels. Some common elements seem to be emerging. Amino acid sequence similarities have revealed the existence of structurally related superfamilies of ion channels (Table 8.2). Electron microscopy image reconstruction techniques have been particularly valuable because it has been possible to visualize not only the hole in the membrane formed by the larger pores, but also the symmetrical arrangement of subunits around the central hole (see Table 8.4). A barrel-like hole formed by amphipathic α-helices from several associated subunits or from separate domains of a single subunit may be a common structural motif for these proteins (1479). The structures of alamethicin and gramicidin A, however, point out that polar amino acid side chains are not required to create a suitably polar lining for a water-filled channel. In no case is it known with certainty which amino acid residues are directly involved in forming a pore. Porins appear to be an important exception to the α-helical structural motif since, in these proteins, the pore appears to be formed by ß-sheet rather than α-helices.

Table 8.4. Pseudosymmetry of some pores and channels.

Pore/channel	Structural symmetry	Comments
Porins	3-fold	3 Identical subunits
Sodium channel	4-fold	4 Homologous domains of a single subunit
Acetylcholine receptor channel	5-fold	$\alpha_2\beta\gamma\delta$ Arrangement of 4 homologous subunits
Gap junction	6-fold	12 Identical subunits, 6 in each membrane
Nuclear pore complex	8-fold	Unknown subunit composition

8.21 Gap Junctions (see 871 for review)

Gap junctions are clusters of membrane channels that connect the interiors of adjacent cells in organized tissues. The channels allow the passage of small molecules such as metabolites and inorganic ions and have a diameter in the range of 12-20 Å in mammalian cells (871, 1515). These channels mediate both chemical and electrical flux between cells. The channels, thus, span two plasma membranes. Cloning experiments (273) and biochemical reconstitution studies have demonstrated that the channel is simply composed of an oligomer of a single polypeptide which, in liver cells, has a molecular weight of 32,000 (871). The amino acid sequence suggests that four transmembrane α-helices may be present in each subunit (1631). Gap junction proteins, sometimes called connexins, have been characterized from several tissues and appear to constitute a diverse family of related proteins (e.g., 871, 916). These channels are usually in an open state but will close (1515) when the cells become leaky or when the metabolic rate is depressed. The primary signal for channel closure appears to be a rise in Ca^{2+}, though a drop in transmembrane potential or acidification of the medium will also result in channel closure. It is possible that the Ca^{2+} effect may be, in part, indirect rather than a direct effect on the protein. Phosphorylation of the channel may be another mode of regulation (871).

The basic model for this channel is provided by the work of Unwin and Ennis (1482), who used electron microscopy to reveal the channel structure. Each channel is composed of 12 subunits, 6 from each cell. The channel through each membrane has a hexameric structure forming a central pore. Each subunit is a rod-shaped molecule traversing the bilayer. Two hexameric units from adjacent membranes are associated in an axial end-to-end manner, forming a continuous channel spanning both membranes. The structure of the gap junction channel is different in the presence or absence of Ca^{2+}. When Ca^{2+} is present, the subunits appear to be more nearly parallel to the central axis through the channel, whereas they appear to be slightly tilted in the absence of Ca^{2+} (open state). This suggests that channel opening and closing may be likened to an iris of a camera where the subunits can slide over one another in response to a gating signal to open and close the central pore. The mechanism is not at all clear.

8.22 Nuclear Pore Complex

As mentioned in Chapter 1, the nuclear envelope is composed of a double membrane, about which little is known. Nuclear pore complexes mediate the transport processes between the nucleus interior and the cytoplasm (1217). These pores, like those of the gap junction, must cross two membranes. Low-resolution electron microscopy (1483) indicates that the pore has octagonal symmetry, but with a more complicated structure than is apparent for the gap junction channel. The nuclear pore is basically two octagonal barrels or annuli, one from each membrane, joined together to span the entire envelope, in a structure somewhat similar to that of the gap junction. The polypeptide structure of the nuclear envelope pore has not been clearly defined, though at least one subunit has been shown to be part of the complex (294).

The nuclear envelope pore is very large, with a functional radius of about 90 Å, able to pass all small solutes as well as many large molecules. There are specific mechanisms whereby macromolecules are transported into and out of the nucleus, but little is known about these mechanisms. See Chapter 10 for a discussion of the sorting of proteins synthesized in the cytoplasm.

8.23 Porins (see 579, 86, 1053 for reviews)

Porins form pores which act as molecular sieves to allow the diffusion of small, hydrophilic solutes across the outer membrane of gram-negative bacteria. Over 40 different porins have been examined (57) and common features are clear. Porins range in molecular weight from 28,000 to 48,000 and usually exist in the membrane as trimers. Porins have a high content (~60%) of ß-sheet (1523). The best characterized porins are those from *E. coli*: OmpF (matrix porin), OmpC, PhoE, and LamB (maltoporin). The structures have been discussed in Section 3.53. The major point is that these proteins form a water-filled transmembrane channel which is almost certainly made of ß-sheet. Figure 8.7 illustrates one possible model based on amphipathic ß-strands. Although porins are strongly associated with both lipopolysaccharide and the peptidoglycan, they do not require either of these to function as transmembrane pores through a phospholipid bilayer.

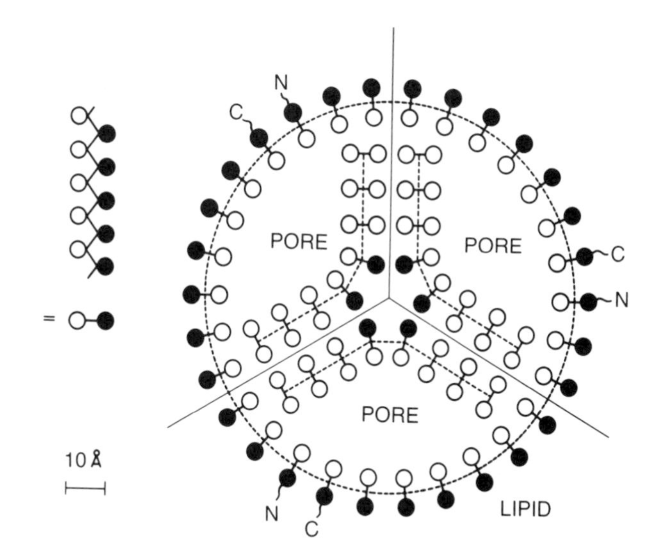

Figure 8.7. Schematic model of a porin trimer where each subunit contains a pore which is lined by hydrophilic amino acid residues. This is a view from above, where each dumbbell represents a ß-strand with alternating hydrophilic (open circles) and hydrophobic (filled circles) residues. Continuous lines designate the boundaries between individual subunits. Adapted from ref. 1523.

Overall, porin channels exhibit a wide range of pore size, 6 Å to 23 Å diameter, and selectivity, 3-fold anion selective to 40-fold cation selective (579). Selectivity appears to result from charged amino acid residues within the channel or near the channel entrance (580). Single-channel conductance measurements, permeability studies, and electron microscopy indicate that the pore structures are not the same for all porins. Some trimeric porins appear to have one large channel; others appear to have three independent channels, apparently associated with the three subunits (e.g., Figure 8.7). The OmpF channel appears to have three entrances on the outside of the cell which merge to form a single exit in the periplasm (579).

Three of the four *E. coli* porins which have been well characterized are closely related, with considerable sequence homology (OmpF, OmpC, and PhoE) and estimated pore sizes of 10 to 12 Å. The fourth porin, LamB (maltoporin), is clearly different, though it exhibits some sequence similarities with the other porins, and is also mostly ß-structure (579). LamB is part of the maltose uptake system, and binds tightly to maltose derivatives. In the absence of maltose, LamB forms small channels which are totally blocked upon maltose binding (87). The main function of LamB appears to be to facilitate maltose uptake and not to act as a general diffusion pore for small solutes. The topography of LamB has been investigated using a number of methods (see Chapter 4), including genetics, resulting in a detailed model for the folding pattern in the membrane (127).

The possibility that porins are gated channels and can exist in a closed state has been investigated by numerous laboratories (see 813). Although porins (e.g., OmpF) can exhibit voltage gating *in vitro*, the physiological relevance is unclear at this time (813).

Finally, it should be noted that a voltage-gated porin has been characterized from the outer membrane of mitochondria, called the voltage-dependent anion-selective channel or VDAC (451, 884). This protein appears to be unrelated to the bacterial porins, but has also been proposed to form a ß-barrel type of pore (451).

In summary, porins are of particular interest because they illustrate the use of ß-strands rather than α-helices in forming transmembrane segments and because, as discussed in Section 3.53, it is anticipated that a high-resolution structure will be forthcoming from crystallographic studies presently in progress.

8.24 The Nicotinic Acetylcholine Receptor Channel (nAChR Channel) (see 1012, 663, 1399, 564 for reviews)

The nAChR channel is an example of a neurotransmitter-stimulated channel and is the best understood of all biological channels. It is located primarily in the postsynaptic membranes of the neuromuscular junction of skeletal muscle, which is called the endplate. When the neuron is electrically excited, the neurotransmitter, acetylcholine, is secreted in a burst at this junction, where it diffuses from the presynaptic neuronal membrane to the skeletal muscle membrane. The nAChR is localized in densely packed clusters in the endplate structure in the muscle cell plasma membrane, and, upon interaction with acetylcholine, the channels open, allowing the selective passage of cations. This changes the transmembrane potential and electrically excites the muscle cell, eventually resulting in muscle contraction.

The nAChR channel is a chemically gated channel in which the chemical signal acts directly on the channel. The chemical signal, acetylcholine, is used for excitatory synaptic transmission between nerve and muscle. The nAChR has been shown to be structurally related by comparison of amino acid sequences to at least two other neurotransmitter receptors: the glycine receptor and the γ-aminobutyric acid (GABA) receptor (1302, 549). These constitute a superfamily of chemically regulated channels (Table 8.2). Despite sequence similarities, these channels do not appear to share a common quaternary structure. Receptors for glutamate, a major neurotransmitter between neurons on the brain, have not been biochemically characterized (1390).

Before proceeding, it is useful to note that the nomenclature and classification of neurotransmitter receptors are largely based on pharmacological studies of drug responses, particularly those drugs which act as agonists, which stimulate in the manner of the physiological neurotransmitter, or antagonists, which block the stimulation by agonists. For example, the effects of acetylcholine were differentiated in the early part of this century as being nicotinic and muscarinic, based on pharmacological distinctions. The nicotinic actions of acetylcholine are mediated by a family of related receptor proteins, one of which is the nAChR channel (523), whereas the muscarinic effects are mediated by a different family of receptor proteins. The muscarinic receptor from brain cells is structurally unrelated to the nAChR channel (1336, 793), and, indeed, is not a channel at all, but acts indirectly by mechanisms which will be discussed in the next chapter. It should be pointed out that nicotinic acetylcholine receptors from brain tissue have been characterized which are structurally and functionally quite different from the endplate channels (1572). In this section, however, we will focus on the properties of the endplate channels.

The reason that the nAChR channel is so well characterized in comparison to other channel proteins is that there is a readily available source of tissue which is enriched in this protein, thus facilitating detailed biochemical studies. Most biochemical work has been accomplished using the receptor purified from the electric organ of the electric eels *Electrophorus* and *Torpedo*, and these "fish" channels have been clearly shown to be very closely related to the endplate channels, and can be considered identical in most properties. Hybrid proteins consisting of portions of the *Torpedo* and bovine subunits have been made by gene fusion technology and shown to be fully functional (677).

The purified nAChR channel consists of five polypeptide subunits of four different types with a stoichiometry $\alpha_2\beta\delta\gamma$. The two copies of the α subunit present in the complex appear to be functionally differentiated (1200). The four subunits are closely related by sequence homologies and range in apparent molecular weight from 40,000 to 68,000 by SDS-PAGE analysis. All of the subunits are phosphorylated and glycosylated, and two of the subunits (α, β) contain covalently bound lipid (1094) (see Chapter 3).

Electron microscopy image reconstruction has revealed that the channel has the structure illustrated schematically in Figure 8.8 (see 663). Negative staining clearly shows a central pit with a diameter of 30 Å at the extracellular (synaptic) end and considerably more narrow on the cytoplasmic side. A pentameric structure is evident in the images, indicating that the five subunits are arranged in a circle around the central pore. Nearest neighbor analyses using crosslinking reagents (see Chapter 4) indicate that the sequence of the subunits is ß-α-δ-γ-α, so the two α-subunits are not adjacent. Numerous studies (see 663) clearly show that the binding sites for acetylcholine, other agonists, and competitive antagonists are located on the α-subunits. Hence, there are two binding sites for the chemical gating agents. Antagonists prevent the activation by agonists either by directly competing for binding at the same sites (competitive antagonists, e.g., α-bungarotoxin) or by binding elsewhere and possibly interacting with the channel per se (noncompetitive antagonists, e.g., $TPMP^+$).

As illustrated in Figure 8.8, the nAChR channel is about 140 Å long, extending over 70 Å beyond the bilayer surface on the extracytoplasmic surface. This forms the large mouth of the channel. In *Torpedo*, the channel exists as dimeric units, held together by a disulfide bond formed between the δ-subunits on the extracel-

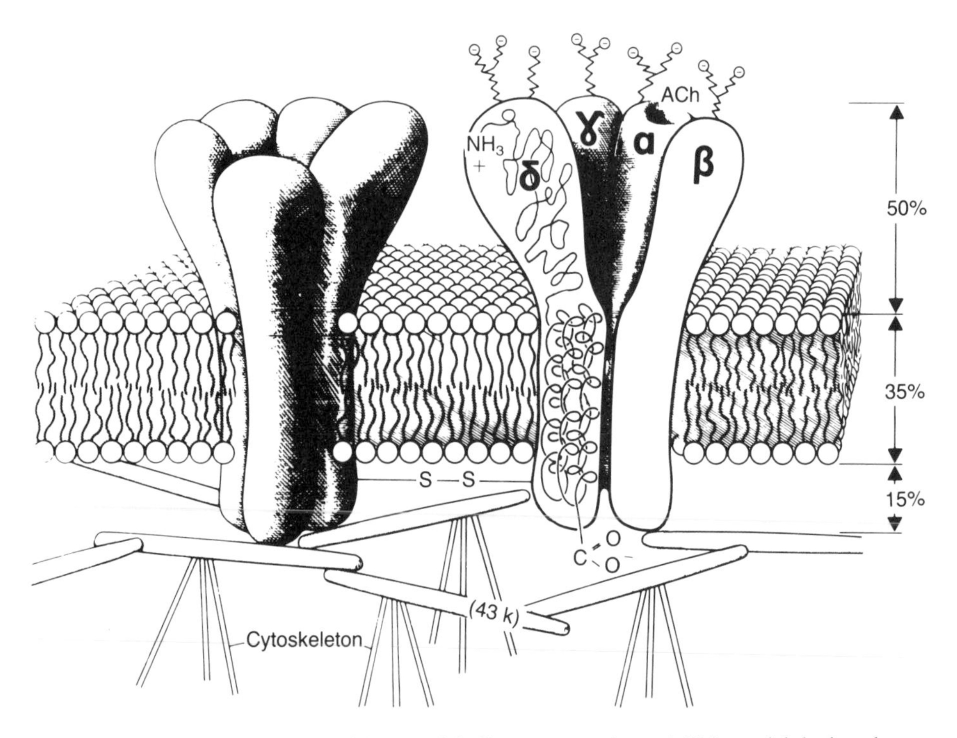

Figure 8.8. Model of the nicotinic acetylcholine receptor channel. This model depicts the general appearance and dimensions of the channel, including the glycosylation sites on the outside and the acetylcholine binding sites on the α-subunits, as well as the putative association with the cytoskeleton on the inside. Adapted from ref. 663.

lular side (356). The extension of the channel on the cytoplasmic side is much less prominent, and this part of the channel presumably interacts with cytoskeletal elements required for forming the densely packed clusters of channel proteins in the endplate. A specific protein (43,000 Da) has been suggested as mediating these interactions (110), as shown in Figure 8.8. In the endplate, the channels can be clustered at a density as high as 10,000 molecules per μm^2, which is close to the theoretical limit (see 633). Interaction with acetylcholine or other agonists results in a rapid, dramatic current through this region of the membrane, depolarizing the postsynaptic membrane.

The kinetic behavior of this channel can be successfully analyzed as that of an allosteric enzyme, capable of existing in several different conformations (see 1477). There are a minimum of three distinct states in which the channel exists: open, closed, and desensitized. The cooperative binding of two acetylcholine molecules is required to induce the channel to go from the closed to the open configuration, where it remains open for about 1 msec (see Figure 8.4). In the desensitized state, the channel remains closed even in the presence of acetylcholine. Single-channel measurements suggest more than one open state (1012), and the minimal kinetic mechanism is quite complicated (see 1477 for review). Reconstitution studies have demonstrated that the lipid environment can influence the ability of the channel to shift conformations (445).

In the open configuration, the channel is permeable to cations and small nonelectrolytes, but not anions. The size limitation of permeable molecules can be used to estimate the dimensions of the narrowest part of the channel, indicated in Figure 8.9 along with data on two other channels. The size of the ionic (i.e., nonhydrated) radius of the largest ion passed by the channel is a reasonable

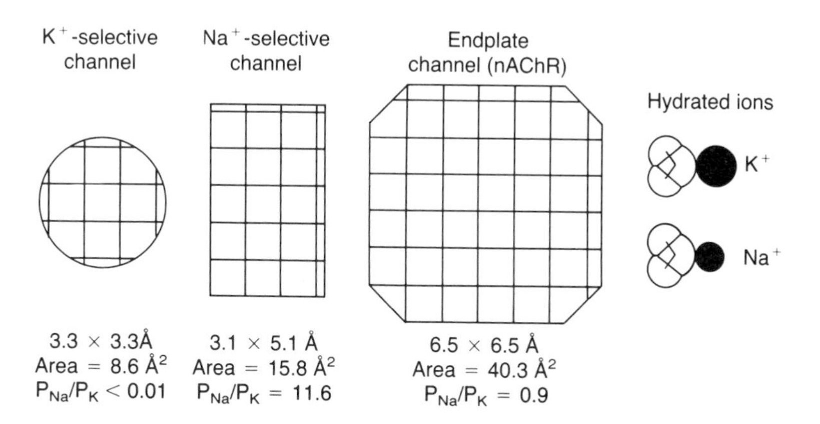

Figure 8.9. Dimensions of three channels from frog nerve and muscle deduced from molecular sieve experiments. Grid marks represent 1 Å steps. The ratios of the permeability coefficients for Na^+ and K^+ are also shown, along with the relative sizes of these ions, drawn to scale with the channel cross sections. In bulk solution, the ratio of the diffusion coefficients for Na^+ and K^+ is 0.7. Adapted from ref. 362a. Reprinted from the *Journal of General Physiology*, *75*, p. 488, Copyright © 1980 the Rockefeller University Press.

estimate of the limiting dimensions of the channel. Electrostatic interactions, possibly due to dipolar or negatively charged groups in the mouth of the channel, are sufficient to explain the fact that the channel is not permeable to anions and, in fact, is three-fold more permeable to cations than to uncharged species. The selectivity of the channel within the series of monovalent and divalent cations is small, but is consistent with a model in which the ions interact predominantly with water, even within the channel proper, and interact only very weakly with the channel itself (852). Although the nAChR channel has relatively large dimensions, it is still too small to allow the passage of ions which are fully hydrated. The channel interior must contain groups poised to easily replace any lost water of hydration.

More is known about the molecular structure of this channel than of any other. Yet, unfortunately, there is no consensus on how this channel is built, despite many quite detailed models that have been put forth (see 663, 1399, 564). Molecular genetics has made an impact, but further experiments are required. The amino acid sequences have been deduced from the cloned genes of each subunit from several sources, showing highly conserved regions. Up to seven different segments have been identified as potential membrane-spanning α-helices (1198). One of these, denoted M5 or MA, has been a focus of attention because it is a potential amphipathic helix with charged groups lined up on one side (see Section 3.6). It is assumed by most that the narrow portion of the channel extending through the bilayer is made up by the subunits contributing one α-helix each to form the central pore. The amphipathic helix is a prime candidate, though the model studies with gramicidin A make it quite clear that charged groups are not necessary to construct a water-filled, cation-selective channel. There is no experimental evidence that this amphipathic helix is directly involved in forming the channel, and, in fact, the experimental evidence that this putative helix extends across the membrane is in dispute (1198, 776). Some evidence points to another putative helix, M2, as being involved in channel formation (664, 677, 1022, 1456). These data come from covalent affinity labeling with noncompetitive antagonists (664, 1022) and from molecular genetics studies (677, 1456). This helix, though not amphipathic from a side-chain analysis, could form a channel lining more like that of gramicidin A or alamethicin. It is noteworthy that the other members of the superfamily of neurotransmitter receptors (Table 8.2) do not have a segment which is equivalent to the amphipathic helix of the nicotinic acetylcholine receptor.

The pioneering work of Numa and his colleagues (677, 996, 997, 1456) has demonstrated the power of site-directed mutagenesis in addressing problems in this and related systems. The genes encoding each of the four subunits of the channel have been injected into COS monkey cells and *Xenopus* oocytes, resulting in the expression of functional channels in the plasma membranes. The use of single-cell voltage clamping and single-channel recordings has been essential for the evaluation of both wild-type and genetically altered channels. It can safely be expected that as these experiments become increasingly fine-tuned, the answers to some long-standing questions about channel structure will be obtained. In combination with the use of well-characterized polyclonal, monoclonal, and peptide-directed antibodies, all the required tools should be available to define those portions of the polypeptides which are involved in acetylcholine (or toxin) binding (e.g., 997, 1060, 71) as well as those portions which constitute the channel itself. The general belief is that this channel is formed from a cluster of α-helices, like alamethicin. The same concept has influenced models of the Na^+ channel, described below.

8.25 The Voltage-Sensitive Sodium Channel (see 193, 79 for reviews)

The voltage-sensitive sodium channel mediates the rapid increase in sodium permeability which causes the depolarizing phase of the action potential in nerve and muscle cells. This channel has been purified to homogeneity from several sources, including rat brain, mammalian skeletal muscle, chicken heart, and the electric eel *Electrophorus electricus.*

The purified channels from *Electrophorus* and chicken heart contain only a single glycoprotein subunit, with a molecular weight around 260,000, while the mammalian channels contain, in addition, one (1277) or two (975) smaller glycoprotein subunits with molecular weights in the range 33,000 to 38,000 (ß1- and ß2-subunits). Most of these purified channel preparations have been successfully reconstituted in planar membranes and shown to retain the electrophysiological and pharmacological properties of the channel in vivo. The mammalian channel appears to require at least one of the small associated proteins for successful reconstitution (975).

The sodium channels interact with a set of toxins, including tetrodotoxin, saxitoxin, and α-scorpion toxin, which bind very tightly to the channel proteins and can be used for convenient, quantitative biochemical assays. The availability of these toxins has been essential for the biochemical purifications of these channels, as well as for localization studies in vivo. Like the endplate (nAChR) channel, the sodium channels are not randomly distributed in the plasma membrane, but are clustered in particular locations (see Chapter 4). In muscle cells, the Na^+ channels are concentrated in the endplate region along with the nAChR channel (349).

Numa and his colleagues have cloned and sequenced the genes encoding the sodium channels from *Electrophorus* (1074), as well as the principal subunit from rat brain (1076, 1075), revealing a family of closely related sodium channels (1076). The *Electrophorus* sodium channel polypeptide sequence was deduced to contain 1820 amino acids, arranged in four repeated homology units (see Figure 8.10). Each homology unit has been postulated to contain 4, 6, or 8 potential transmembrane α-helices in different models (see 1076, 546, 565). Some of these proposed membrane-spanning helices (S4 segments) are amphipathic, similar to that proposed for the nAChR channel (see previous section). There is experimental evidence that the carboxyl terminus is located on the cytoplasmic surface

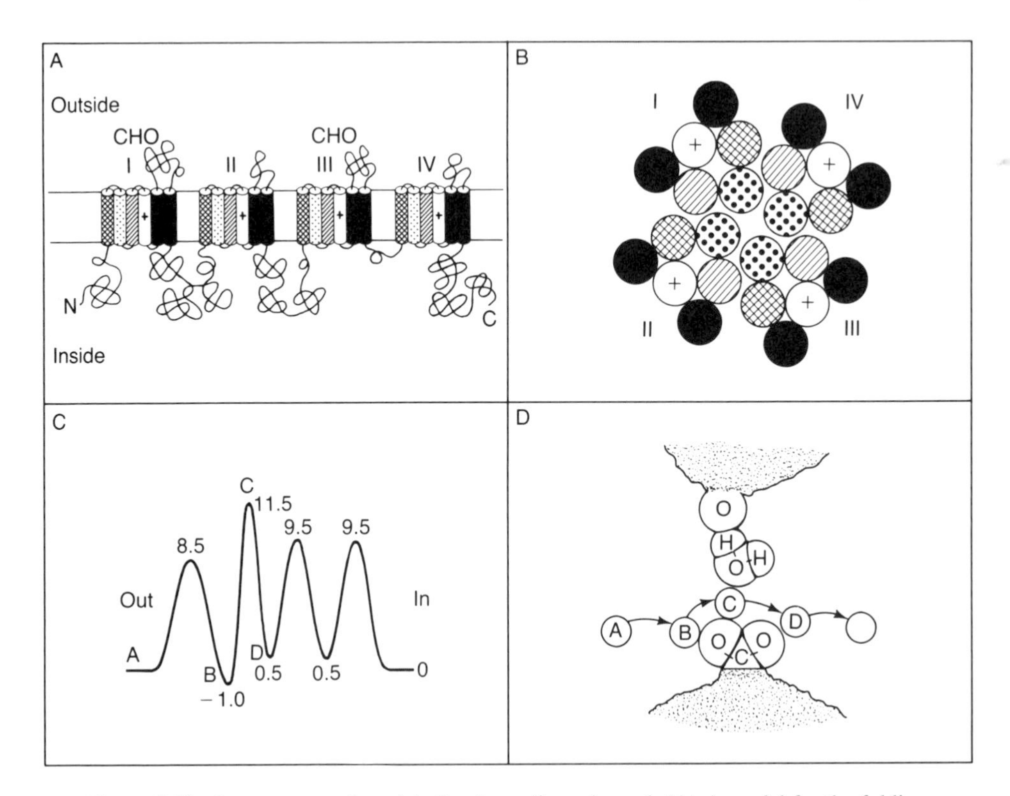

Figure 8.10. Some proposed models for the sodium channel. (A) A model for the folding of the polypeptide chain indicating four homologous domains, each with six membrane-spanning helices. The amphipathic segment S4 is indicated by a plus sign. (B) Speculative arrangement of the six helices from each domain forming a central channel, viewed end-on. (C) Free energy profile consistent with flux data for Na^+ passing through the channel. Numbers indicate energy levels in units of RT relative to the bulk solution. Three binding sites and four barriers are postulated. (D) Speculative model of the narrow "selectivity filter" responsible for the height of barrier "C." The hydrated ion must pass through positions "A," "B," "C," and "D" in sequence, and the height of the kinetic barrier will depend critically on the distance between the postulated carboxyl and carbonyl groups relative to ion size. Adapted from refs. 1076 (A, B) (Reprinted by permission from *Nature*, vol. 320, p. 191, Copyright © 1986 Macmillan Magazines Limited.) and 633a (C, D) Reprinted from the *Journal of General Physiology*, *66*, p. 557, Copyright © 1975 the Rockefeller University press.

(531), and it has been reported that the amphipathic S4 helix is extracytoplasmic (961), but no other information is available to verify the proposed topological models. Similarly, there are no data to indicate which portions of the polypeptide are directly involved in forming the pore. The competing models for the sodium channel are conceptually similar to those of the endplate channel in that the channel per se is made up of a cluster of α-helices with a central pore. In this case, however, instead of 5 closely related subunits, we have four homologous domains of a single subunit. Presumably, specific groups are arranged in the channel to provide the "selectivity filter."

Although molecular sieve experiments result in a picture (see Figure 8.9) showing the limiting channel size, molecular size alone cannot explain the selectivity among ions which are small enough to penetrate. Figure 8.10 diagrams a free energy profile calculated for the diffusion of Na^+ through the channel. If the rate-limiting step is the loss of water, the precise geometry of those components of the channel which temporarily replace the water will be critical to the height of the energy barrier (peak C in Figure 8.10C). For K^+, the groups would not be ideally positioned, since this is a slightly larger ion, and the energy barrier will consequently be higher. Conversely, for the K^+-selective channels, the equivalent energy barrier is minimized for the K^+ ion, and the smaller Na^+ ion is discriminated against.

The manner in which the transmembrane potential regulates the channel gating has also been modeled extensively (see 546, 565, 368) and certainly involves the net movement of charge on the protein in response to the imposed electric field. No data are available to indicate which regions of the polypeptide constitute the gate or respond to the voltage, though the positively charged, amphipathic S4 helix is considered a prime candidate.

On the basis of amino acid sequence similarities, the sodium channel has been shown to be part of a superfamily of voltage-regulated channels (Table 8.2). Other channels in the group are the potassium channel from *Drosophila* (1443) and the dihydropyridine calcium channel from rabbit skeletal muscle (1431). The sequences of these channels include an amphipathic, highly positively charged segment equivalent to the S4 segment of the sodium channel. This strengthens the deduction that this segment is involved in voltage sensing and voltage regulation of these channels. Note that both the mammalian sodium channel and the calcium channel (next section) have subunits of unknown function in addition to those whose sequence is known.

8.26 The Calcium Channel (see 992, 1470 for reviews)

Ca^{2+}-selective channels are very widespread in excitable cells such as nerve and muscle, as well as in most other cells. Some Ca^{2+} channels are voltage-sensitive, whereas others are regulated indirectly via receptor-mediated events (see Chapter 9). Typically, the cytoplasmic $[Ca^{2+}]$ is less than 10^{-7} *M*, which is 10,000-fold lower than the extracellular $[Ca^{2+}]$. Because of these conditions, and unlike the Na^+ and K^+ channels, opening a Ca^{2+} channel can cause significant changes in the cytoplasmic ion concentration. Changes in internal $[Ca^{2+}]$ are known to initiate numerous biochemical events. For example, the voltage-stimulated increase in cytoplasmic Ca^{2+} results in the secretion of neurotransmitters such as acetylcholine from presynaptic membranes.

Several classes of Ca^{2+} channels have been defined pharmacologically, and high-affinity organic blocking agents have proved very valuable for channel purification as well as characterization. The only class of Ca^{2+} channel to be characterized biochemically is the voltage-sensitive channel, which is blocked by a variety of organic compounds, including dihydropyridine derivatives. These have been purified from both cardiac and skeletal muscle, and they appear similar (241, 1422). These channels appear to consist minimally of two subunits with molecular weights of 140,000 and 30,000. The large subunit has been cloned and sequenced, revealing it to be structurally related to the voltage-sensitive sodium channel (1431).

8.27 Summary

The growing body of information gives one the hope that the structures of a variety of channels and pores will have common features. Basically, all these systems must have a water-filled pore with a lining which has polar character, and this can be provided by either α-helix or ß-sheet arrangements. To attain the high rates of transport, high-affinity binding sites for solutes are generally not present: selectivity is usually determined by the heights of the free energy barriers and not by the depth of the wells (see Figure 8.1). The principles behind channel selectivity are reasonably well understood and can be modeled in most cases by placing a few amino acid residues in crucial positions. Similarly, it is possible that gating mechanisms for opening and closing various channels share common structural features, though nothing is really known about this.

As is shown in subsequent sections, similar structural features might also be found in other kinds of transport proteins.

8.3 Several Examples of Uniporters, Symporters, and Antiporters

Several systems which catalyze the transport of one or more solutes have been reasonably well characterized. All of these systems operate at transport velocities considerably less than even the slowest channel. Included in this section are the glucose transporter and anion transporter (Band 3) from the erythrocyte, the lactose permease from *E. coli*, and the mitochondrial transporter family. The transport functions of this collection of proteins are quite diverse, including one-solute facilitated diffusion, H^+-sugar symport for the purpose of sugar accumulation, and solute antiport for the purpose of exchange. Some common features are worth pointing out.

1. In several cases, these transport proteins appear to be oligomers, usually dimers (763). However, only in the case of the mitochondrial transporters (e.g., ATP/ADP exchange system) is it likely that there might be a single transport channel formed from elements of each monomer. It seems probable that, in other cases, each subunit can act independently, even if it is part of an oligomer.
2. Sequence homologies indicate close structural relationships between transport proteins which have very different solute specificities and functions (see Table

8.2). This strongly suggests common mechanisms for a broad group of functionally diverse transporters.

3. Alternating conformation models, similar to that pictured schematically in Figure 8.2, can be successfully applied in most cases. The rate-limiting step in transport is the conformational change required to alter the side of the membrane which is accessible to the binding site.
4. In most cases, the affinity of the transporter for the solute appears to be same regardless of which side the binding site is facing. This is not so for primary active transport systems (see Section 8.4).
5. Transporters in this collection are generally sensitive to sulfhydryl-directed reagents. This is probably fortuitous and not reflective of a profound common structural or mechanistic feature. For example, it is now known that none of the 8 cysteines in the lactose permease is directly involved in the transport mechanism. It has also been suggested that membrane-buried prolines are disproportionately present in transport proteins (133; Section 3.64), but there are no data to confirm speculations on the possible meaning of this observation.

8.31 The Erythrocyte Glucose Transporter (also see Section 6.5)

This transporter is the best characterized system of those which catalyze the diffusion of a single solute across the membrane. It allows D-glucose to pass across the red blood cell membrane for use in glycolysis. The same or very closely related glucose transporters are present in a variety of other animal cell types, and a major contribution in this area has been the sequencing of the cDNA encoding the glucose transporters from human hepatoma cells (1027) and from rat brain (100). The purified erythrocyte transporter is a glycoprotein with an apparent molecular weight of 55,000, which probably exists as at least a dimer in the membrane. The sequence suggests 12 putative membrane-spanning α-helical segments, but experimental evidence for this model is quite limited (1027, 202, 287). The purified transporter has been successfully reconstituted in phospholipid vesicles and shown to be oriented asymmetrically (202) (see Chapter 6).

As is the case with channel proteins, the availability of specific high-affinity inhibitors has been extremely valuable for studying this transporter. These include cytochalasin B and phloretin, which bind with 1:1 stoichiometry to the transporter (e.g., 790). Some inhibitors bind only when presented to the cytoplasmic side (e.g., cytochalasin B), whereas others bind specifically to the outside (e.g., phloretin) (790). Photolabeling experiments (642) have localized the binding sites for the two inhibitors to regions of the carboxyl-terminal half of the polypeptide.

Most, but not all (615), of the kinetic and binding data are consistent with a simple four-state model for the transport cycle (Figure 8.2), in which there is a single binding site for D-glucose located within a channel. NMR (1549) and fluorescence stopped flow (42) methods have been used to measure individual rate constants. The loaded carrier alternates conformations with a rate constant (2,000 sec^{-1}) which is about 7 times that of the unloaded carrier (300 sec^{-1} at

23°C). Hence, glucose exchange is catalyzed at a faster rate than net transport, which requires the unloaded carrier to return across the membrane. The nature of the conformational change is unknown, though it has been detected by Fourier transfer infrared spectroscopy (28) and is speculated to be a "sliding barrier" created by α-helices sliding with respect to each other (1549).

The amino acid sequence of the mammalian glucose transporter has strong homology with several bacterial sugar transport systems (906, 290). Surprisingly, the homologous enzymes are symport systems, which catalyze the cotransport of H^+-arabinose and H^+-xylose. These systems, like the H^+-lactose system described below, utilize the electrochemical proton gradient for the purpose of accumulating these sugars. The erythrocyte glucose transporter does not transport H^+ and is not capable of transport against a glucose gradient. The structural and mechanistic relationships revealed in future studies will certainly be of great interest.

8.32 The Lactose Permease from *E. coli* (see 1608, 717, 716 for reviews)

This is by far the best understood symport protein. It is encoded by the *lacY* gene, which is part of the *lac* operon, and is often referred to as the *lac* permease. The gene has been cloned and sequenced, and the protein has been purified from strains which overproduce it. The permease can be studied in *E. coli* cytoplasmic membrane vesicles (Kaback vesicles), in intact cells, or in reconstituted proteoliposomes. The protein has a molecular weight of 46,500 as deduced from the amino acid sequence, though it migrates abnormally on SDS-polyacrylamide gels (see Chapter 3). It is nearly certain that the functional unit in the membrane is the monomer (253), although some evidence suggests the possibility of a dimeric form of the permease in vivo (1608, 716).

The sequence of the *lac* permease has been used to construct models which have 12 or 14 transmembrane α-helices (716, 1525). These models are generally consistent with spectroscopic data showing a high α-helical content, and with limited topological data. The model proposed by Kaback is shown in Figure 8.11, along with an indication of which regions have been experimentally confirmed as being either cytoplasmic or periplasmic (e.g., 1314).

The kinetics of the permease are examined using the assays indicated in Section 8.13: (1) downhill efflux or influx into vesicles (zero *trans*); (2) active transport, coupling uphill transport of lactose to a proton electrochemical gradient; (3) equilibrium exchange; (4) counterflow (infinite *cis*). The kinetic scheme shown in Figure 8.11 can be used to analyze all the data (1607), though there are significant points of disagreement between the two main contributing groups (see 1607). The major features are clear, however.

1. The permease appears to have one or more sites for H^+ binding and one for lactose binding. These sites are alternately exposed to the periplasmic and cytoplasmic sides of the membrane, and the rate of this conformational change

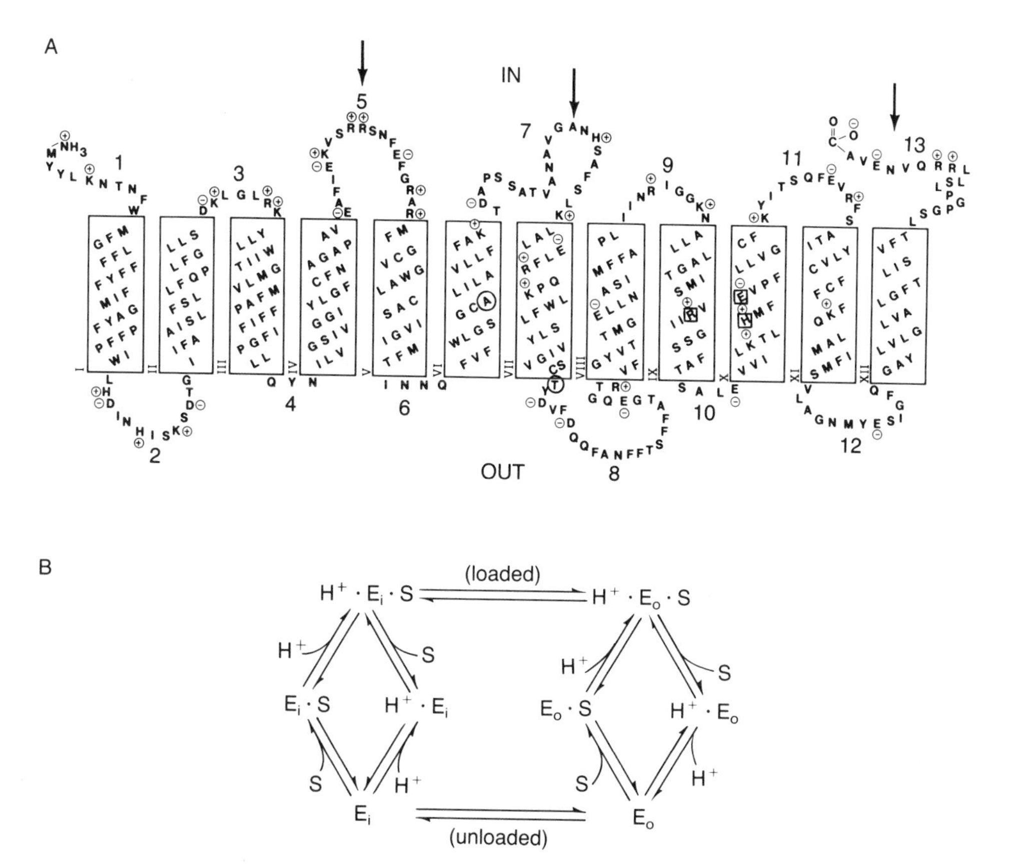

Figure 8.11. Sequence of the H^+-lactose permease of *E. coli*, along with a kinetic scheme used for analysis. Circled residues are implicated in sugar binding specificity and the residues in squares appear to be involved in the mechanism by which protonation is coupled to the net transport of lactose. Dark arrows indicate regions experimentally verified as exposed to the cytoplasm. E_i and E_o refer to the conformations of the permease in which the solute binding sites face the cytoplasm (E_i) or periplasm (E_o). Adapted from ref. 716. Reproduced with permission from the *Annual Review of Biophysics and Biophysical Chemistry*, vol. 15. © 1986 by Annual Reviews Inc. Structural model kindly provided by Dr. H. R. Kaback.

is rate limiting (see 1607). The maximal rate of transport in cells is around 25 to 50 sec^{-1}. In the presence of a transmembrane gradient in the proton electrochemical potential ($\Delta\bar{\mu}_{H^+}$), the K_m for lactose is about 80 μM and the proton binding site probably has a high pK_a value and is, thus, protonated much of the time. In the absence of $\Delta\bar{\mu}_{H^+}$, the K_m for lactose is much higher, 15 to 20 mM.

2. The cotransport of the H^+ and lactose is obligatory and proceeds with a 1:1 stoichiometry.

3. Net transport involves a positive charge moving across the bilayer. Hence, the transmembrane potential ($\Delta\psi$) and the proton chemical potential difference (ΔpH) are important in determining the equilibrium condition as well as the rate of transport (see 1608, 716). (See Section 7.42.)

Box 8.4 Genetics and Site-Directed Mutagenesis Are Powerful Techniques in Studying Lactose Permease

The most exciting developments are coming from studies using genetics to examine this transport protein (see 717, 151, 970). These studies are in their early days, but they clearly are a preview of much more to come, both in this system and in other transport systems. Using random mutagenesis, it was determined that amino acid substitutions for alanine-177 or tyrosine-236 permit the permease to transport maltose. These sites, shown in the model in Figure 8.11, may be involved in sugar binding.

Site-directed mutagenesis is being applied by Kaback and his colleagues to elucidate the mechanism of the permease. Many amino acid substitutions have been found which do not markedly affect the transport kinetics. Not only does this rule out direct involvement in the transport mechanism, it also shows that they do not lead to deleterious "global" conformational changes or prevent proper assembly in the membrane.

Three important mutants have been located which selectively impede parts of the transport mechanism (see 717, 790). Histidine-322, glutamate-325, and arginine-302 appear to be essential for net transport, and are thought possibly to be part of a "charge relay" system within the membrane. It is proposed that they form a hydrogen-bonded group between helices 9 and 10 (Figure 8.11). The substitution of an alanine for a glutamate at position 325 results in a permease which cannot catalyze active transport or efflux of lactose, but which carries out a normal lactose exchange reaction (see Section 8.13). Hence, the loaded form of the carrier can undergo normal conformational alteration, but some other step in the complete cycle is blocked. This is speculated to be deprotonation of the permease. It is through experiments such as these that mechanistic features will be revealed. Obviously, however, a three-dimensional structure would be a tremendous advantage in interpreting these data.

There are other sugar-cation transport systems in *E. coli* as well as in higher organisms (see 1608 for review). The *lac* permease has no sequence homology with the H^+-xylose or H^+-arabinose transporters or with the mammalian glucose transporter described in the previous section. Nor does the *lac* permease have any sequence homology with the melibiose permease from *E. coli* (see 1608, 717). The melibiose permease not only utilizes H^+ symport to drive melibiose uptake, but will also use Na^+. The ability to use Na^+ would seem to rule out mechanisms relying on proton hopping within hydrogen bonding networks through the membrane-spanning portion of the transporter. There is also a Na^+-glucose carrier which catalyzes glucose uptake in the intestinal brush border (see 1608, 1135, 1610).

8.33 Band 3, the Erythrocyte Anion Transporter (see 703, 876 for reviews)

Band 3 comprises about 25% of the total membrane protein in the human erythrocyte, and similar proteins are apparently also present in non-erythroid cells (322). This protein has several distinct functions which can be assigned to two major domains. The amino-terminal half (41,000 Da) is hydrophilic and is located on the cytoplasm side of the erythrocyte membrane. This portion of the protein contains binding sites for cytoskeletal components (ankyrin), as well as for glycolytic enzymes and hemoglobin (see 876 and Sections 4.34 and 6.72). This domain can be removed by proteolysis, leaving behind a carboxyl-terminal domain (52,000 Da) which remains bound to the membrane and which is responsible for mediating Cl^-/HCO_3^- exchange (see Section 8.11) as well as providing a channel for water permeability (see Section 7.21). In addition, the extracytoplasmic portion of this part of the protein carries the carbohydrate antigenic determinants for several blood groups. In the membrane, Band 3 is either a dimer or a tetramer.

The cDNA encoding Band 3 from the mouse erythrocyte has been cloned and sequenced (774), and the deduced amino acid sequence used as the basis for speculative models (775, 774). It is suggested that there are 12 membrane-spanning α-helices, several of which are amphipathic. Experimental evidence in support of this folding pattern exists for only a few regions of the polypeptide, primarily from proteolysis studies and from the location of the attached carbohydrate.

Extensive kinetic studies are consistent with a single-site alternating conformational model (see Figure 8.2). However, the transporter catalyzes anion equilibrium exchange at a rate that is at least 10^4-fold more rapid than net transport. Hence, the unloaded transporter does not rapidly undergo the conformational change required to switch the membrane surface accessible to the anion binding site. NMR experiments utilizing ^{35}Cl have demonstrated the existence of a single binding site within the transporter, and have also shown that this single site can be inward-facing or outward-facing (408, 411). Studies with transport inhibitors (412, 410, 409) also support a model in which the transporter is viewed as a channel with a single anion binding site located somewhere in the middle, and whose access to either side is blocked by a "sliding barrier" which moves from one side to the other as a result of a conformational change. The conformational alteration of the loaded transporter is rate limiting, but proceeds quite rapidly, 10^5 sec^{-1} at 37°C. Presumably, such a fast rate would seem to preclude very major conformational changes in the protein. The nature of this conformational change and the actual structure of the channel have not been experimentally defined.

The rate-limiting conformational change of the loaded transporter is only very slightly dependent on the membrane potential. This is consistent with the conformational transition from the inward to outward orientation of the transporter resulting in the net movement of 0.1 protein-bound charge across the membrane (556). If there is any movement of the anionic solute during this transition it must be accompanied by a counterion, such as a charged amino acid group. In contrast, the voltage-induced conformational change which gates the sodium channel is associated with the net shift of six protein-bound charges across the membrane (193).

8.34 A Family of Mitochondrial Solute Transporters (see 45 for review)

Several transport proteins present in the mitochondrial inner membrane have been shown by amino acid sequence homology to be closely related, presumably derived from a common ancestor by divergent evolution (Table 8.2). There are at least three members of this family: (1) the ADP/ATP translocase, (2) the phosphate transporter (1260), and (3) the uncoupling protein. Although these proteins evidently have related structures, their solute specificities are very different. The ADP/ATP translocator catalyzes the transport of either ADP or ATP across the bilayer. Under physiological conditions, ATP is transported out of the mitochondrion and ADP is transported into the matrix. The mechanism is probably similar to that of Band 3, except that ATP has one more negative charge than ADP, and this exchange is sensitive to the transmembrane potential across the mitochondrial membrane. The phosphate transporter is an H^+-phosphate symport system, and, presumably, has a mechanism similar to the H^+-lactose permease of *E. coli* (Section 8.32). This protein catalyzes net phosphate flux into the mitochondria. Because of the H^+, the net transfer is electroneutral and not sensitive to the transmembrane potential. The uncoupling protein (e.g., 420) is found in the mitochondria of mammalian brown adipose tissue and its function is to allow the proton electrochemical gradient created by the respiratory chain to be dissipated, thus generating heat. The uncoupling protein can also catalyze the transport of anions such as Cl^-, so perhaps the transporter actually catalyzes OH^- transport, which would be indistinguishable experimentally from H^+ transport. This transporter binds to nucleotides, which block transport, and may be regulated by fatty acids in vivo.

All three transporters, and possibly the α-ketoglutarate/malate translocator as well (1260), have a common structure as deduced from their amino acid sequences. They all have similar molecular weights (~33,000), and the polypeptides each contain three homologous domains of about 100 amino acids each. Presumably, this originated from a gene triplication. The sequence suggests a model where each homologous domain crosses the membrane twice, and the subunit is proposed to have 6 transmembrane α-helices. Chemical modification experiments are compatible with this model. Note that this kind of structure is similar to that of the Na^+ channel, where a simple polypeptide was built up of four related homologous domains. The ADP/ATP transporter is the best characterized of the mitochondrial transporters and has been shown to be a dimer, probably with a single channel for transport (see 45). This has been defined by quantifying the binding of high-affinity inhibitors (e.g., carboxyatractyloside). Clearly, this set of transport proteins will provide a rich experimental system for detailed studies in the future.

The ADP/ATP transporter will be discussed further in Chapter 10 as regards the mechanism by which this protein is directed from the cytoplasm, where it is synthesized, to its proper location within the inner mitochondrial membrane.

8.4 Several Examples of Active Transporters Driven by ATP or PEP

General comments concerning primary active transporters: A large number of systems have been characterized which couple solute transport across the membrane to the hydrolysis of a high-energy phosphate bond or another energy-yielding reaction. Our purpose is not to attempt to review this voluminous literature, but rather to point out some of the more salient structural and mechanistic features of these transport systems in the context of what we have learned about other kinds of transport proteins.

As noted previously, the alternating conformation concept, such as that illustrated in Figure 8.2, is the basis for the kinetic models for many primary active transport systems. It is not necessary that the energy input, e.g., ATP hydrolysis, be coupled directly to the conformational change required for transport as shown in Figure 8.2, but it is essential that one or more steps in the kinetic cycle be coupled efficiently to the energy-yielding reaction. The structural requirements for the transmembrane "channel" of a primary active transporter should be the same as for the other transporters which were discussed in previous sections, except that now there is an added complexity of coupling the solute transport to a chemical reaction in a controlled fashion.

Even less is known about the molecular structures of primary active transport systems than about some of the systems discussed earlier in this chapter. The growing number of amino acid sequences has helped identify families of related transporters which probably have similar structures and mechanisms (Table 8.5), but there are few hard data on what the transport sites really look like. Certainly, the dominant structural theme in model building is that the transmembrane channel through which solutes are transported is formed from a cluster of transmembrane amphipathic α-helices. Similarly, although models abound for how chemical reactions are coupled to solute transport, few data distinguish them. Coupling models are of two broad types. *Direct coupling* views the chemical reaction as directly influencing the transported solute in a way which does not require complex protein-mediated conformational changes. For example, the protons involved in ATP hydrolysis could be those which are actually transported (see 1286). This requires the site of the chemical reaction, e.g., ATP hydrolysis, and the site of solute binding to be very close. In contrast, *indirect coupling* views protein-mediated conformational changes as being the manner in which the solute is influenced by the chemical reaction. Indirect coupling models can even have

Table 8.5. Some examples of ion-motive ATPases.[l]

I. *E_1E_2-type*

Ion(s) transported	*Source*	*Membrane*
H^+	Lower eukaryotes (yeast, fungi)	Plasma
H^+	Higher eukaryotes	
	Plants	Plasma
	Animals	Plasma (bladder, tumor)
K^+	*E. coli, S. faecalis*	Cytoplasmic
H^+/K^+	Higher eukaryotes (animals)	Plasma
Na^+/K^+	Higher eukaryotes (animals)	Plasma
Ca^{2+}	Higher eukaryotes (animals)	Plasma
Ca^{2+}	Higher eukaryotes (animals)	Sarcoplasmic

II. *F_1F_0-type*

Ion transported	*Source*	*Membrane*
H^+	Most bacteria	Cytoplasmic
H^+	Eukaryotes	
	Animals, plants	Mitochondrial inner
H^+	Plants	Chloroplast thylakoid

III. *Vacuolar type*

Ion transported	*Source*	*Membrane*
H^+	Lower eukaryotes (yeast, fungi)	Vacuoles
H^+	Higher eukaryotes (plants)	Tonoplasts
H^+	Higher eukaryotes (animals)	Lysosomes
H^+	Higher eukaryotes (animals)	Endosomes
H^+	Higher eukaryotes (animals)	Secretory granules
H^+	Higher eukaryotes (animals)	Storage granules
H^+	Higher eukaryotes (animals)	Clathrin-coated vesicles

[l]Data taken from ref. 1134.

the chemical reaction and active transport occurring on separate subunits within a complex, which does seem to be the case in some systems. These models have the advantage that it is easy to reconcile the fact that different solutes (e.g., Na^+, H^+) can be transported by the same or closely related proteins by a common mechanism.

Active transport systems function at rates typical for many enzymes, i.e., 10^2-10^3 sec^{-1} at saturation. Unlike channel proteins, selectivity is primarily due to the affinity of the transported solute for the active transporter. Furthermore, most primary active transporters normally function under conditions where the solute concentration on the *cis* side is near or above the K_m, and the rate-limiting step for transport may be related to a protein conformation change or to the chemical reaction driving the system.

Finally, the role of a primary active transport system is to move solute unidirectionally (*cis* to *trans*) across the membrane against a concentration gradient. Since the transporter must change its orientation back from *trans* to *cis* once the solute has been delivered, there must be some mechanism to ensure that the transporter does not simply carry the solute back out. Hence, we expect the affinity of the active transporter for the solute to be much higher on the *cis* side, where the solute is picked up at low concentration, than on the *trans* side, when the solute dissociates into a pool at much higher concentration. If this were not the case, the transporter would be very inefficient at high concentrations of solute on the *trans* side. The energy input is used mainly to change the affinity of the transporter for the transported solutes. For example, the phosphorylation by ATP of the Na^+/K^+-ATPase or of the Ca^{2+}-ATPase results in the stabilization of conformations of those transporters which have their ion binding sites facing outward and which have a low affinity for Na^+ or Ca^{2+}, respectively. The binding of ATP, on the other hand, appears to stabilize a form of the Na^+/K^+-ATPase in which the ion binding sites facing the cytoplasm have a low affinity for K^+ (see next section and Figure 8.12). Hence, the mechanism by which energy coupling influences these active transporters is often best understood in thermodynamic terms, i.e., the transient stabilization of particular conformations and the alteration of solute affinity.

What follows is a broad outline of the five classes of transporters which utilize the free energy from a high-energy phosphate bond to drive solute transport. Clearly, nature has found several ways to accomplish this.

8.41 Plasma Membrane (E_1E_2-Type) Cation Transporters: ATP-Linked Ion Pumps (See 1134, 1386 for reviews)

A number of prokaryotic and eukaryotic ion-motive ATPases are members of a single family, related by mechanistic similarities as well as amino acid homologies (see Table 8.5). The best characterized members of this class are the animal plasma membrane Na^+/K^+-ATPase and the Ca^{2+}-ATPase from the sarcoplasmic reticulum (see Section 6.5). Most of the enzymes in this class consist of a single polypeptide with a molecular weight around 100,000, the most notable exception being the Na^+/K^+-ATPase purified from several sources which has a second, smaller subunit whose function is not known. These transporters are diagnostically inhibited by vanadate and are directly phosphorylated by ATP to form a phosphoprotein intermediate that is essential for transport (see Figure 8.12). As

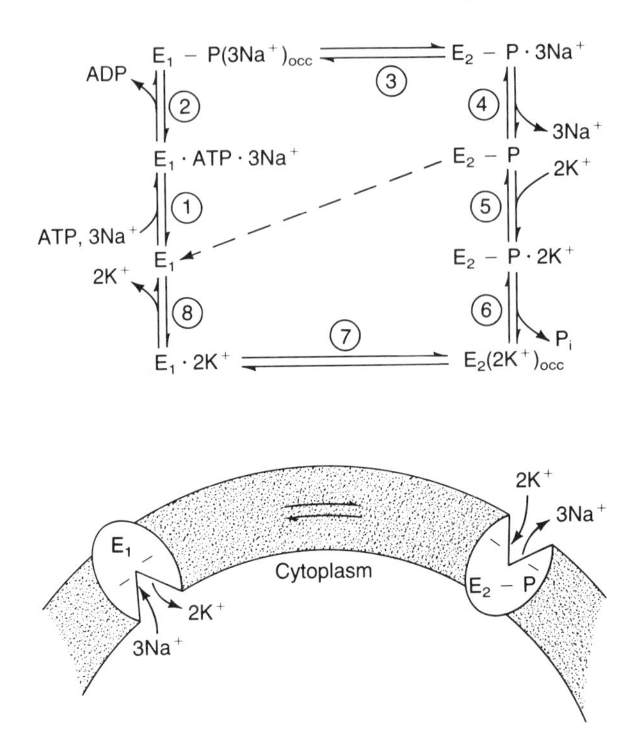

Figure 8.12. Kinetic scheme of the Na^+/K^+-ATPase. For Ca^{2+}-ATPase, the same mechanism applies except that following dephosphorylation of the E_2-P form, the transporter returns to the E_1 conformation in an unloaded form, indicated by the dashed line. Occluded forms (occ) are also indicated.

can be seen in Table 8.5, this superfamily of cation transporters varies considerably in ion specificity. Furthermore, the transport stoichiometries are not identical. For example, the Ca^{2+}-ATPase transports 2 Ca^{2+}/ATP into the lumen of the sarcoplasmic reticulum, whereas the Na^+/K^+-ATPase transports 3 Na^+ out of and 2 K^+ into the cytoplasm across the plasma membrane. Not only are these ions and stoichiometries different, but also the Ca^{2+}-ATPase appears to transport ions only in one direction, while the Na^+/K^+-ATPase does so in both directions.

The name "E_1E_2-type" derives from work on the Na^+/K^+-ATPase, in which it was shown very clearly that this protein exists in at least two distinguishable conformations which not only have different binding properties, but also are cleaved differently by mild proteolysis (see 39, 714). Form E_1 corresponds to the conformation in which the ion binding sites are accessible from the cytoplasmic side (high affinity for Na^+, low affinity for K^+) and which has a high affinity for ATP. The phosphorylated form of E_2 has the ion binding sites exposed to the outside of the cell (high affinity for K^+, low affinity for Na^+). Figure 8.12 describes the catalytic transport cycle involving these two "empty" forms of the transporter as well as two "loaded" forms, E_2 $(2K^+)_{occ}$ and E_1 $(3Na^+)_{occ}$, in which the ions are bound to and "occluded" within the pump molecule. The occlusion of K^+, for example, by the phosphorylated form of the pump (E_2-P) has been examined and shown to occur at two distinct ion binding sites (446). In the occluded form, the ions are inaccessible from either side of the membrane.

Some features of the catalytic cycle to note follow:

1. The E_1 form binds to three Na^+ ions from the cytoplasmic side and then reacts with ATP to form a phosphorylated enzyme. The site of phosphorylation is a specific aspartate which is conserved in this family of enzymes.
2. ADP dissociates from the enzyme and the ions become occluded.
3. The phosphorylation of the protein stabilizes a conformation with a low affinity for Na^+ with the binding sites facing outward. This facilitates the conversion from E_1-P to E_2-P, resulting in transport.
4. The E_2-P form has ion binding sites facing the extracellular medium with a high affinity for K^+, which binds, catalyzes the dephosphorylation of the protein, and becomes occluded. Note that Na^+ is required for rapid phosphorylation and K^+ for rapid dephosphorylation. Vanadate binds to the E_2 form, probably as a phosphate transition state analogue. For other E_1E_2-type enzymes, such as the Ca^{2+}-ATPase (see 1376), the E_2-P form dephosphorylates and returns to the E_1 form, essentially returning in an unloaded form.
5. The rate-limiting step in the catalytic cycle is probably the "de-occlusion" or release of K^+ from the enzyme. This is accelerated by the binding of ATP at a low-affinity binding site on the enzyme (447). Hence, ATP has two distinct roles, as a substrate and as an allosteric effector. It is not clear whether there is more than one ATP binding site.

The amino acid sequences have been deduced for several of the E_1E_2-type ATPases, including the Na^+/K^+-ATPase (α-subunit) from several sources (1337, 733, 734), the Ca^{2+}-ATPase (902, 134), the plasma membrane H^+-ATPase from yeast (1327) and *Neurospora crassa* (572, 3), and the K^+-ATPase from *S. faecalis* (1362). Models based on hydropathy profiles have been proposed containing 6, 8, or 10 transmembrane α-helical segments (see Figure 8.13). Substantial sequence homology exists in several portions of the polypeptides, including the region around the phosphorylation site. The sequences all have a large hydrophilic loop with domains associated with nucleotide binding and phosphorylation (see 902).

In the Ca^{2+}-ATPase, one trypsin cleavage site which is sensitive to the E_1-E_2 conformational transition has been located in the "transduction" domain (Figure 8.13) (35). Occluded forms of Ca^{2+} bound to this enzyme have also been characterized, and it has been suggested that translocation is preceded by the sequential addition of two Ca^{2+} ions to distinct sites within the pump molecule (1433, 678). It has been suggested that the high-affinity Ca^{2+} sites are located a substantial distance from the nucleotide binding site in the Ca^{2+}-ATPase (1373), but this needs to be confirmed. The general structural features of the Ca^{2+}-ATPase implied in the model in Figure 8.13 are consistent with electron microscopic image reconstruction studies showing a large cytoplasmic region extending well above the bilayer (1440). In vivo, it is likely that E_1E_2-type ATPases are aggregated to at least dimers but, as discussed in Chapter 3, this is particularly hard to determine unambiguously. Nevertheless, it is clear that the monomers are capable of catalytic function in at least some cases (e.g., 528, 1518).

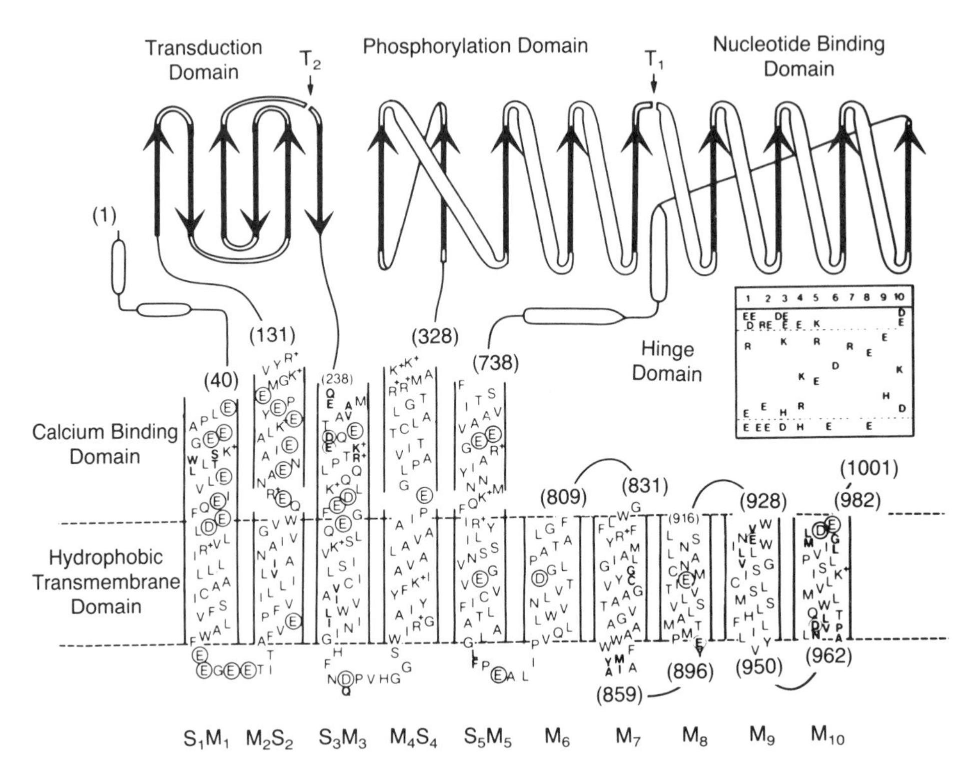

Figure 8.13. Proposed structured model for the Ca^{2+}-ATPase based on theoretical considerations. This model has 10 membrane-spanning α-helices, 5 additional α-helices forming a stalk, and a large globular structure on the cytoplasmic side, made of 3 domains. Proteolysis cleavage sites are indicated (T_1, T_2). The inset shows the distribution of charged residues proposed to be in or near the membrane: negative (D, E) and positive (K, R, H). From ref. 134, © 1986 Cell Press. Kindly provided by Dr. D. MacLennen.

The amino acid sequence of the small ß-subunit of the Na^+/K^+-ATPase has also been determined. It has been suggested that this subunit has 1 or 4 transmembrane α-helices (see 1630, 1111).

In summary, the availability of clones of these plasma membrane ATPases, coupled to the large number of experimental possibilities, makes these excellent systems for the application of site-directed mutagenesis and other genetics studies. Such methods are already being applied to the F_1F_0-ATPase as described in the next section.

8.42 The Mitochondrial, Chloroplast, and Bacterial F_1F_0-Type ATPases (see 1134, 128, 654)

Most bacteria, as well as mitochondria and chloroplasts, have closely related F_1F_0-type ATPases whose function is to utilize the proton electrochemical gradient across the membrane to synthesize ATP from ADP and inorganic phosphate.

Physiologically, these enzymes are ATP synthases. These enzymes contain a minimum of 8 different subunits (*E. coli*) to as many as 13 (bovine heart) and are, therefore, much more complicated structures than the E_1E_2-type plasma membrane ATPases. The enzymes consist of two parts: (1) a hydrophilic globular F_1 portion which contains the nucleotide binding sites and functions as an ATPase and (2) a membrane-bound F_0 portion (654) which functions as an H^+-conducting "channel." The F_1 and F_0 portions can be dissociated from each other and functionally reconstituted after purification.

The F_1 moiety of *E. coli* contains 5 subunits with stoichiometry $\alpha_3\beta_3\gamma\delta\varepsilon$. There appears to be a natural asymmetry among the three $\alpha\beta$ pairs in F_1, perhaps resulting from asymmetric contacts with the other subunits. There are three catalytic sites per F_1, and rapid turnover requires that more than one site be bound to ATP. Alternating-site models of catalysis have been proposed, including one involving physical rotation of portions of the enzyme (256). Probably the δ and ε are involved in binding F_1 to the F_0 portion of the enzyme in the membrane.

The F_0 portion of the *E. coli* enzyme has the simplest composition, containing 3 subunits with a stoichiometry of $a_1b_2c_{10}$. All three subunits are required to make the H^+-conducting channel (1301). Whether F_0 can be viewed as a channel in the sense we have used it in this chapter is uncertain (see 859). It is thought that the H^+-conducting portion of F_0 is essentially made up of α-helices from the multiple copies of the *c*-subunit. This subunit is proposed to contain 5 transmembrane α-helices, while the *a* and *b* subunits probably contain one each. Genetics studies discussed below are consistent with the membrane-buried portions of all three subunits being important to H^+ permeability.

This system is an important proving ground for the successful use of genetics in providing structural information about a membrane protein in the absence of high-resolution crystallographic data. The crucial question is whether single amino acid substitutions will result in artifacts due to protein-mediated conformational changes which influence functions occurring elsewhere in the protein. This may be very system-dependent. So far, for example, the experience obtained in the *lac* permease is very encouraging, as described previously. It is evident that the next decade will witness a large increase in the applications of genetics and mutagenesis to studying structure–function relationships in membrane proteins.

Results obtained so far with the F_1F_0-ATPase from *E. coli* are summarized below.

1. Numerous mutants in the ß-subunit define a region involved in what is likely to be the catalytic nucleotide binding domain (1123, 1122). Conformational coupling between the α- and ß-subunits has also been demonstrated by the use of mutants (1377).
2. Mutations in the ε-subunit have been shown to be defective in coupling F_1-catalyzed ATPase activity to H^+ translocation (257), and in binding F_1 to F_0 (764).
3. Residues in subunits *a* (178), *b* (701), and *c* (424) have been implicated in H^+ permeability, and genetic data suggest that subunit *b* is in direct contact with subunits *a* and *c* in the membrane (801, 800).

Future experiments will reveal whether these data reflect localized perturbations or "global" conformational changes. The goal is to learn about H^+ permeability and the coupling mechanism. So far, little can be said.

A potentially important observation is that the anaerobic bacterium *Propionigenium modestum* contains a Na^+-stimulated F_1F_0-type ATPase in its cytoplasmic membrane (822). If this enzyme functions as an ATP-driven Na^+ pump, then it is reasonable to presume that the mechanisms by which the H^+-ATPases function would also apply to this Na^+-ATPase. This would eliminate from consideration direct coupling mechanisms, for example.

8.43 Three Other Classes of Transporters

In addition to the E_1E_2-type and F_1F_0-type ATP-dependent active transporters, there are three other classes of active transporters which are driven by the free energy available in high-energy phosphate bonds. Even less is known about the actual mechanisms of solute transport or energy coupling for these systems than for those previously discussed. Some interesting variations are presented by these systems, however.

(1) *The Phosphotransferase (PTS) System* (see 548 for review): This system is found only in bacteria and catalyzes the transport of sugars such as glucose and mannitol. The unique feature of this system is that sugar transport is concomitant with its phosphorylation. The sugar–phosphate product is not recognized as a substrate by the transporter in the bacterial membrane, and reverse flux through the system is prevented. Recall that in the E_1E_2-type transporters, the phosphorylation of the transporter stabilizes the form with a lower affinity for the transported solute (Ca^{2+} or Na^+, for example), thereby preventing reverse flux.

The ultimate donor for the phosphate in the PTS system is phosphoenolpyruvate, and the phosphate is passed by a specific sequence of soluble phosphoprotein intermediates in the cytoplasm to a membrane-bound transporter protein, called enzyme II or EII. There is a set of EII enzymes which are sugar specific but whose primary structures are related (132). These EII species are apparently dimers within the membrane (1388), and it is suggested that they form channels (1232). These proteins are sensitive to sulfhydryl-directed reagents and oxidation, and this may be physiologically relevant. Little is known about the actual manner in which the phosphorylated EII species transport and phosphorylate the sugar, but a modified single-site alternating conformational model is a reasonable initial hypothesis.

(2) *The Bacterial Periplasmic Transport Systems* (see 31 for review): In addition to the cation–sugar cotransport system and the PTS system, bacteria have a third system for solute active transport which is used for a variety of amino acids and sugars. The best characterized systems are specific for histidine (*S. typhimurium*) and maltose (*E. coli*). Unfortunately, there are few biochemical data to complement extensive genetics on the membrane-bound components of these systems. What is unique to these systems is that they rely on solute-specific

binding proteins within the periplasmic space to bind to and then "deliver" the solute to the actual transport system within the cytoplasmic membrane. Both the binding proteins and the membrane-bound components contribute to the specificity of the system. The cytoplasmic membrane component contains three subunits, two of which are transmembrane proteins and the third is apparently tightly associated on the cytoplasmic side. This third component of the transport complex appears to have a nucleotide binding site, which is presumed to be the site where ATP or some related compound is hydrolyzed to drive the active transport. However, this has never been directly demonstrated. Virtually nothing is known about the mechanism of transport itself, except for the unique requirement for a soluble binding protein.

This system appears to be similar to a mammalian system used to "pump" out unwanted drugs and, thus, provide multiple drug resistance to cells (see 30, 1272, 508). Possibly, a set of binding proteins allows a single transport system to pump out several different drugs. A similar proposal has been made for the plasmid-encoded bacterial transport system which is used to export arsenicals, conferring resistance to the cells by this mechanism (203). These systems are all thought to be driven by ATP hydrolysis, but this has not been directly demonstrated.

(3) *Vacuolar H^+-ATPases* (see 1134, 128 for reviews): In eukaryotic cells, there are H^+-ATPases located in the membranes of a number of internal organelles besides the mitochondrion (see Table 8.3). These appear to be distinct from either the F_1F_0-type or the E_1E_2-type ATPases. In several cases, it is clear that the function of the H^+-ATPase is to drive protons into the organelle compartment in order to acidify it. The low pH in endocytic vesicles, for example, is known to be essential for the dissociation of ligands from their receptors following endocytosis (see Chapter 9). It is not certain that all these H^+-ATPases are structurally related, but so far it is known that (1) they all appear to contain more than one subunit, (2) they do not form a phosphoprotein intermediate, (3) they are not sensitive to vanadate, and (4) they all transport protons. More biochemical and structural studies are required at this point.

8.5 Active Transport Systems Driven by Electron Transfer or Light

There are several interesting examples of active transport systems where the energy input is provided either by a redox change in a protein-bound prosthetic group (e.g., heme) or by the absorption of a photon by a prosthetic group (retinal). The principles of transport are basically the same as previously discussed. In order to obtain net solute transfer, it is necessary to temporarily stabilize an otherwise unstable form of the protein which is a required intermediate in the transport cycle. This can be accomplished by a number of means including, as discussed previously, protein phosphorylation or even, in principle, ATP binding. The examples in this section demonstrate two other methods.

8.51 Cytochrome *c* Oxidase, a Redox-Linked H^+-Pump (see 502)

This protein is the best example of an enzyme which couples ion transport (H^+) to an electron transfer reaction by functioning as an ion pump. The oxidase catalyzes the oxidation of cytochrome *c*, located in eukaryotes in the intermembrane mitochondrial space, and reduces oxygen to water. The reduction utilizes 4 H^+ per O_2, which are taken from the inside space (matrix) of the mitochondrion. Coincident with this, for every electron passing through the system, at least 1 proton is transported from inside the mitochondrion to the outside (78). The enzyme contains 2 heme *a* prosthetic groups, along with 2 or 3 coppers and 1 zinc (1384). The mitochondrial oxidase is highly conserved through evolution and very similar enzymes are found in many bacteria. The subunit composition is variable, ranging from 2 in some bacteria to 12 distinct subunits in the beef heart oxidase. One subunit (III) has been specifically implicated in H^+ translocation, but there is no consensus about this (1183). The structure of the oxidase has been determined to low resolution by electron microscopy (Figure 3.3), demonstrating that it extends well beyond the plane of the bilayer on the outer surface. Unfortunately, the locations of the prosthetic groups are uncertain, let alone the mechanism by which the ion transport is catalyzed.

A very simple four-state model can be used to demonstrate some of the features of the oxidase (105). This model simply postulates that an H^+ binding site is accessible to one side of the membrane when a metal center is reduced and faces the opposite side when the metal center is oxidized. The pK values (i.e., H^+ binding constant) can also be different in the oxidized or reduced states. Essentially, the electron transfer reaction cycle alternately stabilizes the two conformations of the enzyme in which the H^+ binding site faces the two sides of the membrane. The molecular basis for this has yet to be experimentally deciphered.

Finally, it should be noted that there is at least one example of redox-linked transport of Na^+ (1473), demonstrating that such systems are not limited to H^+ pumping.

8.52 Bacteriorhodopsin, a Light-Driven H^+-Pump

The structure of bacteriorhodopsin has been discussed at length in Chapter 3. The function of this protein is to pump protons across the bacterial cytoplasmic membrane. The absorption of light by the bound retinal prosthetic group initiates a sequence of events which can be optically monitored by observing the retinal, and in some manner this is coupled to the translocation of 2 H^+ per cycle (781). Halorhodopsin, a protein from the same organism, utilizes what appears to be the same mechanism for active Cl^- transport (819). The transient high-energy state of the protein resulting from the absorption of a photon may facilitate the change in the accessibility and affinity of the H^+ binding sites. Most likely, the monomeric form of bacteriorhodopsin is the functional unit, though it exists as a trimer within the membrane. The retinal has been located near the middle of the membrane, and direct coupling mechanisms have been postulated (e.g., 578). The structure of the protein, with 7 parallel α-helices extending through the membrane, does not reveal any channel. How does the H^+ gain access to the retinal? This is not known, but this is certainly one of the questions which can be addressed by utilizing site-directed mutagenesis.

The Khorana laboratory has generated numerous mutational variants of bacteriorhodopsin by site-directed mutagenesis (1002, 570). Remarkably, most of the reported mutants retain normal or nearly normal proton pumping activity. These include amino acid substitutions of each of the 11 tyrosine residues (1002). This represents another interesting case where the ability to use genetics to obtain structural and mechanistic information is being tested.

8.6 Membrane Pores Created by Exogenous Agents

In the previous sections we have reviewed how cells regulate solute flux across their membranes through various endogenous transport systems. Nature, as well as scientific researchers, has also devised ways to create pores in biomembranes through exogenous agents and treatments which act on the membrane. There are many cytolytic peptides (such as gramicidin A and melittin) and proteins which function by creating nonselective pores within the plasma membrane of a targeted cell, thereby killing the cell (see 92 for review). Laboratory procedures also often require the introduction of solutes into cells, preferably in the least disruptive manner. The purpose of this section is to demonstrate the variety of ways in which holes can be made in biomembrane through the use of specific agents or by the use of physical stress.

8.61 Toxins and Cytolytic Proteins

Nature has designed a number of water-soluble toxins whose purpose is to kill particular targeted cells. Minimally, toxins must recognize some receptor on the outer surface of the membrane, and somehow gain entry to the cytoplasm. Some toxins are designed to deliver a lethal enzyme into the cytoplasm (e.g., diphtheria toxin). Other toxins simply form non-selective pores in the membrane, which kills the cells by allowing metabolites, ions, and, in some cases, macromolecules to leak out. The most interesting aspect of all this is how these water-soluble, hydrophilic proteins spontaneously insert into the membrane bilayer. This question is addressed briefly in Section 10.31, where we consider it in the context of how proteins are transported out of cells.

The following summary indicates some of the better characterized toxins which function by forming pores through biomembranes.

(1) *Colicins* (see 235, 1061): These are single polypeptides (~60,000 Da), some of which form nonselective channels in the cytoplasmic membrane of sensitive strains of *E. coli*. Colicins E1 and A are the best characterized. The purified proteins spontaneously form voltage-gated channels in planar membranes

which are large enough to pass NAD^+. The channel-forming portion from the carboxyl terminus of colicin E1 (146 amino acids) has been isolated and characterized (1203, 1146). The amino acid sequence does not resemble that of an intrinsic membrane protein. Remarkably, the transition from the open to the closed state of the colicin E1 pore upon imposition of a voltage results in the translocation across the membrane of a substantial amount of the protein (1203). The pore that is formed involves only a single colicin molecule (1203, 1146).

(2) *Diphtheria toxin* (see 1061): This is a single polypeptide, also produced by bacteria, which, upon binding to specific cell surface receptors, forms a channel in the membrane and delivers a lethal enzyme to the cytoplasm. One fragment of the toxin has been shown to form pores in the membrane which are large enough to allow the translocation of the enzymatically active portion of the toxin (638), though it has yet to be proved that this is the mechanism of translocation. Similar results have been obtained for the botulinum and tetanus toxins (638). Cholera toxin contains two subunits, of which one, the B subunit, forms an oligomer in membranes, apparently with a central channel which is utilized to translocate the enzymatically active A subunit (883). The structure of another toxin, the *Pseudomonas* exotoxin A, which is similar to the diphtheria toxin, has been solved by X-ray crystallography (22). So far, however, the structure has not revealed any clues to how this soluble protein spontaneously inserts into the membrane to form a pore.

(3) *The membrane attack complex of complement and the pore-forming protein from cytotoxic T cells (perforin)*: These related serum proteins aggregate to form large pores (100 Å to 160 Å internal diameter) in the membrane of target cells (1624, 1341). The C9 protein of complement is the primary component which, at the end of a series of self-assembly events involving other components of the complement system, forms a large pore, resulting in cell lysis. This protein has been extensively studied in planar membranes and in liposomes (1341, 88). The molecular details of how the pore forms are not, however, well understood.

Another toxin which aggregates in the membrane to form a large pore is the α-toxin of *S. aureus*. This is known to form a hexameric structure following a conformational change upon membrane binding (676).

8.62 Permeabilization by Detergents

Saponin and digitonin interact with cholesterol within membranes and form aggregates in the form of large pores or tubules (991). These are widely used to form large holes in the plasma membranes of cells to allow small molecules and macromolecules (e.g., antibodies) to enter the cytoplasm (e.g., 991, 516, 820).

8.63 Permeabilization by Osmotic Stress

The physical pressure resulting from osmotic stress results in membrane damage. This has been characterized in the erythrocyte, where a single large hole (diameter > 1 μm) is transiently formed which subsequently reduces in size to a diameter between 14 Å and 280 Å (857). The final size is dependent on the buffer composition. The molecular details of this hemolytic hole are not known.

8.64 Permeabilization by Electroporation

Pores in the plasma membranes of animal cells are formed transiently by the application of an electric field. Studies with erythrocyte ghosts showed large pores (168 Å diameter) formed transiently near the cathode, rapidly reducing in size to 10 Å (1364). Commercial devices are available for electroporation which are used, in part, because this treatment induces cell fusion. Whether the observed pore formation is related to the structural perturbations which result in membrane fusion is unclear.

8.65 Permeabilization by Creating Packing Defects in the Membrane

Any treatment which negatively influences the packing of the lipids and proteins in the membrane can result in destruction of the barrier to hydrophilic solutes. For example, liposomes at the phase transition temperature can have markedly higher permeabilities, measured in one case to be equivalent to those of pores with a diameter of 18 Å (1489). Protein aggregation, either in reconstituted vesicles or induced by the crosslinking of membrane proteins within biomembranes, can also result in large increases in permeability (327, 1488). These are clearly phenomena which are not physiologically relevant, but they do point out that there are limits to the organization of membrane components which are incompatible with membrane function.

8.7 Chapter Summary

The passage of most solute molecules across biological membranes is mediated by transporters or by channel proteins. Channels are designed to facilitate the passage of ions across the membrane and are distinguished by extremely rapid rates (10^6-10^8 ions/sec per molecule). Such rapid rates of ion passage are possible because the proteins need not undergo any conformational changes associated with an ion passing from one side of the membrane to the other. These rapid rates result in such high conductance values that the transport of ions through an individual channel protein can be easily measured. In contrast, transporters undergo a transport cycle in which there is a protein conformational alteration where the solute binding site is exposed first to one side and then to the opposite side of the membrane. As a rule, solute flux through transporters is many orders of magnitude slower than through channels.

Passive transporters simply facilitate diffusion of the solute across the membrane, whereas active transporters use free energy to drive solute transport against a concentration gradient. There are examples of active transporters which couple solute flux to electron transport, to ATP (or PEP) hydrolysis, to the absorption of light, or to the cotransport of an ion.

There are superfamilies of channels and transporters which comprise functionally distinct proteins that have similar structures as revealed by their amino acid sequences. These sequence similarities suggest that there are similar functional mechanisms within each superfamily despite differences in solute specificity. Channels appear to be constructed by having a central pore in the center of a cluster containing several copies (3 to 8) of the same or closely related polypeptides or domains of a single polypeptide. These pores can be opened or closed in response to chemical or electrical signals, depending on the channel.

Chapter 9
The Cell Surface: Receptors, Membrane Recycling and Signal Transduction

9.1 Overview

Whereas the previous chapter focused primarily on mass transport across the membrane, this chapter is largely concerned with the transfer of information across the bilayer. All cells need mechanisms to monitor their environment and to be able to respond to changes which occur. There is a myriad of specialized receptor molecules present in the cytoplasmic membranes of bacterial, plant, and animal cells which interact with extracytoplasmic components and elicit specific cellular responses. Some receptors bind to nutrients or metabolites, others to hormones or neurotransmitters, and still others are involved in cell–cell recognition and adhesion either to other cells or to insoluble matrix components in the environment. Most of the response systems involve a series of steps: (1) binding of a ligand or agonist to a receptor located outside the cell, (2) transmittal of the information concerning receptor occupancy to the inside of the cell, and (3) the cellular response, which can be further subdivided into primary and secondary responses. This is an extremely active area of research which, largely due to the increasing impact of molecular biology, is advancing at a startling rate. It is now becoming clear that numerous seemingly unrelated response systems are, in fact, quite closely related by underlying common features. Several families of sequence-related, homologous receptor proteins have been identified in which individual members each respond to different ligands and elicit unique cellular responses. These are "superfamilies" in that they consist of structurally related but functionally diverse proteins (see Table 9.1). This pattern suggests a modular construction not only of the receptor proteins within each family (e.g., see 984), but also of the other portions of the response system, so that variants of the same basic receptor protein structure can meet a variety of demands in different cell

Table 9.1. Some superfamilies of structurally related receptors in eukaryotes.[1]

(1) *Immunoglobulin superfamily* — many diverse functions
- T cell receptor (α, ß subunits)
- Major histocompatibility complex II MHC (α, ß subunits)
- Lymphocyte function associated antigen-3 (LFA-3)
- CD2 (T cell LFA-2)
- IgA/IgM receptor
- IgG F_c receptor
- High affinity IgE receptor (α subunit)
- Surface immunoglobulins (heavy, light chains)
- N-CAM
- Myelin-associated glycoprotein

(2) *Integrins* — bind to extracellular matrix and adhesion proteins
- Fibronectin receptors
- Vitronectin receptors
- Platelet glycoprotein complex (IIb/IIIa)
- Leukocyte adhesion proteins (LFA-1, Mac1, p150/95)
- T cell very late antigens (VLA family)

(3) *Mitogen/growth factor receptors with tyrosine kinase activity* — stimulate cell growth
- Epidermal growth factor (EGF) receptor
- Platelet-derived growth factor (PDGF) receptor
- Insulin receptor
- Insulin-like growth factor-1 (IGF-1) receptor (1476)
- Colony stimulating factor-1 (CSF-1) receptor

(4) *Neurotransmitter receptors/ion channels* — receptor-operated channels
- Nicotinic acetylcholine receptor (nAChR)
- γ-Aminobutyric acid (GABA) receptor
- Glycine receptor

(5) *Receptors that activate G proteins*
- ß-Adrenergic receptors ($ß_1$, $ß_2$)
- α-Adrenergic receptors (α_1, α_2) (767)
- Opsins (rhodopsin)
- Muscarinic acetylcholine receptors (M1, M2)
- Yeast mating-type (α, a) receptors[2] (624)
- Substance K receptor[2] (935)

(6) *Miscellaneous*
- (A) Asialoglycoprotein receptors
  - Low-affinity (lymphocyte) IgE receptor (unknown function)
- (B) Insulin-like growth factor-2 (IGF-2) receptor (1017)
  - Cation-independent mannose-6-phosphate receptor
  - Cation-dependent mannose-6-phosphate receptor (extracytoplasmic domain) (274)

[1]See text for additional references.

[2]Presumed to activate G proteins.

types and with different effectors. The central roles played by GTP binding proteins (G proteins) and of phosphatidylinositol degradation products in a wide variety of systems are very evident. The main purpose of this chapter is to point out the conceptual framework which has emerged in recent years.

A second, closely related subject discussed in this chapter involves the dynamics of the cell surface itself, in particular the animal cell surface. Whereas in Chapter 5 we saw that individual protein and lipid molecules can often diffuse within the plane of the membrane, we will now describe how the animal cell plasma membrane is in dynamic equilibrium with a pool of internal membrane vesicles, called endosomes or receptosomes, which are formed by pinching off invaginations from the plasma membrane and which can recycle to the plasma membrane. This is part of a complex pattern of intracellular membrane "traffic" which also involves other membranous organelles within the cell, such as the Golgi and lysosomes. Specfic plasma membrane proteins can be sequestered in a cryptic form in an internal membrane pool and then be rapidly delivered to the surface in an active form in response to an appropriate signal [e.g., glucose transporter in response to insulin (712)]. Some macromolecules, such as low density lipoproteins (LDLs), are taken up by cells via the internalization of plasma membrane-derived vesicles in a process called receptor–mediated endocytosis. In the next chapter we will see that newly synthesized plasma membrane proteins and proteins to be secreted travel from their point of initial biosynthesis in the endoplasmic reticulum to their final destinations by a membrane traffic system mediated by similar intracellular vesicles. This is called the exocytic pathway (Section 10.2).

9.2 A View of the Animal Cell Surface

Before proceeding to the discussion of the structures of various membrane-bound receptors and the mechanisms of inducing a cellular response, it will be useful to consider the nature of the animal cell surface, because most of the research in this area concerns animal cells. There are two structural features that differentiate the plasma membrane from other cellular membranes:

1. A relatively high cholesterol content (see Table 1.3). The fraction of cellular cholesterol which is within the plasma membrane has been difficult to quantify (see 1493), but it is clearly substantial. Other membranes in the endocytic pathway (i.e., lysosomes, endosomes) also have a high cholesterol content. Sphingomyelin is another lipid which is found primarily in these membranes.
2. A relatively high content of glycoconjugates (glycolipids and glycoproteins). Although glycoconjugates are not uniquely located in the plasma membrane, they are present to a larger extent in this membrane. The carbohydrate groups of glycoproteins in the plasma membrane are all on the extracellular surface, and, as indicated in Chapter 4, the attachment sites can be used to identify amino acid residues in membrane proteins which must be facing outside. Most,

but not all, plasma membrane proteins are glycosylated. It is also useful to keep in mind that some plasma membrane proteins are anchored to the membrane by linkages to phosphatidylinositol and can be released by the action of phospholipase C (e.g., 675), and others are anchored by covalently bound fatty acids (see Section 3.8).

Although the physical chemistry of cholesterol within model membranes has been a topic of intensive study (see Section 2.42), the functional role of cholesterol in the plasma membrane is not known. Similarly, our understanding of the biological functions of the carbohydrate portions of glycolipids and glycoproteins is poor.

As a model, consider the surface of the red blood cell (1519). The two most abundant proteins, Band 3 and glycophorin A (see Table 4.2, Figure 1.16), are well characterized glycoproteins. Altogether, about 75% of the total monosaccharides on the cell surface are on glycoproteins. The remaining 25% are in glycolipids, which consist of simple globosides and long-chain polylactosamine ceramides (see Chapter 1 for structural definitions). These glycolipids represent only a few mole percent of the total lipid content. The major glycoconjugates of the red blood cell are schematically shown in Figure 9.1, which illustrates the longest reach of the carbohydrate chains. Figure 9.2 is a properly scaled surface view in which the distances between various glycoconjugates are indicated. Although the carbohydrate chains undoubtedly do not stick straight out from the cell surface as pictured in Figures 9.1 and 9.2 (see 539, 891), these structures are likely to play a significant role in the manner in which the cells interact with their environment.

The glycoconjugates provide the cell surface with a very hydrophilic, often highly negatively charged character. Negative charges are largely provided by sialic acid residues, which constitute a family of derivatives of neuraminic acid (e.g., see Figure 1.14). In several examples, the presence of sialic acid has been shown to mask specific recognition sites on cell surface molecules (see 1290). The main point to note is that the glycoconjugates determine the character of a region extending perhaps well beyond 100 Å from the cell surface. In addition, specific components have unique biological functions.

The carbohydrate structures on both glycolipids and glycoproteins have been shown to change during cell development and differentiation and can be used as antigenic markers associated with tumors (415). Gangliosides, which are sialic acid-containing glycosphingolipids (see Figure 1.14), have been identified as the binding sites for cholera and tetanus toxins (GM_1) (see 434, 481). More interestingly, gangliosides have been directly implicated in the modulation of cell growth (1367, 143) and differentiation (1078) and appear to modulate the phosphorylation of the epidermal growth factor receptor, possibly by direct interaction (143). The role of the carbohydrate component of glycoproteins and, specifically, of receptor proteins, is not at all clear. As shown in the next section, glycosylation may play a modulating or regulatory role in some receptors required for cell–cell adhesion.

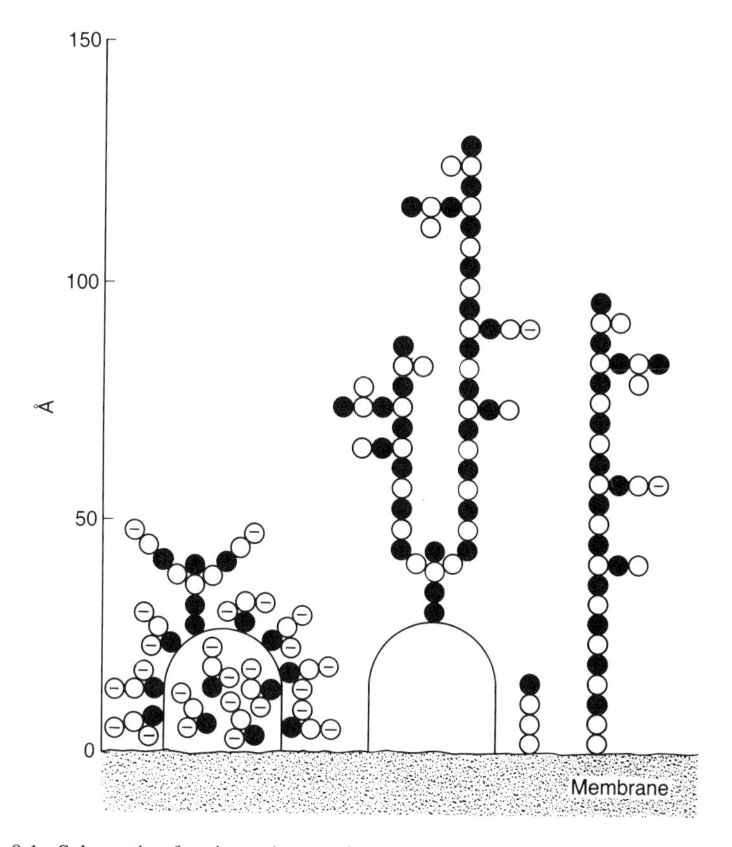

Figure 9.1. Schematic of various glycoconjugates on the surface of the erythrocyte. From left to right: glycophorin A (see Figure 3.17), Band 3, globoside, and polylactosamine ceramide. Open circles represent hexoses, filled circles are *N*-acetylhexosamines, and the circles with the negative sign represent sialic acid residues. The maximum extension away from the membrane surfaces is shown. Adapted from ref. 1519.

9.3 Receptors for Cell Adhesion

A major class of cell surface receptors is involved in cell adhesion. This class includes receptors which are required for recognition and adhesion to other cells, as well as receptors for adhesion to insoluble extracellular matrix components, such as fibronectin or collagen in the case of animal cells. Both protein–protein and protein–sugar interactions are clearly important in the binding interactions. The ability of cells to recognize and specifically adhere to other cells is obviously of primary importance in embryonic development, and some of the cell surface components required for this are now known. In the adult animal, cell–cell and cell–matrix adhesion continue to be essential for tissue stability. Specific systems which have been examined include the cell–cell recognition involved in the immune response, in lymphocyte "homing," and in the induction of the adhesive

Figure 9.2. Representation of an area of the erythrocyte surface 350 Å × 350 Å. Adapted from ref. 1519.

properties of platelets during blood clot formation. The impact of molecular biology has been substantial in each of these areas, revealing structural relationships between receptors. Illustrative examples of adhesion receptors are discussed below.

9.31 Bacterial Adhesion to a Glycolipid

Many bacteria colonize on solid substrates and in some cases adhere to specific surfaces. Among the best described structures involved in this process are the bacterial pili (see 1330). These are hair-like appendages that extend far beyond the cell surface, and they are not analogous to any structures found in higher organisms. One example is the pappili found in the pathogenic strains of *E. coli* which cause infections in the human urinary tract (860). The pappilus is composed of about 1000 protein molecules arranged in a helix at the tip of which is a specialized receptor protein (adhesin) which binds specifically to digalactoside-containing glycolipids. These lipids are present on the surface of the epithelial cells which line the urinary tract, which is where the bacterium colonizes. Note that adhesin, in order to function, must be located well beyond the bacterial cell surface, normally covered with lipopolysaccharide (see Figure 4.6), and that it functions as a lectin, i.e., sugar-binding protein.

9.32 "Homing" of Lymphocytes and Hemopoietic Stem Cells Also Requires Glycoconjugates

Lymphocytes are constantly recirculating between the blood and lymphoid organs (e.g., spleen, adenoids, tonsils). It is in these organs that antigen is collected from all the intercellular spaces, is processed, and is presented to antigen-specific lymphocytes, resulting in the immune response. The lymphocyte migrates between the bloodstream and lymphoid organs via lymphatic vessels called high-walled endothelial venules, and of all the cells in the bloodstream, only recirculating T and B lymphocytes adhere to the luminal walls of these vessels and migrate to the lymphoid organs. This migration is termed "homing," and specific cell surface receptors are required for this (484, 1339). The homing receptor is a protein with a molecular weight around 90,000 which, remarkably, is covalently attached to ubiquitin. Ubiquitin is a small polypeptide (molecular weight 8,451) which has been found covalently linked to numerous cytoplasmic proteins and has been implicated in intracellular protein degradation (see 1339). The role of ubiquitin, which is attached to the extracytoplasmic domain of this cell surface receptor, is not known. However, it appears that other cell surface proteins are similarly modified, so this may be of general importance (see 1339).

The homing receptor also appears to be a lectin, recognizing carbohydrate portions of glycoconjugates on the surface of the endothelial cells (484). Mannose-6-phosphate, specifically, has been implicated in this recognition.

Hemopoietic stem cells also manifest homing (see 8). These cells, which differentiate into macrophages and other cell types, have an affinity for the stromal cells of bone marrow and spleen. It is here that these cells proliferate and differentiate. Although no receptor has been isolated, it appears that the components recognized are galactosyl and mannosyl residues on glycoconjugates in the target tissue. Hence, protein–sugar interactions are important in the adhesion.

9.33 Cell Adhesion Molecules (CAMs) (see 268, 270 for reviews)

The use of monoclonal antibodies to identify surface proteins required for cell–cell interactions has been very productive in revealing a set of proteins which function in intercellular adhesion. These are called cell adhesion molecules or CAMs. The best characterized are a family of glycoproteins involved in neuronal cell adhesion (N-CAMs). Similar molecules have been identified in mouse, chicken, and human brain. Within a particular species, the N-CAMs are expressed during development in several forms at different times in different tissues. It seems possible that the complex patterns of neuronal cell adhesion during development of the brain may be due to the differential expression and post-translational modifications of a small number of molecules (CAMs) rather than a large number of highly specific adhesion proteins. N-CAMs are homophilic, meaning that an N-CAM on one cell binds to a like N-CAM on another cell. Although these adhesion proteins are glycosylated, the carbohydrate is not necessary for the binding interaction. However, the glycosylation may modulate the interaction, and one of the alterations which occurs during the change from an embryonic to adult brain is that the amount of sialic acid on the N-CAMs is reduced by a factor of three. A decreased sialic acid content correlates with enhanced adhesion properties. This is presumed to play a regulatory role in N-CAM interactions (1412).

Other post-translational modifications of N-CAMs include phosphorylation of serine and threonine residues in the cytoplasmic domain, fatty acid acylation, and sulfation of the N-linked oligosaccharides. Presumably, these are all modes of regulating the biological functions of the N-CAMs. In addition, N-CAMs are produced in three different forms as a result of alternative mRNA splicing (180,000 Da, 140,000 Da, 120,000 Da). These N-CAM variants are expressed at different times and places. Two of the forms contain a putative hydrophobic membrane-spanning helix and substantial carboxyl-terminal domains on the cytoplasmic side of the membrane. These two forms differ by the insertion of a 261-amino-acid segment in the cytoplasmic domain. The third form lacks the hydrophobic domain but appears to be associated with the extracytoplasmic surface of the membrane by a phosphatidylinositol linkage (see Section 3.8). The specific biological roles of these molecular variants are not known.

The complete amino acid sequence of N-CAM, deduced from the cDNA, has revealed that it is part of the immunoglobulin gene superfamily. The N-CAM protein contains five contiguous domains of about 100 amino acids each which are not only homologous to each other but also homologous to immunoglobulins. Hence, N-CAMs are composed of immunoglobulin-like domains and appear to be phylogenetically related to immunoglobulins. This superfamily (665) includes other cell surface proteins which function as receptors, including the T cell receptor and the major histocompatibility complexes (see Figure 9.3). Another member is the myelin-associated glycoprotein (810), which appears to mediate interactions between axons and myelinating cells in the peripheral nervous system. This has a structure very much like N-CAM, suggesting a common structure for one class of cell adhesion molecules.

Homophilic protein–protein interactions between the immunoglobulin-like domains of N-CAM are critical to its function. In addition, however, N-CAM binds to heparin at a domain near the amino terminus (233), and this interaction may also be involved in cell–cell interaction (232). Hence, N-CAM is multifunctional. A schematic representation of an N-CAM is shown in Figure 9.4, based on studies utilizing monoclonal antibodies (464).

Other CAMs have been isolated which are distinct from N-CAM. One is neural–glial or Ng-CAM isolated from the chicken embryo. Whereas N-CAM is involved in adhesion between neural cells (homotypic binding), Ng-CAM is involved in adhesion between neural cells and glial cells (heterotypic binding).

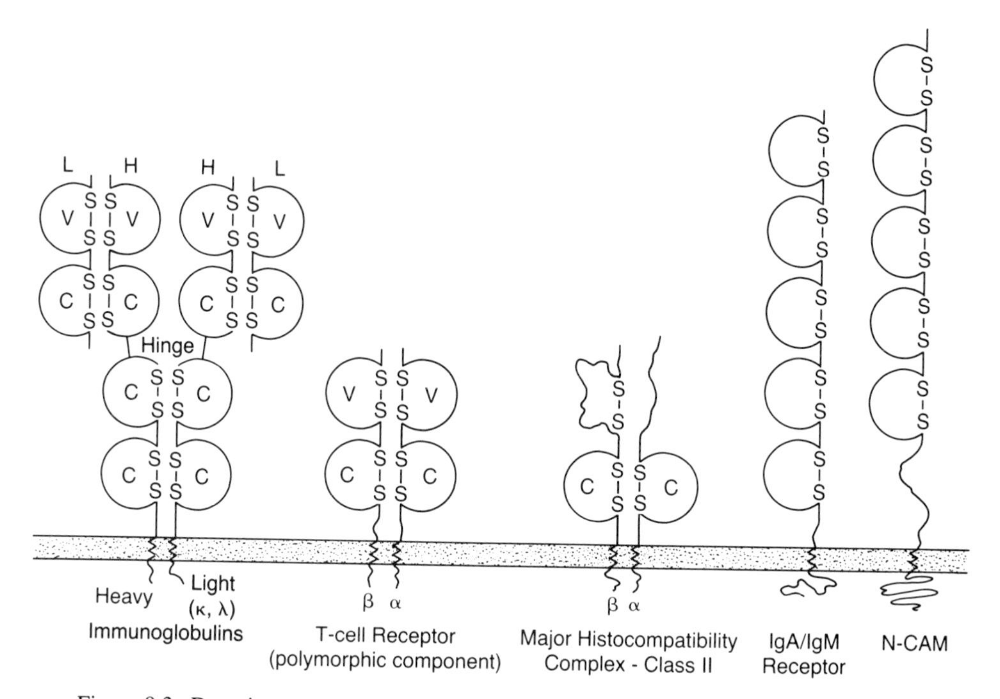

Figure 9.3. Domain structure of several selected members of the immunoglobulin superfamily. The C and V denote constant and variable immunoglobulin domains. In all cases the carboxyl termini are cytoplasmic. Adapted from ref. 665. Reprinted by permission from *Nature*, vol. 323, p. 15, Copyright © 1986 Macmillan Magazines Limited.

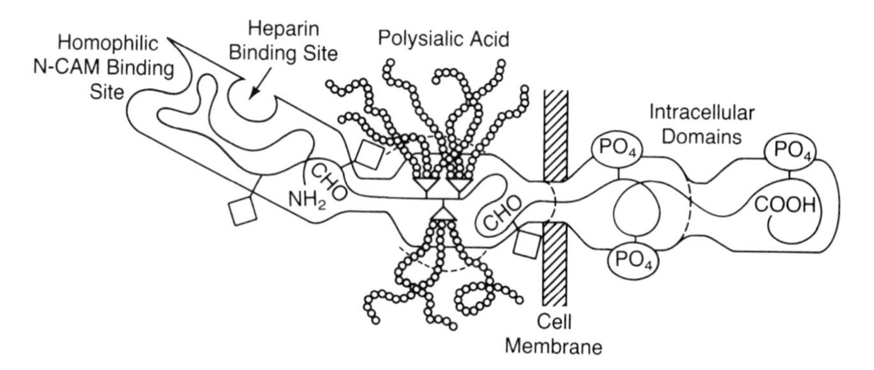

Figure 9.4. Schematic representation of N-CAM in the cell membrane. The dashed lines indicate the three naturally occurring forms of N-CAM which differ in their carboxyl-terminal cytoplasmic extensions.Also, the reduced polysialic acid content of the adult form of N-CAM is indicated. The diamond shapes indicate sites of glycosylation (CHO). The actual folding pattern of the polypeptide chain is unknown. Adapted from ref. 464. Reproduced from the *Journal of Cell Biology*, 1986, vol. 103, p. 1735, by copyright permission of the Rockefeller University Press.

The timing of the appearance and cellular localization of Ng-CAM is distinct from that of N-CAM (1446). Presumably, the programmed appearance of Ng-CAM, along with N-CAM, helps to govern the formation of the neuronal network. The interactions involving these CAMs are Ca^{2+}-independent.

A distinct class of CAMs has been identified which mediates Ca^{2+}-dependent cell–cell interactions (1081, 1220). These CAMs are also referred to as cadherins (1081) and include mouse epithelial (E–) cadherin (1335), placental (P–) cadherin (1081), and uvomorulin (1220). The chicken homologue of E-cadherin is called L-CAM (485), and these glycoproteins mediate interactions between epithelial cells. Similarities in the deduced amino acid sequences show that these Ca^{2+}-dependent CAMs form a distinct superfamily of adhesion molecules (1081, 1220). For example, the complete amino acid sequence deduced from the cDNA reveals that L-CAM is a polypeptide with a molecular weight of 79,700 in the absence of glycosylation. There is no sequence homology to N-CAM, however, and L-CAM is not a member of the immunoglobulin superfamily. However, there are three contiguous homologous segments of 113 amino acids each in L-CAM, suggesting a structural pattern similar to that found in N-CAM, which has five such segments.

It is clear that future studies on CAMs will be prominent in developing concepts of the mechanism of cell–cell recognition and adhesion during embryogenesis, and a variety of other cellular processes.

9.34 Receptors Involved in Cellular Interactions in the Immune Response (see 963, 732, 647, 924 for reviews)

One of the unexpected aspects of the CAM studies is the fact that there appear to be relatively few structurally distinct CAMs involved in mediating cellular interactions during embryogenesis. Quite the opposite is true if one examines the cellular interactions between lymphocytes which are required for the immune response. At least three different cell types are involved in the induction of antibodies directed against a specific antigen: (1) B cells, which produce the antibody; (2) T cells (helper cells), which secrete growth factors; and (3) macrophages or accessory (A) cells. Each of these cells contains cell surface receptors which are required for specific cell–cell interactions and which are essential for the activation of B cells to proliferate and secrete antibody against a specific antigen. The two main receptors known to be involved in this process are (1) the T cell receptor and (2) the class II major histocompatibility complex (class II MHC). These are both members of the immunoglobulin superfamily (665, 647) (Figure 9.3). The specific cell–cell interactions involved in the immune response are primarily determined by protein–protein interactions between these cell surface receptors. A simplified sequence of events is described below.

1. Antigen is first taken up by the A cells, processsed, and presented on the cell surface in a complex with class II MHC.
2. The T cell interacts with the A cell via the T cell receptor. The T cell receptor

is thought to simultaneously recognize the class II MHC and the antigen. Only those T cells with appropriate antigen-specific receptors will interact with the A cell. Although there are many diverse T cell receptor structures among a population of T cells, any individual cell produces only one. One result of this interaction between the A cells and T cells is the production of growth factors which stimulate the proliferation of this subpopulation of T cells as well as other lymphoid cells (B cells, cytotoxic T lymphocytes).

3. B cells can express a large number of distinct membrane-bound immunoglobulins (see Figure 9.3), but, like T cells, they express one type per individual B cell. A subpopulation of B cells will have surface immunoglobulins which bind to the specific antigen presented to the system. To enhance their activation and proliferation, these B cells interact with the activated T cells and the factors that T cells release. The interaction is thought to involve T cell recognition of the antigen which is bound to the class II MHC on the surface of the B cell. This is known as the MHC-restricted immune reaction. Proliferation of these activated B cells results in a population of cells secreting antigen-specific antibody.

The class II MHC is a heterodimer in which each protein has a single transmembrane segment (732). Both the chains are glycoslylated (heavy chain or ß, ~33,000 Da; light chain or α, ~27,000 Da). The structure of a major portion of class II MHC has been determined by X-ray crystallography, revealing a binding site for peptide antigens (104). The interaction of peptide antigens with class II MHC has been demonstrated (559). Also, fluorescence energy transfer has been used to demonstrate that the purified class II MHC, reconstituted in a planar membrane, will form a ternary complex with the T cell (via the T cell receptor) and an antigen, mimicking the antigen-specific interaction between the B cell and T cell (1559). The use of supported artificial membranes appears to be a promising system for studying the interactions between components involved in adhesion of cellular membranes (see 947; Section 2.54).

The T cell receptor contains two portions (see 924 for review). The antigen-specific portion (Ti) is an αß heterodimer which is clonally diverse, like immunoglobulins, and is a member of the immunoglobulin superfamily (see Figure 9.3). This αß dimer forms a complex in the membrane with a structurally invariant (monomorphic) component called the T3 complex. This monomorphic complex has been characterized in both mouse (1275) and human (1567) T cells and consists of four or five polypeptides. Presumably, this portion of the T cell receptor is involved in signal transduction, resulting in the activation of the T cell.

The class II MHC present on the antigen-presenting cells can also interact with another glycoprotein receptor present on the surface of the T cells. This receptor is the CD4 receptor, and its interaction with the class II MHC molecules presumably augments the cell–cell interaction (345).

There is, in addition, a second pair of receptors which mediate the adhesion between the antigen-presenting cell and the helper T cell (see 142). The T lymphocytes contain a membrane glycoprotein called CD2 which specifically interacts with another adhesion protein called LFA-3 (lymphocyte function-associated antigen-3), which is present in the membranes of a variety of cell types (1316, 360). Remarkably, both CD2 (also called LFA-2 and the erythrocyte receptor) and LFA-3 are members of the immunoglobulin superfamily, and are structurally related to N-CAM (see Table 9.1). Like N-CAM, the LFA-3 receptor exists in different forms and can be anchored to the membrane either by a transmembrane helix or by a phosphatidylinositol-mediated linkage (1316, 360). The interaction between the LFA-3 and CD2 receptors is, presumably, important in the activation of the helper T cells.

One consequence of T cell activation is the expression of a receptor for the T cell growth factor, interleukin-2. The gene encoding one subunit of this receptor has been cloned and sequenced (1535), and it appears to have a single transmembrane segment but only a very small carboxyl-terminal cytoplasmic domain. This is unusual for mitogen receptors, which in other cases have a large cytoplasmic domain with tyrosine-specific protein kinase activity. The mechanisms by which animal cells respond to mitogens and other stimuli are discussed in Section 9.7.

9.35 Integrins — A Family of Receptors Which Bind to Extracellular Matrix Components and Adhesive Proteins (see 672, 160, 1261 for reviews)

In addition to the immunoglobulin superfamily of cell surface receptors (Figure 9.3), a second large family of cell adhesion receptors, called integrins, has been identified. Integrins are involved in binding to extracellular matrix proteins and other adhesive proteins. In many cases, the integrins have been shown to recognize and bind to proteins containing the tripeptide Arg-Gly-Asp (RGD in single letter code). Integrins are heterodimers in which each subunit (α, ß) contains one putative transmembrane segment near the carboxyl terminus (Figure 9.5). The subunits each contain a short cytoplasmic domain (about 20–50 amino acids) and a large extracellular domain. The larger α-subunit (~150,000 Da) in many cases is proteolytically processed to two polypeptide chains which are linked by a disulfide bond. The ß-subunit (~110,000 Da) contains four repeats of a 40-amino-acid cysteine-rich motif which appears to contain extensive intrachain disulfide

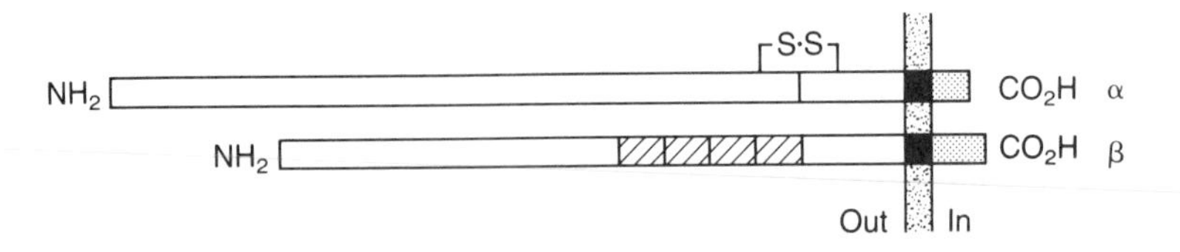

Figure 9.5. Linear representation of the domain structures of the α and ß chains of the integrin family of receptors. Each subunit has a single transmembrane span, with a short cytoplasmic domain and a large amino-terminal extracytoplasmic domain. The ß-subunit contains four repeats of a 40-amino-acid and cysteine-rich motif (crosshatched area). Some of the α subunits are cleaved post-translationally to yield a heavy chain and a light chain linked by disulfide binding. Adapted from ref. 672 © 1987 by Cell Press.

linkages. Both subunits are glycosylated. Approximately 20 different members of this family of receptors have been identified so far, in a variety of cell types. These are related by immunocrossreactivity and/or amino acid sequence homology. Examples are listed in Table 9.1. The integrins have been shown to bind to a variety of matrix or adhesive proteins. However, individual integrins have definite specificity. The RGD sites are critical to these interactions, but obviously other determinants must be present to explain the specificities.

An important function of integrins is to convey information to the cell interior, since binding to extracellular matrix proteins is often essential for determining cell shape and migration and is critical to morphogenesis and differentiation (see 971). In one case, the integrin from chicken fibroblasts has been shown to bind to a cytoskeletal component, talin. This binding is inhibited by competition using a peptide which corresponds to a consensus tyrosine kinase phosphorylation site which is located in the cytoplasmic domain of the integrin ß-subunit (160). Binding the extracellular matrix proteins via the RGD recognition site(s) is not influenced, however, by this peptide, illustrating the quasi-independence of the intracellular (talin binding) and extracellular (RGD binding) domains. This is the best characterized transmembrane protein which simultaneously interacts with both extracellular and cytoskeletal components.

One interesting feature of integrins is that there are variants which contain common ß subunits, of which at least three types exist ($ß_1$, $ß_2$, $ß_3$). However, there is no clear separation of functions of the α and ß subunits, relating one, for example, to extracellular binding and the other to intracellular binding. It has been shown that all the binding interactions of the chicken integrin are lost upon dissociation of the subunits, and are restored upon reconstitution of the heterodimer (160).

Box 9.1 Several Examples of Integrins

(1) *Avian and mammalian extracellular matrix protein receptors*: These bind to glycoprotein components of the extracellular matrix, including fibronectin, laminin, and vitronectin. Binding to the receptors is competitively blocked in most cases by peptides containing the sequence RGD. Fibronectin has been shown to be essential for cell movement, and it is clear that the fibronectin receptor and its intracellular connections are critical to cellular morphogenesis

(2) *Platelet IIb/IIIa glycoproteins*: IIb and IIIa are platelet membrane glycoproteins which are homologous to the α and ß subunits of other integrins. This surface receptor is involved in platelet aggregation which occurs during blood clotting. Platelets are not adhesive until they are "activated" by any one of a number of agonists, including thrombin, collagen, or epinephrine. The details of this activation are not at all clear, but one result is to increase the accessibility of the IIa/IIIb complex at the platelet surface so that it can interact with circulating macromolecular "adhesive" proteins (1605, 236), including fibrinogen, fibronectin, and von Willebrand factor. The binding is calcium-dependent. Fibrinogen is bivalent and is probably most important for bridging platelets in the aggregation cascade. Plasma von Willebrand factor is a multifunctional glycoprotein which has affinity not only

for the activated integrin but also for collagen. Hence, this adhesive protein helps mediate the adhesion of activated platelets to the vascular subendothelium which becomes exposed when the integrity of the endothelial cells is disrupted by a wound (1244). The von Willebrand factor also binds to another platelet receptor protein, called glycoprotein Ib (581).

(3) *Leukocyte adhesion proteins*: Leukocytes must interact with the vascular endothelial cells to be able to migrate to the site of infection and inflammation. Three heterodimeric receptors have been identified in the leukocyte adhesion function (see Table 9.1), and all are members of the integrin family (759). All three appear to share common ß subunits. It has been shown that one of these integrins, LFA-1, binds to a specific cell surface glycoprotein (ICAM-1) present on fibroblasts. This interaction may mediate the binding of T lymphocytes to fibroblasts during an inflammatory response (923).

(4) *Position specific antigens of Drosophila*: It is likely that a family of receptors which is essential for normal morphogenesis during embryo development of *Drosophila* also belongs to the integrin family (160, 1051, 114). Peptides containing the sequence RGD have been shown to disrupt normal embryogenesis and prevent gastrulation in *Drosophila* (1051).

9.36 Other Modes of Binding to Matrix and Adhesive Proteins

Integrins are not unique in mediating binding to extracellular matrix and adhesive proteins. Laminin, thrombospondin, and von Willebrand factor have all been shown to bind specifically to sulfated glycolipids (1226). The physiological importance of this interaction is not clear. The enzyme 5'-nucleotidase has also been implicated as interacting with both extracellular and intracellular components (332), though this also has not been demonstrated to be important in vivo. There are also laminin receptors which are not part of the integrin family (537, 1355).

The aggregation of dissociated sponge cells has been extensively studied (see 1037). The aggregation requires the binding of a high-molecular-weight aggregation factor to the extracellular domain of an aggregation receptor located in the plasma membrane. This apparently triggers the rapid turnover of phosphatidylinositol and the generation of intracellular second messengers (see Section 9.72) which initiates aggregation mediated by a collagen assembly factor. Collagen bundles serve as a matrix to which the sponge cells can adhere.

9.4 Membrane Recycling and Receptor-Mediated Endocytosis (for reviews see 1584, 1447, 1024, 966, 651)

The picture presented, so far, of the animal cell surface has been rather static, with progressive changes in composition accompanying cellular development or differentation. In reality, the cell surface is an extremely dynamic structure, participating in a complex network of membrane flux with internal membranous organelles. This membrane traffic may be considered in two parts: the *endocytic pathway* and the *exocytic pathway* (see Figure 9.6). Endocytosis refers to the internalization of extracellular fluid and particles by enveloping membranes, whereas exocytosis concerns the processing and delivery of newly synthesized proteins and lipid for secretion, or incorporation into plasma membranes, lysosomes, or vacuoles. Both systems involve the selective transfer of specific membrane components between membranes within the cell mediated by vesicles which pinch off from one membrane and can fuse with another. The acidification of the vesicles and vacuoles by H^+-ATPases (see Section 8.43) is a critical feature regulating these systems (966). The endocytic pathway will be discussed in this section, especially as it relates to receptor function. In the next chapter, the exocytic pathway will be discussed in the context of membrane biogenesis (Section 10.2). The general topic of membrane fusion and how this might be biologically regulated is discussed in Section 9.5.

9.41 General Features of the Endocytic Pathway (see 966)

A distinction may be made between the uptake of large particles, solutes, and fluid. The internalization of large particles is called *phagocytosis*. This is observed in only a few cell types (e.g., amoebae, macrophages), is particle-activated, and is sensitive to cytochalasin. After formation of an intermediate *phagosome*, the

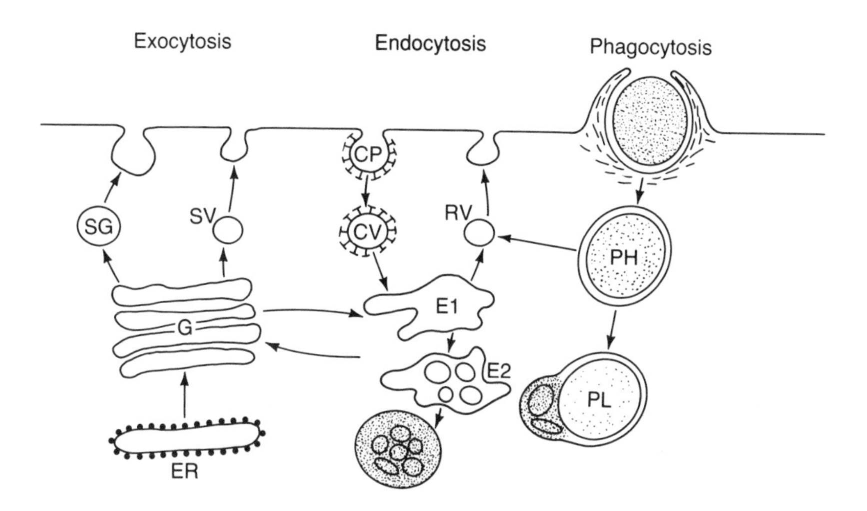

Figure 9.6. Schematic view of membrane traffic between the intracellular vacuoles and the plasma membrane. Abbreviations: ER, endoplasmic reticulum; SG, secretory granule; SV, secretory vesicle; L, lysosome; E1, peripheral endosome; E2, perinuclear endosome; CV, coated vesicle; CP, coated pit; RV, recycling vesicle; PL, phagolysosome; PH, phagosome; G, Golgi apparatus. The exocytic pathway is discussed in Chapter 10. Adapted from ref. 966. Reproduced with permission from the *Annual Review of Biochemistry*, vol. 55, © 1986 by Annual Reviews Inc.

vesicle is acidified and then fuses with a lysosome, which contains degradative enzymes.

The major pathway for both fluid uptake (pinocytosis) and receptor-mediated solute uptake involves coated pits and coated vesicles. The "coating" refers to a morphological feature apparent in electron micrographs due to the presence of a lattice made from a protein called clathrin which is associated with invaginated patches in the plasma membrane (coated pits) and vesicles derived from such invaginations (coated vesicles) (see Section 9.43). Although other endocytic pathways involving uncoated vesicles may exist, little is known about them (see 1466). It should be noted that coated vesicles have also been implicated in parts of the exocytic pathway.

Coated pits typically constitute only about 1-2% of the total area of the plasma membrane surface, and most plasma membrane proteins are excluded from these regions. However, a select few proteins are highly concentrated in the coated pits. For example, about 70% of the receptor for low density lipoprotein (LDL receptor) is present in coated pits. In some cases, the receptor affinity for the coated pits is constitutive (e.g., LDL receptor), but in other cases the receptor only concentrates in the coated pits once the ligand is bound to the receptor (e.g., receptor for the epidermal growth factor, EGF receptor). Table 9.2 lists some plasma membrane receptors which mediate the internalization of specific ligands via coated pits and vesicles.

After the formation of endocytic coated vesicles, the clathrin coating appears to be removed by a specific uncoating protein in an ATP-requiring reaction. This has been demonstrated in vitro, though the role of this protein in vivo has not yet been shown. The stripped vesicles become part of a complex system of tubules and vesicles called peripheral *endosomes*, located adjacent to the plasma membrane. On the basis of functional roles and morphological distinctions (see 651), sub-

Table 9.2. Some receptors which are internalized by endocytosis

Group I:	*Receptor recycled; ligand to lysosomes*
	Mannose receptor
	Asialoglycoprotein (galactose) receptor
	Mannose-6-phosphate receptor
	α_2-Macroglobulin receptor
	LDL receptor
Group II:	*Receptor and ligand to lysosomes*
	EGF receptor
	Insulin receptor
Group III:	*Transcytosis*[1]
	IgA/IgM receptor
	IgG receptor
Group IV:	*Receptor and ligand recycled*
	Transferrin receptor

[1]Transcytosis involves the transfer of the ligand across the cell, from one plasma membrane domain to another. See Section 9.42

populations of endosomes have been assigned various names, such as receptosomes or CURLs (*c*ompartment of *u*ncoupling *r*eceptor and *l*igand). We will refer to all these simply as endosomes. There is also a system of endosomes located close to the centrioles and Golgi called the perinuclear or juxtanuclear endosomes. These are almost certainly also involved in endocytic membrane traffic. The final compartment of the endocytic pathway consists of the secondary lysosomes, where the degradation of selected solutes, such as the low density lipoproteins, occurs.

The rate at which membrane recycling occurs can be astonishing. This can be estimated by using membrane-impermeant, water-soluble tracers to measure the rate at which extracellular fluid is taken up by the cell. Such tracers include, for example, radioactive inulin, dextran, and the dye Lucifer Yellow (see 1287, 1415). Measurements with hepatocytes using inulin indicate that these cells endocytose at least 20% of their volume and 5 times their basolateral plasma membrane surface area each hour (1287)! Other cell types do not undergo such rapid endocytosis, but the rates are still impressive, e.g., macrophages, 2 times the surface area per hour; adipocytes, 0.2 times per hour. Most of the fluid taken up by cells is rapidly regurgitated back out, with a characteristic half-time ranging from 1 min (hepatocytes) to 20 min (adipocytes). Presumably, this represents a rapid dynamic equilibrium between the plasma membrane and the peripheral endosomes, which themselves constitute only about 3% of the total cell volume (1287). In addition, some of the inulin taken up by hepatocytes goes into a much more slowly exchanging internal pool (half-time for exchange ~1 hour), presumably reflecting sequestration into other internal vacuoles, such as lysosomes.

These data reveal a remarkably dynamic membrane surface. Bretscher (146) has pointed out that if the plasma membrane were internalized at one locus on the cell surface and reinserted at another location, there would be a unidirectional bulk membrane flow from the point of insertion to the region of invagination. This has been elaborated into a model to explain amoeboid-like cell locomotion as well as the "capping" of surface antigens (146). Capping is a phenomenon observed in lymphocytes, as well as other cell types, which is initiated by the aggregation of plasma membrane proteins, usually by binding of a multivalent ligand, such as an antibody, to a cell surface component. The bound component aggregates into patches, which then coalesce to form a single, large aggregate called a cap. This is followed by internalization and shedding of the protein aggregates. It is postulated (146) that the initial protein aggregates could be swept up passively by the presumed bulk membrane flow where they collect to form a cap near the locus of membrane internalization. The alternative view, which is more widely held, is that capping is an active process involving the direct participation of cytoskeletal elements (e.g., 744). It is interesting to note that gangliosides can also be induced to form a cap in lymphocytes (1370).

9.42 Receptor-Ligand Sorting

The purpose of the endocytic pathway and the rapid turnover of the plasma membrane is not to internalize fluids. One of the main purposes is to facilitate the internalization of extracellular proteins. Most of the receptors listed in Table 9.2 which concentrate in coated pits bind to macromolecular ligands such as glycoproteins, low density lipoproteins, or immunoglobulins. In addition, there are receptors for growth factors and hormones such as EGF or insulin where membrane recycling rapidly clears the plasma membrane of the hormone–receptor complexes, which are then replaced by newly synthesized receptors. The structural determinants that direct these receptors to bind to the coated pits are not known. The complete primary structures of several of these receptors are known, and no common features are apparent. They all have a single transmembrane segment, but some have their amino termini toward the cytoplasm and others have the reverse. It has been shown in several cases that the cytoplasmic domain of the receptor is required for the receptor to be sequestered in the open pits, but little more is known.

The fates of both the receptor and ligand vary after internalization. There are four general categories of ligand–receptor systems (see Table 9.2 and Figure 9.7): *Group I*: Receptor recycles to cell surface but the ligand is directed to the lysosomes. Examples are the LDL and asialoglycoprotein receptors.

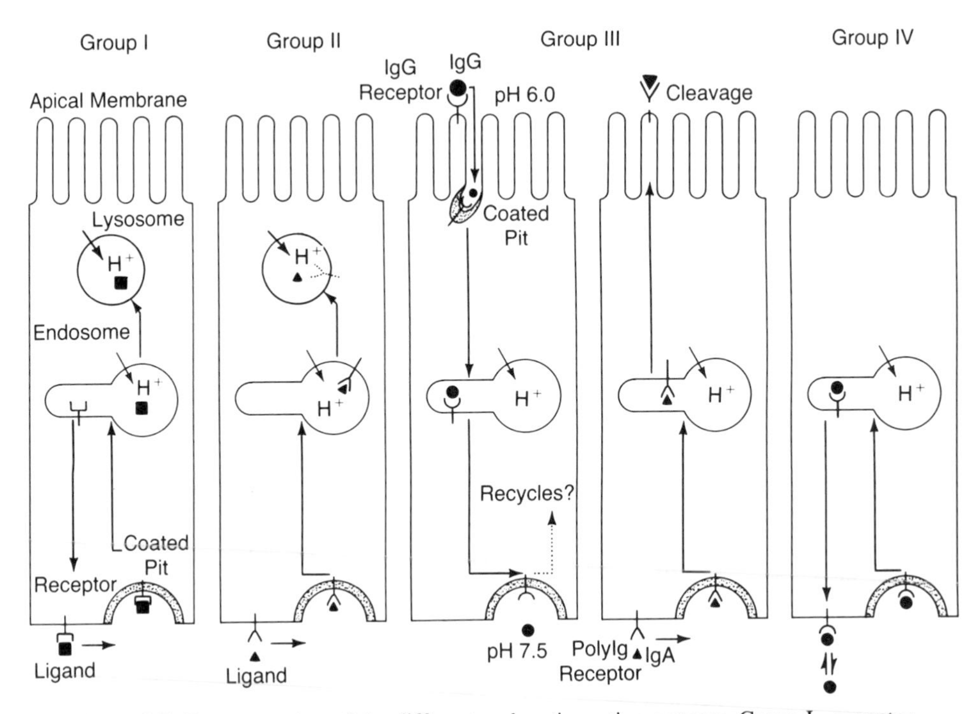

Figure 9.7. Representation of the different endocytic sorting patterns. Group I—receptor recycled, ligand destroyed (e.g., LDL receptor); group II—both receptor and ligand destroyed in the lysosome (e.g., EGF receptor); group III—transcytosis, exemplified by the IgA/IgM receptor and by the IgG receptor, representing opposite directions of ligand transfer; group IV—receptor returns a modified ligand back to the cell surface (e.g., ferritin, apoferritin). Adapted from ref. 1024 ©1985 by Cell Press.

Group II: Receptor and ligand are both directed to lysosomes. Examples are the EGF and insulin receptors.
Group III: Receptor remains bound to the ligand, while the ligand is directed to another plasma membrane domain. This is called *transcytosis* (see 1024) and is observed in polarized epithelial cells with immunoglobulin receptors. The receptor for IgA and IgM is proteolytically cleaved after cell transit, whereas the IgG receptor, under some circumstances, may be recycled.
Group IV: Receptor and ligand are both recycled to the same plasma membrane domain. The best characterized example is the transferrin system, in which a transferrin–iron complex is internalized and, after removal of the iron, the apotransferrin–receptor complex is returned to the cell surface.

The rates at which these events proceed are rather rapid (see 1447). Typically, for receptors recycled to the cell surface, the receptor lifetime on the cell surface is only about 3-6 min, and a full round trip takes about 15-25 min. In the steady state, about 65%-75% of the receptors are found on the cell surface and the remainder are on various intracellular membranes. A complete analysis of the steady-state kinetics of internalization and processing can be made in terms of rate constants for each of the elementary steps (see 1045).

One feature of the sorting mechanism is that it can be remarkably accurate. For example, the transferrin receptor is present almost exclusively on the basolateral membrane and not on the apical membrane of polarized epithelial cells. The error rate of insertion of the transferrin receptor into the apical membrane is only about 0.1%, low enough to maintain the membrane polarity of these cells (471). Error levels as high as 4% have been measured in other systems, which would require an additional sorting step to correct the localization of these receptors (see 471).

All of this requires a sophisticated apparatus for sorting. The various receptor–ligand complexes appear to be taken up together in the same coated vesicle compartments, and the sorting and segregation of the specific receptors and ligands occurs within a few minutes (1395). The receptor for asialoglycoproteins, for example, is observed to segregate laterally within the endosomal system and become highly concentrated in tubules and small vesicle components, but depleted in the large vesicles (511). Presumably, the large endosomal vesicles depleted of this receptor, but containing the glycoprotein ligand, proceed to fuse with the lysosomes. However, the tubules, containing the receptor, return the receptor to the cell surface (Figure 9.7). The sorting clearly occurs within the endosomal (or prelysosomal) organelle complex.

By what mechanism does the rapid sorting of macromolecules occur? Certainly, there must be specific structural "signals" resulting in the compartmentalization of specific receptors and ligands. The only known signal is mannose-6-phosphate, which is present on enzymes which are to be targeted to the lysosome (see 778). Mannose-6-phosphate-specific receptors are present in the Golgi, where they direct newly synthesized proteins to the lysosome (exocytic pathway, Section 10.2), and are also present on the plasma membrane, where they can scavenge such enzymes in the extracellular medium (endocytic pathway).

In addition to specific structural sorting codes, the acidification of the intracellular vesicular compartments is critical to sorting (see 956). The pH in endocytic vesicles is 5.0-6.2 (see 746). It is the low pH which results in the dissociation of the receptor–ligand complexes, including low density lipoproteins, lysosomal enzymes, asialoglycoproteins, EGF, and insulin. The low pH also results in the dissociation of the two iron atoms from the transferrin–receptor complex, though in this case the apotransferrin remains bound to its receptor.

Several opportunistic viruses and toxins have evolved mechanisms for cell entry which utilize this vesicle acidification (see 966). For example, diphtheria toxin binds to an unidentified receptor on the cell surface and is internalized. It is the low pH encountered within the endocytic vesicles which induces a conformational change in the toxin resulting in the penetration of the vesicle membrane and allowing the enzymatically active portion of the toxin to gain access to the cytoplasm. This pH-induced conformational change may result in the formation of a transmembrane pore, as discussed in Section 8.61. Similarly, many enveloped animal viruses gain entry to the cytoplasm after internalization by the endocytic pathway. The low pH encountered in the endocytic vesicles induces a conformational change in the spike glycoproteins in the viral membranes, and this facilitates both the fusion of the viral membrane with that of the endocytic vesicle and delivery of the viral genome into the cytoplasm. Examples are the Semliki Forest virus and the influenza virus. In Section 9.5, aspects of membrane fusion will be discussed.

The pH within intracellular compartments can be raised by the addition of weak bases such as primaquin, chloroquin, methylamine, and ammonium chloride, or by ionophores such as monensin (see 956). The weak bases can cross the bilayer in an uncharged form and become protonated within the acidic compartment. This can raise the internal pH by 1-2 pH units, which effectively blocks many pH-dependent processes. These weak bases are often referred to as "lysosomotropic" or "acidotropic" agents. Monensin will equilibrate the H^+ and K^+ gradients (see Section 7.52), also increasing the vacuolar pH.

The addition of these agents often prevents the recycling of cell surface receptors. The receptor–ligand complexes will not dissociate, and the result is that the receptors accumulate in internal membrane vesicles and are depleted from the cell surface. This is true for the LDL and asialoglycoprotein receptors, for example. The membrane cycling per se, however, appears to continue unabated between the peripheral endosomes and the plasma membrane (1287, 863). Movement into lysosomes, however, may be prevented (863).

Much of the progress in deciphering this complex sorting system has been based on electron microscopy, coupled to the clever use of "tracers," such as horseradish peroxidase, and, more recently, pH-monitoring probes (1100). Genetics has also been useful, taking advantage of the resistance to certain toxins and viruses that results when the endocytic vesicles are not properly acidified (see 956). In this regard, yeast may be a source of useful mutants in future studies (1218). Biochemical studies rely on the isolation and characterization of well-defined subpopulations of endocytic vesicles. Methods have been developed for separating endocytic and exocytic vesicles based on electrophoretic differences

(834, 927), immunoaffinity (555), as well as a cholinesterase-mediated density shift procedure (617, 834). Several laboratories have achieved the fusion of isolated endocytic vesicles in vitro and demonstrated that this requires ATP and occurs only with endocytic vesicles prepared within 5-15 min after internalization (555, 285, 130).

9.43 Clathrin (see 1130 for review)

The structures first encountered in endocytosis are the coated pits and coated vesicles. These are distinguished by the polygonal lattice formed by clathrin that decorates the cytoplasmic surface of these membranes. The properties of clathrin and associated proteins are important because these proteins determine which receptors are sequestered in the coated pits and are somehow involved in the shape change resulting in the pinching off of coated vesicles. The clathrin molecule contains three heavy chains (180,000 Da) and three light chains (30,000-40,000 Da) and has a pinwheel or triskelion shape with three arms, each containing one heavy chain and one light chain subunit. Isolated clathrin will spontaneously self-associate to form polygonal cages, either in the presence or absence of membranes. Reconstitution studies with purified plasma membranes have demonstrated that there is a limited number of high-affinity binding sites for clathrin where coated pits will form (1014). Additional accessory proteins (100,000 Da, 50,000 Da) are also associated with the clathrin lattice.

There are two major classes of clathrin light chains produced by each cell. These are encoded by two different genes, each of which undergoes differential RNA splicing, yielding a total of at least four protein variants (685, 758). Whereas the heavy chains have a clear structural role, the light chains appear to have a regulatory role and contain binding sites for calmodulin and also for the ATP-dependent uncoating protein (1253). In the cell, as much as half of the clathrin is in a soluble pool, in dynamic equilibrium with the membrane cage structures. The removal of clathrin from the coated vesicles is necessary prior to fusion with the endocytic system, and this reaction is catalyzed by the uncoating protein. Although the uncoating protein functions catalytically, in the absence of cages, it remains firmly bound to clathrin triskelions. In the cell, the uncoating protein is present in molar excess to clathrin.

The manner in which clathrin cycles between the monomeric form and membrane-bound cages and the manner in which the cages form on the membrane are obviously critical to understanding receptor-mediated endocytosis and its regulation.

9.44 Some Examples of Receptors Which Become Internalized

Several examples of receptors are briefly discussed. All have a single transmembrane domain, but no other obvious similarities. Examples from groups I, III, and IV are included. The peptide hormone receptors (group II) are described further in Section 9.7.

The Human LDL Receptor (see 152, 1404)

This is the prime example of a group I receptor which recycles while its ligand, serum low density lipoprotein (LDL), is transported to the lysosome. This receptor is especially abundant in liver and in the adrenal gland, where the LDL delivers cholesterol for the purposes of making bile acids and steroid hormones, respectively. The receptor sequence, deduced from the cDNA, suggests five domains, depicted in Figure 9.8. There is no significant homology with other membrane receptors, though the two largest domains of the extracytoplasmic part of the molecule contain different cysteine-rich repeats which are homologous, respectively, to complement factor C9 and to the EGF precursor. There is a single transmembrane spanning domain and a small cytoplasmic carboxyl terminus. Some individuals who suffer from genetic disorders (familial hypercholesterolemia) have been shown to have nonfunctional LDL receptors. Analysis of these

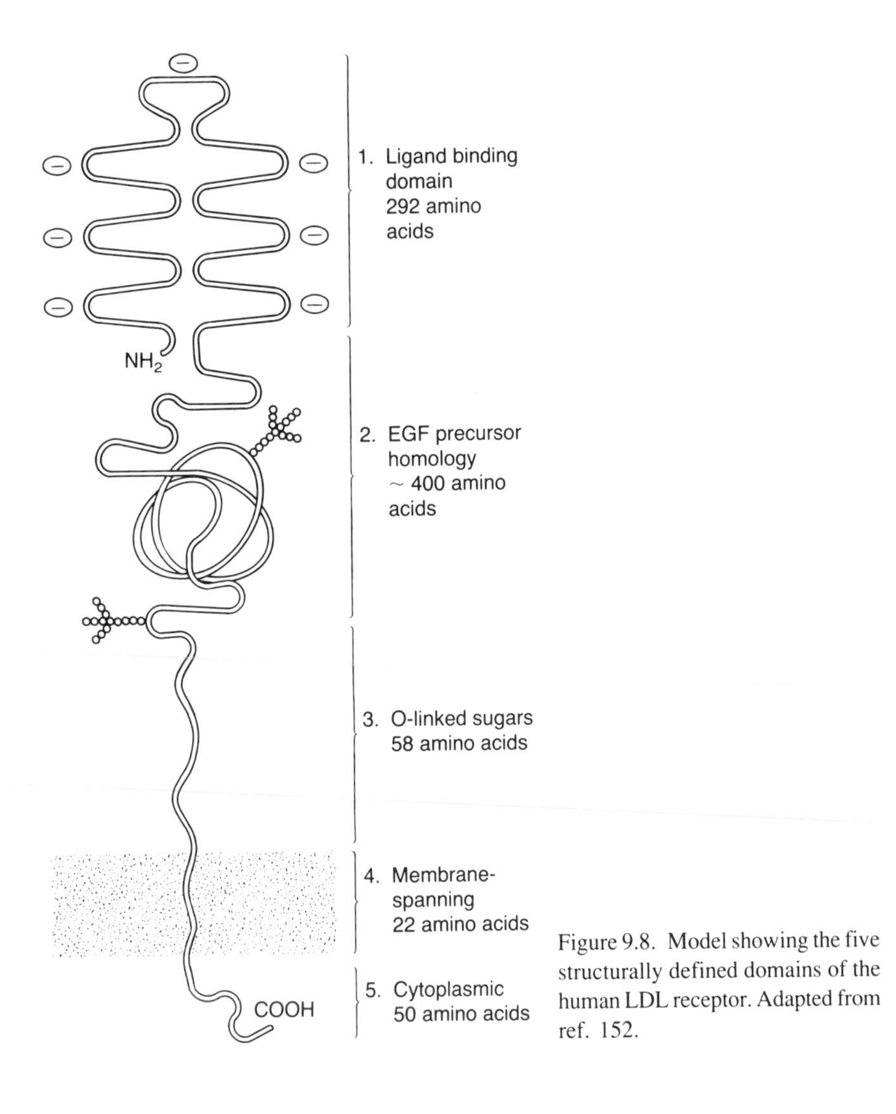

Figure 9.8. Model showing the five structurally defined domains of the human LDL receptor. Adapted from ref. 152.

mutations has been useful in deciphering the functions of the structurally defined domains of the receptor. Mutations within the cytoplasmic domain of the LDL receptor impede binding of the receptor to coated pits, yet do not alter LDL binding to the receptor (833, 289). In addition to the identification of naturally occurring mutations in human subjects, in vitro techniques have been used to generate mutations and test the properties of the expressed gene product. Recombinant DNA methods were used to construct a deletion lacking the domain homologous to the EGF precursor (domain 2). This receptor binds LDL, but no longer releases it upon acidification (288). This results in blocking the recycling of the receptor and leads to its degradation after ligand binding.

Asialoglycoprotein Receptor

This group I receptor, also called hepatic lectin, takes up serum glycoproteins which have been desialated and recognizes glycoproteins with terminal galactose or *N*-acetylgalactose. There are two noteworthy structural features. (1) The receptor has one transmembrane spanning segment and is one of several receptors which have their amino termini facing the cytoplasm (1371, 844; see Figure 10.3). (2) This receptor has been shown to be a hexamer in detergent solution (870), and many exist as such in the membrane. The asialoglycoprotein receptor is homologous to the low-affinity receptor for IgE present on human lymphocytes (747), whose function is unknown.

Polymeric (IgA/IgM) Immunoglobulin Receptor (see 1024, 1023)

This receptor (group III) picks up IgA or IgM from the serum and transports them from the basolateral membrane to the apical surface of hepatic cells (Figure 9.7). During this process, the receptor is cleaved, and a portion is discharged in association with the immunoglobulin molecule into the bile of hepatic cells. Hence, this receptor is used for only one trip and is not recycled. As shown in Figure 9.3, this receptor is a member of the immunoglobulin superfamily and has five extracellular immunoglobulin-like domains. The cytoplasmic domain has been shown to be essential for endocytosis (1023).

Transferrin Receptor

This receptor (group IV) mediates the uptake of iron from serum. The receptor binds to diferric transferrin, is internalized, and remains bound to apotransferrin as the iron dissociates in the acidic endocytic vesicles. Like the LDL receptor, the concentration in the coated pits and internalization proceed in the absence of ligand. The receptor is a disulfide-linked dimer which is glycosylated and is also acylated at at least one cysteine which has been identified by site-directed mutagenesis (710). The subunit is 760 residues in length and has its amino terminus facing the cytoplasm. The transferrin receptor, like those for LDL, IgA (IgM), and EGF, requires the cytoplasmic domain for internalization (1247). This domain contains several serine residues which can be phosphorylated, but these residues have been shown not to be essential for internalization (1247). Nevertheless, it has been demonstrated that phorbol esters, which activate protein kinase C (see Section 9.72), decrease the amount of the transferrin receptor on the cell surface, and concomitantly result in receptor phosphorylation (943, 673). In contrast, insulin and other mitogenic agents cause the redistribution of the receptor to the cell surface, apparently by increasing the rate of receptor externalization (1550, 1437). The mechanisms for this are not known, but the point is that the distribution of this and other receptors between the plasma membrane and internal vesicles is an important regulatory factor.

9.5 Membrane Fusion (see 1594, 881, 1593, 1571 for reviews)

It is evident from the description of membrane flux in the previous section that fusion between selected membranous organelles within the cell is common, and, indeed, an essential cellular process. This membrane fusion must be rapid and highly selective and must also occur without leakage of the vacuolar contents into the cytoplasm. The process must be tightly regulated. The plasma membrane, for example, rapidly fuses with peripheral endocytic vesicles, yet it does not normally fuse with the plasma membranes of the adjacent cells to which it adheres. What is the mechanism of membrane fusion, and how is this regulated? The answers are not yet known, but the minimal requirements for membrane fusion are known from studies with model membranes (1594), and it is also clear that proteins can mediate rapid and selective membrane fusion, as evidenced by studies on the spike proteins of enveloped animal viruses (1571). These studies indicate that membrane fusion can be viewed minimally as a two-step process. First, the two opposing membranes must be brought into close contact. This requires overcoming electrostatic repulsive forces and, most important, requires the dehydration of the polar groups of the lipid molecules. Second, there must be some local packing defect in the bilayers which are in close contact to allow for intermembrane hydrophobic interactions. Each of these steps can be protein mediated, and probably is, in vivo.

9.51 Studies with Lipid Vesicles

Two model systems have been employed: fusion between lipid vesicles and fusion between a planar model membrane and lipid vesicles (see 9). Vesicle fusion can be monitored by light microscopy and electron microscopy (see 718) but is more frequently quantified by directly measuring the coalescing of the internal compartments or by the mixing of the vesicle membrane lipids (see 1594, 362) (Figure 9.9). For example, one population of vesicles can be prepared with dipicolinic acid inside and another population with Tb^{3+} inside. If the vesicles fuse and the internal contents mix, these two reagents rapidly form a Tb^{3+}-dipicolinate complex which is highly fluorescent and, hence, provides a quantitative measure of fusion. Similarly, fluorescence energy transfer between fluorescent lipids (donor and acceptor) is often used to measure mixing of the lipid phases of separate vesicle populations (1343). These assays distinguish between vesicle aggregation and actual fusion (see 362).

A salient feature of unilamellar phospholipid vesicles is that they do not readily fuse spontaneously. Although there is a van der Waals attractive force between

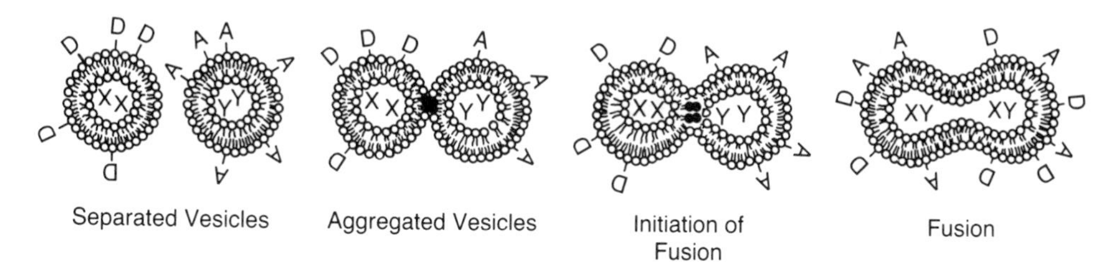

Figure 9.9. Schematic illustrating the fusion of phospholipid vesicles. The lipid phases and internal contents do not mix upon simple vesicle aggregation. The initiation of fusion is depicted as arising from a fluctuating defect in bilayer packing, and no particular structure is implied. Once fusion has occurred, the lipid phases and internal contents mix, providing convenient assays. In many experiments with phospholipid vesicles, fusion is accompanied by leakage of the internal contents.

opposing phospholipid bilayers, there is also a strong repulsive electrostatic interaction, especially if the vesicles contain negatively charged phospholipids. Vesicles which contain phosphatidylserine, for example, will not normally aggregate because there is a substantial free energy barrier preventing the membranes from getting sufficiently close for complex formation. All phospholipids, including zwitterionic lipids such as phosphatidylcholine, bind to water, and this polarized (i.e., oriented) layer of water at the bilayer surface must be eliminated in order for the two apposed bilayers to come into direct contact. Phospholipid dehydration is energetically very costly, and one role of agents which promote membrane fusion is to assist in the dehydration step. Some phospholipids, such as phosphatidylethanolamine, are less hydrated than others, and form vesicles which fuse more readily. For example, vesicles containing a mixture of phosphatidylethanolamine/phosphatidylserine will fuse more readily than vesicles containing phosphatidylcholine/phosphatidylserine (e.g., 718). Vesicles containing phosphatidylethanolamine adhere to each other more strongly than do vesicles containing phosphatidylcholine because of this difference in hydration (718).

Box 9.2 Fusogenic Agents Promote Membrane Fusion

(1) *Calcium* (718, 845): The addition of Ca^{2+} to vesicles containing anionic lipids, such as phosphatidylserine or phosphatidic acid, often leads to vesicle fusion. If the vesicles are purely anionic lipids, the addition of Ca^{2+} results in total vesicle collapse, but when the anionic lipid is mixed (e.g., 1:1) with neutral lipids, such as phosphatidylethanolamine, fusion results. High concentrations (e.g., 10 m*M*) of Ca^{2+} are required for this, and usually Mg^{2+} will not substitute. The Ca^{2+} neutralizes the negative surface potential due to the anionic lipids and facilitates vesicle aggregation. In addition, Ca^{2+} is particularly effective at forming a bridge between anionic phospholipids in the apposing bilayers by virtue of a strong $Ca(phosphatidylserine)_2$ complex (see Section 7.33). This results in dehydration of the lipids, squeezing out the water between the opposed bilayers. Mg^{2+} stabilizes vesicle aggregation, but does not usually stabilize the dehydrated close contact of the opposed bilayers, depending on the lipid content of the vesicles.

The deformation caused by vesicle-vesicle adhesion causes a stress which can be relieved by fusion (718). This can be promoted by lipid bilayer packing defects, which might result from local fluctuations or from phase boundaries resulting from a Ca^{2+}-induced phase separation in the contact region. Generally, there is not a strong correlation between conditions resulting in vesicle fusion and those promoting large-scale phase separations of the lipids (1594). Ca^{2+}-induced fusion is often accompanied by lysis.

Ca^{2+} also promotes the fusion of anionic vesicles to planar membranes (9). However, in this case, there is a need for an osmotic gradient across either the vesicle or planar membrane. This "drives" the fusion, apparently by providing additional mechanical stress. Osmotic swelling has been postulated to be a driving force for membrane fusion in vivo (881), but this is not supported experimentally (1632).

Is Ca^{2+}-induced fusion physiologically relevant? It has frequently been noted that changes in the cytoplasmic Ca^{2+} concentration often precede fusion events in vivo (e.g., the fusion of secretory granules with the plasma membrane). But the Ca^{2+} concentrations required in vitro are far too high to mimic any physiological situation. However, it has been pointed out that if two bilayers are already close to each other, the affinity for Ca^{2+} is very high, and Ca^{2+} bridges between the opposed membrane may be induced by changes in the free Ca^{2+} concentration in a physiologically relevant range (414).

The actual intermediates in bilayer fusion have been the subject of considerable speculation, centering on lipidic particles and nonbilayer forms (see 1594, 1593, and Section 2.2). However, little is known about the infrequently occurring, transient, intermediate structures involved in bilayer fusion.

(2) *Polyethylene glycol and dextran*: These are commonly used fusogenic agents, but they have been less well characterized than Ca^{2+}. It is generally believed that these agents cause vesicle dehydration, resulting in both aggregation and actual fusion (897).

(3) *Electrofusion* (see 1405): Phospholipid vesicles as well as cells can be induced to fuse by the application of short electric impulses. The high fields cause pore formation (see Section 8.64) and can also induce tight contact between opposing bilayers. Long-lived membrane packing defects may also result from the imposition of the large electric field, and these defects can presumably facilitate hydrophobic contacts between closely opposed bilayers.

(4) *Proteins and peptides*: A number of soluble proteins (750) and amphipathic peptides (1016, 1040) have been shown to promote vesicle fusion. In many cases, this is pH-dependent and occurs only upon protonation of groups on the protein (pH < 6.0). For example, α-lactalbumin under acidic conditions appears to undergo a conformational change, presumably resulting in exposure of a hydrophobic loop, which facilitates binding to phospholipid vesicles (750). This interaction somehow promotes vesicle fusion. Melittin (1040, see Section 3.71) and δ-haemolysin (1016) from *S. aureus* are two similar amphipathic peptides which promote the fusion of phosphatidylcholine vesicles. Each of these has a hydrophobic portion which can clearly interact with membranes. These interactions in some way overcome the electrostatic and hydration barriers to vesicle aggregation and fusion, presumably by local disruptions of the bilayer. It has been suggested that hydrophobic peptides might be generated in vivo to promote fusion when and where needed (879). However, there is no strong evidence for this, and it is known that portions of proteins can cause the same effects without being cleaved to form peptides.

9.52 Studies with Viral Spike Proteins (see 1571, 1585)

Certainly, the fusogenic activity of proteins such as α-lactalbumin, described above, is an artifact unrelated to its physiological role. However, one set of proteins has been characterized which are designed specifically to promote membrane fusion. These are the spike glycoproteins of enveloped animal viruses. These viruses contain lipid bilayer shells with several virally encoded proteins (see Section 4.53), and they enter cells by fusing with the cell membrane (see Figure 9.10). Some viruses, such as the Sendai virus, fuse with the plasma membrane. Other viruses, such as the influenza virus, bind to receptors in the plasma membrane (sialoglycoproteins) and are taken up by receptor-mediated endocytosis. These viruses fuse with the membrane of the endocytic vesicle only after the vesicle interior is acidified (pH 5–6).

The spike proteins serve two functions: (1) they attach the virus to the animal cell membrane, usually to a glycoprotein or glycolipid, and (2) they presumably interact directly with the bilayer of the target membrane in a way that brings the membranes in close contact and promotes fusion. In some viruses, the attachment and fusogenic activities are carried out by separate proteins, but in other cases a single protein carries out both functions. For example, the G protein of vesicular stomatitis virus and the hemagglutinin (HA) protein of influenza virus mediate both functions. Each of these proteins is a homotrimer, containing three identical subunits. These proteins have each been purified and reconstituted into liposomes which manifest both attachment and fusogenic activities (e.g., 977). The fusogenic activity in both these cases is observed only under mildly acidic conditions, mimicking the interior pH of endocytic vesicles. Fusogenic activity appears to reside in small segments of these proteins located near their amino termini. A

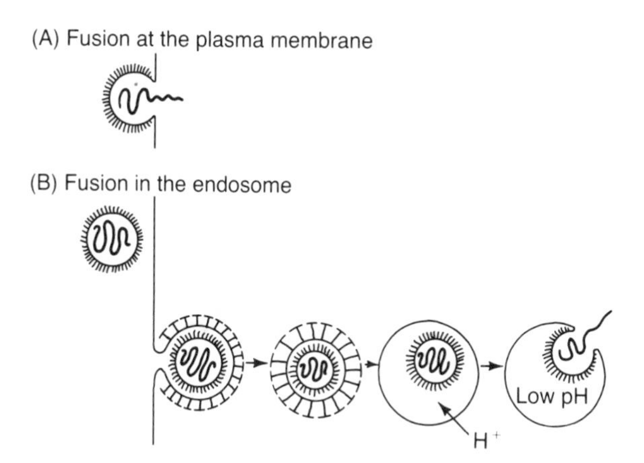

Figure 9.10. Pathways for the entry of enveloped viruses into the cell. Part A shows the pathway used by paramyxoviruses (e.g., Sendai virus) and occurs at neutral pH. Part B shows the pathway used by many other viruses, including the influenza virus. The virus is endocytosed through coated pits into endocytic vesicles. Upon vacuolar acidification, fusion of the viral membrane and endocytic membrane is induced. Adapted from ref. 1571.

25-amino-acid synthetic peptide corresponding to the amino terminus of the G protein of vesicular stomatitus virus has pH-dependent hemolytic activity similar to that of the virus (1295). The relevance of this observation to the fusogenic activity of the G protein, however, has been questioned (1604).

Box 9.3 The Best Characterized Spike Protein is the Hemagglutinin Protein of Influenza Virus

This protein is bound to the viral membrane by a short transmembrane domain at the carboxyl terminus. It is synthesized as a single polypeptide but is proteolytically cleaved during maturation to yield two halves, HA_1 and HA_2, linked by a disulfide bond. The fusogenic portion of the protein is located at the new amino terminus of HA_2, formed as a result of this proteolytic reaction. This corresponds to the amino terminus of the G protein of vesicular stomatitis virus.

The three-dimensional structure of the water-soluble domain of the hemagglutinin spike protein has been determined by X-ray crystallography (1595) (Figure 9.11). This was obtained as a bromelain cleavage product. The protein is a trimer of HA_1-HA_2 which has a rod shape, extending 135 Å above the surface of the bilayer. Each subunit has an α-helical "stem" with a globular "top" which contains the receptor binding site for sialoglycoconjugates in the target membrane.

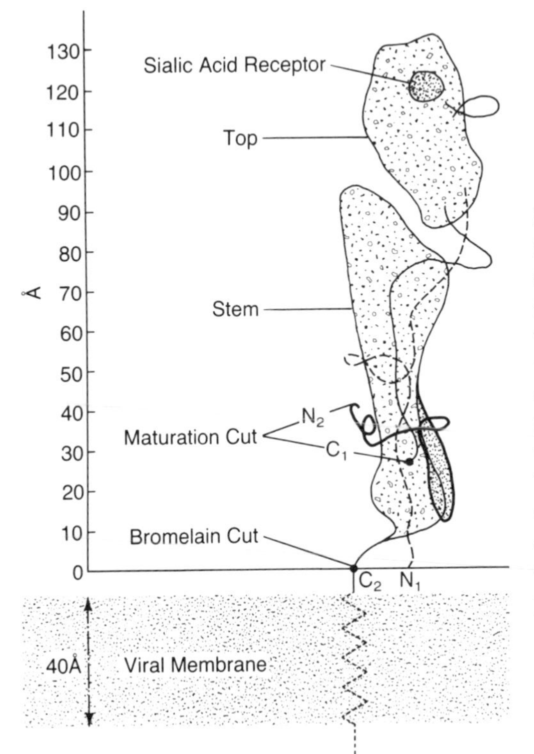

Figure 9.11. Schematic showing the general features of the monomeric unit of the hemagglutinin protein fragment released from the viral membrane by bromelain, which leaves 46 amino acid residues behind in the membrane. Note that in the mature form, the amino terminus of HA_2 (N_2) and the carboxyl terminus of HA_1 (C_1) are separated by 21 Å. Proteolytic cleavage between these residues is essential to maturation. The "fusion peptide" is located at the amino terminus of HA_2. The protein is actually a trimer of three of these subunits. The detailed structure is known from X-ray crystallography studies. Adapted from ref. 1259.

The hydrophobic fusion peptide is buried in the interface between subunits of the trimer, located about 30 Å above the surface of the virus membrane. At low pH, the protein is known to undergo an irreversible conformational change in the tertiary and/or quaternary structure (1259). At pH 5.0, the protein acquires an ability to bind lipids and detergents and will self-aggregate, suggesting exposure of a hydrophobic domain. This presumably correlates with the exposure of the fusion peptide, which becomes accessible to binding to the target membrane, bringing the two membranes close together and facilitating fusion. Mutants have been isolated (280) and produced in vitro (510) which have altered pH optima and fusogenic properties. These confirm the importance of subunit contacts in the pH-dependent conformational change (280) and of the fusion peptide (510). Additionally, the hemagglutinin protein is fatty acylated, and these bound fatty acids appear to be important for fusogenic activity (817).

In summary, although the physical chemical mechanisms are far from clear, it is evident that the specific membranes can be fused selectively within the cell by a protein-mediated process. This requires specific attachment interactions and a membrane protein which becomes fusogenic when required. The viral spike proteins provide a reasonable guide to the likely properties of such proteins, though other models are also possible. Possibly, a family of such proteins direct the intracellular membrane fusions in the exocytic and endocytic pathways.

9.6 Bacterial Signal Response Systems Demonstrate Some Features Found in Higher Organisms

In the first half of this chapter, we have discussed receptor systems specifically designed for adhesion and for the uptake of macromolecules and have pointed out the dynamics of the plasma membrane as a component of the endocytic pathway. In the second half of this chapter we will direct our attention to the question of how cells respond to external chemical stimuli, such as metabolites, hormones, or neurotransmitters. These responses are all mediated by signal transduction systems, in that they convert an extracellular event, the binding of a ligand to a receptor, to a complex intracellular response. In this section we will discuss bacterial systems, and in the next section animal cell signal transduction will be summarized.

Bacteria respond to changes in the concentrations of various solutes in their environment. Free-swimming bacteria such as *E. coli*, for example, exhibit chemotaxis and will respond when they sense an increase in the concentration of particular nutrients in such a way that the cells will swim "up" a concentration gradient toward the nutrient source. This is mediated by receptors in the cytoplasmic membrane that bind to the "attractant" solute and induce a series of events in the cytoplasm affecting flagellar motion. Similarly, *E. coli* will respond to a phosphate or nitrogen limitation by synthesizing proteins that enable the cells to scavenge these compounds from the environment. This response is also mediated by specific receptors in the cytoplasmic membrane.

These sensory response systems have been the subject of considerable study, primarily using genetics and molecular biology. Amino acid sequence homologies have identified two families of prokaryotic receptor proteins:

1. Receptors involved in chemotaxis (*E. coli*, *S. typhimurium*), which influence flagellar rotation.
2. Receptors involved in sensory responses that influence the transcriptional apparatus and turn genes on.

The patterns that have emerged are striking in their similarities to those observed in higher organisms. A brief description will introduce themes which will recur when we discuss more complex systems from animal cells in Section 9.7.

9.61 Chemotaxis Receptors of *E. coli*

There are four receptors in this protein family, also known as transducers or methyl-accepting chemotaxis proteins (MCPs). They are often referred to as the gene products from the four genes *tsr*, *tar*, *tap*, and *trg*. The *tsr* protein, for example, binds to the attractant serine and is also required for the chemotactic response away from the repellent leucine. The best studied is the aspartate receptor (the *tar* gene product), which binds to the attractant aspartate and also to the maltose-binding protein (see Section 8.43). All four receptors contain single polypeptide chains (~60,000 Da). The deduced amino acid sequences suggest that these are transmembrane proteins, each with two membrane-spanning α-helices. The folding pattern of the aspartate receptor shown in Figure 9.12 is consistent with gene fusion experiments (918), as well as genetic (1264) and biochemical data (454). Biochemical characterization of the aspartate receptor (454) suggests that it may be a tetramer within the cytoplasmic membrane. All four chemotaxis receptor proteins have carboxyl-terminal halves, corresponding

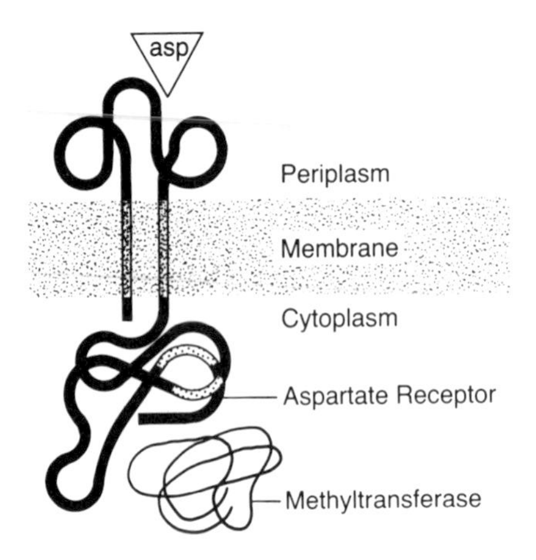

Figure 9.12. Model of the aspartate chemotactic receptor of *E. coli*. The location of the two hydrophobic segments is shown, as are the regions where the protein becomes covalently modified by methylation and deamination (stippling). The methyltransferase is also schematically pictured. The actual folding pattern of the polypeptide chains is not known. Adapted from ref. 454.

to the cytoplasmic domains, which are highly conserved, whereas the amino-terminal halves, the periplasmic domains, show much less sequence similarity.

The aspartate receptor has three functions that can be shown to be separable.

1. *Aspartate binding*, which occurs in the amino-terminal domain, facing the periplasm. Truncated proteins lacking portions of the cytoplasmic domain still bind normally to aspartate (918).
2. *Signal transduction*, which certainly involves residues in the transmembrane helices (1095), as shown by the deleterious effect of the substitution of a lysine for an alanine at position 19, in the first transmembrane helix (see Figure 9.12). This mutant binds aspartate but elicits no response, presumably due to an alteration in protein conformation. It has also been shown that the receptor can bind to aspartate and the maltose binding protein simultaneously and that the responses are, remarkably, additive (1026). The nature of the conformational change in the receptor induced by ligand binding is not known but it does appear to extend to a large portion of the polypeptide (408). Nor is the mechanism known by which the receptor influences the flagellar motor, though it does involve the carboxyl-terminal domain of the receptor (1353) and may be mediated by the phosphorylation of one of the other protein components of the system (626).
3. *Adaptation*, which refers to the fact that the response of the system to an increase in aspartate concentration is only transient, and within a few minutes the receptor becomes "desensitized" or adapts to the new level of attractant. The receptor "resets" itself to respond again to further changes in concentration of the attractant. This function is, in part, related to methylation and demethylation of the receptor at several glutamyl residues located in the cytoplasmic domain (1264, but see 1393). Adaptation is selectively eliminated in a truncated form of the receptor lacking a portion of the carboxyl terminus (1264), resulting in a mutant in which the response to aspartate is continuous as long as aspartate is present.

Numerous experiments have demonstrated the quasi-independence of the functional domains located in the amino-terminal and carboxyl-terminal domains (e.g., 1264, 788). In fact, chimeric gene products have been examined in which the amino-terminal portion of the *tsr* protein (serine receptor) is fused to the carboxyl terminal portion of the aspartate receptor (788). These are chemotactically active to serine. This kind of modular construction of signal-response receptors is also observed in the family of peptide hormone receptors in animal cells (see Section 9.7).

In summary, the main points to note are

1. A family of sequence-related transmembrane receptors.
2. Separable functions, suggesting a receptor structure composed of distinct structural/functional domains.
3. Signal transduction involving other proteins in the cytoplasm, presumably modified in some way due to a conformational change in the receptor upon ligand binding.

4. Adaptation, resulting in a transient response. This is common in many sensory response systems and is also known as desensitization (see Section 9.74).

9.62 Receptors Coupled to Transcriptional Activation (see 1241 for review)

A second receptor family has been identified in a number of bacteria, in which the cellular response is apparently transcriptional activation of particular genes. In all cases, there are two protein components: (1) a *sensor* or *receptor* and (2) a *regulator*. The receptors all appear to have a similar structure, with two putative transmembrane helices in the amino-terminal half of the molecule. The carboxyl termini show substantial sequence homology within this family of proteins and are presumed, by analogy to the chemotactic receptors, to be on the cytoplasmic side of the membrane (Figure 9.12). The regulator proteins appear to be a soluble species within the cytoplasm. Presumably, the receptors modify the regulators, which then, either directly or indirectly, act as transcriptional activators. Possibly, receptors activate the regulator by phosphorylation, as suggested by some evidence for the nitrogen limitation response system. In Section 9.74, the critical roles of protein phosphorylation in animal cell signal transduction will be discussed.

The bacterial receptor families illustrate how basic structural motifs can become elaborated to accommodate a variety of ligands and a variety of responses. This is also exemplified by some of the animal cell receptors (984; Table 9.1). Whereas little is known about the details of the signal response mechanisms in bacterial systems, much more is known in animal cells, as described in the next section.

9.7 Signal Transduction in Animal Cells

Animal cells respond to a wide variety of external signal molecules. The first step is always the binding of the ligand (e.g., neurotransmitter or hormone) to a specific receptor on the external surface of plasma membrane. The manner in which this ligand binding initiates the cytoplasmic cascade of cell-specific and receptor-specific events that follow is a very fertile and rapidly maturing area of research. This is a field where new concepts will be required in the near future to accommodate the rapidly accumulating mass of data. Nevertheless, broad unifying themes are already apparent, even if many of the details are somewhat murky or unknown. The purpose of this section is to introduce these central themes and those membrane-associated biochemical events that are well characterized.

Table 9.1 includes a number of animal cell receptors which are involved in signal transduction, many of which have been cloned and whose primary structures are known. The table has been organized to emphasize structural similarities. Functionally, most of the signal tranducing receptors bind to either mitogenic agents or neurotransmitters. Mitogenic agents include growth factors such as EGF, as well as peptide hormones and regulatory peptides (e.g., insulin). These agents regulate cell growth in a variety of circumstances, such as during embryogenesis, cell maturation, or the cell proliferation that is part of the immune response. Neurotransmitters include epinephrine, norepinephrine, acetylcholine, glycine, and a number of other small molecules used to elicit a cellular response from target cells.

A few additional words on receptor classifications (see Section 8.24) will be useful at this point, especially as regards neurotransmitters. These receptors are named and classified according to their responsiveness to various agonists and antagonists (pharmacological properties), physiological functions, and anatomical locations. In many cases, there are multiple receptors which respond to the same agonist, and the pharmacological and anatomical classifications may not be adequate to define the different receptor proteins. The adrenergic receptors, for example, which all respond to catecholamine agonists such as norepinephrine and epinephrine were first subdivided into α and ß types and subsequently into α_1, α_2, $ß_1$, and $ß_2$ subtypes. Receptor proteins have been isolated corresponding to each subclassification, though they do not all fit neatly into the pharmacologically defined systematics (see 342). Similarly, there are two classes of pharmacologically defined muscarinic cholinergic (acetylcholine) receptors (M1 and M2) and at least four cloned genes (121). These can be distinguished pharmacologically from the nicotinic acetylcholine receptor (Section 8.24) on the basis of drug responses. There are also multiple receptor classes for histamine, dopamine, opiates, and other agonists. The continued application of biochemical and molecular cloning techniques can be expected to clarify the identity of the receptor proteins and define the coupling mechanisms that are involved in eliciting the cellular responses. This will provide further information for modifying the pharmacologically based classifications of receptors.

9.71 The Primary Response and Receptor Families

Upon ligand binding, there must be a conformational change induced in the receptor resulting in functional alteration elsewhere in the molecule. In no case is the structural nature of the conformational change known. In at least one case, the high-affinity IgE receptor, aggregation of the receptor within the plane of the membrane is induced by ligand binding (see Section 9.75), but there is no indication that this is generally true for all signal-transducing receptors. Although the structural details are not known, the functional consequences of hormone or neurotransmitter binding are becoming clear. So far, three classes of primary events have been associated with agonist binding (see Figure 9.13):

1. The receptor is a channel, and agonist binding opens the channel. Examples are the nicotinic acetylcholine receptor (nAChR), the γ-aminobutyric acid (GABA) receptor, and the glycine receptor. These are all neurotransmitter receptors which are structurally related as a superfamily (1302, 549; Table 9.1). The primary structures of these receptors/ion channels suggest a model with four transmembrane segments for each polypeptide, though the experi-

mental studies on the topology of the nAChR are not conclusive (see Section 8.24).

2. The cytoplasmic domain of the receptor is a tyrosine-specific protein kinase which becomes activated upon ligand binding. The receptor itself is usually one target (autophosphorylation), but the details of what proteins are phosphorylated and how they affect the cell are generally not known. Mitogenic peptide hormone and growth factor receptors utilize this mechanism, and many of these receptors are structurally related (1614) (Figure 9.14; Table 9.1). These receptor polypeptides have one transmembrane segment each.
3. The receptor forms a complex with one of a group of membrane-bound GTP-binding proteins, called "G proteins." Upon ligand binding to the receptor, the conformational change in the receptor–G protein complex results in facilitating an exchange of bound GDP for GTP on the G protein. The sequence of events that follows is depicted in Figure 9.15. The G protein is transiently activated when bound to GTP, and, in this state, it can dissociate from the receptor and one or more subunits of the G protein can bind to and influence other membrane proteins, labeled "targets" in Figure 9.15. These targets include ion channels, adenylate cyclase, cGMP phosphodiesterase, and phospholipase C. A more detailed description is given below in Section 9.72. Several receptors whose effects are mediated through G proteins also belong to a receptor superfamily (342), including the ß-adrenergic receptors, the muscarinic acetylcholine receptors, and the opsins (light receptors). These receptors each have seven transmembrane segments. Figure 4.1 shows the proposed structure of rhodopsin, the best characterized member of this group.

9.72 G Proteins (for reviews see 1400, 515)

The guanine nucleotide-binding regulatory proteins, or G proteins, are responsible for the transduction of a variety of hormone or neurotransmitter signals to an equally diverse set of targets within the cell. Examples are shown in Table 9.3. Four G proteins have been purified to homogeneity and characterized biochemically: G_t (transducin), G_s, G_i, and G_o. These each appear to have unique targets or effector proteins (see Figure 9.15). G_t activates the cGMP-specific phosphodiesterase in outer segments of retinal rod cells; G_s and G_i, respectively, stimulate and inhibit adenylate cyclase and are present in nearly all cells; G_o is present in large quantity in brain cells and has been shown to inhibit the voltage-sensitive Ca^{2+} channel in neurons (see 625). In addition, other G proteins are strongly suggested to exist, though they have not yet been isolated. A G protein, G_p (see 228), is presumed to have as its target the phosphatidylinositol-specific phospholipase C, which initiates rapid turnover of phosphatidylinositol in the plasma membrane and the generation of several second messengers (see Section 9.73). Note that there is also a G protein named G_p which has been isolated from human placental membranes but whose function is not known (401). Another G protein, G_k, is presumed to open the K^+-specific channels in cardiac muscle and other cells (see 1283); G proteins (G_e?) have also been implicated in the control of exocytosis (167). In some cases, a single G protein within a cell can respond to ligand binding

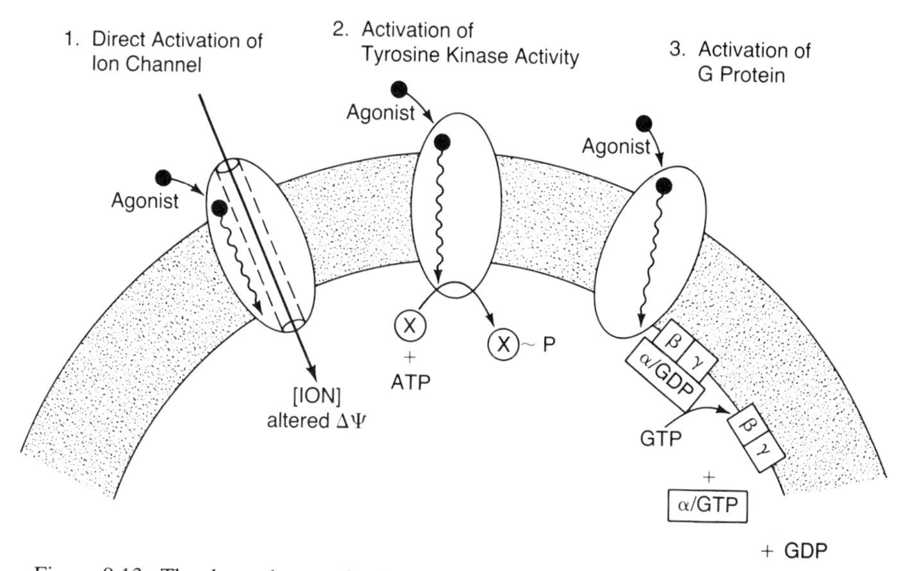

Figure 9.13. The three characterized primary events that are induced by the binding of an external agonist to its receptor in the plasma membrane. The wavy arrow is meant to depict a conformational change induced upon agonist binding. The substrates of the tyrosine-specific protein kinase activities are primarily proteins.The G protein is represented as a three-subunit αßγ heterotrimer. See text for details.

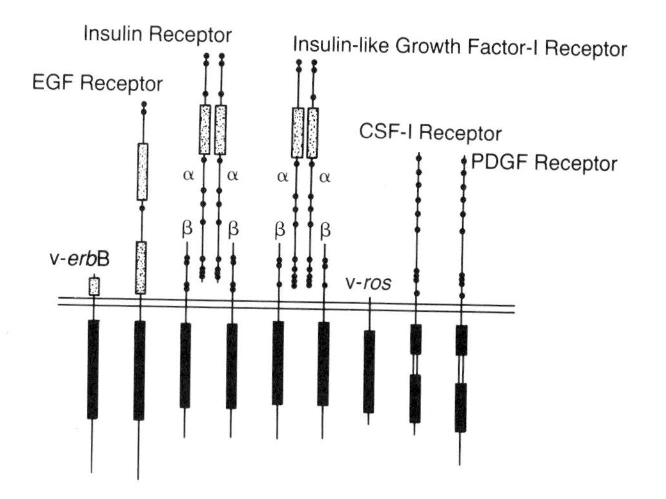

Figure 9.14. Schematic view of structurally similar receptors for mitogenic peptides and some related viral oncogenes. Note that the receptors for insulin and the insulin-like growth factor-1 contain two subunits, whereas others have only a single subunit. Some receptors have cysteine-rich domains, indicated by the stippled boxes. Filled circles represent other cysteines located in the extracellular domains. The tyrosine kinase domains (shaded boxes) are divided in some cases and continuous in others. The products of the two oncogenes v-*erb*B and v-*ros* are related, respectively, to the EGF receptor and the ß-subunit of the insulin receptor. CSF-1 is the colony stimulating factor-1, PDGF is the platelet-derived growth factor, and EGF is the epidermal growth factor. Adapted from ref. 1476.

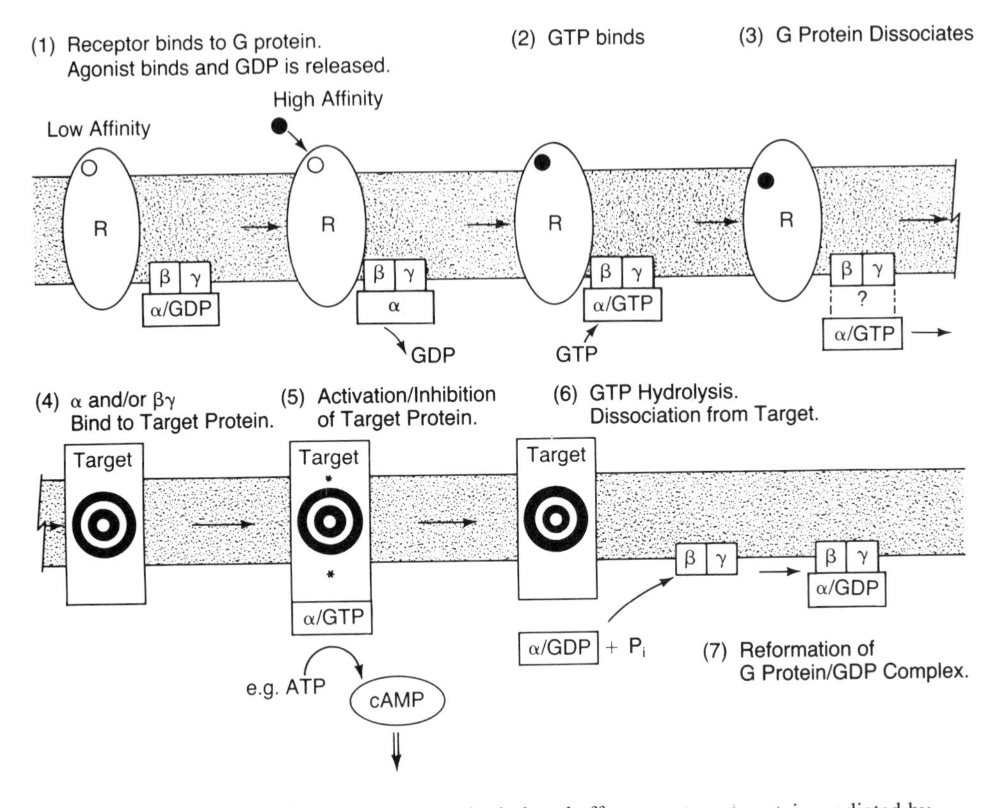

Figure 9.15. Representation of the agonist-induced effect on a target protein mediated by a G protein. An example would be the ß-adrenergic receptor (R) coupled to adenylate cyclase (Target) by a specific G protein (G_s), which is shown as a three-subunit species (αßγ). (• represents agonist.) See text for details.

to any one of several different receptors, which is strongly suggested for G_k, for example, in *Aplysia* ganglion cells (1283). In addition, G proteins can have more than one target. One example is G_t (transducin), which activates both the cGMP phosphodiesterase and phospholipase A2 activity in bovine rod outer segments (704). These activations appear to be accomplished by different subunits of G_t.

The four G proteins which have been isolated are all αßγ heterotrimers. Biochemical and cDNA sequence analysis indicates considerable diversity among the subunits (e.g., 3 α_i subunits) (77), at least two distinct ß-subunits (489, 444), and evidence for diversity among the γ subunits as well (see 444). The α subunits bind GDP and have GTPase activity when dissociated from the ßγ subunit pair. The α subunit also contains a target site for ADP-ribosylation by bacterial exotoxins. G_i, G_o, and G_t (transducin) are modified by pertussis toxin, whereas G_s and G_t are modified by cholera toxin. This covalent modification blocks G protein function and serves as one of the diagnostics for the involvement of G proteins as intermediates in cellular responses. Note that the modifications by cholera and

Table 9.3. Functions and properties of G proteins (see 228, 515, 1400)

	G_s(purified)	G_i(purified)	G_t(purified)	G_o(purified)[2]	G_p(putative)[2]	G_k(putative)[2]
Receptor protein(s)[1]	ß-Adrenergic receptors ($ß_1$, $ß_2$)	α_2-Adrenergic receptor Muscarinic cholinergic receptor A_1 adenosine receptor D_2 dopamine receptor Opiate receptor	Rhodopsin	Opiate receptor (brain)	Muscarinic cholinergic α_1-Adrenergic receptor Vasopressin receptor Angiotensin receptor High-affinity IgE receptor	Muscarinic cholinergic receptor Dopamine receptor Histamine receptor
Target protein(s)	Adenylate cylase	Adenylate cylase	cGMP-specific phosphodiesterase; Phospholipase A2	Voltage-sensitive Ca^{2+} channel	Phosphatidylinositol-specific Phospholipase C	K^+ channel (?)
Function	Stimulate	Inhibit	Stimulate	Close channel	Stimulate	Open channel
Subunit mol. wt.						
α	45,000 and 52,000	41,000	39,000	39,000	Unknown	Unknown
ß	35,000 and 36,000	35,000 and 36,000	36,000	35,000 and 36,000		
γ	8,000	8,000	8,000	8,000		
Toxin suscepti-bility	Cholera	Pertussis	Pertussis and cholera	Pertussis (?)		

[1]There may be more than one receptor for the indicated hormones and neurotransmitters. These lists are not meant to be all-inclusive.

[2]Some of these conclusions are speculative inferences from physiological data. The function of G_o and the existence of G_p and G_k have yet to be firmly demonstrated. The G_p referred to here is not the one isolated from placental membranes (401). It appears likely that G_k is one of the variants of G_i.

pertussis toxins have very different effects. Cholera toxin leads to the permanent activation of G_s in the presence of ATP, whereas pertussis toxin uncouples G_i, G_o, and G_t from the receptor required for activation.

G proteins can be stimulated in the absence of a hormone by the addition into the cytosol of a nonhydrolyzable GTP analogue, GTPγS, which binds to the α subunit. This provides a second criterion for implicating G proteins in a physiological response (see 1400). The addition of fluoride also can stimulate G protein function, apparently by forming a fluoro–aluminate complex with trace amounts of aluminum. This $Al(F)_4^-$ binds to and activates the G protein/GDP complex, apparently as an analogue of the γ phosphate of GTP (97).

G proteins are all firmly bound to the plasma membrane, with the exception of transducin, which can easily be dissociated *in vitro*. In at least two cases, G_i and G_o, it has been demonstrated that the α subunits require the ßγ pair to be firmly anchored to the membrane, as depicted in Figure 9.15 (1389). None of the subunits appears to be a transmembrane protein. However, in at least some cases, the α subunit is myristylated (Table 3.8) and can be tethered to the membrane by the covalently bound fatty acid. Hormone or neurotransmitter occupancy of the receptor (or light activation of rhodopsin) promotes the rapid exchange of GDP, bound on the α subunit, for GTP. The G protein dissociates from the receptor, and eventually α dissociates from ßγ. The G protein or the dissociated α or ßγ subunits diffuse either along the membrane surface or through the cytoplasm to the target(s). The modes of binding of the G protein subunits to the membrane and to their target and receptor partners are uncertain (see 196). However, α and ßγ can have distinct functions once dissociated, as shown by the subunits of transducin, where α_t stimulates the phosphodiesterase and $\beta\gamma_t$ reportedly stimulates phospholipase A_2 (704).

Functionally, the key points about the system are that it provides both amplification as well as limited response duration. The occupied receptor can activate many G proteins. For example, each light-activated rhodopsin molecule can activate about 500 G_t molecules. Each of these activates a molecule of phosphodiesterase, which, during the lifetime of the activated state, hydrolyzes about 1000 molecules of cGMP. Hence, there is a 10^6-fold amplification. In the case of the rhodopsin system, the reduced level of cGMP in the cytoplasm results in closing a cGMP-dependent K^+ channel, which hyperpolarizes the plasma membrane, generating the neural response (43). It takes only a matter of seconds for the bound GTP to be hydrolyzed by the α subunit, requiring a return to the occupied receptor (or light-activated rhodopsin) to be activated once again. Control of the response duration can, thus, be exercised at the level of the receptor, and numerous mechanisms have been elucidated by which such receptors become "desensitized" (see Section 9.74).

The first G protein (transducin) was discovered and characterized in the rhodopsin-linked system, where it is present in relative abundance. It is now recognized that G proteins are a fundamental and versatile mechanism used to transduce information about receptor occupancy on the outside of a cell to an amplified, specifically directed biochemical response in the cytoplasm. One can easily anticipate the characterization of additional G proteins as well as considerable efforts to determine how specificity is determined, especially when a single cell contains several different G proteins.

9.73 Phosphatidylinositol Turnover and Second Messengers
(for reviews see 907, 1124, 908, 930, 80, 94)

Two of the most widespread targets of G proteins are adenylate cyclase (G_s, G_i) and the phospholipase C which is responsible for the hydrolysis of phosphatidylinositol (G_p). Modulation of adenylate cyclase results in altering the intracellular concentration of cyclic AMP (cAMP), which has long been recognized as a "second messenger," influencing a number of intracellular processes. One consequence of increasing cAMP is, for example, the stimulation of the cAMP-dependent protein kinase (protein kinase A), which, in turn, phosphorylates specific protein substrates (e.g., 239). Cells also contain two types of Ca^{2+}-dependent protein kinases, activated, respectively, by Ca^{2+}-calmodulin and by Ca^{2+} in conjunction with diacylglycerol and phosphatidylserine (protein kinase C). The activity of both of these protein kinases is regulated by second messengers generated by the breakdown of phosphatidylinositol, which is initiated in many cells by the G protein-dependent activation of a specific phospholipase C. The discovery and elucidation of the pathway for phosphatidylinositol turnover and of the physiological effects of the breakdown products has been a major advance in our understanding of how hormones and neurotransmitters influence cellular functions.

Phosphatidylinositol (PI) constitutes only about 2%–8% of the phospholipid in eukaryotic cell membranes (908). The configuration of the polar headgroup is shown in Figure 9.16, and this stereo-isomer is referred to as *myo*-inositol, since it was first isolated from muscle digests. A small fraction of the phosphatidylinositol is phosphorylated at the 4 position or at both the 4 and 5 positions. About 1% to 10% of the phosphatidylinositol present in the membrane is phosphatidylinositol(4, 5)bisphosphate, abbreviated as PIP_2 or PI(4, 5)P_2. This compound is likely the initial target for the phosphatidylinositol-specific phospholipase C which is activated in many cells by a G protein. The resulting hydrolysis leads to a rapid turnover of phosphatidylinositol in the plasma membrane and the transient increase in breakdown products. During platelet activation, for example, half of the total pool of phosphatidylinositol is degraded within 90 sec (908). It is the breakdown products that act as second messengers and are linked to numerous cellular processes. The complexities of this system are still being discovered, but the basic scheme is illustrated in Figure 9.16. The sequence of events is as follows.

(1) The initial products of PIP_2 hydrolysis are diacylglycerol and inositol(1, 4, 5) trisphosphate [I(1, 4, 5)P_3]. Diacylglycerol is membrane-bound, whereas I(1, 4, 5)P_3 is a soluble compound. Typically, the fatty acids of the phosphatidylinositol are stearate at position 1 and arachidonate at position 2 of the glycerol. Two different phosphatidylinositol-specific phospholipase Cs have been isolated from bovine brain (1266, 1204), but their activation by a G protein has not been demonstrated in vitro.

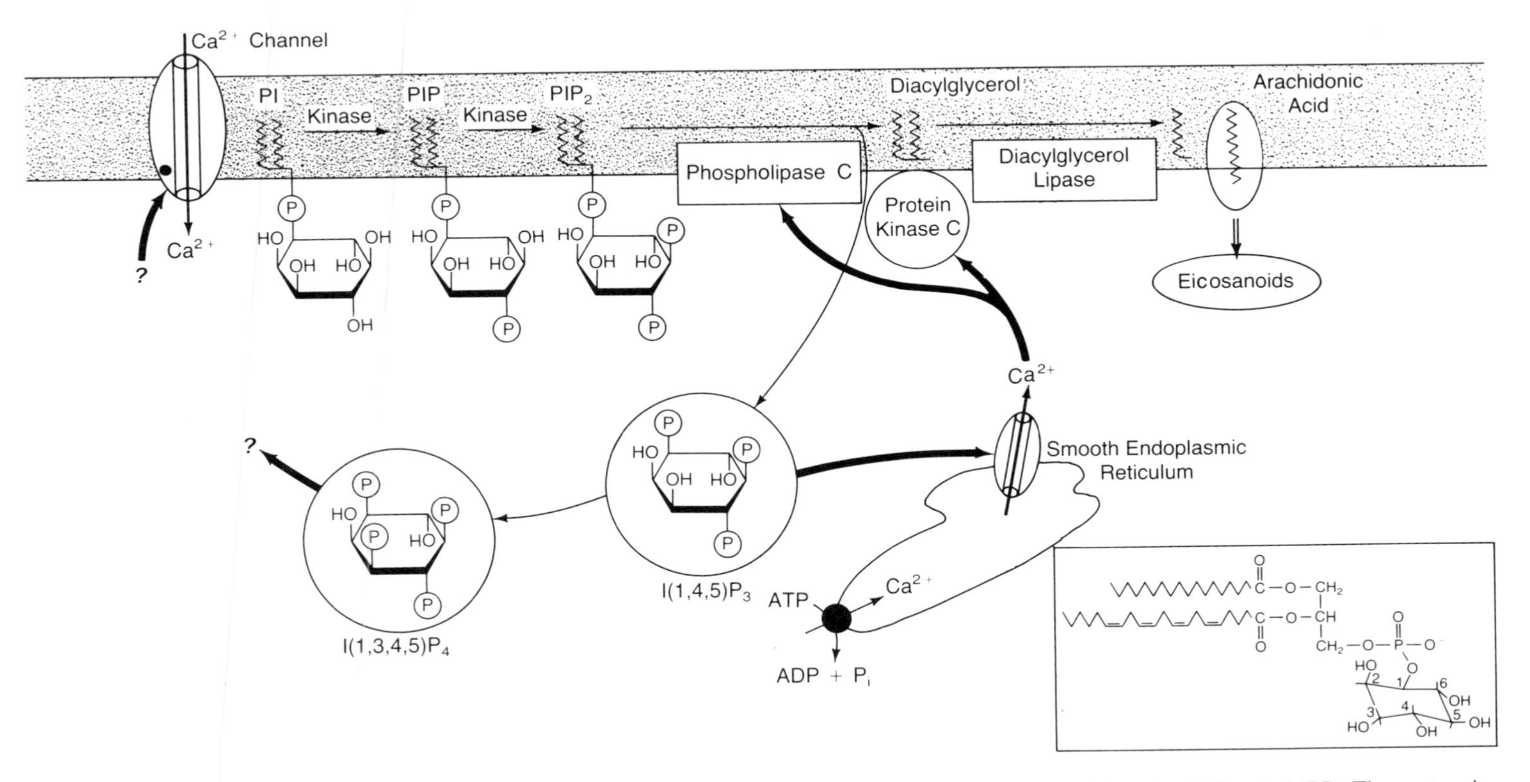

Figure 9.16. Representation of the generation of second messengers by the agonist-induced breakdown of phosphatidylinositol (PI). The system is activated in many cells by a G protein, which stimulates the PI-specific phospholipase C. IP_3 is known to result in the mobilization of Ca^{2+} sequestered in the endoplasmic reticulum, and it is possible that IP_3 and IP_4 act together to allow Ca^{2+} to enter the cell across the plasma membrane. The inset shows the structure of a typical PI species, enriched in stearate (*sn*-1) and arachidonate (*sn*-2). See text for details.

(2) The I(1, 4, 5)P_3 is a second messenger whose major function appears to be to mobilize the Ca^{2+} sequestered in the endoplasmic reticulum. This compound, possibly by direct binding, opens Ca^{2+}-specific channels in the endoplasmic reticulum, which results in increasing the concentration of Ca^{2+} in the cytoplasm several-fold. Typically, free Ca^{2+} is present at a concentration of 0.1 μM in the cytoplasm.

(3) A specific kinase converts some of the I(1, 4, 5)P_3 to the tetraphosphorylated compound I(1, 3, 4, 5)P_4 (see 657). Other phosphoinositides, including cyclic phosphoinositides, are also formed, and some of these may also have physiological functions (see 908).

(4) One of the enzymes influenced by Ca^{2+} is phospholipase C, which prefers PIP_2 as a substrate at low Ca^{2+} but, in vitro at least, utilizes the unphosphorylated phosphatidylinositol at higher Ca^{2+}. Possibly, this facilitates the rapid hydrolysis of the bulk of the phosphatidylinositol directly, though this is not clear (see 908).

(5) The enzyme of central importance that is activated by Ca^{2+} is protein kinase C, which has been discussed in Section 6.71. This enzyme is located predominantly in the cytosol until diacylglycerol and Ca^{2+} are generated. It then binds to the plasma membrane, dependent on the presence of phosphatidylserine, and is activated (see 80). Phorbol esters (see Figure 9.17) mimic the effects of diacylglycerol, and protein kinase C is also referred to as the phorbol ester receptor (see 50). One diagnostic implicating protein kinase C and the phosphoinositide system in a cellular response to an agonist is to duplicate the effects of the agonist by the addition of phorbol esters, which activate protein kinase C directly (see 50).

In the activated state, protein kinase C is a serine- and threonine-specific protein kinase, which phosphorylates specific targets, as well as itself (1063). The basis for its substrate specificity and the physiological consequences of the protein phosphorylation are not clearly known. However, molecular cloning has revealed the existence of a family of protein kinase C species, three, for example, from rabbit brain (see 186). Perhaps these have different specificities, like the family of G proteins.

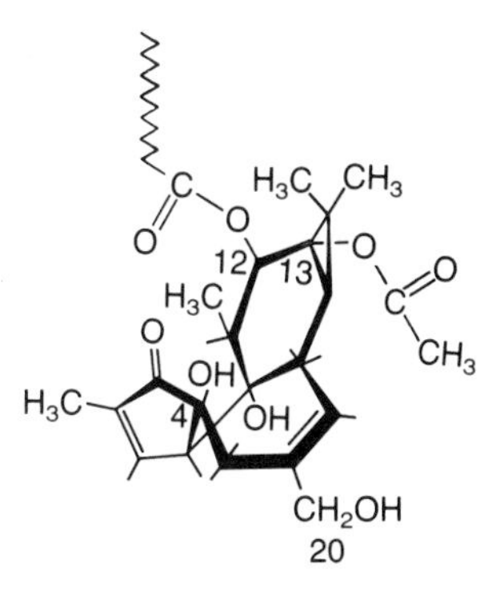

Figure 9.17. Structure of the most potent phorbol diester tumor-promoting agent 12-O-tetradecanoylphorbol-13-acetate (TPA or PMA). This compound acts as a diacylglycerol analogue and binds to and activates protein kinase C. Adapted from ref. 50.

The regulation of protein kinase C is not understood, but it has been demonstrated that gangliosides (786) and lysosphingolipids (587) inhibit the enzyme, and a protein inhibitor has also been isolated (892).

(6) Diacylglycerol can be further degraded by diacylglycerol lipase to yield arachidonic acid, which is oxidized to a number of biologically active metabolites called eicosanoids, which include the prostaglandins. Although arachidonic acid has been implicated itself as a second messenger (1167), it is not clear how significant the phosphatidylinositol-specific phospholipase C-dependent pathway is in its formation. Alternatively, arachidonic acid can be formed from a number of phospholipids by the action of phospholipase A_2. In several cell types, the phospholipase A_2 pathway has been shown to be more important (e.g., 163, 317).

In summary, this complex system can generate at least three known second messengers: diacylglycerol, IP_3, and arachidonic acid. Each of these has specific functions, including the increase in intracellular Ca^{2+} and activation of the Ca^{2+}-dependent protein kinases. The system is very versatile, though the molecular details concerning the manner in which specific cellular functions are influenced have yet to be elucidated.

9.74 Receptor Phosphorylation and Desensitization (see 1340 for review)

The physiological response to the presence of a hormone or neurotransmitter usually is transient, even in the continued presence of the agonist. This phenomenon is called desensitization, and it is observed in numerous eukaryotic as well as prokaryotic response systems. As discussed in Section 9.6, the desensitization of the *E. coli* chemotactic system is due, in part, to the covalent modification of the receptors by methylation. In animal cells, receptor phosphorylation plays a critical role in desensitization.

Ligand-induced desensitization can be considered in two categories. *Homologous desensitization* is when the response to a specific agonist is reduced, but the response to other agonists operating through different receptors is unimpaired. *Heterologous desensitization* is when the application of one agonist reduces the response of the cell to multiple agonists operating through different receptors. Both kinds of mechanisms involve phosphorylation reactions, in large part directed at the receptor proteins. Two kinases which have been implicated in these reactions are protein kinase C and the cAMP-dependent protein kinase, also called protein kinase A. In addition, receptors (e.g., rhodopsin and the ß-adrenergic receptors) can be modified by receptor-specific kinases, and there are also autophosphorylation reactions catalyzed by those receptors which are tyrosine-specific protein kinases.

A major mechanism for heterologous desensitization is by a classical feedback regulation. Agonists that stimulate adenylate cyclase (e.g., prostaglandin E_1 via the ß-adrenergic receptor) activate protein kinase A, which, in turn, phosphorylates the receptor and causes its desensitization. The phosphorylated ß-adrenergic receptor, for example, is less able to activate the G_s protein. Similarly, agonists that stimulate the phosphatidylinositol system activate protein kinase C, which phosphorylates and reduces the response of the receptor. An example of this is the α_1-adrenergic receptor. There is also the possibility of "crosstalk" whereby receptors coupled to the phosphatidylinositol system can be modulated by protein kinase A, stimulated by cAMP. An example of this is the muscarinic acetylcholine receptor (see 166).

A mechanism for homologous desensitization has been described for the ß-adrenergic receptors which involves receptor phosphorylation by a receptor-specific kinase. This kinase will only phosphorylate the receptor–agonist complex, and the consequence is that the phosphorylated receptor is internalized by an endocytic pathway. The endocytic vesicles have a phosphatase which removes the phosphate group(s) and regenerates the active form of the receptor, which is then returned to the plasma membrane. A similar kinase has been characterized for the "light receptor" rhodopsin, where the substrate is the bleached form of rhodopsin.

Phosphorylation at specific sites may be one signal for receptor internalization. For example, phosphorylation of the EGF receptor by protein kinase C results in internalization of the receptor. This kind of internalization is blocked by replacing a specific threonine (no. 654) with an alanine residue by site-directed mutagenesis. However, EGF-stimulated internalization of the receptor is unabated, indicating multiple signals for internalization. As already mentioned, phorbol esters result in phosphorylation and internalization of the transferrin receptor (673, 943).

Internalization is not the only mechanism by which phosphorylation attenuates the receptor response. The activity of the receptor in the plasma membrane can be modulated. For example, the phosphorylated EGF receptor has a lower tyrosine kinase activity as well as a lower affinity for the agonist EGF. The phosphorylated muscarinic acetylcholine receptor is also deactivated in the plasma membrane (166).

The autophosphorylation reactions of the receptors which possess tyrosine kinase activity generally are stimulatory. This has been demonstrated for insulin receptor, where autophosphorylation enhances the ability of the receptor to phosphorylate other proteins and renders the receptor activity insulin-independent. Specific tyrosines located in the cytoplasmic domain of the ß-subunit have been implicated by site-directed mutagenesis in mediating these effects.

9.75 Several Examples of Receptors Involved in Signal Transduction in Animal Cells

The purpose of this section is to point out some of the most salient features of three representative examples of animal cell signal transduction systems. The examples selected are relatively well defined and illustrate interesting features of how animal cells respond to external stimuli. Key features to note are the central, though not fully understood, roles played by protein kinases and by alterations in ion fluxes.

The Epidermal Growth Factor (EGF) Receptor (see 187 for review)

Epidermal growth factor is a potent mitogenic protein which binds to the specific EGF receptor located on the surface of various epithelial, epidermal, and fibroblast cells. The receptor is a single glycosylated polypeptide with a molecular weight of about 170,000 and is one of a family of similar mitogen receptors (187) (Figure 9.14). The amino-terminal half is extracellular and is the EGF-binding domain. There are a single membrane-spanning segment and a large intracellular carboxyl terminus that has protein tyrosine kinase activity. The EGF receptor contains extracellular cysteine-rich domains, as do other receptor proteins, but their function is not known.

The EGF receptor is present on the plasma membrane in high-affinity and low-affinity forms, whose origins are not understood. Possibly, these result from different states of association (monomer/dimer) within the membrane. It has been shown that EGF binding stabilizes a dimeric form of the purified receptor in detergent, suggesting the possibility that receptor aggregation may be important in signal transduction (1614, 120). Biochemically, it is clear that EGF binding induces tyrosine kinase activity, including the autophosphorylation of the receptor at several tyrosines in the cytoplasmic domain. The relationship of the kinase activity to the cellular response is not known (see 1185, 867). In some cell types there is an independent mechanism by which EGF influences the phosphatidylinositol system, possibly by stimulating the enzyme that phosphorylates phosphatidylinositol to form the monophosphorylated derivative (1158).

It has already been pointed out in the previous section that the EGF receptor is phosphorylated by protein kinase C, which desensitizes the receptor and induces its internalization (see 1340). One of the early events associated with EGF stimulation of cells is the activation of a Na^+/H^+-antiporter in the plasma membrane. This exchanges extracellular Na^+ for intracellular H^+ and raises the pH of the cytoplasm slightly (1246). This is an early response with many mitogenic agents (1258), and the altered pH may be enough to trigger other enzyme activities within the cell. The mitogenic response, i.e., the induction of DNA synthesis, is a long-term response, requiring the presence of EGF in the extracellular medium for several hours (1185). The mechanism by which this occurs is not understood.

The EGF receptor is internalized by receptor-mediated endocytosis, as described in Section 9.5. Initially, the receptors are distributed randomly in the plasma membrane, and, upon binding EGF, the receptor–EGF complex is sequestered in coated pits, internalized, and delivered to the lysosome where the complex is degraded (357). This effectively clears the surface of about 80% of the receptors within about 20 minutes. The lateral mobility of the receptor has been measured by the FRAP technique (see Section 5.4) to be 1.5×10^{-10} cm^2/sec, indicating that it is impeded somehow in its lateral motion. Mutational variants lacking most of the cytoplasmic domain still diffuse slowly, suggesting that this domain is not required for the restricted lateral mobility (866). Possibly, interactions with extracellular matrix components are important for this. The cytoplasmic domain is, however, necessary for endocytosis (868,1185), as also shown with other receptors. Deletion of 63 amino acids from the carboxyl terminus results in a receptor which lacks the autophosphorylation sites and which does not have a high affinity for EGF. Yet this receptor still undergoes endocytosis and is mitogenic (868).

Although mutagenesis will be very helpful for elucidating the mechanism by which the receptor functions, an important unknown factor is the state of association of the receptor within the membrane. This is an especially interesting point for those receptors that apparently have a single helical transmembrane segment, because somehow a conformational change must be transmitted from the extracellular agonist binding domain to the intracellular tyrosine kinase domain via this single helix. It is unlikely that EGF binding would alter the helical geometry, and there is no possibility for the helix to "slide" along another helix within the transmembrane domain if the receptor is a monomer within the membrane. Possibly, EGF binding alters the protein–lipid interactions which can push or pull the helix within the bilayer. Alternatively, the state of receptor aggregation might be altered by EGF binding, as suggested by studies with the isolated protein (1614, 120).

The ß-Adrenergic Receptor (see 342 for review)

Many of the important features of the ß-adrenergic receptors have already been discussed. Several of these receptors have been cloned, including those from hamster, human, and turkey, and have been shown to have structural similarity with the family of opsins, including rhodopsin, and with the muscarinic cholinergic receptors (see 342; Table 9.1). Presumably, all these proteins have seven transmembrane helices (341), as has been demonstrated experimentally for rhodopsin (see Figure 4.1). The conserved residues among these receptors are primarily within the hydrophobic regions and not in the hydrophilic loops connecting the transmembrane segments. Several polar residues within the transmembrane segments are among those conserved. Note that these proteins are not homologous to bacteriorhodopsin.

The retinal binding site of rhodopsin is known to be in the hydrophobic core, making a Schiff base with a lysine in helix VII (Figure 4.1). It is tempting to speculate by analogy that the agonist binding site in the ß-adrenergic receptor might also be in the hydrophobic core. Experiments to test this by site-directed mutagenesis of the hamster ß-adrenergic receptor have been performed (1396, 388, 339). Results, indeed, suggest that the hydrophilic loops are not essential for agonist or antagonist binding, and a specific aspartate (no. 113), located in helix III, appears to be required for antagonist binding. Not surprisingly, these studies show that a hydrophilic loop is probably involved in the interaction with the G protein (1397).

The IgE Receptor of Mast Cells and Basophils

Mast cells and basophils contain secretory granules loaded with histamine. Upon appropriate stimulation, these exocytic vesicles fuse with the plasma membrane and release the stored histamine (1632). The triggering event has been shown to be the aggregation of cell surface receptors for IgE, which is induced either by binding to IgE or by binding to other ligands which crosslink the receptors. The

IgE system is of particular interest because of the documented role that receptor aggregation plays in signal transduction (973, 972) and because it is an example of a system linked to the breakdown of phosphatidylinositol initiated by a G protein (65, 1054).

The IgE receptor is an $\alpha\beta\gamma_2$ tetramer which has been purified (1222), and the α-subunit gene has been cloned and sequenced (754). This subunit has sequence homology to the IgG F_c receptor (1202), which is part of the immunoglobulin superfamily (see Table 9.1). In the absence of a crosslinking ligand, the IgE receptors are uniformly distributed (10^3 receptors/cm^2) and show restricted mobility (lateral diffusion coefficient ~3×10^{-10} cm^2/sec). Each receptor can bind one IgE, and the cellular response is only elicited by multimeric forms of IgE which are capable of binding several receptors simultaneously. Oligomers containing only a few IgE molecules cause the rapid immobilization of the receptor clusters, which appears to be a consequence of interactions with cytoskeletal elements and cannot be explained by simple hydrodynamics (973). The result is to trigger the cellular response, which includes production of phosphoinositides and arachidonic acid, Ca^{2+} mobilization, and then histamine secretion (65, 1054, 1177). This response is inhibited by pertussis toxin, suggesting mediation by a G protein.

9.8 Oncogenes and Signal Transduction (for reviews see 1036, 1563, 101, 102, 350)

The receptors and associated signal transduction machinery described in the previous section play essential roles in regulating cell growth and differentiation. Studies on the molecular genetics of cancer have clearly implicated a number of these components in various forms of neoplasia. The best experimental models have been provided by retroviral oncogenes (101). Twenty different retroviral genes, termed oncogenes, have been shown to be responsible for the neoplastic transformation of infected cells. These oncogenes have cellular counterparts that have been identified in a number of cases. For example, the protein encoded by the viral oncogene v-*erb*B is a truncated version of the EGF receptor (see Figure 9.14). Hence, in this case, the viral oncogene is related to the cellular gene encoding the EGF receptor. These cellular genes are called proto-oncogenes. Oncogenes have also been isolated from the DNA of tumor cells by their ability to transform cells in culture to neoplastic growth (102). These oncogenes result from genetic lesions in proto-oncogenes. About half of the proto-oncogenes identified by analogy with viral oncogenes have also emerged from the examination of tumors. There is no simple correlation between the genetic lesion or oncogene and the type of cancer or cellular malfunction. However, there is an emerging pattern identifying proto-oncogenes as encoding proteins that are involved either in the membrane-associated signal transduction apparatus or the targets influenced by second messengers (102, 1563).

Several of the oncogenes which have been identified have tyrosine-specific protein kinase activity and are analogues of membrane receptors for mitogenic peptides and growth hormones. The pathology may result from overexpression of the oncogene, or altered specificity for protein substrates, or altered regulation of its activity. Simple point mutations can convert a proto-oncogene to an oncogene. Not all of the oncogenes with tyrosine kinase activity are transmembrane proteins, but some are peripheral membrane proteins (see 101).

One set of oncogenes, v-*ras*, is homologous to the α subunit of G proteins and may comprise altered versions of a gene encoding the G protein which stimulates the phosphatidylinositol system. It has been demonstrated that the introduction of these oncogenes results in elevated levels of diacylglycerol, although the soluble phosphoinositides were not increased (1603). How this relates to the induction of the transformed state is not known. It should be noted that phorbol esters are tumor-promoting agents, and this function is probably related to the activation of protein kinase C.

The main point is that a derangement of components of the signal transduction apparatus is clearly associated with cancer. This realization not only may be useful in elucidating the mechanisms behind the pathological state but also should be useful for tracing the complex network of interactions in the signal transduction apparatus responsible for normal cell proliferation and differentiation.

9.9 Chapter Summary

Many interactions of a cell with its environment are mediated through cell surface receptors. These receptors range widely in function. Some determine the adhesive properties of cells, either to other cells or to extracellular matrix components. Other receptors are involved in signal–response systems or are involved in the import of macromolecules into the cytoplasm. Amino acid sequence comparisons have revealed that a number of receptors can be grouped within superfamilies of proteins which are structurally related but functionally diverse. A number of receptors involved in mediating cell–cell adhesion, for example, are members of the immunoglobulin superfamily. The integrins constitute another superfamily which includes a number of receptors for extracellular matrix components.

Receptors that bind to an extracellular ligand, such as a hormone or neurotransmitter, and then elicit a cellular response do so in one of several ways. The receptor, in some cases, is itself a protein kinase and, in other cases, may be an ion channel. Binding to the external site on the receptor alters these functions, initiating a cascade of events in the cytoplasm. In still other examples, the receptor binds to and activates a guanine nucleotide binding protein (G protein) in the cytoplasm in response to the extracellular signal. This G protein goes on to alter other cellular processes, such as the breakdown of phophatidylinositol in the cell plasma membrane. Receptors that interact with G proteins form one of several superfamilies of receptors involved in signal–response systems.

An important point to recall is that the animal cell surface is very dynamic and that the plasma membrane participates in an extraordinary phenomenon called membrane recycling. Intracellular vesicles are constantly fusing with the plasma membrane and, conversely, portions of the plasma membrane are pinching off to form intracellular vesicles. These events constitute part of the endocytic and the exocytic pathways.

Chapter 10
Membrane Biogenesis

10.1 Overview

In this chapter we examine how membranes are made. First, the individual protein and lipid components must be synthesized, and then they must be delivered to their final destination. Considering all the distinct membranes within a typical eukaryotic cell, this is a prodigious task, requiring highly accurate and remarkable mechanisms.

In the first part of this chapter, membrane protein biogenesis will be discussed. Conceptually, there are two major problems to consider relating to membrane protein assembly.

(1) Since all nuclear-encoded proteins are synthesized by a common pool of ribosomes, how are individual membrane proteins directed to their proper destination? What distinguishes a plasma membrane protein from one which resides in the inner mitochondrial membrane or one one which is resident in the endoplasmic reticulum? This is a complex sorting problem, and must require distinct signals within each polypeptide as well as recognition apparatus.

(2) How are membrane proteins actually inserted into the membrane to attain the proper topology with respect to the membrane bilayer? Do insertion and orientation also require special signals and apparatus and, if so, what are they? What are the requirements for membrane proteins to attain their proper tertiary folding and their quaternary structures in the case of multisubunit assemblies?

The last decade of research has seen rapid progress relating to all of these questions, and this is proceeding at an accelerating pace, in large measure due to the ability to utilize recombinant DNA technology to explore the roles of specific polypeptide sequences as signals in these processes. Although not all the answers are known, there has been a satisfying trend toward the realization that seemingly quite different systems actually have substantial common features. For example, it was realized rather early (see 1268) that the mechanism of how proteins are secreted had much in common with the mechanism of how plasma membrane proteins are synthesized. More recently, the common aspects of how proteins are translocated across the mitochondrial membranes, the endoplasmic reticulum, and the membranes of gram-negative bacteria have been more fully appreciated (1576, 1136, 1288). These are the experimental systems which have been most intensively studied. Although there are significant differences between these processes, there are important common features:

(1) There is an identifiable portion of the polypeptide sequence which serves as a recognition site or "signal," directing the individual polypeptide to the membrane into which it is inserted. Often, these signals are at the amino terminus of the newly synthesized polypeptide, and they are often cleaved off by specific signal peptidases following insertion into or translocation across the target membrane. The terms *signal peptide*, *signal sequence*, *transit peptide*, *leader peptide*, and *pre-sequence* have all been used to refer to that amino-terminal signal.

(2) The processes of translation and of insertion into the membrane are experimentally separable. The assembly of membrane proteins, in most cases, requires energy input distinct from that required by the ribosome during translation. It is noted that in vivo translation and translocation are often closely coupled temporally.

(3) Once associated with the target membrane, the polypeptide must still be *competent* for translocation or insertion in order for the process to proceed. Protein translocation across membranes appears to proceed, in many cases, from the amino terminus to the carboxyl terminus and requires that the protein be at least partially unfolded or loosely folded during this process. The polypeptide may be translocated in an extended form in an energy-requiring step (1252).

The first set of questions we will address deals with protein sorting during biogenesis and assembly. Figure 10.1 summarizes the complexity of this problem for eukaryotic cells and for gram-negative bacteria. A hierarchy of signals encoded within each polypeptide must be involved in directing each protein to its correct destination. For example, most proteins which are directed to either the endoplasmic reticulum or the mitochondrion are synthesized as larger precursors (preproteins) with amino-terminal extensions which are subsequently removed by proteolytic enzymes within these organelles. These *primary signals* are very different and are required for these polypeptides to be recognized by the translocation machinery via specific receptors in the organelles. The association with the mitochondrion occurs after the completion of translation, but this is not the case for most proteins directed to the mammalian endoplasmic reticulum. As shown in Figure 10.1, further sorting is necessary once the proteins have become associated with the appropriate organelle. This requires additional information which must also be encoded within each polypeptide sequence, and these can be termed *secondary signals*. In some cases, these have been defined and are physically separable from the primary signal, though this may not always be the case. Specific examples will be described later in this chapter to illustrate how these signals are utilized for protein sorting.

Our concerns are specifically directed at how membrane proteins are as-

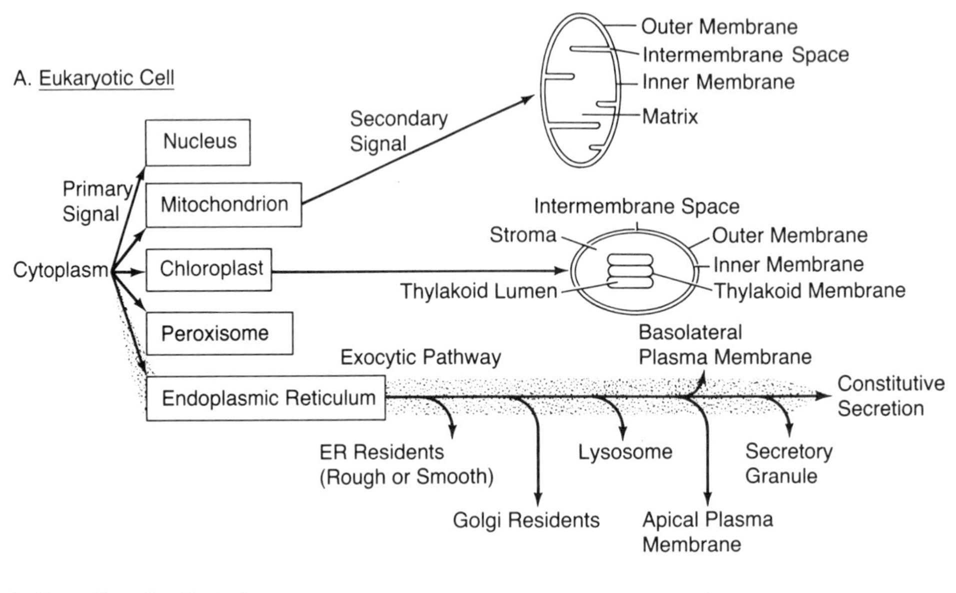

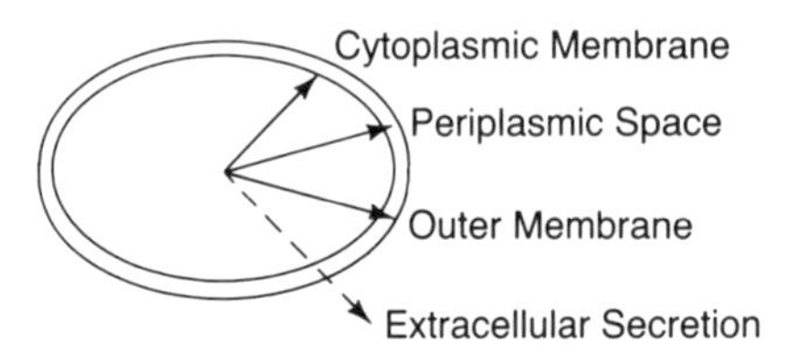

Figure 10.1. Protein sorting in eukaryotic cells and in gram-negative bacteria. Primary sorting signals direct the polypeptide to the correct organelle and are necessary for translocation. Secondary signals determine the final destination. The exocytic pathway is indicated.

sembled, but this must be examined in the context of protein sorting. Figure 10.2 shows schematically three commonly invoked mechanisms for the insertion of a precursor polypeptide into a membrane. Mechanisms A and B are variants of a linear extrusion scheme in which it is imagined that the signal directs the polypeptide to the translocation apparatus, which includes a water-filled channel. The signal might go directly through (model A) or remain tethered, as indicated, to form a loop (model B). In the absence of any signal to stop the translocation process, the entire polypeptide will be transported across the membrane. However, if a second signal, called a *stop-transfer* signal, is located within the polypeptide, the process is halted and the stop-transfer signal becomes a transmembrane segment of the mature membrane protein. By fixing the protein in the membrane, a stop-transfer signal acts as a sorting signal (1491). Additional transmembrane segments would be formed if there were additional *start-transfer* and *stop-transfer* signals within the protein. Figure 10.3 illustrates how a small number of kinds of signals can instruct the translocation apparatus in a sequential

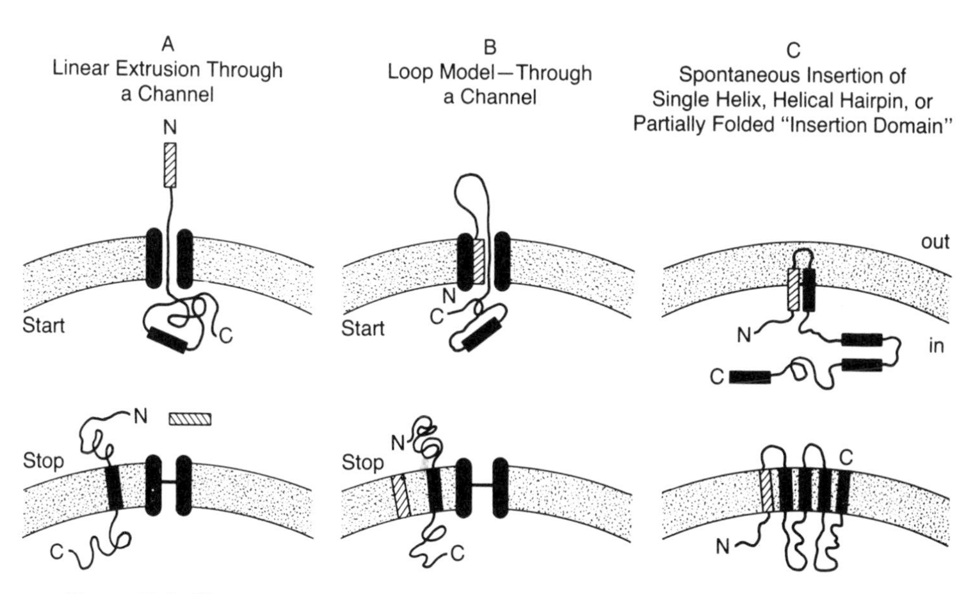

Figure 10.2. Three general models of how proteins might assemble in a membrane. The first two models (A and B) assume that the protein is transported in a linear fashion through a protein channel. The presence of a stop-transfer signal halts the process; otherwise the protein is translocated entirely across the membrane. Model C assumes that hydrophobic elements of the polypeptide spontaneously insert into the lipid bilayer. The hydrophobic element may be a single helix or a higher-order structure. This process could also be protein-mediated.

manner to produce a wide variety of protein folding patterns across the endoplasmic reticulum membrane. Note that signals that are not removed proteolytically usually remain as transmembrane segments and can be used to initiate the transport of the flanking polypeptide domains on either the amino or carboxyl sides. Unfortunately, this simple view is insufficient, and there are examples where it appears that the signals change their function depending on their context (51, 1000) or that interactions between putative signals within a polypeptide may be important to direct proper insertion into the membrane (365, 275).

Model C in Figure 10.2 illustrates a possible role for the spontaneous insertion into the membrane bilayer of hydrophobic elements of a precursor polypeptide. This mechanism can be invoked only when membrane insertion occurs after the translation of the polypeptide. The prime example supporting this mechanism is the M13 procoat protein, which is discussed in Section 10.31. The spontaneous insertion model can be invoked to explain how amphipathic α-helices or ß-sheet can be inserted across a membrane. This could, of course, also be a protein-mediated process.

The next two sections in this chapter review the exocytic pathway in eukaryotic cells and discuss how and where membrane proteins are assembled and transported within the cell. Little is known about how the biosynthesis of membrane lipids and proteins is coordinated in either prokaryotic or eukaryotic cells, al-

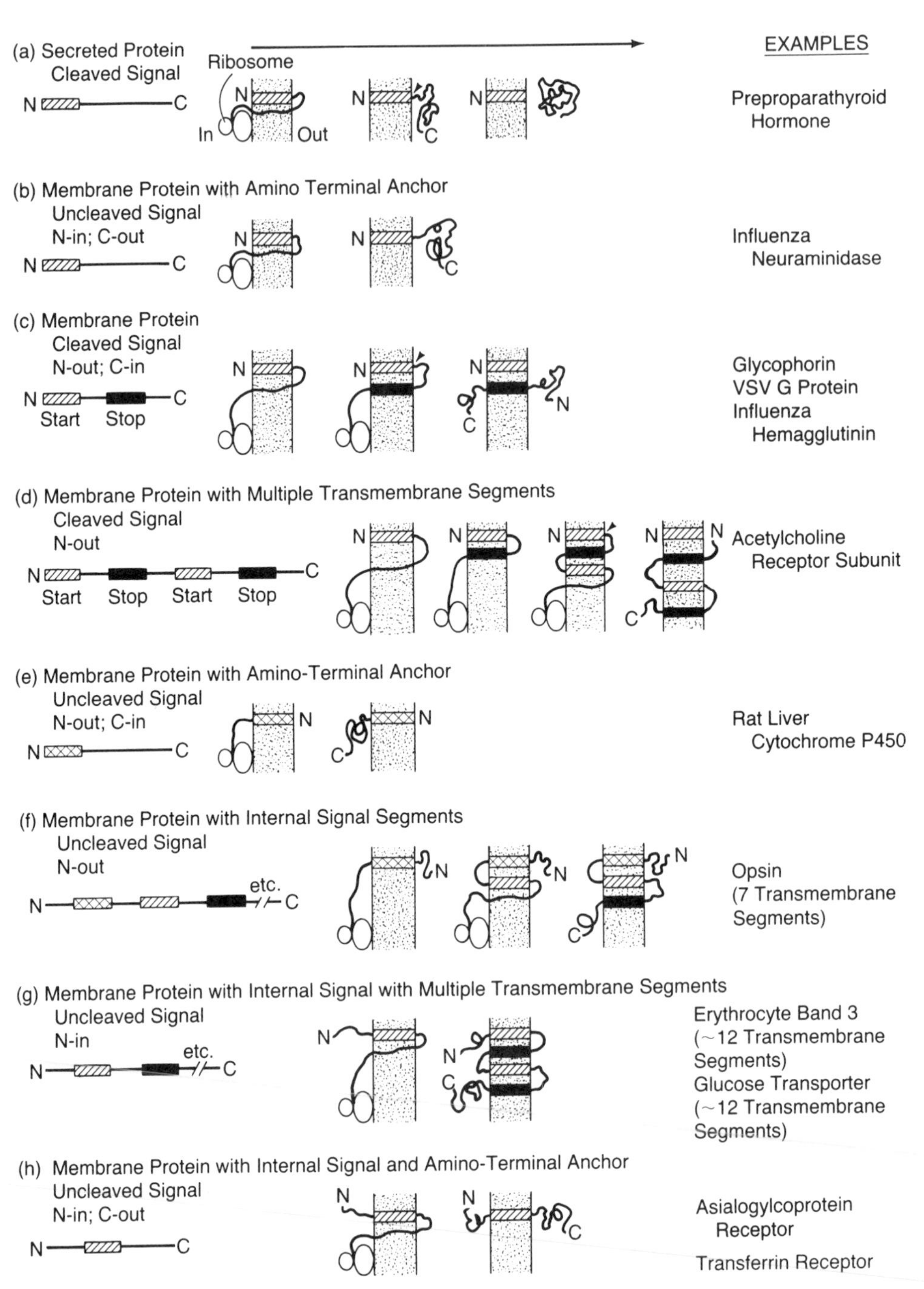

Figure 10.3. Models used to explain a variety of folding patterns of membrane proteins with cleaved signals, uncleaved amino-terminal signals, and internal signals. If assembly is directed by a linear sequence of signals, then three types are required: (1) a signal causing the translocation of the carboxyl-terminal flanking domain (a, b, c, d, g, h), (2) a signal causing the translocation of the amino-terminal flanking domain (e, f), and (3) stop-transfer signals (b–h).

though there are several examples where the overproduction of a particular membrane protein results in the elaboration of intracellular membranes containing lipids and, predominantly, the overproduced protein (1588, 383, 768). The last portion of this chapter focuses on lipid biogenesis. The fact that the lipid compositions of each of the various membranes within a eukaryotic cell are very distinct (see Chapter 1) raises the question of how these different compositions are created and maintained. One aspect considered is how rapidly lipid species are exchanged between different membranes and what mechanisms facilitate the transfer of lipids from their site of synthesis to their destinations. Finally, the lipid composition of the membranes of numerous organisms have been shown to vary in an adaptive response to environmental alterations. The nature of this adaptive response in membrane lipid biogenesis is discussed in the final section.

10.2 General Features of the Exocytic Pathway (for reviews, see 966, 738, 552, 358, 1153, 634)

The exocytic pathway in eukaryotic cells defines the trail followed by proteins which are secreted or are incorporated in the cytoplasmic membrane. Secreted proteins are synthesized by membrane-bound ribosomes on the cytoplasmic surface of the rough endoplasmic reticulum and are translocated into the lumen by the same mechanism used to generate membrane proteins in the endoplasmic reticulum (Figure 10.3). If a water-soluble protein in the lumen lacks any secondary sorting signals, it is transported to the cell surface and secreted by the "constitutive" pathway (1580). This pathway for exocytosis has been characterized by a combination of cytochemical methods, genetics, and cell-free biochemical techniques. A protein following this pathway proceeds sequentially from the endoplasmic reticulum, through the various compartments of the Golgi apparatus, and eventually is delivered to the cell surface (Figure 10.4). Membrane proteins following this pathway can become components of the cytoplasmic membrane or, due to the presence of secondary sorting signals, remain behind as either residents of the endoplasmic reticulum (e.g., ribophorin, cytochrome P450) or residents of the Golgi apparatus (e.g., various glycosyltransferases). In the Golgi, further protein sorting distinguishes proteins to be secreted by the constitutive pathway (the default pathway) from those which are directed to the lysosome or those which are concentrated in secretory granules, which then are used to secrete the protein upon appropriate cellular stimulation (i.e., the regulated secretory pathway). It has been noted, in addition, that the outer membrane of the nuclear envelope also may be a locus for the synthesis of membrane glycoproteins which subsequently are transported via the exocytic pathways (1465).

As proteins are transported through the exocytic pathway, they are subject to a series of post-translational modifications, in particular, glycosylation reactions (see 358, 1245). The compartmentalization of N-linked oligosaccharide processing is well defined (Figures 10.4 and 10.5) and has been very useful in defining distinct compartments constituting the Golgi (*cis*, *medial*, and *trans* cisternae). The high-mannose oligosaccharide precursor (see Figure 3.16) is assembled at the glycosidation sites on the polypeptide (asparagine residues) while the protein

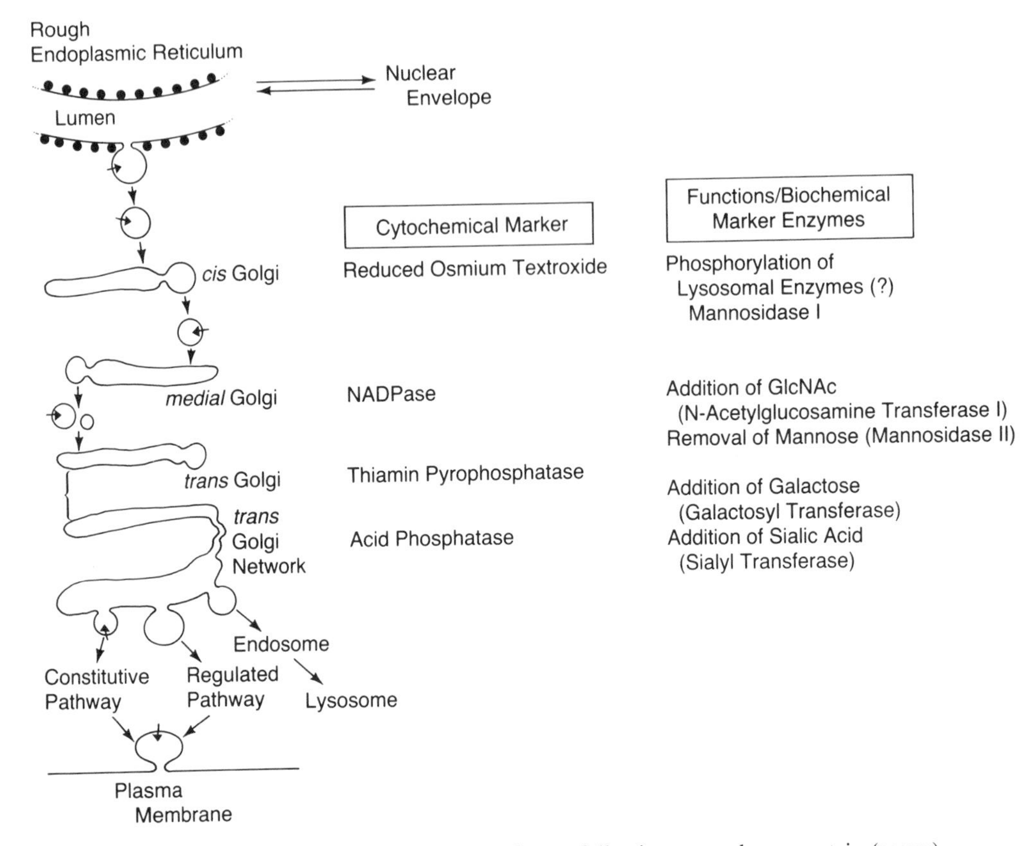

Figure 10.4. Schematic of the exocytic pathway, following a membrane protein (arrow) from its insertion in the endoplasmic reticulum to the plasma membrane. The distinction between the *trans* Golgi and the *trans* Golgi network is tentative (see 11). It is assumed that transport is mediated by vesicles. GlcNAc, *N*-acetylglucosamine. Adapted from ref. 552. Copyright 1986 by the AAAS.

is within the endoplasmic reticulum. This is then processed by a series of enzymes that are compartmentalized (552, 358, 1153, 634, 1245), allowing one to monitor the progress of proteins through the system by analyzing their glycosylation status. Figure 10.5 illustrates some pathways for the processing of the high-mannose precursor oligosaccharide.

Studies on the exocytic pathway have benefited from the use of enveloped viruses, in particular, vesicular stomatitis virus (VSV). The mode of entry of the enveloped viral genome by membrane fusion with the animal cell membrane has already been discussed (Section 9.52), as has the budding of newly synthesized viral particles from the animal cell membrane (Section 4.53). The viral spike glycoprotein, called the G protein for the vesicular stomatitis virus, is a trans-membrane protein which is synthesized by membrane-bound ribosomes on the endoplasmic reticulum following infection. The G protein utilizes the cellular exocytic pathway and is brought to the plasma membrane, where it helps to form

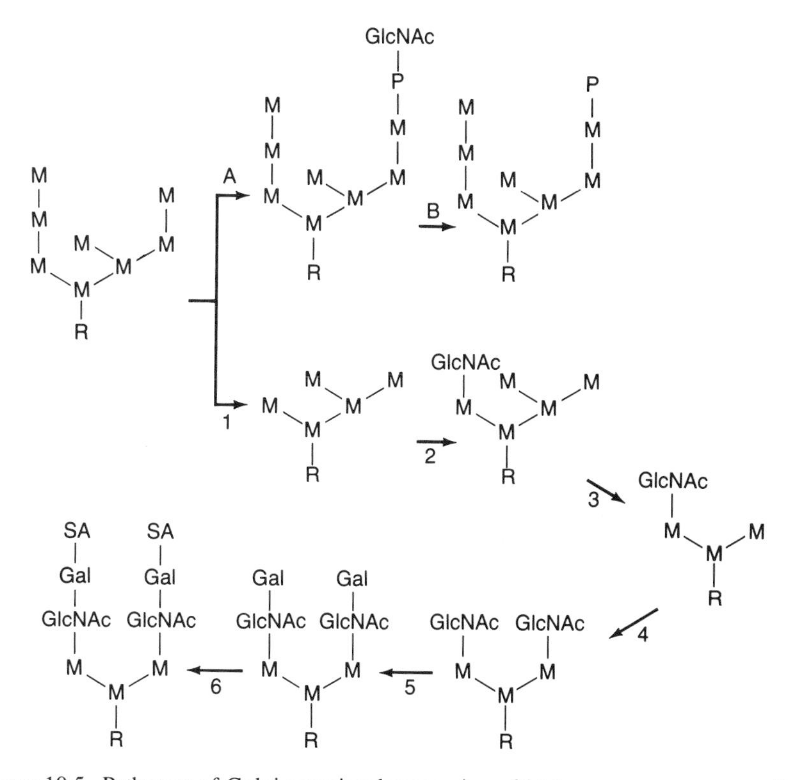

Figure 10.5. Pathways of Golgi-associated processing of high-mannose precursor oligosaccharides. Three glucose residues and at least one mannose residue are removed in the endoplasmic reticulum to yield the precursor at the top left. Steps A and B show how a mannose residue is phosphorylated, as occurs for lysosomal enzymes. Steps 1-6 show how a two-branched (biantennary) oligosaccharide is made. The enzymes are (1) Golgi mannosidase I, (2) GlcNAc transferase, (3) mannosidase II, (4) GlcNAc transferase II, (5) galactosyltransferase, and (6) sialyltransferase. M, mannose; GlcNAc, *N*-acetylglucosamine; Gal, galactose; SA, sialic acid; P, phosphate; R, $(GlcNAc)_2$-asparagine-(protein). Adapted from ref. 358. © 1985 by Cell Press.

the framework of the budding virus (Figure 4.9). By infecting cells with this virus, one can follow the progress of the virally encoded G protein to examine the exocytic pathway (e.g., 1250, 61, 1099, 168, 60). This approach has been exploited to define vesicles involved in intracellular transport (1099), to examine the effects of glycosylation on delivery to the cell surface (168), to examine the energetics of exocytic transport (60), and, most important, to construct a cell-free system for vesicle-mediated intracellular transport (1250, 61).

Several salient features of the exocytic transport system have emerged from these studies.

(1) The transport between various organelle compartments is mediated by vesicles which bud off the "donor" and subsequently fuse with the "acceptor" membranes. The vesicles mediating transport between Golgi compartments have been shown to be "coated" vesicles, but the protein coat is not clathrin, as in

Box 10.1 An In Vitro Assay for Vesicle-Mediated Intracellular Transport

Figure 10.6 illustrates an in vitro assay for transport of the viral G protein between successive Golgi compartments. This kind of assay is important in that it allows one to define biochemically the components required for the process. Membrane fractions containing Golgi are prepared from two cells, termed the "donor" and "acceptor." The donor fraction is prepared from mutant cells which lack the enzyme UDP-GlcNAc glycosyltransferase I and which have been infected with vesicular stomatitis virus. Processing of the oligosaccharide on the G protein in these cells is, thus, blocked. The acceptor Golgi fraction is prepared from wild-type, uninfected cells. The addition of [^{3}H]GlcNAc to the newly synthesized G protein requires the transfer of the G protein from the donor Golgi to that of the acceptor which contains the required enzyme. Incorporation of the [^{3}H]GlcNAc is quantified after immunoprecipitation. This particular assay measures transport from the *cis* to the *medial* Golgi compartment (61), and similar studies observing the addition of sialic acid have demonstrated transport *in vitro* between *trans* Golgi elements (1250).

endocytic vesicles (1099). Considerable evidence indicates that clathrin is not an essential component of the exocytic pathway (344, 1254), although it does appear to be essential for the normal growth of some yeast strains (832). The extent to which the endocytic and exocytic pathways share components is not clear, though some vesicles appear to function as parts of both systems (433).

(2) Intracellular transport requires ATP (1250, 61, 60) as well as cytosolic protein components (1558). It has also been shown that fatty acyl-CoA is a required cofactor for vesicle-mediated transport between Golgi cisternae (520, 1250). The specific roles these play in the mechanism of vesicle budding and fusion are not known.

(3) The role of the oligosaccharide as a sorting signal for the newly synthesized glycoproteins appears to vary. In some cases, processing of the N-linked carbohydrate is clearly not required for a plasma membrane protein to be expressed at the cell surface (285, 168). However, in some examples, glycosylation is essential for delivery to the cell surface (557) and, for example, is required for opsin to be incorporated in the retinal rod outer segment membrane (440). In many mammalian cells, mannose-6-phosphate acts as a sorting signal for protein to be directed to the lysosomes (see 778 for review). However, this is not true for yeast (1485, 711). In those mammalian cells where mannose-6-phosphate acts as a sorting signal, two membrane-bound receptors with a high affinity for glycoproteins containing mannose-6-phosphate have been characterized and the genes cloned (1385, 869, 1168). These are evidently critical to targeting particular polypeptides to the lysosomes, and one of these receptors appears to function in both the endocytic as well as the exocytic pathway (1385, 139).

As shown by the cell-free assay depicted in Figure 10.6, the availability of specific mutants can be particularly valuable in the study of such a complex system. Numerous other mutants defective in various stages of the exocytic

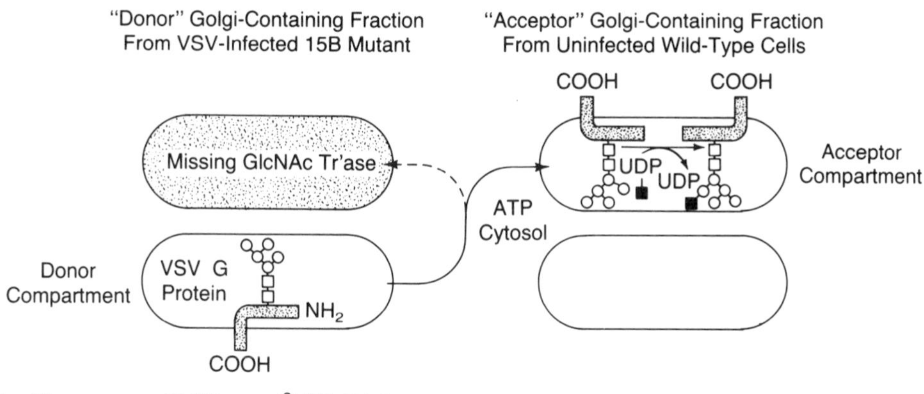

Figure 10.6. Schematic of the assay used to monitor transport of protein between Golgi compartments in vitro. In order for [^{3}H]GlcNAc to be added to the core oligosaccharide, the viral G protein from the Golgi from infected donor cells must be transferred to the Golgi obtained from cells which have an active *N*-acetylglucosamine transferase I. GlcNAc, *N*-acetylglucosamine. Adapted from ref. 61. © 1984 by Cell Press.

pathway have been characterized in yeast (326, 421) and, to some extent, exploited to develop a cell-free interorganelle transport assay (599). The yeast and mammalian systems appear to be quite similar (359), and one can anticipate further progress utilizing the yeast mutants. Finally, it is noted that numerous enveloped viruses bud from membranes other than the plasma membrane and, thus, are potentially useful for studying other aspects of intracellular protein sorting and membrane traffic (1240).

10.3 General Features of Membrane Protein Biosynthesis (for reviews see 1268, 1576, 1195)

The essential question of interest to us is how membrane proteins are assembled. As we will see in this section, this does not typically occur as a spontaneous event due to the interaction between a nascent polypeptide and a lipid bilayer. Rather, this process is energy-requiring and is mediated by protein structures which are not yet well defined. The experimental data justify the view that the translocation of proteins entirely across a particular membrane (e.g., into the lumen of the endoplasmic reticulum) and the assembly of intrinsic membrane proteins are closely related variations of the same process. It would seem logical that there might be a universal solution to the problem of how to transport proteins across or into membranes. Before proceeding to summarize those features which the membrane assembly processes in various systems have in common, it will be useful to review how this is examined experimentally.

Those systems which are best defined have advanced due to the availability of cell-free systems to observe and quantify protein translocation and proteolytic processing. These systems all involve the use of membrane vesicle or organelle preparations in which the cytosolic surface is facing outward, since protein translocation proceeds from the cytoplasm. Microsomes prepared from the endoplasmic reticulum of secreting cells provide such a system, as do mitochondria and chloroplasts. Inverted vesicles from *E. coli* membrane preparations can be prepared which have their cytosolic surface facing outside and which are suitable for cell-free studies of protein translocation (1034, 204, 500).

When a precursor polypeptide is present in the external medium, and all the conditions are properly adjusted, the polypeptide will be translocated into or across the membrane vesicle or organelle. This is most frequently monitored by adding proteases to the external medium. The degree of protection against proteolysis indicates the extent to which the polypeptide has been transported to the inside of the vesicle or organelle. As shown schematically in Figure 10.7, proteolytic processing by the signal peptidase, if it occurs, can also be observed by SDS-PAGE. If one is studying the translocation into microsomal vesicles, glycosylation can also be monitored in some cases. One can distinguish proteins which have become assembled in the membrane by an alkaline extraction. Presumably, proteins which bind to the membrane surface will be removed by this treatment. However, this need not always be the case, so results obtained by alkaline extraction should be interpreted with caution.

With these cell-free systems one can examine the biochemical requirements for protein translocation and identify required soluble components. In addition, the nature of the polypeptide "substrate" which is translocated can be varied. Similar studies can also often be done in vivo by pulse-label techniques, and these methods avoid the possibility of artifacts due to the artificial conditions of the cell-free systems (e.g., 1345, 1497, 275).

Several important conclusions about the requirements for protein translocation have been derived from cell-free assay systems.

(1) *Post-translational vs. co-translational translocation*: It is now accepted that in all the systems examined, the process of translocation into or across membranes can be carried out independently of translational elongation (see 1288). This has always been clearly demonstrable for chloroplast and mitochondrial protein translocation and, though long debated, for translocation across the bacterial membrane as well. Although protein translocation into or across the endoplasmic reticulum of eukaryotes was thought for many years to be an obligately co-translational process, this has been clearly demonstrated not to be the case. Post-translational transport which is ribosome-independent has been demonstrated into yeast microsomes in at least one case (prepro-α factor) (1555, 589). It has also been shown that, although protein elongation is not required for translocation across the endoplasmic reticulum in higher eukaryotes (1136, 1029, 1028), in most cases, protein translocation is still ribosome-dependent and must occur while the nascent polypeptide is still tethered to the ribosome (1136). The important conclusion is that the energy required for the translocation process is not derived from the ribosomal biosynthetic apparatus.

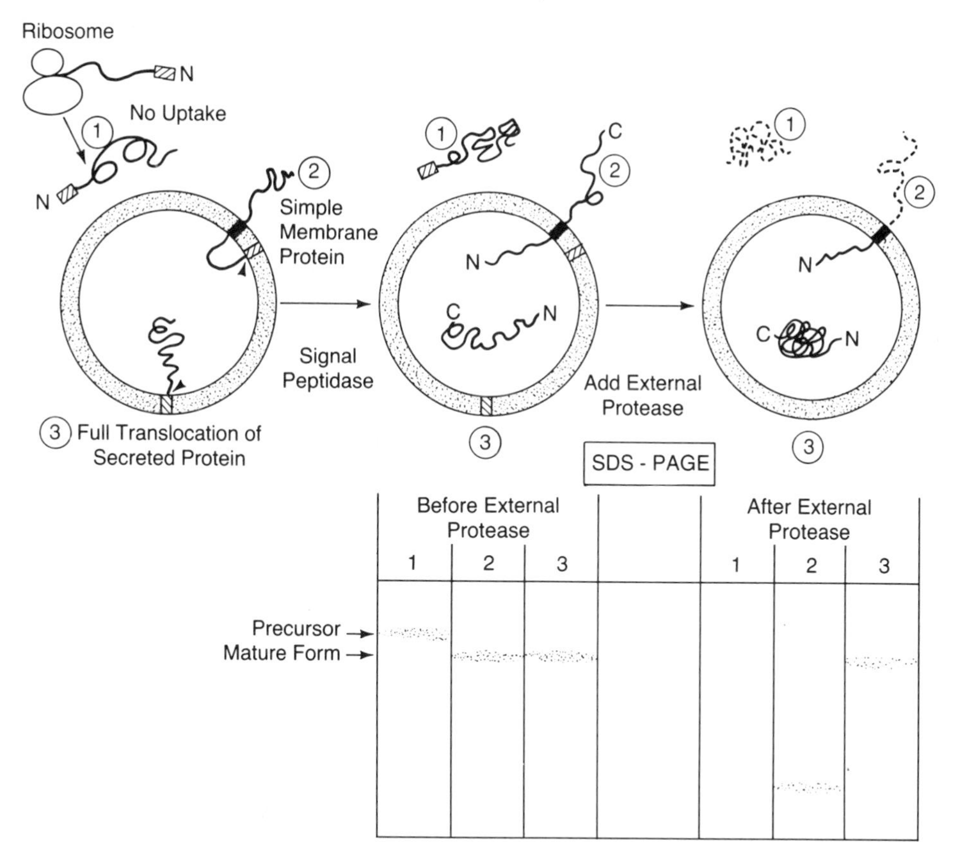

Figure 10.7. Schematic showing an in vitro assay for protein insertion into or across a vesicle membrane. The assay is based on protection against proteolytic degradation, assuming that protection is due to import of part or all of the polypeptide into the lumen of the vesicle. The proteins are usually prepared in radioactive form, immunoprecipitated, and analyzed by SDS-polyacrylamide gel electrophoresis. An SDS-PAGE profile is shown for each of the three examples. The signal peptide will not be observed by this method.

Note that these data merely state that translocation and translation are experimentally separable. In the cell, translocation is closely coupled to translation, at least for proteins directed into the endoplasmic reticulum of mammalian cells and for many *E. coli* proteins (83).

(2) *Energy requirements for translocation*: As a general rule, protein translocation into or across membranes is energy-requiring. The common feature in both the prokaryotic and eukaryotic systems appears to be a requirement for ATP (or nucleoside triphosphate) hydrolysis. This has been demonstrated for (a) import of proteins into the chloroplast stroma (1117, 442); (b) import of proteins into the mitochondrial matrix (371, 1149, 207), inner membrane (1149), and outer membrane (76); (c) translocation of a protein across the yeast endoplasmic reticulum (589, 1555) and post-translational insertion of a membrane protein into the mammalian endoplasmic reticulum (1028); and (d) translocation of proteins across

the *E. coli* cytoplasmic membrane (204, 1612, 500, 1151). The ATPase activity required is not due to the F_1F_0-type ATPase in either the mitochondria or *E. coli* membranes, and the role of the ATPase activity is not to generate a potential across the membranes.

An independent requirement has been demonstrated for a transmembrane potential across the mitochondrial inner membrane in order to import proteins into the matrix (371, 1149, 207) and into the inner mitochondrial membrane (1149). This potential is apparently required for an early step in the import process, probably relating to the initial binding to the mitochondrion. This does not appear to be the case for the import of at least some proteins into the chloroplast (1117). However, optimal translocation across the *E. coli* cytoplasmic membrane also has an independent requirement for a proton motive force across the membrane (1612, 500, 204, 308). Note that the direction of protein translocation with respect to the polarity of $\Delta\psi$ is opposite in *E. coli* in comparison to the mitochondrion, whereas the endoplasmic reticulum membrane is not known to have a transmembrane potential.

(3) *Competence of the precursor for translocation (see 1252)*: Considerable evidence indicates that the tertiary conformation of the protein to be translocated is a critical factor for successful translocation. The most obvious explanation is that the signal sequence(s) recognized by the translocation apparatus must be accessible. Hence, the protein must be in a "loosely folded" or partially unfolded conformation in order to be *competent* for translocation. In addition, if proteins are translocated in an extended conformation across the membrane, then the translocation apparatus must be capable of unfolding proteins during the actual transport process. Some precursor proteins, if allowed to acquire a stable tertiary conformation, would be difficult to unfold and, hence, would not be translocation competent.

The best evidence that proteins are translocated in an extended conformation comes from the work of Schleyer and Neupert (1298), who trapped intermediates in the translocation of two different proteins into the mitochondrial matrix. These intermediates were shown to have their amino termini exposed to the matrix, while the bulk of each polypeptide was outside of the mitochondrion. Hence, the intermediates must span both the inner and outer mitochondrial membranes, strongly suggesting that their site of entry is located at the junction of these two membranes (see 1307).

Box 10.2 The Transport of Proteins Across a Membrane Requires an Unfolded Conformation

Schatz and his co-workers (370, 1514) have studied an artificial fusion protein in which a mitochondrial signal sequence was linked to a protein which is not normally imported into the mitochondrion, tetrahydrofolate reductase. The resulting protein is imported into the mitochondrial matrix, where the signal sequence is removed by a signal peptidase. In order to test whether the protein could be translocated in a folded conformation across the mitochondrial membranes, import was measured in the presence of methotrexate, an inhibitor which binds with high affinity to the native form of tetrahydrofolate reductase. It was found that the binding of methotrexate abolished import into the mitochondrion, presumably because it trapped the protein in a tightly folded conformation, stabilized by the binding of the inhibitor. Unfolding of the precursor of the ß-subunit of the F_1F_0-ATPase has also been shown to be required for its import into the mitochondrial matrix (206).

Truncated versions of the mitochondrially directed tetrahydrofolate reductase precursor have also been studied (1514). These have the mitochondrial signal sequence, but translation is interrupted before the synthesis of the full polypeptide is complete. These truncated precursors cannot bind methotrexate or, presumably, fold into a native-like conformation. However, they are imported into mitochondria. Of particular interest is the fact that the transport of these incomplete precursors does not require ATP, whereas import of the completed precursor is ATP-dependent. This strongly suggests that the ATP is needed to unfold the polypeptide, prior to or concomitant with its transport into the mitochondrion. Figure 10.8 schematically illustrates a working model of the steps involved in mitochondrial import, showing distinct points requiring a transmembrane potential and ATP.

Studies on the import of porin into the outer mitochondrial membrane have yielded similar conclusions regarding the role of ATP (1612, 1147). This protein does not have a cleaved signal sequence, and the mature protein has encoded within it all the required signal information for mitochondrial import. The protein has been isolated in a water-soluble form which is presumably partially denatured, and this form is competent for translocation. Import of this water-soluble precursor does not require ATP, in contrast to the import of the protein when it is imported directly after synthesis in an in vitro expression system. Again, it is most reasonable that the ATP is required in some active process to unfold the protein during the translocation process.

The translocation of the maltose-binding protein across the *E. coli* cytoplasmic membrane into the periplasm has also been shown to be dependent on the conformation of the precursor (1191). A mutant with an altered signal sequence which is not translocated was shown to exist as a precursor which is less susceptible to proteolytic degradation, i.e., more "tightly" folded. A conformation more accessible to proteases is competent for translocation, consistent with the mitochondrial examples. Presumably, the presence of the amino-terminal signal peptide retards the folding into a native-like conformation. Of interest is the fact that the mutation in the signal peptide which prevents translocation can be suppressed by a second mutation within the mature protein (254). The precursor with both mutations was considerably less stable in the cytoplasm than the species with only the signal sequence mutation, possibly reflecting a more open conformation.

The possibility that a soluble protein cofactor may be required to prevent the precursor from folding into a native-like conformation has also been addressed. The precursor for the *E. coli* outer membrane protein OmpA has been isolated in a water-soluble form (259), similar to the mitochondrial porin. However, in the case of the *E. coli* protein, it is not efficiently translocated across the cytoplasmic membrane in the absence of a cytosolic protein, termed a "trigger factor." It has long been known that the translocation of proteins across or into the mammalian endoplasmic reticulum requires a soluble cofactor, the *signal recognition particle* (1312) (see Section 10.34). Possibly a function of this factor is to prevent folding of the precursor polypeptide and maintain competence for translocation (1547).

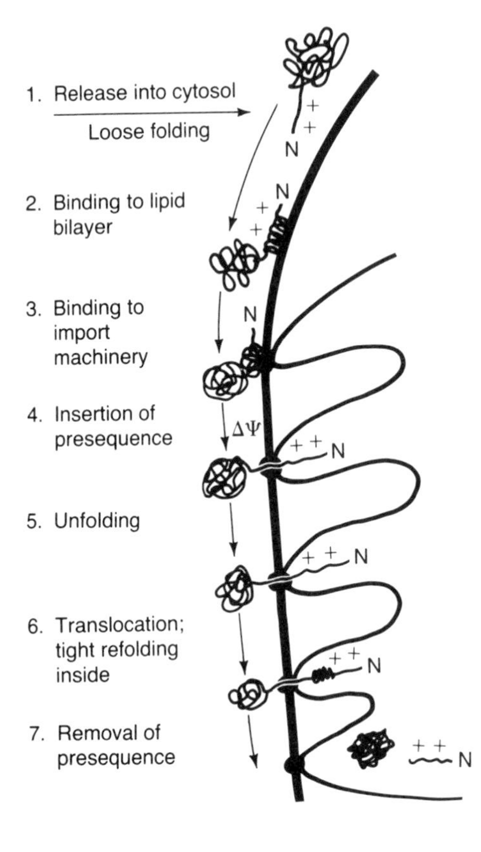

Figure 10.8. Working model for post-translational protein transport into the mitochondrial matrix. The signal or presequence is indicated by the three positive charges. Some steps may occur simultaneously (e.g., 3 and 4 or 5, 6 and 7). Step 5 can be bypassed if the precursor is incompletely folded. Adapted from ref. 1514.

10.31 Are Channels Needed for Protein Translocation?

There are no experimental data absolutely proving the existence of channels mediating the assembly of membrane proteins or the complete translocation of proteins across a membrane. Certainly the existence of membrane-bound receptors which specifically recognize proteins to be translocated has been demonstrated both on the mitochondrial surface (1147) and in the endoplasmic reticulum (1578). Conceivably, these proteins could be part of a complex translocation apparatus which could include a channel through which the protein is extruded (1350).

In order for protein translocation to proceed at a rate approximating polypeptide synthesis (1-10 residues per sec) the kinetic free energy barrier can be no greater than about 18 kcal/mol (388). Engelman and Steitz (388) estimated that two nearby helices could be expected to partition spontaneously into the bilayer in the form of a helical hairpin with a favorable free energy component as large as 60 kcal/mol and that this hydrophobic driving force could essentially be used to “drag” polar and even charged groups into the lipid bilayer to some extent. However, as discussed in Chapter 3, the transfer of ionizable and polar groups from an aqueous environment into the lipid bilayer has a considerable free energy cost (389, 646). It is not reasonable, based on simple energetic considerations, to expect the spontaneous insertion model be generally applicable, because in the assembly of many membrane proteins, it is necessary to transport long, often highly charged, stretches of polypeptide across the membrane.

However, several small membrane proteins have been shown to insert spontaneously into lipid bilayers, including cytochrome b_5, which has a single hydrophobic anchor at the carboxyl terminus (see Section 4.22), and the bacteriophage M13 procoat protein, which has a pair of putative transmembrane helices which could potentially penetrate the bilayer as a helical hairpin or loop. The procoat protein has a 23-residue signal sequence which is normally cleaved off during assembly in the *E. coli* cytoplasmic membrane. The mature coat protein (50 amino acid residues) contains an acidic amino terminus which is periplasmic, a transmembrane segment, and a basic carboxyl terminus which faces the cytoplasm (Figure 10.9). The procoat inserts spontaneously into phospholipid liposomes (501). Furthermore, the rate of assembly in vivo has been shown to be grossly retarded if mutations are located either within the transmembrane region or in the carboxyl terminus of the mature protein (798, 797), consistent with a model in which the two hydrophobic segments spontaneously insert as a hairpin or loop (388, 796). Figure 10.9 summarizes a model of the membrane insertion of this protein. Interestingly, the M13 procoat also is assembled by mammalian microsomes, and this assembly requires ATP, possibly to maintain a transport-competent conformation (1577). It should be pointed out that the M13 procoat is not typical of proteins assembled in the *E. coli* cytoplasmic membrane in that it has a cleaved signal sequence, and assembly is independent of the functions of genes [*secA*, *secY* (*prlA*)] which are required for the translocation of proteins into or across the cytoplasmic membrane (1600; see 84).

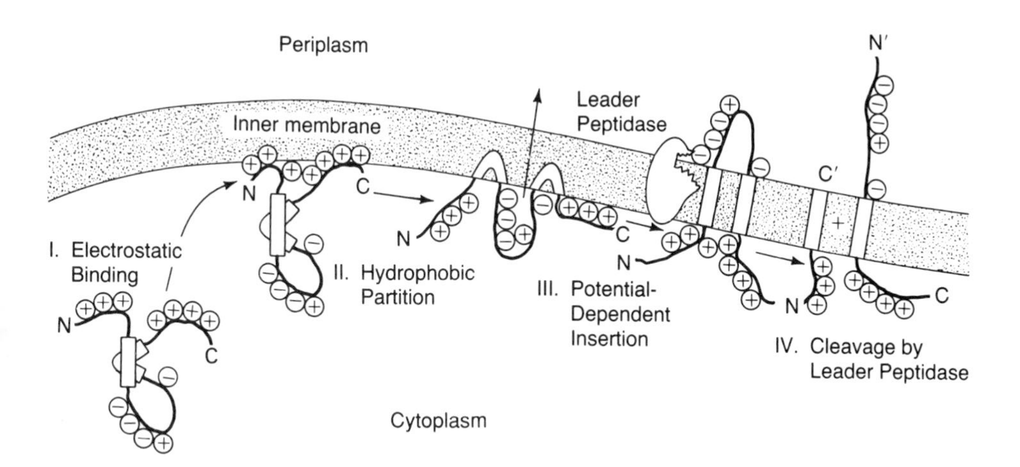

Figure 10.9. Working model for the spontaneous insertion of the M13 procoat in the inner membrane of *E. coli*. Adapted from ref. 798. Reprinted by permission from *Nature*, vol. 322, p. 336. Copyright © 1986 Macmillan Magazines Limited.

The M13 procoat studies provide the strongest supporting evidence for a spontaneous insertion mechanism requiring no protein-mediated assistance. The water-soluble precursor is presumed to fold in a conformation competent for membrane insertion. Insertion requires a major conformational change in the precursor which occurs upon interaction with the bilayer. This general model has been proposed as part of the "membrane trigger hypothesis" (1575), and a similar mechanism has been proposed for the assembly of at least portions of more complicated membrane proteins such as the glucose transporter (1028). Note that spontaneous insert mechanisms involving loops or helical hairpins can be invoked only in cases where translocation is not tightly coupled to polypeptide translation.

Apocytochrome *c*, which is the precursor for mitochondrial cytochrome *c*, provides another example where spontaneous insertion into the lipid bilayer has been proposed to play an important role in translocation (90, 1067). This precursor does not differ in length from mature cytochrome *c*, but it lacks the covalently attached heme *c*, which is added only after the protein has been transported across the mitochondrial outer membrane. The mature protein is located in the intermembrane space of the mitochondrion (see Figure 10.1). It has been shown that apocytochrome *c* binds to anionic lipids in pure phospholipid vesicles, can penetrate into the lipid bilayer, and can actually traverse a pure phospholipid bilayer (90). The mechanism for this remarkable activity is totally unknown but one would expect substantial rearrangement of the lipids. The polypeptide has no significant stretch of contiguous hydrophobic residues and 40% of its amino acids are charged!

It is not clear whether the observations with artificial phospholipid vesicles are relevant to the physiological mechanism of import, but one proposed model requires that apocytochrome *c* penetrate the outer membrane sufficiently to bind to a specific protein receptor on the inner surface of the outer membrane (1067). Conceivably, this could involve a channel, however. In vivo, the covalent attachment of heme effectively traps the mature protein in the intermembrane space and, perhaps, induces a conformational change needed for complete translocation (1067).

There are many other examples of water-soluble proteins which spontaneously insert into lipid bilayers and biomembranes, though in no case is the mechanism known. The predominant examples are provided by toxins and pore-forming proteins, mentioned in Section 8.61. The conceptual models offered usually involve postulated conformational changes in which hydrophobic residues which are facing the protein interior in the water-soluble conformation became exposed to the lipid bilayer interior upon contact with the bilayer. Examples include the *Staphylococcus aureus* α-toxin (1457), complement component C9 (1380), and colicin A (75). In many cases, such as *Pseudomonas* exotoxin A (413) and diphtheria toxin (343), a low pH is required for the conformational transition to be triggered. Presumably, this delays membrane penetration until the toxin is within the acidic endocytic vesicles following endocytosis (see Section 9.4). It is doubtful that these toxins and pore-forming proteins are relevant models for the assembly of many membrane proteins, but they do provide clearcut examples where water-soluble precursors can spontaneously assemble within a bilayer to form complex, transmembrane, biochemically active species.

In summary, if translocation occurs by transporting the extended polypeptide linearly from the amino terminus to the carboxyl terminus, then energetic considerations would demand a water-filled pore or channel to provide a hydrophilic environment for the charged or polar groups (1350). This is a reasonable working model for most proteins, though there are numerous examples, like the M13 procoat, where spontaneous insertion of individual helices or of domains into the lipid bilayer can occur. If specific channels for protein translocation do exist, they must be remarkable structures, insofar as they allow the passage of virtually any polypeptide chain, without being leaky to ions and small metabolites. No attempts to observe such pores by patch-clamping techniques have been reported.

10.32 Polypeptide Signals Which Direct Protein Sorting and Membrane Insertion (see 1528, 1289, 237, 1352)

Although very little is known about the translocation apparatus or mechanism, quite a bit is known about the codes or signals present within the polypeptides which direct each protein to its proper location. Much of the progress has resulted from the use of recombinant DNA techniques to generate hybrid or chimeric polypeptides in which a particular amino acid sequence to be tested is taken from one protein and fused to another protein. Hence, one can examine the effect of a putative signal sequence on the localization of a "passenger protein." This approach can be used to great advantage only if the information determining the localization of the final product is entirely contained in the primary sequence of the signal and if the passenger protein is just a neutral participant and, essentially, will go where it is instructed. In many examples, this is precisely what is observed (see 1291), but there are examples where translocation efficiency is dependent on the passenger protein (1497, 1554, 798) or the final localization may even depend on the passenger protein (689). Poor translocation of a chimeric protein could result if the passenger protein is in a conformation that is not competent for translocation. In addition, the function of some signal sequences is dependent on their position within the polypeptide or on interactions with other portions of the polypeptide (e.g., 234, 51, 1345, 365). Despite these complications, much has been learned about a variety of signal sequences. (see Table 10.1, p. 393).

The Signal Sequence for Insertion into the Endoplasmic Reticulum (see 1528 for review)

Most proteins which are inserted into or across the membrane of the endoplasmic reticulum are guided there by a transient signal peptide (15 to 30 residues) located at the amino terminus of the polypeptide. This signal sequence directly interacts with at least two receptors, one soluble (signal recognition particle) and the other in the membrane (see 1312, 1547, 1578) (see Section 10.34). It might be expected that this signal peptide would have a highly conserved sequence shared by all the translocated proteins, but this is not the case. Neither the length nor the amino acid sequence is conserved among these signal sequences, and numerous mutagenesis studies have demonstrated that considerable structural variations are tolerated. That signal peptides contain all the information required to direct proteins across or into the endoplasmic reticulum membrane is demonstrated by the generation of chimeric polypeptides in which the fused amino-terminal signal sequence results in the import of normally cytoplasmic passenger proteins, such as globin, into the endoplasmic reticulum lumen (e.g., 1345).

The "comparative anatomy" of the amino-terminal signal sequences suggests three structurally distinct regions: (i) a positively charged amino-terminal region (n-region); (ii) a central hydrophobic core of 7 to 15 residues (h-region); (iii) a carboxyl-terminal region which is polar and defines the cleavage site recognized by the signal peptidase which is on the lumenal side of the endoplasmic reticulum (c-region) (1528). It has been demonstrated that numerous randomly generated sequences are able to replace the normal signal sequence of yeast invertase and direct its export from the cell (720). Analysis of the random sequences which can function as signal sequences in this experiment indicated that overall hydrophobicity is a critical structural feature. Figure 10.10 compares the hydrophobicity and length of the hydrophobic segments of known eukaryotic signal peptides to the most hydrophobic segments found in eukaryotic cytosolic proteins, many being at the amino terminus, and to known transmembrane anchor segments of membrane proteins. This plot shows that the characteristics of the h-region are

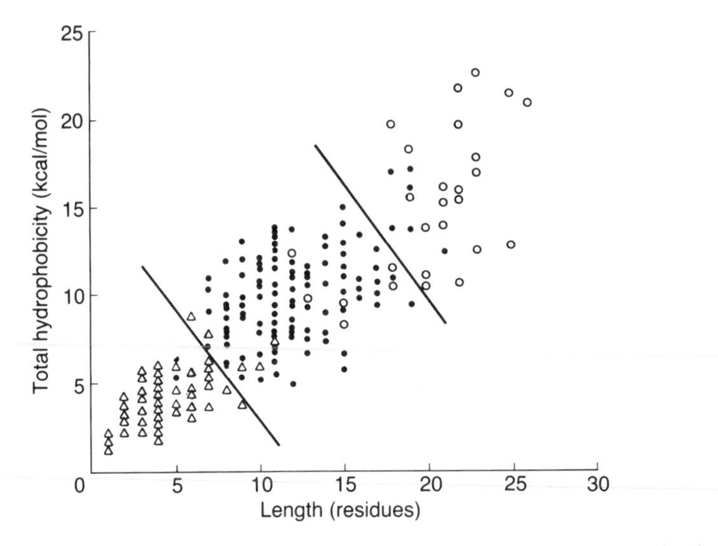

Figure 10.10. Scatter plot of total hydrophobicity versus length. (1) Eukaryotic signal peptides (filled circles); (2) carboxyl-terminal transmembrane anchor segments (open circles); (3) the most hydrophobic segments of a set of cytosolic proteins (triangles). The two lines encompass 92% of the signal peptides, 20% of the anchor segments, and 4% of the cytosolic sequences. Adapted from ref. 1528.

intermediate between those of the segments in cytosolic proteins and those of typical transmembrane segments.

Obviously, the structural specificity for recognition by the translocation machinery is quite low. Nevertheless, it should be remembered that a factor of 1,000-fold in affinity corresponds to a free energy difference of less than 5 kcal/mol, which is approximately the stabilization provided by a single hydrogen bond. Rather subtle differences between functional and nonfunctional signal sequences can easily amount to such a difference in affinity. A model for what the signal peptide receptor might look like is provided by the structure of the soluble fragment of the class I histocompatibility antigen, HLA-A2, whose three-dimensional structure is known (104). This protein binds to peptides derived from foreign antigens as part of the immune response (see Section 9.34). The peptide binding site is a large groove, which is open on one end and which can accommodate a 20-residue peptide if it is α-helical. Although little is known about the peptides that can bind to HLA-A2, the closely related class II histocompatibility antigen shows a high affinity for a wide variety of peptide sequences. Secondary structure and amphipathic character appear to be the most important common features of the peptides that can bind with high affinity. Numerous interactions at loci within the binding site can contribute to stability of the complex.

Relatively subtle differences among signal sequences are known to result in quite different behavior. For example, if the signal sequence is not recognized by the signal peptidase, the protein usually remains bound to the membrane rather than secreted (Figure 10.3b) (861), though this is not always the case (1345, 608). Generally, the signal sequences that also serve as amino-terminal anchors of this type have a longer hydrophobic h-segment, around 20 residues long, which must be necessary to stop the translocation and/or form a stable anchor in the membrane bilayer (1528). An example of one such signal/anchor sequence is in the transferrin receptor (Figure 10.11). Note that in this example, the signal sequence is not located at the amino terminus but is internal, located more than 50 amino acid residues from the amino end of the polypeptide.

There are also several examples of proteins where the signal sequence anchors the mature protein with the opposite orientation, i.e., amino terminus facing the extracytoplasmic surface (1528). These include rat microsomal cytochrome P450 (108; see Figure 3.11), the invariant chain of mouse class II histocompatibility antigens (862), several viral proteins (1589), and the H-subunit of the *R. viridis* reaction center (Figure 3.6). In some way, these signal/anchor sequences have flipped their amino terminus across the membrane and function to stop translocation so that the bulk of the protein remains cytoplasmic (Figure 10.3e). It has been noted (1528) that in several of these examples, the "start/stop" signals have at least one negative charge in the n-region (see Figure 10.11 for example). However, the same translocation apparatus (signal recognition particle) appears to be required for insertion of this class of membrane protein as for the more usual type (1274, 862). Possibly, the negative charge facilitates a spontaneous or protein-mediated translocation of the amino-terminal residues across the membrane.

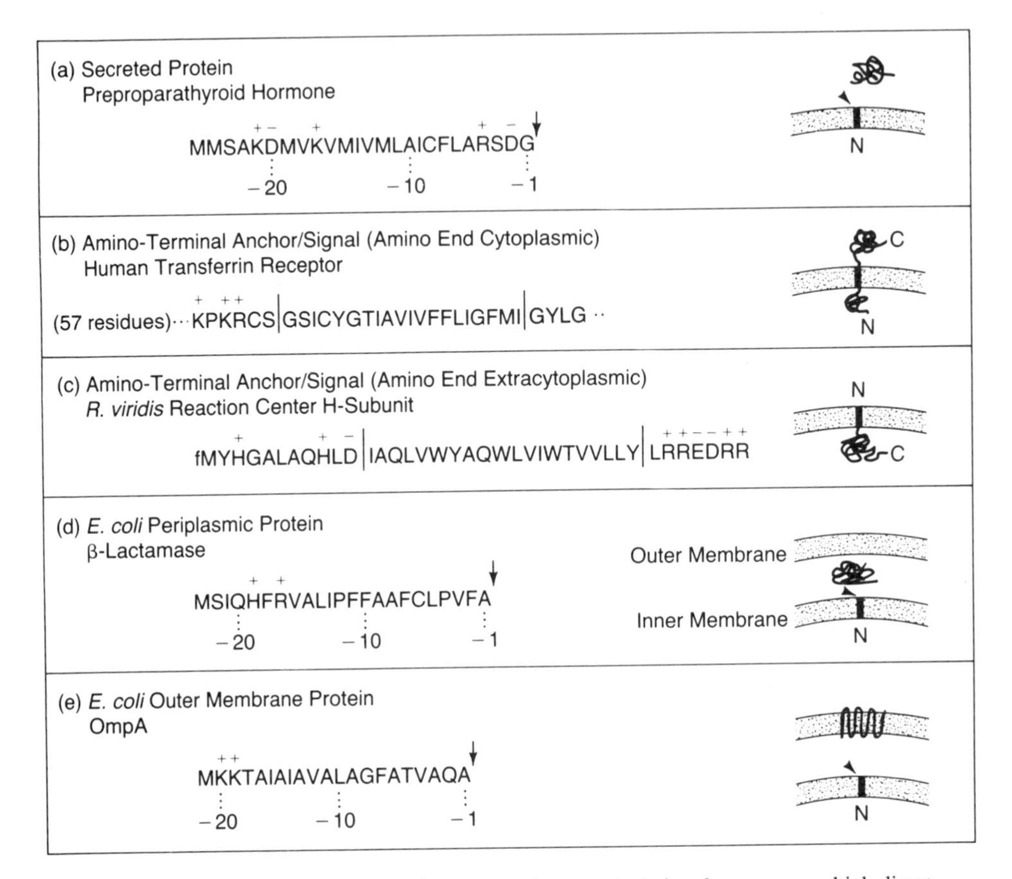

Figure 10.11. Selected examples of several amino-terminal signal sequences which direct protein transport across or into the eukaryotic endoplasmic reticulum (a and b) or the cytoplasmic membrane of a gram-negative bacterium. The signal sequences in b and c also act as stop-transfer signals and membrane anchors, but they have the opposite orientation in the bilayer.

Signal sequences, as pointed out above, need not be located at the amino terminus and, in fact, can direct the translocation of both of the flanking protein domains, at least in an artificial hybrid (1345, 1137). Ovalbumin is a unique example where an uncleaved, interior signal appears to direct secretion of the entire polypeptide (423). There are many examples of membrane proteins assembled in the endoplasmic reticulum which have uncleaved signal sequences which are internal and which also function as transmembrane anchors (Figure 10.3h). The asialoglycoprotein receptor is one example (641, 1372). The internal signal sequence of this protein requires the same translocation apparatus as do the amino-terminal signal sequences, and, in fact, the internal signal sequence will function as a normal amino-terminal signal sequence in artificial constructs (1372). The glucose transporter and Band 3 anion transporter are examples of proteins with internal, uncleaved signals that have multiple transmembrane segments and have their amino termini on the cytoplasmic side of the membrane (see 1268, 1576, 1028) (Figure 10.3g). In contrast, opsin is an example of a membrane protein with an internal, uncleaved signal peptide with the amino terminus on the extracytoplasmic side of the membrane (51) (Figure 10.3f). The internal signal (presumably, the first transmembrane segment) directs the hydrophilic amino-terminal domain (36 amino acids) across the membrane and, thus, has the opposite orientation to that in the more common examples where translocation of the polypeptide to the carboxyl side is mediated by the signal. The reason why opsin assembles in this manner is not known, though the nature of the amino-terminal peptide may be important (51).

In summary, subtle variations in signal sequences determine whether the "passenger protein" is secreted into the lumen of the endoplasmic reticulum or remains anchored in the membrane and determine the orientation of the amino terminus of the membrane protein. All possible topogenic variations have been shown to exist (Figure 10.3). One important point is that the assembly appears to be mediated by the same translocation apparatus.

Stop-Transfer Sequences

As pointed out in the previous section, those signal sequences which are not cleaved and which serve as amino-terminal anchors of the resulting membrane protein tend to have longer hydrophobic regions (1528). This suggests that the requirement to stop translocation may simply be a matter of having a sufficiently long hydrophobic segment which is capable of forming an α-helix across the membrane. Other data support this notion. Recombinant DNA techniques were used to introduce a hydrophobic segment into the middle of an *E. coli* protein which is normally secreted across the cytoplasmic membrane (295). Artificial hydrophobic domains, at least 16 residues long, successfully halted translocation and anchored the protein in the cytoplasmic membrane. This is a bacterial system, but, presumably, the translocation mechanisms in the prokaryotic and eukaryotic systems are related (see section on bacterial signals below). In contrast, variants of the vesicular stomatitis G protein were made with altered membrane-spanning domains. The length of the hydrophobic segment could be reduced from 20 to as small as eight residues and still produce a transmembrane polypeptide, though transport to the plasma membrane was blocked (2). Hence, what constitutes a stop transfer element is not well defined. Two issues need to be clarified: (1) Is the stop-transfer event mediated by specific proteins in the translocation apparatus? (2) Is the stop-transfer signal itself simply a matter of hydrophobicity or are more subtle factors involved? It has been shown that stop-transfer segments that halt translocation across the endoplasmic reticulum do not necessarily halt translocation across the chloroplast envelope (878). This suggests significant differences in these two processes.

The definition of "start transfer" and "stop transfer" assumes a linear scheme of translocation starting at the amino terminus, and this is justified by the behavior of simple systems. However, it has been shown that sequences which function to halt transfer in one context can initiate translocation in another (51, 1000).

Hence, not only is the nature of the stop or start transfer sequences themselves important, but also the surrounding polypeptide can be equally important.

Secondary Signals of the Exocytic System

The function of the signal peptide is to direct proteins to the endoplasmic reticulum and to initiate translocation. As shown in Figure 10.1, there are several final destinations for membrane proteins as well as soluble proteins which are imported into the lumen. Somehow information is encoded in the mature polypeptide which directs proper localization. In the absence of any secondary signal, water-soluble proteins are secreted by the "constitutive" secretory system (1249). Progress is being made to determine the signals responsible for directing other soluble proteins to lysosomes or to secretory granules or retaining such proteins in the endoplasmic reticulum or the Golgi (see Table 10.1). It is likely that regions of these polypeptides are involved in interactions with membrane-bound receptors so that their export through the system is retarded (1249).

Little is known about the signals that are responsible for the localization of integral membrane proteins within the exocytic system. A short sequence at the carboxyl terminus of the E19 protein of adenovirus appears to be entirely responsible for the retention of this protein in the endoplasmic reticulum membrane (1115). This protein has a single transmembrane segment and a 15-residue cytoplasmic tail at the carboxyl terminus. Shortening this cytoplasmic tail by only 8 amino acids results in the intracellular transport of this protein out of the endoplasmic reticulum. Presumably the signal region binds either directly or indirectly to some cytoplasmic structure, thus maintaining the E19 protein as a resident of the endoplasmic reticulum. Work on the E1 glycoprotein of coronavirus indicates that the signal that is required for this protein to remain in the Golgi is located in one of the three putative transmembrane helices (899).

A related sorting problem is to elucidate the secondary signals responsible for directing membrane proteins to the correct plasma membrane domain in polarized epithelial cells (see 937). Most of this work has utilized enveloped viruses which bud specifically from either the apical or basolateral surfaces of cultured epithelial cells. For example, the G protein of vesicular stomatitis virus is localized exclusively in the basolateral membrane from which the virus buds, whereas the hemagglutinin (HA) glycoprotein is transported to the apical membrane. A chimeric hybrid consisting of the extracytoplasmic domain of the HA protein and the membrane-spanning segment plus cytoplasmic tail of the G protein was localized exclusively in the apical membrane. These and other experiments suggest a critical role for the extracytoplasmic domain in localization (958,959). However, evidence also suggests that the cytoplasmic domain may also have important sorting determinants (1186, 958). Sorting of plasma membrane proteins in polarized epithelial cells appears to take place in the Golgi, but this may not be the case in rat hepatocytes, where all the plasma membrane proteins appear to go first to the basolateral domain (72).

An important point is that the secondary sorting determinants are contained within the mature polypeptide and do not appear to be related to or even contiguous with the primary signal peptide responsible for the initial localization to the endoplasmic reticulum. This is quite different from most of the mitochondrial and chloroplast secondary sorting signals (237) and, perhaps, sorting in the bacterial envelope.

Bacterial Signals for Translocation and Sorting (see 1192, 1352, 83)

Most studies on prokaryotic membrane protein assembly and translocation have been with *E. coli*. Protein synthesized in the cytoplasm can be directed into the cytoplasmic membrane, the periplasmic intermembrane space, or the outer

Table 10.1. Signals for protein sorting in eukaryotic cells.

(A) Primary signals (eukaryotic cells)

Organelle or membrane	Signal	References
(1) *Endoplasmic reticulum*	(1) Usually at the extreme amino terminus and usually removed by signal peptidase. (2) No conserved sequence per se, but overall hydrophobic character is critical. (3) *Three regions*: *n-region* — short, positively charged at amino terminus *h-region* — 7 to 15 residues, hydrophobic core *c-region* — defines the cleavage site by the signal peptidase, at carboxyl terminus (4) Signal peptide can also serve as an amino-terminal anchor for mature protein with the amino end either cytoplasmic or extracytoplasmic.	1528, 720
(2) *Mitchondrion*	(1) Usually at the extreme amino terminus and usually removed by proteolysis, unless the final destination is the outer membrane. (2) No conserved sequence per se, but conserved characteristic is positively charged, amphiphilic helix. Can be as short as 9 residues.	1289, 237
(3) *Chloroplast*	Similar to mitochondrial signal.	237, 1356
(4) *Nucleus*	Short sequence ProLysLysLysArgLysValGlu identified in SV40 large T antigen responsible for nuclear localization. Original site includes Lys-128, but it also works when at the amino terminus.	1225, 523
(5) *Microbodies* (peroxisomes, glyoxysomes, glycosomes)	No cleaved signal sequence. Nature of the signal is not known.	124

Table 10.1 continued

(B) Secondary signals in the exocytic pathway

Destination	Signal	References
(1) *Lysosome*	Mannose-6-phosphate in most mammalian cells but not in yeast. Amino acid sequence determinants are not known.	1385, 869, 1168
(2) *Endoplasmic reticulum residents*	Short sequence(s) in the carboxyl terminus causes retention of some lumenal and membrane components of the endoplasmic reticulum.	1038, 1115
(3) *Golgi residents*	Signal is within one of the transmembrane segments of E1 glycoprotein of coronavirus.	899
(4) *Apical vs. basolateral plasma membrane*	Extracytoplasmic (lumenal) domain in some cases contains a signal determining final localization but cytoplasmic domain can also play a role.	937, 958, 1186
(5) *Secretory granules*	Signal for trypsinogen shown not to require the signal peptide (primary signal) or first 12 amino acid residues at the amino terminus of the native protein.	1249
(6) *Constitutive secretion*	Default pathway if no secondary signals are present.	165, 738

membrane (see Figure 10.1). The translocation of proteins across the inner membrane to either the periplasm or the outer membrane is often referred to as "export." In addition, several proteins are translocated across both membranes and either are secreted to the external medium or become components of the pili (1352). Most of the research activity has focused on bacterial export (1192), which has many features in common with import into the endoplasmic reticulum.

As a rule, proteins translocated into the periplasm or to the outer membrane have transient amino-terminal signal peptides which are very similar to those found on secreted proteins imported into the endoplasmic reticulum of eukaryotic cells. In fact, to a certain extent, the prokaryotic and eukaryotic signal peptides are interchangeable and recognized in heterologous systems (493). For example, the signal peptide of the outer membrane lipoprotein of *E. coli* promotes protein translocation across mammalian microsomes (493). As shown for the eukaryotic signal peptides, there is no evident sequence homology between known prokaryotic signal peptides (1352). With only a few exceptions, cytoplasmic membrane proteins do not have a cleaved signal peptide (1352). The exceptions include the M13 coat protein (798), which is not indigenous to *E. coli*, and penicillin binding proteins, which apparently are not localized uniquely in the cytoplasmic membrane (67). Genetic evidence suggests that, in general, the translocation of exported proteins and the assembly of cytoplasmic membrane proteins share some common biochemical apparatus (1600) and have similar mechanistic features (1601).

In contrast, the secretion of proteins across both the cytoplasmic and outer membranes can be quite different. In the case of haemolysin, for example, it has been shown that the secretion signal is within the last 27 amino acids at the carboxyl terminus of the polypeptide and not at the amino terminus (901). Remarkably, the secretion of the cholera toxin across the outer membrane of the gram-negative *Vibrio cholerae* is claimed to occur after the polypeptides have folded into their tertiary and quaternary conformation within the periplasm (635).

Genetic analysis indicates at least four genes whose products are required for the translocation of most, but not all, envelope proteins across the cytoplasmic membrane: *secA*, *secB*, *secY* (or *prlA*), and *secD* (83, 495). The *secA* and *secY* gene products are also required for the assembly of at least some cytoplasmic membrane proteins such as the leader peptidase, but not the M13 coat protein (1600). The functions of these gene products are not known, but it seems likely that they are directly involved in mediating protein translocation. Biochemical studies (1034) have proved to be much more difficult than the genetics. Although the available mutants which are defective in protein secretion have not resulted in any mechanistic insights, the data do suggest a close coupling in vivo between protein secretion and translation (1398, 864). Biochemical studies also suggest that membrane translocation and translation are often closely coupled in vivo, though in some cases, including the M13 coat protein, post-translational insertion into the membrane occurs in vivo (1192). However, the *secY* gene product has been shown to be able to function post-translationally (55). This gene product (10) may be a membrane-bound receptor or channel protein which interacts with the bacterial signal peptide.

Additional Primary Signal Determinants (see 417 for review)

The signal peptide of exported proteins has been shown in several cases to be sufficient to direct the translation of passenger proteins across the cytoplasmic membrane. An example is the OmpA signal sequence (900). On the other hand, transport of the LamB outer membrane protein appears to require a portion of the mature polypeptide (85). Randall and her colleagues showed that the maltose-binding protein, though synthesized on membrane-bound ribosomes, is not translocated into the periplasm until translation is about 80% complete (see 1192). This suggests a possible role for parts of the mature sequence in initiating translation, though other explanations are possible.

Studies on two proteins in the cytoplasmic membrane, the M13 coat protein and leader peptidase, show that they presumably require structural determinants in addition to the amino-terminal signal sequence. The assembly of the M13 coat protein (Figure 10.9) has already been discussed (Section 10.32). More representative of proteins in the cytoplasmic membrane is the leader peptidase, whose proposed topology is shown in Figure 10.12. This is the enzyme responsible for the proteolytic cleavage of the signal peptide from most exported proteins in *E. coli*, and its active site is located on the periplasmic surface of the cytoplasmic membrane. This protein (323 amino acids) has two putative transmembrane

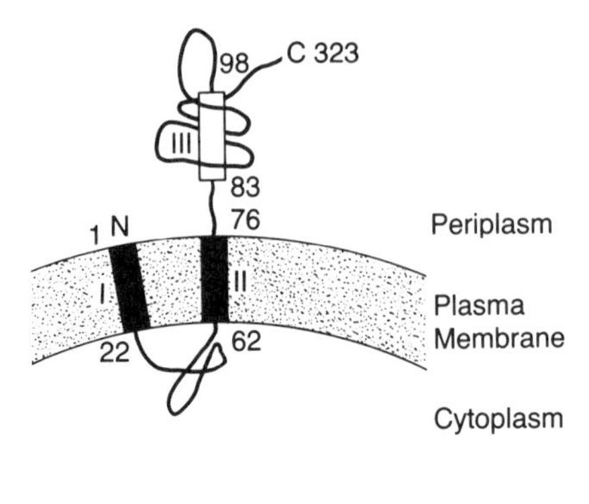

Figure 10.12. A model of the topography of the leader peptidase from *E. coli*. Amino acid residues are numbered. The rectangles represent hydrophobic segments. The first hydrophobic span has not been demonstrated experimentally, but is surmised from its hydrophobicity. The existence and orientation of the second hydrophobic span have been experimentally verified and this segment appears to function as an internal signal sequence. The third hydrophobic segment is only weakly hydrophobic and is part of the periplasmic domain. Adapted from ref. 1013.

segments and a large carboxyl-terminal domain in the periplasm (275, 1013). Deletion of residues 142-323 prevents translocation of the truncated polypeptide across the cytoplasmic membrane (277), consistent with some important role of the carboxyl half of the molecule for translocation (see Figure 10.12). In addition, a construct in which residues 4-50, containing the first transmembrane segment, are deleted assembles normally in the membrane, and the second transmembrane segment has been shown to act as a signal peptide (275).

In summary, studies on several *E. coli* proteins suggest that information encoded in the structure of the protein beyond the recognized signal sequence is required for translocation across the cytoplasmic membrane. This may simply reflect the fact that the artificial constructs used in these studies are in a conformation incompetent for translocation, e.g., occluding the signal peptide, or it may indicate specific sequence determinants needed to interact with parts of the translocation machinery.

Secondary Signals

What guides proteins to the inner membrane, periplasm or outer membrane? The answer is not known. Genetic studies (83, 1600) as well as competition studies (192, 62) suggest that many of these proteins share some common biochemical translocation apparatus. An analysis of the signal peptides suggests that subtle variations of hydrophobicity and size may result in different localization of the translocated protein (1352). On the other hand, the signal sequence of a periplasmic protein can successfully replace that of an outer membrane protein (1462), suggesting that the assembly information is incorporated in the mature outer membrane protein. Perhaps sorting to the cytoplasmic membrane is simply due to the presence of a stop-transfer sequence. Recall that the major outer membrane proteins, including the porins, lack hydrophobic transmembrane segments (see Figure 8.7).

Studies with Synthetic Signal Peptides and Lipid Partitioning

Peptides corresponding to the wild-type as well as mutated signal peptides of the outer membrane LamB protein have been synthesized, and their interactions with model phospholipid membranes and *E. coli* vesicles have been examined (205, 148, 147). The peptide corresponding to the wild-type signal sequence effectively inhibits the translocation of precursors of both a periplasmic and outer membrane protein in vitro. A peptide corresponding to a mutated signal sequence which is export-defective does not inhibit translocation in the cell-free assay. This suggests that the signal peptides recognize a common receptor in either the cytosolic or membrane fraction (205). In addition, the ability of these peptides to bind to model membranes (monolayer and bilayer) also correlates to their effectiveness to function as translocation signals. Studies on the maltose-binding protein precursor also indicate a correlation between the hydrophobic partitioning of the signal and the ability to act effectively to initiate translocation (334).

These data are consistent with a model where the primary signal sequence serves to localize the precursor polypeptide in the membrane by nonspecific interactions with the lipid bilayer, followed by a more specific interaction with a protein receptor. A similar model has been proposed for amphipathic peptide hormones (see Section 3.71). Note, however, that the signal peptide in mammalian cells is known to interact first with a soluble receptor before it interacts with the membrane (see Section 10.34). The significance of the lipid-binding properties of the signal peptide is unclear.

Other Bacterial Signal Sequences

Some data are available for bacterial signals other than from *E. coli*. Of particular interest is the unique kind of signal present in bacteriorhodopsin, synthesized in *H. halobium*. This protein (Figure 3.8) is synthesized in vivo with an amino-terminal signal which is 13 residues long and which has the potential to form an amphipathic α-helix (1531). This is very different from the *E. coli* signal sequence, and little is known about how it functions.

Signals for Mitochondrial Import and Sorting (see 1289, 237)

Nuclear-encoded mitochondrial and chloroplast proteins are synthesized as water-soluble precursors on free ribosomes, and are translocated post-translationally in vivo. Sorting places mitochondrial proteins either in the outer membrane, the intermembrane space, the inner membrane, or the matrix (Figure 10.1). Many experiments with chimeric constructs or with gene deletions have demonstrated that, in general, sufficient information for both import and sorting is present in the amino terminus. Most mitochondrial and chloroplast proteins are synthesized with amino-terminal pre-sequences which are removed by signal peptidases during or after the translocation steps. These pre-sequences contain the information required for import and sorting. For example, the first 22 amino acids of cytochrome c oxidase subunit IV bound to mouse dihydrofolate reductase result in the import of this cytosolic enzyme into the mitochondrial matrix (371). In general, if no additional signals are present, the mitochondrial import signal will direct the passenger protein into the matrix. This is the "default" pathway, analogous to constitutive secretion for proteins imported into the endoplasmic reticulum. When additional sorting signals are present, they are, as a rule, also in the pre-sequence, just downstream of the signal for mitochondrial import. For example, if the pre-sequence of cytochrome c_1 (Figure 10.13) is fused to tetrahydrofolate reductase, the passenger protein is localized in the intermembrane space (1289). Even in those few examples where there is no cleaved signal (ADP/ATP translocator) import and sorting signals are located at the amino terminus of the polypeptide. It should be noted, however, that several studies do suggest a role in the import of mitochondrial proteins for sequences other than those near the amino terminus (1148, 1150).

Figure 10.13 summarizes the signal sequence structure of representative mitochondrial proteins. All have a matrix-targeting sequence at the extreme amino terminus and additional signal information as required.

Outer Membrane Proteins

The best studied example is the 70-kDa outer membrane protein of yeast (see 1289). As is the case with other outer membrane proteins, there is no cleaved signal peptide, but the amino terminus of the polypeptide is a matrix-targeting signal. This was demonstrated by fusing the first 12 amino acids of the 70-kDa protein to a cytosolic protein and demonstrating that the passenger protein is imported into the mitochondrial matrix. The sorting signal that causes the 70-kDa proteins to localize in the outer membrane is apparently a stretch of 28 uncharged amino acids which is adjacent to the matrix-targeting signal. Deletion of as few as two of these uncharged residues results in translocation into the matrix. Note that mitochondrial import is thought to occur via channels located at the junctions of the inner and outer membranes (1298,1307), as shown in the model in Figure 10.8. Unlike the example shown in Figure 10.8, the assembly of the 70-kDa protein does not require a transmembrane potential, which is typical for outer membrane proteins.

The simplest model for the assembly of the 70-kDa protein is that the hydrophobic stretch which causes retention of the protein in the outer membrane acts as a stop-transfer signal, leaving the bulk of the protein in the outer surface of the mitochondrion. Another outer membrane protein which has been studied, porin from *Neurospora crassa*, is probably sorted by a different mechanism. This protein has no stretch of hydrophobic amino acids and, like bacterial porins, may be predominantly transmembrane ß-sheet (761).

Intermembrane Space

Cytochrome b_2 is a representative of a class of proteins within the intermembrane space (596). The pre-sequence is 80 residues long and contains two distinct

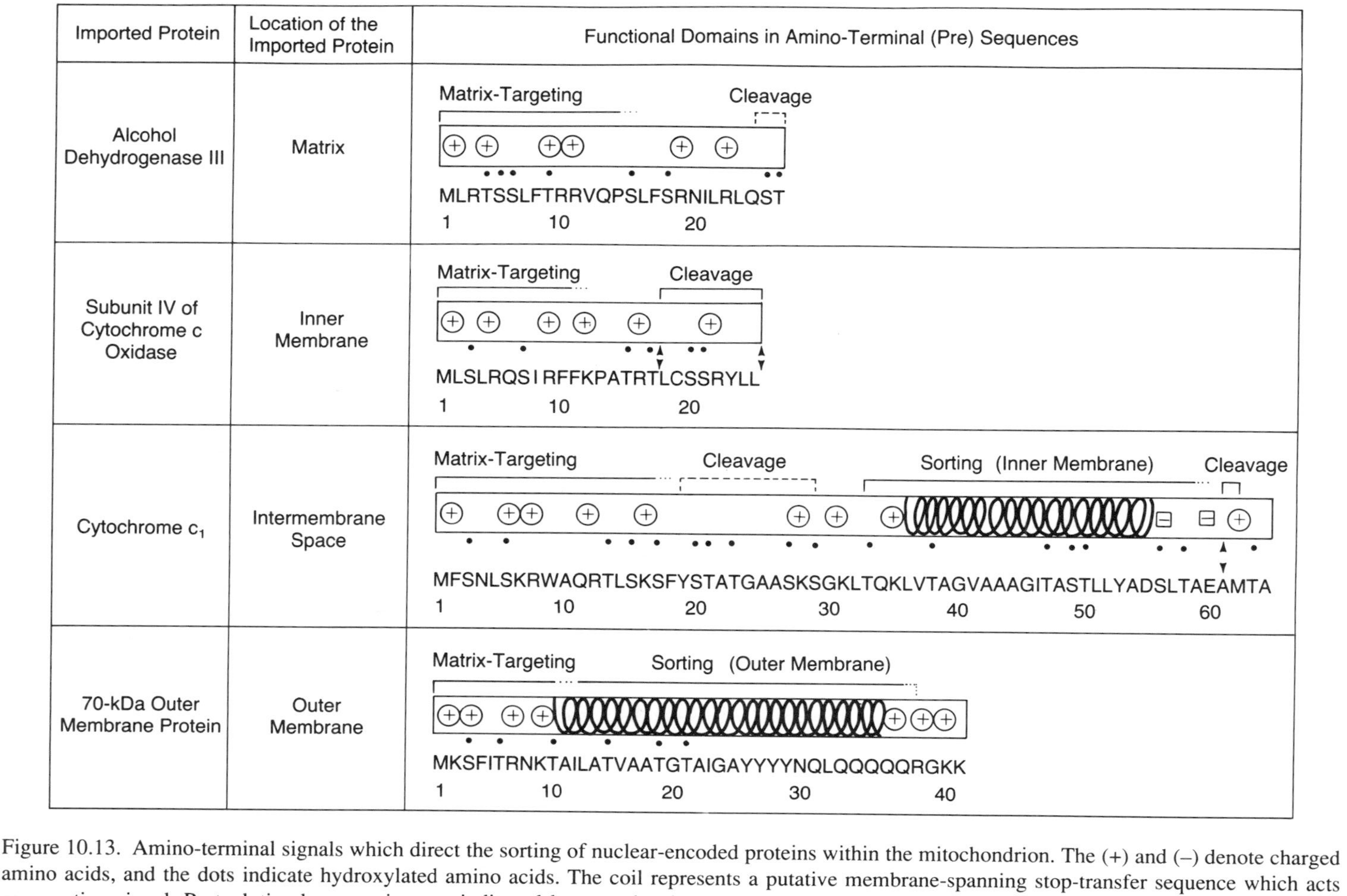

Figure 10.13. Amino-terminal signals which direct the sorting of nuclear-encoded proteins within the mitochondrion. The (+) and (–) denote charged amino acids, and the dots indicate hydroxylated amino acids. The coil represents a putative membrane-spanning stop-transfer sequence which acts as a sorting signal. Proteolytic cleavage sites are indicated by arrowheads. Adapted from ref. 1289.

portions. The amino-terminal part of the pre-sequence is a matrix-targeting signal, and this directs the entire polypeptide into the matrix. This step is dependent on a membrane potential across the inner membrane. Proteolytic processing by the signal peptidase within the matrix removes the amino-terminal portion of the pre-sequence and reveals the second part of the signal, which directs the polypeptide back across the inner membrane and into the intermembrane space. This step does not require a membrane potential. A second proteolytic event occurs at the outer surface of the inner membrane, yielding the water-soluble, mature form of the protein. The import of the Rieske iron-sulfur subunit of the bc_1 complex follows a similar pathway, but in this case all the proteolytic processing occurs inside the matrix (Figure 10.14) (595). Note that the translocation of proteins out of the matrix is very similar to bacterial export of proteins (596).

Two mechanisms have been proposed for the import of cytochrome c_1, another subunit of the bc_1 complex. One mechanism is similar to that described above for cytochrome b_2 (596). A different mechanism, illustrated in Figure 10.14, has also been proposed (1491, 1490). The pre-sequence is long (61 residues, Figure 10.13) and very similar to that of cytochrome b_2. It has been proposed that the

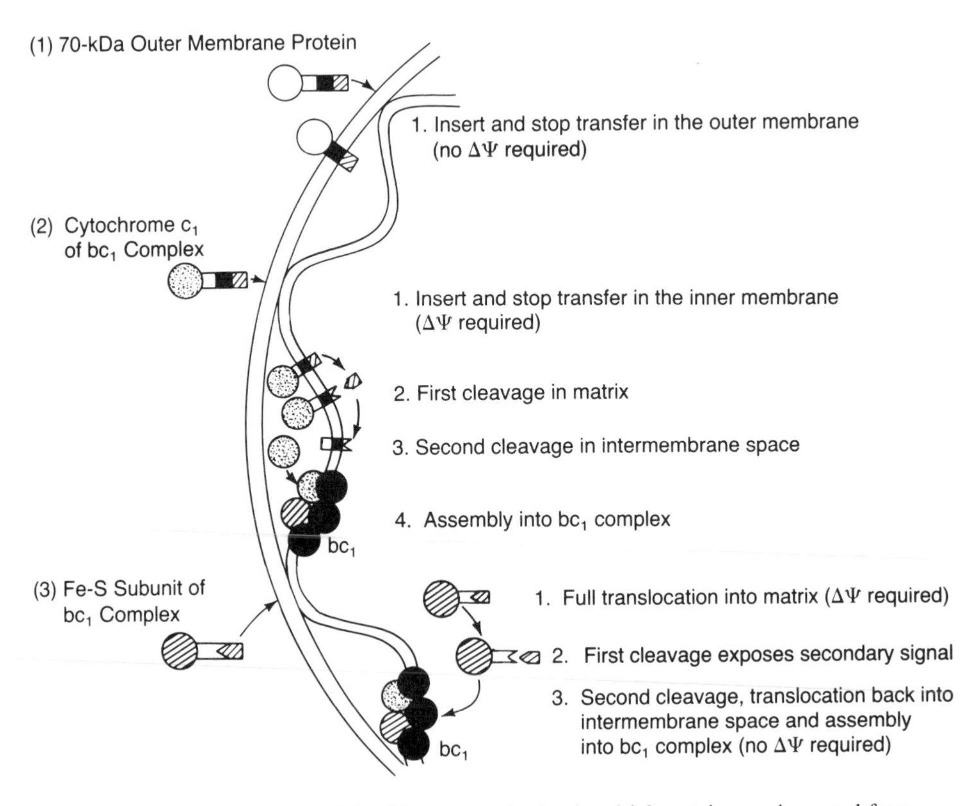

Figure 10.14. Working models of how several mitochondrial proteins are imported from the cytoplasm and sorted. Note that the mechanism proposed for cytochrome c_1 is in dispute (596). Adapted from ref. 1491.

matrix-targeting signal at the amino terminus directs the protein into the matrix until translocation is apparently stopped by a hydrophobic segment containing 19 uncharged residues at the carboxyl terminus of the pre-sequence. This fixes the protein in the inner membrane, as shown schematically in Figure 10.14. Once cytochrome c_1 translocation has been halted, two signal peptidases, one in the matrix and the second in the intermembrane space, cleave the pre-sequence, resulting in the mature protein, which assembles in the multisubunit bc_1 complex. The mature protein is probably anchored to the inner membrane by a carboxyl-terminal hydrophobic helix.

Note that the import mechanism of cytochrome *c* to the intermembrane space appears to be unique. There is no cleaved pre-sequence and this protein appears to have its own specific receptor (1067).

Inner Membrane

It is possible that the sorting signal for many inner membrane proteins which are imported from the cytoplasm may be an appropriately placed stretch of hydrophobic residues that can serve as a stop-transfer signal. This is suggested by the localization of a hybrid protein in which a matrix-targeting signal was fused to the carboxyl half of the vesicular stomatitis G protein (1064). The protein is imported into mitochondria in vitro and anchored in the inner membrane by the transmembrane domain of the G protein. The ADP/ATP translocator also contains a segment within its amino-terminal end which functions as a stop-transfer signal in the inner but not the outer membrane (4).

Chloroplast Signals

The primary sorting signals of nuclear-encoded chloroplast proteins are similar to those for mitochondrial proteins (237, 1356). In fact, in one case, the pre-sequence of a chloroplast protein was shown to direct passenger proteins into yeast mitochondria (237). It is not clear what features allow the cells to route proteins to the proper organelle when both chloroplasts and mitochondria are present. The sorting of chloroplast proteins within the organelle is an even more complicated problem than for the mitochondrion. Chloroplast proteins can be localized on either of the two membranes of the envelope or in the stroma and can also incorporate in the thylakoid membrane or cross it completely to the thylakoid lumen (Figure 10.1). The secondary signals appear to be localized within the amino-terminal pre-sequence in a series of separable domains, similar to the mitochondrial sorting signals (237). Thus, for a nuclear-encoded protein to get to the thylakoid lumen, at least two signals are required, one to cross the envelope to form a soluble precursor in the stroma and a second signal for transport across thylakoid membrane.

Structural Characteristics of the Matrix-Targeting Sequence

A comparison of mitochondrial targeting sequences reveals no specific sequence conservation but rather a common structural motif which suggests that these signals can form positively charged amphipathic helices (1527). Typically, these signals have no or few negative charges and have lysines and arginines spaced so that if an α-helix were formed they would be largely on one side, resulting in a large hydrophobic moment (see Section 3.62). Such amphipathic sequences are fairly common within soluble proteins. It has been shown that internal sequences from the cytosolic enzyme dihydrofolate reductase, as well as sequences generated randomly from the *E. coli* genome, when fused to a passenger protein, result in transport of that protein into the mitochondrion (667, 59). A variety of artificial sequences were also shown to function as matrix-targeting signals (21). It is interesting to note the similarity between the mitochondrial targeting signal and the putative amphipathic helix which is postulated to form the voltage sensor in voltage-regulated ion channels (see Section 8.25). Evidently, some features of the secondary structure, rather than specific amino acid residues, are sufficient for recognition by the receptors mediating import into mitochondria. As discussed previously, the histocompatibility antigen HLA-A2 provides a good model of a receptor which can bind a wide variety of different peptides (104).

Chemically synthesized peptides corresponding to mitochondrial signals have been shown to interact with lipid bilayers and monolayers and have been shown to form a helix in the presence of some phospholipids and detergents (1239, 394, 1046). In addition, a synthetic peptide will block the import of mitochondrial precursor proteins, presumably by interaction with specific receptors (514). Possibly, the signal sequence results first in concentrating the precursor in membranes and, second, binding to a specific receptor in the outer membrane. As depicted in Figure 10.8, the binding step requires a transmembrane potential across the inner membrane. One model which has been proposed (1527, 1239) is that the amphipathic signal peptide initially binds to the surface of the membrane and then is moved into or across the membrane by virtue of the electric potential, which is negative inside. This proposed transient structure of the positively charged helix buried in the membrane is certainly unstable. However, if the free energy barrier to forming this state is less than about 18 kcal/mol, this still might cross the membrane at a reasonable rate (1 amino acid per sec). This is unlikely, especially considering the high density of hydroxylated amino acids typically present in the pre-sequences in addition to the positive charges. An alternative model is that the transmembrane voltage influences the receptor proteins or putative channel within the membrane which mediates translocation. It should be noted that the translocation into the chloroplast stroma does not require a transmembrane potential, though the organelle targeting signals are similar.

10.33 Signal Peptidases

Specific proteins are required to remove the transient amino-terminal signal peptides. The signal proteases from *E. coli* are the best characterized. Most exported proteins in *E. coli* have their signal peptide cleaved at the periplasmic surface of the inner membrane by the *leader peptidase*, whose structure is illustrated in Figure 10.12. This peptidase is not required for protein translocation across the inner membrane but is necessary to release the exported protein from the cytoplasmic membrane (276). The enzyme has been purified and functions in vitro when incorporated into liposomes (1090). The specificity of cleavage is largely but not entirely determined by residues near the cleavage site (333, 1530). The signal peptidase in the endoplasmic reticulum has the same specificity as that of *E. coli* (1530), which is not surprising considering the similarity of the signal sequences. The eukaryotic microsomal signal peptidase has also been purified and shown to be associated with other polypeptides, possibly related to the translocation machinery (400).

E. coli has a second signal peptidase which processes prolipoproteins. These polypeptide components of the *E. coli* envelope are unique in that their amino termini are modified by a glyceride during maturation (see Section 3.8). The prolipoprotein signal peptidase is also located in the cytoplasmic membrane (679). After cleavage, the signal peptide remains in the cytoplasmic membrane, where it is degraded by a second membrane-bound enzyme, protease IV (674).

More than one signal peptidase must be present within mitochondria and chloroplasts since processing occurs in more than one compartment (Figure 10.14). A soluble peptidase has been partially purified from the mitochondrial matrix but has not been extensively characterized (see 1289).

10.34 Soluble and Membrane-Bound Proteins Required for Translocation

Several cytosolic and membrane-bound protein components which are required for translocation have been identified. Best characterized by far are the protein factors required for translocation across and incorporation into the mammalian endoplasmic reticulum (see Figure 10.15).

(1) *Signal recognition particle* (SRP): This is a soluble ribonucleoprotein complex consisting of six different proteins plus a 7S RNA molecule (207, 1548). The SRP is required to initiate translocation. It binds to the signal sequence of the nascent polypeptide as it is being synthesized by the ribosome. For preprolactin, for example, the dissociation constant is estimated to be 1 n*M* (1196). Photochemical crosslinking has identified one of the polypeptides (54 kDa) as directly interacting with the signal sequence of the precursor (1579). In some cell-free assays, the SRP-signal interaction inhibits translation or causes a translational pause (elongation arrest), but this might be an experimental artifact (see 1288) and, in any event, has been shown in modeling studies to be unnecessary to explain the kinetics of protein translocation in vivo (1196). One likely role of the SRP is to prevent the nascent polypeptide from folding in such a way as to become incompetent for translocation, by, for example, occluding the signal sequences. The retardation of translational elongation would, presumably, reduce the chances of such aberrant folding and, thus, increase the efficiency of protein translocation.

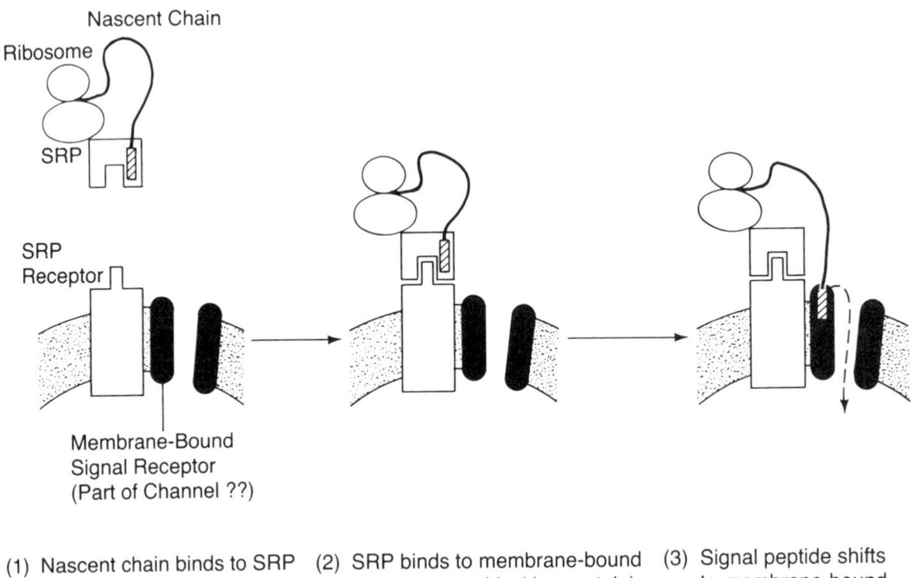

Figure 10.15. Schematic showing the early steps in the co-translational translocation of a polypeptide across the mammalian endoplasmic reticulum. The signal recognition particle (SRP) and SRP receptor (docking protein) are well characterized. The membrane-bound signal receptor has been drawn as a component of a channel across the membrane, but the existence of such channels and the role of the signal receptor are both speculative. After the formation of the complex between the signal peptide and the membrane-bound receptor in step 3, the SRP and docking protein can dissociate and recycle, leaving a membrane-bound ribosome and the nascent chain bound to the translocation apparatus.

Some small proteins (<8.5 kDa) will be translocated into the endoplasmic reticulum independently of the SRP. These include frog prepropeptide GLa (1297), prepromelittin (1033), which are both precursors of secreted proteins, and the M13 procoat protein (1577). In all these examples, the conformation of the precursor is such that they must remain competent for translocation even in the absence of the SRP and ribosomes (1033).

(2) *SRP receptor or docking protein*: The complex containing the SRP/ribosome/nascent chain is targeted to the rough endoplasmic reticulum through a strong interaction between the SRP and a membrane-bound SRP receptor, also called the docking protein (656, 823 1419). The SRP receptor contains a 73-kDa subunit which has an amino-terminal anchor in the membrane. Presumably, the ribosome also binds to specific receptors present in the membrane (1620).

(3) *Signal-sequence receptor*: The signal sequence on the nascent polypeptide chain moves from the SRP to a second receptor present in the membrane called

the signal-sequence receptor (1578, 1547). This has tentatively been identified by photochemical crosslinking using a label attached to the preprolactin signal sequence. The putative membrane-bound receptor is a glycoprotein with a molecular weight of 35,000. Possibly, this forms part of the translocation channel, but this is not known. Another candidate for a signal sequence receptor (45,000 Da) has also been found by the same crosslinking approach, but using a synthetic signal peptide (1233). The relationship between the two proteins is not known, nor have the functions of these polypeptides been demonstrated. Once the nascent chain is associated with the membrane-bound receptor, the SRP and SRP receptor can be released from the membrane-bound ribosome and can be recycled. Nothing is known about the putative channel which mediates translocation, and the isolation of such a species is the goal of considerable research effort.

Soluble factors have also been shown to be required for translocation across the yeast endoplasmic reticulum (589, 1556), though they have not been shown to be like SRP. Translocation of various secreted proteins in yeast can occur post-translationally, so possibly the nascent chains, in these cases, can interact directly with the membrane-bound signal sequence receptor (1547). Soluble protein factors appear to be required for the export of proteins in *E. coli* (1034) and also for the import of proteins into the mitochondrion (1152), but these have not been well characterized. The *secY* gene product, required for protein export in *E. coli*, has been shown to be membrane-bound, but its function is unknown (10).

Some progress has been made identifying a receptor in the mitochondrial outer membrane that is required for protein import. Binding studies with porin have shown that there is a high-affinity receptor (dissociation constant $\sim 10^{-8}\,M$) which is also required for the import of the ADP/ATP transporter, an inner membrane protein (1147). Two different polypeptides have been implicated as being components of an outer membrane signal sequence receptor (513, 1089). These have been found by using antibodies which block protein import (1089) or by using chemical crosslinking to a synthetic signal peptide (513). Interestingly, import into mitochondria can reportedly bypass the outer membrane entirely if it is stripped off to form mitoplasts. This observation suggests that another signal sequence receptor is in the inner membrane (1088).

10.35 Assembly of Multisubunit Complexes and Membrane Protein Turnover

Once a membrane polypeptide has been inserted into the membrane, it still must attain the proper conformation required for biological activity and, in the case of multisubunit complexes, bind to other proteins. In eukaryotes, in particular, these processes require covalent modifications such as glycosylation, acylation, sulfation, or the formation of disulfide bonds. Even when such modifications are not necessary, the conformational maturation process can be slow and temporally separable from insertion into the membrane.

In *E. coli*, for example, the assembly of stable trimers of both proteins LamB (1533) and OmpF (66) can be observed well after the insertion of monomers into the outer membrane. For LamB, this maturation takes about 5 min. In eukaryotic cells, it has been shown that in order to be transported from the endoplasmic reticulum to the Golgi, the influenza hemagglutinin glycoprotein must attain the properly folded trimeric quaternary structure corresponding to the mature form (509). Unfolded molecules of hemagglutinin are retained in the endoplasmic reticulum. Trimer formation takes about 7-10 min (509). A similar requirement for oligomerization has also been shown for the vesicular stomatitis G protein (785).

The assembly of many multisubunit complexes containing different kinds of subunits also appears to occur within the endoplasmic reticulum. An example is the nicotinic acetylcholine receptor (1359), which contains two copies of an α subunit and one copy each of ß, γ, and δ subunits (see Figure 8.8). Antibodies can distinguish distinct forms of the α subunit: (1) the initial product inserted into the endoplasmic reticulum which cannot bind the antagonist α-bungarotoxin; (2) a form of the α subunit that can bind to α-bungarotoxin that forms several minutes after translation within the endoplasmic reticulum; (3) assembled receptor, containing all the subunits ($\alpha_2\beta\gamma\delta$), which can be observed 15 minutes after translation within the endoplasmic reticulum; (4) the final receptor on the cell surface, which appears about 2 hours later. Complete maturation includes the formation of disulfide bonds, oligosaccharide processing, and fatty acylation. Phosphorylation of the subunits appears to play a role in subunit assembly, which has also been observed to take place in the Golgi (1243).

The stability of stoichiometric complexes such as the acetylcholine receptor appears to be a critical aspect of the assembly process. In several systems, it appears that individual subunits are synthesized in substantial excess, and that subunits that are not assembled into stable complexes are proteolytically degraded (see 887 for review). This has been shown for several multisubunit complexes, including the T cell antigen receptor (994).

Maturation and assembly of the sodium channel has also been studied (1300). The mature channel requires a disulfide linkage between the α and ß2 subunits (see Section 8.25). The formation of this covalent link, however, occurs about 1 hour after translation and occurs after the subunits have been transported to the Golgi. The receptor appears on the cell surface 4 hours after translation. In this case, free α subunits are not rapidly degraded but are maintained in an intracellular pool, possibly to be used as precursors for channel assembly in growing neurons.

Once membrane proteins have fully matured, they undergo turnover at varying rates. The sodium channel, for example, has a half-life of about 30 hr (1300), which is typical for surface proteins. In growing cultured cells, the large subunit of the Na/K-ATPase has a half-life of about 20 to 40 hr (727, 1429). It is likely that at least some, if not all, membrane protein degradation occurs in the lysosome. This has been shown for cytochrome P450, for example, which is located in the endoplasmic reticulum (931).

Box 10.3 Two Interesting Examples of Membrane Protein Turnover

Many different factors determine the stability of individual proteins within a cell (see 1205). A particularly interesting example of membrane protein degradation is HMG CoA reductase (512, 768). This enzyme is in the smooth endoplasmic reticulum, and it regulates the endogenous synthesis of cholesterol. It has been shown that degradation of this protein is rapid (2-4 hr) and that it is accelerated by an interaction with cholesterol. The membrane-bound domain of the enzyme is required for this acceleration of degradation (512). Under certain conditions, a strain of Chinese hamster ovary cells will be induced to overproduce this enzyme 500 times above the normal levels. This causes the formation of membrane tubules occupying 15% of the cell volume and containing, in addition, lipids as well as other membrane proteins (768). The addition of cholesterol causes the rapid degradation of this excess membrane, indicating both coordinate synthesis and degradation of the membrane components.

Another interesting example of selective turnover of a membrane protein is the 32-kDa herbicide-binding protein (also called the D1 and Q_B protein) from chloroplast thylakoids. This protein is synthesized as a precursor, processed within the unstacked thylakoid lamellae, and assembled as part of photosystem II in the stacked thylakoid lamellae (940). In the presence of light, the rate of turnover of this polypeptide is much faster than other proteins in the membrane (630). The reason for this is not known, but it may be caused by light-induced damage within photosystem II.

10.4 Membrane Lipid Biosynthesis and Distribution
(see 298 for review)

As pointed out in Chapter 1, an enormous number of different lipids are found in the membranes of a single eukaryotic cell. These lipids are not uniformly distributed among the various cell membranes, however. This applies to the distribution in terms of the polar headgroup (Table 1.3) as well as the acyl chains (Table 1.6). For example, the degree of unsaturation of the phospholipids generally decreases as one proceeds from the endoplasmic reticulum to the Golgi and then to the plasma membrane (298). Whereas the average molar ratio of cholesterol/phospholipid is 0.3-0.4 for an animal cell, the ratio is much higher (0.8-0.9) in the plasma membrane than in other membranes, such as the rough endoplasmic reticulum (cholesterol/phospholipid ~0.1). In addition to this obvious uneven distribution of lipid components among the various membranes, in many membranes the two halves of the bilayer are asymmetric with respect to lipid composition, as discussed in Chapter 4.

The problem of how these compositional differences are established and maintained is extraordinarily challenging. It must be recalled that many of the membranes in eukaryotic cells participate in the endocytic and exocytic pathways, in which vesicles are constantly budding off one membrane and fusing with another. This process is selective for those proteins which are transported by the vesicles, but somehow the system does not result in identical lipid compositions of the various membranes involved. The way in which this is accomplished is not known, but at least some aspects of the problem have been clarified. To start, let us first see where membrane lipids are synthesized and then discuss how they may be transported to their destinations.

10.41 Where are Membrane Lipids Synthesized? (see 298, 1498, 1235)

Prokaryotes

In *E. coli*, all the phospholipid synthesis occurs in the cytoplasmic membrane (see 1235). The general features of the biosynthetic pathway are shown in Figure 10.16. Fatty acids are synthesized as precursors covalently attached to the acyl carrier protein and are then incorporated in the membrane by acylation reactions to form CDP-diacylglycerol. The two acylating enzymes (Figure 10.16) have substrate specificities resulting in the preponderance of unsaturated fatty acyl groups being in the *sn*-2 position. The CDP-diacylglycerol is a branchpoint, from which either phosphatidylserine or phosphatidylglycerol is made, as shown in Figure 10.16. Regulation of the balance of phospholipids in the membrane is poorly understood. Strains that overproduce the enzyme required for cardiolipin biosynthesis by 10-fold have only a marginal increase in the amount of this phospholipid in the membrane (1091). Similar results have been obtained for the phosphatidylserine synthase (step 4, Figure 10.16) and phosphatidylglycerol-phosphate synthase (step 5, Figure 10.16) (see 687). Overproduction of these enzymes has little influence on the phospholipid composition of the membrane. On the other hand, the complete elimination of phosphatidylglycerolphosphate synthase is lethal, probably because the acidic phospholipids are required to provide precursors for other biosynthetic pathways or because they are important for regulation (610). Strains can be produced in which phosphatidylethanolamine is the only major phospholipid in mutants where the level of phosphatidylglycerolphosphate synthase is greatly reduced (999). Several enzymes in the biosynthetic pathway have been purified, including CDP-diglyceride synthase (step 3) (1366) and phosphatidylserine synthase (step 4) (1188). The former, not unexpectedly, is a membrane-bound enzyme, but phosphatidylserine synthase is associated with ribosomes in cell-free extracts of *E.coli*.

Eukaryotes (see 1498)

Figure 10.17 summarizes the more complex biosynthetic network for phospholipid biosynthesis in eukaryotic cells. Fatty acid elongation occurs in both the endoplasmic reticulum and the mitochondrion. Fatty acyl CoA is used as a substrate, since animal cells do not have an equivalent of the acyl carrier protein. Most of the enzymes for phospholipid biosynthesis are located on the cytoplasmic

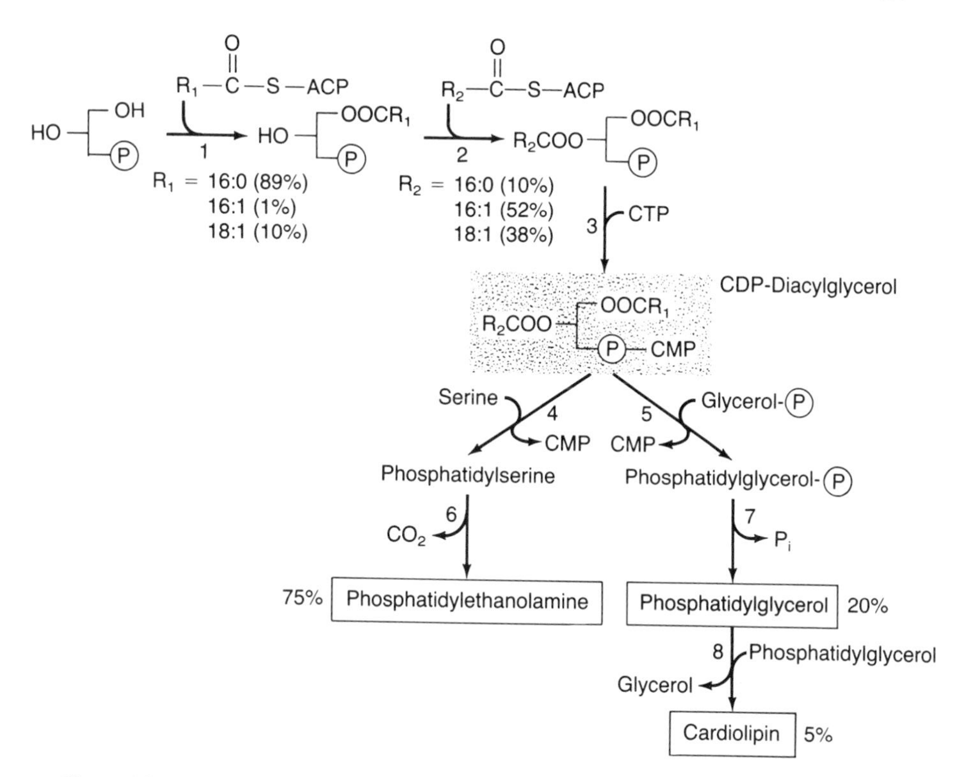

Figure 10.16. General scheme for phospholipid biosynthesis in *E. coli*. CDP-diacylglycerol is a branchpoint, leading to the synthesis of either phosphatidylethanolamine or phosphatidylglycerol and cardiolipin. The percentages of the lipids in the membrane are indicated. Note that phosphatidic acid and phosphatidylserine are necessary intermediates but do not accumulate in wild-type cells. The enzymes which catalyze each step are as follows: *step 1*, *sn*-glycerol-3-phosphate acyltransferase; *step 2*, 1-acyl-*sn*-glycerol-3-phosphate acyltransferase; *step 3*, CDP-diacylglycerol synthase; *step 4*, phosphatidylserine synthase; *step 5*, phosphatidylglycerol phosphate synthase; *step 6*, phosphatidylserine decarboxylase; *step 7*, phosphatidylglycerol phosphate phosphatase; step 8, cardiolipin synthase. See ref. 1235.

side of the endoplasmic reticulum membrane, but there are interesting exceptions. Eukaryotic cells can make phosphatidylglycerol and cardiolipin via the CDP-diacylglycerol pathway, as in *E. coli*. However, the enzymes required for this are located in the mitochondrial inner membrane and these phospholipids are uniquely found in the mitochondrion. Lower eukaryotes, such as yeast, can also make phosphatidylserine by the same pathway as found in *E. coli*, but this enzyme (1069) is found in the endoplasmic reticulum as well as in the mitochondrion. In yeast, phosphatidylserine constitutes only about 5% of the total phospholipid, but it is the precursor for both phosphatidylethanolamine and phosphatidylcholine.

Eukaryotes, including yeast, can make phosphatidylethanolamine and phosphati-

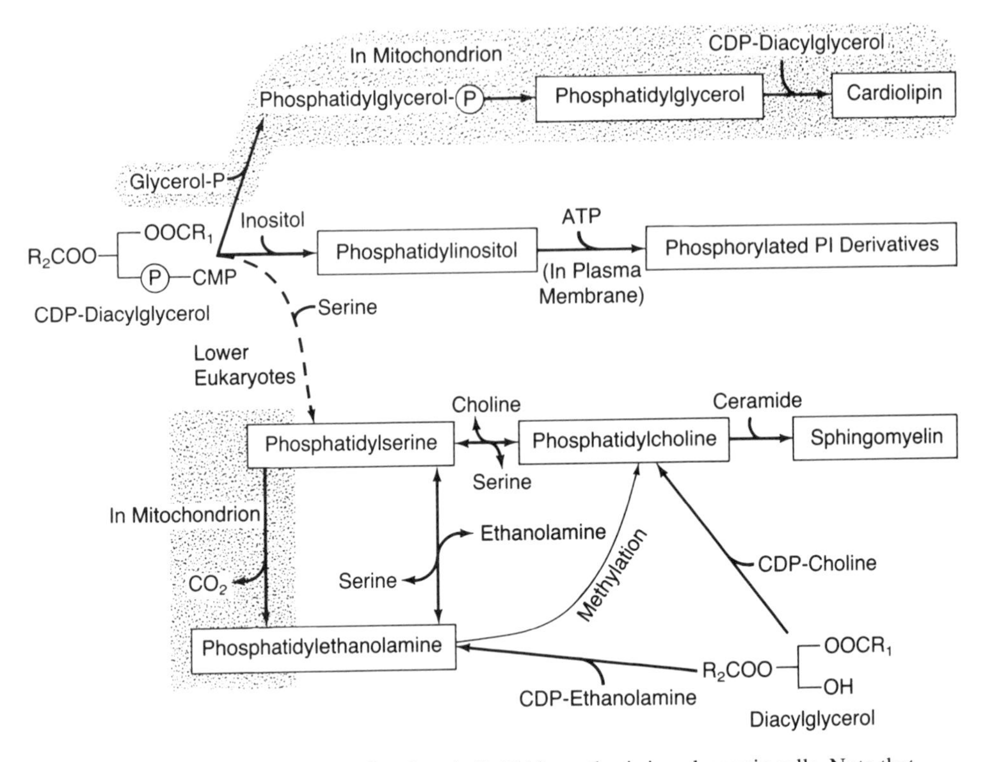

Figure 10.17. General scheme for phospholipid biosynthesis in eukaryotic cells. Note that there are essentially two pathways into the network of reactions. CDP-diacylglycerol, on the left, and diacylglycerol, on the lower right, are precursors. Although most of the enzymes catalyzing these reactions are in the endoplasmic reticulum, the enzymes required for the synthesis of phosphatidylglycerol and cardiolipin, as well as phosphatidylserine decarboxylase, are in the mitochondrion. In yeast, the synthesis of phosphatidylserine is from CDP-diacylglycerol, as in bacteria. In higher eukaryotes, phosphatidylserine is made primary from either phosphatidylethanolamine or phosphatidylcholine by the base-exchange enzymes.

dylcholine by a reaction with l, 2-diacylglycerol (see Figure 10.17). This is the major pathway for phospholipid synthesis in animal cells. Base-exchange enzymes (794) located in the endoplasmic reticulum of higher eukaryotes (but not yeast) then catalyze a headgroup exchange to form phosphatidylserine. Higher eukaryotes require this exchange reaction to make phosphatidylserine. Once formed in the endoplasmic reticulum, phosphatidylserine is then decarboxylated, at least in some cells, to reform phosphatidylethanolamine. The enzyme that catalyzes this reaction, phosphatidylserine decarboxylase, is located in the mitochondrion. Hence, phosphatidylserine can be a major precursor for all of the phosphatidylethanolamine in the cell (794, 1521). This series of reactions requires transfer of the lipids between the endoplasmic reticulum and the mitochondrion.

Phosphatidylethanolamine can be converted to phosphatidylcholine by one or two methylases located in the endoplasmic reticulum (1118, 769), but this is usually not a major pathway.

Two lipid components which are predominately localized in the plasma membrane are sphingomyelin and cholesterol. Sphingomyelin appears to be made in some cells by the transfer of the phosphorylcholine group from phosphatidylcholine to ceramide by an enzyme located in the plasma membrane. In contrast, even though cholesterol is most concentrated in the plasma membrane, it is synthesized by enzymes located in the smooth endoplasmic reticulum and, possibly, in peroxisomes (1451).

10.42 Transport of Lipids from Their Site of Synthesis (see 298, 1520)

In order for membrane lipids to get from their place of synthesis to their destination, two processes are implied: (1) transbilayer flip-flop and (2) interbilayer transport. The rate of flip-flop across a bilayer is discussed in Section 4.43 in the context of membrane asymmetry. The rate of phospholipid flip-flop is particularly fast across membranes where lipid biosynthesis occurs, with half-times on the order of several minutes. There is some indication that this may be protein-mediated and possibly even an active process, driven by ATP hydrolysis. Cholesterol has also been shown to be able to spontaneously flip-flop very rapidly across membranes (see 298). Hence, moving from the cytosolic side of the endoplasmic reticulum to the lumenal side is a rapid process.

In considering the movement of lipids from one cellular membrane to another, several possible processes need to be considered. Each may contribute in particular situations.

1. Spontaneous lipid transfer by diffusion of monomeric lipid species through the aqueous phase.
2. Lipid diffusion through direct contact of the two membranes via either permanent or transient junctions.
3. Protein-mediated transport, catalyzed either by proteins which facilitate the desorption of lipids out of the donor membrane or by lipid-binding proteins.
4. Vesicle-mediated transport in which lipids are moved between membranes much as membrane proteins are thought to be transported by continuous budding and fusion of intracellular vesicles. This could be energy requiring.

Let us first consider what is known about the movement of membrane lipids by spontaneous diffusion between membranes (see 1154, 298). Numerous studies have demonstrated that lipids can spontaneously transfer between unilamellar vesicles or between phospholipid vesicles and biomembranes. In most cases, but not all (1044), the mechanism appears to involve the desorption of monomeric lipids from the surface of the donor membrane, and free diffusion through the aqueous medium to the acceptor membrane. The rate-limiting step is the removal from the donor membrane, at least under conditions where the acceptor membrane is in excess. The half-times for transfer are, under these conditions, dependent on the height of the free energy barrier for desorption (Figure 10.18). Not surprisingly, lipids which are less water-soluble (i.e., low critical micelle concentration), as a rule, have a higher barrier to desorption and, thus, a longer half-time for transfer. Transfer rates depend not only on the hydrophobicity of the lipid being transferred but also on the composition and physical state of the donor bilayer (153, 63, 1618). For example, the ganglioside GM_1 exists in a monodisperse state when present in phosphatidylcholine vesicles. Due to the hydrophilic polar groups, GM_1 will not flip-flop across the vesicle membrane, but it will transfer between vesicles with a half-time of about 40 hr at 45°C (153). In contrast, neutral gangliosides lacking the sialic acid residues, such as asialo-GM_1, form a gel-like cluster in these vesicles, and their half-time for transfer is about 500 hours. Vesicles containing mixtures of cholesterol and phospholipids also form complex phases (see Section 2.42) and this phenomenon might also be important in determining the transfer kinetics of cholesterol (63). The relative stability of cholesterol within a membrane could be determined by favorable interactions with specific phospholipids (1618, 1557) such as sphingomyelin.

Table 10.2 lists some values for the interbilayer transfer half-times of several membrane lipids. At one extreme are cholesterol esters which are very nonpolar and, essentially, do not transfer between membranes by the monomer diffusion mechanism. At the other extreme are lysophospholipids which very rapidly move between membranes. Cholesterol, typically, has a half-life for transfer of 1-2 hr. Hence, a pertinent question concerning cholesterol is how the unequal distribution in the different membranes is maintained.

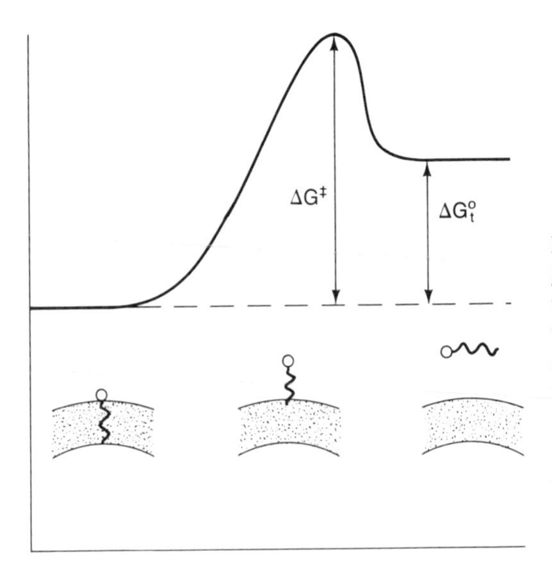

Figure 10.18. Free energy diagram for the rate-limiting desorption of a lipid from a bilayer. Note that other things being equal, the activation barrier to solubilization will increase in proportion to the standard state transfer free energy into water, which is related to the critical micelle concentration (see Section 2.32). Adapted from ref. 1154.

Table 10.2. Half-times for phospholipid transfer from phospholipid vesicles[l] (see 1154, 153)

Compound	Half-time
Cholesterol oleate	10^7 hr (calculated)
Cholesterol	1–2 hr
Dipalmitoyl phosphatidylcholine	83 hr
1-Palmitoyl-2-oleoyl phosphatidylcholine	63 hr
Dimyristoyl phosphatidylcholine	2 hr
Palmitoyl lysophosphatidylcholine	< 0.05 hr
Ganglioside GM_1	40 hr (45°C)
Asialo-GM_1	580 hr (45°C)

[l]Experiments were performed at 37°C unless otherwise noted. The compositions of the donor vesicles are not identical for these measurements. The half-times are characteristic of the rate-limiting desorption of lipid from the donor vesicle.

Transit times as short as 10 min have been reported for the transfer of newly synthesized cholesterol from the endoplasmic reticulum to the plasma membrane (see 298). The process is sensitive to reagents such as cyanide which block bioenergetic reactions within the cell. These and other data suggest an energy-requiring, vesicle-mediated intracellular transport mechanism for cholesterol. In principle, this could overcome any scrambling due to spontaneous transfer. However, there is no consensus about this. One serious problem is that there are large inconsistencies in the literature in estimating the fraction of the total cellular cholesterol which is present in the plasma membrane (25% to 95%) (1493). Although it seems likely that the flux of cholesterol into and out of some cells can be understood using spontaneous monomeric diffusion rates (1154), the relevance of these rates and of the spontaneous transfer mechanism inside the cell is not known.

The transfer half-times of phospholipids from phospholipid vesicles are much longer than for cholesterol (Table 10.2). For example, dipalmitoyl phosphatidylcholine has a half-time for transfer of 83 hr at 37° from vesicles prepared from dimyristoyl phosphatidylcholine (1154). The rate of transfer of phospholipids by this mechanism is much too slow to be relevant for the interbilayer transport that occurs in vivo.

Prokaryotes

Let us first discuss phospholipid transfer in gram-negative bacteria, where the problem is relatively simple since there are only two membranes (see 1235). The phospholipids are synthesized in the cytoplasmic membrane and must be transported to the outer membrane (for review see 1520). Several experiments indicate rapid exchange of the lipids in these two membranes. Phosphatidylethanolamine gets to the outer membrane with a half-time of about 3 min. This process does not depend on protein, lipid, or ATP synthesis but is dependent on a proton motive force in an unknown way. Lipids fused into the outer membrane can also rapidly move to the inner membrane, even phosphatidylcholine, which is not normally found in *E. coli*. Hence, the process seems nonspecific. However, in a mutant which has a defective diacylglycerol kinase, diacylglycerol accumulates in the cytoplasmic membrane and is not transported to the outer membrane.

The mechanism of lipid exchange between these two membranes is not known, but it is generally believed that there are junctions or zones of adhesion between the inner and outer membranes. These junctions may serve as sites for diffusion of lipids between the two membranes. Conceivably, proteins exported to the outer membrane could also be assembled at these junctions.

Eukaryotes

The mechanism by which phospholipids are distributed in eukaryotic cells presents a much more complicated situation. Cardiolipin is the only phospholipid which is localized uniquely at its site of synthesis in the mitochondrion. Intracellular transport of phospholipids has been shown in numerous experiments to be considerably faster than can be accounted for by spontaneous diffusion through the aqueous medium (see 298, 1520). For example, the transfer of newly synthesized phospholipid from the endoplasmic reticulum to the mitochondrion in rat liver takes only minutes. Since the decarboxylase that converts phosphatidylserine to phosphatidylethanolamine is located in the mitochondrion (Figure 10.17), the rate of this reaction can be used to monitor the net transfer of phosphatidylserine from the endoplasmic reticulum to the mitochondrion. Results show not only that transfer is rapid but also that it is inhibited by azide and agents that block bioenergetic reactions. Studies have also indicated rapid transfer of newly synthesized phospholipids to the plasma membrane in just a few minutes, though this is very dependent on the lipid and the cell type (see 298, 1520). In some examples, lipid transport is blocked by inhibitors of bioenergetic reactions (energy poisons) and/or by agents which disrupt the cytoskeleton. The main point of agreement is that the rates are fast relative to spontaneous diffusion.

One possible mechanism to facilitate phospholipid transfer between membranes is for proteins to mediate the process. A number of water-soluble proteins called phospholipid transfer proteins have been isolated which mediate the intermembrane transfer of phospholipids in vitro (see 200 for review). These proteins all have lipid binding sites and form water-soluble phospholipid-protein complexes. They facilitate the one-to-one exchange of lipids between membranes by carrying monomeric species back and forth. In some cases, such as the phosphatidylcholine-specific transfer protein from bovine liver, the lipid binding is specific and the affinity high (1492). In other cases, the binding affinity and specificity are relatively low (1066). A low affinity, in principle, would facilitate the ability of the protein to catalyze the net transfer of lipids from one membrane to another, because the protein could return to the donor membrane with an empty lipid binding site. Phospholipid transfer proteins have been isolated from bacteria

(1418) and plants (126), in addition to animal cells. Similar proteins have been isolated that bind to glycolipids (1284) and to long-chain fatty acids (141).

Unfortunately, the function of phospholipid transfer proteins has never been demonstrated in vivo, so it remains an open question whether they are physiologically important for the intracellular flux of phospholipids. Alternatively, vesicle-mediated processes are likely candidates, though both mechanisms may be operative.

It is important to realize that, ultimately, the distribution of phospholipids among the various cellular membranes must rely on proteins. If the distribution of lipids represents equilibrium, then their distribution is determined by the lipid-binding affinities of membrane proteins which are present in the various membranes. A bias in the distribution of some lipids due to interaction with proteins could influence the partitioning of other lipids, such as cholesterol (1557, 1618). If the lipid distribution is not at equilibrium, then the different compositions among the various membranes must be determined by different rates of delivery and removal of the lipid species from each membrane (see 1006). In this case, proteins once again must be responsible for this differential kinetics.

10.43 Phospholipid Turnover (see 298, 1235)

Little is known about phospholipid turnover, except that it is rapid. In animal cells, half of the total phospholipid is turned over every one or two cell divisions. The rates of turnover for the polar and nonpolar portions of phospholipids are different. This appears to be a manifestation of the reactions of extralysosomal phospholipases and acyltransferases which function to remodel intracellular phospholipids in situ. Sphingolipids are degraded in the lysosomes. Disease states result when sphingolipid degradation is impaired. For example, in Niemann-Pick disease there is an accumulation of sphingomyelin due to the lack of a lysosomal enzyme, sphingomyelinase. The degradation of glycosphingolipids also occurs in the lysosome in vivo and is thought to be mediated by proteins called sphingolipid activator proteins (SAPs) which bind to and presumably solubilize specific lipids such as GM_1 ganglioside (1611).

In *E. coli*, there are at least ten degradative enzymes which could play roles in phospholipid turnover (1235). Little is known about the functions in vivo, however. During exponential growth, the phospholipids exhibit little turnover of their glycerol backbone or fatty acids, and acyl exchange also appears to be quite slow (1235, 1234). The turnover of the glycerolphosphate headgroup of phosphatidylglycerol, however, has been demonstrated to be unusually rapid. This results from the utilization of phosphatidylglycerol to donate *sn*-glycerol-1-phosphate for the synthesis of membrane-derived oligosaccharides (686). These oligosaccharides are located in the periplasm and play a role in osmotic adaptation of the organism. The reaction results in the conversion of phosphatidylglycerol to diacylglycerol, which is then salvaged by diacylglycerol kinase (1545), which converts it to phosphatidic acid.

10.5 Adaptation of Membrane Lipid Composition in Response to Environmental Changes (see 252 for review)

From the previous discussion, it is clear that little is known about how the lipid composition of individual cellular membranes is controlled. Obviously, one minimal requirement is that the membrane lipids must form a stable bilayer in a liquid crystalline (fluid) state. A number of organisms have adaptive mechanisms to alter their lipid composition in response to changes in the environment, most notably temperature and pressure. Several examples illustrate the phenomenon. The mechanism for the thermal adaptation of *E. coli* is best understood.

(1) *Thermal adaptation of E. coli* (see 1235). When *E. coli* is grown at low temperatures, the fatty acid composition changes to favor a higher proportion of unsaturated fatty acids. This maintains the overall fluidity of the membrane and permits the cells to survive at temperatures which would otherwise be lethal. This has been termed homeoviscous adaptation. The predominant fatty acids in *E. coli* (see Figure 10.16) are palmitoyl (16:0), palmitoleoyl (16:1), and *cis*-vaccenoyl (18:1). At low temperatures, more *cis*-vaccenic acid is incorporated in the bilayer, whereas the amount of palmitoleic acid is constant. The explanation lies in the activity of one of the two enzymes which catalyze the elongation of fatty acids. At low temperatures the enzyme 3-ketoacyl-ACP synthase II is more active in the conversion of palmitoleic acid to *cis*-vaccenic acid, resulting in an increase in the pool size of unsaturated fatty acids to be incorporated into the phospholipids.

(2) *Organisms at high pressure*: The fatty acid composition of the barophilic marine bacterium (NPT3) varies as a function of pressure (321). Pressure can also induce a deleterious phase transition in membranes, and this adaptation response, as in the previous example, allows the organism to survive. At high pressures, the proportion of unsaturated fatty acids increases, as one would expect. The packing properties of the unsaturated fatty acids maintain the membrane in the liquid crystalline state. Similar results have been obtained in examining the liver mitochondrial phospholipids of deep-sea fish obtained from different ocean depths (251).

(3) *Studies with Acholeplasma laidlawii*: This organism is a mycoplasma which has no cell wall and a single plasma membrane. It is particularly useful for laboratory study because it will readily incorporate exogenous fatty acids into its membrane, which can be easily isolated and examined. The major lipid components are monoglucosyldiglycerides (MGDG) and diglucosyldiglycerides (DGDG). As explained in Chapter 2, because of the difference in the size of the polar headgroup, isolated MGDG forms a reverse hexagonal phase, whereas DGDG forms a stable bilayer. The phase properties of *A. laidlawii* membranes are determined primarily by the MGDG/DGDG ratio (1582). The lipid composition has been examined when the organism is grown in particular fatty acids and temperatures (1219) or in the presence of membrane perturbants such as

alcohols and detergents (1582). In these examples, a simple consideration of membrane viscosity is not sufficient to explain the adaptive response, since the equilibrium between the lamellar and nonlamellar phases must also be important. The results can generally be interpreted in terms of lipid packing parameters as they influence the phase properties of the membrane lipid mixture (see Chapter 2). The mechanisms by which the organism adapts to its environment by altering its lipid composition are not known.

10.6 Chapter Summary

A eukaryotic cell contains many membranous organelles and many distinct intracellular membranes, each with a unique protein and lipid composition. Any membrane protein which is nuclear-encoded must be properly directed from its site of synthesis on a ribosome in the cytoplasm to its final destination. This requires a complex series of signals contained within either the mature form of the polypeptide or a precursor, as well as receptors within the cell to recognize these signals. Although some membrane proteins have been shown to spontaneously insert into a lipid bilayer, in most cases proper assembly of a protein within a cellular membrane is an energy-requiring process which is mediated by specialized apparatus. It is likely that proteins cannot be inserted into a cellular membrane unless they are in a partially unfolded conformation. ATP hydrolysis as well as the binding of specific proteins in the cytoplasm may help to attain or maintain the unfolded conformations that are competent for membrane translocation.

In a eukaryotic cell, most of the lipids are synthesized in the endoplasmic reticulum. Interbilayer transport is required to get the lipids to their final destinations. The nature of the transport mechanisms is not known, although vesicle-mediated transport and protein-mediated transport via soluble lipid-binding proteins are two possibilities that are prominently discussed. The manner by which the distinct compositions of the various intracellular membranes are maintained is not known.

Appendix 1
Single-Letter Codes for Amino Acids

Amino acid	Three-letter abbreviation	One-letter symbol
Alanine	Ala	A
Arginine	Arg	R
Asparagine	Asn	N
Aspartic acid	Asp	D
Cysteine	Cys	C
Glutamine	Gln	Q
Glutamic acid	Glu	E
Glycine	Gly	G
Histidine	His	H
Isoleucine	Ile	I
Leucine	Leu	L
Lysine	Lys	K
Methionine	Met	M
Phenylalanine	Phe	F
Proline	Pro	P
Serine	Ser	S
Threonine	Thr	T
Tryptophan	Trp	W
Tyrosine	Tyr	Y
Valine	Val	V

References

1. Abeywardena, M.Y. and J.S. Charnock, (1983), Modulation of Cardiac Glycoside Inhibition of ($Na^+ + K^+$)-ATPase by Membrane Lipids: Differences between Species. Biochim. Biophys. Acta 729, 75-84.
2. Adams, G.A. and J.K. Rose, (1985), Structural Requirements of a Membrane-Spanning Domain for Protein Anchoring and Cell Surface Transport. Cell 41, 1007-1015.
3. Addison, R. (1986), Primary Structure of the Neurospora Plasma Membrane H^+-ATPase Deduced from the Gene Sequence: Homology to Na^+/K^+-, Ca^{2+}, and K^+-ATPases. J. Biol. Chem. 261, 14896-14901.
4. Adrian, G.S., M.T. McCammon, D.L. Montgomery, and M.G. Douglas, (1986), Sequences Required for Delivery and Localization of the ADP/ATP Translocator to the Mitochondrial Inner Membrane. Mol. Cell. Biol. 6, 626-634.
5. Agard, D.A. and R.M. Stroud, (1981), Linking Regions Between Helices in Bacteriorhodopsin Revealed. Biophys. J. 37, 589-602.
6. Aggeler, R., Y.-Z. Zhang, and R.A. Capaldi, (1987), Labeling of the ATP Synthase of Escherichia coli from the Head-Group Region of the Lipid Bilayer. Biochemistry 26, 7107-7113.
7. Aiyer, R.A. (1983), Structural Characterization of Insulin Receptors: I. Hydrodynamic Properties of Receptors from Turkey Erythrocytes. J. Biol. Chem. 258, 14992-14999.
8. Aizawa, S. and M. Tavassoli, (1987), In Vitro Homing of Hemopoietic Stem Cells Is Mediated by a Recognition System with Galactosyl and Mannosyl Specificities. Proc. Natl. Acad. Sci. USA 84, 4485-4489.
9. Akabas, M.H., F.S. Cohen, and A. Finkelstein, (1984), Separation of the Osmotically Driven Fusion Event from Vesicle-Planar Membrane Attachment in a Model System for Exocytosis. J. Cell Biol. 98, 1063-1071.
10. Akiyama, Y. and K. Ito, (1986), Overproduction, Isolation and Determination of the Amino-terminal Sequence of the SecY Protein, a Membrane Protein Involved in Protein Export in Escherichia coli. Eur. J. Biochem. 159, 263-266.
11. Akiyama, Y. and K. Ito, (1987), Topology Analysis of the SecY Protein, an Integral

Membrane Protein Involved in Protein Export in Escherichia coli. The EMBO Journal 6, 3465-3470.

12. Akutsu, H., Y. Suezaki, W. Yoshikawa, and Y. Kyogoku, (1986), Influence of Metal Ions and a Local Anesthetic on the Conformation of the Choline Group of Phosphatidylcholine Bilayers Studied by Raman Spectroscopy. Biochim. Biophys. Acta 854, 213-218.
13. Albert, A.D., S.A. Lane, and P.L. Yeagle, (1985), ^{2}H and ^{31}P Nuclear Magnetic Resonance Studies of Membranes Containing Bovine Rhodopsin. J. Membrane Biol. 87, 211-215.
14. Albertsson, P.-A. (1974), Countercurrent Distribution of Cells and Cell Organelles. Methods in Enz. 31, 761-769.
15. Albertsson, P.-A. (1979), Phase Partition of Cells and Subcellular Particles. Separation of Cells and Subcellular Elements (H. Peeters, Ed.), pp. 17-22. Pergamon Press, New York.
16. Albon, N. and J.M. Sturtevant, (1978), Nature of the Gel to Liquid Crystal Transition of Synthetic Phosphatidylcholines. Proc. Natl. Acad. Sci. USA 75, 2258-2260.
17. Allen, J.P., G. Feher, T. O. Yeates, H. Komiya, and D.C. Rees, (1987), Structure of the Reaction Center from Rhodobacter sphaeroides R-26: The Cofactors. Proc. Natl. Acad. Sci. USA 84, 5730-5734.
18. Allen, J.P., G. Feher, T.O. Yeates, D.C. Rees, J. Deisenhofer, H. Michel, and R. Huber, (1986), Structural Homology of Reaction Centers from Rhodopseudomonas sphaeroides and Rhodopseudomonas viridis as Determined by X-Ray Diffraction. Proc. Natl. Acad. Sci. USA 83, 8589-8593.
19. Allen, P. and G. Feher, (1984), Crystallization of Reaction Center from Rhodopseudomonas sphaeroides: Preliminary Characterization. Proc. Natl. Acad. Sci. USA 81, 4795-4799.
20. Allen, T.M., A.Y. Romans, H. Kercret, and J.P. Segrest, (1980), Detergent Removal During Membrane Reconstitution. Biochim. Biophys. Acta 601, 328-342.
21. Allison, D.S. and G. Schatz, (1986), Artificial Mitochondrial Presequences. Proc. Natl. Acad. Sci. USA 83, 9011-9015.
22. Allured, V.S., R.J. Collier, S.F. Carroll, and D.B. McKay, (1986), Structure of Exotoxin A of Pseudomonas aeruginosa at 3.0-Angstrom Resolution. Proc. Natl. Acad. Sci. USA 83, 1320-1324.
23. Almog, S., T. Kushnir, S. Nir, and D. Lichtenberg, (1986), Kinetic and Structural Aspects of Reconstitution of Phosphatidylcholine Vesicles by Dilution of Phosphatidylcholine-Sodium Cholate Mixed Micelles. Biochemistry 25, 2597-2605.
24. Alonso, A., C.J. Restall, M. Turner, J.C. Gomez-Fernandez, F.M. Goni, and D. Chapman, (1982), Protein-Lipid Interactions and Differential Scanning Calorimetric Studies of Bacteriorhodopsin Reconstituted Lipid-Water Systems. Biochim. Biophys. Acta 689, 283-289.
25. Alpes, H., K. Allman, H. Plattner, J. Reichert, R. Riek, and S. Schulz, (1986), Formation of Large Unilamellar Vesicles Using Alkyl Maltoside Detergents. Biochim. Biophys. Acta 862, 294-302.
26. Altenbach, C. and J. Seelig, (1985), Binding of the Lipophilic Cation Tetraphenylphosphonium to Phosphatidylcholine Membranes. Biochim. Biophys. Acta 818, 410-415.
27. Altenbach, C. and J. Seelig, (1984), Ca^{2+} Binding to Phosphatidylcholine Bilayers As Studied by Deuterium Magnetic Resonance. Evidence for the Formation of a Ca^{2+} Complex with Two Phospholipid Molecules. Biochemistry 23, 3913-3920.

28. Alvarez, J., D.C. Lee, S.A. Baldwin, and D. Chapman, (1987), Fourier Transform Infrared Spectroscopic Study of the Structure and Conformational Changes of the Human Erythrocyte Glucose Transporter. J. Biol. Chem. 262, 3502-3509.
29. Ameloot, M., H. Hendrickx, W. Herreman, H. Pottel, F. Van Cauwelaert, and W. Van Der Meer, (1984), Effect of Orientational Order on the Decay of the Fluorescence Anisotropy in Membrane Suspensions. Biophys. J. 46, 525-539.
30. Ames, G.F.-L. (1986), The Basis of Multidrug Resistance in Mammalian Cells: Homology with Bacterial Transport. Cell 47, 323-324.
31. Ames, G. Ferro-Luzzi (1986), Bacterial Periplasmic Transport Systems: Structure, Mechanism, and Evolution. Ann. Rev. Biochem. 55, 397-425.
32. Amey, R.L. and D. Chapman, (1984), Infrared Spectroscopic Studies of Model and Natural Biomembranes. Biomembrane Structure and Function (D. Chapman, Ed.), pp. 199-256. Verlag Chemie, Basel.
33. Anantharamaiah, G.M., J.L. Jones, C.G. Brouillette, C.F. Schmidt, B.H. Chung, T.A. Hughes, A.S. Bhown, and J.P. Segrest, (1985), Studies of Synthetic Peptide Analogs of the Amphipathic Helix: Structures of Complexes with Dimyristoyl Phosphatidylcholine. J. Biol. Chem. 260, 10248-10255.
34. Anderle, G. and R. Mendelsohn, (1986), Fourier-Transform Infrared Studies of Ca ATPase/Phospholipid Interaction: Survey of Lipid Classes. Biochemistry 25, 2174-2179.
35. Andersen, J.P., B. Vilsen, J.H. Collins, and P.L. Jorgensen, (1986), Localization of E1-E2 Conformational Transitions of Sarcoplasmic Reticulum Ca-ATPase by Tryptic Cleavage and Hydrophobic Labeling. J. Membrane Biol. 93, 85-92.
36. Anderson, R.A. and R.E. Lovrien, (1984), Glycophorin Is Linked by Band 4.1 Protein to the Human Erythrocyte Membrane Skeleton. Nature 307, 655-658.
37. Anderson, R.A. and V.T. Marchesi, (1985), Regulation of the Association of Membrane Skeletal Protein 4.1 with Glycophorin by a Polyphosphoinositide. Nature 318, 295-298.
38. Andersson, K. and R. Hjorth, (1984), Isopycnic Centrifugation of Rat-Liver Microsomes in Iso-Osmotic Gradients of Percoll and Release of Microsomal Material by Low Concentrations of Sodium Deoxycholate. Biochim. Biophys. Acta 770, 97-100.
39. Anner, B.M. (1985), Interaction of (Na^+ + K^+)-ATPase with Artificial Membranes. I. Formation and Structure of (Na^+ + K^+)-ATPase-Liposomes. Biochim. Biophys. Acta 822, 319-334.
40. Anraku, Y. and R.B. Gennis, (1987), The Aerobic Respiratory Chain of Escherichia coli. TIBS 12, 262-266.
41. Apell, H.-J. and P. Lauger, (1986), Quantitative Analysis of Pump-mediated Fluxes in Reconstituted Lipid Vesicles. Biochim. Biophys. Acta 861, 302-310.
42. Appleman, J.R. and G.E. Lienhard, (1985), Rapid Kinetics of the Glucose Transporter from Human Erythrocytes. J. Biol. Chem. 260, 4575-4578.
43. Applebury, M.L. (1987), Biochemical Puzzles about the Cyclic-GMP-dependent Channel. Nature 326, 546-547.
44. Applebury, M.L. and P.A. Hargrave, (1986), Molecular Biology of the Visual Pigments. Vision Res. 26, 1881-1895.
45. Aquila, H., T.A. Link, and M. Klingenberg, (1987), Solute Carriers Involved in Energy Transfer of Mitochondria Form a Homologous Protein Family. FEBS Lett. 212, 1-9.
46. Aquila, H., T.A. Link, and M. Klingenberg, (1985), The Uncoupling Protein from Brown Fat Mitochondria Is Related to the Mitochondrial ADP/ATP Carrier. Analysis

of Sequence Homologies and of Folding of the Protein in the Membrane. The EMBO Journal 4, 2369-2376.

47. Argos, P., J.K.M. Rao, and P.A. Hargrave, (1982), Structural Prediction of Membrane-bound Proteins. Eur. J. Bioch. 128, 565-575.
48. Arinc, E., L.M. Rzepecki, and P. Strittmatter, (1987), Topography of the C Terminus of Cytochrome b_5 Tightly Bound to Dimyristoylphosphatidylcholine Vesicles. J. Biol. Chem. 262, 15563-15567.
49. Aris, J.P. and R.D. Simoni, (1983), Crosslinking and Labeling of the Escherichia coli F_1F_0-ATP Synthase Reveal a Compact Hydrophilic Portion of F_0 Close to an F_1 Catalytic Subunit. J. Biol. Chem. 258, 14599-14609.
50. Ashendel, C.L. (1985), The Phorbol Ester Receptor: A Phospholipid-Regulated Protein Kinase. Biochim. Biophys. Acta 822, 219-242.
51. Audinger, Y., M. Friedlander, and G. Blobel, (1987), Multiple Topogenic Sequences in Bovine Opsin. Proc. Natl. Acad. Sci. USA 84, 5783-5787.
52. Avers, C.J., A. Szabo, and C.A. Price, (1969), Size-Separation of Yeast Mitochondria in the Zonal Centrifuge. J. Bacteriology 100, 1044-1048.
53. Aveyard, R. and D.A. Haydon, (1973), An Introduction to the Principles of Surface Chemistry. Cambridge University Press, Cambridge.
54. Azzone, G.F., D. Pietrobon, and M. Zoratti, (1984), Determination of the Proton Electrochemical Gradient across Biological Membranes. Current Topics in Bioenergetics (C.P. Lee, Ed.), Vol. 13, pp. 1-77. Academic Press, Inc., New York.
55. Bacallao, R., E. Crooke, K. Shiba, W. Wickner, and K. Ito, (1986), The SecY Protein Can Act Post-translationally to Promote Bacterial Protein Export. J. Biol. Chem. 261, 12907-12910.
56. Bach, D. (1984), Calorimetric Studies of Model and Natural Biomembranes. Biomembrane Structure and Function (D. Chapman, Ed.), pp. 1-41. Verlag Chemie, Basel.
57. Baines, A.J. and V. Bennett, (1985), Synapsin I is a Spectrin-Binding Protein Immunologically Related to Erythrocyte Protein 4.1. Nature 315, 410-413.
58. Baines, A.J. and V. Bennett, (1986), Synapsin I is a Microtubule-Bundling Protein. Nature 319, 145-147.
59. Baker, A. and G. Schatz, (1987), Sequences from a Prokaryotic Genome or the Mouse Dihydrofolate Reductase Gene Can Restore the Import of a Truncated Precursor Protein into Yeast Mitochondria. Proc. Natl. Acad. Sci. USA 84, 3117-3121.
60. Balch, W.E. and D.S. Keller, (1986), ATP-coupled Transport of Vesicular Stomatitis Virus G Protein between the Endoplasmic Reticulum and the Golgi. J. Biol. Chem. 261, 14690-14696.
61. Balch, W.E., W.G. Dunphy, W.A. Braell, and J.E. Rothman, (1984), Reconstitution of the Transport of Protein between Successive Compartments of the Golgi Measured by the Coupled Incorporation of N-Acetylglucosamine. Cell 39, 405-416.
62. Bankaitis, V.A. and P.J. Bassford, (1984), The Synthesis of Export-defective Proteins Can Interfere with Normal Protein Export in Escherichia coli. J. Biol. Chem. 259, 12193-12200.
63. Bar, L.K., Y. Barenholz, and T.E. Thompson, (1987), Dependence on Phospholipid Composition of the Fraction of Cholesterol Undergoing Spontaneous Exchange between Small Unilamellar Vesicles. Biochemistry 26, 5460-5465.
64. Bar, R.S., D.W. Deamer, and D.G. Cornwell, (1966), Surface Area of Human Erthyrocyte Lipids: Reinvestigation of Experiments on Plasma Membrane. Science 153, 1010-1012.

65. Baranes, D., F.T. Liu, and E. Razin, (1986), Thrombin and IgE Antigen Induce Formation of Inositol Phosphates by Mouse E-mast Cells. FEBS Lett. 206, 64-68.
66. Barbas, J.A., D. Vazquez, and A. Rodriguez-Tebar, (1985), Final Steps of the Maturation of Omp F, a Major Protein from the Outer Membrane of Escherichia coli. FEBS Lett. 188, 73-76.
67. Barbas, J.A., J. Diaz, A. Rodriguez-Tebar, and D. Vazquez, (1986), Specific Location of Penicillin-Binding Proteins within the Cell Envelope of Escherichia coli. J. Bacteriology 165, 269-275.
68. Barber, J. (1980), Membrane Surface Charges and Potentials in Relation to Photosynthesis. Biochim. Biophys. Acta 594, 253-308.
69. Barber, J. (1987), Photosynthetic Reaction Centres: A Common Link. TIBS 12, 321-326.
70. Barchi, R.L. (1984), Voltage-sensitive Na^+ Ion Channels: Molecular Properties and Functional Reconstitution. TIBS 9, 358-361.
71. Barkas, T., A. Mauron, B. Roth, C. Alliod, S.J. Tzartos, and M. Ballivet, (1987), Mapping the Main Immunogenic Region and Toxin-Binding Site of the Nicotinic Acetylcholine Receptor. Science 235, 77-80.
72. Bartles, J.R., H.M. Feracci, B. Stieger, and A.L. Hubbard, (1987), Biogenesis of the Rat Hepatocyte Plasma Membrane In Vivo: Comparison of the Pathways Taken by Apical and Basolateral Proteins Using Subcellular Fractionation. J. Cell Biol. 105, 1241-1251.
73. Barton, P.G. and F.D. Gunstone, (1975), Hydrocarbon Chain Packing and Molecular Motion in Phospholipid Bilayers Formed from Unsaturated Lecithins. J. Biol. Chem. 250, 4470-4476.
74. Batenburg, A.M., P.E. Bougis, H. Rochat, A.J. Verkleij, and B. de Kruijff, (1985), Penetration of a Cardiotoxin into Cardiolipin Model Membranes and Its Implications on Lipid Organization. Biochemistry 24, 7101-7110.
75. Baty, D., M. Knibiehler, H. Verheij, F. Pattus, D. Shire, A. Bernadae, and C. Lazdunski, (1987), Site-directed Mutagenesis of the COOH-terminal Region of Colicin A: Effect on Secretion and Voltage-dependent Channel Activity. Proc. Natl. Acad. Sci. USA 84, 1152-1156.
76. Bazzi, M.D. and G.L. Nelsestuen, (1987), Role of Substrate in Determining the Phospholipid Specificity of Protein Kinase C Activation. Biochemistry 26, 5002-5008.
77. Beals, C.R., C.B. Wilson, and R.M. Perlmutter, (1987), A Small Multigene Family Encodes G_i Signal-Transduction Proteins. Proc. Natl. Acad. Sci. USA 84, 7886-7890.
78. Beavis, A.D. (1987), Upper and Lower Limits of the Charge Translocation Stoichiometry of Cytochrome c Oxidase. J. Biol. Chem. 262, 6174-6181.
79. Begenisich, T. (1987), Molecular Properties of Ion Permeation Through Sodium Channels. Ann. Rev. Biophys. Biophys. Chem. 16, 247-263.
80. Bell, R.M. (1986), Protein Kinase C Activation by Diacylglycerol Second Messengers. Cell 45, 631-632.
81. Benga, G., O. Popescu, and V.I. Pop, (1986), p-(Chloromercuri)-benzenesulfonate Binding by Membrane Proteins and the Inhibition of Water Transport in Human Erythrocytes. Biochemistry 25, 1535-1538.
82. Bennett, V. (1985), The Membrane Skeleton of Human Erythrocytes and Its Implications for More Complex Cells. Ann. Rev. Biochem. 54, 273-304.
83. Benson, S.A., M.N. Hall, and B.A. Rasmussen, (1987), Signal Sequence Mutations

That Alter Coupling of Secretion and Translation of an Escherichia Coli Outer Membrane Protein. J. Bacteriology 169, 4686-4691.

84. Benson, S.A., M.N. Hall, and T.J. Silhavy, (1985), Genetic Analysis of Protein Transport in Escherichia coli K12. Ann. Rev. Biochem. 54, 101-134.
85. Benson, S.A., E. Bremer, and T.J. Silhavy, (1984), Intragenic Regions Required for LamB Export. Proc. Natl. Acad. Sci. USA 81, 3830-3834.
86. Benz, R. (1986), Porin from Bacterial and Mitochondrial Outer Membranes. CRC Critical Reviews in Biochemistry 19, 145-185.
87. Benz, R., A. Schmid, T. Nakae, and G.H. Vos-Scheperkeuter, (1986), Pore Formation by LamB of Escherichia coli in Lipid Bilayer Membranes. J. Bacteriology 165, 978-986.
88. Benz, R., A. Schmidt, T. Wiedmer, and P.J. Sims, (1986), Single-Channel Analysis of the Conductance Fluctuations Induced in Lipid Bilayer Membranes by Complement Proteins C5b-9. J. Membrane Biol. 94, 37-45.
89. Bergmann, W.L., V. Dressler, C.W.M. Haest, and B. Deuticke, (1984), Reorientation Rates and Asymmetry of Distribution of Lysophospholipids Between the Inner and Outer Leaflet of the Erythrocyte Membrane. Biochim. Biophys. Acta 772, 328-336.
90. Berkhout, T.A., A. Rietveld, and B. de Kruijff, (1987), Preferential Lipid Association and Mode of Penetration of Apocytochrome c in Mixed Model Membranes as Monitored by Tryptophanyl Fluorescence Quenching Using Brominated Phospholipids. Biochim. Biophys. Acta 897, 1-4.
91. Berliner, L.J. and K. Koga, (1987), α-Lactalbumin Binding to Membranes: Evidence for a Partially Buried Protein. Biochemistry 26, 3006-3009.
92. Bernheimer, A.W. and B. Rudy, (1986), Interactions between Membranes and Cytolytic Peptides. Biochim. Biophys. Acta 864, 123-141.
93. Bernier-Valentin, F., D. Aunis, and B. Rousset, (1983), Evidence for Tubulin-binding Sites on Cellular Membranes: Plasma Membranes, Mitochondrial Membranes and Secretory Granule Membranes. J. Cell Biol. 97, 209-216.
94. Berridge, M.J. (1987), Inositol Trisphosphate and Diacylglycerol: Two Interacting Second Messengers. Ann. Rev. Biochem. 56, 159-193.
95. Besterman, J.M., R.S. Pollenz, E.L. Booker, and P. Cuatrecasas, (1986), Diacylglycerol-induced Translocation of Diacylglycerol Kinase: Use of Affinity-purified Enzyme in a Reconstitution System. Proc. Natl. Acad. Sci. USA 83, 9378-9382.
96. Beth, A.H., T.E. Conturo, S.D. Venkataramu, and J.V. Staros, (1986), Dynamics and Interactions of the Anion Channel in Intact Human Erythrocytes: An Electron Paramagnetic Resonance Spectroscopic Study Employing a New Membrane-Impermeant Bifunctional Spin-Label. Biochemistry 25, 3824-3832.
97. Bigay, J., P. Deterre, C. Pfister, and M. Chabre, (1987), Fluoride Complexes of Aluminium or Beryllium Act on G-Proteins as Reversibly Bound Analogues of the gamma Phosphate of GTP. The EMBO Journal 6, 2907-2913.
98. Bigelow, D.J. and D.D. Thomas, (1987), Rotational Dynamics of Lipid and the Ca-ATPase in Sarcoplasmic Reticulum. J. Biol. Chem. 262, 13449-13456.
99. Birchmeier, W. (1984), Cytoskeleton Structure and Function. TIBS 9, 192-195.
100. Birnbaum, M.J., H.C. Haspel, and O.M. Rosen, (1986), Cloning and Characterization of a cDNA Encoding the Rat Brain Glucose-transporter Protein. Proc. Natl. Acad. Sci. USA 83, 5784-5788.
101. Bishop, J.M. (1985), Viral Oncogenes. Cell 42, 23-38.
102. Bishop, J.M. (1987), The Molecular Genetics of Cancer. Science 235, 305-311.

103. Bishop, W.R. and R.M. Bell, (1985), Assembly of the Endoplasmic Reticulum Phospholipid Bilayer: The Phosphatidylcholine Transporter. Cell 42, 51-60.
104. Bjorkman, P.J., M.A. Saper, B. Samraoui, W.S. Bennett, J.L. Strominger, and D.C. Wiley, (1987), The Foreign Antigen Binding Site and T Cell Recognition Regions of Class I Histocompatibility Antigens. Nature 329, 512-518.
105. Blair, D.F., J. Gelles, and S.I. Chan, (1986), Redox-linked Proton Translocation in Cytochrome Oxidase: The Importance of Gating Electron Flow: The Effects of Slip in a Model Transducer. Biophys. J. 50, 713-733.
106. Blake, R., L.P. Hager, and R.B. Gennis, (1978), Activation of Pyruvate Oxidase by Monomeric and Micellar Amphiphiles. J. Biol. Chem. 253, 1963-1971.
107. Blanc, J.P. and E.T. Kaiser, (1984), Biological and Physical Properties of a beta-Endorphin Analog Containing Only D-Amino Acids in the Amphiphilic Helical Segment 13-31. J. Biol. Chem. 259, 9549-9556.
108. Blaurock, A.E. and M.H.F. Wilkins, (1969), Structure of Frog Photoreceptor Membranes. Nature 223, 906-909.
109. Blaurock, A.F. and C.R. Worthington, (1966), Treatment of Low Angle X-ray Data from Planar and Concentric Multilayer Structures. Biophys. J. 6, 305-312.
110. Bloch, R.J. and S.C. Froehner, (1987), The Relationship of the Postsynaptic ^{43}K Protein to Acetylcholine Reseptors in Receptor Clusters Isolated from Cultured Rat Myotubes. J. Cell Biol. 104, 645-654.
111. Blocher, D., L. Six, R. Gutermann, B. Henkel, and K. Ring, (1985), Physicochemical Characterization of Tetraether Lipids from Thermoplasma acidophilum. Calorimetric Studies on Miscibility with Diether Model Lipids Carrying Branched or Unbranched Alkyl Chains. Biochim. Biophys. Acta 818, 333-342.
112. Blume, A. (1983), Apparent Molar Heat Capacities of Phospholipids in Aqueous Dispersion. Effects of Chain Length and Head Group Structure. Biochemistry 22, 5436-5442.
113. Boekema, E.J., M.G. VanHeel, and E.F.J. VanBruggen, (1984), Three-dimensional Structure of Bovine NADH:Ubiquinone Oxidoreductase of the Mitochondrial Respiratory Chain. Biochim. Biophys. Acta 787, 19-26.
114. Bogaert, T., N. Brown, and M. Wilcox, (1987), The Drosophila PS2 Antigen Is an Invertebrate Integrin That, Like the Fibronectin Receptor, Becomes Localized to Muscle Attachments. Cell 51, 929-940.
115. Boggs, J.M. (1987), Lipid Intermolecular Hydrogen Bonding: Influence on Structural Organization and Membrane Function. Biochim. Biophys. Acta 906, 353-404.
116. Boggs, J.M. (1983), The Hydrophobic and Electrostatic Effects of Proteins on Lipid Fluidity and Organization. Membrane Fluidity in Biology (R.C. Aloia, Ed.), Vol. 2, pp. 89-130. Academic Press, New York.
117. Bohnenberger, E. and H. Sandermann, (1983), Lipid Dependence of Diacylglycerol Kinase from Escherichia coli. Eur. J. Bioch. 132, 645-650.
118. Bolard, J. (1986), How Do the Polyene Macrolide Antibiotics Affect the Cellular Membrane Properties? Biochim. Biophys. Acta 864, 257-304.
119. Bon, F., E. Lebrun, J. Gomel, R. VanRapenbusch, J. Cartaud, J.-L. Popot, and J.-P. Changeux, (1984), Image Analysis of the Heavy Form of the Acetylcholine Receptor from Torpedo marmorata. J. Mol. Biol. 176, 205-237.
120. Boni-Schnetzler, M. and P.F. Pilch, (1987), Mechanism of Epidermal Growth Factor Receptor Autophosphorylation and High-Affinity Binding. Proc. Natl. Acad. Sci. USA 84, 7832-7836.
121. Bonner, T.I., N.J. Buckley, A.C. Young, and M.R. Brann, (1987), Identification of a Family of Muscarinic Acetylcholine Receptor Genes. Science 237, 527-532.

122. Borle, F. and J. Seelig, (1985), Ca^{2+} Binding to Phosphatidylglycerol Bilayers as Studied by Differential Scanning Calorimetry and ^{2}H- and ^{31}P-Nuclear Magnetic Resonance. Chem. Phys. Lipids 36, 263-283.
123. Borle, F. and J. Seelig, (1983), Hydration of Escherichia coli Lipids: Deuterium T_1 Relaxation Time Studies of Phosphatidylglycerol, Phosphatidylethanolamine and Phosphatidylcholine. Biochim. Biophys. Acta 735, 131-136.
124. Borst, P. (1986), How Proteins Get into Microbodies (Peroxisomes, Glyoxysomes, Glycosomes). Biochim. Biophys. Acta 866, 179-203.
125. Bosterling, B. and J.R. Trudell, (1982), Association of Cytochrome b_5 and Cytochrome P-450 Reductase with Cytochrome P-450 in the Membrane of Reconstituted Vesicles. J. Biol. Chem. 257, 4783-4787.
126. Bouillon, P., C. Drischel, C. Vergnolle, H. Duranton, and J.-C. Kader, (1987), The Primary Structure of Spinach-leaf Phospholipid-transfer Protein. Eur. J. Biochem. 166, 387-391.
127. Boulain, J.C., A. Charbit, and M. Hofnung, (1986), Mutagenesis by Random Linker Insertion into the lamB Gene of Escherichia coli K_{12}. Mol. Gen. Genet. 205, 339-348.
128. Bowman, B.J. and E.J. Bowman, (1986), H^+-ATPase from Mitochondria, Plasma Membranes, and Vacuoles of Fungal Cells. J. Membrane Biol. 94, 83-97.
129. Boyd, D., C. Manoil, and J. Beckwith, (1987), Determinants of Membrane Protein Topology. Proc. Natl. Acad. Sci. USA 84, 8525-8529.
130. Braell, W.A. (1987), Fusion between Endocytic Vesicles in a Cell-Free System. Proc. Natl. Acad. Sci. USA 84, 1137-1141.
131. Bramhall, J. (1986), Use of the Fluorescent Weak Acid Dansylglycine to Measure Transmembrane Proton Concentration Gradients. Biochemistry 25, 3958-3962.
132. Bramley, H.F. and H.L. Kornberg, (1987), Sequence Homologies between Proteins of Bacterial Phosphoenolpyruvate-dependent Sugar Phosphotransferase Systems: Identification of Possible Phosphate-carrying Histidine Residues. Proc. Natl. Acad. Sci. USA 84, 4777-4780.
133. Brandl, C.J. and C. M. Deber, (1986), Hypothesis about the Function of Membrane-buried Proline Residues in Transport Proteins. Proc. Natl. Acad. Sci. USA 83, 917-921.
134. Brandl, C.J., N.M. Green, B. Korczak, and D.H. MacLennan, (1986), Two Ca^{2+} ATPase Genes: Homologies and Mechanistic Implications of Deduced Amino Acid Sequences. Cell 44, 597-607.
135. Branton, D. (1966), Fracture Faces of Frozen Membranes. Proc. Natl. Acad. Sci. USA 55, 1048-1056.
136. Branton, D. (1971), Freeze-etching Studies of Membrane Structure. Phil. Trans. Roy. Soc. London B 261, 133-138.
137. Brasseur, R., J.A. Killian, B. de Kruijff, and J.M. Ruysschaert, (1987), Conformational Analysis of Gramacidin-Gramacidin Interactions at the Air/Water Interface Suggests that Gramacidin Aggregates into Tube-like Structures Similar as Found in the Gramacidin-induced Hexagonal H_{II} Phase. Biochim. Biophys. Acta 903, 11-17.
138. Brasseur, R. and J.M. Ruysschaert, (1986), Conformation and Mode of Organization of Amphiphilic Membrane Components: A Conformational Analysis. Biochem. J. 238, 1-11.
139. Braulke, T., C. Gartung, A. Hasilik, and K. von Figura, (1987), Is Movement of Mannose 6-Phosphate-specific Receptor Triggered by Binding of Lysosomal Enzymes? J. Cell Biol. 104, 1735-1742.

140. Brauner, T., D.F. Hulser, and R.J. Strasser, (1984), Comparative Measurements of Membrane Potentials with Microelectrodes and Voltage-sensitive Dyes. Biochim. Biophys. Acta 771, 208-216.
141. Brecher, P., R. Saouaf, J.M. Sugarman, D. Eisenberg, and K. LaRosa, (1984), Fatty Acid Transfer between Multilamellar Liposomes and Fatty Acid-binding Proteins. J. Biol. Chem. 259, 13395-13401.
142. Breitmeyer, J.B. (1987), How T Cells Communicate. Nature 329, 760-761.
143. Bremer, E.G., J. Schlessinger, and S. Hakomori, (1986), Ganglioside-mediated Modulation of Cell Growth: Specific Effects of GM_3 on Tyrosine Phosphorylation of the Epidermal Growth Factor Receptor. J. Biol. Chem. 261, 2434-2440.
144. Brent, L.G. and P.A. Srere, (1987), The Interaction of Yeast Citrate Synthase with Yeast Mitochondrial Inner Membranes. J. Biol. Chem. 262, 319-325.
145. Bretcher, M.S. (1972), Phosphatidylethanolamine: Differential Labelling in Intact Cells and Ghosts of Human Erythrocytes by a Membrane-impermeable Reagent. J. Mol. Biol. 71, 523-528.
146. Bretscher, M.S. (1984), Endocytosis: Relation to Capping and Cell Locomotion. Science 224, 681-686.
147. Briggs, M.S., D.G. Cornell, R.A. Dluhy, and L.M. Gierasch, (1986), Conformations of Signal Peptides Induced by Lipids Suggest Initial Steps in Protein Export. Science 233, 206-208.
148. Briggs, M.S., L.M. Gierasch, A. Zlotnick, J.D. Lear, and W.F. DeGrado, (1985), In Vivo Function and Membrane Binding Properties Are Correlated for Escherichia coli LamB Signal Peptides. Science 228, 1096-1099.
149. Brisson, A. and P.N.T. Unwin, (1985), Quarternary Structure of the Acetylcholine Receptor. Nature 315, 474-477.
150. Brooker, R.J., K. Fiebig, and T.H. Wilson, (1985), Characterization of Lactose Carrier Mutants Which Transport Maltose. J. Biol. Chem. 260, 16181-16186.
151. Brothers, J.R., O.H. Griffith, M.O. Brothers, P.C. Jost, J.R., Silvius, and L.E. Hokin, (1981), Lipid-Protein Multiple Binding Equilibria in Membranes. Biochemistry 20, 5261-5267.
152. Brown, M.S. and J.L. Goldstein, (1986), A Receptor-Mediated Pathway for Cholesterol Homeostasis. Science 232, 34-47.
153. Brown, R.E. and T.E. Thompson, (1987), Spontaneous Transfer of Ganglioside GM_1 Between Phospholipid Vesicles. Biochemistry 26, 5454-5460.
154. Bruckdorfer, N.R. and M.K. Sherry, (1984), The Solubility of Cholesterol and Its Exchange Between Membranes. Biochim. Biophys. Acta 769, 187-196.
155. Brugge, J.S. (1986), The p35/p36 Substrates of Protein-Tyrosine Kinases as Inhibitors of Phospholipase A2. Cell 46, 149-150.
156. Brunner, J., A.J. Franzusoff, B. Luscher, C. Zugliani, and G. Semenza, (1985), Membrane Protein Topology: Amino Acid Residues in a Putative Transmembrane Alpha-Helix of Bacteriorhodopsin Labeled with the Hydrophobic Carbene-Generating Reagent 3-(Trifluoromethyl)-3-(m-[^{125}I]iodophenyl)diazirine. Biochemistry 24, 5422-5430.
157. Brunner, J., A.J. Franzusoff, B. Luscher, C. Zugliani, and G. Semenza, (1985), Membrane Protein Topology: Amino Acid Residues in a Putative Transmembrane alpha-Helix of Bacteriorhodopsin Labeled with the Hydrophobic Carbene-Generating Reagent 3-(Trifluoromethyl)-3-(m[^{125}I] iodophenyl)diazirine. Biochemistry 24, 5422-5430.
158. Brunner, J., M. Spiess, R. Aggeler, P. Huber, and G. Semenza, (1983), Hydrophobic

Labeling of a Single Leaflet of the Human Erythrocyte Membrane. Biochemistry 22, 3812-3820.

159. Buchel, D.E., B. Gronenborn, and B. Muller-Hill, (1980), Sequence of the Lactose Permease Gene. Nature 283, 541-545.
160. Buck, C.A. and A.F. Horwitz, (1987), Cell Surface Receptors for Extracellular Matrix Molecules. Ann. Rev. Cell Biol. 3, 179-205.
161. Buldt, G., H.U. Gally, A. Seelig, J. Seelig, and G. Zaccai, (1978), Neutron Diffraction Studies on Selectively Deuterated Phospholipid Bilayers. Nature 271, 182-184.
162. Bullivant, S. (1974), Freeze-etching Techniques Applied to Biological Membranes. Phil. Trans. Roy. Soc. London B 268, 5-14.
163. Burch, R.M., A. Luini, and J. Axelrod, (1986), Phospholipase A2 and Phospholipase C Are Activated by Distinct GTP-Binding Proteins in Response to α_1-Adrenergic Stimulation in FRTL5 Thyroid Cells. Proc. Natl. Acad. Sci. USA 83, 7201-7205.
164. Burdette, R.A. and D.M. Quinn, (1986), Interfacial Reaction Dynamics and Acyl-enzyme Mechanism for Lipoprotein Lipase-catalyzed Hydrolysis of Lipid p-Nitrophenyl Esters. J. Biol. Chem. 261, 12016-12021.
165. Burgess, T.L., C.S. Craik, L. Matsuuchi, and R.B. Kelly, (1987), In Vitro Mutagenesis of Trypsinogen: Role of the Amino Terminus in Intracellular Protein Targeting to Secretory Granules. J. Cell Biol. 105, 659-668.
166. Burgoyne, R.D. (1987), Phosphorylation of the Muscarinic Receptor. TIBS 12, 208-209.
167. Burgoyne, R.D. (1987), Control of Exocytosis. Nature 328, 112-113.
168. Burke, B., K. Matlin, E. Bause, G. Legler, N. Peyrieras, and H. Ploegh, (1984), Inhibition of N-linked Oligosaccharide Trimming Does Not Interfere with Surface Expression of Certain Integral Membrane Proteins. The EMBO Journal 3, 551-556.
169. Burn, P., A. Rotman, R.K. Meyer, and M.M. Burger, (1985), Diacylglycerol in Large α-Actinin/Actin Complexes and in the Cytoskeleton of Activated Platelets. Nature 314, 469-472.
170. Burnett, B.K., R.J. Robson, Y. Takagaki, R. Radhakrishnan, and H.G. Khorana, (1985), Synthesis of Phospholipids Containing Photoactivatible Carbene Precursors in the Headgroups and Their Crosslinking with Membrane Proteins. Biochim. Biophys. Acta 815, 57-67.
171. Bush, S.F., R.G. Adams, and I.W. Levin, (1980), Structural Reorganizations in Lipid Bilayer Systems: Effect of Hydration and Sterol Addition on Raman Spectra of Dipalmitoylphosphatidylcholine Multilayers. Biochemistry 19, 4429-4436.
172. Buss, J.E., S.M. Mumby, P.J. Casey, A.G. Gilman, and B.M. Sefton, (1987), Myristoylated α Subunits of Guanine Nucleotide-binding Regulatory Proteins. Proc. Natl. Acad. Sci. USA 84, 7493-7497.
173. Byers, T.J. and D. Branton, (1985), Visualization of the Protein Association in the Erythrocyte Membrane Skeleton. Proc. Natl. Acad. Sci. USA 82, 6153-6157.
174. Caffrey, M. and G.W. Feigenson, (1981), Fluoresence Quenching in Model Membranes. 3. The Relationship Between Calcium Adenosine Triphosphatase Enzyme Activity and the Affinity of the Protein for Phosphatidylcholine with Different Acyl Chain Characteristics. Biochemistry 20, 1949-1961.
175. Caffrey, M. and G.W. Feigenson, (1984), Influence of Metal Ions on the Phase Properties of Phosphatidic Acid in Combination with Natural and Synthetic Phosphatidylcholines: An X-ray Diffraction Study Using Synchrotron Radiation. Biochemistry 23, 323-331.

176. Cafiso, D.S. and W.L. Hubbell, (1981), Investigation of Electrical Phenomena in Membranes: New Approaches with Spin Labelling. Molecular Basis of Drug Action (Singer and Ondarza, Eds.), pp. 254-269. Elsevier, New York.
177. Cafiso, D.S. and W.L. Hubbell, (1978), Estimation of Transmembrane Potentials from Phase Equilibria of Hydrophobic Paramagnetic Ions. Biochemistry 17, 187-194.
178. Cain, B.D. and R.D. Simoni, (1986), Impaired Proton Conductivity Resulting from Mutations in the α Subunit of F_1F_0 ATPase in Escherichia coli. J. Biol. Chem. 261, 10043-10050.
179. Callender, R. and B. Honig, (1977), Resonance Raman Studies of Visual Pigments. Annu. Rev. Biophys. Bioeng. 6, 33-55.
180. Cameron, D.G. and H.H. Mantsch, (1978), The Phase Transition of 1,2-Dipalmitoyl-sn-glycerophosphocholine as Seen by Fourier Transform Infrared Difference Spectroscopy. Bioch. Biophys. Res. Comm. 83, 886-892.
181. Cameron, D.G., H.L. Casal, and H.H. Mantsch, (1980), Characterization of the Pretransition in 1,2-Dipalmitoyl-sn-glycero-3-phosphocholine by Fourier Transform Infrared Spectroscopy. Biochemistry 19, 3665-3672.
182. Campbell, I.D. and R.A. Dwek, (1984), Biological Spectroscopy. Benjamin/ Cummings Publishing Co., Menlo Park, California.
183. Cantor, C.R. and P.R. Schimmel, (1980), Biophysical Chemistry. W.H. Freeman and Co., San Francisco.
184. Caras, I.W., G.N. Weddell, M.A. Davitz, V. Nussenzweig, and D.W. Martin, (1987), Signal for Attachment of a Phospholipid Membrane Anchor in Decay Accelerating Factor. Science 238, 1280-1283.
185. Cardoza, J.D., A.M. Kleinfeld, K.C. Stallcup, and M.F. Mescher, (1984), Hairpin Configuration of H-2K in Liposomes Formed by Detergent Dialysis. Biochemistry 23, 4401-4409.
186. Carpenter, D., T. Jackson, and M.R. Hanley, (1987), Coping with a Growing Family. Nature 325, 107-108.
187. Carpenter, G. (1987), Receptors for Epidermal Growth Factor and Other Polypeptide Mitogens. Ann. Rev. Biochem. 56, 881-914.
188. Carrier, D. and M. Pezolet, (1986), Investigation of Polylysine-Dipalmitoylphosphatidylglycerol Interactions in Model Membranes. Biochemistry 25, 4167-4174.
189. Carruthers, A. and D.L. Melchior, (1986), How Bilayer Lipids Affect Membrane Protein Activity. TIBS 11, 331-335.
190. Casal, H.L. and H.H. Mantsch, (1984), Polymorphic Phase Behavior of Phospholipid Membranes Studied by Infrared Spectroscopy. Biochim. Biophys. Acta 779, 381-401.
191. Casey, R.P. (1984), Membrane Reconstitution of the Energy-Conserving Enzymes of Oxidative Phosphorylation. Biochim. Biophys. Acta 768, 319-347.
192. Catron, K.M. and C.A. Schnaitman, (1987), Export of Protein in Escherichia coli: a Novel Mutation in ompC Affects Expression of Other Major Outer Membrane Proteins. J. Bacteriology 169, 4327-4334.
193. Catterall, W.A. (1986), Molecular Properties of Voltage-Sensitive Sodium Channels. Ann. Rev. Biochem. 55, 953-985.
194. Cevc, G. (1987), How Membrane Chain Melting Properties Are Regulated by the Polar Surface of the Lipid Bilayer. Biochemistry 26, 6305-6310.
195. Cevc, G. and D. Marsh, (1985), Hydration of Noncharged Lipid Bilayer Mem-

branes: Theory and Experiments with Phosphatidylethanolamines. Biophys. J. 47, 21-31.

196. Chabre, M. (1987), The G Protein Connection: Is It in the Membrane or the Cytoplasm? TIBS 12, 213-215.
197. Chang, Y.-Y. and J.E. Cronan, (1986), Molecular Cloning, DNA Sequencing, and Enzymatic Analyses of Two Escherichia coli Pyruvate Oxidase Mutants Defective in Activation by Lipids. J. Bacteriology 167, 312-318.
198. Chapman, D. (1982), Protein-Lipid Interactions in Model and Natural Biomembranes. Biological Membranes (D. Chapman, Ed.), Vol. 4, pp. 179-229. Academic Press, New York.
199. Chapman, D. and G. Benga, (1984), Biomembrane Fluidity—Studies of Model and Natural Biomembranes. Biological Membranes (D. Chapman, Ed.), Vol. 5, pp. 1-56. Academic Press, New York.
200. Charbit, A., J.C. Boulain, A. Ryter, and M. Hofnung, (1986), Probing the Topology of a Bacterial Membrane Protein by Genetic Insertion of a Foreign Epitope; Expression at the Cell Surface. The EMBO Journal 5, 3029-3037.
201. Cheifetz, S., J.M. Boggs, and M.A. Moscarello, (1985), Increase in Vesicle Permeability Mediated by Myelin Basic Protein: Effect of Phosphorylation of Basic Protein. Biochemistry 24, 5170-5175.
202. Chen, C.-C., T. Kurokawa, S.-Y. Shaw, L.G. Tillotson, S. Kalled, and K.J. Isselbacher, (1986), Human Erythrocyte Glucose Transporter: Normal Asymmetric Orientation and Function in Liposomes. Proc. Natl. Acad. Sci. USA 83, 2652-2656.
203. Chen, C.-M., T.K. Misra, S. Silver, and B.P. Rosen, (1986), Nucleotide Sequence of the Structural Genes for an Anion Pump: The Plasmid-Encoded Arsenical Resistance Operon. J. Biol. Chem. 261, 15030-15038.
204. Chen, L. and P.C. Tai, (1986), Roles of H^+-ATPase and Proton Motive Force in ATP-Dependent Protein Translocation In Vitro. J. Bacteriology 167, 389-392.
205. Chen, L., P.C. Tai, M.S. Briggs, and L.M. Gierasch, (1987), Protein Translocation into Escherichia coli Membrane Vesicles Is Inhibited by Functional Synthetic Signal Peptides. J. Biol. Chem. 262, 1427-1429.
206. Chen, W.-J. and M.G. Douglas, (1987), The Role of Protein Structure in the Mitochondrial Import Pathway. J. Biol. Chem. 262, 15605-15609.
207. Chen, W.-J. and M.G. Douglas, (1987), Phosphodiester Bond Cleavage Outside Mitochondria Is Required for the Completion of Protein Import into the Mitochondrial Matrix. Cell 49, 651-658.
208. Cheneval, D., M. Muller, R. Toni, S. Ruetz, and E. Carafoli, (1985), Adriamycin as a Probe for the Transversal Distribution of Cardiolipin in the Inner Mitochondrial Membrane. J. Biol. Chem. 260, 13003-13007.
209. Cheng, K.-H., J.R. Lepock, S.W. Hui, and P.L. Yeagle, (1986), The Role of Cholesterol in the Activity of Reconstituted Ca-ATPase Vesicles Containing Unsaturated Phosphatidylethanolamine. J. Biol. Chem. 261, 5081-5087.
210. Cheresh, D.A., R. Pytela, M.D. Pierschbacher, F.G. Klier, E. Ruoslahti, and R.A. Reisfeld, (1987), An Arg-Gly-Asp-directed Receptor on the Surface of Human Melanoma Cells Exists in a Divalent Cation-dependent Functional Complex with the Disialoganglioside GD_2. J. Cell Biol. 105, 1163-1173.
211. Cherry, R.J. (1979), Rotational and Lateral Diffusion of Membrane Proteins. Biochim. Biophys. Acta 559, 289-327.
212. Cherry, R.J., U. Muller, R. Henderson, and M.P. Heyn, (1978), Temperature-

dependent Aggregation of Bacteriorhodopsin in Dipalmitoyl- and Dimyristoylphophatidylcholine Vesicles. J. Mol. Biol. 121, 283-298.

213. Chicken, C.A. and F.J. Sharom, (1984), Lipid-Protein Interactions of the Human Erythrocyte Concanavalin. A Receptor in Phospholipid Bilayers. Biochim. Biophys. Acta 774, 110-118.

214. Chung, B.H., G.M. Anatharamaiah, C.G. Brouillette, T. Nishida, and J.P. Segrest, (1985), Studies of Synthetic Peptide Analogs of the Amphipathic Helix: Correlation of Structure with Function. J. Biol. Chem. 260, 10256-10262.

215. Chong, P.L.-G. and G. Weber, (1983), Pressure Dependence of 1,6-Diphenyl-1,3,5-hexatriene Fluorescence in Single-component Phosphatidylcholine Liposomes. Biochemistry 22, 5544-5550.

216. Chong, P.L.-G., P.A.G. Fortes, and D.M. Jameson, (1985), Mechanisms of Inhibition of (Na,K)-ATPase by Hydrostatic Pressure Studied with Fluorescent Probes. J. Biol. Chem. 260, 14484-14490.

217. Chothia, C. (1976), The Nature of the Accessible and Buried Surfaces in Proteins. J. Mol. Biol. 105, 1-14.

218. Chou, P.Y. and G.D. Fasman, (1974), Prediction of Protein Conformation. Biochemistry 13, 222-245.

219. Christiansen, K. and J. Carlsen, (1985), Reconstitution of Cytochrome b_5 into Lipid Vesicles in a Form Which Is Nonsusceptible to Attack by Carboxypeptidase Y. Biochim. Biophys. Acta 815, 215-222.

220. Chung, L., G. Kaloyanides, R. McDaniel, A. McLaughlin, and S. McLaughlin, (1985), Interaction of Gentamicin and Spermine with Bilayer Membranes Containing Negatively Charged Phospholipids. Biochemistry 24, 442-452.

221. Clancy, R.M., A.R. Wissenberg, and M. Glaser, (1981), Use of Phospholipase D to Alter the Surface Charge of Membranes and Its Effect on the Enzymatic Activity of D-ß-Hydroxybutyrate Dehydrogenase. Biochemistry 20, 6060-6065.

222. Clancy, R.M., L.H. McPherson, and M. Glaser, (1983), Effects of Change in the Phospholipid Composition on the Enzyme Activity of D-ß-Hydroxybutyrate Dehydrogenase in Rat Hepatocytes. Biochemistry 22, 2358-2364.

223. Clarke, S. (1975), The Size and Detergent Binding of Membrane Proteins. J. Biol. Chem. 250, 5459-5469.

224. Classen, J., C.W.M. Haest, H. Tournois, and B. Deuticke, (1987), Gramacidin-Induced Enhancement of Transbilayer Reorientation of Lipids in the Erythrocyte Membrane. Biochemistry 26, 6604-6612.

225. Clejan, S. and R. Bittman, (1985), Rates of Amphotericin B and Filipin Association with Sterols. J. Biol. Chem. 260, 2884-2889.

226. Clejan, S. and R. Bittman, (1984), Distribution and Movement of Sterols with Different Side Chain Structures Between the Two Leaflets of the Membrane Bilayer of Mycoplasma Cells. J. Biol. Chem. 259, 449-455.

227. Clement, J.M., E. Lepouce, C. Marchal, and M. Hofnung, (1983), Genetic study of a Membrane Protein: DNA Sequence Alterations due to 17 lamB Point Mutations Affecting Adsorption of Phage lambda. The EMBO Journal 2, 77-80.

228. Cockcroft, S. (1987), Polyphosphoinositide Phosphodiesterase: Regulation by a Novel Guanine Nucleotide Binding Protein, G_p. TIBS 12, 75-78.

229. Caffrey, M. and G.W. Feigenson, (1981), Fluorescence Quenching in Model Membranes. 3. Relationships Between Calcium Adenosine Triphosphatase Enzyme Activity and the Affinity of the Protein for Phosphatidylcholine with Different Acyl Chain Characteristics. Biochemistry 20, 1949-1961.

230. Cohen, L.B. and B.M. Salzberg, (1978), Optical Measurement of Membrane Potential. Rev. Physiol. Biochem. Pharmacol. 83, 35-88.
231. Cohn, E.J. and J.T. Edsall, (1943), Proteins, Amino Acids, and Peptides. Reinhold, New York.
232. Cole, G.J., A. Loewy, and L. Glaser, (1986), Neuronal Cell-Cell Adhesion Depends on Interactions of N-CAM with Heparin-like Molecules. Nature 320, 445-447.
233. Cole, G.J., A. Loewy, N.V. Cross, R. Akeson, and L. Glaser, (1986), Topographic Localization of the Heparin-binding Domain of the Neural Cell Adhesion Molecule N-CAM. J. Cell Biol. 103, 1739-1744.
234. Coleman, J., M. Inukai, and M. Inouye, (1985), Dual Functions of the Signal Peptide in Protein Transfer Across the Membrane. Cell 43, 351-360.
235. Collarini, M., G. Amblard, C. Lazdunski, and F. Pattus, (1987), Gating Processes of Channels Induced by Colicin A, Its C-terminal Fragment and Colicin E1 in Planar Lipid Bilayers. Eur. Biophys. J. 14, 147-153.
236. Coller, B.S. (1986), Activation Affects Access to the Platelet Receptor for Adhesive Glycoproteins. J. Cell Biol. 103, 451-456.
237. Colman, A. and C. Robison, (1986), Protein Import into Organelles: Hierarchical Targeting Signals. Cell 46, 321-322.
238. Condon, C., R. Cammack, D.S. Patil, and P. Owen, (1985), The Succinate Dehydrogenase of Escherichia coli: Immunochemical Resolution and Biophysical Characterization of a 4-Subunit Enzyme Complex. J. Biol. Chem. 260, 9427-9434.
239. Connelly, P.A., R.B. Sisk, H. Schulman, and J.C. Garrison, (1987), Evidence for the Activation of the Multifunctional Ca^{2+}/Calmodulin-dependent Protein Kinase in Response to Hormones That Increase Intracellular Ca^{2+}. J. Biol. Chem. 262, 10154-10163.
240. Connolly, T.J., A. Carruthers, and D.L. Melchior, (1985), Effect of Bilayer Cholesterol Content on Reconstituted Human Erythrocyte Sugar Transporter Activity. J. Biol. Chem. 260, 2617-2620.
241. Cooper, C.L., S. Vandaele, J. Barhanin, M. Fosset, M. Lazdunski, and M.M. Hosey, (1987), Purification and Characterization of the Dihydropyridine-sensitive Voltage-dependent Calcium Channel from Cardiac Tissue. J. Biol. Chem. 262, 509-512.
242. Copeland, B.R. and H.M. McConnel, (1980), The Rippled Structure in Bilayer Membranes of Phosphatidylcholine and Binary Mixtures of Phosphatidylcholine and Cholesterol. Biochim. Biophys. Acta 599, 95-109.
243. Corey, D.P. (1983), Patch Clamp: Current Excitement in Membrane Physiology. Neuroscience Commentaries 1, 99-110.
244. Cornell, R. and D.E. Vance, (1987), Translocation of CTP:Phosphocholine Cytidylyltransferase from Cytosol to Membranes in HeLa Cells: Stimulation by Fatty Acid, Fatty Alcohol, Mono- and Diacylglycerol. Biochim. Biophys. Acta 919, 26-36.
245. Cornette, J.L., K.B. Cease, H. Margalit, J.L. Spouge, J.A. Berzofsky, and C. DeLisi, (1987), Hydrophobicity Scales and Computational Techniques for Detecting Amphipathic Structures in Proteins. J. Mol. Biol. 195, 659-685.
246. Coronado, R. (1986), Recent Advances in Planar Phospholipid Bilayer Techniques for Monitoring Ion Channels. Ann. Rev. Biophys. Biophys. Chem. 15, 259-277.
247. Correas, I., D.W. Speicher, and V.T. Marchesi, (1986), Structure of the Spectrin-Actin Binding Site of Erythrocyte Protein 4.1. J. Biol. Chem. 261, 13362-13366.
248. Cortese, J.D., J.C. Vidal, P. Churchill, J.O. McIntyre, and S. Fleischer, (1982), Reactivation of D-ß-Hydroxybutyrate Dehydrogenase with Short-chain Lecithins: Stoichiometry and Kinetic Mechanism. Biochemistry 21, 3899-3908.

249. Cortese, J.D. and S. Fleischer, (1987), Noncooperative vs. Cooperative Reactivation of D-ß-Hydroxybutyrate Dehydrogenase: Multiple Equilibria for Lecithin Binding Are Determined by the Physical State (Soluble vs. Bilayer) and Composition of the Phospholipids. Biochemistry 26, 5283-5293.
250. Cortijo, M., A. Alonzo, J.C. Gomez-Fernandez, and D. Chapman, (1982), Intrinsic Protein-Lipid Interactions: Infrared Spectroscopic Studies of Gramacidin A, Bacteriorhodopsin, and Ca^{2+}-ATPase in Biomembranes and Reconstituted Systems. J. Mol. Biol. 157, 597-618.
251. Cossins, A.R. and A.G. Macdonald, (1986), Homeoviscous Adaptation Under Pressure. III. The Fatty Acid Composition of Liver Mitochondrial Phospholipids of Deep-sea Fish. Biochim. Biophys. Acta 860, 325-335.
252. Cossins, A.R. and M. Sinensky, (1984), Adaptation of Membranes to Temperature, Pressure, and Exogenous Lipids. Physiology of Membrane Fluidity (M. Shinitzky, Ed.), Vol. 2, pp. 1-20. CRC Press, Inc., Boca Raton, Florida.
253. Costello, M.J., J. Escaig, K. Matsushita, P.V. Viitanen, D.R. Menick, and H.R. Kaback, (1987), Purified lac Permease and Cytochrome c Oxidase Are Functional as Monomers. J. Biol. Chem. 262, 17072-17082.
254. Cover, W.H., J.P. Ryan, P.J. Bassford, K.A. Walsh, J. Bollinger, and L.L. Randall, (1987), Suppression of a Signal Sequence Mutation by an Amino Acid Substitution in the Mature Portion of the Maltose-Binding Protein. J. Bacteriology 169, 1794-1800.
255. Cowan, A.E., D.G. Myles, and D.E. Koppel, (1987), Lateral Diffusion of the PH-20 Protein on Guinea Pig Sperm: Evidence That Barriers to Diffusion Maintain Plasma Membrane Domains in Mammalian Sperm. J. Cell Biol. 104, 917-923.
256. Cox, G.B., D.A. Jans, A.L. Fimmel, F. Gibson, and L. Hatch, (1984), The Mechanism of ATP Synthase: Conformational Change by Rotation of the ß-Subunit. Biochim. Biophys. Acta 768, 201-208.
257. Cox. G.B., L. Hatch, D. Webb, A.L. Fimmel, Z.-H. Lin, A.E. Senior, and F. Gibson, (1987), Amino Acid Substitutions in the Epsilon-Subunit of the F_1F_0-ATPase of Escherichia coli. Biochim. Biophys. Acta 890, 195-204.
258. Criado, M., S. Hochschwender, V. Sarin, J.L. Fox, and J. Lindstrom, (1985), Evidence for Unpredicted Transmembrane Domains in Acetylcholine Receptor Subunits. Proc. Natl. Acad. Sci. USA 82, 2004-2008.
259. Crooke, E. and W. Wickner, (1987), Trigger Factor: A Soluble Protein That Folds pro-OmpA into a Membrane-assembly-competent Form. Proc. Natl. Acad. Sci. USA 84, 5216-5220.
260. Cross, G.A.M. (1987), Eukaryotic Protein Modification and Membrane Attachment via Phosphatidylinositol. Cell 48, 179-181.
261. Crowther, R.A. and A. Klug, (1975), Structural Analysis of Macromolecular Assemblies by Image Reconstruction from Electron Micrographs. Ann. Rev. Biochem. 44, 161-182.
262. Cullen, J., M.C. Phillips, and G.G. Shipley, (1971), The Effects of Temperature on the Composition and Physical Properties of the Lipids of Pseudomonas fluorescens. Biochem. J. 125, 733-742.
263. Cullis, P.R., B. de Kruijff, M.J. Hope, A.J. Verkleij, R. Nayar, S.B. Farren, C. Tilcock, T.D. Madden, and M.B. Bally, (1983), Structural Properties of Lipids and Their Functional Roles in Biological Membranes. Membrane Fluidity in Biology (R.C. Aloia, Ed.), Vol. 1, pp. 39-81. Academic Press, New York.
264. Cullis, P.R. and B. de Kruijff, (1979), Lipid Polymorphism and the Functional Roles of Lipids in Biological Membranes. Biochim. Biophys. Acta 559, 399-420.

265. Cullis, P.R., B. de Kruijff, M.J. Hope, A.J. Verkleij, R. Nayar, S.B. Farren, C. Tilcock, T.D. Madden, and M.B. Bally, (1983), Structural Properties of Lipids and Their Functional Roles in Biological Membranes. Membrane Fluidity in Biology (R.C. Aloia, Ed.), Vol. 1, pp. 39-81. Academic Press, New York.
266. Cullis, P.R., B. de Kruijff, M.J. Hope, R. Nayar, R. Rietveld, and A.J. Verkleij, (1980), Structural Properties of Phospholipids in the Rat Liver Inner Mitochondrial Membrane: A ^{31}P-NMR Study. Biochim. Biophys. Acta 600, 625-635.
267. Cullis, P.R., M.J. Hope, and C.P.S. Tilcock, (1986), Lipid Polymorphism and the Roles of Lipids in Membranes. Chem. Phys. Lipids 40, 127-144.
268. Cunningham, B.A. (1986), Cell Adhesion Molecules: A New Perspective on Molecular Embryology. TIBS 11, 423-426.
269. Cunningham, B.A., J.E. Shimotake, W. Tamura-Lis, T. Mastran, W.-M. Kwok, J.W. Kauffman, and L.J. Lis, (1986), The Influence of Ion Species on Phosphatidylcholine Bilayer Structure and Packing. Chem. Phys. Lipids 39, 135-143.
270. Cunningham, B.A., J.J. Hemperly, B.A. Murray, E.A. Prediger, R. Brackenbury, and G.M. Edelman, (1987), Neural Cell Adhesion Molecule: Structure, Immunoglobulin-like Domains, Cell Surface Modulation, and Alternative RNA Splicing. Science 236, 799-806.
271. Curatolo, W. (1987), Glycolipid Function. Biochim. Biophys. Acta 906, 137-160.
272. Curatolo, W. and L.J. Neuringer, (1986), The Effects of Cerebrosides on Model Membrane Shape. J. Biol. Chem. 261, 17177-17182.
273. Dahl, G., T. Miller, D. Paul, R. Voellmy, and R. Werner, (1987), Expression of Functional Cell-Cell Channels from Cloned Rat Liver Gap Junction Complementary DNA. Science 236, 1290-1293.
274. Dahms, N.M., P. Lobel, J. Breitmeyer, J.M. Chirgwin, and S. Kornfeld, (1987), 46-kd Mannose 6-Phosphate Receptor: Cloning, Expression, and Homology to the 215-kd Mannose 6-Phosphate Receptor. Cell 50, 181-192.
275. Dalbey, R.E., A. Kuhn, and W. Wickner, (1987), The Internal Signal Sequence of Escherichia coli Leader Peptidase Is Necessary, but Not Sufficient, for Its Rapid Membrane Assembly. J. Biol. Chem. 262, 13241-13245.
276. Dalbey, R.E. and W. Wickner, (1985), Leader Peptidase Catalyzes the Release of Exported Proteins from the Outer Surface of the Escherichia coli Plasma Membrane. J. Biol. Chem. 260, 15925-15931.
277. Dalbey, R.E. and W. Wickner, (1986), The Role of the Polar, Carboxyl-terminal Domain of Escherichia coli Leader Peptidase in Its Translocation across the Plasma Membrane. J. Biol. Chem. 261, 13844-13849.
278. Dandliker, W.B. and V.A. de Saussure, (1971), Stabilization of Macromolecules by Hydrophobic Bonding: Role of Water Structure and of Chaotropic Ions. The Chemistry of Biosurfaces (M.L. Hair, Ed.), pp. 1-43. Marcel Dekker, New York.
279. Danielli, J.F. and H. Davson, (1935), A Contribution to the Theory of Permeability of Thin Films. J. Cell. Comp. Physiol. 5, 495-508.
280. Daniels, R.S., J.C. Downie, A.J. Hay, M. Knossow, J.J. Skehel, M.L. Wang, and D.C. Wiley, (1985), Fusion Mutants of the Influenza Virus Hemagglutinin Glycoprotein. Cell 40, 431-439.
281. Danielsen, E.M. and G.M. Cowell, (1986), Biosynthesis of Intestinal Microvillar Proteins. Biochem. J. 240, 777-782.
282. Darszon, A. (1983), Strategies in the Reassembly of Membrane Proteins into Lipid Bilayer Systems and Their Functional Assay. J. Bioenerg. and Biomemb. 15, 321-334.

283. Dasseux, J.-I., J.-F. Faucon, M. Lafleur, M. Pezolet, and J. Dufourcq, (1984), A Restatement of Melittin-induced Effects on the Thermotropism of Zwitterionic Phospholipid. Biochim. Biophys. Acta 775, 7-50.
284. Daum, G. (1985), Lipids of Mitochondria. Biochim. Biophys. Acta 822, 1-42.
285. Davey, J., S.M. Hurtley, and G. Warren, (1985), Reconstitution of an Endocytic Fusion Event in a Cell-Free System. Cell 43, 643-652.
286. Davidson, F.F., E.A. Dennis, M. Powell, and J.R. Glenney, (1987), Inhibition of Phospholipase A2 by "Lipocortins" and Calpactins. J. Biol. Chem. 262, 1698-1705.
287. Davies, A., K. Meeran, M.T. Cairns, and S.A. Baldwin, (1987), Peptide-specific Antibodies as Probes of the Orientation of the Glucose Transporter in the Human Erythrocyte Membrane. J. Biol. Chem. 262, 9347-9352.
288. Davis, C.G., J.L. Goldstein, T.C. Sudhof, R.G.W. Anderson, D.W. Russell, and M.S. Brown, (1987), Acid-dependent Ligand Dissociation and Recycling of LDL Receptor Mediated by Growth Factor Homology Region. Nature 326, 760-765.
289. Davis, C.G., M.A. Lehrman, D.W. Russell, R.G.W. Anderson, M.S. Brown, and J.L. Goldstein, (1986), The J.D. Mutation in Familial Hypercholesterolemia: Amino Acid Substitution in the Cytoplasmic Domain Impedes Internalization of LDL Receptors. Cell 45, 15-24.
290. Davis, E.O. and P.J.F. Henderson, (1987), The Cloning and DNA Sequence of the Gene xylE for Xylose-Proton Symport in Escherichia coli K12. J. Biol. Chem. 262, 13928-13932.
291. Davis, J.H. (1983), The Description of Membrane Lipid Conformation, Order and Dynamics by ^{2}H-NMR. Biochim. Biophys. Acta 737, 117-171.
292. Davis, J.Q. and V. Bennett, (1984), Brain Ankyrin: A Membrane-Associated Protein with Binding Sites for Spectrin, Tubulin, and the Cytoplasmic Domain of the Erythrocyte Anion Channel. J. Biol. Chem. 259, 13550-13559.
293. Davies, J.T. and E.K. Rideal, (1963), Interfacial Phenomena. Academic Press, New York.
294. Davis, L.I. and G. Blobel, (1987), Nuclear Pore Complex Contains a Family of Glycoproteins That Includes p62: Glycosylation Through a Previously Unidentified Cellular Pathway. Proc. Natl. Acad. Sci. USA 84, 7552-7556.
295. Davis, N.G. and P. Model, (1985), An Artificial Anchor Domain: Hydrophobicity Suffices to Stop Transfer. Cell 41, 607-614.
296. Davison, M.D. and J.B.C. Findlay, (1986), Modification of Ovine Opsin with the Photosensitive Hydrophobic Probe 1-Azido-4-[^{125}I] Iodobenzene. Biochem. J. 234, 413-420.
297. Davoust, J. and P.F. Devaux, (1982), Simulation of Electron Spin Resonance Spectra of Spin-Labeled Fatty Acids Covalently Attached to the Boundary of an Intrinsic Membrane Protein. A Chemical Exchange Model. J. Mag. Res. 48, 475-494.
298. Dawidowicz, E.A. (1987), Dynamics of Membrane Lipid Metabolism and Turnover. Ann. Rev. Biochem. 56, 43-61.
299. Dawson, C.R., A.F. Drake, J. Helliwell, and R.C. Hider, (1978), The Interaction of Bee Melittin with Lipid Bilayer Membranes. Biochim. Biophys. Acta 510, 75-86.
300. de Curtis, I., G. Fumagalli, and N. Borgese, (1986), Purification and Characterization of Two Plasma Membrane Domains from Ejaculated Bull Spermatozoa. J. Cell Biol. 102, 1813-1825.
301. De Cuyper, M. and M. Joniau, (1985), Spontaneous Intervesicular Transfer of

Anionic Phospholipids Differing in the Nature of Their Polar Head Group. Biochim. Biophys. Acta 814, 374-380.

302. De Grado, W.F., F.J. Kezdy, and E.T. Kaiser, (1981), Design, Synthesis and Characterization of a Cytotoxic Peptide with Melittin-like Activity. J. Amer. Chem. Soc. 103, 679-681.
303. De Kruijff, B. (1987), Polymorphic Regulation of Membrane Lipid Composition. Nature 329, 587-588.
304. de Kruijff, B., A. Rietveld, N. Telders, and B. Vaandrager, (1985), Molecular Aspects of the Bilayer Stabilization Induced by Poly(L-lysines) of Varying Size in Cardiolipin Liposomes. Biochim. Biophys. Acta 820, 295-304.
305. de Leij, L. and B. Witholt, (1977), Structural Heterogeneity of the Cytoplasmic and Outer Membranes of Escherichia coli. Biochim. Biophys. Acta 471, 92-104.
306. De Lemos-Chiarandini, C., A.B. Frey, D.D. Sabatini, and G. Kreibich, (1987), Determination of the Membrane Topology of the Phenobarbital-inducible Rat Liver Cytochrome P-450 Isoenzyme PB-4 Using Site-specific Antibodies. J. Cell Biol. 104, 209-219.
307. de Rosa, M., A. Gambacorta, and A. Gliozzi, (1986), Structure, Biosynthesis, and Physicochemical Properties of Archaebacterial Lipids. Microbiological Reviews 50, 70-80.
308. De Vrije, T., J. Tommassen, and B. De Kruijff, (1987), Optimal Posttranslational Translocation of the Precursor of PhoE Protein across Escherichia coli Membrane Vesicles Requires both ATP and the Protonmotive Force. Biochim. Biophys. Acta 900, 63-72.
309. de Wolf, F.A., B.H. Groen, L.P.A. van Houte, F.A.L.J. Peters, K. Krab, and R. Kraayenhof, (1985), Studies on Well-coupled Photosystem I-enriched Subchloroplast Vesicles. Neutral Red as Probe for External Surface Charge Rather than Internal Protonation. Biochim. Biophys. Acta 809, 204-214.
310. Deamer, D.W. (1987), Proton Permeation of Lipid Bilayers. J. Bioenerg. and Biomemb. 19, 457-479.
311. Deamer, D.W. and J. Bramhall, (1986), Permeability of Lipid Bilayers to Water and Ionic Solutes. Chem. Phys. Lipids 40, 167-188.
312. Deamer, D.W. and P.S. Uster, (1983), Liposome Preparation: Methods and Mechanism. Liposomes (M.J. Ostro, Ed.), pp. 27-51. Marcel Dekker, New York.
313. Deamer, D.W., R. Leonard, A. Tardieu, and D. Branton, (1970), Lamellar and Hexagonal Lipid Phases Visualized by Freeze-Etching. Biochim. Biophys. Acta 219, 47-60.
314. Deatherage, J.F., R. Henderson, and R.A. Capaldi, (1982), Three-dimensional Structure of Cytochrome c Oxidase Vesicle Crystals in Negative Stain. J. Mol. Biol. 158, 487-499.
315. Deber, C.M., C.J. Brandl, R.B. Deber, L.C. Hsu, and X.K. Young, (1986), Amino Acid Composition of the Membrane and Aqueous Domains of Integral Membrane Proteins. Arch. Biochem. Biophys. 251, 68-76.
316. de Duve, C. (1983), Lysosomes Revisited. Eur. J. Bioch. 137, 391-397.
317. DeGeorge, J.J., A.H. Ousley, K.D. McCarthy, P. Morell, and E.G. Lapetina, (1987), Glucocorticoids Inhibit the Liberation of Arachidonate but Not the Rapid Production of Phospholipase C-dependent Metabolites in Acetylcholine-stimulated C62B Glioma Cells. J. Biol. Chem. 262, 9979-9983.
318. Deisenhofer, J., H. Michel, and R. Huber, (1985), The Structural Basis of Photosynthetic Light Reactions in Bacteria. TIBS 10, 243-248.

319. Deisenhofer, J., O. Epp, K. Miki, R. Huber, and H. Michel, (1985), Structure of the Protein Subunits in the Photosynthetic Reaction Centre of Rhodopseudomonas viridis at 3 Å Resolution. Nature 318, 618-621.
320. Dejter-Juszynski, M., N. Harpaz, H.M. Flowers, and N. Sharon, (1978), Blood-Group ABH-Specific Macroglycolipids of Human Erythrocytes: Isolation in High Yield from a Crude Membrane Glycoprotein Fraction. Eur. J. Bioch. 83, 363- 378.
321. DeLong, E.F. and A.A. Yayanos, (1985), Adaptation of the Membrane Lipids of a Deep-Sea Bacterium to Changes in Hydrostatic Pressure. Science 228, 1101-1103.
322. Demuth, D.R., L.C. Showe, M. Ballantine, A. Palumbo, P.J. Fraser, L. Cioe, G. Rovera, and P.J. Curtis, (1986), Cloning and Structural Characterization of a Human Non-erythroid Band 3-like Protein. The EMBO Journal 5, 1205-1214.
323. Dencher, N.A. (1986), Spontaneous Transmembrane Insertion of Membrane Proteins into Lipid Vesicles Facilitated by Short-Chain Lecithins. Biochemistry 25, 1195-1200.
324. Dennis, E.A. (1983), Phospholipases. The Enzymes (P.D. Boyer, Ed.), Vol. 16, pp. 307-353. Academic Press, New York.
325. Desaymard, C., M. Debarbouille, M. Jolit, and M. Schwartz, (1986), Mutations Affecting Antigenic Determinants of an Outer Membrane Protein of Escherichia coli. The EMBO Journal 5, 1383-1388.
326. Deshaies, R.J. and R. Schekman, (1987), A Yeast Mutant Defective at an Early Stage in Import of Secretory Protein Precursors into the Endoplasmic Reticulum. J. Cell Biol. 105, 633-645.
327. Deuticke, B., P. Lutkemeier, and M. Sistemich, (1984), Ion Selectivity of Aqueous Leaks Induced in the Erythrocyte Membranes by Crosslinking of Membrane Proteins. Biochim. Biophys. Acta 775, 150-160.
328. Devaux, P.F. (1983), ESR and NMR Studies of Lipid-Protein Interactions in Membranes. Biological Magnetic Resonance (L.J. Berliner and J. Reuben, Eds.), pp. 183-299. Plenum Press, New York.
329. Devaux, P.F., G.L Hoatson, E. Favre, P. Fellmann, B. Farren, A.L. MacKay, and M. Bloom, (1986), Interaction of Cytochrome c with Mixed Dimyristoylphosphatidylcholine-Dimyristoylphosphatidylserine Bilayers: A Deuterium Nuclear Magnetic Resonance Study. Biochemistry 25, 3804-3812.
330. Devaux, P.F. and M. Seigneuret, (1985), Specificity of Lipid-Protein Interaction as Determined by Spectroscopic Techniques. Biochim. Biophys. Acta 822, 63-125.
331. Diamond, J.M. and Y. Katz, (1974), Interpretation of Nonelectrolyte Partition Coefficients between Dimyristoyl Lecithin and Water. J. Membrane Biol. 17, 121-154.
332. Dieckhoff, J., B. Niggemeyer, R. Lietzke, and H.G. Mannherz, (1987), Reconstitution of Purified Chicken Gizzard 5'-Nucleotidase in Phospholipid Vesicles: Evidence for Its Transmembranous Character and the Existence of Functional Domains on Both Sides of the Phospholipid Bilayer. Eur. J. Biochem. 162, 451-459.
333. Dierstein, R. and W. Wickner, (1986), Requirements for Substrate Recognition by Bacterial Leader Peptidase. The EMBO Journal 5, 427-431.
334. Dierstein, R. and W. Wickner, (1985), The Leader Region of Pre-maltose Binding Protein Binds Amphiphiles: A Model for Self-Assembly in Protein Export. J. Biol. Chem. 260, 15919-15924.
335. Dijkstra, B.W., K.H. Kalk, J. Drenth, G.H. de Haas, M.R. Egmond, and A.J. Slotboom, (1984), Role of the N-Terminus in the Interaction of Pancreatic Phospholi-

pase A2 with Aggregated Substrates. Properties and Crystal Structure of Transaminated Phospholipase A2. Biochemistry 23, 2759-2766.

336. Dill, K.A. and P.J. Flory, (1981), Molecular Organization in Micelles and Vesicles. Proc. Natl. Acad. Sci. USA 78, 676-680.
337. Divakar, S. and K.R.K. Easwaran, (1987), Conformational Studies of A23187 with Mono-, Di- and Trivalent Metal Ions by Circular Dichroism Spectroscopy. Biophys. Chem. 27, 139-147.
338. Dixon, R.A.F., I.S. Sigal, E. Rands, R.B. Register, M.R. Candelore, A.D. Blake, and C.D. Strader, (1987), Ligand Binding to the ß-Adrenergic Receptor Involves Its Rhodopsin-like Core. Nature 326, 75-77.
339. Dixon, R.A.F., I.S. Sigal, M.R. Candelore, R.B. Register, W. Scattergood, E. Rands, and C.D. Strader, (1987), Structural Features Required for Ligand Binding to the ß-Adrenergic Receptor. The EMBO Journal 6, 3269-3275.
340. Dluhy, R.A., D.G. Cameron, H.H. Mantsch, and R. Mendelsohn, (1983), Fourier Transform Infrared Spectroscopic Studies of the Effects of Calcium Ions on Phosphatidylserine. Biochemistry 22, 6318-6325.
341. Dohlman, H.G., M. Bouvier, J.L. Benovic, M.G. Caron, and R.J. Lefkowitz, (1987), The Multiple Membrane Spanning Topography of the $ß_2$-Adrenergic Receptor. J. Biol. Chem. 262, 14282-14288.
342. Dohlman, H.G., M.G. Caron, and R.J. Lefkowitz, (1987), A Family of Receptors Coupled to Guanine Nucleotide Regulatory Proteins. Biochemistry 26, 2657-2664.
343. Donovan, J.J., M.I. Simon, and M. Montal, (1985), Requirements for the Translocation of Diphtheria Toxin Fragment A across Lipid Membranes. J. Biol. Chem. 260, 8817-8823.
344. Doxsey, S.J., F.M. Brodsky, G.S. Blank, and A. Helenius, (1987), Inhibition of Endocytosis by Anti-Clathrin Antibodies. Cell 50, 453-463.
345. Doyle, C. and J.L. Strominger, (1987), Interaction between CD4 and Class II MHC Molecules Mediates Cell Adhesion. Nature 330, 256-259.
346. Dratz, E.A., J.F.L. Van Breemen, K.M.P. Kamps, W. Keegstra, E.F.J. Van Bruggen, (1985), Two-dimensional Crystallization of Bovine Rhodopsin. Biochim. Biophys. Acta 832, 337-342.
347. Dressler, V., C.W.M. Haest, G. Plasa, B. Deuticke, and J.D. Erusalimsky, (1984), Stabilizing Factors of Phospholipid Asymmetry in the Erythrocyte Membrane. Biochim. Biophys. Acta 775, 189-196.
348. Drews, G. (1985), Structure and Functional Organization of Light-Harvesting Complexes and Photochemical Reaction Centers in Membranes of Phototrophic Bacteria. Microbiol. Rev. 49, 59-70.
349. Dreyfus, P., F. Rieger, M. Murawsky, L. Garcia, A. Lombet, M. Fosset, D. Pauron, J. Barhanin, and M. Lazdunski, (1986), The Voltage-dependent Sodium Channel Is Co-localized with the Acetylcholine Receptor at the Vertebrate Neuromuscular Junction. Biochem. Biophys. Res. Comm. 139, 196-201.
350. Duesberg, P.H. (1987), Cancer Genes: Rare Recombinants Instead of Activated Oncogenes (A Review). Proc. Natl. Acad. Sci. USA 84, 2117-2124.
351. Dufourc, E.J. and I.C. Smith, (1985), 2H NMR Evidence for Antibiotic-Induced Cholesterol Immobilization in Biological Model Membranes. Biochemistry 24, 2420-2424.
352. Dufourc, E.J., I.C.P. Smith, and H.C. Jarrell, (1984), Role of Cyclopropane Moieties in the Lipid Properties of Biological Membranes: A ^{2}H-NMR Structural and Dynamical Approach. Biochemistry 23, 2300-2309.
353. Dufourcq, J., J.-F. Faucon, G. Fourche, J.-L. Dasseux, M. Le Marie, and T. Gulik-

Krzywicki, (1986), Morphological Changes of Phosphatidylcholine Bilayers Induced by Melittin: Vesicularization, Fusion, Discoidal Particles. Biochim. Biophys. Acta 859, 33-48.

354. Dunn, R., J. McCoy, M. Simsek, A. Majumdar, S.H. Chang, U.L. RajBhandary, and H.G. Khorana, (1981), The Bacteriorhodopsin Gene. Proc. Natl. Acad. Sci. USA 78, 6744-6748.

355. Dunn, R.J., N.R. Hackett, J.M. McCoy, B.H. Chao, K. Kimura, and H.G. Khorana, (1987), Structure-Function Studies on Bacteriorhodopsin. J. Biol. Chem. 262, 9246-9254.

356. Dunn, S.M.J., B.M. Conti-Tronconi, and M.A. Raftery, (1986), Acetylcholine Receptor Dimers Are Stabilized by Extracellular Disulfide Bonding. Biochemical and Biophysical Research Communications 139, 830-837.

357. Dunn, W.A. and A.L. Hubbard, (1984), Receptor-mediated Endocytosis of Epidermal Growth Factor by Hepatocytes in the Perfused Rat Liver: Ligand and Receptor Dynamics. J. Cell Biol. 98, 2148-2159.

358. Dunphy, W.G. and J.E. Rothman, (1985), Compartmental Organization of the Golgi Stack. Cell 42, 13-21.

359. Dunphy, W.G., S.R. Pfeffer, D.O. Clary, B.W. Wattenberg, B.S. Glick, and J.E. Rothman, (1986), Yeast and Mammals Utilize Similar Cytosolic Components to Drive Protein Transport through the Golgi Complex. Proc. Natl. Acad. Sci. USA 83, 1622-1626.

360. Dustin, M.L., P. Selvarai, R.J. Mattaliano, and T.A. Springer, (1987), Anchoring Mechanisms for LFA-3 Cell Adhesion Glycoprotein at Membrane Surface. Nature 329, 846-848.

361. Dux, L. and A. Martonosi, (1983), The Regulation of ATPase-ATPase Interactions in Sarcoplasmic Reticulum Membrane. J. Biol. Chem. 258, 11903-11907.

362. Duzgunes, N., T.M. Allen, J. Fedor, and D. Papahadjopoulos, (1987), Lipid Mixing During Membrane Aggregation and Fusion: Why Fusion Assays Disagree. Biochemistry 26, 8435-8442.

362a. Dwyer, T.M., D.J. Adams, and B. Hille, (1980), The Permeability and the Endplate Channel to Organic Cations in Frog Muscle. J. Gen. Physiol. 75, 469-492.

363. East, J.M. and A.G. Lee, (1982), Lipid Selectivity of the Calcium and Magnesium Ion Dependent Adenosinetriphosphatase, Studied with Fluorescence Quenching by a Brominated Phospholipid. Biochemistry 21, 4144-4151.

364. East, J.M., O.T. Jones, A.C. Simmonds, and A.G. Lee, (1984), Membrane Fluidity Is Not an Important Physiological Regulator of the (Ca^{2+}-Mg^{2+})-dependent ATPase of Sarcoplasmic Reticulum. J. Biol. Chem. 259, 8070-8071.

365. Eble, B.E., D.R. MacRae, V.R. Lingappa, and D. Ganem, (1987), Multiple Topogenic Sequences Determine the Transmembrane Orientation of Hepatitis B Surface Antigen. Mol. Cell Biol. 7, 3591-3601.

366. Edidin, M. (1981), Molecular Motions and Membrane Organization and Function. Membrane Structure (J.B. Finean and R.H. Michell, Eds.), pp. 37-82. Elsevier, New York.

367. Edidin, M. and M. Zuniga, (1984), Lateral Diffusion of Wild-type and Mutant Ld Antigens in L Cells. J. Cell Biol. 99, 2333-2335.

368. Edmonds, D.T. (1987), A Physical Model of Sodium Channel Gating. Eur. Biophys. J. 14, 195-201.

369. Edwards, H.C. and A.G. Booth, (1987), Calcium-sensitive, Lipid-binding Cytoskeletal Proteins of the Human Placental Microvillar Region. J. Cell Biol. 105, 303-311.

370. Eilers, M. and G. Schatz, (1986), Binding of a Specific Ligand Inhibits Import of a Purified Precursor Protein into Mitochondria. Nature 322, 228-232.
371. Eilers, M., W. Oppliger, and G. Schatz, (1987), Both ATP and an Energized Inner Membrane Are Required to Import a Purified Precursor Protein into Mitochondria. The EMBO Journal 6, 1073-1077.
372. Eisenberg, D. (1984), Three-dimensional Structure of Membrane and Surface Proteins. Annu. Rev. Biochem. 53, 595-623.
373. Eisenberg, D., E. Schwarz, M. Komaromy, and R. Wall, (1984), Analyses of Membrane and Surface Protein Sequences with the Hydrophobic Moment Plot. J. Mol. Biol. 179, 125-142.
374. Eisenberg, D., R.M. Weiss, and T.C. Terwilliger, (1984), The Hydrophobic Moment Detects Periodicity in Protein Hydrophobicity. Proc. Natl. Acad. Sci. USA 81, 140-144.
375. Eisenberg, D., R.M. Weiss, and T.C. Terwilliger, (1984), The Helical Hydrophobic Moment: A Measure of the Amphiphilicity of a Helix. Nature 299, 371-374.
376. Eisenberg, M., T. Gresalfi, T. Riccio, and S. McLaughlin, (1979), Adsorption of Monovalent Cations to Bilayer Membranes Containing Negative Phospholipids. Biochemistry 18, 5213-5223.
377. Eisenman, G. and J.A. Dani, (1987), An Introduction to Molecular Architecture and Permeability of Ion Channels. Ann. Rev. Biophys. Biophys. Chem. 16, 205-226.
378. Eisinger, J., J. Flores, and W.P. Petersen, (1986), A Milling Crowd Model for Local and Long-Range Obstructed Lateral Diffusion. Biophys. J. 49, 987-1001.
379. El Kebbaj, M.S., N. Latruffe, M. Monsigny, and A. Obrenovitch, (1986), Interactions between apo-(D-ß-Hydroxybutyrate dehydrogenase) and Phospholipids Studied by Intrinsic and Extrinsic Fluorescence. Biochem. J. 237, 359-364.
380. Elgsaeter, A., B.T. Stokke, A. Mikkelsen, and D. Branton, (1986), The Molecular Basis of Erythrocyte Shape. Science 234, 1217-1221.
381. Ellena, J.F., R.D. Pates, and M.F. Brown, (1986), ^{31}P NMR Spectra of Rod Outer Segment and Sarcoplasmic Reticulum Membranes Show No Evidence of Immobilized Components due to Lipid-Protein Interactions. Biochemistry 25, 3742-3748.
382. Elliott, J.R., R.D. Murrell, and D.A. Haydon, (1987), Local Anesthetic Action of Carboxylic Esters: Evidence for the Significance of Molecular Volume and for the Number of Sites Involved. J. Memb. Biol. 95, 143-149.
383. Elmes, M.L., D.G. Scraba, and J.H. Weiner, (1986), Isolation and Characterization of the Tubular Organelles Induced by Fumarate Reductase Overproduction in Escherichia coli. Journal of General Microbiology 132, 1429-1439.
384. Engel, A., A. Massalski, H. Schindler, D.L. Dorset, and J.P. Rosenbusch, (1985), Porin Channel Triplets Merge into Single Outlets in Escherichia coli Outer Membranes. Nature 317, 643-645.
385. Engelman, D.M., A. Goldman, and T.A. Steitz, (1982), The Identification of Helical Segments in the Polypeptide Chain of Bacteriorhodopsin. Methods in Enzymology 88, 81-89.
386. Engelman, D.M. and G. Zaccai, (1980), Bacteriorhodopsin Is an Inside-Out Protein. Proc. Natl. Acad. Sci. USA 77, 5894-5898.
387. Engelman, D.M., R. Henderson, A.D. McLachlan, and B.A. Wallace, (1980), Path of the Polypeptide in Bacteriorhodopsin. Proc. Natl. Acad. Sci. USA 77, 2023-2027.
388. Engelman, D.M. and T.A. Steitz, (1981), The Spontaneous Insertion of Proteins into and across Membranes: The Helical Hairpin Hypothesis. Cell 23, 411-422.
389. Engelman, D.M., T.A. Steitz, and A. Goldman, (1986), Identifying Nonpolar

Transbilayer Helices in Amino Acid Sequences of Membrane Proteins. Ann. Rev. Biophys. Biophys. Chem. 15, 321-353.

390. Enoch, H.G. and P. Strittmatter, (1979), Formation and Properties of 1000-Å Diameter, Single-bilayer Phospholipid Vesicles. Proc. Natl. Acad. Sci. USA 76, 145-149.
391. Enoch, H.G., A. Catala, and P. Strittmatter, (1976), Mechanism of Rat Liver Microsomal Stearyl-CoA Desaturase: Studies of the Substrate Specificity, Enzyme-Substrate Interactions and the Function of Lipid. J. Biol. Chem. 251, 5095-5103.
392. Epand, R.M., J.K. Seyler, and R.C. Orlowski, (1986), The Hydrophobic Moment of the Amphipathic Helix of Salmon Calcitonin and Biological Potency. Eur. J. Biochem. 159, 125-127.
393. Epand, R.M. and J.M. Sturtevant, (1981), A Calorimetric Study of Peptide-Phospholipid Interactions: The Glucagon-Dimyristoyl Phosphatidylcholine Complex. Biochemistry 20, 4603-4606.
394. Epand, R.M., S.-W. Hui, C. Argan, L.L. Gillespie, and G.C. Shore, (1986), Structural Analysis and Amphiphilic Properties of a Chemically Synthesized Mitochondrial Signal Peptide. J. Biol. Chem. 261, 10017-10020.
395. Esmann, M., A. Watts, and D. Marsh, (1985), Spin-Label Studies of Lipid-Protein Interactions in (Na^+, K^+)-ATPase Membranes from Rectal Glands of Squalus acanthias. Biochemistry 24, 1386-1393.
396. Esmann, M. and D. Marsh, (1985), Spin-Label Studies on the Origin of the Specificity of Lipid-Protein Interactions in (Na^+, K^+)-ATPase Membranes from Squalus acanthias. Biochemistry 24, 3572-3578.
397. Esposito, G., J.A. Carver, J. Boyd, and I.D. Campbell, (1987), High-Resolution ^1H NMR Study of the Solution Structure of Alamethicin. Biochemistry 26, 1043-1050.
398. Estep, T.N., D.B. Mountcastle, Y. Barenholz, R.L. Biltonen, and T.E. Thompson, (1979), Thermal Behavior of Synthetic Sphingomyelin-Cholesterol Dispersions. Biochemistry 18, 2112-2117.
399. Etemadi, A.-H. (1980), Membrane Asymmetry. A Survey and Critical Appraisal of the Methodology. I. Methods for Assessing the Asymmetric Orientation and Distribution of Proteins. Biochim. Biophys. Acta 604, 347-422.
400. Evans, E.A., R. Gilmore, and G. Blobel, (1986), Purification of Microsomal Signal Peptidase as a Complex. Proc. Natl. Acad. Sci. USA 83, 581-585.
401. Evans, T., M.L. Brown, E.D. Fraser, and J.K. Northup, (1986), Purification of the Major GTP-binding Proteins from Human Placental Membranes. J. Biol. Chem. 261, 7052-7059.
402. Evans, W.H. (1980), A Biochemical Dissection of the Functional Polarity of the Plasma Membrane of the Hepatocyte. Biochim. Biophys. Acta 604, 27-64.
403. Evans, W.H. and N. Flint, (1985), Subfractionation of Hepatic Endosomes in Nycodenz Gradients and by Free-flow Electrophoresis. Biochem. J. 232, 25-32.
404. Evans, W.H. and W.G.M. Hardison, (1985), Phospholipid, Cholesterol, Polypeptide and Glycoprotein Composition of Hepatic Endosome Subfractions. Biochem. J. 232, 33-36.
405. Eytan, G.D. (1982), The Use of Liposomes for Reconstitution of Biological Functions. Biochim. Biophys. Acta 694, 185-202.
406. Fagan, M.H. and T.G. Dewey, (1986), Resonance Energy Transfer Study of Membrane-bound Aggregates of the Sarcoplasmic Reticulum Calcium ATPase. J. Biol. Chem. 261, 3654-3660.

407. Falke, J.J. and D.E. Koshland, (1987), Global Flexibility in a Sensory Receptor: A Site-Directed Cross-Linking Approach. Science 237, 1596-1600.
408. Falke, J.J., R.J. Pace, and S.I. Chan, (1984), Direct Observation of the Transmembrane Recruitment of Band 3 Transport Sites by Competitive Inhibitors. J. Biol. Chem. 259, 6481-6491.
409. Falke, J.J. and S.I. Chan, (1986), Molecular Mechanisms of Band 3 Inhibitors. 2. Channel Blockers. Biochemistry 25, 7895-7898.
410. Falke, J.J. and S.I. Chan, (1986), Molecular Mechanisms of Band 3 Inhibitors. 1. Transport Site Inhibitors. Biochemistry 25, 7888-7894.
411. Falke, J.J. and S.I. Chan, (1985), Evidence that Anion Transport by Band 3 Proceeds via a Ping-Pong Mechanism Involving a Single Transport Site. J. Biol. Chem. 260, 9537-9544.
412. Falke, J.J. and S.I. Chan, (1986), Molecular Mechanisms of Band 3 Inhibitors. 3. Translocation Inhibitors. Biochemistry 25, 7899-7906.
413. Farahbakhsh, Z.T., R.L. Baldwin, and B.J. Wisnieski, (1987), Effect of Low pH on the Conformation of Pseudomonas Exotoxin A. J. Biol. Chem. 262, 2256-2261.
414. Feigenson, G.W. (1986), On the Nature of Calcium Ion Binding between Phosphatidylserine Lamellae. Biochemistry 25, 5819-5825.
415. Feizi, T. and R.A. Childs, (1985), Carbohydrate Structures of Glycoproteins and Glycolipids as Differentiation Antigens, Tumour-Associated Antigens and Components of Receptor Systems. TIBS 10, 24-29.
416. Feller, D.J., J.A. Talvenheimo, and W.A. Catterall, (1985), The Sodium Channel from Rat Brain. J. Biol. Chem. 260, 11542-11547.
417. Ferenci, T. and T.J. Silhavy, (1987), Sequence Information Required for Protein Translocation from the Cytoplasm. J. Bacteriology 169, 5339-5342.
418. Ferguson, M.A.J., M.G. Low, and G.A.M. Cross, (1985), Glycosyl-sn-1,2-dimyristoylphosphatidylinositol Is Covalently Linked to Trypanosoma brucei Variant Surface Glycoprotein. J. Biol. Chem. 260, 14547-14555.
419. Ferguson, S.J. (1985), Fully Delocalised Chemiosmotic or Localised Proton Flow Pathways in Energy Coupling? A Scrutiny of Experimental Evidence. Biochim. Biophys. Acta 811, 47-95.
420. Fernandez, M., D.G. Nicholls, and E. Rial, (1987), The Uncoupling Protein from Brown-adipose-tissue Mitochondria: Chymotrypsin-induced Structural and Functional Modifications. Eur. J. Biochem. 164, 675-680.
421. Ferro-Novick, S., P. Novick, C. Field, and R. Schekman, (1984), Yeast Secretory Mutants That Block the Formation of Active Cell Surface Enzymes. J. Cell Biol. 98, 35-43.
422. Freudl, R., S. MacIntyre, M. Degen, and U. Henning, (1986), Cell Surface Exposure of the Outer Membrane Protein OmpA of Escherichia coli K-12. J. Mol. Biol. 188, 491-494.
423. Fidelio, G.D., B.M. Austen, D. Chapman, and J.A. Lucy, (1987), Interactions of Ovalbumin and of Its Putative Signal Sequence with Phospholipid Monolayers. Biochem. J. 244, 295-301.
424. Fillingame, R.H., L.K. Peters, L.K. White, M.E. Mosher, and C.R. Paule, (1984), Mutations Altering Aspartyl-61 of the Omega Subunit (uncE Protein) of Escherichia coli H^+-ATPase Differ in Effect on Coupled ATP Hydrolysis. J. Bacteriology 158, 1078-1083.
425a. Finean, J.B., R. Coleman, and R.H. Michell, (1978), Membranes and Their Cellular Functions, 2nd ed. Blackwell, Oxford, UK.

425. Finean, J.B. (1962), The Nature and Stability of the Plasma Membrane. Circulation 26, 1151-1162.
426. Finean, J.B., R. Coleman, W.G. Green, and A.R. Limbrick, (1966), Low Angle X-ray Diffraction and Electron Microscope Studies of Isolated Cell Membranes. J. Cell Sci. 1, 287-296.
427. Finean, J.B. and R.E. Burge, (1963), The Determination of the Fourier Transform of the Myelin Layer from a Study of Swelling Phenomena. J. Mol. Biol. 7, 672-682.
428. Finean, J.B. and R.H. Michell, (1981), Isolation, Composition and General Structure of Membranes. Membrane Structure (J.B. Finean and R.H. Michell, Eds.), pp. 1-36. Elsevier, New York.
429. Finer-Moore, J. and R.M. Stroud, (1984), Amphipathic Analysis and Possible Formation of the Ion Channel in an Acetylcholine Receptor. Proc. Natl. Acad. Sci. USA 81, 155-159.
430. Finkelstein, A. and O.S. Andersen, (1981), The Gramicidin A Channel: A Review of Its Permeability Characteristics with Special Reference to the Single-file Aspect of Transport. J. Membrane Biol. 59, 155-171.
431. Fischl, A.S., M.J. Homann, M.A. Poole, and G.M. Carman, (1986), Phosphatidylinositol Synthase from Saccharomyces cerevisiae: Reconstitution, Characterization, and Regulation of Activity. J. Biol. Chem. 261, 3178-3183.
432. Fisher, K. and D. Branton, (1974), Application of the Freeze-Fracture Technique to Natural Membranes. Methods in Enz. 32, 35-44.
433. Fishman, J.B. and R.E. Fine, (1987), A Trans Golgi-Derived Exocytic Coated Vesicle Can Contain Both Newly Synthesized Cholinesterase and Internalized Transferrin. Cell 48, 157-164.
434. Fishman, P.H. (1986), Recent Advances in Identifying the Functions of Gangliosides. Chem. Phys. Lipids 42, 137-151.
435. Fleischer, S. and L. Packer, (1974), Methods in Enzymology, Vol. 31. Academic Press, New York.
436. Fleischer, S. and M. Kervina, (1974), Subcellular Fractionation of Rat Liver. Methods in Enz. 31, 6-41.
437. Rogers, M.J. and P. Strittmatter, (1975), The Interaction of NADH-Cytochrome b_5 Reductase and Cytochrome b_5 Bound to Egg Lecithin Liposomes. J. Biol. Chem. 250, 5713-5718.
438. Flewelling, R.F. and W.L. Hubbell, (1986), The Membrane Dipole Potential in a Total Membrane Potential Model. Biophys. J. 49, 541-552.
439. Flewelling, R.F. and W.L. Hubbell, (1986), Hydrophobic Ion Interactions with Membranes: Thermodynamic Analysis of Tetraphenylphosphonium Binding to Vesicles. Biophys. J. 49, 531-540.
440. Fliesler, S.J. and S.F. Basinger, (1985), Tunicamycin Blocks the Incorporation of Opsin into Retinal Rod Outer Segment Membranes. Proc. Natl. Acad. Sci. USA 82, 1116-1120.
441. Flinta, C., G. von Heijne, and J. Johansson, (1983), Helical Sidedness and the Distribution of Polar Residues in Trans-membrane Helices. J. Mol. Biol. 168, 193-196.
442. Flugge, U.I. and G. Hinz, (1986), Energy Dependence of Protein Translocation into Chloroplasts. Eur. J. Biochem. 160, 563-570.
443. Fluhler, E., V.G. Burnham, and L.M. Loew, (1985), Spectra, Membrane Binding, and Potentiometric Responses of New Charge Shift Probes. Biochemistry 24, 5749-5755.

444. Fong, H.K.W., T.T. Amatruda III, B.W. Birren, and M.I. Simon, (1987), Distinct Forms of the ß Subunit of GTP-Binding Regulatory Proteins Identified by Molecular Cloning. Proc. Natl. Acad. Sci. USA 84, 792-3796.
445. Fong, T.M. and M.G. McNamee, (1986), Correlation between Acetylcholine Receptor Function and Structural Properties of Membranes. Biochemistry 25, 830-840.
446. Forbush, B. III (1987), Rapid Release of $^{42}K^+$ or $^{86}Rb^+$ from Two Distinct Transport Sites on the Na, K-Pump in the Presence of P_i or Vanadate. J. Biol. Chem. 262, 11116-11127.
447. Forbush, B. III (1987), Rapid Release of $^{42}K^+$ and $^{86}Rb^+$ from an Occluded State of the Na, K-Pump in the Presence of ATP or ADP. J. Biol. Chem. 262, 11104-11115.
448. Ford, R.C., D. Picot, and R.M. Garavito, (1987), Crystallization of the Photosystem I Reaction Centre. The EMBO Journal 6, 1581-1586.
449. Forman, S.D. and Y. Nemerson, (1986), Membrane-dependent Coagulation Reaction Is Independent of the Concentration of Phospholipid-bound Substrate: Fluid Phase Factor X Regulates the Extrinsic System. Proc. Natl. Acad. Sci. USA 83, 4675-4679.
450. Forrest, B.J. and J. Mattai, (1985), ^2H and ^{31}P NMR Study of the Interaction of General Anesthetics with Phosphatidylcholine Membranes. Biochemistry 24, 7148-7153.
451. Forte, M., H.R. Guy, and C.A. Mannella, (1987), Molecular Genetics of the VDAC Ion Channel: Structural Model and Sequence Analysis J. Bioenerg. Biomemb. 19, 341-350.
452. Foster, D.L. and R.H. Fillingame, (1982), Stoichiometry of Subunits in the H^+-ATPase Complex of Escherichia coli. J. Biol. Chem. 257, 2009-2015.
453. Foster, D.L., M. Boublik, and H.R. Kaback, (1983), Structure of the lac Carrier Protein of Escherichia coli. J. Biol. Chem. 258, 31-34.
454. Foster, D.L., S.L. Mowbray, B.K. Jap, and D.E. Koshland, (1985), Purification and Characterization of the Asparate Chemoreceptor. J. Biol. Chem. 260, 11706-11710.
455. Fox, J.F.B. (1985), Identification of Actin-Binding Protein as the Protein Linking Membrane Skeleton to Glycoproteins on Platelet Plasma Membrane. J. Biol. Chem. 260, 11970-11977.
456. Fox, R.O. and F.M. Richards, (1982), A Voltage-gated Ion Channel Model Inferred from the Crystal Structure of Alamethicin at 1.5 Å Resolution. Nature 300, 325-330.
457. Franck, P.F.H., J.A.F. Op den Kamp, B. Roelofsen, and L.L.M. van Deenen, (1986), Does Diamide Treatment of Intact Human Erythrocytes Cause Loss of Phospholipid Asymmetry? Biochim. Biophys. Acta 857, 127-130.
458. Franke, W.W. (1987), Nuclear Lamins and Cytoplasmic Intermediate Filament Proteins: A Growing Multigene Family. Cell 48, 3-4.
459. Franke, W.W. and U. Scheer, (1970), The Ultrastructure of the Nuclear Envelope of Amphibian Oocytes: A Reinvestigation. 1. The Mature Oocyte. J. Ultrastruc. Res. 30, 288-316.
460. Franks, F. and D. Eagland, (1975), The Role of Solvent Interactions in Protein Conformation. CRC Critical Reviews on Biochemistry 3, 165-219.
461. Franks, N.P. (1976), Structural Analysis of Hydrated Egg Lecithin in Cholesterol Bilayers I. X-ray Diffraction. J. Mol. Biol. 100, 345-358.
462. Franks, N.P. and W.R. Lieb, (1984), Do General Anaesthetics Act by Competitive Binding to Specific Receptors? Nature 310, 599-601.
463. Freedman, R.B. (1981), Membrane-bound Enzymes. Membrane Structure (J.B. Finean and R.H. Michell, Eds.), pp. 161-214. Elsevier, New York.

464. Frelinger, A.L. and U. Rutishauser, (1986), Topography of N-CAM Structural and Functional Determinants. II. Placement of Monoclonal Antibody Epitopes. J. Cell Biol. 103, 1729-1737.
465. French, J.S., F.P. Guengerich, and M.J. Coon, (1980), Interactions of Cytochrome P-450, NADPH-Cytochrome P-450 Reductase, Phospholipid, and Substrate in the Reconstituted Liver Microsomal Enzyme System. J. Biol. Chem. 255, 4112-4119.
466. Frey, T.G., M.J. Costello, B. Karlsson, J.C. Haselgrove, and J.S. Leigh, (1982), Structure of the Cytochrome Oxidase Dimer: Electron Microscopy of Two-dimensional Crystals. J. Mol. Biol. 162, 113-130.
467. Frolich, O. and R.B. Gunn, (1986), Erythrocyte Anion Transport: The Kinetics of a Single-site Obligatory Exchange System. Biochim. Biophys. Acta 864, 169-194.
468. Frye, L.D. and M. Edidin, (1970), The Rapid Intermixing of Cell Surface Antigens After Formation of Mouse-Human Heterokaryons. J. Cell Sci. 7, 319-335.
469. Fukuda, M., M. Lauffenburger, H. Sasaki, M.E. Rogers, and A. Dell, (1987), Structures of Novel Sialylated O-Linked Oligosaccharides Isolated from Human Erythrocyte Glycophorins. J. Biol. Chem. 262, 11952-11957.
470. Fukushima, D., J.P. Kupferberg, S. Yokoyama, D.J. Kroon, E.T. Kaiser, and F.J. Kezdy, (1979), A Synthetic Amphiphilic Helical Docosapeptide with the Surface Properties of Plasma Apolipoprotein A-I. J. Am. Chem. Soc. 101, 3703-3704.
471. Fuller, S.D. and K. Simons, (1986), Transferrin Receptor Polarity and Recycling Accuracy in "Tight" and "Leaky" Strains of Madin-Darby Canine Kidney Cells. J. Cell Biol. 103, 1767-1779.
472. Furst, P. and M. Solioz, (1985), The Vandate-sensitive ATPase of Streptococcus faecalis Pumps Potassium in a Reconstituted System. J. Biol. Chem. 261, 4302-4308.
473. Furth, A.J. (1980), Removing Unbound Detergent from Hydrophobic Proteins. Anal. Bioch. 109, 207-215.
474. Furthmayr, H. and V.T. Marchesi, (1976), Subunit Structure of Human Erythrocyte Glycophorin A. Biochemistry 15, 1137-1143.
475. Furthmayr, H. (1978), Structural Comparison of Glycophorins and Immunochemical Analysis of Genetic Variants. Nature (London) 271, 519-524.
476. Futai, M. and H. Kanazawa, (1983), Structure and Function of Proton Translocating Adenosine Triphosphatase: Biochemical and Molecular Biological Approaches. Microbiol. Rev. 47, 285-312.
477. Gaber, B.P. and W.L. Peticolas, (1977), On the Quantitative Interpretation of Biomembrane Structure by Raman Spectroscopy. Biochim. Biophys. Acta 465, 260-274.
478. Gabriel, N.E. and M.F. Roberts, (1984), Spontaneous Formation of Stable Unilamellar Vesicles. Biochemistry 23, 4011-4015.
479. Gabriel, N.E. and M.F. Roberts, (1986), Interaction of Short-Chain Lecithin with Long-Chain Phospholipids: Characterization of Vesicles That Form Spontaneously. Biochemistry 25, 2812-2821.
480. Gaffney, B.J. (1985), Chemical and Biochemical Crosslinking of Membrane Components. Biochim. Biophys. Acta 822, 289-317.
481. Gahmberg, C.G. (1981), Membrane Glycoproteins and Glycolipids: Structure, Localization and Function of the Carbohydrate. Membrane Structure (J.B. Finean and R.H. Michell, Eds.), pp. 127-160. Elsevier, New York.
482. Gaines, G.L. (1966), Insoluble Monolayers at Liquid-Gas Interfaces. Interscience Publishers, a division of John Wiley and Sons, Inc., New York.
483. Galla, H.J., W. Hartman, U. Theilen, and E. Sackmann, (1979), On Two-dimen-

sional Passive Random Walls in Lipid Bilayers and Fluid Pathways in Biomembranes. J. Membrane Biol. 48, 215-236.

484. Gallatin, M., T.P. St. John, M. Siegelman, R. Reichert, E.C. Butcher, and I.L. Weissman, (1986), Lymphocyte Homing Receptors. Cell 44, 673-680.

485. Gallin, W.J., B.C. Sorkin, G.M. Edelman, and B.A. Cunningham, (1987), Sequence Analysis of a cDNA Clone Encoding the Liver Cell Adhesion Molecule, L-CAM. Proc. Natl. Acad. Sci. USA 84, 2808-2812.

486. Gally, H.U., G. Pluschke, P. Overath, and J. Seelig, (1980), Structure of Escherichia coli Membranes. Fatty Acyl Chain Order Parameters of Inner and Outer Membranes and Derived Liposomes. Biochemistry 19, 1638-1634.

487. Gally, H.U., G. Pluschke, P. Overath, and J. Seelig, (1979), Structure of Escherichia coli Membranes. Phospholipid Conformation in Model Membranes and Cells as Studied by Deuterium Magnetic Resonance. Biochemistry 18, 5605-5610.

488. Ganong, B.R. and R.M. Bell, (1984), Transmembrane Movement of Phosphatidylglycerol and Diacylglycerol Sulfhydryl Analogues. Biochemistry 23, 4977-4983.

489. Gao, B., S. Mumby, and A.G. Gilman, (1987), The G Protein $ß_2$ Complementary DNA Encodes the $ß_{35}$ Subunit. J. Biol. Chem. 262, 17254-17257.

490. Garavito, R.M., J. Jenkins, J.N. Jansonius, R. Karlsson, and J.P. Rosenbusch, (1983), X-ray Diffraction Analysis of Matrix Porin, an Integral Membrane Protein from Escherichia coli Outer Membrane. J. Mol. Biol. 164, 313-327.

491. Garavito, R.M., U. Hinz, and J.-M. Neuhaus, (1984), The Crystallization of Outer Membrane Proteins from Escherichia coli: Studies on lamB and ompA Gene Products. J. Biol. Chem. 259, 4254-4257.

492. Garcia, M.L., M. Kitada, H.C. Eisenstein, and T.A. Krulwich, (1984), Voltage-Dependent Proton Fluxes in Liposomes. Biochim. Biophys. Acta 766, 109-115.

493. Garcia, P.D., J. Ghrayeb, M. Inouye, and P. Walter, (1987), Wild Type and Mutant Signal Peptides of Escherichia coli Outer Membrane Lipoprotein Interact with Equal Efficiency with Mammalian Signal Recognition Particle. J. Biol. Chem. 262, 9463-9468.

494. Gardas, A. (1976), A Structural Study on a Macroglycolipid Containing 22 Sugars Isolated from Human Erythrocytes. Eur. J. Bioch. 68, 177-183.

495. Gardel, C., S. Benson, J. Hunt, S. Michaelis, and J. Beckwith, (1987), secD, a New Gene Involved in Protein Export in Escherichia coli. J. Bacteriology 169, 1286-1290.

496. Garland, P.B. (1984), Optical Probes and the Detection of Conformational Changes in Membrane Proteins. Biological Membranes (D. Chapman, Ed.), Vol. 5, pp. 279-288. Academic Press, New York.

497. Geiger, B. (1983), Membrane-Cytoskeleton Interaction. Biochim. Biophys. Acta 737, 305-341.

498. Geiger, B. (1985), Microfilament-Membrane Interaction. TIBS 10, 456-461.

499. Geisow, M.J. (1986), Common Domain Structure of Ca^{2+} and Lipid-binding Proteins. FEBS Lett. 203, 99-103.

500. Geller, B.L., N.R. Movva, and W. Wickner, (1986), Both ATP and the Electrochemical Potential Are Required for Optimal Assembly of pro-OmpA into Escherichia coli Inner Membrane Vesicles. Proc. Natl. Acad. Sci. USA 83, 4219-4222.

501. Geller, B.L. and W. Wickner, (1985), M13 Procoat Inserts into Liposomes in the Absence of Other Membrane Proteins. J. Biol. Chem. 260, 13281-13285.

502. Gelles, J., D.F. Blair, and S.I. Chan, (1987), The Proton-pumping Site of Cytochrome c Oxidase: A Model of Its Structure and Mechanism. Biochim. Biophys. Acta 853, 205-236.

503. Gennis, R.B. and A. Jonas, (1977), Protein Lipid Interactions. Ann. Rev. of Biophys. and Bioeng. 6, 195-238.
504. Gennis, R.B., M. Sinensky, and J.L. Strominger, (1976), Activation of C55-Isoprenoid Alcohol Phosphokinase from Staphylococcus aureus. J. Biol. Chem. 251, 1270-1276.
505. Georgallas, A., D.L. Hunter, T. Lookman, M.J. Zuckermann, and D.A. Pink, (1984), Interactions Between Two Sheets of a Bilayer Membrane and Its Internal Lateral Pressure. Eur. Biophys. J. 11, 79-86.
506. Gerber, G.E., C.P. Gray, D. Wildenauer, and H.G. Khorana, (1977), Orientation of Bacteriorhodopsin in Halobacterium halobium as Studied by Selective Proteolysis. Proc. Natl. Acad. Sci. USA 74, 5426-5430.
507. Georgatos, S.D., K. Weber, N. Geisler, and G. Blobel, (1987), Binding of Two Desmin Derivatives to the Plasma Membrane and the Nuclear Envelope of Avian Erythrocytes: Evidence for a Conserved Site-specificity in Intermediate Filament-Membrane Interactions. Proc. Natl. Acad. Sci. USA 84, 6780-6784.
508. Gerlach, J.H., J.A. Endicott, P.F. Juranka, G. Henderson, F. Sarangi, K.L. Deuchars, and V. Ling, (1986), Homology between P-glycoprotein and a Bacterial Haemolysin Transport Protein Suggests a Model for Multidrug Resistance. Nature 324, 485-489.
509. Gething, M.-J., K. McCammon, and J. Sambrook, (1986), Expression of Wild-Type and Mutant Forms of Influenza Hemagglutinin: The Role of Folding in Intracellular Transport. Cell 46, 939-950.
510. Gething, M.-J., R.W. Doms, D. York, and J. White, (1986), Studies on the Mechanism of Membrane Fusion: Site-specific Mutagenesis of the Haemagglutinin of Influenza Virus. J. Cell Biol. 102, 11-23.
511. Geuze, H.J., J.W. Slot, and A.L. Schwartz, (1987), Membranes of Sorting Organelles Display Lateral Heterogeneity in Receptor Distribution. J. Cell Biol. 104, 1715-1723.
512. Gil, G., J.R. Faust, D.J. Chin, J.L. Goldstein, and M.S. Brown, (1985), Membrane-Bound Domain of HMG CoA Reductase Is Required for Sterol-Enhanced Degradation of the Enzyme. Cell 41, 249-258.
513. Gillespie, L.L. (1987), Identification of an Outer Mitochondrial Membrane Protein That Interacts with a Synthetic Signal Peptide. J. Biol. Chem. 262, 7939-7942.
514. Gillespie, L.L., C. Argan, A.T. Taneja, R.S. Hodges, K.B. Freeman, and G.C. Shore, (1985), A Synthetic Signal Peptide Blocks Import of Precursor Proteins Destined for the Mitochondrial Inner Membrane or Matrix. J. Biol. Chem. 260, 16045-16048.
515. Gilman, A.G. (1987), G Proteins: Transducers of Receptor-Generated Signals. Ann Rev. Biochem. 56, 615-649.
516. Giugni, T.D., L.C. James, and H.T. Haigler, (1985), Epidermal Growth Factor Stimulates Tyrosine Phosphorylation of Specific Proteins in Permeabilized Human Fibroblasts. J. Biol. Chem. 260, 15081-15090.
517. Giugni, T.D., D.L. Braslau, and H.T. Haigler, (1987), Electric Field-induced Redistribution and Postfield Relaxation of Epidermal Growth Factor Receptors on A431 Cells. J. Cell Biol. 104, 1291-1297.
518. Glaeser, R.M. and B.K. Jap, (1985), Absorption Flattening in the Circular Dichroism Spectra of Small Membrane Fragments. Biochemistry 24, 6398-6401.
519. Glenney, J.R. and P. Glenney, (1983), Fodrin Is the General Spectrin-Like Protein Found in Most Cells Whereas Spectrin and the TW Protein Have a Restricted Distribution. Cell 34, 503-512.
520. Glick, B.S. and J.E. Rothman, (1987), Possible Role for Fatty Acyl-Coenzyme A in Intracellular Protein Transport. Nature 326, 309-312.

521. Goins, B., M. Masserini, B.G. Barisas, and E. Freire, (1986), Lateral Diffusion of Ganglioside GM_1 in Phospholipid Bilayer Membranes. Biophys. J. 49, 849-856.
522. Golan, D.E., M.R. Alecio, W.R. Veatch, and R.R. Rando, (1984), Lateral Mobility of Phospholipid and Cholesterol in the Human Erythrocyte Membrane: Effects of Protein-Lipid Interactions. Biochemistry 23, 332-339.
523. Goldfarb, D.S., J. Gariepy, G. Schoolnik, and R.D. Kornberg, (1986), Synthetic Peptides as Nuclear Localization Signals. Nature 322, 641-644.
524. Goldkorn, T., G. Rimon, E.S. Kempner, and H.R. Kaback, (1984), Functional Molecular Weight of the lac Carrier Protein from Escherichia coli as Studied by Radiation Inactivation Analysis. Proc. Natl. Acad. Sci. USA 81, 1021-1025.
525. Goldman, D., E. Deneris, W. Luyten, A. Kochhar, J. Patrick, and S. Heinemann, (1987), Members of a Nicotinic Acetylcholine Receptor Gene Family Are Expressed in Different Regions of the Mammalian Central Nervous System. Cell 48, 965-973.
526. Gomez-Fernandez, J.C., F.M. Goni, D. Bach, C. Restall, and D. Chapman, (1980), Protein Lipid Interactions: Biophysical Studies of (Ca^{2+} + Mg^{2+})-ATPase Reconstituted Systems. Biochim. Biophys. Acta 598, 502-516.
527. Gonenne, A. and R. Ernst, (1978), Solubilization of Membrane Proteins by Sulfobetaines, Novel Zwitterionic Surfactants. Anal. Bioch. 87, 28-38.
528. Goormaghtigh, E., C. Chadwick, and G.A. Scarborough, (1986), Monomers of the Neurospora Plasma Membrane H^+-ATPase Catalyze Efficient Proton Translocation. J. Biol. Chem. 261, 7466-7471.
529. Goormaghtigh, E. and G.A. Scarborough, (1986), Density-Based Separation of Liposomes by Glycerol Gradient Centrifugation. Analytical Biochemistry 159, 122-131.
530. Goormaghtigh, E., R. Brasseur, P. Huart, and J.M. Ruysschaert, (1987), Study of the Adriamycin-Cardiolipin Complex Structure Using Attenuated Total Reflection Infrared Spectroscopy. Biochemistry 26, 1789-1794.
531. Gordon, R.D., W.E. Fieles, D.L. Schotland, R. Hogue-Angeletti, and R.L. Barchi, (1987), Topographical Localization of the C-terminal Region of the Voltage-dependent Sodium Channel from Electrophorus electricus Using Antibodies Raised Against a Synthetic Peptide. Proc. Natl. Acad. Sci. USA 84, 308-312.
532. Gorter, E. and F. Grendel, (1925), On Biomolecular Layers of Lipid on the Chromacytes of the Blood. J. Exp. Med. 41, 439-443.
533. Gounaris, K. and J. Barber, (1983), Monogalactosyldiacylglycerol: The Most Abundant Polar Lipid in Nature. TIBS 8, 378-381.
534. Gounaris, K., J. Barber, and J.L. Harwood, (1986), The Thylakoid Membranes of Higher Plant Chloroplasts. Biochem. J. 237, 313-326.
535. Govil, G. and R.V. Hosur, (1982), Conformation of Biological Molecules: New Results for NMR. Springer-Verlag, New York.
536. Grabau, C. and J.E. Cronan, (1986), In Vivo Function of Escherichia coli Pyruvate Oxidase Specifically Requires a Functional Lipid Binding Site. Biochemistry 25, 3748-3751.
537. Graf, J., R.C. Ogle, F.A. Robey, M. Sasaki, G.R. Martin, Y. Yamada, and H.K. Kleinman, (1987), A Pentapeptide from the Laminin B1 Chain Mediates Cell Adhesion and Binds the 67000 Laminin Receptor. Biochemistry 26, 6896-6900.
538. Graham, I., J. Gagne, and J.R. Silvius, (1985), Kinetics and Thermodynamics of Calcium-Induced Lateral Phase Separations in Phosphatidic Acid Containing Bilayers. Biochemistry 24, 7123-7131.

539. Grant, C.W.M. (1984), Cell Surface Structural Implications of Some Experiments with Isolated Glycolipids and Glycoproteins. Can. J. Biochem. Cell Biol. 62, 1151-1157.

540. Grasberger, B., A.P. Minton, C. DeLisi, and H. Metzger, (1986), Interaction between Proteins Localized in Membranes. Proc. Natl. Acad. Sci. USA 83, 6258-6262.

541. Gray, T.M. and B.W. Mathews, (1984), Intrahelical Hydrogen Binding of Serine, Threonine and Cysteine Residues within alpha-Helices and Its Relevance to Membrane-bound Proteins. J. Mol. Biol. 175, 75-81.

542. Green, D.E. and J.F. Perdue, (1966), Correlation of Mitochondrial Structure and Function. Ann. N.Y. Acad. Sci. 137, 667-684.

543. Green, D.E. and S. Fleischer, (1963), The Role of Lipids in Mitochondrial Electron Transfers and Oxidative Phosphorylation. Biochim. Biophys. Acta 70, 554-582.

544. Green, P.R. and R.M. Bell, (1984), Asymmetric Reconstitution of Homogeneous Escherichia coli sn-Glycerol-3-phosphate Acyltransferase into Phospholipid Vesicles. J. Biol. Chem. 259, 14688-14694.

545. Greenberg, M. and T.Y. Tsong, (1984), Detergent Solubilization and Affinity Purification of a Local Anesthetic Binding Protein from Mammalian Axonal Membranes. J. Biol. Chem. 259, 13241-13245.

546. Greenblatt, R.E., Y. Blatt, and M. Montal, (1985), The Structure of the Voltage-sensitive Sodium Channel. FEBS Lett. 193, 125-134.

547. Greenhut, S.F., V.R. Bourgeois, and M.A. Roseman, (1986), Distribution of Cytochrome b_5 between Small and Large Unilamellar Phospholipid Vesicles. J. Biol. Chem. 261, 3670-3675.

548. Grenier, F.C., E.B. Waygood, and M.H. Saier, (1986), The Bacterial Phosphotransferase System: Kinetic Characterization of the Glucose, Mannitol, Glucitol, and N-Acetylglucosamine Systems. J. Cell. Biochem. 31, 97-105.

549. Grenningloh, G., A. Rienitz, B. Schmitt, C. Methfessel, M. Zensen, K. Beyreuther, E.D. Gundelfinger, and H. Betz, (1987), The Strychnine-binding Subunit of the Glycine Receptor Shows Homology with Nicotinic Acetylcholine Receptors. Nature 328, 215-220.

550. Griffith, O.H., D.A. McMillen, J.F.W. Keana, and P.C. Jost, (1986), Lipid-Protein Interactions in Cytochrome c Oxidase. A Comparison of Covalently Attached Phospholipid Photo-Spin-Label with Label Free to Diffuse in the Bilayer. Biochemistry 25, 574-584.

551. Griffith, O.H., P.J. Dehlinger, and S.P. Van, (1974), Shape of the Hydrophobic Barrier of Phospholipid Bilayers: Evidence for Water Penetration in Biological Membranes. J. Membrane Biol. 15, 159-192.

552. Griffiths, G. and K. Simons, (1986), The trans Golgi Network: Sorting at the Exit Site of the Golgi Complex. Science 234, 438-443.

553. Grimes, H.D., N.M. Watanabe, and R.W. Breidenbach, (1986), Plasma Membrane Isolated with a Defined Orientation Used to Investigate Protein Topography. Biochim. Biophys. Acta 862, 165-177.

554. Gross, R.W. (1985), Identification of Plasmalogen as the Major Phospholipid Constituent of Cardiac Sarcoplasmic Reticulum. Biochemistry 24, 1662-1668.

555. Gruenberg, J.E. and K.E. Howell, (1986), Reconstitution of Vesicle Fusions Occurring in Endocytosis with a Cell-Free System. The EMBO Journal 5, 3091-3101.

556. Grygorczyk, R., W. Schwarz, and H. Passow, (1987), Potential Dependence of the "Electrically Silent" Anion Exchange across the Plasma Membrane of Xenopus Oocytes Mediated by the Band-3 Protein. J. Membrane Biol. 99, 127-136.

557. Guan, J., C.E. Machamer, and J.K. Rose, (1985), Glycosylation Allows Cell-Surface Transport of an Anchored Secretory Protein. Cell 42, 489-496.

558. Guidotti, G. (1976), The Structure of Membrane Transport System. TIBS 1, 11-13.

559. Guillet, J.-G., M.-Z. Lai, T.J. Briner, J.A. Smith, and M.L. Gefter, (1986), Interaction of Peptide Antigens and Class II Major Histocompatibility Complex Antigens. Nature 324, 260-262.

560. Gumbiner, B. and D. Louvard, (1985), Localized Barriers in the Plasma Membrane: A Common Way to Form Domains. TIBS 10, 435-438.

561. Gut, J., C. Richter, R.J. Cherry, K.H. Winterhalter, and S. Kawato, (1982), Rotation of Cytochrome P-450 II. Specific Interactions of Cytochrome P-450 with NADPH-Cytochrome P-450 Reductase in Phospholipid Vesicles. J. Biol. Chem. 257, 7030-7036.

562. Gutknecht, J. (1987), Proton Conductance Through Phospholipid Bilayers: Water Wires or Weak Acids? J. Bioenerg. and Biomemb. 19, 427-442.

563. Gutknecht, J. (1984), Proton/Hydroxide Conductance through Lipid Bilayer Membranes. J. Membrane Biol. 82, 105-112.

564. Guy, H.R. and F. Hucho, (1987), The Ion Channel of the Nicotinic Acetylcholine Receptor. TINS 10, 318-321.

565. Guy, H.R. and P. Seetharamulu, (1986), Molecular Model of the Action Potential Sodium Channel. Proc. Natl. Acad. Sci. USA 83, 508-512.

566. Habermann, V.E. and J. Jentsch, (1967), Sequenzanalyse des Melittins aus den tryptischen und Peptischen Spaltstücken. Hoppe-Seyler's Z. Physiol. Chem. 348, 37-50.

567. Hackenberg, H. and M. Klingenberg, (1980), Molecular Weight and Hydrodynamic Parameters of the Adenosine 5'-Diphosphate-Adenosine 5'-Triphosphate Carrier in Triton X-100. Biochemistry 19, 548-555.

568. Hackenbrock, C.R., B. Chazotte, and S.S. Gupte, (1986), The Random Collision Model and a Critical Assessment of Diffusion and Collision in Mitochondrial Electron Transport. J. Bioenerg. and Biomemb. 18, 331-368.

569. Hackett, C.S. and P. Strittmatter, (1984), Covalent Cross-linking of the Active Sites of Vesicle-bound Cytochrome b_5 and NADPH-Cytochrome b_5 Reductase. J. Biol. Chem. 259, 3275-3282.

570. Hackett, N.R., L.J. Stern, B.H. Chao, K.A. Kronis, and H.G. Khorana, (1987), Structure-Function Studies on Bacteriorhodopsin. J. Biol. Chem. 262, 9277-9284.

571. Haest, C.W.M. (1982), Interactions Between Membrane Skeleton Proteins and the Erythrocyte Membrane. Biochim. Biophys. Acta 694, 331-352.

572. Hager, K.M., S.M. Mandala, J.W. Davenport, D.W. Speicher, E.J. Benz, Jr., and C.W. Slayman, (1986), Amino Acid Sequence of the Plasma Membrane ATPase of Neurospora crassa: Deduction from Genomic and cDNA Sequences. Proc. Natl. Acad. Sci. USA 83, 7693-7697.

573. Hakomori, S. (1984), Glycosphingolipids as Differentiation-Dependent, Tumor-Associated Markers and as Regulators of Cell Proliferation. TIBS 9, 453-458.

574. Hall, P.F. (1984), The Role of the Cytoskeleton in Hormone Action. Can. J. Biochem. Cell Biol. 62, 653-665.

575. Hall, T.G. and V. Bennett, (1987), Regulatory Domains of Erythrocyte Ankyrin. J. Biol. Chem. 262, 10537-10545.

576. Hanahan, D.J. (1986), Platelet Activating Factor: A Biologically Active Phosphoglyceride. Ann. Rev. Biochem. 55, 483-509.

577. Hanahan, D.J. and D.R. Nelson, (1984), Phospholipids as Dynamic Participants in Biological Processes. J. Lipid Res. 25, 1528-1535.
578. Hanamoto, J.H., P. Dupuis, and M.A. El-Sayed, (1984), On the Protein (Tyrosine)-Chromophore (Protonated Schiff Base) Coupling in Bacteriorhodopsin. Proc. Natl. Acad. Sci. USA 81, 7083-7087.
579. Hancock, R.E.W. (1987), Role of Porins in Outer Membrane Permeability. J. Bacteriology 169, 929-933.
580. Hancock, R.E.W., A. Schmidt, K. Bauer, and R. Benz, (1986), Role of Lysines in Ion Selectivity of Bacterial Outer Membrane Porins. Biochim. Biophys. Acta 860, 263-267.
581. Handa, M., K. Titani, L.Z. Holland, J.R. Roberts, and Z.M. Ruggeri, (1986), The von Willebrand Factor-binding Domain of Platelet Membrane Glycoprotein Ib: Characterization by Monoclonal Antibodies and Partial Amino Acid Sequence Analysis of Proteolytic Fragments. J. Biol. Chem. 261, 12579-12585.
582. Hanke, W. (1986), Reconstitution of Ion Channels. CRC Critical Reviews in Biochem. 19, 1-44.
583. Hannig, K. (1979), Continuous Free-flow Electrophoresis. Separation of Cells and Subcellular Elements (H. Peeters, Ed.), pp. 23-30. Pergamon Press, New York.
584. Hannig, K. and H.-G. Heidrich, (1974), The Use of Continuous Preparative Free-Flow Electrophoresis for Dissociating Cell Fractions and Isolation of Membranous Components. Methods in Enz. 31, 746-761.
585. Hannun, Y.A., C.R. Loomis, A.H. Merrill, Jr., and R.M. Bell, (1986), Sphingosine Inhibition of Protein Kinase C Activity and of Phorbol Dibutyrate Binding in Vitro and in Human Platelets. J. Biol. Chem. 261, 12604-12609.
586. Hannun, Y.A., C.R. Loomis, and R.M. Bell, (1986), Protein Kinase C Activation in Mixed Micelles: Mechanistic Implications of Phospholipid, Diacylglycerol, and Calcium Interdependencies. J. Biol. Chem. 261, 7184-7190.
587. Hannun, Y.A. and R.M. Bell, (1987), Lysophingolipids Inhibit Protein Kinase C: Implications for the Sphingolipidoses. Science 235, 670-674.
588. Hannun, Y.A. and R.M. Bell, (1986), Phorbol Ester Binding and Activation of Protein Kinase C in Triton X-100 Mixed Micelles Containing Phosphatidylserine. J. Biol. Chem. 261, 9341-9347.
589. Hansen, W., P.D. Garcia, and P. Walter, (1986), In Vitro Protein Translocation across the Yeast Endoplasmic Reticulum: ATP-Dependent Posttranslational Translocation of the Prepro-α-Factor. Cell 45, 397-406.
590. Hanson, B.A. and R.L. Lester, (1980), The Extraction of Inositol-containing Phospholipids and Phosphatidylcholine from Saccharomyces cerevisiae and Neurospora crassa. J. Lipid Res. 21, 309-315.
591. Hantke, K. and V. Braun, (1973), Covalent Binding of Lipid to Protein: Diglyceride and Amide-Linked Fatty Acid at the N-Terminal End of the Murein-Lipoprotein of the E. coli Outer Membrane. Eur. J. Bioch. 34, 284-296.
592. Hanukoglu, I. and Z. Hanukoglu, (1986), Stoichiometry of Mitochondrial Cytochromes P-450, Adrenodoxin and Adrenodoxin Reductase in Adrenal Cortex and Corpus Luteum: Implications for Membrane Organization and Gene Regulation. Eur. J. Biochem. 157, 27-31.
593. Hargrave, P.A. (1986), Topography of Membrane Proteins. Determination of Regions Exposed to the Aqueous Phase. Techniques for the Analysis of Membrane Proteins (C.I. Ragan and R.J. Cherry, Eds.), pp. 129-151. Chapman and Hall, New York.

593a. Hargrave, P.A., J.H. McDowell, D.R. Curtis, J.K. Wang, E. Juszczak, S.-L. Fong, J.K.M. Rao, and P. Argos, (1983), The Structure of Bovine Rhodopsin. Biophys. Struct. Mech. 9, 235-244.

594. Harlos, K., H. Eibl, I. Pascher, and S. Sundell, (1984), Conformation and Packing Properties of Phosphatidic Acid: The Crystal Structure of Monosodium Dimyristoylphosphatidate. Chem. Phys. Lipids 34, 115-126.

595. Hartl, F.-U., B. Schmidt, E. Wachter, H. Weiss, and W. Neupert, (1986), Transport into Mitochondria and Intramitochondrial Sorting of the Fe/S Protein of Ubiquinol-Cytochrome c Reductase. Cell 47, 939-951.

596. Hartl, F.-U., J. Ostermann, B. Guiard, and W. Neupert, (1987), Successive Translocation into and out of the Mitochondrial Matrix: Targeting of Proteins to the Intermembrane Space by a Bipartite Signal Peptide. Cell 51, 1027-1037.

597. Hartsel, S.C. and D.S. Cafiso, (1986), A Test of Discreteness-of-Charge Effects in Phospholipid Vesicles: Measurements Using Paramagnetic Amphiphiles. Biochemistry 25, 8214-8219.

598. Harwood, J.L. (1987), Lung Surfactant. Prog. Lipid Res. 25, 211-256.

599. Haselbeck, A. and R. Scheckman, (1986), Interorganelle Transfer and Glycosylation of Yeast Invertase In Vitro. Proc. Natl. Acad. Sci. USA 83, 2017-2021.

600. Hauser, H. (1984), Some Aspects of the Phase Behavior of Charged Lipids. Biochim. Biophys. Acta 772, 37-50.

601. Hauser, H., D. Oldani, and M.C. Phillips, (1973), Mechanisms of Ion Escape from Phosphatidylcholine and Phosphatidylserine Single Bilayer Vesicles. Biochemistry 12, 4507-4517.

602. Hauser, H. and G.G. Shipley, (1984), Interactions of Divalent Cations with Phosphatidylserine Bilayer Membranes. Biochemistry 23, 34-41.

603. Hauser, H., G.P. Hazlewood, and R.M.C. Dawson, (1985), Characterization of Membrane Lipids of a General Fatty Acid Auxotrophic Bacterium by Electron Spin Resonance Spectroscopy and Differential Scanning Calorimetry. Biochemistry 24, 5247-5253.

604. Hauser, H., I. Pascher, R.H. Pearson, and S. Sundell, (1981), Preferred Conformation and Molecular Packing of Phosphatidylethanolamine and Phosphatidylcholine. Biochim. Biophys. Acta 650, 21-51.

605. Hauser, H., I. Pascher, and S. Sundell, (1980), Conformation of Phospholipids: Crystal Structure of a Lysophosphatidylcholine Analogue. J. Mol. Biol 137, 249-264.

606. Hauser, H., N. Gains, and M. Muller, (1983), Vesiculation of Unsonicated Phospholipid Dispersions Containing Phosphatidic Acid by pH Adjustment: Physicochemical Properties of the Resulting Unilamellar Vesicles. Biochemistry 22, 4775-4781.

607. Haverstick, D.M. and M. Glaser, (1987), Visualization of Ca^{2+}-induced Phospholipid Domains. Proc. Natl. Acad. Sci. USA 84, 4475-4479.

608. Hay, B., S.B. Prusiner, and V.R. Lingappa , (1987), Evidence for a Secretory Form of the Cellular Prion Protein. Biochemistry 26, 8110-8115.

609. Hayward, S.B. and R.M. Stroud, (1981), Projected Structure of Purple Membrane Determined to 3.7 Å Resolution by Low Temperature Electron Microscopy. J. Mol. Biol. 151, 491-517.

610. Heacock, P.N. and W. Dowhan, (1987), Construction of a Lethal Mutation in the Synthesis of the Major Acidic Phospholipids of Escherichia coli. J. Biol. Chem. 262, 13044-13049.

611. Hedrick, J.L. and A.J. Smith, (1968), Size and Charge Isomer Separation and

Estimation of Molecular Weights of Proteins by Disc Gel Electrophoresis. Arch. Bioch. Biophys. 126, 155-164.

612. Helenius, A., D.R. McCaslin, E. Fries, and C. Tanford, (1979), Properties of Detergents. Methods in Enz. 56, 734-749.
613. Helenius, A. and K. Simons, (1972), The Binding of Detergents to Lipophilic and Hydrophilic Proteins. J. Biol. Chem. 247, 3656-3661.
614. Helenius, A. and K. Simons, (1975), Solubilization of Membranes by Detergents. Biochim. Biophys. Acta 415, 29-79.
615. Helgerson, A.L. and A. Carruthers, (1987), Equilibrium Ligand Binding to the Human Erythrocyte Sugar Transporter. J. Biol. Chem. 262, 5464-5475.
616. Helmkamp, G.M. (1986), Phospholipid Transfer Proteins: Mechanism of Action. J. Bioenerg. and Biomemb. 18, 71-91.
617. Helmy, S., K. Porter-Jordan, E.A. Dawidowicz, P. Pilch, A.L. Schwartz, and R.E. Fine, (1986), Separation of Endocytic from Exocytic Coated Vesicles Using a Novel Cholinesterase Mediated Density Shift Technique. Cell 44, 497-506.
618. Henderson, R. (1985), Structure of a Bacterial Photosynthetic Reaction Centre. Nature 318, 598-599.
619. Henderson, R. and P.N.T. Unwin, (1975), Three-dimensional Model of Purple Membrane Obtained by Electron Microscopy. Nature (London) 257, 28-32.
620. Henderson, R. and P.N.T. Unwin, (1975), Three-dimensional Model of Purple Membrane Obtained by Electron Microscopy. Nature 257, 28-32.
621. Henry, G.D., J.H. Weiner, and B.D. Sykes, (1987), Backbone Dynamics of a Model Membrane Protein: Assignment of the Carbonyl Carbon ^{13}C NMR Resonances in Detergent-Solubilized M13 Coat Protein. Biochemistry 26, 3619-3626.
622. Herbette, L., C.A. Napolitano, and R.V. McDaniel, (1984), Direct Determination of the Calcium Profile Structure for Dipalmitoyllecithin Multilayers Using Neutron Diffraction. Biophys. J. 46, 677-685.
623. Hereld, D., J.L. Krakow, J.D. Bangs, G.W. Hart, and P.T. Englund, (1986), A Phospholipase C from Trypanosoma brucei Which Selectively Cleaves the Glycolipid on the Variant Surface Glycoprotein. J. Biol. Chem. 261, 13813-13819.
624. Herskowitz, I. and L. Marsh, (1987), Conservation of a Receptor/Signal Transduction System. Cell 50, 995-996.
625. Hescheler, J., W. Rosenthal, W. Trautwein, and G. Schultz, (1987), The GTP-Binding Protein, G_o, Regulates Neuronal Calcium Channels. Nature 325, 445-477.
626. Hess, J.F., K. Oosawa, P. Matsumura, and M.I. Simon, (1987), Protein Phosphorylation Is Involved in Bacterial Chemotaxis. Proc. Natl. Acad. Sci. USA 84, 7609-7613.
627. Heyn, M.P. (1979), Determination of Lipid Order Parameters and Rotational Correlation Times from Fluorescence Depolarization Experiments. FEBS Lett. 108, 359-364.
628. Heyn, M.P., R.J. Cherry, and U. Muller, (1977), Transient and Linear Dichroism Studies on Bacteriorhodopsin: Determination of the Orientation of the 568 nm All-trans Retinal Chromophore. J. Mol. Biol. 117, 607-620.
629. Hidalgo, C. (1987), Lipid-Protein Interactions and the Function of the Ca^{2+} ATPase of Sarcoplasmic Reticulum. CRC Critical Reviews in Biochemistry 21, 319-347.
630. Hilditch, P.H. Thomas and L.J. Rogers, (1986), Two Processes for the Breakdown of the Q_b Protein of Chloroplasts. FEBS Lett. 208, 313-317.
631. Hilhorst, R., R. Spruijt, C. Laane, and C. Veeger, (1984), Rules for the Regulation

of Enzyme Activity in Reversed Micelles as Illustrated by the Conversion of Apolar Steroids by 20 ß-Hydroxysteroid Dehydrogenase. Eur. J. Bioch. 144, 459-466.

632. Hill, T.L. (1977), Free Energy Transduction in Biology. Academic Press, New York.
633. Hille, B. (1984), Ionic Channels of Excitable Membranes. Sinauer Associates, Inc., Sunderland, Massachusetts.
633a. Hille, B. (1975), Ionic Selectivity, Saturation, and Block in Sodium Channels. J. Gen. Physiol. 66, 535-560.
634. Hirschberg, C.B. and M.D. Snider, (1987), Topography of Glycosylation in the Rough Endoplasmic Reticulum and Golgi Apparatus. Ann. Rev. Biochem. 56, 63-87.
635. Hirst, T.R. and J. Holmgren, (1987), Conformation of Protein Secreted across Bacterial Outer Membranes: A Study of Enterotoxin Translocation from Vibrio cholerae. Proc. Natl. Acad. Sci. USA 84, 7418-7422.
636. Hitchcock, P.B., R. Mason, K.M. Thomas, and G.G. Shipley, (1974), Structural Chemistry of 1,2-Dilauroyl-DL-phosphatidylethanolamine: Molecular Conformation and Intermolecular Packing of Phospholipids. Proc. Natl. Acad. Sci. USA 71, 3036-3040.
637. Hjelmeland, L.M. (1980), A Nondenaturing Zwitterionic Detergent for Membrane Biochemistry: Design and Synthesis. Proc. Natl. Acad. Sci. USA 77, 6368-6370.
638. Hoch, D.H., M. Romero-Mira, B.E. Ehrlich, A. Finkelstein, B.R. DasGupta, and L.L. Simpson, (1985), Channels Formed by Botulinum, Tetanus, and Diphtheria Toxins in Planar Lipid Bilayers: Relevance to Translocation of Proteins across Membranes. Proc. Natl. Acad. Sci. USA 82, 1692-1696.
639. Hochman, Y. and D. Zakim, (1983), Evidence that UDP-glucuronyl-transferase in Liver Microsomes at 37° C Is in a Gel Phase Lipid Environment. J. Biol. Chem. 258, 11758-11762.
640. Hochman, Y., M. Kelley, and D. Zakim, (1983), Modulation of the Number of Ligand Binding Sites of UDP-glucuronyltransferase by the Gel to Liquid-Crystal Phase Transition of Phosphatidylcholines. J. Biol. Chem. 258, 6509-6516.
641. Holland, E.C. and K. Drickamer, (1985), Signal Recognition Particle Mediates the Insertion of a Transmembrane Protein Which Has a Cytoplasmic NH_2 Terminus. J. Biol. Chem. 261, 1286-1292.
642. Holman, G.D. and W.D. Rees, (1987), Photolabelling of the Hexose Transporter at External and Internal Sites: Fragmentation Patterns and Evidence for a Conformational Change. Biochim. Biophys. Acta 897, 395-405.
643. Holt, G.D., R.S. Haltiwanger, C.-R. Torres, and G.W. Hart, (1987), Erythrocytes Contain Cytoplasmic Glycoproteins. J. Biol. Chem. 262, 14847-14850.
644. Wu, S.H.-W. and H.M. McConnell, (1975), Phase Separations in Phospholipid Membranes. Biochemistry 14, 847-854.
645. Honig, B.H. and W.L. Hubbell, (1984), Stability of "Salt Bridges" in Membrane Proteins. Proc. Natl. Acad. Sci. USA 81, 5412-5416.
646. Honig, B.H., W.L. Hubbell, and R.F. Flewelling, (1986), Electrostatic Interactions in Membranes and Proteins. Ann. Rev. Biophys. and Biophys. Chem. 15, 163-193.
647. Hood, L., M. Kronenberg, and T. Hunkapiller, (1985), T Cell Antigen Receptors and the Immunoglobulin Supergene Family. Cell 40, 225-229.
648. Hope, M.J., M.B. Bally, G. Webb, and P.R. Cullis, (1985), Production of Large Unilamellar Vesicles by a Rapid Extrusion Procedure. Characterization of Size

Distribution, Trapped Volume and Ability to Maintain a Membrane Potential. Biochim. Biophys. Acta 812, 55-65.

649. Hope, M.J., M.B. Bally, L.D. Mayer, A.S. Janoff, and P.R. Cullis, (1986), Generation of Multilamellar and Unilamellar Phospholipid Vesicles. Chem. Phys. Lipids 40, 89-107.

650. Hope, M.J. and P.R. Cullis, (1987), Lipid Asymmetry Induced by Transmembrane pH Gradients in Large Unilamellar Vesicles. J. Biol. Chem. 262, 4360-4366.

651. Hopkins, C.R. (1986), Membrane Boundaries Involved in the Uptake and Intracellular Processing of Cell Surface Receptors. TIBS 11, 473-477.

652. Hoppe, J., D. Gatti, H. Weber, and W. Sebald, (1986), Labeling of Individual Amino Acid Residues in the Membrane-embedded F_0 Part of the F_1F_0 ATP Synthase from Neurospora crassa: Influence of Oligomycin and Dicyclohexylcarbodiimide. Eur. J. Biochem. 155, 259-264.

653. Hoppe, J., J. Brunner, and B.B. Jorgensen, (1984), Structure of the Membrane-Embedded F_0 Part of F_1F_0 ATP Synthase from Escherichia coli as Inferred from Labelling with 3-(trifluoromethyl)-3-(m-[^{125}I]iodophenyl) diazirine. Biochemistry 23, 5610-5616.

654. Hoppe, J. and W. Sebald, (1984), The Proton Conducting F_0-Part of Bacterial ATP Synthases. Biochim. Biophys. Acta 768, 1-27.

655. Horne, W.A., G.A. Weiland, and R.E. Oswald, (1986), Solubilization and Hydrodynamic Characterization of the Dihydropyridine Receptor from Rat Ventricular Muscle. J. Biol. Chem. 261, 3588-3594.

656. Hortsch, M., D. Avossa, and D.I. Meyer, (1985), A Structural and Functional Analysis of the Docking Protein. J. Biol. Chem. 260, 9137-9145.

657. Houslay, M.D. (1987), Egg Activation Unscrambles a Potential Role for IP4. TIBS 12, 1-2.

658. Houslay, M.D. and K.K. Stanley, (1982), Dynamics of Biological Membranes: Influence on Synthesis, Structure and Function. John Wiley and Sons, New York.

659. Huang, C.-H. (1969), Studies on Phosphatidylcholine Vesicles. Formation and Physical Characteristics. Biochemistry 8, 344-351.

660. Huang, C. and J.T. Mason, (1978), Geometric Packing Constraints in Egg Phosphatidylcholine Vesicles. Proc. Natl. Acad. Sci. USA 75, 308-310.

661. Huang, K.-S., B.P. Wallner, R.J. Mattaliano, R. Tizard, C. Burne, A. Frey, C. Hession, P. McGray, L.K. Sinclair, E.P. Chow, J.L. Browning, K.L. Ramachandran, J. Tang, J.E. Smart, and R.B. Pepinsky, (1986), Two Human 35 kd Inhibitors of Phospholipase A2 Are Related to Substrates of pp60v-src and of the Epidermal Growth Factor Receptor/Kinase. Cell 46, 191-199.

662. Huang, K.-S., R. Radhakrishnan, H. Bayley, and H.G. Khorana, (1982), Orientation of Retinal in Bacteriorhodopsin as Studied by Cross-linking Using a Photosensitive Analog of Retinal. J. Biol. Chem. 257, 13616-13623.

663. Hucho, F. (1986), The Nicotinic Acetylcholine Receptor and Its Ion Channel. Eur. J. Biochem. 158, 211-226.

664. Hucho, F., W. Oberthur, and F. Lottspeich, (1986), The Ion Channel of the Nicotinic Acetylcholine Receptor Is Formed by the Homologous Helices M II of the Receptor Subunits. FEBS Lett. 205, 137-141.

665. Hunkapiller, T. and L. Hood, (1986), The Growing Immunoglobulin Gene Superfamily. Nature 323, 15-16.

666. Hunziker, W., M. Spiess, G. Semenza, and H.F. Lodish, (1986), The Sucrase-Isomaltase Complex: Primary Structure, Membrane-Orientation, and Evolution of a Stalked, Intrinsic Brush Border Protein. Cell 46, 227-234.

667. Hurt, E.C. and G. Schatz, (1987), A Cytosolic Protein Contains a Cryptic Mitochondrial Targeting Signal. Nature 325, 499-503.
668. Husebye, E.S. and T. Flatmark, (1984), The Content of Long-Chain Free Fatty Acids and Their Effect on Energy Transduction in Chromaffin Granule Ghosts. J. Biol. Chem. 259, 15272-15276.
669. Husten, E.J., C.T. Esmon, and A.E. Johnson, (1987), The Active Site of Blood Coagulation Factor Xa. J. Biol. Chem. 262, 12953-12961.
670. Hutson, J.L. and J.A. Higgins, (1982), Asymmetric Synthesis Followed by Transmembrane Movement of Phosphatidylethanolamine in Rat Liver Endoplasmic Reticulum. Biochim. Biophys. Acta 687, 247-256.
671. Hymel, L., A. Maurer, C. Berenski, C.Y. Jung, and S. Fleischer, (1984), Target Size of Calcium Pump Protein from Skeletal Muscle Sarcoplasmic Reticulum. J. Biol. Chem. 259, 4890-4895.
672. Hynes, R.O. (1987), Integrins: A Family of Cell Surface Receptors. Cell 48, 549-554.
673. Iacopetta, B., J.-L. Carpentier, T. Pozzan, D.P. Lew, P. Gorden, and L. Orci, (1986), Role of Intracellular Calcium and Protein Kinase C in the Endocytosis of Transferrin and Insulin by HL60 Cells. J. Cell Biol. 103, 851-856.
674. Ichihara, S., T. Suzuki, M. Suzuki, and S. Mizushima, (1986), Molecular Cloning and Sequencing of the sppA Gene and Characterization of the Encoded Protease IV, a Signal Peptide Peptidase, of Escherichia coli. J. Biol. Chem. 261, 9405-9411.
675. Ikehara, Y., Y. Hayashi, S. Ogata, A. Miki, and T. Kominami, (1987), Purification and Characterization of a Major Glycoprotein in Rat Hepatoma Plasma Membranes. Biochem. J. 241, 63-70.
676. Ikigai, H. and T. Nakae, (1987), Assembly of the α-Toxin-Hexamer of Staphylococcus aureus in the Liposome Membrane. J. Biol. Chem. 262, 2156-2160.
677. Imoto, K., C. Methfessel, B. Sakmann, M. Mishina, Y. Mori, T. Konno, K. Fukuda, M. Kurasaki, H. Bujo, Y. Fujita, and S. Numa, (1986), Location of a δ-Subunit Region Determining Ion Transport Through the Acetylcholine Receptor Channel. Nature 324, 670-674.
678. Inesi, G. (1987), Sequential Mechanism of Calcium Binding and Translocation in Sarcoplasmic Reticulum Adenosine Triphosphatase. J. Biol. Chem. 262, 16338-16342.
679. Innis, M.A., M. Tokunaga, M.E. Williams, J.M. Loranger, S.-Y. Chang, S. Chang, and H.C. Wu, (1984), Nucleotide Sequence of the Escherichia coli Prolipoprotein Signal Peptidase (lsp) Gene. Proc. Natl. Acad. Sci. USA 81, 3708-3712.
680. Inui, K., T. Okano, M. Takano, S. Kitazawa, and R. Hori, (1981), A Simple Method for the Isolation of Basolateral Plasma Membrane Vesicles from Rat Kidney Cortex. Biochim. Biophys. Acta 647, 150-154.
681. Ishidate, K., E.S. Creeger, J. Zrike, S. Deb, B. Glauner, T.J. MacAlister, and L.I. Rothfield, (1986), Isolation of Differentiated Membrane Domains from Escherichia coli and Salmonella typhimurium, Including a Fraction Containing Attachment Sites Between the Inner and Outer Membranes and the Murein Skeleton and the Cell Envelope. J. Biol. Chem. 261, 428-443.
682. Israelachvili, J.N., D.J. Mitchell, and B.W. Ninham, (1976), Theory of Self-assembly of Hydrocarbon Amphiphiles into Micelles and Bilayers. J. Chem. Soc. Faraday Trans. II. 72, 1525-1568.
683. Israelachvili, S., S. Marcelja, and R.G. Horn, (1980), Physical Principles of Membrane Organization. Quart. Rev. Biophys. 13, 121-200.

684. Issacs, B.S., E.J. Husten, C.T. Esmon, and A.E. Johnson, (1986), A Domain of Membrane-Bound Blood Coagulation Factor Va Is Located Far from the Phospholipid Surface. Biochemistry 25, 4958-4969.
685. Jackson, A.P., H.-F. Seow, N. Holmes, K. Drickamer, and P. Parham, (1987), Clathrin Light Chains Contain Brain-specific Insertion Sequences and a Region of Homology with Intermediate Filaments. Nature 326, 154-159.
686. Jackson, B.J., J.-P. Bohin, and E.P. Kennedy, (1984), Biosynthesis of Membrane-Derived Oligosaccharides: Characterization of mdoB Mutants Defective in Phosphoglycerol Transferase I Activity. J. Bacteriology 160, 976-981.
687. Jackson, B.J., J.M. Gennity, and E.P. Kennedy, (1986), Regulation of the Balanced Synthesis of Membrane Phospholipids: Experimental Test Models for Regulation in Escherichia coli. J. Biol. Chem. 261, 13464-13468.
688. Jackson, C.M. and Y. Nemerson, (1980), Blood Coagulation. Ann. Rev. Biochem. 49, 765-811.
689. Jackson, M.E., J.M. Pratt, N.G. Stoker, and I.B. Holland, (1985), An Inner Membrane Protein N-Terminal Signal Sequence Is Able to Promote Efficient Localisation of an Outer Membrane Protein in Escherichia coli. The EMBO Journal 4, 2377-2383.
690. Jacobson, K. (1983), Lateral Diffusion in Membranes. Cell Motility 3, 367-373.
691. Jacobson, K., A. Ishihara, and R. Inman, (1987), Lateral Diffusion of Proteins in Membranes. Ann. Rev. Physiol. 49, 163-175.
692. Jacobson, K., D. O'Dell, and J.T. August, (1984), Lateral Diffusion of an 80,000-dalton Glycoprotein in the Plasma Membrane of Murine Fibroblasts: Relationships to Cell Structure and Function. J. Cell. Biol. 99, 1624-1633.
693. Jain, M.K. (1983), Nonrandom Lateral Organization in Bilayers and Biomembranes. Membrane Fluidity in Biology (R.C. Aloia, Ed.), Vol. 1, pp. 1-37. Academic Press, New York.
694. Jain, M.K. and D. Zakim, (1987), The Spontaneous Incorporation of Proteins into Preformed Bilayers. Biochim. Biophys. Acta 906, 33-68.
695. Jain, M.K., G.H. DeHaas, J.F. Marecek, and F. Ramirez, (1986), The Affinity of Phospholipase A2 for the Interface of the Substrate and Analogs. Biochim. Biophys. Acta 860, 475-483.
696. Jain, M.K., J. Rogers, D.V. Jahagirdar, J.F. Marecek, and F. Ramirez, (1986), Kinetics of Interfacial Catalysis by Phospholipase A2 in Intravesicle Scooting Mode, and Heterofusion of Anionic and Zwitterionic Vesicles. Biochim. Biophys. Acta 860, 435-447.
697. Jain, M.K., J. Rogers, J.F. Marecek, F. Ramirez, and H. Eibl, (1986), Effect of the Structure of Phospholipid on the Kinetics of Intravesicle Scooting of Phospholipase A2. Biochim. Biophys. Acta 860, 462-474.
698. Jain, M.K. and R.C. Wagner, (1980), Introduction to Biological Membranes. John Wiley and Sons, New York.
699. Jaken, S. and S.C. Kiley, (1987), Purification and Characterization of Three Types of Protein Kinase C from Rabbit Brain Cytosol. Proc. Natl. Acad. Sci. USA 84, 4418-4422.
700. Janiak, M.J., D.M. Small, and G.G. Shipley, (1976), Nature of the Thermal Pretransition of Synthetic Phospholipids: Dimyristoyl- and Dipalmitoyllecithin. Biochemistry 15, 4575-4580.
701. Jans, D.A., L. Hatch, A.L. Fimmel, F. Gibson, and G.B. Cox, (1984), An Acidic or Basic Amino Acid at Position 26 of the ß Subunit of Escherichia coli F_1F_0-

ATPase Impairs Membrane Proton Permeability: Suppression of the uncF469 Nonsense Mutation. J. Bacteriology 160, 764-770.

702. Jaworsky, M. and R. Mendelsohn, (1987), Unusual Partitioning Behavior of CaATPase in Dipalmitoylphosphatidylethanolamine/Dielaidoylphosphatidylcholine Mixtures. Biophys. J. 52, 241-248.
703. Jay, D. and L. Cantley, (1986), Structural Aspects of the Red Cell Anion Exchange Protein. Ann. Rev. Biochem. 55, 511-538.
704. Jelsema, C.L. and J. Axelrod, (1987), Stimulation of Phospholipase A2 Activity in Bovine Rod Outer Segments by the ß/γ Subunits of Transducin and Its Inhibition by the α Subunit. Proc. Natl. Acad. Sci. USA 84, 3623-3627.
705. Jenkins, J.D., D.P. Madden, and T.L. Steck, (1984), Association of Phosphofructokinase and Aldolase with the Membrane of the Intact Erythrocyte. J. Biol. Chem. 259, 9374-9378.
706. Jenkins, J.D., F.J. Kezdy, and T.L. Steck, (1985), Mode of Interaction of Phosphofructokinase with the Erythrocyte Membrane. J. Biol. Chem. 260, 10426-10433.
707. Jennings, M.L., M. Adams-Lackey, and G.H. Denney, (1984), Peptides of Human Erythrocyte Band 3 Produced by Extracellular Papain Cleavage. J. Biol. Chem. 259, 4652-4660.
708. Jennings, M.L., M.P. Anderson, and R. Monaghan, (1986), Monoclonal Antibodies Against Human Erythrocyte Band 3 Protein: Localization of Proteolytic Cleavage Sites and Stilbenedisulfonate-binding Lysine Residues. J. Biol. Chem. 261, 9002-9010.
709. Jensen, J.W. and J.S. Schutzbach, (1984), Activation of Mannosyltransferase II by Nonbilayer Phospholipids. Biochemistry 23, 1115-1119.
710. Jing, S. and I.S. Trowbridge, (1987), Identification of the Intermolecular Disulfide Bonds of the Human Transferrin Receptor and Its Lipid-attachment Site. The EMBO Journal 6, 327-331.
711. Johnson, L.M., V.A. Bankaitis, and S.D. Emr, (1987), Distinct Sequence Determinants Direct Intracellular Sorting and Modification of a Yeast Vacuolar Protease. Cell 48, 875-885.
712. Joost, H.G., T.M. Weber, S.W. Cushman, and I.A. Simpson, (1987), Activity and Phosphorylation State of Glucose Transporters in Plasma Membranes from Insulin-, Isoproterenol-, and Phorbol Ester-treated Rat Adipose Cells. J. Biol. Chem. 262, 11261-11267.
713. Jordan, P.C. (1987), How Pore Mouth Charge Distributions Alter the Permeability of Transmembrane Ionic Channels. Biophys. J. 51, 297-311.
714. Jorgensen, P.L. (1982), Mechanism of the Na^+, K^+ Pump: Protein Structure and Conformation of the Pure (Na^+, K^+)-ATPase. Biochim. Biophys. Acta 694, 27-68.
715. Jubb, J.S., D.L. Worcester, H.L. Crespi, and G. Zaccai, (1984), Retinal Location in Purple Membrane of Halobacterium halobium: A Neutron Diffraction Study of Membranes Labelled In Vivo with Deuterated Retinal. The EMBO Journal 3, 1455-1461.
716. Kaback, H.R. (1986), Active Transport in Escherichia coli: Passage to Permease. Ann. Rev. Biophys. Biophys. Chem. 15, 279-319.
717. Kaback, H.R. (1987), Use of Site-Directed Mutagenesis to Study the Mechanism of a Membrane Transport Protein. Biochemistry 26, 2071-2076.
718. Kachar, B., N. Fuller, and R.P. Rand, (1986), Morphological Responses to Calcium-

Induced Interaction of Phosphatidylserine-Containing Vesicles. Biophys. J. 50, 779-788.

719. Kagawa, Y. (1980), Energy Transducing Proteins in Thermophilic Biomembranes. J. Memb. Biol. 55, 1-8.
720. Kaiser, C.A., D. Preuss, P. Grisafi, and D. Botstein, (1987), Many Random Sequences Functionally Replace the Secretion Signal Sequence of Yeast Invertase. Science 235, 312-317.
721. Kaiser, E.T. and F.J. Kezdy, (1987), Peptides with Affinity for Membranes. Ann. Rev. Biophys. Biophys. Chem. 16, 561-581.
722. Kakitani, T., B. Honig, and A.R. Crofts, (1982), Theoretical Studies of the Electrochromic Response of Carotenoids in Photosynthetic Membranes. Biophys. J. 39, 57-63.
723. Kamegai, J., S. Kimura, and Y. Imanishi, (1986), Conformation of Sequential Polypeptide Poly(Leu-Leu-D-Phe-Pro) and Formation of Ion Channel Across Bilayer Lipid Membrane. Biophys. J. 49, 1101-1108.
724. Kantor, H.L. and J.H. Prestegard, (1978), Fusion of Phosphatidylcholine Bilayer Vesicles: Role of Free Fatty Acid. Biochemistry 17, 3592-3597.
725. Kapitza, H.G., D.A. Ruppel, H.J. Galla, and E. Sackmann, (1984), Lateral Diffusion of Lipids and Glycophorin in Solid Phosphatidylcholine Bilayers. Biophys. J. 45, 577-587.
726. Kapitza, H.G., G. McGregor, and K.A. Jacobson, (1985), Direct Measurement of Lateral Transport in Membranes by Using Time-resolved Spatial Photometry. Proc. Natl. Acad. Sci. USA 82, 4122-4126.
727. Karin, N.J. and J.S. Cook, (1986), Turnover of the Catalytic Subunit of Na,K-ATPase in HTC Cells. J. Biol. Chem. 261, 10422-10428.
728. Karnovsky, M.J., A.M. Kleinfeld, R.L. Hoover, and R.D. Klausner, (1982), The Concept of Lipid Domains in Membranes. J. Cell Biol. 94, 1-6.
729. Kasianowicz, J., R. Benz, and S. McLaughlin, (1984), The Kinetic Mechanism by which CCCP (Carbonyl Cyanide m-Chlorophenylhydrazone) Transports Protons across Membranes. J. Membrane Biol. 82, 179-190.
730. Kasianowicz, J., R. Benz, and S. McLaughlin, (1987), How Do Protons Cross the Membrane-Solution Interface? Kinetic Studies on Bilayer Membranes Exposed to the Protonophore S-13 (5-chloro-3-tert-butyl-2'-chloro-4' nitrosalicylanilide). J. Memb. Biol. 95, 73-89.
731. Katre, N.V., J. Finer-Moore, R.M. Stroud, and S.B. Hayward, (1984), Location of an Extrinsic Label in the Primary and Tertiary Structure of Bacteriorhodopsin. Biophys. J. 46, 195-204.
732. Kaufman, J.F., C. Auffray, A.J. Korman, D.A. Shackelford, and J. Strominger, (1984), The Class II Molecules of the Human and Murine Major Histocompatibility Complex. Cell 36, 1-13.
733. Kawakami, K., S. Noguchi, M. Noda, H. Takahashi, T. Ohta, M. Kawamura, H. Nojima, K. Nagano, T. Hirose, S. Inayana, H. Hayashida, T. Miyata, and S. Numa, (1985), Primary Structure of the α-Subunit of Torpedo californica ($Na^+ + K^+$)ATPase Deduced from cDNA Sequence. Nature 316, 733-736.
734. Kawakami, K., T. Ohta, H. Nojima, and K. Nagano, (1986), Primary Structure of the α-Subunit of Human Na,K-ATPase Deduced from cDNA Sequence. J. Biochem. 100, 389-397.
735. Kawashima, Y. and R.M. Bell, (1987), Assembly of the Endoplasmic Reticulum Phospholipid Bilayer. J. Biol. Chem. 262, 16495-16502.

736. Kawato, S., E. Sigel, E. Carafoli, and R.J. Cherry, (1981), Rotation of Cytochrome Oxidase in Phospholipid Vesicles. J. Biol. Chem. 256, 7518-7527.
737. Kleffel, B., R.M. Garavito, W. Baumeister, and J.P. Rosenbusch, (1985), Secondary Structure of a Channel Forming Protein: Porin from E. coli Outer Membranes. The EMBO Journal 4, 1589-1592.
738. Kelly, R.B. (1985), Pathways of Protein Secretion in Eukaryotes. Science 230, 25-32.
739. Kelusky, E.C. and I.C.P. Smith, (1983), Characterization of the Binding of the Local Anesthetics Procaine and Tetracaine to Model Membranes of Phosphatidylethanolamine: A Deuterium Nuclear Magnetic Resonance Study. Biochemistry 22, 6011-6017.
740. Kempf, C., R.D. Klausner, J.N. Weinstein, J. Van Renswonde, M. Pincus, and R. Blumenthal, (1982), Voltage-dependent Trans-bilayer Orientation of Melittin. J. Biol. Chem. 257, 2469-2476.
741. Kempner, E.S. and H.T. Haigler, (1985), Target Analysis of Growth Factor Receptors. Growth and Maturation Factors (G. Guroff, Ed.), Vol. 3, pp. 149-173. John Wiley, New York.
742. Kempner, E.S. and W. Schlegel, (1979), Size Determination of Enzymes by Radiation Inactivation. Anal. Bioch. 92, 2-10.
743. Kensil, C.R. and P. Strittmatter, (1986), Binding and Fluorescence Properties of the Membrane Domain of NADH-Cytochrome-b_5 Reductase: Determination of the Depth of Trp-16 in the Bilayer. J. Biol. Chem. 261, 7316-7321.
744. Kerrick, W.G.L. and L.Y.W. Bourguignon, (1984), Regulation of Receptor Capping in Mouse Lymphoma T Cells by Ca^{2+}-Activated Myosin Light Chain Kinase. Proc. Natl. Acad. Sci. USA 81, 165-169.
745. Khorana, H.G., G.E. Gerber, W.C. Herlihy, C.P. Gray, R.J. Anderegg, K. Nihei, and K. Biemann, (1979), Amino Acid Sequence of Bacteriorhodopsin. Proc. Natl. Acad. Sci. USA 76, 5046-5050.
746. Kielian, M.C., M. Marsh, and A. Helenius, (1986), Kinetics of Endosome Acidification Detected by Mutant and Wild-type Semliki Forest Virus. The EMBO Journal 5, 3103-3109.
747. Kikutani, H., S. Inui, R. Sato, E.L. Barsumian, H. Owaki, K. Yamasaki, T. Kaisho, N. Uchibayashi, R.R. Hardy, T. Hirano, S. Tsunasawa, F. Sakiyama, M. Suemura, and T. Kishimoto, (1986), Molecular Structure of Human Lymphocyte Receptor for Immunoglobulin E. Cell 47, 657-665.
748. Killian, J.A., A.J. Verkleij, J. Leunissen-Bijvelt, and B. de Kruijff, (1985), External Addition of Gramicidin Induces H_{II} Phase in Dioleoylphosphatidylcholine Model Membranes. Biochim. Biophys. Acta 812, 21-26.
749. Killian, J.A. and B. de Kruijff, (1985), Importance of Hydration for Gramicidin-Induced Hexagonal H_{II} Phase Formation in Dioleoylphosphatidylcholine Model Membranes. Biochemistry 24, 7890-7898.
750. Kim, J. and H. Kim, (1986), Fusion of Phospholipid Vesicles Induced by α-Lactalbumin at Acidic pH. Biochemistry 25, 7867-7874.
751. Kim, K.S., D.P. Vercauteren, M. Welti, S. Chin, and E. Clementi, (1985), Interaction of K^+ Ion with the Solvated Gramicidin A Transmembrane Channel. Biophys. J. 47, 327-335.
752. Kimelberg, H.K. and D. Papahadjopoulos, (1972), Phospholipid Requirements for (Na^+ + K^+)-ATPase Activity: Head-Group Specificity and Fatty Acid Fluidity. Biochim. Biophys. Acta 282, 277-292.

753. Kimura, K., T.L. Mason, and H.G. Khorana, (1982), Immunological Probes for Bacteriorhodopsin: Identification of Three Distinct Antigenic Sites on the Cytoplasmic Surface. J. Biol. Chem. 257, 2859-2867.
754. Kinet, J.-P., H. Metzger, J. Hakimi, and J. Kochan, (1987), A cDNA Presumptively Coding for the α Subunit of the Receptor with High Affinity for Immunoglobulin E. Biochemistry 26, 4605-4610.
755. King, G.I., R.E. Jacobs, and S.H. White, (1985), Hexane Dissolved in Dioleoyllecithin Bilayers Has a Partial Molar Volume of Approximately Zero. Biochemistry 24, 4637-4645.
756. King, G.I., W. Stoeckenius, H.L. Crespi, and B.P. Schoenborn, (1979), The Location of Retinal in the Purple Membrane Profile by Neutron Diffraction. J. Mol. Biol. 130, 395-404.
757. Kinosita, K., S. Kawato, and A. Ikegami, (1977), A Theory of Fluorescence Polarization Decay in Membranes. Biophys. J. 20, 289-305.
758. Kirchhausen, T., P. Scarmato, S.C. Harrison, J.J. Monroe, E.P. Chow, R.J. Mattaliano, K.L. Ramachandran, J.E. Smart, A.H. Ahn, and J. Brosius, (1987), Clathrin Light Chains LCA and LCB Are Similar, Polymorphic, and Share Repeated Heptad Motifs. Science 236, 320-324.
759. Kishimoto, T.K., K. O'Connor, A. Lee, T.M. Roberts, and T.A. Springer, (1987), Cloning of the ß Subunit of the Leukocyte Adhesion Proteins: Homology to an Extracellular Matrix Receptor Defines a Novel Supergene Family. Cell 48, 681-690.
760. Kistler, J. and R.M. Stroud, (1981), Crystalline Arrays of Membrane-bound Acetylcholine Receptor. Proc. Natl. Acad. Sci. USA 78, 3678-3682.
761. Kleene, R., N. Pfanner, R. Pfaller, T.A. Link, W. Sebald, W. Neupert, and M. Tropschug, (1987), Mitochondrial Porin of Neurospora crassa: cDNA Cloning, In Vitro Expression and Import into Mitochondria. The EMBO Journal 6, 2627-2634.
762. Kleinberg, M.E. and A. Finkelstein, (1984), Single-Length and Double-Length Channels Formed by Nystatin in Lipid Bilayer Membranes. J. Membrane Biol. 80, 257-269.
763. Klingenberg, M. (1981), Membrane Protein Oligomeric Structure and Transport Function. Nature 290, 449-453.
764. Klionsky, D.J., W.S.A. Brusilow, and R.D. Simoni, (1984), In Vivo Evidence for the Role of the Epsilon Subunit as an Inhibitor of the Proton-Translocating ATPase of Escherichia coli. J. Bacteriology 160, 1055-1060.
765. Knoll, W., G. Schmidt, K. Ibel, and E. Sackmann, (1985), Small-Angle Neutron Scattering Study of Lateral Phase Separation in Dimyristoylphosphatidylcholine-Cholesterol Mixed Membranes. Biochemistry 24, 5240-5246.
766. Knowles, P.F., A. Watts, and D. Marsh, (1979), Spin-Label Studies of Lipid Immobilization in Dimyristoylphosphatidylcholine-Substituted Cytochrome Oxidase. Biochemistry 18, 4480-4487.
767. Koblika, B.K., H. Matsui, T.S. Koblika, T.L. Yang-Feng, U. Francke, M.G. Caron, R.J. Leftkowitz, and J.W. Regan, (1987), Cloning, Sequencing, and Expression of the Gene Coding for the Human Platelet α_2-Adrenergic Receptor. Science 238, 650-656.
768. Kochevar, D.T. and R.G.W. Anderson, (1987), Purified Crystalloid Endoplasmic Reticulum from UT-1 Cells Contains Multiple Proteins in Addition to 3-Hydroxy-3-methylglutaryl Coenzyme A Reductase. J. Biol. Chem. 262, 10321-10326.

769. Kodaki, T. and S. Yamashita, (1987), Yeast Phosphatidylethanolamine Methylation Pathway. J. Biol. Chem. 262, 15428-15435.
770. Koland, J.G., M.J. Miller, and R.B. Gennis, (1984), Reconstitution of the Membrane-Bound, Ubiquinone-Dependent Pyruvate Oxidase Respiratory Chain of E. coli with the Cytochrome d Terminal Oxidase. Biochemistry 445-453.
771. Kominami, S., H. Hara, T. Ogishima, and S. Takemori, (1984), Interaction between Cytochrome P-450 (P-450c21) and NADPH-Cytochrome P-450 Reductase from Adrenocortical Microsomes in a Reconstituted System. J. Biol. Chem. 259, 2991-2999.
772. Kominami, S., Y. Itoh, and S. Takemori, (1986), Studies on the Interaction of Steroid Substrates with Adrenal Microsomal Cytochrome P-450 (P-450c21) in Liposome Membranes. J. Biol. Chem. 261, 2077-2083.
773. Kop, J.M.M., P.A. Cuypers, T. Lindhout, H.C. Hemker, and W.T. Hermens, (1984), The Adsorption of Prothrombin to Phospholipid Monolayers Quantitated by Ellipsometry. J. Biol. Chem. 259, 13993-13998.
774. Kopito, R.R. and H.F. Lodish, (1985), Primary Structure and Transmembrane Orientation of the Murine Anion Exchange Protein. Nature 316, 234-238.
775. Kopito, R.R. and H.F. Lodish, (1985), Structure of the Murine Anion Exchange Protein. J. Cell. Biochem. 29, 1-17.
776. Kordossi, A.A. and S.J. Tzartos, (1987), Conformation of Cytoplasmic Segments of Acetylcholine Receptor α- and ß-Subunits Probed by Monoclonal Antibodies: Sensitivity of the Antibody Competition Approach. The EMBO Journal 6, 1605-1610.
777. Kornacki, J.A. and W. Firshein, (1986), Replication of Plasmid RK2 In Vitro by a DNA-Membrane Complex: Evidence for Initiation of Replication and Its Coupling to Transcription and Translation. J. Bacteriology 167, 319-326.
778. Kornfeld, S. (1987), Trafficking of Lysosomal Enzymes. FASEB J. 1, 462-468.
779. Korsgren, C. and C.M. Cohen, (1986), Purification and Properties of Human Erythrocyte Band 4.2: Association with the Cytoplasmic Domain of Band 3. J. Biol. Chem. 261, 5536-5543.
780. Kosower, E.M. (1985), A Structural and Dynamic Molecular Model for the Sodium Channel of Electrophorus electricus. FEBS Lett. 182, 234-242.
781. Kouyama, T., A.N.-Kouyama, and A. Ikegami, (1987), Bacteriorhodopsin Is a Powerful Light-driven Proton Pump. Biophys. J. 51, 839-841.
782. Krab, K., H.S. van Walraven, M.J.C. Scholts, and R. Kraayenhof, (1985), Measurement of Diffusion Potentials in Liposomes. Origin and Properties of the Threshold Level in the Oxonol VI Response. Biochim. Biophys. Acta 809, 228-235.
783. Krbecek, R., C. Gebhardt, H. Gruler, and E. Sackmann, (1979), Three Dimensional Microscopic Surface Profiles of Membranes Reconstituted from Freeze-Etching Electron Micrographs. Biochim. Biophys. Acta 554, 1-22.
784. Krebs, K.E. and M.C. Phillips, (1983), The Helical Hydrophobic Moments and Surface Activities of Serum Apoliproteins. Biochim. Biophys. Acta 754, 227-230.
785. Kreis, T.E. and H.F. Lodish, (1986), Oligomerization Is Essential for Transport of Vesicular Stomatitis Viral Glycoprotein to the Cell Surface. Cell 46, 929-937.
786. Kreutter, D., J.Y.H. Kim, J.R. Goldenring, H. Rasmussen, C. Ukomadu, R.J. DeLorenzo, and R.K. Yu, (1987), Regulation of Protein Kinase C Activity by Gangliosides. J. Biol. Chem. 262, 1633-1637.
787. Krieg, U.C., B.S. Issacs, S.S. Yemul, C.T. Esmon, H. Bayley, and A.E. Johnson,

(1987), Interaction of Blood Coagulation Factor Va with Phospholipid Vesicles Examined by Using Lipophilic Photoreagents. Biochemistry 26, 103-109.

788. Krikos, A., M.P. Conley, A. Boyd, H.C. Berg, and M.I. Simon, (1985), Chimeric Chemosensory Transducers of Escherichia coli. Proc. Natl. Acad. Sci. USA 82, 1326-1330.

789. Krishnamoorthy, G. and P.C. Hinkle, (1984), Non-Ohmic Proton Conductance of Mitochondria and Liposomes. Biochemistry 23, 1640-1645.

790. Krupka, R.M. (1985), Asymmetrical Binding of Phloretin to the Glucose Transport System of Human Erythrocytes. J. Membrane Biol. 83, 71-80.

791. Krupka, R.M. and R. Deves, (1983), Kinetics of Inhibition of Transport Systems. Int. Rev. of Cyt. 84, 303-352.

792. Kubesch, P., J. Boggs, L. Luciano, G. Maass, and B. Tummler, (1987), Interaction of Polymyxin B Nonapeptide with Anionic Phospholipids. Biochemistry 26, 2139-2149.

793. Kubo, T., K. Fukuda, A. Mikami, A. Maeda, H. Takahashi, M. Mishina, T. Haga, K. Haga, A. Ichiyama, K. Kangawa, M. Kojima, H. Matsuo, T. Hirose, and S. Numa, (1986), Cloning, Sequencing and Expression of Complementary DNA Encoding the Muscarinic Acetylcholine Receptor. Nature 323, 411-416.

794. Kuge, O., M. Nishijima, and Y. Akamatsu, (1986), Phosphatidylserine Biosynthesis in Cultured Chinese Hamster Ovary Cells: Genetic Evidence for Utilization of Phosphatidylcholine and Phosphatidylethanolamine as Precursors. J. Biol. Chem. 261, 5795-5798.

795. Kuhlbrandt, W. (1984), Three-dimensional Structure of the Light Harvesting Chlorophyll a/b-Protein Complex. Nature 307, 478-480.

796. Kuhn, A. (1987), Bacteriophage M13 Procoat Protein Inserts into the Plasma Membrane as a Loop Structure. Science 238, 1413-1415.

797. Kuhn, A., G. Kreil, and W. Wickner, (1986), Both Hydrophobic Domains of M13 Procoat Are Required to Initiate Membrane Insertion. The EMBO Journal 5, 3681-3685.

798. Kuhn, A., W. Wickner, and G. Kreil, (1986), The Cytoplasmic Carboxy Terminus of M13 Procoat Is Required for the Membrane Insertion of Its Central Domain. Nature 322, 335-339.

799. Kuhn, L.A. and J.S. Leigh, (1985), A Statistical Technique for Predicting Membrane Protein Structure. Biochim. Biophys. Acta 828, 351-361.

800. Kumamoto, C.A. and R.D. Simoni, (1986), Genetic Evidence for Interaction between the a and b Subunits of the F_0 Portion of the Escherichia coli Proton Translocating ATPase. J. Biol. Chem. 261, 10037-10042.

801. Kumamoto, C.A. and R.D. Simoni, (1987), A Mutation of the c Subunit of the Escherichia coli Proton-translocating ATPase That Suppresses the Effects of a Mutant b Subunit. J. Biol. Chem. 262, 3060-3064.

802. Kumar, A. and C.M. Gupta, (1985), Transbilayer Phosphatidylcholine Distribution in Small Unilamellar Sphingomyelin-phosphatidylcholine Vesicles: Effect of Altered Polar Head Group. Biochemistry 24, 5157-5163.

803. Kusano, T., D. Steinmetz, W.G. Hendrickson, J. Murchie, M. King, A. Benson, and M. Schaechter, (1984), Direct Evidence for Specific Binding of the Replicative Origin of the Escherichia coli Chromosome to the Membrane. J. Bacteriology 158, 313-316.

804. Kyte, J. and R.F. Doolittle, (1982), A Simple Method for Displaying the Hydrophobic Character of a Protein. J. Mol. Biol. 157, 105-132.

805. Labarca, P., M.S. Montal, J.M. Lindstrom, and M. Montal, (1985), The Occurrence of Long Openings in the Purified Cholinergic Receptor Channel Increases with Acetylcholine Concentration. The Journal of Neuroscience 5, 3409-3413.
806. Labischinski, H., G. Barnickel, H. Bradaczek, D. Naumann, E.T. Rietchel, and P. Giesbrecht, (1985), High State of Order of Isolated Bacterial Lipopolysaccharide and Its Possible Contribution to the Permeation Barrier Property of the Outer Membrane. J. Bacteriology 162, 9-20.
807. Laffan, J. and W. Firshein, (1987), DNA Replication by a DNA-Membrane Complex Extracted from Bacillus subtilis: Site of Initiation In Vitro and Initiation Potential of Subcomplexes. J. Bacteriology 169, 2819-2827.
808. Lagaly, G., A. Weiss, and E. Stuke, (1977), Effect of Double-Bonds on Bimolecular Films in Membrane Models. Biochim. Biophys. Acta 470, 331-341.
809. Lai, C.-S. and J.S. Schutzbach, (1986), Localization of Dolichols in Phospholipid Membranes. FEBS Lett. 203, 153-156.
810. Lai, C., M.A. Brow, K.-A. Nave, A.B. Noronna, R.H. Quarles, F.E. Bloom, R.J. Milner, and J.G. Sutcliffe, (1987), Two Forms of 1B236/Myelin-associated Glycoprotein, a Cell Adhesion Molecule for Postnatal Neural Development, Are Produced by Alternative Splicing. Proc. Natl. Acad. Sci. USA 84, 4337-4341.
811. Lai, J.-S., M. Sarvas, W.J. Brammar, K. Neugebauer, and H.C. Wu, (1981), Bacillus licheniformis Penicillinase Synthesized in E. coli Contains Covalently Linked Fatty Acid and Glyceride. Proc. Natl. Acad. Sci. USA 78, 3506-3510.
812. Laitinen, J., R. Lopponen, J. Merenmies, and H. Rauvala, (1987), Binding of Laminin to Brain Gangliosides and Inhibition of Laminin-Neuron Interaction by the Gangliosides. FEBS Lett. 217, 94-100.
813. Lakey, J.H. (1987), Voltage Gating in Porin Channels. FEBS Lett. 211, 1-4.
814. Lambeth, J.D., D.W. Seybert, and H. Kamin, (1980), Phospholipid Vesicle-reconstituted Cytochrome P-450scc: Mutually Facilitated Binding of Cholesterol and Adrenodoxin. J. Biol. Chem. 255, 138-143.
815. Lambeth, J.D., L.M. Geren, and F. Millett, (1984), Adrenodoxin Interaction with Adrenodoxin Reductase and Cytochrome P-450scc: Cross-linking of Protein Complexes and Effects of Adrenodoxin Modification by 1-Ethyl-3-(3-Dimethylaminopropyl)Carbodiimide. J. Biol. Chem. 259, 10025-10029.
816. Lambeth, J.D. and S.O. Pember, (1983), Cytochrome P-450scc-Adrenodoxin Complex. J. Biol. Chem. 258, 5596-5602.
817. Lambrecht, B. and M.F.G. Schmidt, (1986), Membrane Fusion Induced by Influenza Virus Hemagglutinin Requires Protein Bound Fatty Acids. FEBS Lett. 202, 127-132.
818. Lampe, P.D., M.L. Pusey, G.J. Wei, and G.L. Nelsestuen, (1984), Electron Microscopy and Hydrodynamic Properties of Blood Clotting Factor V and Activation Fragments of Factor V with Phospholipid Vesicles. J. Biol. Chem. 259, 9959-9964.
819. Lanyi, J.K. (1986), Halorhodopsin: A Light-driven Chloride Ion Pump. Ann. Rev. Biophys. Biophys. Chem. 15, 11-28.
820. Lapetina, E.G., S.P. Watson, and P. Cuatrecasas, (1984), myo-Inositol 1,4,5-trisphosphate Stimulates Protein Phosphorylation in Saponin-permeabilized Human Platelets. Proc. Natl. Acad. Sci. USA 81, 7431-7435.
821. Lau, S.H., J. Rivier, W. Vale, E.T. Kaiser, and F.J. Kezdy, (1983), Surface Properties of an Amphiphilic Peptide Hormone and of Its Analog: Corticotropin-releasing Factor and Sauvagine. Proc. Natl. Acad. Sci. USA 80, 7070-7074.
822. Laubinger, W. and P. Dimroth, (1987), Characterization of the Na^+-stimulated

ATPase of Propionigenium modestum as an Enzyme of the F_1F_0 Type. Eur. J. Biochem. 168, 475-480.

823. Lauffer, L., P.D. Garcia, R.N. Harkins, L. Coussens, A. Ullrich, and P. Walter, (1985), Topology of Signal Recognition Particle Receptor in Endoplasmic Reticulum Membrane. Nature 318, 334-338.
824. Lauger, P. (1973), Ion Transport through Pores: A Rate Theory Analysis. Biochim. Biophys. Acta 311, 423-441.
825. Lauger, P. (1984), Thermodynamic and Kinetic Properties of Electrogenic Ion Pumps. Biochim. Biophys. Acta 779, 307-341.
826. Laursen, R.A., M. Samiullah, and M.B. Lees, (1984), The Structure of Bovine Brain Myelin Proteolipid and Its Organization in Myelin. Proc. Natl. Acad. Sci. USA 81, 2912-2916.
827. Lecompte, M.-F., S. Krishnaswamy, K.G. Mann, M.E. Nesheim, and C. Gilter, (1987), Membrane Penetration of Bovine Factor V and Va Detected by Labelling with 5-Iodonaphthalene-1-Azide. J. Biol. Chem. 262, 1935-1937.
828. Lee, A.G. (1987), Interactions of Lipids and Proteins: Some General Principles. J. Bioenerg. and Biomemb. 19, 581-603.
829. Lee, A.G. (1983), Lipid Phase Transitions and Mixtures. Membrane Fluidity in Biology (R.C. Aloia, Ed.), Vol. 2, pp. 43-88. Academic Press, New York.
830. Lee, A.G. (1978), Calculation of Phase Diagrams for Non-Ideal Mixtures of Lipids, and a Possible Non-Random Distribution of Lipids in Lipid Mixtures in the Lipid Crystalline Phase. Biochim. Biophys. Acta 507, 433-444.
831. Lee, D.C., A.A. Durrani, and D. Chapman, (1984), A Difference Infrared Spectroscopic Study of Gramicidin A, Alamethicin, and Bacteriorhodopsin in Perdeuterated Dimyristoylphosphatidylcholine. Biochim. Biophys. Acta 769, 49-56.
832. Lee, D.C., J.A. Hayward, C.J. Restall, and D. Chapman, (1985), Second-derivative Infrared Spectroscopic Studies of the Secondary Structures of Bacteriorhodopsin and Ca^{2+}-ATPase. Biochemistry 24, 4364-4373.
833. Lehrman, M.A., J.L. Goldstein, M.S. Brown, D.W. Russell, and W.J. Schneider, (1985), Internalization-Defective LDL Receptors Produced by Genes with Nonsense and Frameshift Mutations That Truncate the Cytoplasmic Domain. Cell 41, 735-743.
834. Lemansky, P., A. Hasilik, K. von Figura, S. Helmy, J. Fishman, R.E. Fine, N. L. Kedersha, and L.H. Rome, (1987), Lysosomal Enzyme Precursors in Coated Vesicles Derived from the Exocytic and Endocytic Pathways. J. Cell Biol. 104, 1743-1748.
835. Lemmon, S.K. and E.W. Jones, (1987), Clathrin Requirement for Normal Growth of Yeast. Science 238, 504-509.
836. Lenard, J. and S.J. Singer, (1966), Protein Conformation in Cell Membrane Preparations as Studied by Optical Rotatory Dispersion and Circular Dichroism. Proc. Natl. Acad. Sci. USA 56, 1828-1835.
837. Lenaz, G. and R. Fato, (1986), Is Ubiquinone Diffusion Rate Limiting for Electron Transfer? J. Bioenerg. and Biomemb. 18, 369-401.
838. Lemke, H.-D. and D. Oesterhelt, (1981), Lysine 216 Is a Binding Site of the Retinyl Moiety in Bacteriorhodopsin. FEBS Lett. 128, 255-260.
839. Lentz, B.R., D.A. Barrow, and M. Hoechli, (1980), Cholesterol-Phosphatidylcholine Interactions in Multilamellar Vesicles. Biochemistry 19, 1943-1954.
840. Lentz, B.R., K.W. Clubb, D.R. Alford, M. Hochli, and G. Meissner, (1985), Phase Behavior of Membranes Reconstituted from Dipentadecanoyl Phosphatidylcholine

and the Mg^{2+}-Dependent, Ca^{2+}-Stimulated Adenosinetriphosphatase of Sarcoplasmic Reticulum: Evidence for a Disrupted Lipid Domain Surrounding Protein. Biochemistry 24, 433-442.

841. Lentz, B.R., Y. Barenholz, and T.E. Thompson, (1976), Fluorescence Depolarization Studies of Phase Transitions and Fluidity in Phospholipid Bilayers 2. Two-Component Phosphatidylcholine Liposomes. Biochemistry 15, 4529-4537.
842. Leo, G.C., L.A. Colnago, K.G. Valentine, and S.J. Opella, (1987), Dynamics of fd Coat Protein in Lipid Bilayers. Biochemistry 26, 854-862.
843. Leonard, K., P. Wingfield, T. Arad, and H. Weiss, (1981), Three-dimensional Structure of Ubiquinol: Cytochrome c Reductase from Neurospora Mitochondria Determined by Electron Microscopy of Membrane Crystals. J. Mol. Biol. 149, 259-274.
844. Leung, J.O., E.C. Holland, and K. Drickamer, (1985), Characterization of the Gene Encoding the Major Rat Liver Asialoglycoprotein Receptor. J. Biol. Chem. 260, 12523-12527.
845. Leventis, R., J. Gagne, N. Fuller, R.P. Rand, and J.R. Silvius, (1986), Divalent Cation Induced Fusion and Lipid Lateral Segregation in Phosphatidylcholine-Phosphatidic Acid Vesicles. Biochemistry 25, 6978-6987.
846. Levitt, D.G. (1986), Interpretation of Biological Ion Channel Flux Data—Reaction-Rate versus Continuum Theory. Ann. Rev. Biophys. Biophys. Chem. 15, 29-57.
847. Levitzki, A. (1985), Reconstitution of Membrane Receptor Systems. Biochim. Biophys. Acta 822, 127-153.
848. Lewis, B.A. and D.M. Engelman, (1983), Lipid Bilayer Thickness Varies Linearly with Acyl Chain Length in Fluid Phosphatidylcholine Vesicles. J. Mol. Biol. 166, 211-217.
849. Lewis, B.A., G.M. Gray, R. Coleman, and R.H. Michell, (1975), Differences in the Enzymic, Polypeptide, Glycopeptide, Glycolipid, and Phospholipid Compositions of Plasma Membranes from the Two Surfaces of Intestinal Epithelial Cells. Bioch. Soc. Trans. 3, 752-753.
850. Lewis, B.A., G.S. Harbison, J. Herzfeld, and R.C. Griffin, (1985), NMR Structural Analysis of a Membrane Protein: Bacteriorhodopsin Peptide Backbone Orientation and Motion. Biochemistry 24, 4671-4679.
851. Lewis, B.A. and P.M. Engelman, (1983), Bacteriorhodopsin Remains Dispersed in Fluid Phospholipid Bilayers Over a Wide Range of Bilayer Thickness. J. Mol. Biol. 166, 203-210.
852. Lewis, C.A. and C.F. Stevens, (1983), Acetylcholine Receptor Channel Ionic Selectivity: Ions Experience an Aqueous Environment. Proc. Natl. Acad. Sci. USA 80, 6110-6113.
853. Liao, M.-J. and H.G. Khorana, (1984), Removal of the Carboxyl-terminal Peptide Does Not Affect Refolding or Function of Bacteriorhodopsin as a Light-dependent Proton Pump. J. Biol. Chem. 259, 4194-4199.
854. Lichtenberg, D., G. Romero, M. Menashe, and R.L. Biltonen, (1986), Hydrolysis of Dipalmitoylphosphatidylcholine Large Unilamellar Vesicles by Porcine Pancreatic Phospholipase A2. J. Biol. Chem. 261, 5334-5340.
855. Lichtenberg, D., M. Menasche, S. Donaldson, and R.L. Biltonen, (1984), Thermodynamic Characterization of the Pretransition of Unilamellar Dipalmitoylphosphatidylcholine Vesicles. Lipids 19, 395-400.
856. Lieb, W.R. and W.D. Stein, (1986), Non-Stokesian Nature of Transverse Diffusion within Human Red Cell Membranes. J. Membrane Biol. 92, 111-119.

857. Lieber, M.R. and T.L. Steck, (1982), Dynamics of the Holes in Human Erythrocyte Membrane Ghosts. J. Biol. Chem. 257, 11660-11666.
858. Lieber, M.R., Y. Lange, R.S. Weinstein, and T.L. Steck, (1984), Interaction of Chlorpromazine with the Human Erythrocyte Membrane. J. Biol. Chem. 259, 9225-9234.
859. Lill, H., S. Engelbrecht, G. Schonknecht, and W. Junge, (1986), The Proton Channel, CF_0, in Thylakoid Membranes: Only a Low Proportion of CF_1-lacking CF_0 Is Active with a High Unit Conductance (169 fS). Eur. J. Biochem. 160, 627-634.
860. Lindberg, F., B. Lund, L. Johansson, and S. Normark, (1987), Localization of the Receptor-Binding Protein Adhesin at the Tip of the Bacterial Pilus. Nature 328, 84-87.
861. Lipp, J. and B. Dobberstein, (1986), The Membrane-Spanning Segment of Invariant Chain (I γ) Contains a Potentially Cleavable Signal Sequence. Cell 46, 1103-1112.
862. Lipp, J. and B. Dobberstein, (1986), Signal Recognition Particle-dependent Membrane Insertion of Mouse Invariant Chain: A Membrane-spanning Protein with a Cytoplasmically Exposed Amino Terminus. J. Cell Biol. 102, 2169-2175.
863. Lippincott-Schwartz, J. and D.M. Fambrough, (1987), Cycling of the Integral Membrane Glycoprotein LEP100, between Plasma Membrane and Lysosomes: Kinetic and Morphological Analysis. Cell 49, 669-677.
864. Liss, L.R. and D.B. Oliver, (1986), Effects of secA Mutations on the Synthesis and Secretion of Proteins in Escherichia coli: Evidence for a Major Export System for Cell Envelope Proteins. J. Biol. Chem. 261, 2299-2303.
865. Liu, S.-C., L.H. Derick, and J. Palek, (1987), Visualization of the Hexagonal Lattice in the Erythrocyte Membrane Skeleton. J. Cell Biol. 104, 527-536.
866. Livneh, E., M. Benveniste, R. Prywes, S. Felder, Z. Kam, and J. Schlessinger, (1986), Large Deletions in the Cytoplasmic Kinase Domain of the Epidermal Growth Factor Receptor Do Not Affect Its Lateral Mobility. J. Cell Biol. 103, 327-331.
867. Livneh, E., N. Reiss, E. Berent, A. Ullrich, and J. Schlessinger, (1987), An Insertional Mutant of Epidermal Growth Factor Receptor Allows Dissection of Diverse Receptor Functions. The EMBO Journal 6, 2669-2676.
868. Livneh, E., R. Prywes, O. Kashles, N. Reiss, I. Sasson, Y. Mory, A. Ullrich, and J. Schlessinger, (1986), Reconstitution of Human Epidermal Growth Factor Receptor and Its Deletion Mutants in Cultured Hamster Cells. J. Biol. Chem. 261, 12490-12497.
869. Lobel, P., N.M. Dahms, J. Breitmeyer, J.M. Chirgwin, and S. Kornfeld, (1987), Cloning of the Bovine 215-kDa Cation-independent Mannose 6-phosphate Receptor. Proc. Natl. Acad. Sci. USA 84, 2233-2237.
870. Loeb, J.A. and K. Drickamer, (1987), The Chicken Receptor for Endocytosis of Glycoproteins Contains a Cluster of N-Acetylglucosamine-binding Sites. J. Biol. Chem. 262, 3022-3029.
871. Loewenstein, W.R. (1987), The Cell-to-Cell Channel of Gap Junctions. Cell 48, 725-726.
872. London, E. and G.W. Feigenson, (1981), Fluorescence Quenching in Model Membranes 2. Determination of the Local Lipid Environment of the Calcium Adenosine-triphosphatase from Sarcoplasmic Reticulum. Biochemistry 20, 1939-1948.
873. Lorenzina-Fiallo, M.M. and A. Garnier-Suillerot, (1986), Interaction of Adriamycin

with Cardiolipin-containing Vesicles. Evidence of an Embedded Site for the Dihydroanthraquinone Moiety. Biochim. Biophys. Acta 854, 143-146.
874. Louie, K., Y.-C. Chen, and W. Dowhan, (1986), Substrate-Induced Membrane Association of Phosphatidylserine Synthase from Escherichia coli. J. Bacteriology 165, 805-812.
875. Low, M.G. (1987), Biochemistry of the Glycosyl-phosphatidylinositol Membrane Protein Anchors. Biochem. J. 244, 1-13.
876. Low, P.S. (1986), Structure and Function of the Cytoplasmic Domain of Band 3: Center of Erythrocyte Membrane-Peripheral Protein Interactions. Biochim. Biophys. Acta 864, 145-167.
877. Lu, P.-W., C.-J. Soong, and M. Tao, (1985), Phosphorylation of Ankyrin Decreases Its Affinity for Spectrin Tetramer. J. Biol. Chem. 260, 14958-14964.
878. Lubben, T.H., J. Bansberg, and K. Keegstra, (1987), Stop-Transfer Regions Do Not Halt Translocation of Proteins into Chloroplasts. Science 238, 1112-1114.
879. Lucy, J.A. (1984), Do Hydrophobic Sequences Cleaved from Cellular Polypeptides Induce Membrane Fusion Reactions In Vivo? FEBS Lett. 166, 223-231.
880. Lucy, J.A. and A.M. Glauert, (1964), Structure and Assembly of Macromolecular Lipid Complexes Composed of Globular Micelles. J. Mol. Biol. 8, 727-748.
881. Lucy, J.A. and Q.F. Ahkong, (1986), An Osmotic Model for the Fusion of Biological Membranes. FEBS Lett. 199, 1-11.
882. Luderitz, O., M.A. Freudenberg, C. Galanos, V. Lehmann, E.T. Rietschel, and D.H. Shaw, (1982), Lipopolysaccharides of Gram Negative Bacteria. Current Topics in Membranes and Transport 17, 79-151.
883. Ludwig, D.S., H.O. Ribi, G.K. Schoolnik, and R.D. Kornberg, (1986), Two-dimensional Crystals of Cholera Toxin B-subunit-receptor Complexes: Projected Structure at 17-Å Resolution. Proc. Natl. Acad. Sci. USA 83, 8585-8588.
884. Ludwig, O., V. De Pinto, F. Palmieri, and R. Benz, (1986), Pore Formation by the Mitochondrial Porin of Rat Brain in Lipid Bilayer Membranes. Biochim. Biophys. Acta 860, 268-276.
885. Lugtenberg, B. and L. Van Alphen, (1983), Molecular Architecture and Functioning of the Outer Membrane of Escherichia coli and Other Gram-negative Bacteria. Biochim. Biophys. Acta 737, 51-115.
886. Luxnat, M. and H.-J. Galla, (1986), Partition of Chlorpromazine into Lipid Bilayer Membranes: The Effect of Membrane Structure and Composition. Biochim. Biophys. Acta 856, 274-282.
887. Luzikov, V.N. (1986), Proteolytic Control Over Topogenesis of Membrane Proteins. FEBS Lett. 200, 259-264.
888. Luzzati, V., M. DeRosz, A. Gulik, and A. Gambacorta, (1987), Polar Lipids of Thermophilic Prokaryotic Organisms: Chemical and Physical Structure. Ann. Rev. Biophys. Biophys. Chem. 16, 25-47.
889. Mabrey, S. and J.M. Sturtevant, (1978), Incorporation of Saturated Fatty Acids into Phosphatidylcholine Bilayers. Biochim. Biophys. Acta 486, 444-450.
890. Mabrey, S., P.L. Mateo, and J.M. Sturtevant, (1978), High-sensitivity Calorimetric Study of Mixtures of Cholesterol with Dimyristoyl- and Dipalmitoylphosphatidylcholines. Biochemistry 17, 2464-2468.
891. MacDonald, A.L. and D.A. Pink, (1987), Thermodynamics of Glycophorin in Phospholipid Bilayer Membranes. Biochemistry 26, 1909-1917.
892. MacDonald, J.R., U. Groschel-Stewart, and M.P. Walsh, (1987), Properties and Distribution of the Protein Inhibitor (M_r 17000) of Protein Kinase C. Biochem. J. 242, 695-705.

893. Macdonald, P.M., B.D. Sykes, and R.N. McElhaney, (1984), ^{19}F NMR Studies of Lipid Fatty Acyl Chain Order and Dynamics in Acholeplasma laidlawii B Membranes. ^{19}F NMR Line Shape and Orientation Order in the Gel State. Biochemistry 23, 4496-4502.
894. Macdonald, P.M. and J. Seelig, (1987), Calcium Binding to Mixed Phosphatidylglycerol-Phosphatidylcholine Bilayers as Studied by Deuterium Nuclear Magnetic Resonance. Biochemistry 26, 1231-1240.
895. Macdonald, P.M. and J. Seelig, (1987), Calcium Binding to Mixed Cardiolipin-Phosphatidylcholine Bilayers as Studied by Deuterium Nuclear Magnetic Resonance. Biochemistry 26, 6292-6298.
896. MacDonald, R.C. and S.A. Simon, (1987), Lipid Monolayer States and Their Relationships to Bilayers. Proc. Natl. Acad. Sci. USA 84, 4089-4093.
897. MacDonald, R.I. (1985), Membrane Fusion Due to Dehydration by Polyethylene Glycol, Dextran, or Sucrose. Biochemistry 24, 4058-4066.
898. MacDonald, R.I. and R.C. MacDonald, (1983), Lipid Mixing During Freeze-Thawing of Liposome Membranes as Monitored by Fluorescence Energy Transfer. Biochim. Biophys. Acta 735, 243-251.
899. Machamer, C.E. and J.K. Rose, (1987), A Specific Transmembrane Domain of a Coronavirus E1 Glycoprotein Is Required for Its Retention in the Golgi Region. J. Cell Biol. 105, 1205-1214.
900. MacIntyre, S., R. Freudl, M. Degen, I. Hindennach, and U. Henning, (1987), The Signal Sequence of an Escherichia coli Outer Membrane Protein Can Mediate Translocation of a Not Normally Secreted Protein across the Plasma Membrane. J. Biol. Chem. 262, 8416-8422.
901. Mackman, N., K. Baker, L. Gray, R. Haigh, J.-M. Nicaud, and I.B. Holland, (1987), Release of a Chimeric Protein into the Medium from Escherichia coli Using the C-terminal Secretion Signal of Hemolysin. The EMBO Journal 6, 2835-2841.
902. MacLennan, D.H., C.J. Brandl, B. Korczak, and N.M. Green, (1985), Amino-acid Sequence of a Ca^{2+} + Mg^{2+}-dependent ATPase from Rabbit Muscle Sarcoplasmic Reticulum, Deduced from Its Complementary DNA Sequence. Nature 316, 696-700.
903. Madden, T.D., M.J. Hope, and P.R. Cullis, (1984), Influence of Vesicle Size and Oxidase Content on Respiratory Control in Reconstituted Cytochrome Oxidase Vesicles. Biochemistry 23, 1413-1418.
904. Maher, P. and S.J. Singer, (1984), Structural Changes in Membranes Produced by the Binding of Small Amphiphilic Molecules. Biochemistry 23, 232-240.
905. Maher, P.A. and S.J. Singer, (1985), Anomalous Interaction of the Acetylcholine Receptor Protein with the Nonionic Detergent Triton X-114. Proc. Natl. Acad. Sci. USA 82, 958-962.
906. Maiden, M.C.J., E.O. Davis, S.A. Baldwin, D.C.M. Moore, and P.J.F. Henderson, (1987), Mammalian and Bacterial Sugar Transport Proteins Are Homologous. Nature 325, 641-643.
907. Majerus, P.W., D.B. Wilson, T.M. Connolly, T.E. Bross, and E.J. Neufeld, (1985), Phosphoinositide Turnover Provides a Link in Stimulus-Response Coupling. TIBS 10, 168-171.
908. Majerus, P.W., T.M. Connolly, H. Deckmyn, T.S. Ross, T.E. Bross, H. Ishii, V.S. Bansal, and D.B. Wilson, (1986), The Metabolism of Phosphoinositide-Derived Messenger Molecules. Science 234, 1519-1526.
909. Makino, S., J.A. Reynolds, and C. Tanford, (1973), The Binding of Deoxycholate and Triton X-100 to Proteins. J. Biol. Chem. 248, 4926-4932.

910. Makowski, L. and J. Li, (1984), X-Ray Diffraction and Electron Microscope Studies of the Molecular Structure of Biological Membranes. Biomembrane Structure and Function (D. Chapman, Ed.), pp. 43-166. Verlag Chemie, Basel.
911. Makriyannis, A., D.J. Siminovitch, S.K. Das Gupta, and R.G. Griffin, (1986), Studies on the Interaction of Anesthetic Steroids with Phosphatidylcholine Using ^{2}H and ^{13}C Solid State NMR. Biochim. Biophys. Acta 859, 49-55.
912. Maksymiw, R., S. Sui, H. Gaub, and E. Sackmann, (1987), Electrostatic Coupling of Spectrin Dimers to Phosphatidylserine Containing Lipid Lamellae. Biochemistry 26, 2983-2990.
913. Malatesta, F., V. Darley-Usmar, C. de Jong, L.J. Prochaska, R. Bisson, R.A. Capaldi, G.C.M. Steffens, and G. Buse, (1983), Arrangement of Subunit IV in Beef Heart Cytochrome c Oxidase Probed by Chemical Labeling and Protease Digestion Experiments. Biochemistry 22, 4405-4411.
914. Mamelok, R.D., D.F. Groth, and S.B. Prusiner, (1980), Separation of Membrane-bound γ-Glutamyl Transpeptidase from Brush Border Transport and Enzyme Activities. Biochemistry 19, 2367-2373.
915. Mangeat, P. and K. Burridge, (1984), Actin-Membrane Interactions in Fibroblasts: What Proteins Are Involved in This Association? J. Cell Biol. 99, 95s-103s.
916. Manjunath, C.K., B.J. Nicholson, D. Teplow, L. Hood, E. Page, and J.-P. Revel, (1987), The Cardiac Gap Junction Protein (M_r 47,000) Has a Tissue-Specific Cytoplasmic Domain of M_r 17,000 at its Carboxy-Terminus. Biochemical and Biophysical Research Communications 142, 228-234.
917. Mann, K.G. (1987), The Assembly of Blood Clotting Complexes on Membranes. TIBS 12, 229-233.
918. Manoil, C. and J. Beckwith, (1986), A Genetic Approach to Analyzing Membrane Protein Topology. Science 233, 1403-1408.
919. Mantsch, H.H., A. Martin, and D.G. Cameron, (1981), Characterization by Infrared Spectroscopy of the Bilayer to Nonbilayer Phase Transition of Phosphatidylethanolamines. Biochemistry 20, 3138-3145.
920. Maraganore, J.M. (1987), Structural Elements for Protein-Phospholipid Interactions May Be Shared in Protein Kinase C and Phospholipase A_2. TIBS 12, 176-177.
921. Marcelja, S. (1974), Chain Ordering in Lipid Crystals. II. Structure of Bilayer Membranes. Biochim. Biophys. Acta 367, 165-176.
922. Marchesi, V.T., H. Furthmayr, and M. Tomita, (1976), The Red Cell Membrane. Ann. Rev. Bioch. 45, 667-698.
923. Marlin, S.D. and T.A. Springer, (1987), Purified Intercellular Adhesion Molecule-1 (ICAM-1) Is a Ligand for Lymphocyte Function-Association Antigen 1 (LFA-1). Cell 51, 813-819.
924. Marrack, P. and J. Kappler, (1987), The T Cell Receptor. Science 238, 1073-1079.
925. Marsh, D. (1987), Selectivity of Lipid-Protein Interactions. J. Bioenerg. and Biomemb. 19, 677-689.
926. Marsh, D., A. Watts, R.D. Pates, R. Uhl, P.F. Knowles, and M. Esmann, (1982), ESR Spin-Label Studies of Lipid-Protein Interactions in Membranes. Biophys. J. 37, 265-274.
927. Marsh, M., S. Schmid, H. Kern, E. Harms, P. Male, I. Mellman, and A. Helenius, (1987), Rapid Analytical and Preparative Isolation of Functional Endosomes by Free Flow Electrophoresis. J. Cell Biol. 104, 875-886.
928. Martin, D.W., C. Tanford, and J.A. Reynolds, (1984), Monomeric Solubilized Sarcoplasmic Reticulum Ca Pump Protein: Demonstration of Ca Binding and

Dissociation Coupled to ATP Hydrolysis. Proc. Natl. Acad. Sci. USA 81, 6623-6626.

929. Martin, O.C. and R.E. Pagano, (1987), Transbilayer Movement of Fluorescent Analogs of Phosphatidylserine and Phosphatidylethanolamine at the Plasma Membrane of Cultured Cells. J. Biol. Chem. 262, 5890-5898.
930. Marx, J.L. (1987), Polyphosphoinositide Research Updated. Science 235, 974-976.
931. Masaki, R., A. Yamamoto, and Y. Tashiro, (1987), Cytochrome P-450 and NADPH-Cytochrome P-450 Reductase Are Degraded in the Autolysosomes in Rat Liver. J. Cell Biol. 104, 1207-1215.
932. Masamoto, K. (1984), Dependence on Surface pH and Surface Substrate Concentration of Activity of Microsome-Bound Arylsulfatase C and the Surface Charge Density in the Vicinity of the Enzyme. J. Biochem. 95, 715-719.
933. Masters, C. and D. Crane, (1984), The Role of Peroxisomes in Lipid Metabolism. TIBS 9, 314-317.
934. Masters, C.J. (1981), Interactions Between Soluble Enzymes and Subcellular Structure. CRC Crit. Rev. Bioch. 11, 105-143.
935. Masu, Y., K. Nakayama, H. Tamaki, Y. Harada, M. Kuno, and S. Nakanishi, (1987), cDNA Cloning of Bovine Substance-K Receptor through Oocyte Expression System. Nature 329, 836-838.
936. Mather, M.W. and R.B. Gennis, (1985), Spectroscopic Studies of Pyruvate Oxidase Flavoprotein from Escherichia coli Trapped in the Lipid-Activated Form by Crosslinking. J. Biol. Chem. 260, 10395-10397.
937. Matlin, K.S. (1986), The Sorting of Proteins to the Plasma Membrane in Epithelial Cells. J. Cell Biol. 103, 2565-2568.
938. Matsushita, K., L. Patel, R.B. Gennis, and H.R. Kaback, (1983), Reconstitution of Active Transport in Proteoliposomes Containing Cytochrome c Oxidase and lac Carrier Protein Purified from Escherichia coli. Proc. Natl. Acad. Sci. USA 80, 4889-4893.
939. Matteoni, R. and T.E. Kreis, (1987), Translocation and Clustering of Endosomes and Lysosomes Depends on Microtubules. J. Cell Biol. 105, 1253-1265.
940. Mattoo, A.K. and M. Edelman, (1987), Intramembrane Translocation and Posttranslational Palmitoylation of the Chloroplast 32-kDa Herbicide-binding Protein. Proc. Natl. Acad. Sci. USA 84, 1497-1501.
941. Maurer, A., J.O. McIntyre, S. Churchill, and S. Fleischer, (1985), Phospholipid Protection against Proteolysis of D-ß-Hydroxybutyrate Dehydrogenase, a Lecithin-requiring Enzyme. J. Biol. Chem. 260, 1661-1669.
942. May, D.C., E.M. Ross, A.G. Gilman, and M.D. Smigel, (1985), Reconstitution of Catecholamine-stimulated Adenylate Cyclase Activity Using Three Purified Proteins. J. Biol. Chem. 260, 15829-15833.
943. May, W.S., N. Sahyoun, S. Jacobs, M. Wolf, and P. Cuatrecasas, (1985), Mechanism of Phorbol Diester-induced Regulation of Surface Transferrin Receptor Involves the Action of Activated Protein Kinase C and an Intact Cytoskeleton. J. Biol. Chem. 260, 9419-9426.
944. Mayer, L.D., M.B. Bally, M.J. Hope, and P.R. Cullis, (1985), Uptake of Dibucaine into Large Unilamellar Vesicles in Response to a Membrane Potential. J. Biol. Chem. 260, 802-808.
945. Mayer, L.D., M.L. Pusey, M.A. Griep, and G.L. Nelsestuen, (1983), Association of Blood Coagulation Factors V and X with Phospholipid Monolayers. Biochemistry 22, 6226-6232.

946. McConnell, H.M., L.K. Tamm, and R.M. Weis, (1984), Periodic Structures in Lipid Monolayer Phase Transitions. Proc. Natl. Acad. Sci. USA 81, 3249-3253.
947. McConnell, H.M., T.H. Watts, R.M. Weis, and A.A. Brian, (1986), Supported Planar Membranes in Studies of Cell-Cell Recognition in the Immune System. Biochim. Biophys. Acta 864, 95-106.
948. McDaniel, R.V., K. Sharp, D. Brooks, A.C. McLaughlin, A.P. Winiski, D. Cafiso, and S. McLaughlin, (1986), Electrokinetic and Electrostatic Properties of Bilayers Containing Gangliosides GM_1, GD_{1a}, or GT_1. Biophys. J. 49, 741-752.
949. McEhlaney, R.N. (1986), Differential Scanning Calorimetric Studies of Lipid-Protein Interactions in Model Membrane Systems. Biochim. Biophys. Acta 864, 361-421.
950. McIntosh, T.J., S.A. Simon, and R.A. MacDonald, (1980), The Organization of n-Alkanes in Lipid Bilayers. Biochim. Biophys. Acta 597, 445-463.
951. McLaughlin, S. (1983), Experimental Tests of the Assumptions Inherent in the Gouy-Chapman-Stern Theory of the Aqueous Diffuse Double Layer. Physical Chemistry of Transmembrane Ion Motions (G. Spach, Ed.), pp. 69-76. Elsevier, Amsterdam.
952. McLaughlin, S. (1977), Electrostatic Potentials at Membrane-Solution Interfaces. Current Topics in Membranes and Transport 9, 71-144.
953. McLaughlin, S. (1982), Divalent Cations, Electrostatic Potentials, Bilayer Membranes. Membranes and Transport (A.N. Martonosi, Ed.), Vol. 1, pp. 51-55. Plenum Press, New York.
954. McLaughlin, S., N. Mulrine, T. Gresalfi, G. Vaio, and A. McLaughlin, (1981), Adsorption of Divalent Cations to Bilayer Membranes Containing Phosphatidylserine. J. Gen. Physiol. 77, 445-473.
955a. McLean, L.R. and M.C. Phillips, (1984), Kinetics of Phosphatidylcholine and Lysophosphatidylcholine Exchange between Unilamellar Vesicles. Biochemistry 23, 4624-4630.
955. McLean, L.R. and R.L. Jackson, (1985), Interaction of Lipoprotein Lipase and Apolipoprotein C-II with Sonicated Vesicles of 1,2-Ditetradecylphosphatidylcholine: Comparison of Binding Constants. Biochemistry 24, 4196-4201.
956. McMillan, P.N. and R.B. Luftig, (1973), Preservation of Erythrocyte Ghost Ultrastructure Achieved by Various Fixatives. Proc. Natl. Acad. Sci. USA 70, 3060-3064.
957. McMillen, D.A., J.J. Volwerk, J. Ohishi, M. Erion, J.F.W. Keana, P.C. Jost, and O. Hayes Griffith, (1986), Identifying Regions of Membrane Proteins in Contact with Phospholipid Head Groups: Covalent Attachment of a New Class of Aldehyde Lipid Labels to Cytochrome c Oxidase. Biochemistry 25, 182-193.
958. McQueen, N., D.P. Nayak, E.B. Stephens, and R.W. Compans, (1986), Polarized Expression of a Chimeric Protein in Which the Transmembrane and Cytoplasmic Domains of the Influenza Virus Hemagglutinin Have Been Replaced by Those of the Vesicular Stomatitis Virus G Protein. Proc. Natl. Acad. Sci. USA 83, 9318-9322.
959. McQueen, N.L., D.P. Nayak, E.B. Stephens, and R.W. Compans, (1987), Basolateral Expression of a Chimeric Protein in Which the Transmembrane and Cytoplasmic Domains of Vesicular Stomatitis Virus G Protein Have Been Replaced by Those of the Influenza Virus Hemagglutinin. J. Biol. Chem. 262, 16233-16240.
959a. Medoff, G. and G.A. Kobayashi, (1980), The Polyenes. Antifungal Chemotherapy (D.C.E. Speller, Ed.), pp. 3-33. John Wiley and Sons, New York.

960. Meier, P., J.-H. Sachse, P.J. Brophy, D. Marsh, and G. Kothe, (1987), Integral Membrane Proteins Significantly Decrease the Molecular Motion in Lipid Bilayers: A Deuteron NMR Relaxation Study of Membranes Containing Myelin Proteolipid Apoprotein. Proc. Natl. Acad. Sci. USA 84, 3704-3708.
961. Meiri, H., G. Spira, M. Sammar, M. Namir, A. Schwartz, A. Komoriya, E.M. Kosower, and Y. Palti, (1987), Mapping a Region Associated with Na Channel Inactivation Using Antibodies to a Synthetic Peptide Corresponding to a Part of the Channel. Proc. Natl. Acad. Sci. USA 84, 5058-5062.
962. Meister, H., R. Bachofen, G. Semenza, and J. Brunner, (1985), Membrane Topology of Light-harvesting Protein B870-alpha of Rhodopsirillum rubrum G-9. J. Biol. Chem. 260, 16326-16331.
963. Melchers, F. and J. Andersson, (1984), B Cell Activation: Three Steps and Their Variations. Cell 37, 715-720.
964. Melchior, D.L. (1986), Lipid Domains in Fluid Membranes: A Quick-Freeze Differential Scanning Calorimetry Study. Science 234, 1577-1580.
965. Melis, A., P. Svensson, and P.-A. Albertsson, (1986), The Domain Organization of the Chloroplast Thylakoid Membrane. Localization of Photosystem I and of the Cytochrome b_6-f Complex. Biochim. Biophys. Acta 850, 402-412.
966. Mellman, I., R. Fuchs, and A. Helenius, (1986), Acidification of the Endocytic and Exocytic Pathways. Ann. Rev. Biochem. 55, 663-700.
967. Menashe, M., G. Romero, R.L. Biltonen, and D. Lichtenberg, (1986), Hydrolysis of Dipalmitoylphosphatidylcholine Small Unilamellar Vesicles by Porcine Pancreatic Phospholipase A2. J. Biol. Chem. 261, 5328-5333.
968. Mendelsohn, R., R.A. Dluhy, T. Crawford, and H.H. Mantsch, (1984), Interaction of Glycophorin with Phosphatidylserine: A Fourier Transform Infrared Investigation. Biochemistry 23, 1498-1504.
969. Menestrina, G., K.-P. Voges, G. Jung, and G. Boheim, (1986), Voltage-Dependent Channel Formation by Rods of Helical Polypeptides. J. Membrane Biol. 93, 111-132.
970. Menick, D.R., N. Carrasco, L. Antes, L. Patel, and H.R. Kaback, (1987), lac Permease of Escherichia coli: Arginine-302 as a Component of the Postulated Proton Relay. Biochemistry 26, 6638-6644.
971. Menko, A.S. and D. Boettiger, (1987), Occupation of the Extracellular Matrix Receptor, Integrin, Is a Control Point for Myogenic Differentiation. Cell 51, 51-57.
972. Menon, A.K., D. Holowka, W.W. Webb, and B. Baird, (1986), Clustering, Mobility, and Triggering Activity of Small Oligomers of Immunoglobulin E on Rat Basophilic Leukemia Cells. J. Cell Biol. 102, 534-540.
973. Menon, A.K., D. Holowka, W.W. Webb, and B. Baird, (1986), Cross-linking of Receptor-bound IgE to Aggregates Larger than Dimers Leads to Rapid Immobilization. J. Cell Biol. 102, 541-550.
974. Mertens, K. and R.M. Bertina, (1984), The Contribution of Ca^{2+} and Phospholipids to the Activation of Human Blood-coagulation Factor X by Activated Factor IX. Biochem. J. 223, 607-615.
975. Messner, D.J., D.J. Feller, T. Scheuer, and W.A. Catterall, (1986), Functional Properties of Rat Brain Sodium Channels Lacking the $ß_1$ or $ß_2$ Subunit. J. Biol. Chem. 261, 14882-14890.
976. Metcalf, T.N., III, J.L. Wang, and M. Schindler, (1986), Lateral Diffusion of Phospholipids in the Plasma Membrane of Soybean Protoplasts: Evidence for Membrane Lipid Domains. Proc. Natl. Acad. Sci. USA 83, 95-99.

977. Metsikko, K., G. van Meer, and K. Simons, (1986), Reconstitution of the Fusogenic Activity of Vesicular Stomatitis Virus. The EMBO Journal 5, 3429-3435.
978. Michel, H. (1982), Characterization and Crystal Packing of Three-dimensional Bacteriorhodopsin Crystals. The EMBO Journal 1, 1267-1271.
979. Michel, H. (1983), Crystallization of Membrane Proteins. TIBS 8, 56-59.
980. Michel, H. (1982), Three-Dimensional Crystals of a Membrane Protein Complex: The Photosynthetic Reaction Centre from Rhodopseudomonas viridis. J. Mol. Biol. 158, 567-572.
981. Michel, H., D. Oesterhelt, and R. Henderson, (1980), Orthorhombic Two-dimensional Crystal Form of Purple Membrane. Proc. Natl. Acad. Sci. USA 77, 338-342.
982. Michel, H., K.A. Weyer, H. Gruenberg, I. Dunger, D. Oesterhelt, and F. Lottspeich, (1986), The 'Light' and 'Medium' Subunits of the Photosynthetic Reaction Centre from Rhodopseudomonas viridis: Isolation of the Genes, Nucleotide and Amino Acid Sequence. The EMBO Journal 5, 1149-1158.
983. Michel, H., O. Epp, and J. Deisenhofer, (1986), Pigment-Protein Interactions in the Photosynthetic Reaction Centre from Rhodopseudomonas viridis. The EMBO Journal 5, 2445-2451.
984. Michell, B. and M. Houslay, (1986), Pleiotypic Responses: Regulation by Programmable Messengers or by Multiple Receptors? TIBS 11, 239-241.
985. Middelkoop, E., B.H. Lubin, J.A.F. Op den Kamp, and B. Roelofsen, (1986), Flip-Flop Rates of Individual Molecular Species of Phosphatidylcholine in the Human Red Cell Membrane. Biochim. Biophys. Acta 855, 421-424.
986. Middlekoop, E., B.H. Lubin, J.A.F. Op den Kamp, and B. Roelofsen, (1986), Flip-Flop Rates of Individual Molecular Species of Phosphatidylcholine in the Human Red Cell Membrane. Biochim. Biophys. Acta 855, 421-424.
987. Miller, C. and M.M. White, (1980), A Voltage-dependent Chloride Conductance Channel from Torpedo Electroplax Membrane. Ann. N.Y. Acad. Sci. 341, 534-551.
988. Miller, C. and M.M. White, (1984), Dimeric Structure of Single Chloride Channels from Torpedo Electroplax. Proc. Natl. Acad. Sci. USA 81, 2772-2775.
989. Miller, K.R. and M.K. Lyon, (1985), Do We Really Know Why Chloroplast Membranes Stack? TIBS 10, 219-222.
990. Miller, M.J. and R.B. Gennis, (1983), The Purification and Characterization of the Cytochrome d Terminal Oxidase Complex of the E. coli Aerobic Respiratory Chain. J. Biol. Chem. 258, 9159-9165.
991. Miller, R.G. (1984), Interactions Between Digitonin and Bilayer Membranes. Biochim. Biophys. Acta 774, 151-157.
992. Miller, R.J. (1987), Multiple Calcium Channels and Neuronal Function. Science 235, 46-52.
993. Mimms, L.T., G. Zampighi, Y. Nozaki, C. Tanford, and J.A. Reynolds, (1981), Phospholipid Vesicle Formation and Transmembrane Protein Incorporation Using Octyl Glucoside. Biochemistry 20, 833-840.
994. Minami, Y., L.E. Samelson, A.M. Weissman, and R.D. Klausner, (1987), Building a Multichain Receptor: Synthesis, Degradation, and Assembly of the T-cell Antigen Receptor. Proc. Natl. Acad. Sci. USA 84, 2688-2692.
995. Mischel, M., J. Seelig, L.F. Braganza, and G. Buldt, (1987), A Neutron Diffraction Study of the Headgroup Conformation of Phosphatidylglycerol from Escherichia coli Membranes. Chem. Phys. Lipids 43, 237-246.
996. Mishina, M., T. Kurosaki, T. Tobimatsu, Y. Morimoto, M. Noda, T. Yamamoto, M. Terao, J. Lindstrom, T. Takahashi, M. Kuno, and S. Numa, (1984), Expres-

sion of Functional Acetylcholine Receptor from Cloned cDNAs. Nature 307, 604-608.

997. Mishina, M., T. Tobimatsu, K. Imoto, K. Tanaka, Y. Fujita, K. Fukuda, M. Kurasaki, H. Takahashi, Y. Morimoto, T. Hirose, S. Inayama, T. Takahashi, M. Kuno, and S. Numa, (1985), Location of Functional Regions of Acetylcholine Receptor α-Subunit by Site-directed Mutagenesis. Nature 313, 364-369.
998. Mitaku, S., S. Hoshi, T. Abe, and R. Kataoka, (1984), Spectral Analysis of Amino Acid Sequence I: Intrinsic Membrane Proteins. J. Phys. Soc. Japan 53, 4083-4090.
999. Miyazaki, C., M. Kuroda, A. Ohta, and I. Shibuya, (1985), Genetic Manipulation of Membrane Phospholipid Composition in Escherichia coli: pgsA Mutants Defective in Phosphatidylglycerol Synthesis. Proc. Natl. Acad. Sci. USA 82, 7530-7534.
1000. Mize, N.K., D.W. Andrews, and V.R. Lingappa, (1986), A Stop Transfer Sequence Recognizes Receptors for Nascent Chain Translocation across the Endoplasmic Reticulum Membrane. Cell 47, 711-719.
1001. Moe, G.R., R.J. Miller, and E.T. Kaiser, (1983), Design of a Peptide Hormone: Synthesis and Characterization of a Model Peptide with Calcitonin-like Activity. J. Am. Chem. Soc. 105, 4100-4102.
1002. Mogi, T., L.J. Stern, N.R. Hackett, and H.G. Khorana, (1987), Bacteriorhodopsin Mutants Containing Single Tyrosine to Phenylalanine Substitutions Are All Active in Proton Translocation. Proc. Natl. Acad. Sci. USA 84, 5595-5599.
1003. Mohamed, A.H. and T.L. Steck, (1986), Band 3 Tyrosine Kinase: Association with the Human Erythrocyte Membrane. J. Biol. Chem. 261, 2804-2809.
1004. Mohandas, N., M. Rossi, S. Bernstein, S. Ballas, Y. Ravindranath, J. Wyatt, and W. Mentzer, (1985), The Structural Organization of Skeletal Proteins Influences Lipid Translocation Across Erythrocyte Membrane. J. Biol. Chem. 260, 14264-14268.
1005. Mohraz, M., M.V. Simpson, and P.R. Smith, (1987), The Three-dimensional Structure of the Na, K-ATPase from Electron Microscopy. J. Cell Biol. 105, 1-8.
1006. Molitoris, B.A. and F.R. Simon, (1986), Maintenance of Epithelial Membrane Lipid Polarity: A Role for Differing Phospholipid Translocation Rates. J. Membrane Biol. 94, 47-53.
1007. Mancuso, C.A., G. Odham, G. Westerdahl, J.N. Reeve, and D.C. White, (1985), C15, C20, and C25 Isoprenoid Homologues in Glycerol Diether Phospholipids of Methanogenic Archaebacteria. J. Lipid Res. 26, 1120-1125.
1008. Mondat, M., A. Georgallas, D.A. Pink, and M.J. Zuckermann, (1984), The Thermodynamic Properties of Mixed Phospholipid Bilayers: A Theoretical Analysis. Can. J. Biochem. Cell Biol. 62, 796-802.
1009. Montal, M. (1974), Formation of Bimolecular Membranes from Lipid Monolayers. Methods in Enz. 32, 545-556.
1010. Montal, M. (1987), Reconstitution of Channel Proteins from Excitable Cells in Planar Lipid Bilayer Membranes. J. Membrane Biol. 98, 101-115.
1011. Montal, M., A. Darzon, and H. Schindler, (1981), Functional Reassembly of Membrane Proteins in Planar Lipid Membranes. Quart. Rev. Biophys. 14, 1-79.
1012. Montal, M., R. Anholt, and P. Labarca, (1986), The Reconstituted Acetylcholine Receptor. Ion Channel Reconstitution (C. Miller, Ed.), pp. 157-204. Plenum Publishing Corporation, New York.
1013. Moore, K.E. and S. Miura, (1987), A Small Hydrophobic Domain Anchors Leader Peptidase to the Cytoplasmic Membrane of Escherichia coli. J. Biol. Chem. 262, 8806-8813.

1014. Moore, M.S., D.T. Mahaffey, F.M. Brodsky, and R.G.W. Anderson, (1987), Assembly of Clathrin-Coated Pits onto Purified Plasma Membranes. Science 236, 558-563.
1015. Mooseker, M.S. (1983), Actin Binding Proteins of the Brush Border. Cell 35, 11-13.
1016. Morgan, C.G., J.E. Fitton, and Y.P. Yianni, (1986), Fusogenic Activity of δ-Haemolysin from Staphylococcus aureus in Phospholipid Vesicles in the Liquid-Crystalline Phase. Biochim. Biophys. Acta 863, 129-138.
1017. Morgan, D.O., J.C. Edman, D.N. Standring, V.A. Fried, M.C. Smith, R.A. Roth, and W.J. Rutter, (1987), Insulin-like Growth Factor II Receptor as a Multifunctional Binding Protein. Nature 329, 301-307.
1018. Morita, T., B.S. Issacs, C.T. Esmon, and A.E. Johnson, (1984), Derivatives of Blood Coagulation Factor IX Contain a High Affinity Ca^{2+}-binding Site That Lacks γ-Carboxyglutamic Acid. J. Biol. Chem. 259, 5698-5704.
1019. Morona, R., C. Kramer, and U. Henning, (1985), Bacteriophage Receptor Area of Outer Membrane Protein OmpA of Escherichia coli K-12. J. Bacteriology 164, 539-543.
1020. Morris, S.J. and I. Schovanka, (1977), Some Physical Properties of Adrenal Medulla Chromaffin Granules Isolated by a New Continuous Iso-osmotic Density Gradient Method. Biochim. Biophys. Acta 464, 53-64.
1021. Morrow, M.R., J.C. Huschilt, and J.H. Davis, (1985), Simultaneous Modeling of Phase and Calorimetric Behavior in an Amphiphilic Peptide/Phospholipid Model Membrane. Biochemistry 24, 5396-5406.
1022. Mosckovitz, R., R. Haring, J.M. Gershoni, Y. Kloog, and M. Sokolovsky, (1987), Localization of Azidophencyclidine-binding Site on the Nicotinic Acetylcholine Receptor α-Subunit. Biochemical and Biophysical Research Communications 145, 810-816.
1023. Mostov, K.E., A. de Bruyn Kops, and D.L. Deitcher, (1986), Deletion of the Cytoplasmic Domain of the Polymeric Immunoglobulin Receptor Prevents Baso-lateral Localization and Endocytosis. Cell 47, 359-364.
1024. Mostov, K.E. and N.E. Simister, (1985), Transcytosis. Cell 43, 389-390.
1025. Mouritsen, O.G. and M. Bloom, (1984), Mattress Model of Lipid-Protein Interactions in Membranes. Biophys. J. 46, 141-153.
1026. Mowbray, S.L. and D.E. Koshland, (1987), Additive and Independent Responses in a Single Receptor: Asparate and Maltose Stimuli on the Tar Protein. Cell 50, 171-180.
1027. Mueckler, M., C. Caruso, S.A. Baldwin, M. Panico, I. Blench, H.R. Morris, W.J. Allard, G.E. Lienhard, and H.F. Lodish, (1985), Sequence and Structure of a Human Glucose Transporter. Science 229, 941-945.
1028. Mueckler, M. and H.F. Lodish, (1986), The Human Glucose Transporter Can Insert Posttranslationally into Microsomes. Cell 44, 629-637.
1029. Mueckler, M. and H.F. Lodish, (1986), Post-Translational Insertion of a Fragment of the Glucose Transporter into Microsomes Requires Phosphoanhydride Bond Cleavage. Nature 322, 549-552.
1030. Mueller, P. and P.O. Rudin, (1968), Action Potentials Induced in Biomolecular Lipid Membranes. Nature 217, 713-719.
1031. Mueller, P., T.F. Chien, and B. Rudy, (1983), Formation and Properties of Cell-Size Lipid Bilayer Vesicles. Biophys. J. 44, 375-381.
1032. Muhlebach, T. and R.J. Cherry, (1985), Rotational Diffusion and Self-Association of Band 3 in Reconstituted Lipid Vesicles. Biochemistry 24, 975-983.
1033. Muller, G. and R. Zimmermann, (1987), Import of Honeybee Prepromelittin into

the Endoplasmic Reticulum: Structural Basis for Independence of SRP and Docking Protein. The EMBO Journal 6, 2099-2107.

1034. Muller, M. and G. Blobel, (1984), Protein Export in Escherichia coli Requires a Soluble Activity. Proc. Natl. Acad. Sci. USA 81, 7737-7741.

1035. Muller, M., R. Moser, D. Cheneval, and E. Carafoli, (1985), Cardiolipin Is the Membrane Receptor for Mitochondrial Creatine Phosphokinase. J. Biol. Chem. 260, 3839-3843.

1036. Muller, R. (1986), Proto-Oncogenes and Differentiation. TIBS 11, 129-132.

1037. Muller, W.E.G., M. Rottmann, B. Diehl-Seifert, B. de Kruijff, G. Uhlenbruck, and H.C. Schroder, (1987), Role of Aggregation Factor in the Regulation of Phosphoinositide Metabolism in Sponges. J. Biol. Chem. 262, 9805-9858.

1038. Munro, S. and H.R.B. Pelham, (1987), A C-Terminal Signal Prevents Secretion of Luminal ER Proteins. Cell 48, 899-907.

1039. Munthe-Kaas, A.C. and P.O. Seglen, (1974), The Use of Metrizamide as a Gradient Medium for Isopycnic Separation of Rat Liver Cells. FEBS Lett. 43, 252-256.

1040. Murata, M., K. Nagayama, and S. Ohnishi, (1987), Membrane Fusion Activity of Succinylated Melittin Is Triggered by Protonation of Its Carboxyl Groups. Biochemistry 26, 4056-4062.

1041. Murphy, D.J. (1982), The Importance of Non-planar Bilayer Regions in Photosynthetic Membranes and Their Stabilization by Galactolipids. FEBS Lett. 150, 19-26.

1042. Murphy, D.J. (1986), The Molecular Organization of the Photosynthetic Membranes of Higher Plants. Biochim. Biophys. Acta 864, 33-94.

1043. Murray, J.M. (1983), Three-Dimensional Structure of a Membrane-Microtubule Complex. J. Cell Biol. 98, 283-295.

1044. Mutsch, B., N. Gains, and H. Hauser, (1986), Interaction of Intestinal Brush Border Membrane Vesicles with Small Unilamellar Phospholipid Vesicles. Exchange of Lipids between Membranes Is Mediated by Collisional Contact. Biochemistry 25, 2134-2140.

1045. Myers, A.C., J.S. Kovach, and S. Vuk-Pavlovic, (1987), Binding, Internalization, and Intracellular Processing of Protein Ligands: Derivation of Rate Constants by Computer Modeling. J. Biol. Chem. 262, 6494-6499.

1046. Myers, M., O.L. Mayorga, J. Emtage, and E. Freire, (1987), Thermodynamic Characterization of Interactions between Ornithine Transcarbamylase Leader Peptide and Phospholipid Bilayer Membranes. Biochemistry 26, 4309-4315.

1047. Nabedryk, E., A.M. Bardin, and J. Breton, (1985), Further Characterization of Protein Secondary Structures in Purple Membrane by Circular Dichroism and Polarized Infrared Spectroscopies. Biophys. J. 48, 873-876.

1048. Nagle, J.F. (1980), Theory of the Main Lipid Bilayer Phase Transition. Annu. Rev. Phys. Chem. 31, 157-195.

1049. Nagle, J.F. (1987), Theory of Passive Proton Conductance in Lipid Bilayers. J. Bioenerg. Biomemb. 19, 413-426.

1050. Nagle, J.F. and H.J. Morowitz, (1978), Molecular Mechanisms for Proton Transport in Membranes. Proc. Natl. Acad. Sci. USA 75, 298-302.

1051. Naidet, C., M. Semeriva, K.M. Yamada, and J.P. Thiery, (1987), Peptides Containing the Cell-Attachment Recognition Signal Arg-Gly-Asp Prevent Gastrulation in Drosophila Embryos. Nature 325, 348-350.

1052. Nakache, M., A.B. Schreiber, H. Gaub, and H.M. McConnell, (1985), Heterogeneity of Membrane Phospholipid Mobility in Endothelial Cells Depends on Cell Substrate. Nature 317, 75-77.

1053. Nakae, T. (1986), Outer-Membrane Permeability of Bacteria. CRC Critical Reviews in Microbiology 13, 1-62.

1054. Nakamura, T. and M. Ui, (1985), Simultaneous Inhibitions of Inositol Phospholipid Breakdown, Arachidonic Acid Release, and Histamine Secretion in Mast Cells by Islet-activating Protein, Pertussis Toxin. J. Biol. Chem. 260, 3584-3593.

1055. Nakashima, R.A., P.S. Mangan, M. Colombini, and P.L. Pederson, (1986), Hexokinase Receptor Complex in Hepatoma Mitochondria: Evidence from N,N'-Dicyclohexylcarbodiimide-Labeling Studies for the Involvement of the Pore-Forming Protein VDAC. Biochemistry 25, 1015-1021.

1056. Nalecz, M.J., J. Zborowski, K.S. Famulski, and L. Wojtczak, (1980), Effect of Phospholipid Composition on the Surface Potential of Liposomes and the Activity of Enzymes Incorporated into Liposomes. Eur. J. Bioch. 112, 75-80.

1057. Napier, R.M., J.M. East, and A.G. Lee, (1987), State of Aggregation of the (Ca^{2+} + Mg^{2+})-ATPase Studied Using Saturation-Transfer Electron Spin Resonance. Biochim. Biophys. Acta 903, 365-373.

1057a. Napier, R.M., J.M. East, and A.G. Lee, (1987), State of Aggregation of the (Ca^{2+} + Mg^{2+})-ATPase Studied Using Chemical Cross-linking. Biochim. Biophys. Acta 903, 374-380.

1057b. Nelander, J.C. and A.E. Blaurock, (1978), Disorder in Nerve Myelin: Phasing the Higher Order Reflections by Means of the Diffuse Scatter. J. Mol. Biol. 118, 497-532.

1058. Nelson, W.J. and P.J. Veshnock, (1987), Ankyrin Binding to (Na^+ + K^+)ATPase and Implications for the Organization of Membrane Domains in Polarized Cells. Nature 328, 533-536.

1059. Nemes, P.P., G.P. Miljanich, D.L. White, and E.A. Dratz, (1980), Covalent Modification of Rhodopsin with Imidoesters: Evidence for Transmembrane Arrangement of Rhodopsin in Rod Outer Segment Disk Membranes. Biochemistry 19, 2067-2074.

1060. Neumann, D., J.M. Gershoni, M. Fridkin, and S. Fuchs, (1985), Antibodies to Synthetic Peptides as Probes for the Binding Site on the α Subunit of the Acetylcholine Receptor. Proc. Natl. Acad. Sci. USA 82, 3490-3493.

1061. Neville, Jr, D.M., and T.H. Hudson, (1986), Transmembrane Transport of Diphtheria Toxin, Related Toxins, and Colicins. Ann. Rev. Biochem. 55, 195-224.

1062. Newman, M.J., D.L. Foster, T.H. Wilson, and H.R. Kaback, (1981), Purification and Reconstitution of Functional Lactose Carrier from Escherichia coli. J. Biol. Chem. 256, 11804-11808.

1063. Newton, A.C. and D.E. Koshland, (1987), Protein Kinase C Autophosphorylates by an Intrapeptide Reaction. J. Biol. Chem. 262, 10185-10188.

1064. Nguyen, M. and G.C. Shore, (1987), Import of Hybrid Vesicular Stomatitis G Protein to the Mitochondrial Inner Membrane. J. Biol. Chem. 262, 3929-3931.

1065. Nicholls, D.G. (1982), Bioenergetics. Academic Press, New York.

1066. Nichols, J.W. (1987), Binding of Fluorescent-labeled Phosphatidylcholine to Rat Liver Nonspecific Lipid Transfer Protein. J. Biol. Chem. 262, 14172-14177.

1067. Nicholson, D.W., H. Kohler, and W. Neupert, (1987), Import of Cytochrome c into Mitochondria: Cytochrome c Heme Lyase. Eur. J. Biochem. 164, 147-157.

1068. Nikaido, H. and M. Vaara, (1985), Molecular Basis of Bacterial Outer Membrane Permeability. Microbiol. Rev. 49, 1-32.

1069. Nikawa, J., Y. Tsukagoshi, T. Kodaki, and S. Yamashita, (1987), Nucleotide Sequence and Characterization of the Yeast PSS Gene Encoding Phosphatidylserine Synthase. Eur. J. Biochem. 167, 7-12.

1070. Nishikawa, M., S. Nojima, T. Akiyama, U. Sankawa, and K. Inoue, (1984), Interaction of Digitonin and Its Analogs with Membrane Cholesterol. J. Biochem. 96, 1231-1239.

1071. Nishizuka, Y. (1984), The Role of Protein Kinase C in Cell Surface Signal Transduction and Tumour Promotion. Nature 308, 693-398.

1072. Nisimoto, Y. and D. Lambeth, (1985), NADPH-Cytochrome P-450 Reductase-Cytochrome b_5 Interactions: Crosslinking of the Phospholipid Vesicle-Associated Proteins by a Water-Soluble Carbodiimide. Arch. Biochem. Biophys. 241, 386-396.

1073. Noda, M., K. Yoon, G.A. Rodan, and D.E. Koppel, (1987), High Lateral Mobility of Endogenous and Transfected Alkaline Phosphatase: A Phosphatidylinositol-anchored Membrane Protein. J. Cell Biol. 105, 1671-1677.

1074. Noda, M., S. Shimizu, T. Tanabe, T. Takai, T. Kayano, T. Ikeda, H. Takahashi, H. Nakayama, Y. Kanaoka, N. Minamino, K. Kangawa, H. Matsuo, M.A. Raftery, T. Hirose, S. Inayama, H. Hayashida, T. Miyata, and S. Numa, (1984), Primary Structure of Electrophorus electricus Sodium Channel Deduced from cDNA Sequence. Nature 312, 121-127.

1075. Noda, M., T. Ikeda, H. Suzuki, H. Takeshima, T. Takahashi, M. Kuno, and S. Numa, (1986), Expression of Functional Sodium Channels from Cloned cDNA. Nature 322, 826-828.

1076. Noda, M., T. Ikeda, T. Kayano, H. Suzuki, H. Takeshima, M. Kurasaki, H. Takahashi, and S. Numa, (1986), Existence of Distinct Sodium Channel Messenger RNAs in Rat Brain. Nature 320, 188-192.

1077. Noguchi, S., M. Noda, H. Takahashi, K. Kawakami, T. Ohta, K. Nagano, T. Hirose, S. Inayama, M. Kawamura, and S. Numa, (1986), Primary Structure of the ß-Subunit of Torpedo californica (Na^+ + K^+)-ATPase Deduced from the cDNA Sequence. FEBS Lett. 196, 315-320.

1078. Nojiri, H., F. Takaku, Y. Terui, Y. Miura, and M. Saito, (1986), Ganglioside GM_3: An Acidic Membrane Component That Increases During Macrophage-Like Cell Differentiation Can Induce Monocytic Differentiation of Human Myeloid and Monocytoid Leukemic Cell Lines HL-60 and U937. Proc. Natl. Acad. Sci. USA 83, 782-786.

1079. Noordam, P.C., A. Killian, R.F.M. Oude Elferink, and J. de Gier, (1982), Comparative Study on the Properties of Saturated Phosphatidylethanolamine and Phosphatidylcholine Bilayers: Barrier Characteristics and Susceptibility to Phospholipase A2 Degradation. Chem. Phys. Lipids 31, 191-204.

1080. Noren, O., H. Sjostrom, G.M. Cowell, J. Tranum-Jensen, O.C. Hansen, and K.G. Welinder, (1986), Pig Intestinal Microvillar Maltase-glucoamylase: Structure and Membrane Insertion. J. Biol. Chem. 261, 12306-12309.

1081. Nose, A., A. Nagafuchi, and M. Takeichi, (1987), Isolation of Placental Cadherin cDNA: Identification of a Novel Gene Family of Cell-Cell Adhesion Molecules. The EMBO Journal 6, 3655-3662.

1082. Nozaki, Y., J.A. Reynolds, and C. Tanford, (1978), Conformational States of a Hydrophobic Protein: The Coat Protein of fd Bacteriophage. Biochemistry 17, 1239-1246.

1083. O'Brien, T.A., R. Blake, and R.B. Gennis, (1977), Regulation by Lipids of Cofactor Binding to a Peripheral Membrane Enzyme: Binding of Thiamin Pyrophosphate to Pyruvate Oxidase. Biochemistry 16, 3105-3109.

1084. O'Leary, T.J. and I.W. Levin, (1984), Raman Spectroscopic Study of an Interdigitated Lipid Bilayer Dipalmitoylphosphatidylcholine Dispersed in Glycerol. Biochim. Biophys. Acta 776, 185-189.

1085. O'Neill, L.J., J.G. Miller, and N.O. Peterson, (1986), Evidence for Nystatin Micelles in L-Cell Membranes from Fluorescence Photobleaching Measurements of Diffusion. Biochemistry 25, 177-181.
1086. O'Shea, P.S., G. Petrone, R.P. Casey, and A. Azzi, (1984), The Current-Voltage Relationships of Liposomes and Mitochondria. Biochem. J. 219, 719-726.
1087. O'Shea, P.S., S. Feuerstein-Thelen, and A. Azzi, (1984), Membrane-potential-dependent Changes of the Lipid Microviscosity of Mitochondria and Phospholipid Vesicles. Biochem. J. 220, 795-801.
1088. Ohba, M. and G. Schatz, (1987), Disruption of the Outer Membrane Restores Protein Import to Trypsin-treated Yeast Mitochondria. The EMBO Journal 6, 2117-2122.
1089. Ohba, M. and G. Schatz, (1987), Protein Import into Yeast Mitochondria Is Inhibited by Antibodies Raised against 45-kd Proteins of the Outer Membrane. The EMBO Journal 6, 2109-2115.
1090. Ohno-Iwashita, Y., P. Wolfe, K. Ito, and W. Wickner, (1984), Processing of Preproteins by Liposomes Bearing Leader Peptidase. Biochemistry 23, 6178-6184.
1091. Ohta, A., T. Obara, Y. Asami, and I. Shibuya, (1985), Molecular Cloning of the cls Gene Responsible for Cardiolipin Synthesis in Escherichia coli and Phenotypic Consequences of Its Amplification. J. Bacteriology 163, 506-514.
1092. Oldfield, E. and D. Chapman, (1972), Dynamics of Lipids in Membranes: Heterogeneity and the Role of Cholesterol. FEBS Lett. 23, 285-297.
1093. Oldfield, E., J.L. Bowers, and J. Forbes, (1987), High-Resolution Proton and Carbon-13 NMR of Membranes: Why Sonicate? Biochemistry 26, 6919-6923.
1094. Olson, E.N., L. Glaser, and J.P. Merlie, (1984), α and ß Subunits of the Nicotinic Acetylcholine Receptor Contain Covalently Bound Lipid. J. Biol. Chem. 259, 5364-5367.
1095. Oosawa, K. and M. Simon, (1986), Analysis of Mutations in the Transmembrane Region of the Asparate Chemoreceptor in Escherichia coli. Proc. Natl. Acad. Sci. USA 83, 6930-6934.
1096. Op den Kamp, J.A.F. (1981), The Asymmetric Architecture of Membranes. Membrane Structure (J.B. Finean and R.H. Michell, Eds.), pp. 83-126. Elsevier, New York.
1097. Op den Kamp, J.A.F. (1979), Lipid Asymmetry in Membranes. Ann. Rev. Biochem. 48, 47-71.
1098. Op den Kamp, J.A.F., B. Roelofsen, and L.L.M. Van Deenen, (1985), Structural and Dynamic Aspects of Phosphatidylcholine in the Human Erythrocyte Membrane. TIBS 10, 320-323.
1099. Orci, L., B.S. Glick, and J.E. Rothman, (1986), A New Type of Coated Vesicular Carrier That Appears Not to Contain Clathrin: Its Possible Role in Protein Transport within the Golgi Stack. Cell 46, 171-184.
1100. Orci, L., M. Ravazzola, and R.G.W. Anderson, (1987), The Condensing Vacuole of Exocrine Cells Is More Acidic Than the Mature Secretory Vesicle. Nature 326, 77-79.
1101. Osborn, M. and K. Weber, (1986), Intermediate Filament Proteins: A Multigene Family Distinguishing Major Cell Lineages. TIBS 11, 469-472.
1102. Osborn, M.J. and R. Munson, (1974), Separation of the Inner (Cytoplasmic) and Outer Membranes of Gram-Negative Bacteria. Methods in Enz. 31, 642-653.
1103. Osterman, D.G. and E.T. Kaiser, (1985), Design and Characterization of Peptides with Amphiphilic beta-Strand Structures. J. Cellular Biochem. 29, 57-72.

1104. Ott, P. (1985), Membrane Acetylcholinesterases: Purification, Molecular Properties and Interactions with Amphiphilic Environments. Biochim. Biophys. Acta 822, 375-392.

1105. Ottolenghi, P. (1979), The Relipidation of Delipidated Na,K-ATPase: An Analysis of Complex Formation with Dioleoylphosphatidylcholine and with Dioleoylphosphatidylethanolamine. Eur. J. Biochem. 99, 113-131.

1106. Ovchinnikov, Y.A., N.G. Abdulaev, and A.V. Kiselev, (1984), Bacteriorhodopsin Topography in Purple Membrane. Biological Membranes (D. Chapman, Ed.), Vol. 5, pp. 197-220. Academic Press, New York.

1107. Ovchinnikov, Y.A., N.G. Abdulaev, M.Y. Feigina, A.V. Kiselev, and N.A. Lobanov, (1979), The Structural Basis of the Functioning of Bacteriorhodopsin: An Overview. FEBS Lett. 100, 219-224.

1108. Ovchinnikov, Y.A., N.G. Abdulaev, R.G. Vasilov, I.Y. Vturina, A.B. Kuryatov, and A.V. Kiselev, (1985), The Antigenic Structure and Topography of Bacteriorhodopsin in Purple Membranes as Determined by Interaction with Monoclonal Antibodies. FEBS Lett. 179, 343-350.

1109. Ovchinnikov, Y.A., V.V. Demin, A.N. Barnakov, A.P. Kuzin, A.V. Lunev, N.N. Modyanov, and K.N. Dzhandzhugazyan, (1985), Three-dimensional Structure of $(Na^+ + K^+)$-ATPase Revealed by Electron Microscopy of Two-dimensional Crystals. FEBS Lett. 190, 73-76.

1110. Ovchinnikov, Y.A. (1987), Probing the Folding of Membrane Proteins. TIBS 12, 434-438.

1111. Ovchinnikov, Y.A., N.M. Arzamazova, E.A. Arystarkhova, N.M. Gevondyan, N.A. Aldanova, and N.N. Modyanov, (1987), Detailed Structural Analysis of Exposed Domains of Membrane-bound Na^+, K^+-ATPase. FEBS Lett. 217, 269-274.

1112. Overton, E. (1895), Über die osmotischen Eigenschaften der lebenden Pflanzen und Tierzelle. Vjsch. Naturf. Ges. Zurich 40, 159-201.

1113. Ozols, J. (1972), Cytochrome b_5 from a Normal Human Liver. J. Biol. Chem. 247, 2242-2244.

1114. Ozols, J., F.S. Heinemann, and E.F. Johnson, (1985), The Complete Amino Acid Sequence of a Constitutive Form of Liver Microsomal Cytochrome P-450. J. Biol. Chem. 260, 5427-5434.

1115. Paabo, S., B.M. Bhat, W.S.M. Wold, and P.A. Peterson, (1987), A Short Sequence in the COOH-Terminus Makes an Adenovirus Membrane Glycoprotein a Resident of the Endoplasmic Reticulum. Cell 50, 311-317.

1116. Page, M.G.P. and J.P. Rosenbusch, (1986), Topographic Labelling of Pore-forming Proteins from the Outer Membrane of Escherichia coli. Biochem. J. 235, 651-661.

1117. Pain, D. and G. Blobel, (1987), Protein Import into Chloroplasts Requires a Chloroplast ATPase. Proc. Natl. Acad. Sci. USA 84, 3288-3292.

1118. Pajares, M.A., S. Alemany, I. Varela, D. Marin Cao, and J.M. Mato, (1984), Purification and Photo-Affinity Labelling of Lipid Methyltransferase from Rat Liver. Biochem. J. 223, 61-66.

1119. Papahadjopoulos, D., W.J. Vail, K. Jacobson, and G. Poste, (1975), Cochleate Lipid Cylinders: Formation by Fusion of Unilamellar Lipid Vesicles. Biochim. Biophys. Acta 394, 483-491.

1120. Parente, R.A. and B.R. Lentz, (1984), Phase Behavior of Large Unilamellar Vesicles Composed of Synthetic Phospholipids. Biochemistry 23, 2353-2362.

1121. Parker, P.J., L. Coussens, N. Totty, L. Rhee, S. Young, E. Chen, S. Stabel, M.D. Waterfield, and A. Ullrich, (1986), The Complete Primary Structure of Protein Kinase C— the Major Phorbol Ester Receptor. Science 233, 853-859.
1122. Parsonage, D., S. Wilke-Mounts, and A.E. Senior, (1987), Directed Mutagenesis of the ß-Subunit of F_1-ATPase from Escherichia coli. J. Biol. Chem. 262, 8022-8026.
1123. Parsonage, D., T.M. Duncan, S. Wilke-Mounts, F.A.S. Kironde, L. Hatch, and A.E. Senior, (1987), The Defective Proton-ATPase of uncD Mutants of Escherichia coli: Identification by DNA-Sequencing of Residues in the ß-Subunit Which Are Essential for Catalysis or Normal Assembly. J. Biol. Chem. 262, 6301-6307.
1124. Parthasarathy, R. and F. Eisenberg, (1986), The Inositol Phospholipids: A Stereochemical View of Biological Activity. Biochem. J. 235, 313-322.
1125. Pascher, I. and S. Sundell, (1977), Molecular Arrangements in Sphingolipids. The Crystal Structure of Cerebroside. Chem. Phys. Lipids 20, 175-191.
1126. Pascher, I., S. Sundell, K. Harlos, and H. Eibl, (1987), Conformation and Packing Properties of Membrane Lipids: The Crystal Structure of Sodium Dimyristoylphosphatidylglycerol. Biochim. Biophys. Acta 896, 77-88.
1127. Pasternack, G.R., R.A. Anderson, T.L. Leto, and V.T. Marchesi, (1985), Interactions Between Protein 4.1 and Band 3: An Alternative Binding Site for an Element of the Membrane Skeleton. J. Biol. Chem. 260, 3676-3683.
1128. Pates, R.D. and D. Marsh, (1987), Lipid Mobility and Order in Bovine Rod Outer Segment Disk Membranes. A Spin-Label Study of Lipid-Protein Interactions. Biochemistry 26, 29-39.
1129. Paul, C. and J.P. Rosenbusch, (1985), Folding Patterns of Porin and Bacteriorhodopsin. The EMBO Journal 4, 1593-1597.
1130. Pearse, B.M.F. and R.A. Crowther, (1987), Structure and Assembly of Coated Vesicles. Ann. Rev. Biophys. Biophys. Chem. 16, 49-68.
1131. Pearson, L.T., J. Edelman, and S.I. Chan, (1984), Statistical Mechanisms of Lipid Membranes: Protein Correlation Functions and Lipid Ordering. Biophys. J. 45, 863-871.
1132. Pearson, L.T., S.I. Chan, B.A. Lewis, and D.M. Engelman, (1983), Pair Distribution Functions of Bacteriorhodopsin and Rhodopsin in Model Bilayers. Biophys. J. 43, 167-174.
1133. Pearson, R.H. and I. Pascher, (1979), The Molecular Structure of Lecithin Hydrates. Nature 281, 499-501.
1134. Pedersen, P.L. and E. Carafoli, (1987), Ion Motive ATPases. I. Ubiquity, Properties, and Significance to Cell Function. TIBS 12, 146-150.
1135. Peerce, B.E. and E.M. Wright, (1987), Examination of the Na^+-Induced Conformational Change of the Intestinal Brush Border Sodium/Glucose Symporter Using Fluorescent Probes. Biochemistry 26, 4272-4279.
1136. Perara, E., R.E. Rothman, and V.R. Lingappa, (1986), Uncoupling Translocation from Translation: Implications for Transport of Proteins Across Membranes. Science 232, 348-352.
1137. Perara, E. and V.R. Lingappa, (1985), A Former Amino Terminal Signal Sequence Engineered to an Internal Location Directs Translocation of Both Flanking Protein Domains. J. Cell Biol. 101, 2292-2301.
1138. Perides, G., C. Harter, and P. Traub, (1987), Electrostatic and Hydrophobic Interactions of the Intermediate Filament Protein Vimentin and Its Amino Terminus with Lipid Bilayers. J. Biol. Chem. 262, 13742-13749.

1139. Perkins, W.R. and D.S. Cafiso, (1987), Characterization of H^+/OH-Currents in Phospholipid Vesicles. J. Bioenerg. Biomemb. 19, 443-455.
1140. Pertoft, H., T.C. Laurent, R. Seljelid, G. Akerstrom, L. Kagedal, and M. Hirtenstein, (1979), The Use of Density Gradients of Percoll for the Separation of Biological Particles. Separation of Cells and Subcellular Elements (H. Peeters, Ed.), pp. 67-72. Pergamon Press, New York.
1141. Peschke, J., J. Riegler, and H. Mohwald, (1987), Quantitative Analysis of Membrane Distortions Induced by Mismatch of Protein and Lipid Hydrophobic Thickness. Eur. Biophys. J. 14, 385-391.
1142. Pessin, J.E. and M. Glaser, (1980), Budding of Rous Sarcoma Virus and Vesicular Stomatitis Virus from Localized Lipid Regions in the Plasma Membrane of Chicken Embryo Fibroblasts. J. Biol. Chem. 255, 9044-9050.
1143. Peters, J., J. Takemoto, and G. Drews, (1983), Spatial Relationships Between the Photochemical Reaction Center and the Light Harvesting Complexes in the Membrane of Rhodopseudomonas capsulata. Biochemistry 22, 5660-5667.
1144. Peters, K.-R., G.E. Palade, B.G. Schneider, and D.S. Papermaster, 1983), Fine Structure of a Periciliary Ridge Complex of Frog Retinal Cells Revealed by Ultrahigh Resolution Scanning Electron Microscopy. J. Cell Biol. 96, 265-276.
1145. Peters, R. and K. Beck, (1983), Translational Diffusion in Phospholipid Monolayers Measured by Fluorescence Microphotolysis. Proc. Natl. Acad. Sci. USA 80, 7183-7187.
1146. Peterson, A.A. and W.A. Cramer, (1987), Voltage-Dependent, Monomeric Channel Activity of Colicin E1 in Artificial Membrane Vesicles. J. Membrane Biol. 99, 197-204.
1147. Pfaller, R. and W. Neupert, (1987), High-Affinity Binding Sites Involved in the Import of Porin into Mitochondria. The EMBO Journal 6, 2635-2642.
1148. Pfanner, N., H.K. Muller, M.A. Harmey, and W. Neupert, (1987), Mitochondrial Protein Import: Involvement of the Mature Part of a Cleavable Precursor Protein in the Binding to Receptor Sites. The EMBO Journal 6, 3449-3454.
1149. Pfanner, N., M. Tropschug, and W. Neupert, (1987), Mitochondrial Protein Import: Nucleoside Triphosphates Are Involved in Conferring Import-Competence to Precursors. Cell 49, 815-823.
1150. Pfanner, N., P. Hoeben, M. Tropschug, and W. Neupert, (1987), The Carboxyl-Terminal Two-Thirds of the ADP/ATP Carrier Polypeptide Contains Sufficient Information to Direct Translocation into Mitochondria. J. Biol. Chem. 262, 14851-14854.
1151. Pfanner, N. and W. Neupert, (1985), Transport of Proteins into Mitochondria: A Potassium Diffusion Potential Is Able to Drive the Import of ADP/ATP Carrier. The EMBO Journal 4, 2819-2825.
1152. Pfanner, N. and W. Neupert, (1987), Distinct Steps in the Import of ADP/ATP Carrier into Mitochondria. J. Biol. Chem. 262, 7528-7536.
1153. Pfeffer, S.R. and J.E. Rothman, (1987), Biosynthetic Protein Transport and Sorting by the Endoplasmic Reticulum and Golgi. Ann. Rev. Biochem. 56, 829-852.
1154. Phillips, M.C., W.J. Johnson, and G.H. Rothblat, (1987), Mechanisms and Consequences of Cellular Cholesterol Exchange and Transfer. Biochim. Biophys. Acta 906, 223-276.
1155. Piela, L., G. Nemethy, and H.A. Scheraga, (1987), Proline-Induced Constraints in α-Helices. Biopolymers 26, 1587-1600.
1156. Pierce, G.N. and K.D. Philipson, (1985), Binding of Glycolytic Enzymes to Cardiac

Sarcolemmal and Sarcoplasmic Reticular Membranes. J. Biol. Chem. 260, 6862-6870.

1157. Pieterson, W.A., J.C. Vidal, J.J. Volwerk, and G.H. de Haas, (1974), Zymogen-Catalyzed Hydrolysis of Monomeric Substrates and the Presence of a Recognition Site for Lipid-Water Interfaces in Phospholipase A2. Biochemistry 13, 1455-1460.

1158. Pike, L.J. and A.T. Eakes, (1987), Epidermal Growth Factor Stimulates the Production of Phosphatidylinositol Monophosphate and the Breakdown of Polyphosphoinositides in A431 Cells. J. Biol. Chem. 262, 1644-1651.

1159. Pilch, P.F. and M.P. Czech, (1979), Interaction of Cross-linking Agents with the Insulin Effector Systems of Isolated Fat Cells. J. Biol. Chem. 254, 3375-3381.

1160. Pink, D.A. (1984), Theoretical Studies of Phospholipid Bilayers and Monolayers. Perturbing Probes, Monolayer Phase Transitions, and Computer Simulation of Lipid-Protein Bilayers. Can. J. Bioch. Cell Biol. 62, 760-777.

1161. Pink, D.A. (1984), Theoretical Models of Monolayers, Bilayers and Biological Membranes. Biomembrane Structure and Function (D. Chapman, Ed.), pp. 319-354. Verlag Chemie, Basel.

1162. Pink, D.A. (1985), Protein Lateral Movement in Lipid Bilayers. Stimulation Studies of Its Dependence upon Protein Concentration. Biochim. Biophys. Acta 818, 200-204.

1163. Pink, D.A. (1982), Theoretical Models of Phase Changes in One- and Two-Component Lipid Bilayers. Biological Membranes (D. Chapman, Ed.), Vol. 4, pp. 131-178. Academic Press, New York.

1164. Pink, D.A., D J. Laidlaw, and D.M. Chisholm, (1986), Protein Lateral Movement in Lipid Bilayers. Monte Carlo Simulation Studies of Its Dependence upon Attractive Protein-Protein Interactions. Biochim. Biophys. Acta 863, 9-17.

1165. Pink, D.A. and A.L. MacDonald, (1986), Theoretical Studies of Phospholipid-Glycophorin Bilayer Membranes Using Electron Spin Resonance and Fluorescent Probes. Biochim. Biophys. Acta 863, 243-252.

1166. Pink, D.A., D. Chapman, D.J. Laidlaw, and T. Wiedmer, (1984), Electron Spin Resonance and Steady State Fluorescence Polarization Studies of Lipid Bilayers Containing Integral Proteins. Biochemistry 23, 4051-4058.

1167. Piomelli, D., A. Volterra, N. Dale, S.A. Siegelbaum, E.R. Kandel, J.H. Schwartz, and F. Belardetti, (1987), Lipoxygenase Metabolites of Arachidonic Acid as Second Messengers for Presynaptic Inhibition of Aplysia Sensory Cells. Nature 328, 38-43.

1168. Pohlman, R., G. Nagel, B. Schmidt, M. Stein, G. Lorkowski, C. Krentler, J. Cully, H.E. Meyer, K.-H. Grzeschik, G. Mersmann, A. Hasilik, and K. vonFigura, (1987), Cloning of a cDNA Encoding the Human Cation-dependent Mannose 6-Phosphate-Specific Receptor. Proc. Natl. Acad. Sci. USA 84, 5575-5579.

1169. Pollet, R.J., B.A. Haase, and M.L. Standaert, (1981), Characterization of Detergent-solubilized Membrane Proteins: Hydrodynamic and Sedimentation Equilibrium Properties of the Insulin Receptor of the Cultured Lymphoblastoid Cell. J. Biol. Chem. 256, 12118-12126.

1170. Porpaczy, Z., B. Sumegi, and I. Alkonyi, (1987), Interaction between NAD-dependent Isocitrate Dehydrogenase, α-Ketoglutarate Dehydrogenase Complex, and NADH:Ubiquinone Oxidoreductase. J. Biol. Chem. 262, 9509-9514.

1171. Powell, G.L., P.F. Knowles, and D. Marsh, (1985), Association of Spin-labelled Cardiolipin with Dimyristoylphosphatidylcholine-Substituted Bovine Heart Cytochrome c Oxidase. A Generalized Specificity Increase Rather than Highly Specific Binding Sites. Biochim. Biophys. Acta 816, 191-194.

1172. Pownall, H.J., R.D. Knapp, A.M. Gotto, and J.B. Massey, (1983), Helical Amphipathic Moment: Application to Plasma Lipoproteins. FEBS Lett. 159, 17-23.
1173. Prasad, P.V., M. Yamaguchi, and Y. Hatefi, (1986), Role of Phospholipids in Activation of Mitochondrial D(-)-ß-Hydroxybutyrate Dehydrogenase. Biochem. Int. 12, 641-648.
1174. Prats, M., J. Teissie, and J.-F. Tocanne, (1986), Lateral Proton Conduction at Lipid-Water Interfaces and Its Implications for the Chemiosmotic-Coupling Hypothesis. Nature 322, 756-758.
1175. Prats, M., J.-F. Tocanne, and J. Teissie, (1987), Lateral Proton Conduction at a Lipid/Water Interface. Effects of Lipid Nature and Ionic Content of the Aqueous Phase. Eur. J. Biochem. 162, 379-385.
1176. Pressman, B.C. (1973), Properties of Ionophores with Broad Range of Cation Selectivity. Fed. Proc. 32, 1698-1703.
1177. Pribluda, V.S. and H. Metzger, (1987), Calcium-independent Phosphoinositide Breakdown in Rat Basophilic Leukemia Cells: Evidence for an Early Rise in Inositol 1,4,5-Triphosphate Which Precedes the Rise in Other Inositol Phosphates and in Cytoplasmic Calcium. J. Biol. Chem. 262, 11449-11454.
1178 Price, C.A. (1982), Centrifugation in Density Gradients. Academic Press, New York.
1179. Price, C.A. (1974), Plant Cell Fractionation. Methods in Enz. 31, 501-519.
1180. Prigent-Dachary, J., J.-F. Faucon, M.-R. Boisseau, and J. Dufourcq, (1986), Topology of the Binding Site of Blood-clotting Factors in Model Membranes: A Fluorescence Study. Eur. J. Biochem. 155, 133-140.
1181. Prince, R.C. (1987), Hopanoids: The World's Most Abundant Biomolecules? TIBS 12, 455-456.
1182. Prochaska, L., R. Bisson, and R.A. Capaldi, (1980), Structure of the Cytochrome c Oxidase Complex: Labelling by Hydrophilic and Hydrophobic Protein Modifying Reagents. Biochemistry 19, 3174-3179.
1183. Prochaska, L.J. and P.S. Fink, (1987), On the Role of Subunit III in Proton Translocation in Cytochrome c Oxidase. J. Bioenerg. and Biomemb. 19, 143-166.
1184. Pryde, J.G. (1986), Triton X-114: A Detergent That Has Come in from the Cold. TIBS 11, 160-163.
1185. Prywes, R., E. Livneh, A. Ullrich, and J. Schlessinger, (1986), Mutations in the Cytoplasmic Domain of EGF Receptor Affect EGF Binding and Receptor Internalization. The EMBO Journal 5, 2179-2190.
1186. Puddington, L., C. Woodgett, and J.K. Rose, (1987), Replacement of the Cytoplasmic Domain Alters Sorting of a Viral Glycoprotein in Polarized Cells. Proc. Natl. Acad. Sci. USA 84, 2756-2760.
1187. Quiocho, F.A., J.S. Sack, and N.K. Vyas, (1987), Stabilization of Charges on Isolated Ionic Groups Sequestered in Proteins by Polarized Peptide Units. Nature 329, 561-564.
1188. Raetz, C.R.H., G.M. Carman, W. Dowhan, R.-T. Jiang, W. Waszkuc, W. Loffredo, and M.-D. Tsai, (1987), Phospholipids Chiral at Phosphorus. Steric Course of the Reactions Catalyzed by Phosphatidylserine Synthase from Escherichia coli and Yeast. Biochemistry 26, 4022-4027.
1189. Raftery, M.A., M.W. Hunkapiller, C.D. Strader, and L.E. Hood, (1980), Acetylcholine Receptor: Complex of Homologous Subunits. Science 208, 1454-1457.

1190. Ragan, C.I. and J.J. Reed, (1986), Regulation of Electron Transfer by the Quinone Pool. J. Bioenerg. and Biomemb. 18, 403-418.
1191. Randall, L.L. and S.J.S. Hardy, (1986), Correlation of Competence for Export with Lack of Tertiary Structure of the Mature Species: A Study In Vivo of Maltose-Binding Protein in E. coli. Cell 46, 921-928.
1192. Randall, L.L. and S.J.S. Hardy, (1984), Export of Protein in Bacteria. Microbiological Reviews 48, 290-298.
1193. Rand, R.P. and V.A. Parsegian, (1984), Physical Force Considerations in Model and Biological Membranes. Can. J. Biochem. Cell Biol. 62, 752-759.
1194. Rao, J.K.M. and P. Argos, (1986), A Conformational Preference Parameter to Predict Helices in Integral Membrane Proteins. Biochim. Biophys. Acta 869, 197-214.
1195. Rapoport, T.A. (1986), Protein Translocation Across and Integration Into Membranes. CRC Critical Reviews in Biochemistry 20, 73-137.
1196. Rapoport. T.A., R. Heinrich, P. Walter, and T. Schulmeister, (1987), Mathematical Modeling of the Effects of the Signal Recognition Particle on Translation and Translocation of Proteins Across the Endoplasmic Reticulum Membrane. J. Mol. Biol. 195, 621-636.
1197. Rashin, A.A. and B. Honig, (1984), On the Environment of Ionizable Groups in Globular Proteins. J. Mol. Biol. 173, 515-521.
1198. Ratnam, M., D. Le Nguyen, J. Rivier, P.B. Sargent, and J. Lindstrom, (1986), Transmembrane Topography of Nicotinic Acetylcholine Receptor: Immunochemical Tests Contradict Theoretical Predictions Based on Hydrophobicity Profiles. Biochemistry 25, 2633-2643.
1199. Ratnam, M., P.B. Sargent, V. Sarin, J.L. Fox, D. Le Nguyen, J. Rivier, M. Criado, and J. Lindstrom, (1986), Location of Antigenic Determinants on Primary Sequences of Subunits of Nicotinic Acetylcholine Receptor by Peptide Mapping. Biochemistry 25, 2621-2632.
1200. Ratnam, M., W. Gullick, J. Spiess, K. Wan, M. Criado, and J. Lindstrom, (1986), Structural Heterogeneity of the α Subunits of the Nicotinic Acetylcholine Receptor in Relation to Agonist Affinity Alkylation and Antagonist Binding. Biochemistry 25, 4268-4275.
1201. Raudino, A. and D. Mauzerall, (1986), Dielectric Properties of the Polar Head Group Region of Zwitterionic Lipid Bilayers. Biophys. J. 50, 441-449.
1202. Ravetch, J.V., A.D. Luster, R. Weinshank, J. Kochan, A. Pavlovec, D.A. Portnoy, J. Hulmes, Y.-C. E. Pan, and J.C. Unkeless, (1986), Structural Heterogeneity and Functional Domains of Murine Immunoglobulin G F_c Receptors. Science 234, 718-725.
1203. Raymond, L., S.L. Slatin, A. Finkelstein, Q.-R. Liu, and C. Levinthal, (1986), Gating of a Voltage-Dependent Channel (Colicin E1) in Planar Lipid Bilayers: Translocation of Regions Outside the Channel-Forming Domain. J. Membrane Biol. 92, 255-268.
1204 Rebecchi, M.J. and O.M. Rosen, (1987), Purification of a Phosphoinositide-specific Phospholipase C from Bovine Brain. J. Biol. Chem. 262, 12526-12532.
1205. Rechsteiner, M., S. Rogers, and K. Rote, (1987), Protein Structure and Intracellular Stability. TIBS 12, 390-394.
1206. Recny, M.A., C. Grabau, J.E. Cronan, Jr., and L.P. Hager, (1985), Characterization of the α-Peptide Released upon Protease Activation of Pyruvate Oxidase. J. Biol. Chem. 260, 14287-14291.

1207. Reed, P.W. (1979), Ionophores. Methods in Enz. 55, 435-454.

1208. Rehorek, M., N.A. Dencher, and M.P. Heyn, (1985), Long-Range Lipid-Protein Interactions. Evidence from Time-Resolved Fluorescence Depolarization and Energy-Transfer Experiments with Bacteriorhodopsin-Dimyristoylphosphatidylcholine Vesicles. Biochemistry 24, 5980-5988.

1209. Reid, E. and R. Williamson, (1974), Centrifugation. Methods in Enz. 31, 713-733.

1210. Renetseder, R., S. Brunie, B.W. Dijkstra, J. Drenth, and P.B. Sigler, (1985), A Comparison of the Crystal Structures of Phospholipase A2 from Bovine Pancreas and Crotalus atrox Venom. J. Biol. Chem. 260, 11627-11634.

1211. Reynolds, J.A. and C. Tanford, (1970), Binding of Dodecyl Sulfate to Proteins at High Binding Ratios. Possible Implications for the State of Proteins in Biological Membranes. Proc. Natl. Acad. Sci. USA 66, 1002-1007.

1212. Reynolds, J.A., D.B. Gilbert, and C. Tanford, (1974), Empirical Correlation Between Hydrophobic Free Energy and Aqueous Cavity Surface Area. Proc. Natl. Acad. Sci. USA 71, 2925-2927.

1213. Reynolds, J.A. and D.R. McCaslin, (1985), Determination of Protein Molecular Weights in Complexes with Detergents without Knowledge of Binding. Methods in Enz. 117, 41-53.

1214. Reynolds, J.A., Y. Nozaki, and C. Tanford, (1983), Gel-Exclusion Chromatography on S1000 Sephacryl: Application to Phospholipid Vesicles. Anal. Biochem. 130, 471-474.

1215. Rich, G.T., J.S. Pryor, and A.P. Dawson, (1985), Lack of Binding of Glyceraldehyde-3-phosphate Dehydrogenase to Erythrocyte Membranes under In Vivo Conditions. Biochim. Biophys. Acta 817, 61-66.

1216. Richardson, S.H., H.O. Hultin, and S. Fleischer, (1964), Interactions of Mitochondrial Structural Protein with Phospholipids. Arch. Biochem. Biophys. 105, 254-260.

1217. Riedel, N. and H. Fasold, (1987), Nuclear-Envelope Vesicles as a Model System to Study Nucleocytoplasmic Transport. Biochem. J. 241, 213-219.

1218. Riezman, H. (1985), Endocytosis in Yeast: Several of the Yeast Secretory Mutants Are Defective in Endocytosis. Cell 40, 1001-1009.

1219. Rilfors, L. (1985), Difference in Packing Properties between Iso and Anteiso Methyl-branched Fatty Acids as Revealed by Incorporation into the Membrane Lipids of Acholeplasma laidlawii Strain A. Biochim. Biophys. Acta 813, 151-160.

1220. Ringwald, M., R. Schuh, D. Vestweber, H. Eistetter, F. Lottspeich, J. Engel, R. Dolz, F. Jahnig, J. Epplen, S. Mayer, C. Muller, and R. Kemler, (1987), The Structure of Cell Adhesion Molecule Uvomorulin. Insights into the Molecular Mechanism of Ca^{2+}-dependent Cell Adhesion. The EMBO Journal 6, 3647-3653.

1221. Rivnay, B. and G. Fischer, (1986), Phospholipid Distribution in the Microenvironment of the Immunoglobulin E-Receptor from Rat Basophilic Leukemia Cell Membrane. Biochemistry 25, 5686-5693.

1222. Rivnay, B., G. Rossi, M. Henkart, and H. Metzger, (1984), Reconstitution of the Receptor for Immunoglobulin E into Liposomes. J. Biol. Chem. 259, 1212-1217.

1223. Rizzo, V., S. Stankowski, and G. Schwarz, (1987), Alamethicin Incorporation in Lipid Bilayers: A Thermodynamic Study. Biochemistry 26, 2751-2759.

1224. Roman, L.M. and H. Garoff, (1985), Revelation through Exploitation: The Viral Model for Intracellular Traffic. TIBS 10, 428-432.

1225. Roberts, B.L., W.D. Richardson, and A.E. Smith, (1987), The Effect of Protein Context on Nuclear Location Signal Function. Cell 50, 465-475.

1226. Roberts, D.D., C.N. Rao, L.A. Liotta, H.R. Gralnick, and V. Ginsburg, (1986),

Comparison of the Specificities of Laminin, Thrombospondin, and von Willebrand Factor for Binding to Sulfated Glycolipids. J. Biol. Chem. 261, 6872-6877.

1227. Roberts, R.H. and R.L. Barchi, (1987), The Voltage-sensitive Sodium Channel from Rabbit Skeletal Muscle: Chemical Characterization of Subunits. J. Biol. Chem. 262, 2298-2303.

1228. Robertson, D.E. and H. Rottenberg, (1983), Membrane Potential and Surface Potential in Mitochondria. J. Biol. Chem. 258, 11039-11048.

1229. Robertson, J.D. (1960), The Molecular Structure and Contact Relationships of Cell Membranes. Progress in Biophysics 343-418.

1230. Robertson, J.D. (1966), Granulo-Fibrillar and Globular Substructure in Unit Membranes. Ann. N.Y. Acad. Sci. 137, 421-440.

1231. Robertson, J.D. (1959), The Ultrastructure of Cell Membranes and Their Derivatives. Biochemical Society Symposium 16, 3-43.

1232. Robillard, G.T. and R.B. Beechey, (1986), Evidence for the Existence of a Channel in the Glucose-Specific Carrier EIIGlc of the Salmonella typhimurium Phosphoenolpyruvate-Dependent Phosphotransferase System. Biochemistry 25, 1346-1354.

1233. Robinson, A., M.A. Kaderbhai, and B.M. Austen, (1987), Identification of Signal Sequence Binding Proteins Integrated into the Rough Endoplasmic Reticulum Membrane. Biochem. J. 242, 767-777.

1234. Rock, C.O. (1984), Turnover of Fatty Acids in the 1-Position of Phosphatidylethanolamine in Escherichia coli. J. Biol. Chem. 259, 6188-6194.

1235. Rock, C.O. and J.E. Cronan, (1985), Lipid Metabolism in Procaryotes. Biochemistry of Lipids and Membranes (D.E. Vance and J.E. Vance, Eds.), pp. 73-115. The Benjamin/Cummings Publishing Company, Inc., Menlo Park, California.

1236. Rogalski, A.A. and S.J. Singer, (1985), An Integral Glycoprotein Associated with the Membrane Attachment Sites of Actin Microfilaments. J. Cell Biol. 101, 785-801.

1237. Rogers, M.J. and P. Stritmatter, (1974), Evidence for Random Distribution and Translational Movement of Cytochrome b_5 in Endoplasmic Reticulum. J. Biol. Chem. 249, 895-900.

1238. Rogers, M.J. and P. Stritmatter, (1975), The Interaction of NADH-Cytochrome b_5 Reductase and Cytochrome b_5 Bound to Egg Lecithin Liposomes. J. Biol. Chem. 250, 5713-5718.

1239. Roise, D., S.J. Horvath, J.M. Tomich, J.H. Richards, and G. Schatz, (1986), A Chemically Synthesized Pre-sequence of an Imported Mitochondrial Protein Can Form an Amphiphilic Helix and Perturb Natural and Artificial Phospholipid Bilayers. The EMBO Journal 5, 1327-1334.

1240. Roman, L.M. and H. Garoff, (1985), Revelation through Exploitation: The Viral Model for Intracellular Traffic. TIBS 10, 428-432.

1241. Ronson, C.W., B.T. Nixon, and F.M. Ausubel, (1987), Conserved Domains in Bacterial Regulatory Proteins That Respond to Environmental Stimuli. Cell 49, 579-581.

1242. Rosevear, P., T. VanAken, J. Baxter, and S. Ferguson-Miller, (1980), Alkyl Glycoside Detergents: A Simpler Synthesis and Their Effects on Kinetic and Physical Properties of Cytochrome c Oxidase. Biochemistry 19, 4108-4115.

1243. Ross, A.F., M. Rapuano, J.H. Schmidt, and J.M. Prives, (1987), Phosphorylation and Assembly of Nicotinic Acetylcholine Receptor Subunits in Cultured Chick Muscle Cells. J. Biol. Chem. 262, 14640-14647.

1244. Roth, G.J., K. Titani, L.W. Hoyer, and M.J. Hickey, (1986), Localization of Binding Sites within Human von Willebrand Factor for Monomeric Type III Collagen. Biochemistry 25, 8357-8361.

1245. Roth, J. (1987), Subcellular Organization of Glycosylation in Mammalian Cells. Biochim. Biophys. Acta 906, 405-436.

1246. Rothenberg, P., L. Glaser, P. Schlesinger, and D. Cassel, (1983), Activation of Na^+/H^+ Exchange by Epidermal Growth Factor Elevates Intracellular pH in A431 Cells. J. Biol. Chem. 258, 12644-12653.

1247. Rothenberger, S., B.J. Iacopetta, and L.C. Kuhn, (1987), Endocytosis of the Transferrin Receptor Requires the Cytoplasmic Domain but Not Its Phosphorylation Site. Cell 49, 423-431.

1248. Rothman, J. (1981), The Golgi Apparatus: Two Organelles in Tandem. Science 213, 1212-1219.

1249. Rothman, J.E. (1987), Protein Sorting by Selective Retention in the Endoplasmic Reticulum and Golgi Stack. Cell 50, 521-522.

1250. Rothman, J.E. (1987), Transport of the Vesicular Stomatitis Glycoprotein to trans Golgi Membranes in a Cell-free System. J. Biol. Chem. 262, 12502-12510.

1251. Rothman, J.E. and E.P. Kennedy, (1977), Rapid Transmembrane Movement of Newly Synthesized Phospholipids During Membrane Assembly. Proc. Natl. Acad. Sci. USA 74, 1821-1825.

1252. Rothman, J.E. and R.D. Kornberg, (1986), An Unfolding Story of Protein Translocation. Nature 322, 209-210.

1253. Rothman, J.E. and S.L. Schmid, (1986), Enzymatic Recycling of Clathrin from Coated Vesicles. Cell 46, 5-9.

1254. Rothman, J.E. (1986), Life without Clathrin. Nature 319, 96-97.

1255. Rottenberg, H. (1984), Membrane Potential and Surface Potential in Mitochondria: Uptake and Binding of Lipophilic Cations. J. Membrane Biol. 81, 127-138.

1256. Rottenberg, H. and S. Steiner-Mordoch, (1986), Free Fatty Acids Decouple Oxidative Phosphorylation by Dissipating Intramembranal Protons without Inhibiting ATP Synthesis Driven by the Proton Electrochemical Gradient. FEBS Lett. 202, 314-318.

1257. Rouser, G. (1983), Membrane Composition, Structure, and Function. Membrane Fluidity in Biology (R.C. Aloia, Ed.), Vol. 1, pp. 235-289. Academic Press, NewYork.

1258. Rozengurt, E. (1986), Early Signals in the Mitogenic Response. Science 234, 161-166.

1259. Ruigrok, R.W.H., N.G. Wrigley, L.J. Calder, S. Cusack, S.A. Wharton, E.B. Brown, and J.J. Skehel, (1986), Electron Microscopy of the Low pH Structure of Influenza Virus Haemagglutinin. The EMBO Journal 5, 41-49.

1260. Runswick, M.J., S.J. Powell, P. Nyren, and J.E. Walker, (1987), Sequence of the Bovine Mitochondrial Phosphate Carrier Protein: Structural Relationship to ADP/ATP Translocase and the Brown Fat Mitochondria Uncoupling Protein. The EMBO Journal 6, 1367-1373.

1261. Ruoslahti, E. and M.D. Pierschbacher, (1987), New Perspectives in Cell Adhesion: RGD and Integrins. Science 238, 491-497.

1262. Russell, N.J. (1984), Mechanisms of Thermal Adaptation in Bacteria: Blueprints for Survival. TIBS 9, 108-112.

1263. Russell, P., H.L. Schrock, and R.B. Gennis, (1977), Lipid Activation and Protease Activation of Pyruvate Oxidase: Evidence Suggesting a Common Site of Interaction on the Protein. J. Biol. Chem. 252, 7883-7887.

1264. Russo, A.F. and D.E. Koshland, (1983), Separation of Signal Transduction and Adaptation Functions of the Asparate Receptor in Bacterial Sensing. Science 220, 1016-1020.

1265. Ryba, N.J.P., L.I. Horvath, A. Watts, and D. Marsh, (1987), Molecular Exchange at the Lipid-Rhodopsin Interface: Spin-Label Electron Spin Resonance Studies of Rhodopsin-Dimyristoylphosphatidylcholine Recombinants. Biochemistry 26, 3234-3240.

1266. Ryu, S.H., K.S. Cho, K.-Y. Lee, P.-G. Suh, and S.G. Rhee, (1987), Purification and Characterization of Two Immunologically Distinct Phosphoinositide-specific Phospholipase C from Bovine Brain. J. Biol. Chem. 262, 12511-12518.

1267. Rzepecki, L.M., P. Strittmatter, and L.G. Herbette, (1986), X-ray Diffraction Analysis of Cytochrome b_5 Reconstituted in Egg Phosphatidylcholine Vesicles. Biophys. J. 49, 829-838.

1268. Sabatini, D.D., G. Kreibich, T. Morimoto, and M. Adesnik, (1982), Mechanisms for the Incorporation of Proteins in Membranes and Organelles. J. Cell Biol. 92, 1-22.

1269. Sackmann, E. (1983), Physical Foundations of the Molecular Organization and Dynamics of Membranes. Biophysics (Walter Hoppe, Wolfgang Lohmann, Hubert Markl, and Hubert Ziegler, Eds.), pp. 425-457. Springer-Verlag, New York.

1270. Sackmann, E. (1984), Physical Basis of Trigger Processes and Membrane Structures. Biological Membranes (D. Chapman, Ed.), Vol. 5, pp. 105-143. Academic Press, New York.

1271. Sackmann, E., R. Kotulla, and F.-J. Heiszler, (1984), On the Role of Lipid Bilayer Elasticity for the Lipid-Protein Interaction and the Indirect Protein-Protein Coupling. Can. J. Biochem. Cell Biol. 62, 778-788.

1272. Safa, A.R., C.J. Glover, J.L. Sewell, M.B. Meyers, J.L. Biedler, and R.L. Felsted, (1987), Identification of the Multidrug Resistance-related Membrane Glycoprotein as an Acceptor for Calcium Channel Blockers. J. Biol. Chem. 262, 7884-7888.

1273. Saffman, P.G. and M. Delbrück, (1975), Brownian Motion in Biological Membranes. Proc. Natl. Acad. Sci. USA 72, 3111-3113.

1274. Sakaguchi, M., K. Mihara, and R. Sato, (1987), A Short Amino-Terminal Segment of Microsomal Cytochrome P-450 Functions Both as an Insertion Signal and as a Stop-Transfer Sequence. The EMBO Journal 6, 2425-2431.

1275. Samelson, L.E., J.B. Harford, and R.D. Klausner, (1985), Identification of the Components of the Murine T Cell Antigen Receptor Complex. Cell 43, 223-231.

1276. Sanderman, H. and B.A. Gottwald, (1983), Cooperative Lipid Activation of (Na^+ + K^+)-ATPase as a Consequence of Non-Cooperative Lipid-Protein Interactions. Biochim. Biophys. Acta 732, 332-335.

1277. Sandermann, H., J.O. McIntyre, and S. Fleischer, (1986), Site-Site Interaction in the Phospholipid Activation of D-ß-Hydroxybutyrate Dehydrogenase. J. Biol. Chem. 261, 6201-6208.

1278. Sandermann, H. (1978), Regulation of Membrane Enzymes by Lipids. Biochim. Biophys. Acta 515, 209-237.

1279. Sandermann, H. (1982), Lipid-Dependent Membrane Enzymes: A Kinetic Model for Cooperative Activation in the Absence of Cooperativity in Lipid Binding. Eur. J. Biochem. 127, 123-128.

1280. Sandermann, H. (1983), Lipid Solvation and Kinetic Cooperativity of Functional Membrane Proteins. TIBS 8, 408-411.

1281. Saraste, M., T. Penttila, and M. Wikstrom, (1981), Quaternary Structure of Bovine Cytochrome Oxidase. Eur. J. Biochem. 115, 261-268.

1282. Sarges, R. and B. Witkop, (1965), Gramicidin A. V. The Structure of Valine- and Isoleucine-Gramicidin A. J. Amer. Chem. Soc. 87, 2011-2020.
1283. Sasaki, K. and M. Sato, (1987), A Single GTP-binding Protein Regulates K^+-Channels Coupled with Dopamine, Histamine and Acetylcholine Receptors. Nature 325, 259-262.
1284. Sasaki, T. (1985), Glycolipid-binding Proteins. Chem. Phys. Lipids 38, 63-77.
1285. Saxton, M.J. (1982), Lateral Diffusion in an Archipelago: Effects of Impermeable Patches on Diffusion in a Cell Membrane. Biophys. J. 39, 165-173.
1286. Scarborough, G.A. (1985), Binding Energy, Conformational Change, and the Mechanisms of Transmembrane Solute Movements. Microbiological Reviews 49, 214-231.
1287. Scharschmidt, B.F., J.R. Lake, E.L. Renner, V. Licko, and R.W. Van Dyke, (1986), Fluid Phase Endocytosis by Cultured Rat Hepatocytes and Perfused Rat Liver: Implications for Plasma Membrane Turnover and Vesicular Trafficking of Fluid Phase Markers. Proc. Natl. Acad. Sci. USA 83, 9488-9492.
1288. Schatz, G. (1986), A Common Mechanism for Different Membrane Systems? Nature 321, 108-109.
1289. Schatz, G. (1987), Signals Guiding Proteins to Their Correct Locations in Mitochondria. Eur. J. Biochem. 165, 1-6.
1290. Schauer, R. (1985), Sialic Acids and Their Role as Biological Masks. TIBS 10, 357-360.
1291. Schekman, R. (1985), Protein Transport — It's What's Up Front that Counts. TIBS 10, 177.
1292. Schenkman, S., A. Tsugita, M. Schwartz, and J.P. Rosenbusch, (1984), Topology of Phage Lambda Receptor Protein: Mapping Targets of Proteolytic Cleavage in Relation to Binding Sets for Phage or Monoclonal Antibodies. J. Biol. Chem. 259, 7570-7576.
1293. Scherer, P.G. and J. Seelig, (1987), Structure and Dynamics of the Phosphatidylcholine and the Phosphatidylethanolamine Head Group in L-M Fibroblasts as Studied by Deuterium Nuclear Magnetic Resonance. The EMBO Journal 6, 2915-2922.
1294. Schiffer, M. and A.B. Edmundson, (1967), Use of Helical Wheels to Represent the Structures of Proteins and to Identify Segments with Helical Potential. Biophys. J. 7, 121-135.
1295. Schlegel, R. and M. Wade, (1984), A Synthetic Peptide Corresponding to the NH_2 Terminus of Vesicular Stomatitis Virus Glycoprotein Is a pH-Dependent Hemolysin. J. Biol. Chem. 259, 4691-4694.
1296. Schleifer, K.H. and E. Stackebrandt, (1983), Molecular Systematics of Prokaryotes. Ann. Rev. Microbiol. 37, 143-187.
1297. Schlenstedt, G. and R. Zimmermann, (1987), Import of Frog Prepropeptide GLa into Microsomes Requires ATP but Does Not Involve Docking Protein or Ribosomes. The EMBO Journal 6, 699-703.
1298. Schleyer, M. and W. Neupert, (1985), Transport of Proteins into Mitochondria: Translocational Intermediates Spanning Contact Sites between Outer and Inner Membranes. Cell 43, 339-350.
1299. Schmidt, F.O., R.S. Bear, and G. Clark, (1935), X-ray Diffraction Studies on Nerve. Radiology 25, 131-151.
1300. Schmidt, J.W. and W.A. Catterall, (1987), Palmitylation, Sulfation, and Glycosylation of the α Subunit of the Sodium Channel. J. Biol. Chem. 262, 13713-13723.

1301. Schneider, E. and K. Altendorf, (1985), All Three Subunits Are Required for the Reconstitution of an Active Proton Channel (F_0) of Escherichia coli ATP Synthase (F_1F_0). The EMBO Journal 4, 515-518.
1302. Schofield, P.R., M.G. Darlison, N. Fujita, D.R. Burt, F.A. Stephenson, H. Rodriguez, L.M. Rhee, J. Ramachandran, V. Reale, T.A. Glencorse, P.H. Seeburg, and E.A. Barnard, (1987), Sequence and Functional Expression of the GABA Receptor Shows a Ligand-gated Receptor Super-family. Nature 328, 221-227.
1303. Schreier, S., C.F. Polnaszek, and I.C. Smith, (1978), Spin Labels in Membranes: Problems in Practice. Biochim. Biophys. Acta 515, 395-436.
1304. Schrock, H.L. and R.B. Gennis, (1977), High Affinity Lipid Binding Sites on the Peripheral Membrane Enzyme Pyruvate Oxidase. J. Biol. Chem. 252, 5990-5995.
1305. Schulz, G.E. and R.H. Schirmer, (1979), Principles of Protein Structure. Springer-Verlag, New York.
1306. Schurtenberger, P. and H. Hauser, (1984), Characterization of the Size Distribution of Unilamellar Vesicles by Gel Filtration, Quasi-elastic Light Scattering and Electron Microscopy. Biochim. Biophys. Acta 778, 470-480.
1307. Schwaiger, M., V. Herzog, and W. Neupert, (1987), Characterization of Translocation Contact Sites Involved in the Import of Mitochondrial Proteins. J. Cell Biol. 105, 235-246.
1308. Scopes, R.K. (1982), Protein Purification: Principles and Practice. Springer-Verlag, New York.
1309. Scott, S., S.A. Pendlebury, and C. Green, (1984), Lipid Organization in Erythrocyte Membrane Microvesicles. Biochem. J. 224, 285-290.
1310. Scotto, A.W. and D. Zakim, (1985), Reconstitution of Membrane Proteins. Spontaneous Association of Integral Membrane Proteins with Preformed Unilamellar Lipid Bilayers. Biochemistry 24, 4066-4075.
1311. Scotto, A.W. and D. Zakim, (1986), Reconstitution of Membrane Proteins: Catalysis by Cholesterol of Insertion of Integral Membrane Proteins into Preformed Lipid Bilayers. Biochemistry 25, 1555-1561.
1312. Scoulica, E., E. Krause, K. Meese, and B. Dobberstein, (1987), Disassembly and Domain Structure of the Proteins in the Signal-recognition Particle. Eur. J. Biochem. 163, 519-528.
1313. Scullion, B.F., Y. Hou, L. Puddington, J.K. Rose, and K. Jacobson, (1987), Effects of Mutations in Three Domains of the Vesicular Stomatitis Viral Glycoprotein on Its Lateral Diffusion in the Plasma Membrane. J. Cell Biol. 105, 69-75.
1314. Seckler, R., T. Moroy, J.K. Wright, and P. Overath, (1986), Anti-Peptide Antibodies and Proteases as Structural Probes for the Lactose/H+ Transporter of Escherichia coli: A Loop around Amino Acid Residue 130 Faces the Cytoplasmic Side of the Membrane. Biochemistry 25, 2403-2409.
1315. Seddon, J.M., G. Cevc, R.D. Kaye, and D. Marsh, (1984), X-ray Diffraction Study of the Polymorphism of Hydrated Diacyl and Dialkylphosphatidylethanolamines. Biochemistry 23, 2634-2644.
1316. Seed, B. (1987), An LFA-3 cDNA Encodes a Phospholipid-linked Membrane Protein Homologous to Its Receptor CD2. Nature 329, 840-842.
1317. Seelig, A. and J. Seelig, (1985), Phospholipid Composition and Organization of Cytochrome c Oxidase Preparations as Determined by 31P-Nuclear Magnetic Resonance. Biochim. Biophys. Acta 815, 153-158.
1318. Seelig, J. (1978), Phosphorus-31 Nuclear Magnetic Resonance and the Headgroup Structure of Phospholipids in Membranes. Biochim. Biophys. Acta 515, 105-141.

1319. Seelig, J. (1977), Deuterium Magnetic Resonance: Theory and Application to Lipid Membranes. Quart. Rev. Biophys. 10, 353-415.
1320. Seelig, J. and A. Seelig, (1980), Lipid Conformation in Model Membranes and Biological Membranes. Quart. Rev. Biophys. 13, 19-61.
1321. Seelig, J. and J.L. Browning, (1978), General Features of Phospholipid Conformation in Membranes. FEBS Lett. 92, 41-44.
1322. Sefton, B.M. and J.E. Buss, (1987), The Covalent Modification of Eukaryotic Proteins with Lipid. J. Cell Biol. 104, 1449-1453.
1323. Segrest, J.P., I. Kahane, R.L. Jackson, and V.T. Marchesi, (1973), Major Glycoprotein of the Human Erythrocyte Membrane: Evidence for an Amphipathic Structure. Arch. Biochem. Biophys. 155, 167-183.
1324. Seiff, F., J. Westerhausen, I Wallat, and M.P. Heyn, (1986), Location of the Cyclohexene Ring of the Chromophore of Bacteriorhodopsin by Neutron Diffraction with Selectively Deuterated Retinal. Proc. Natl. Acad. Sci. USA 83, 7746-7750.
1325. Seigneuret, M. and J.-L. Rigaud, (1986), Analysis of Passive and Light-Driven Ion Movements in Large Bacteriorhodopsin Liposomes Reconstituted by Reverse-Phase Evaporation. 1. Factors Governing the Passive Proton Permeability of the Membrane. Biochemistry 25, 6716-6722.
1326. Semin, B.K., M. Saraste, and M. Wikstrom, (1984), Calorimetric Studies of Cytochrome Oxidase-Phospholipid Interactions. Biochim. Biophys. Acta 769, 15-22.
1327. Serrano, R., M.C. Kielland-Brandt, and G.R. Fink, (1986), Yeast Plasma Membrane ATPase Is Essential for Growth and Has Homology with (Na^+ + K^+), K^+-and Ca^{2+}-ATPases. Nature 319, 689-693.
1328. Sessa, G., J.H. Freer, G. Colacicco, and G. Weissmann, (1969), Interactions of a Lytic Polypeptide Melittin with Lipid Membrane Systems. J. Biol. Chem. 244, 3575-3582.
1329. Severs, N.J. and H. Robenek, (1983), Detection of Microdomains in Biomembranes: An Appraisal of Recent Developments in Freeze-Fracture Cytochemistry. Biochim. Biophys. Acta 737, 373-408.
1330. Sharon, N. (1987), Bacterial Lectins, Cell-Cell Recognition and Infectious Disease. FEBS Lett. 217, 145-157.
1331. Sheetz, M., M. Schindler, and D.E. Koppel, (1980), Lateral Mobility in Integral Membrane Proteins Increased in Spherocytic Erythrocytes. Nature 285, 510-512.
1332. Shimada, K. and N. Murata, (1976), Chemical Modification by Trinitrobenzenesulfonate of a Lipid and Protein of Intracytoplasmic Membranes Isolated from Chromatium vinosum and Azotobacter vinlandii. Biochim. Biophys. Acta 455, 605-620.
1333. Shinitzky, M. (1984), Membrane Fluidity and Cellular Functions. Physiology of Membrane Fluidity (M. Shinitzky, Ed.), Vol. 1, pp. 1-51. CRC Press, Boca Raton.
1334. Shipley, G.G. (1973), Recent X-ray Diffraction Studies of Biological Membrane Components. Biological Membranes (D. Chapman and D.F.H. Wallach, Eds.), Vol. 2, pp. 1-89. Academic Press, New York.
1334a. Shipley, G.G., J.P. Green, and B.W. Nichols, (1973), The Phase Behavior of Monogalactosyl, Digalactosyl, and Sulphoquinovosyl Diglycerides. Biochim. Biophys. Acta 311, 531-544.
1335. Shirayoshi, Y., K. Hatta, M. Hosoda, S. Tsunasawa, F. Sakiyama, and M. Takeichi, (1986), Cadherin Cell Adhesion Molecules with Distinct Binding Specificities Share a Common Structure. The EMBO Journal 5, 2485-2488.

1336. Shreeve, S.M., W.R. Roeske, and J.C. Venter, (1984), Partial Functional Reconstitution of the Cardiac Muscarine Cholinergic Receptor. J. Biol. Chem. 259, 12398-12402.
1337. Shull, G.E., A. Schwarz, and J.B. Lingrel, (1985), Amino-acid Sequence of the Catalytic Subunit of the $(Na^+ + K^+)$ATPase Deduced from a Complementary DNA. Nature 316, 691-695.
1338. Shull, G.E., L.K. Lane, and J.B. Lingrel, (1986), Amino-acid Sequence of the ß-Subunit of the $(Na^+ + K^+)$ATPase Deduced from a cDNA. Nature 321, 429-431.
1339. Siegelman, M., M.W. Bond, W.M. Gallatin, T. St. John, H.T. Smith, V.A. Fried, and I.L. Weissman, (1986), Cell Surface Molecule Associated with Lymphocyte Homing Is a Ubiquitinated Branched-Chain Glycoprotein. Science 231, 823-829.
1340. Sibley, D.R., J.L. Benovic, M.G. Caron, and R.J. Lefkowitz, (1987), Regulation of Transmembrane Signalling by Receptor Phosphorylation. Cell 48, 913-922.
1341. Silversmith, R.E. and G.L. Nelsestuen, (1986), Assembly of the Membrane Attack Complex of Complement on Smaller Unilamellar Phospholipid Vesicles. Biochemistry 25, 852-860.
1342. Silvius, J.R., D.A. McMillen, N.D. Saley, P.C. Jost, and O.H. Griffith, (1984), Competition between Cholesterol and Phosphatidylcholine for the Hydrophobic Surface of Sarcoplasmic Reticulum Ca^{2+}-ATPase. Biochemistry 23, 538-547.
1343. Silvius, J.R., R. Leventis, P.M. Brown, and M. Zuckermann, (1987), Novel Fluorescent Phospholipids for Assays of Lipid Mixing between Membranes. Biochemistry 26, 4279-4287.
1344. Simmonds, A.C., E.K. Rooney, and A.G. Lee, (1984), Interactions of Cholesterol Hemisuccinate with Phospholipids and $(Ca^{2+} - Mg^{2+})$-ATPase. Biochemistry 23, 1432-1441.
1345. Simon, K., E. Perara, and V.R. Lingappa, (1987), Translocation of Globin Fusion Proteins across the Endoplasmic Reticulum Membrane in Xenopus laevis Oocytes. J. Cell Biol. 104, 1165-1172.
1346. Simon, S.A. and T.J. McIntosh, (1984), Interdigitated Hydrocarbon Chain Packing Causes the Biphasic Transition Behavior in Lipid/Alcohol Suspensions. Biochim. Biophys. Acta 773, 169-172.
1347. Simons, K., H. Garoff, and A. Helenius, (1977), The Glycoproteins of the Semliki Forest Virus Membrane. Membrane Proteins and Their Interaction with Lipids (R.A. Capaldi, Ed.), pp. 207-234. Marcel Dekker, Inc., New York.
1348. Singer, S.J. (1974), The Molecular Organization of Membranes. Ann. Rev. Bioch. 43, 805-833.
1349. Singer, S.J. and G.L. Nicolson, (1972), The Fluid Mosaic Model of the Structure of Cell Membranes. Science 175, 720-731.
1350. Singer, S.J., P.A. Maher, and M.P. Yaffe, (1987), On the Transfer of Integral Proteins into Membranes. Proc. Natl. Acad. Sci. USA 84, 1960-1964.
1351. Sixl, F., P.J. Brophy, and A. Watts, (1984), Selective Protein-Lipid Interactions at Membrane Surfaces: A Deuterium and Phosphorus Nuclear Magnetic Resonance Study of the Association of Myelin Basic Protein with the Bilayer Head Groups of Dimyristoylphosphatidylcholine and Dimyristoylphosphatidylglycerol. Biochemistry 23, 2032-2039.
1352. Sjostrom, M., S. Wold, A. Wieslander, and L. Rilfors, (1987), Signal Peptide Amino Acid Sequences in Escherichia coli Contain Information Related to Final Protein Localization. A Multivariate Data Analysis. The EMBO Journal 6, 823-831.
1353. Slocum, M.K., N.F. Halden, and J.S. Parkinson, (1987), Hybrid Escherichia coli

Sensory Transducers with Altered Stimulus Detection and Signaling Properties. J. Bacteriology 169, 2938-2944.

1354. Smaal, E.B., K. Nicolay, J.G. Mandersloot, J. de Gier, and B. de Kruijff, (1987), ^{2}H-NMR, ^{31}P-NMR and DSC Characterization of a Novel Lipid Organization in Calcium-dioleoylphosphatidate Membranes. Implications for the Mechanism of the Phosphatidate Calcium Transmembrane Shuttle. Biochim. Biophys. Acta 897, 453-466.

1355. Smalheiser, N.R. and N.B. Schwartz, (1987), Cranin: A Laminin-binding Protein of Cell Membranes. Proc. Natl. Acad. Sci. USA 84, 6457-6461.

1356. Smeekens, S., C. Bauerle, J. Hageman, K. Keegstra, and P. Wiesbeek, (1986), The Role of the Transit Peptide in the Routing of Precursors toward Different Chloroplast Compartments. Cell 46, 365-375.

1357. Smigel, M. and S. Fleischer, (1977), Characterization of Triton X-100-solubilized Prostaglandin E. Binding Protein of Rat Liver Plasma Membranes. J. Biol. Chem. 252, 3689-3696.

1358. Smith, C., M. Phillips, and C. Miller, (1986), Purification of Charybdotoxin, a Specific Inhibitor of the High-conductance Ca^{2+}-activated K^{+} Channel. J. Biol. Chem. 261, 14607-14613.

1359. Smith, M.M., J. Lindstrom, and J.P. Merlie, (1987), Formation of the α-Bungarotoxin Binding Site and Assembly of the Nicotinic Acetylcholine Receptor Subunits Occur in the Endoplasmic Reticulum. J. Biol. Chem. 262, 4367-4376.

1360. Smith, R.L. and E. Oldfield, (1984), Dynamic Structure of Membranes by Deuterium NMR. Science 225, 280-288.

1361. Snoek, G.T., I. Rosenberg, S.W. de Laat, and C. Gitler, (1986), The Interaction of Protein Kinase C and Other Specific Cytoplasmic Proteins with Phospholipid Bilayers. Biochim. Biophys. Acta 860, 336-344.

1362. Solioz, M., S. Mathews, and P. Furst, (1987), Cloning of the K^{+}-ATPase of Streptococcus faecalis: Structural and Evolutionary Implications of Its Homology to the KdpB-Protein of Escherichia coli. J. Biol. Chem. 262, 7358-7362.

1363. Sowers, A.E. and C.R. Hackenbrock, (1985), Variation in Protein Lateral Diffusion Coefficients Is Related to Variation in Protein Concentration Found in Mitochondrial Inner Membranes. Biochim. Biophys. Acta 821, 85-90.

1364. Sowers, A.E. and M.R. Lieber, (1986), Electropore Diameters, Lifetimes, Numbers, and Locations in Individual Erythrocyte Ghosts. FEBS Lett. 205, 179-184.

1365. Sparrow, C.P., B.R. Ganong, and C.R.H. Raetz, (1984), Escherichia coli Membrane Vesicles with Elevated Phosphatidic Acid Levels. Biochim. Biophys. Acta 796, 373-383.

1366. Sparrow, C.P. and C.R.H. Raetz, (1985), Purification and Properties of the Membrane-bound CDP-diglyceride Synthetase from Escherichia coli. J. Biol. Chem. 260, 12084-12091.

1367. Spiegel, S. and P.H. Fishman, (1985), Direct Evidence That Endogenous GM_1 Ganglioside Can Mediate Thymocyte Proliferation. Science 230, 1285-1287.

1368. Spiegel, S. and P.H. Fishman, (1987), Gangliosides as Bimodal Regulators of Cell Growth. Proc. Natl. Acad. Sci. USA 84, 141-145.

1369. Spiegel, S., R. Blumenthal, P.H. Fishman, and J.S. Handler, (1985), Gangliosides Do Not Move from Apical to Basolateral Plasma Membrane in Cultural Epithelial Cells. Biochim. Biophys. Acta 821, 310-318.

1370. Spiegel, S., S. Kassis, M. Wilchek, and P.H. Fishman, (1984), Direct Visualization of Redistribution and Capping of Fluorescent Gangliosides on Lymphocytes. J. Cell Biol. 99, 1575-1581.

1371. Spiess, M., A.L. Schwartz, and H.F. Lodish, (1985), Sequence of Human Asialoglycoprotein Receptor cDNA: An Internal Signal Sequence for Membrane Insertion. J. Biol. Chem. 260, 1979-1982.
1372. Spiess, M. and H.F. Lodish, (1986), An Internal Signal Sequence: The Asialoglycoprotein Receptor Membrane Anchor. Cell 44, 177-185.
1373. Squier, T.C., D.J. Bigelow, J. Garcia de Ancos, and G. Inesi, (1987), Localization of Site-specific Probes on the Ca-ATPase of Sarcoplasmic Reticulum Using Fluorescence Energy Transfer. J. Biol. Chem. 262, 4748-4754.
1374. Srere, P.A. (1987), Complexes of Sequential Metabolic Enzymes. Ann. Rev. Biochem. 56, 89-124.
1375. Srivastava, D.K. and S.A. Bernhard, (1986), Metabolic Transfer via Enzyme-Enzyme Complexes. Science 234, 1081-1086.
1376. Stahl, N. and W.P. Jencks, (1987), Reactions of the Sarcoplasmic Reticulum Calcium Adenosinetriphosphatase with Adenosine 5'-Triphosphate and Ca^{2+} That Are Not Satisfactorily Described by an E1-E2 Model. Biochemistry 26, 7654-7667.
1377. Stan-Lotter, H. and P.D. Bragg, (1986), Conformational Interactions between α and ß Subunits in the F_1 ATPase of Escherichia coli as Shown by Chemical Modification of uncA401 and uncD412 Mutant Enzymes. Eur. J. Biochem. 160, 169-174.
1378. Stankowski, S. (1984), Large-Ligand Adsorption to Membranes III. Cooperativity and General Ligand Shapes. Biochim. Biophys. Acta 777, 167-182.
1379. Stankowski, S. (1983), Large-Ligand Adsorption to Membranes I. Linear Ligands as a Limiting Case. Biochim. Biophys. Acta 735, 341-351.
1380. Stanley, K.K. and J. Herz, (1987), Topological Mapping of Complement Component C9 by Recombinant DNA Techniques Suggests a Novel Mechanism for Its Insertion into Target Membranes. The EMBO Journal 6, 1951-1957.
1381. Stark, W., W. Kuhlbrandt, I. Wildhaber, E. Wehrli, and K. Muhlethaler, (1984), The Structure of the Photoreceptor Unit of Rhodopseudomonas viridis. The EMBO Journal 3, 777-783.
1382. Staros, J.V. (1980), Aryl Azide Photolabels in Biochemistry. TIBS 5, 320-322.
1383. Steele, J.C.H. Jr, C. Tanford, and J.A. Reynolds, (1978), Determination of Partial Specific Volumes for Lipid Associated Proteins. Methods in Enz. 48, 11-23.
1384. Steffens, G.C.M., R. Biewald, and G. Buse, (1987), Cytochrome c Oxidase is a Three-copper, Two-heme-A Protein. Eur. J. Biochem. 164, 295-300.
1385. Stein, M., J.E. Zijderhand-Bleekemolen, H. Geuze, A. Hasilik, and K. vonFigura, (1987), M_r-46000 Mannose 6-Phosphate Specific Receptor: Its Role in Targeting of Lysosomal Enzymes. The EMBO Journal 6, 2677-2681.
1386. Stein, W.D. (1986), Transport and Diffusion across Cell Membranes. Academic Press, Inc., New York.
1387. Stein, W.D. and J.F. Danielli , (1956), Structure and Function in Red Cell Permeability. Disc. Faraday Soc. 21, 238-251.
1388. Stephan, M.M. and G.R. Jacobson, (1986), Subunit Interactions of the Escherichia coli Mannitol Permease: Correlation with Enzymic Activities. Biochemistry 25, 4046-4051.
1389. Sternweis, P.C. (1986), The Purified α Subunits of G_o and G_i from Bovine Brain Require ß γ for Association with Phospholipid Vesicles. J. Biol. Chem. 261, 631-637.
1390. Stevens, C.F. (1986), Are There Two Functional Classes of Glutamate Receptors? Nature 322, 210-211.
1391. Stidham, M.A., T.J. McIntosh, and J.N. Siedow, (1984), On the Location of

Ubiquinone in Phosphatidylcholine Bilayers. Biochim. Biophys. Acta 767, 423-413.
1392. Stillinger, F.H. (1980), Water Revisited. Science 209, 451-457.
1393. Stock, J., G. Kersulis, and D.E. Koshland, (1985), Neither Methylating nor Demethylating Enzymes Are Required for Bacterial Chemotaxis. Cell 42, 683-690.
1394. Stoeckenius, W.R. and R.A. Bogomolni, (1982), Bacteriorhodopsin and Related Pigments of Halobacteria. Ann. Rev. Biochem. 51, 587-616.
1395. Stoorvogel, W., H.J. Geuze, and G.J. Strous, (1987), Sorting of Endocytosed Transferrin and Asialoglycoprotein Occurs Immediately after Internalization in HepG2 Cells. J. Cell Biol. 104, 1261-1268.
1396. Strader, C.D., I.S. Sigall, R.B. Register, M.R. Candelore, E. Rands, and R.A.F. Dixon, (1987), Identification of Residues Required for Ligand Binding to the ß-Adrenergic Receptor. Proc. Natl. Acad. Sci. USA 84, 4384-4388.
1397. Strader, C.D., R.A.F. Dixon, A.H. Cheung, M.R. Candelore, A.D. Blake, and I.S. Sigal, (1987), Mutations That Uncouple the ß-Adrenergic Receptor from G_s and Increase Agonist Affinity. J. Biol. Chem. 262, 16439-16443.
1398. Strauch, K.L., C.A. Kumamoto, and J. Beckwith, (1986), Does secA Mediate Coupling between Secretion and Translation in Escherichia coli? J. Bacteriology 166, 505-512.
1399. Stroud, R.M. and J. Finer-Moore, (1985), Acetylcholine Receptor Structure, Function, and Evolution. Ann. Rev. Cell Biol. 1, 317-351.
1400. Stryer, L. and H.R. Bourne, (1986), G Proteins: A Family of Signal Transducers. Ann. Rev. Cell Biol. 2, 391-419.
1401. Suarez, M.D., A. Revzin, R. Narlock, E.S. Kempner, D.A. Thompson, and S. Ferguson-Miller, (1984), The Functional and Physical Form of Mammalian Cyto-chrome c Oxidase Determined by Gel Filtration, Radiation Inactivation and Sedimentation Equilibrium Analysis. J. Biol. Chem. 259, 13791-13799.
1402. Suarez-Isla, B.A., K. Wan, J. Lindstrom, and M. Montal, (1983), Single-channel Recordings from Purified Acetylcholine Receptors Reconstituted in Bilayers Formed at the Tip of Patch Pipets. Biochemistry 22, 2319-2323.
1403. Subramaniam, S., M. Seul, and H.M. McConnell, (1986), Lateral Diffusion of Specific Antibodies Bound to Lipid Monolayers on Alkylated Substrates. Proc. Natl. Acad. Sci. USA 83, 1169-1173.
1404. Sudhof, T.C., J.L. Goldstein, M.S. Brown, and D.W. Russell, (1985), The LDL Receptor Gene: A Mosaic of Exons Shared with Different Proteins. Science 228, 815-822.
1405. Sugar, I.P., W. Foster, and E. Neumann, (1987), Model of Cell Electrofusion: Membrane Electroporation, Pore Coalescence and Percolation. Biophys. Chem. 26, 321-335.
1406. Sumegi, B. and P.A. Srere, (1984), Complex I Binds Several Mitochondrial NAD-coupled Dehydrogenases. J. Biol. Chem. 259, 15040-15045.
1407. Sumegi, B. and P.S. Srere, (1984), Binding of the Enzymes of Fatty Acid ß-Oxidation and Some Related Enzymes to Pig Heart Inner Mitochondrial Membrane. J. Biol. Chem. 259, 8748-8752.
1408. Sun, S.-T., A. Milon, T. Tanaka, G. Ourisson, and Y. Nakatani, (1986), Osmotic Swelling of Unilamellar Vesicles by the Stopped-flow Light Scattering Method. Elastic Properties of Vesicles. Biochim. Biophys. Acta 860, 525-530.
1409. Sundberg, S.A. and W.L. Hubbell, (1986), Investigation of Surface Potential Asymmetry in Phospholipid Vesicles by a Spin Label Relaxation Method. Biophys. J. 49, 553-562.

1410. Sundby, C. and C. Larsson, (1985), Transbilayer Organization of the Thylakoid Galactolipids. Biochim. Biophys. Acta 813, 61-67.

1411. Sung, S.-S. and P.C. Jordan, (1987), Why Is Gramicidin Valence Selective? Biophys. J. 51, 661-672.

1412. Sunshine, J., K. Balak, U. Rutishauser, and M. Jacobson, (1987), Changes in Neural Cell Adhesion Molecule (NCAM) Structure During Vertebrate Neural Development. Proc. Natl. Acad. Sci. USA 84, 5986-5990.

1413. Surewicz, W.K., M.A. Moscarello, and H.H. Mantsch, (1987), Fourier Transform Infrared Spectroscopic Investigation of the Interaction between Myelin Basic Protein and Dimyristoylphosphatidylglycerol Bilayers. Biochemistry 26, 3881-3886.

1414. Surewicz, W.K., M.A. Moscarello, and H.H. Mantsch, (1987), Secondary Strucutre of the Hydrophobic Myelin Protein in a Lipid Environment as Determined by Fourier-transform Infrared Spectrometry. J. Biol. Chem. 262, 8598-8602.

1415. Swanson, J.A., B.D. Yirinec, and S.C. Silverstein, (1985), Phorbol Esters and Horseradish Peroxidase Stimulate Pinocytosis and Redirect the Flow of Pinocytosed Fluid in Macrophages. J. Cell Biol. 100, 851-859.

1416. Symington, F.W., W.A. Murray, S.I. Bearman, and S. Hakomori, (1987), Intracellular Localization of Lactosylceramide, the Major Human Neutrophil Glycosphingolipid. J. Biol. Chem. 262, 11356-11363.

1417. Szoka, F. and D. Papahadjopoulos, (1980), Comparative Properties and Methods of Preparation of Lipid Vesicles (Liposomes). Annu. Rev. Biophys. Bioeng. 9, 467-508.

1418. Tai, S.-P. and S. Kaplan, (1985), Intracellular Localization of Phospholipid Transfer Activity in Rhodopseudomonas sphaeroides and a Possible Role in Membrane Biogenesis. J. Bacteriology 164, 181-186.

1419. Tajima, S., L.Lauffer, V.L. Rath, and P. Walter, (1986), The Signal Recognition Particle Receptor Is a Complex That Contains Two Distinct Polypeptide Chains. J. Cell Biol. 103, 1167-1178.

1420. Takagaki, Y., R. Radhakrishnam, K.W.A. Wirtz, and H.G. Khorana, (1983), The Membrane-embedded Segment of Cytochrome b_5 as Studied by Cross-linking with Photoactivatable Phospholipids: II The Non-transformable Form. J. Biol. Chem. 258, 9136-9142.

1421. Takagaki, Y., R. Radhakrishnan, C.M. Gupta, and H.G. Khorana, (1983), The Membrane-embedded Segment of Cytochrome b_5 as Studied by Cross-linking with Photoactivatable Phospholipids. I. The Transferable Form. J. Biol. Chem. 258, 9128-9135.

1422. Takahashi, M., M.J. Seagar, J.F. Jones, B.F.X. Reber, and W.A. Catterall, (1987), Subunit Structure of Dihydropyridine-sensitive Calcium Channels from Skeletal Muscle. Proc. Natl. Acad. Sci. USA 84, 5478-5482.

1423. Takemori, S. and S. Kominami, (1984), The Role of Cytochromes P-450 in Adrenal Steroidogenesis. TIBS 9, 393-396.

1424. Tal, M., A. Silberstein, and E. Nusser, (1985), Why Does Coomassie Brilliant Blue R Interact Differently with Different Proteins?: A Partial Answer. J. Biol. Chem. 260, 9976-9980.

1425. Talvenheimo, J.A. (1985), The Purification of Ion Channels from Excitable Cells. J. Membrane Biol. 87, 77-91.

1426. Tamburini, P.P. and G.G. Gibson, (1983), Thermodynamic Studies of the Protein-Protein Interactions Between Cytochrome P-450 and Cytochrome b_5. J. Biol. Chem. 258, 13444-13452.

1427. Tamburini, P.P. and J.B. Schenkman, (1987), Purification to Homogeneity and

Enzymological Characterization of a Functional Covalent Complex Composed of Cytochromes P-450 Isozyme 2 and b_5 from Rabbit Liver. Proc. Natl. Acad. Sci. USA 84, 11-15.

1428. Tamburini, P.P., S. MacFarquhar, and J.B. Schenkman, (1986), Evidence of Binary Complex Formation Between Cytochrome P-450, Cytochrome b_5, and NADPH-Cytochrome P-450 Reductase of Hepatic Microsomes. Biochem. Biophys. Res. Comm. 134, 519-526.
1429. Tamkun, M.M. and D.M. Fambrough, (1986), The (Na^+ + K^+)-ATPase of Chick Sensory Neurons: Studies on Biosynthesis and Intracellular Transport. J. Biol. Chem. 261, 1009-1019.
1430. Tamm, L.K. and H.M. McConnell, (1985), Supported Phospholipid Bilayers. Biophys. J. 47, 105-113.
1431. Tanabe, T., H. Takeshima, A. Mikami, V. Flockerzi, H. Takahashi, K. Kangawa, M. Kojima, H. Matsuo, T. Hirose, and S. Numa, (1987), Primary Structure of the Receptor for Calcium Channel Blockers from Skeletal Muscle. Nature 328, 313-318.
1432. Tanford, C. (1973), The Hydrophobic Effect: Formation of Micelles in Biological Membranes. John Wiley, New York.
1433. Tanford, C., J.A. Reynolds, and E.A. Johnson, (1987), Sarcoplasmic Reticulum Calcium Pump: A Model for Ca^{2+} Binding and Ca^{2+}-coupled Phosphorylation. Proc. Natl. Acad. Sci. USA 84, 7094-7098.
1434. Tanford, C., Y. Nozaki, and M.F. Rohde, (1977), Size and Shape of Globular Micelles Formed in Aqueous Solution by n-Alkyl Polyoxyethylene Ethers. J. Phys. Chem. 81, 1555-1560.
1435. Taniguchi, H., Y. Imai, and R. Sato, (1987), Protein-Protein and Lipid-Protein Interactions in a Reconstituted Cytochrome P-450 Dependent Microsomal Monooxygenase. Biochemistry 26, 7084-7090.
1436. Tank, D.W., E.-S. Wu, and W.W. Webb, (1982), Enhanced Molecular Diffusibility in Muscle Membrane Blebs: Release of Lateral Constraints. J. Cell Biol. 92, 207-212.
1437. Tanner, L.I. and G.E. Lienhard, (1987), Insulin Elicits a Redistribution of Transferrin Receptors in 3T3-L1 Adipocytes through an Increase in the Rate Constant for Receptor Externalization. J. Biol. Chem. 262, 8975-8980.
1438. Tanner, M.J.A. (1979), Isolation of Integral Membrane Proteins and Criteria for Identifying Carrier Proteins. Curr. Topics Memb. Transp. 12, 1-51.
1439. Taraschi, T.F., B. de Kruijff, A. Verkleij, and C.J.A. Van Echteld, (1982), Effect of Glycophorin on Lipid Polymorphism: A ^{31}P-NMR Study. Biochim. Biophys. Acta 685, 153-161.
1440. Taylor, K.A., L. Dux, and A. Martonosi, (1986), Three-dimensional Reconstruction of Negatively Stained Crystals of the Ca^{2+}-ATPase from Muscle Sarcoplasmic Reticulum. J. Mol. Biol. 187, 417-427.
1441. Taylor, M.G. and I.C.P. Smith, (1983), The Conformations of Nitroxide-Labelled Fatty Acid Probes of Membrane Structure as Studied by ^{2}H-NMR. Biochim. Biophys. Acta 733, 256-263.
1442. Teeters, C.L., J. Eccles, and B.A. Wallace, (1987), A Theoretical Analysis of the Effects of Sonication on Differential Absorption Flattening in Suspensions of Membrane Sheets. Biophys. J. 51, 527-532.
1443. Tempel, B.L., D.M. Papazian, T.L. Schwarz, Y.N. Jan, and L.Y. Jan, (1987), Sequence of a Probable Potassium Channel Component Encoded at Shaker Locus of Drosophila. Science 237, 770-775.

1444. Tenchov, B.G., A.I. Boyanov, and R.D. Koynova, (1984), Lyotropic Polymorphism of Racemic Dipalmitoylphosphatidylethanolamine. A Differential Scanning Calorimetry Study. Biochemistry 23, 3553-3555.
1445. Terwilliger, T.C., L. Weissman, and D. Eisenberg, (1982), The Structure of Melittin in the Form I Crystals and Its Implication for Melittin's Lytic and Surface Activities. Biophys. J. 37, 353-361.
1446. Thiery, J.-P., A. Delouvee, M. Grumet, and G.M. Edelman, (1985), Initial Appearance and Regional Distribution of the Neuron-Glia Cell Adhesion Molecule in the Chick Embryo. J. Cell Biol. 100, 442-456.
1447. Thilo, L. (1985), Quantification of Endocytosis-derived Membrane Traffic. Biochim. Biophys. Acta 822, 243-266.
1448. Thomas, D.D. (1978), Large-Scale Rotational Motions of Proteins Detected by Electron Paramagnetic Resonance and Fluorescence. Biophys. J. 24, 439-462.
1449. Thompson, J.E., R. Coleman, and J.B. Finean, (1968), Comparative X-ray Diffraction and Electron Microscope Studies of Isolated Mitochondrial Membranes. Biochim. Biophys. Acta 150, 405-415.
1450. Thompson, N.L., H.M. McConnell, and T.P. Burghardt, (1984), Order in Supported Phospholipid Monolayers Detected by the Dichroism of Fluorescence Excited with Polarized Evanescent Illumination. Biophys. J. 46, 739-747.
1451. Thompson, S.L., R. Burrows, R.J. Laub, and S.K. Krisans, (1987), Cholesterol Synthesis in Rat Liver Peroxisomes. J. Biol. Chem. 262, 17420-17425.
1452. Thurnhofer, H., N. Gains, B. Mutsch, and H. Hauser, (1986), Cholesterol Oxidase as a Structural Probe of Biological Membranes: Its Application to Brush Border-Membrane. Biochim. Biophys. Acta 856, 174-181.
1453. Tilcock, C.P.S., M.J. Hope, and P.R. Cullis, (1984), Influence of Cholesterol Esters of Varying Unsaturation on the Polymorphic Phase Preferences of Egg Phosphatidylethanolamine. Chem. Phys. Lipid 35, 363-370.
1454. Tilcock, C.P.S., P.R. Cullis, and S.M. Gruner, (1986), On the Validity of ^{31}P-NMR Determinations of Phospholipid Polymorphic Phase Behavior. Chem. Phys. Lipids 40, 47-56.
1455. Tilley, L., S. Cribier, B. Roelofsen, J.A.F. Op den Kamp, and L.L.M. Van Deenen, (1986), ATP-dependent Translocation of Amino Phospholipids Across the Human Erythrocyte Membrane. FEBS Lett. 194, 21-27.
1456. Tobimatsu, T., Y. Fujita, K. Fukuda, K. Tanaka, Y. Mori, T. Konno, M. Mishina, and S. Numa, (1987), Effects of Substitution of Putative Transmembrane Segments on Nicotinic Acetylcholine Receptor Function. FEBS Lett. 222, 56-62.
1457. Tobkes, N., B.A. Wallace, and H. Bayley, (1985), Secondary Structure and Assembly Mechanism of an Oligomeric Channel Protein. Biochemistry 24, 1915-1920.
1458. Tokutomi, S., R. Lew, and S.I. Ohnishi, (1981), Ca^{2+}-Induced Phase Separation in Phosphatidylserine, Phosphatidylethanolamine, and Phosphatidylcholine Mixed Membranes. Biochim. Biophys. Acta 643, 276-282.
1459. Tolbert, N.E. (1974), Isolation of Subcellular Organelles of Metabolism on Isopycnic Sucrose Gradients. Methods in Enz. 31, 734-746.
1460. Tomita, M., H. Furthmayr, and V.T. Marchesi, (1978), Primary Structure of Human Erythrocyte Glycophorin A. Isolation and Characterization of Peptides and Complete Amino Acid Sequence. Biochemistry 17, 4756-4770.
1461. Tomita, M. and V.T. Marchesi, (1975), Amino-acid Sequence and Oligosaccharide Attachment Sites of Human Erythrocyte Glycophorin. Proc. Natl. Acad. Sci. USA 72, 2964-2968.
1462. Tommassen, J., H. van Tol, and B. Lugtenberg, (1983), The Ultimate Localization

of an Outer Membrane Protein of Escherichia coli K-12 Is Not Determined by the Signal Sequence. The EMBO Journal 2, 1275-1279.

1463. Toon, M.R. and A.K. Solomon, (1987), Modulation of Water and Urea Transport in Human Red Cells: Effects of pH and Phloretin. J. Memb. Biol. 99, 157-164.

1464. Toro, M., J. Cerbon, E. Arzt, G. Alegria, R. Alva, Y. Meas, and S. Estrada-O, (1987), Formation of Ion-Translocating Oliogomers by Nigericin. J. Memb. Biol. 95, 1-8.

1465. Torrisi, M.R., L.V. Lotti, A. Pavan, G. Migliaccio, and S. Bonatti, (1987), Free Diffusion to and from the Inner Nuclear Membrane of Newly Synthesized Plasma Membrane Glycoproteins. J. Cell Biol. 104, 733-737.

1466. Tran, D., J.-L. Carpentier, F. Sawano, P. Gorden, and L. Orci, (1987), Ligands Internalized through Coated or Noncoated Invaginations Follow a Common Intracellular Pathway. Proc. Natl. Acad. Sci. USA 84, 7957-7961.

1467. Trauble, H. and H. Eibl, (1974), Electrostatic Effects on Lipid Phase Transitions: Membrane Structure and Ionic Environment. Proc. Natl. Acad. Sci., USA 71, 214-219.

1468. Trewhella, J., J.-L. Popot, G. Zaccai, and D.M. Engelman, (1986), Localization of Two Chymotryptic Fragments in the Structure of Renatured Bacteriorhodopsin by Neutron Diffraction. The EMBO Journal 5, 3045-3049.

1469. Trewhella, J., S. Anderson, R. Fox, E. Gogol, S. Khan, D. Engelman, and C. Zaccai, (1983), Assignment of Segments of the Bacteriorhodopsin Sequence to Positions in the Structural Map. Biophys. J. 42, 233-241.

1470. Tsien, R.W., P. Hess, E.W. McCleskey, and R.L. Rosenberg, (1987), Calcium Channels: Mechanisms of Selectivity, Permeation, and Block. Ann. Rev. Biophys. Biophys. Chem. 16, 265-290.

1471. Tsui, F.C., D.M. Ojcius, and W.L. Hubbell, (1986), The Intrinsic pK_a Values for Phosphatidylserine and Phosphatidylethanolamine in Phosphatidylcholine Host Bilayers. Biophys. J. 49, 459-468.

1472. Tsuji, A. and S. Ohnishi, (1986), Restriction of the Lateral Motion of Band 3 in the Erythrocyte Membrane by the Cytoskeletal Network: Dependence on Spectrin Association State. Biochemistry 25, 6133-6139.

1473. Udagawa, T., T. Unemoto, and H. Tokuda, (1986), Generation of Na^+ Electrochemical Potential by the Na^+-motive NADH Oxidase and Na^+/H^+ Antiport System of a Moderately Halophilic Vibrio costicola. J. Biol. Chem. 261, 2616-2622.

1474. Udgaonkar, J.B. and G.P. Hess, (1986), Acetylcholine Receptor Kinetics: Chemical Kinetics. J. Membrane Biol. 93, 93-109.

1475. Ueno, M., C. Tanford, and J.A. Reynolds, (1984), Phospholipid Vesicle Formation Using Nonionic Detergents with Low Monomer Solubility. Kinetic Factors Determine Vesicle Size and Permeability. Biochemistry 23, 3070-3076.

1476. Ullrich, A., A. Gray, A.E. Tam, T. Yang-Feng, M. Tsubokawa, C. Collins, W. Henzel, T. Le Bon, S. Kathuria, E. Chen, S. Jacobs, U. Francke, J. Ramachandran, and Y. Fujita-Yamaguchi, (1986), Insulin-like Growth Factor I Receptor Primary Structure: Comparison with Insulin Receptor Suggests Structural Determinants That Define Functional Specificity. The EMBO Journal 5, 2503-2512.

1477. Ulrich, E.L., M.E. Girvin, W.A. Cramer, and J.L. Markley, (1985), Location and Mobility of Ubiquinones of Different Chain Lengths in Artificial Membrane Vesicles. Biochemistry 24, 2501-2508.

1478. Umbreit, J.N. and J.L. Strominger, (1973), Relation of Detergent HLB Number to Solubilization and Stabilization of D-alanine Carboxypeptidase from Bacillus subtilis Membranes. Proc. Natl. Acad. Sci. USA 70, 2997-3001.

1479. Unwin, N. (1986), Is There a Common Design for Cell Membrane Channels? Nature 323, 12-13.
1480. Unwin, N. and R. Henderson, (1984), The Structure of Proteins in Biological Membranes. Sci. Am. 250, 78-94.
1481. Unwin, P.N.T. and G. Zampighi, (1980), Structure of the Junction between Communicating Cells. Nature 283, 545-549.
1482. Unwin, P.N.T. and P.D. Ennis, (1984), Two Configurations of a Channel-forming Membrane Protein. Nature 307, 609-613.
1483. Unwin, P.N.T. and R.A. Milligan, (1982), A Large Particle Associated with the Perimeter of the Nuclear Pore Complex. J. Cell Biol. 93, 63-75.
1484. Urry, D.W. (1971), The Gramicidin A Transmembrane Channel: A Proposed Pi-L,D Helix. Proc. Natl. Acad. Sci. USA 68, 672-676.
1485. Valls, L.A., C.P. Hunter, J.H. Rothman, and T.H. Stevens, (1987), Protein Sorting in Yeast: The Localization Determinant of Yeast Vacuolar Carboxypeptidase Y Resides in the Propeptide. Cell 48, 887-897.
1486. Van den Bosch, H. (1980), Intracellular Phospholipases A. Biochim. Biophys. Acta 604, 191-246.
1487. van Heeswijk, M.P.E. and C.H. van Os, (1986), Osmotic Water Permeabilities of Brush Border and Basolateral Membrane Vesicles from Rat Renal Cortex and Small Intestine. J. Membrane Biol. 92, 183-193.
1488. van Hoogevest, P., A.P.M. Du Maine, B. de Kruijff, and L. de Gier, (1984), The Influence of Lipid Composition on the Barrier Properties of Band 3-Containing Lipid Vesicles. Biochim. Biophys. Acta 777, 241-252.
1489. van Hoogevest, P., J. de Gier, and B. de Kruijff, (1984), Determination of the Size of the Packing Defects in Dimyristoylphosphatidylcholine Bilayers Present at the Phase Transition Temperature. FEBS Lett. 171, 160-164.
1490. van Loon, A.P.G.M., A.W. Braendli, B. Pesold-Hurt, D. Blank, and G. Schatz, (1987), Transport of Proteins to the Mitochondrial Intermembrane Space: The 'Matrix-targeting' and the 'Sorting' Domains in the Cytochrome c_1 Presequence. The EMBO Journal 6, 2433-2439.
1491. van Loon, A.P.G.M. and G. Schatz, (1987), Transport of Proteins to the Mitochondrial Intermembrane Space: The 'Sorting' Domain of the Cytochrome c_1 Presequence Is a Stop-Transfer Sequence Specific for the Mitochondrial Inner Membrane. The EMBO Journal 6, 2441-2448.
1492. Van Loon, D., Th.A. Berkhout, R.A. Demel, and K.W.A. Wirtz, (1985), The Lipid Binding Site of the Phosphatidylcholine Transfer Protein from Bovine Liver. Chem. Phys. Lipids 38, 29-39.
1493. van Meer, G. (1987), Plasma Membrane Cholesterol Pools. TIBS 12, 375-376.
1494. Van Meer, G., B.J.H.M. Poorthuis, K.W.A. Wirtz, J.A.F. Op den Kamp, and L.L.M. Van Deenen, (1980), Transbilayer Distribution and Mobility of Phosphatidylcholine in Intact Erythrocyte Membranes. Eur. J. Biochem. 103, 283-288.
1495. van Meer, G. and K. Simons, (1986), The Function of Tight Junctions in Maintaining Differences in Lipid Composition between the Apical and the Basolateral Cell Surface Domains of MDCK Cells. The EMBO Journal 5, 1455-1464.
1496. van Rijn, J.L.M.L., J.W.P. Govers-Riemslag, R.F.A. Zwaal, and J. Rosing, (1984), Kinetic Studies of Prothrombin Activation: Effect of Factor Va and Phospholipids on the Formation of the Enzyme-Substrate Complex. Biochemistry 23, 4557-4564.
1497. Van Steeg, H., P. Oudshoorn, B. Van Hell, J.E.M. Polman, and L.A. Grivell, (1986), Targeting Efficiency of a Mitochondrial Pre-sequence Is Dependent on the Passenger Protein. The EMBO Journal 5, 3643-3650.

1498. Vance, D.E. (1985), Phospholipid Metabolism in Eucaryotes. Biochemistry of Lipids and Membranes (D.E. Vance and J.E. Vance, Eds.), pp. 242-270. The Benjamin/Cummings Publishing Company, Inc., Menlo Park, California.
1499. Vance, D.E. and S.L. Pelech, (1984), Enzyme Translocation in the Regulation of Phosphatidylcholine Biosynthesis. TIBS 9, 17-20.
1500. Vandenheuval, F.A. (1965), Study of Biological Structure at the Molecular Level with Stereomodel Projections. II. The Structure of Myelin in Relation to Other Membrane Systems. J. Am. Oil Chemists' So. 42, 481-492.
1501. Vanderkooi, G. and D.E. Green, (1970), Biological Membrane Structure. I. The Protein Crystal Model for Membranes. Proc. Natl. Acad. Sci. USA 66, 615-621.
1502. Vaz, W.L.C., A. Nicksch, and F. Jahnig, (1978), Electrostatic Interactions at Charged Lipid Membranes: Measurement of Surface pH with Fluorescent Lipoid pH Indicators. Eur. J. Biochem. 83, 299-305.
1503. Vaz, W.L.C., F. Goodsaid-Zalduondo, and K. Jacobson, (1984), Lateral Diffusion of Lipids and Proteins in Bilayer Membranes. FEBS Lett. 174, 199-207.
1504. Vaz, W.L.C., I. Derzko, and K.A. Jacobson, (1982), Photobleaching Measurements of the Lateral Diffusion of Lipids and Proteins in Artificial Phospholipid Bilayer Membranes. Cell Surface Reviews 8, 84-128.
1505. Vaz, W.L.C., R.M. Clegg, and D. Hallmann, (1985), Translational Diffusion of Lipids in Liquid Crystalline Phase Phosphatidylcholine Multibilayers. A Comparison of Experiment with Theory. Biochemistry 24, 781-786.
1506. Verhallen, P.F.J., R.A. Demel, H. Zwiers, and W.H. Gispen, (1984), Adrenocorticotropic Hormone (ACTH)-Lipid Interactions: Implications for Involvement of Amphipathic Helix Formation. Biochim. Biophys. Acta 775, 246-254.
1507. Verkleij, A.J. (1984), Lipidic Intramembranous Particles. Biochim. Biophys. Acta 779, 43-63.
1508. Verkleij, A.J., B. Humbel, D. Studer, and M. Muller, (1985), Lipid Particle Systems Visualized by Thin-section Electron Microscopy. Biochim. Biophys. Acta 812, 591-594.
1509. Verkleij, A.J. and P.H.J.Th. Ververgaert, (1975), The Architecture of Biological and Artificial Membranes as Visualized by Freeze Etching. Ann. Rev. Phys. Chem. 26, 101-121.
1510. Verkleij, A.J., R.F.A. Zwaal, B. Roelofsen, P. Comfurius, D. Kastelijn, and L.L.M. Van Deenen, (1973), The Asymmetric Distribution of Phospholipids in the Human Red Cell Membrane. Biochim. Biophys. Acta 323, 178-193.
1511. Verkman, A.S. (1987), Mechanism and Kinetics of Merocyanine 540 Binding to Phospholipid Membranes. Biochemistry 26, 4050-4056.
1512. Verkman, A.S., K. Skorecki, and D.A. Ausiello, (1984), Radiation Inactivation of Oligomeric Enzyme Systems: Theoretical Considerations. Proc. Natl. Acad. Sci. USA 81, 150-154.
1513. Verma, S.P. and D.F.H. Wallach, (1984), Raman Spectroscopy of Lipids and Biomembranes. Biomembrane Structure and Function (D. Chapman, Ed.), pp. 167-198. Verlag Chemie, Basel.
1514. Verner, K. and G. Schatz, (1987), Import of an Incompletely Folded Precursor Protein into Isolated Mitochondria Requires an Energized Inner Membrane, but No Added ATP. The EMBO Journal 6, 2449-2456.
1515. Verselis, V., R.L. White, D.C. Spray, and M.V.L. Bennett, (1986), Gap Junctional Conductance and Permeability Are Linearly Related. Science 234, 461-464.
1516. Vigo, C., S.H. Grossman, and W. Drost-Hansen, (1984), Interaction of Dolichol

and Dolichol Phosphate with Phospholipid Bilayers. Biochim. Biophys. Acta 774, 221-226.

1517. Viitanen, P.V., P.J. Geiger, S. Erickson-Viitanen, and S.P. Bessman, (1984), Evidence for Functional Hexokinase Compartmentation in Rat Skeletal Muscle Mitochondria. J. Biol. Chem. 259, 9679-9686.

1518. Vilsen, B., J.P. Andersen, J. Petersen, and P.L. Jorgensen, (1987), Occlusion of $^{22}Na^+$ and $^{86}Rb^+$ in Membrane-bound and Soluble Protomeric αß-Units of Na,K-ATPase. J. Biol. Chem. 262, 10511-10517.

1519. Viitala, J. and J. Jarnefelt, (1985), The Red Cell Surface Revisited. TIBS 10, 392-395.

1520. Voelker, D.R. (1985), Lipid Assembly into Cell Membranes. Biochemistry of Lipids and Membranes (D.E. Vance and J.E. Vance, Eds.), pp. 475-502. The Benjamin/Cummings Publishing Company, Inc., Menlo Park, California.

1521. Voelker, D.R. (1984), Phosphatidylserine Functions as the Major Precursor of Phosphatidylethanolamine in Cultured BHK-21 Cells. Proc. Natl. Acad. Sci. USA 81, 2669-2673.

1522. Vogel, H. (1981), Incorporation of Melittin in Phosphatidylcholine Bilayers. FEBS Lett. 134, 37-42.

1523. Vogel, H. and F. Jahnig, (1986), Models for the Structure of Outer-membrane Proteins of Escherichia coli Derived from Raman Spectroscopy and Prediction Methods. J. Mol. Biol. 190, 191-199.

1524. Vogel, H. and F. Jahnig, (1986), The Structure of Melittin in Membranes. Biophys. J. 50, 573-582.

1525. Vogel, H., J.K. Wright, and F. Jahnig, (1985), The Structure of the Lactose Permease Derived from Raman Spectroscopy and Prediction Methods. The EMBO Journal 4, 3625-3631.

1526. Vogel, H. and W. Gartner, (1987), The Secondary Structure of Bacteriorhodopsin Determined by Raman and Circular Dichroism Spectroscopy. J. Biol. Chem. 262, 11464-11469.

1527. von Heijne, G. (1986), Mitochondrial Targeting Sequences May Form Amphiphilic Helices. The EMBO Journal 5, 1335-1342.

1528. von Heijne, G. (1986), Towards a Comparative Anatomy of N-Terminal Topogenic Protein Sequences. J. Mol. Biol. 189, 239-242.

1529. von Heijne, G. (1986), The Distribution of Positively Charged Residues in Bacterial Inner Membrane Proteins Correlates with the Transmembrane Topology. The EMBO Journal 5, 3021-3027.

1530. von Heijne, G. (1986), A New Method for Predicting Signal Sequence Cleavage Sites. Nucleic Acids Research 14, 4683-4690.

1531. von Heijne, G. and J.P. Segrest, (1987), The Leader Peptides from Bacteriorhodopsin and Halorhodopsin Are Potential Membrane-spanning Amphipathic Helices. FEBS Lett. 213, 238-240.

1532. von Jagow, G., T.A. Link, and T. Ohnishi, (1986), Organization and Function of Cytochrome b and Ubiquinone in the Cristae Membrane of Beef Heart Mitochondria. J. Bioenerg. and Biomemb. 18, 157-179.

1533. Vos-Scheperkeuter, G.H. and B. Witholt, (1984), Assembly Pathway of Newly Synthesized LamB Protein, an Outer Membrane Protein of Escherichia coli K-12. J. Mol. Biol. 175, 511-528.

1534. Waggoner, A.S. (1979), Dye Indicators of Membrane Potential. Ann. Rev. Biophys. Bioeng. 8, 47-68.

1535. Waldemann, T.A. (1986), The Structure, Function, and Expression of Interleukin-2 Receptors on Normal and Malignant Lymphocytes. Science 232, 727-732.
1536. Walker, J.E., A.F. Carne, and H.W. Schmitt, (1979), The Topography of the Purple Membrane. Nature 278, 653-654.
1537. Wallace, B.A. (1985), Structure of Gramicidin A. Biophysical Discussions Abstracts 277-286.
1538. Wallace, B.A. and C.L. Teeters, (1987), Differential Absorption Flattening Optical Effects Are Significant in the Circular Dichroism Spectra of Large Membrane Fragments. Biochemistry 26, 65-70.
1539. Wallace, B.A. and D. Mao, (1984), Circular Dichroism Analysis of Membrane Proteins: An Examination of Differential Light Scattering and Absorption Flattening Effects in Large Membrane Vesicles and Membrane Sheets. Anal. Bioch. 142, 317-328.
1540. Wallace, B.A., M. Cascio, and D.L. Mielke, (1986), Evaluation of Methods for the Prediction of Membrane Protein Secondary Structures. Proc. Natl. Acad. Sci. USA 83, 9423-9427.
1541. Wallace, B.A. and N. Kohl, (1984), The C-Terminus of Bacteriorhodopsin is a Random Coil. Biochim. Biophys. Acta 777, 93-98.
1542. Wallace, B.A. and R. Henderson, (1982), Location of Carboxyl Terminus of Bacteriorhodopsin in Purple Membrane. Biophys. J. 39, 233-239.
1543. Wallach, D.F.H. and P.H. Zahler, (1966), Protein Conformations in Cellular Membranes. Proc. Natl. Acad. Sci. USA 56, 1552-1559.
1544. Wallach, D.F.H., S.P. Verma, and J. Fookson, (1979), Application of Lasar Raman and Infrared Spectroscopy to the Analysis of Membrane Structure. Biochim. Biophys. Acta 559, 153-208.
1545. Walsh, J.P. and R.M. Bell, (1986), sn-1,2-Diacylglycerol Kinase of Escherichia coli. J. Biol. Chem. 261, 6239-6247.
1546. Walter, A. and J. Gutknecht, (1986), Permeability of Small Nonelectrolytes through Lipid Bilayer Membranes. J. Membrane Biol. 90, 207-217.
1547. Walter, P. (1987), Signal Recognition: Two Receptors Act Sequentially. Nature 328, 763-764.
1548. Walter, P. and G. Blobel, (1982), Signal Recognition Particle Contains a 7S RNA Essential for Protein Translocation across the Endoplasmic Reticulum. Nature 299, 691-784.
1549. Wang, J.-F., Falke J.J., and S.I. Chan, (1986), A Proton NMR Study of the Mechanism of the Erythrocyte Glucose Transporter. Proc. Natl. Acad. Sci. USA 83, 3277-3281.
1550. Ward, D.M. and J. Kaplan, (1986), Mitogenic Agents Induce Redistribution of Transferrin Receptors from Internal Pools to the Cell Surface. Biochem. J. 238, 721-728.
1551. Warren, G.B., M.D. Houslay, J.C. Metcalfe, and N.J.M. Birdsall, (1975), Cholesterol Is Excluded from the Phospholipid Annulus Surrounding an Active Calcium Transport Protein. Nature 255, 684-687.
1552. Warschel, A. (1981), Electrostatic Basis of Structure-Function Correlation in Proteins. Acc. Chem. Res. 14, 284-290.
1553. Warshel, A., S.T. Russell, and A.K. Churg, (1984), Macroscopic Models for Studies of Electrostatic Interactions in Proteins: Limitations and Applicability. Proc. Natl. Acad. Sci. USA 81, 4785-4789.
1554. Wasmann, C.C., B. Reiss, S.G. Bartlett, and H.J. Bohnert, (1986), The Importance

of the Transit Peptide and the Transported Protein for Protein Import into Chloroplasts. Mol. Gen. Genet. 205, 446-453.

1555. Waters, M.G. and G. Blobel, (1986), Secretory Protein Translocation in a Yeast Cell-free System Can Occur Posttranslationally and Requires ATP Hydrolysis. J. Cell Biol. 102, 1543-1550.
1556. Waters, M.G., W.J. Chirico, and G. Blobel, (1986), Protein Translocation across the Yeast Microsomal Membrane Is Stimulated by a Soluble Factor. J. Cell Biol. 103, 2629-2636.
1557. Wattenberg, B.W. and D.F. Silbert, (1983), Sterol Partitioning among Intracellular Membranes. J. Biol. Chem. 258, 2284-2289.
1558. Wattenberg, B.W. and J.E. Rothman, (1986), Multiple Cytosolic Components Promote Intra-Golgi Protein Transport: Resolution of a Protein Acting at a Late Stage, Prior to Membrane Fusion. J. Biol. Chem. 261, 2208-2213.
1559. Watts, T.H., H.E. Gaub, and H.M. McConnell, (1986), T-Cell-Mediated Association of Peptide Antigen and Major Histocompatibility Complex Protein Detected by Energy Transfer in an Evanescent Wave-Field. Nature 320, 179-181.
1560. Waxman, S.G. and J.M. Ritchie, (1986), Organization of Ion Channels in the Myelinated Nerve Fiber. Science 228, 1502-1507.
1561. Weber, K., N. Johnsson, U. Plessmann, P. Nguyen Van, H.-D. Soling, C. Ampe, and J. Vandekerckhove, (1987), The Amino Acid Sequence of Protein II and its Phosphorylation Site for Protein Kinase C; the Domain Structure Ca^{2+}-modulated Lipid Binding Proteins. The EMBO Journal 6, 1599-1604.
1562. Weckstrom, K. (1985), Aqueous Micellar Systems in Membrane Protein Crystallization. FEBS Lett. 192, 220-224.
1563. Weinberg, R.A. (1985), The Action of Oncogenes in the Cytoplasm and Nucleus. Science 230, 770-776.
1564. Weinstein, S., J.T. Durkin, W.R. Veatch, and E.R. Blout, (1985), Conformation of the Gramicidin A Channel in Phospholipid Vesicles: A Fluorine-19 Nuclear Magnetic Resonance Study. Biochemistry 24, 4374-4382.
1565. Weis, R.M. and H.M. McConnell, (1984), Two-dimensional Chiral Crystals of Phospholipid. Nature 310, 47-49.
1566. Weiss, H. and P. Wingfield, (1979), Enzymology of Ubiquinone-Utilizing Electron Transfer Complexes in Nonionic Detergent. Eur. J. Bioch. 99, 151-160.
1567. Weissman, A.M., L.E. Samelson, and R.D. Klausner, (1986), A New Subunit of the Human T-Cell Antigen Receptor Complex. Nature 324, 480-482.
1568. Welsh, E.J., D. Thom, E.R. Morris, and D.A. Rees, (1985), Molecular Organization of Glycophorin A: Implications for Membrane Interactions. Biopolymers 24, 2301-2332.
1569. Weyer, K.A., W. Schafer, F. Lottspeich, and H. Michel, (1987), The Cytochrome Subunit of the Photosynthetic Reaction Center from Rhodopseudomonas viridis Is a Lipoprotein. Biochemistry 26, 2909-2914.
1570. White, D.A. (1973), The Phospholipid Composition of Mammalian Tissues. Form and Function of Phospholipids (G.B. Ansell, J.N. Hawthorne, and R.M.C. Dawson, Eds.), 2nd Edition, pp. 441-482. Elsevier Scientific Publication Company, New York.
1571. White, J., M. Kielian, and A. Helenius, (1983), Membrane Fusion Proteins of Enveloped Animal Viruses. Quarterly Reviews of Biophysics 16, 151-195.
1572. Whiting, P. and J. Lindstrom, (1987), Affinity Labelling of Neuronal Acetylcholine Receptors Localizes Acetylcholine-binding Sites to Their ß-Subunits. FEBS Lett. 213, 55-60.

1573. Whitmarsh, J. (1986), Mobile Electron Carriers in Thylakoids. Encyclopedia of Plant Physiology, New Series, Volume 19, Photosynthesis III 19, 508-527.

1574. Wickner, W. (1976), Asymmetric Orientation of Phage M13 Coat Protein in Escherichia coli Cytoplasmic Membranes and in Synthetic Lipid Vesicles. Proc. Natl. Acad. Sci. USA 73, 1159-1163.

1575. Wickner, W. (1979), The Assembly of Proteins into Biological Membranes: The Membrane Trigger Hypothesis. Ann. Rev. Biochem. 48, 23-45.

1576. Wickner, W.T. and H.F. Lodish, (1985), Multiple Mechanisms of Protein Insertion Into and Across Membranes. Science 230, 400-407.

1577. Wiech, H., M. Sagstetter, G. Muller, and R. Zimmermann, (1987), The ATP Requiring Step in Assembly of M13 Procoat Protein into Microsomes Is Related to Preservation of Transport Competence of the Precursor Protein. The EMBO Journal 6, 1011-1016.

1578. Wiedmann, M., T.V. Kurzchalia, E. Hartmen, and T.A. Rapoport, (1987), A Signal Sequence Receptor in the Endoplasmic Reticulum Membrane. Nature 328, 830-833.

1579. Wiedmann, M., T.V. Kurzchalia, H. Bielka, and T.A. Rapoport, (1987), Direct Probing of the Interaction between the Signal Sequence of Nascent Preprolactin and the Signal Recognition Particle by Specific Cross-linking. J. Cell Biol. 104, 201-208.

1580. Wieland, F.T., M.L. Gleason, T.A. Serafini, and J.E. Rothman, (1987), The Rate of Bulk Flow from the Endoplasmic Reticulum to the Cell Surface. Cell 50, 289-300.

1581. Weiner, J.R., R. Pal, Y. Barenholz, and R.R. Wagner, (1985), Effect of the Vesicular Stomatitis Virus Matrix Protein on the Lateral Organization of Lipid Bilayers Containing Phosphatidylglycerol: Use of Fluorescent Phospholipid Analogues. Biochemistry 24, 7651-7658.

1582. Wieslander, A., L. Rilfors, and G. Lindblom, (1986), Metabolic Changes of Membrane Lipid Composition in Acholeplasma laidlawii by Hydrocarbons, Alcohols, and Detergents: Arguments for Effects on Lipid Packing. Biochemistry 25, 7511-7517.

1583. Wilcox, C.A. and E.N. Olson, (1987), The Majority of Cellular Fatty Acid Acylated Proteins Are Localized to the Cytoplasmic Surface of the Plasma Membrane. Biochemistry 26, 1029-1036.

1584. Wileman, T., C. Harding, and P. Stahl, (1985), Receptor-mediated Endocytosis. Biochem. J. 232, 1-14.

1585. Wiley, D.C. and J.J. Skehel, (1987), The Structure and Function of the Hemagglutinin Membrane Glycoprotein of Influenza Virus. Ann. Rev. Biochem. 56, 365-394.

1586. Wilkens, M.H.F., A.E. Blaurock, and D.M. Engelman, (1971), Bilayer Structure in Membranes. Nature New Biology 230, 72-76.

1587. Wilkinson, D.A. and J.F. Nagle, (1979), Dilatometric Study of Binary Mixtures of Phosphatidylcholines. Biochemistry 18, 4244-4249.

1588. Wilkison, W.O., J.P. Walsh, J.M. Corless, and R.M. Bell, (1986), Crystalline Arrays of the Escherichia coli sn-Glycerol-3-phosphate Acyltransferase, an Integral Membrane Protein. J. Biol. Chem. 1986, 9951-9958.

1589. Williams, M.A. and R.A. Lamb, (1986), Determination of the Orientation of an Integral Membrane Protein and Sites of Glycosylation by Oligonucleotide-Directed Mutagenesis: Influenza B Virus NB Glycoprotein Lacks a Cleavable Signal Sequence and Has an Extracellular NH2-Terminal Region. Mol. Cell. Biol. 6, 4317-4328.

1590. Williams, R.W. (1983), Estimation of Protein Secondary Structure from the Laser Raman Amide I Spectrum. J. Mol. Biol. 166, 581-603.

1591. Williamson, P., R. Antia, and R.A. Schlegel, (1987), Maintenance of Membrane Phospholipid Asymmetry. FEBS Lett. 219, 316-320.
1592. Wills, N.K. and A. Zweifach, (1987), Recent Advances in the Characterization of Epithelial Ionic Channels. Biochim. Biophys. Acta 906, 1-31.
1593. Wilschut, J. and D. Hoekstra, (1984), Membrane Fusion: From Liposomes to Biological Membranes. TIBS 9, 479-483.
1594. Wilschut, J. and D. Hoekstra, (1986), Membrane Fusion: Lipid Vesicles as a Model System. Chem. Phys. Lipids 40, 145-166.
1595. Wilson, I.A., J.J. Skehel, and D.C. Wiley, (1981), Structure of the Haemagglutinin Membrane Glycoprotein of Influenza Virus at 3 Å Resolution. Nature 289, 366-373.
1596. Winiski, A.P., A.C. McLaughlin, R.V. McDaniel, M. Eisenberg, and S. McLaughlin, (1986), An Experimental Test of the Discreteness-of-Charge Effect in Positive and Negative Lipid Bilayers. Biochemistry 25, 8206-8214.
1597. Witzke, N.M. and R. Bittman, (1984), Dissociation Kinetics and Equilibrium Binding Properties of Polyene Antibiotic Complexes with Phosphatidylcholine/Sterol Vesicles. Biochemistry 23, 1668-1674.
1598. Wolf, D.E. (1985), Determination of the Sidedness of Carbocyanine Dye Labeling of Membranes. Biochemistry 24, 582-586.
1599. Wolf, D.E., B.K. Scott, and C.F. Millette, (1986), The Development of Regionalized Lipid Diffusibility in the Germ Cell Plasma Membrane during Spermatogenesis in the Mouse. J. Cell Biol. 103, 1745-1750.
1600. Wolfe, P.B., M. Rice, and W. Wickner, (1985), Effects of Two sec Genes on Protein Assembly into the Plasma Membrane of Escherichia coli. J. Biol. Chem. 260, 1836-1841.
1601. Wolfe, P.B. and W. Wickner, (1984), Bacterial Leader Peptidase, a Membrane Protein without a Leader Peptide, Uses the Same Export Pathway as Pre-secretory Proteins. Cell 36, 1067-1072.
1602. Wolfenden, R., L. Andersson, P.M. Cullis, and C.C.B. Southgate, (1981), Affinities of Amino Acid Side Chains for Solvent Water. Biochemistry 20, 849-855.
1603. Wolfman, A. and I.G. Macara, (1987), Elevated Levels of Diacylglycerol and Decreased Phorbol Ester Sensitivity in Ras-Transformed Fibroblasts. Nature 325, 359-361.
1604. Woodgett, C. and J.K. Rose, (1986), Amino-Terminal Mutation of the Vesicular Stomatitis Virus Glycoprotein Does Not Affect Its Fusion Activity. J. Virology 59, 486-489.
1605. Woods, Jr., V.L., L.E. Wolff, and D.M. Keller, (1986), Resting Platelets Contain a Substantial Centrally Located Pool of Glycoprotein IIb-IIIa Complex Which May Be Accessible to Some but Not Other Extracellular Proteins. J. Biol. Chem. 261, 15242-15251.
1606. Worcester, D.L. and N.P. Franks, (1976), Structural Analysis of Hydrated Egg Lecithin and Cholesterol Bilayers. II. Neutron Diffraction. J. Mol. Biol. 100, 359-378.
1607. Wright, J.K. (1986), The Kinetic Mechanism of Galactoside/H^+ Cotransport in Escherichia coli. Biochim. Biophys. Acta 855, 391-416.
1608. Wright, J.K., R. Seckler, and P. Overath, (1986), Molecular Aspects of Sugar: Ion Cotransport. Ann. Rev. Biochem. 55, 225-248.
1609. Wright, P.S., J.N. Morand, and C. Kent, (1985), Regulation of Phosphatidylcholine Biosynthesis in Chinese Hamster Ovary Cells by Reversible Membrane Association of CTP: Phosphocholine Cytidylyltransferase. J. Biol. Chem. 260, 7919-7926.

1610. Wu, J.-S.R. and J.E. Lever, (1987), Purification and Reconstitution of a 75-Kilodalton Protein Identified as a Component of the Renal Na^+/Glucose Symporter. Biochemistry 26, 5958-5962.

1611. Wynn, C.H. (1986), A Triple-binding-domain Model Explains the Specificity of the Interaction of a Sphingolipid Activator Protein (SAP-1) with Sulphatide, GM_1-ganglioside and Globotriaosylceramide. Biochem. J. 240, 921-924.

1612. Yamane, K., S. Ichihara, and S. Mizushima, (1987), In Vitro Translocation of Protein across Escherichia coli Membrane Vesicles Requires Both the Proton Motive Force and ATP. J. Biol. Chem. 262, 2358-2362.

1613. Yarden, Y., I. Harari, and J. Schlessinger, (1985), Purification of an Active EGF Receptor Kinase with Monoclonal Antibodies. J. Biol. Chem. 260, 315-319.

1614. Yarden, Y. and J. Schlessinger, (1987), Epidermal Growth Factor Induces Rapid, Reversible Aggregation of the Purified Epidermal Growth Factor Receptor. Biochemistry 26, 1443-1451.

1615. Yarden, Y., J.A. Escobedo, W.-J. Kuang, T.L. Yang-Feng, T.O. Daniel, P.M. Tremble, E.Y. Chen, M.E. Ando, R.N. Harkins, U. Francke, V.A. Fried, A. Ullrich, and L.T. Williams, (1986), Structure of the Receptor for Platelet-derived Growth Factor Helps Define a Family of Closely Related Growth Factor Receptors. Nature 323, 226-232.

1616. Yeagle, P. (1986), Hydration and the Lamellar to Hexagonal II Phase Transition of Phosphatidylethanolamine. Biochemistry 25, 7518-7522.

1617. Yeagle, P.L. (1985), Cholesterol and the Cell Membrane. Biochim. Biophys. Acta 822, 267-287.

1618. Yeagle, P.L. and J.E. Young, (1986), Factors Contributing to the Distribution of Cholesterol among Phospholipid Vesicles. J. Biol. Chem. 261, 8175-8181.

1619. Yeates, T.O., H. Komiya, D.C.Rees, J.P. Allen, and G. Feher, (1987), Structure of the Reaction Center from Rhodobacter sphaeroides R-26: Membrane-Protein Interactions. Proc. Natl. Acad. Sci. USA 84, 6438-6442.

1620. Yoshida, H., N. Tondokoro, Y. Asano, K. Mizusawa, R. Yamagishi, T. Horigome, and H. Sugano, (1987), Studies on Membrane Proteins Involved in Ribosome Binding on the Rough Endoplasmic Reticulum. Biochem. J. 245, 811-819.

1621. Yoshikawa, W., H. Akutsu, and Y. Kyogoku, (1983), Light-scattering Properties of Osmotically Active Liposomes. Biochim. Biophys. Acta 735, 397-406.

1622. Yoshima, H., H. Furthmayr, and A. Kobata, (1980), Structures of the Asparagine-linked Sugar Chains of Glycophorin A. J. Biol. Chem. 255, 9713-9718.

1623. Young, E.F., E. Ralston, J. Blake, J. Ramachandran, Z.W. Hall, and R.M. Stroud, (1985), Topological Mapping of Acetylcholine Receptors: Evidence for a Model with Five Transmembrane Segments and a Cytoplasmic COOH-terminal Peptide. Proc. Natl. Acad. Sci. USA 82, 626-630.

1624. Young, J.D.-E., Z.A. Cohn, and E.R. Podack, (1986), The Ninth Component of Complement and the Pore-Forming Protein (Perforin 1) from Cytotoxic T Cells: Structural, Immunological, and Functional Similarities. Science 233, 184-190.

1625. Yu, J., A. Elgsaeter, and D. Branton, (1977), Intramembrane Particle Aggregation in Erythrocyte Membranes and Band 3-Lipid Recombinants. Cell Shape and Surface Architecture (J.P. Revel, U. Henning, and C.F. Fox, Eds.), 453-458. Alan R. Liss, Inc., New York.

1626. Yu, J. and D. Branton, (1976), Reconstitution of Intramembrane Particles in Recombinants of Erythrocyte Protein Band 3 and Lipid: Effects of Spectrin-Actin Association. Proc. Natl. Acad. Sci. USA 73, 3891-3895.

1627. Zachowski, A., A. Herrmann, A. Paraf, and P.F. Devaux, (1987), Phospholipid

Outside-Inside Translocation in Lymphocyte Plasma Membranes Is a Protein-Mediated Phenomenon. Biochim. Biophys. Acta 897, 197-200.

1628. Zachowski, A., E. Favre, S. Cribier, P. Herve, and P.F. Devaux, (1986), Outside-Inside Translocation of Aminophospholipids in the Human Erythrocyte Membrane Is Mediated by a Specific Enzyme. Biochemistry 25, 2585-2590.

1629. Zhang, Y.-Z., R.A. Capaldi, P.R. Cullis, and T.D. Madden, (1985), Orientation of Cytochrome c Oxidase Molecules in the Two Populations of Reconstituted Vesicles Resolved by Column Chromatography on DEAE-Sephacryl. Biochim. Biophys. Acta 808, 209-211.

1630. Zibirre, R., G. Hippler-Feldtmann, J. Kuhne, P. Poronnik, G. Warnecke, and G. Koch, (1987), Detection and Localization of a Cytoplasmic Domain on the ß-Subunit of Na^+, K^+-ATPase. J. Biol. Chem. 262, 4349-4354.

1631. Zimmer, D.B., C.R. Green, W.H. Evans, and N.B. Gilula, (1987), Topological Analysis of the Major Protein in Isolated Intact Rat Liver Gap Junctions and Gap Junction-derived Single Membrane Structures. J. Biol. Chem. 262, 7751-7763.

1632. Zimmerberg, J., M. Curran, F.S. Cohen, and M. Brodwick, (1987), Simultaneous Electrical and Optical Measurements Show That Membrane Fusion Precedes Secretory Granule Swelling During Exocytosis of Beige Mouse Mast Cells. Proc. Natl. Acad. Sci. USA 84, 1585-1589.

1633. Zingsheim, H.P., D.-C. Neugebauer, J. Frank, W. Hanicke, and F.S. Barrantes, (1982), Dimeric Arrangement and Structure of the Membrane-bound Acetylcholine Receptor Studied by Electron Microscopy. The EMBO Journal 1, 541-547.

Index